BIOLOGICAL, BIOCHEMICAL, AND BIOMEDICAL ASPECTS OF ACTINOMYCETES

Academic Press Rapid Manuscript Reproduction

Proceedings of the V International Symposium
on Actinomycetes Biology
Held in Oaxtepec, Morelos, México
August 16–19, 1982

BIOLOGICAL, BIOCHEMICAL, AND BIOMEDICAL ASPECTS OF ACTINOMYCETES

Edited by

Librado Ortiz-Ortiz

Departamento de Inmunología
Instituto de Investigaciones Biomédicas
Universidad Nacional Autónoma de México
México, D.F., México

Luis F. Bojalil

División de Ciencias
Biológicas y de la Salud Unidad Xochimilco
Universidad Autónoma Metropolitana
México, D.F., México

Veronica Yakoleff

Departamento de Inmunología
Instituto de Investigaciones Biomédicas
Universidad Nacional Autónoma de México
México, D.F., México

1984

AP

ACADEMIC PRESS, INC.
(Harcourt Brace Jovanovich, Publishers)
Orlando San Diego New York London
Toronto Montreal Sydney Tokyo

ACADEMIC PRESS, INC.
Orlando, Florida 32887

United Kingdom Edition published by
ACADEMIC PRESS, INC. (LONDON) LTD.
24/28 Oval Road, London NW1 7DX

Library of Congress Cataloging in Publication Data

Main entry under title:

Biological, biochemical, and biomedical aspects of actinomycetes.

Proceedings of the Fifth International Symposium on Actinomycetes Biology, held in Oaxtepec, Morelos, Mexico on August 16-19, 1982
Includes index.
1. Actinomycetales--Congresses. I. Ortiz-Ortiz, Librado. II. Bojalil, Luis F. III. Yakoleff, Veronica. IV. International Symposium on Actinomycetes Biology (5th : 1982 : Oaxtepec, Mexico)
QR81.B56 1984 589.9'2 84-45221
ISBN 0-12-528620-1 (alk. paper)

PRINTED IN THE UNITED STATES OF AMERICA

84 85 86 87 9 8 7 6 5 4 3 2 1

CONTENTS

CONTRIBUTORS

Numbers in parentheses indicate the pages on which the authors' contributions begin.

N. S. Agre (229), *Institute of Biochemistry and Physiology of Microorganisms, Academy of Sciences of the USSR, Pustchino, USSR*

Alfredo Aguilar (259), *Departamento de Microbiología, Facultad de Biología, Universidad de León, León, Spain, and John Innes Institute, Norwich, United Kingdom*

Y. Aharonowitz (357), *Department of Microbiology, George S. Wise Faculty of Life Sciences, Tel Aviv University, Tel Aviv, Israel 69978*

G. Alderson (537, 597), *School of Studies in Medical Sciences, University of Bradford, Bradford BP7 1DP, United Kingdom*

Maria T. Alegre (273), *Departamento de Microbiología, Facultad de Biología, Universidad de León, León, Spain*

M. Athalye[1] (597), *Department of Bacteriology, Institute of Dermatology, London, United Kingdom*

Richard W. Attwell[2] (441), *Department of Microbiology, University of Maryland, College Park, Maryland 20742*

Dwight Baker (563), *Charles F. Kettering Research Laboratory, Yellow Springs, Ohio 45387*

Blaine L. Beaman (73, 89, 107), *Department of Medical Microbiology and Immunology, School of Medicine, University of California, Davis, California 95616*

Dale C. Birdsell (33), *Department of Oral Biology, University of Washington, Seattle, Washington 98195*

S. Biró (197), *Institute of Biology, University Medical School, Debrecen H-4012, Hungary*

J. S. Bond (315, 381), *Departments of Biochemistry and of Microbiology, and Immunology, Virginia Commonwealth University, Richmond, Virginia 23298*

[1]Present address: Department of Molecular Biology, University of Edinburgh, Edinburgh EH9 3JR, United Kingdom.

[2]Present address: Department of Biological Sciences, Manchester Polytechnic, Manchester, United Kingdom.

G. H. W. Bowden (1), *Department of Oral Biology, Faculty of Dentistry, The University of Manitoba, Winnipeg, Manitoba R3E OW3, Canada*

S. G. Bradley (315, 381), *Departments of Biochemistry and of Microbiology and Immunology, Virginia Commonwealth University, Richmond, Virginia 23298*

Dale Brown (325), *Laboratory of the Biology of Viruses, National Institute of Allergy and Infectious Diseases, National Institutes of Health, Bethesda, Maryland 20205*

Johann J. Burckhardt[3] (47), *Department of Oral Microbiology and General Immunology, Dental Institute, University of Zürich, Zürich, Switzerland*

Donald R. Callihan[4] (33), *Microbiology Department, School of Medicine and Dentistry, University of Rochester Medical Center, Rochester, New York 14642*

Jose M. Castro (273), *Departamento de Microbiología, Facultad de Biología, Universidad de León, León, Spain*

James A. Clagett (61), *Departments of Periodontics and Microbiology–Immunology, Schools of Dentistry and Medicine, University of Washington, Seattle, Washington 98195*

Rita R. Colwell (441), *Department of Microbiology, University of Maryland, College Park, Maryland 20742*

Carmen Conde (119), *Centro de Fijación de Nitrógeno, Universidad Nacional Autónoma de México, Cuernavaca, Morelos, México*

T. Cross (521), *Postgraduate School of Biological Sciences, University of Bradford, Bradford BD7 1DP, United Kingdom*

Eric Cundliffe (303), *Department of Biochemistry, University of Leicester, Leicester LE1 7RH, United Kingdom*

L. David Engel (61), *Departments of Periodontics and Pathology, Schools of Dentistry and Medicine, University of Washington, Seattle, Washington 98195*

Laura Escalante (343), *Instituto de Investigaciones Biomédicas, Universidad Nacional Autónoma de México, México, D.F., 04510 México*

Amelia Farres (343), *Instituto de Investigaciones Biomédicas, Universidad Nacional Autónoma de México, México, D.F., 04510 México*

Gregory A. Filice (107), *Department of Medicine, University of Minnesota, Minneapolis, Minnesota 55455 and Infectious Disease Section, Medical Service, Veterans Administration Medical Center, Minneapolis, Minnesota 55417*

Werner Fischlschweiger (33), *Department of Basic Dental Sciences, College of Dentistry, University of Florida, Gainesville, Florida 32610*

[3]Present address: Research and Development Department, Diagnostica Division, F. Hoffmann–La Roche & Co. AG, CH 4002 Basel, Switzerland.

[4]Present address: Diagnostic Laboratory, New York State College of Veterinary Medicine, Cornell University, Ithaca, New York 14851-0786.

Maria Elena Flores (343), *Facultad de Química, Universidad Autónoma del Estado de México, Toluca, México*

John Foster (325), *Department of Microbiology, Marshall University School of Medicine, Huntington, West Virginia 25701*

M. Goodfellow (453, 537, 583), *Department of Microbiology, The University, Newcastle upon Tyne NE1 7RU, United Kingdom*

J. A. Guijarro (179), *Departamento de Microbiología, Universidad de Oviedo, Oviedo, Spain*

C. Hardisson (179), *Departamento de Microbiología, Universidad de Oviedo, Oviedo, Spain*

J. A. Haynes (453), *Department of Microbiology, The University, Newcastle upon Tyne NE1 7RU, United Kingdom*

Michael J. M. Hitchcock (325), *Bristol Laboratories, Syracuse, New York 13201*

Thuioshi Ioneda (239), *Instituto de Química, Departamento de Bioquímica, Universidade de São Paulo, 01498 São Paulo—SP Brazil*

Edward Katz (325), *Department of Microbiology, Schools of Medicine and Dentistry, Georgetown University, Washington, D.C. 20007*

F. T. Kirnberg (357), *Department of Microbiology, George S. Wise Faculty of Life Sciences, Tel Aviv University, Tel Aviv, Israel 69978*

James A. Krick (107), *Department of Medicine, School of Medicine, University of California, Davis, California 95616*

Viswanath P. Kurup (145), *Department of Medicine, The Medical College of Wisconsin, Milwaukee, Wisconsin 53226, and The Research Service, Veterans Administration Medical Center, Wood, Wisconsin 53193*

S. Lanning[5] (473), *Botany Department, University of Liverpool, Liverpool L69 3BX, United Kingdom*

Hubert A. Lechevalier (575), *Waksman Institute of Microbiology, Rutgers, The State University of New Jersey, Piscataway, New Jersey 08854*

Mary P. Lechevalier (575), *Waksman Institute of Microbiology, Rutgers, The State University of New Jersey, Piscataway, New Jersey 08854*

Paloma Liras (273), *Departamento de Microbiología, Facultad de Biología, Universidad de León, León, Spain*

R. Locci (395), *Department of Mycology, University of Milan, 20133 Milan, Italy*

Hector López (343), *Instituto de Investigaciones Biomédicas, Universidad Nacional Autónoma de México, México, D.F., 04510 México*

N. M. Magal (357), *Department of Microbiology, George S. Wise Faculty of Life Sciences, Tel Aviv University, Tel Aviv, Israel 69978*

Juan F. Martín (273), *Departamento de Microbiología, Facultad de Biología, Universidad de León, León, Spain*

[5]Present address: Biotechnica Ltd., 5 Chiltern Close, Cardiff CF4 5DL, United Kingdom.

Rosa del Carmen Mateos (343), *Instituto de Investigaciones Biomédicas, Universidad Nacional Autónoma de México, México, D.F., 04510, México*

A. J. McCarthy[6] (521), *Postgraduate School of Biological Sciences, University of Bradford, Bradford BD7 1DP, United Kingdom*

Norvel M. McClung (251), *Department of Biology, University of South Florida, Tampa, Florida 33620*

Emma I. Melendro (119), *Departamento de Inmunología y Reumatología, Instituto Nacional de la Nutrición, México, D.F., México*

S. Mendelovitz (357), *Department of Microbiology, George S. Wise Faculty of Life Sciences, Tel Aviv University, Tel Aviv, Israel 69978*

D. E. Minnikin (583), *Department of Organic Chemistry, The University, Newcastle upon Tyne NE1 7RU, United Kingdom*

A. M. Mortimer (537), *Department of Botany, University of Liverpool, Liverpool L69 3BX, United Kingdom*

I. B. Naumova (229), *Biology Department, N. V. Lomonosov State University, Moscow, USSR*

Hiroshi Ogawara (215), *Department of Biochemistry, Meiji College of Pharmacy, Tokyo 154, Japan*

Satoshi Omura (367), *School of Pharmaceutical Sciences, Kitasato University, and The Kitasato Institute, Tokyo 108, Japan*

Librado Ortiz-Ortiz (119), *Departamento de Inmunología, Instituto de Investigaciones Biomédicas, Universidad Nacional Autónoma de México, México, D.F., 04510, México*

A. C. Pier (135), *Division of Microbiology and Veterinary Medicine, University of Wyoming, Laramie, Wyoming 82071*

Burton M. Pogell (289), *Department of Medicinal Chemistry/Pharmacognosy, School of Pharmacy, University of Maryland, Baltimore, Maryland 21201*

James T. Powell (33), *Department of Basic Dental Sciences, College of Dentistry, University of Florida, Gainesville, Florida 32610*

H. Prauser (617), *Zentralinstitut für Mikrobiologie und Experimentelle Therapie, Akademie der Wissenschaften der DDR, DDR-6900 Jena, German Democratic Republic*

G. Pulverer (13, 161), *Institute of Hygiene, University of Cologne, D-5000 Cologne 41, Federal Republic of Germany*

Jack S. Remington (107), *Department of Immunology and Infectious Diseases, Research Institute, Palo Alto Medical Foundation, Palo Alto, California, and Division of Infectious Diseases, Department of Medicine, Stanford University School of Medicine, Stanford, California 94305*

G. W. Ross (553), *Regulatory Affairs Division, Glaxo Group Research Ltd., Greenford, Middlesex, United Kingdom*

[6]Present address: Department of Biochemistry and Applied Molecular Biology, University of Manchester, Institute of Science and Technology, Manchester M60 1QD, United Kingdom.

Julieta Rubio (343), *Instituto de Investigaciones Biomédicas, Universidad Nacional Autónoma de México, México, D.F., 04510, México*

M. J. Sackin (537), *Department of Microbiology, University of Leicester, Leicester LE1 7RH, United Kingdom*

J. A. Salas (179), *Departamento de Microbiología, Universidad de Oviedo, Oviedo, Spain*

Sergio Sanchez (343), *Instituto de Investigaciones Biomédicas, Universidad Nacional Autónoma de México, México, D.F., 04510 México*

K. P. Schaal[7] (13, 161, 505), *Institute of Hygiene, University of Cologne, D5000 Cologne 41, Federal Republic of Germany*

Karl-Heinz Schleifer (485), *Lehrstuhl für Mikrobiologie, Institut Botanik und Mikrobiologie Technischen Universität München, D-8000 Munich 2, Federal Republic of Germany*

Geraldine M. Schofield (505), *Microbiology Safety Reference Laboratory, PHLS Centre for Applied Microbiology and Research, Wiltshire SP4 OJG, United Kingdom*

Ed Seling (563), *Museum of Comparative Zoology, Harvard University, Cambridge, Massachusetts 02138*

J. D. Shannon[8] (381), *Departments of Biochemistry and of Microbiology and Immunology, Virginia Commonwealth University, Richmond, Virginia 23298*

A. S. Shashkov (229), *N. D. Zelinsky Institute of Organic Chemistry, Academy of Sciences of the USSR, Moscow, USSR*

Richard H. Skinner[9] (303), *Department of Biochemistry, University of Leicester, Leicester LE1 7RH, United Kingdom*

N. K. Skoblilova (229), *Institute of Biochemistry and Physiology of Microorganisms, Academy of Sciences of the USSR, Pustchino, USSR*

Richard A. Smucker (171, 409), *Chesapeake Biological Laboratory, Center for Environmental Estuarine Studies, University of Maryland, Solomons, Maryland 20688*

P. H. A. Sneath (537), *Department of Microbiology, University of Leicester, Leicester LE1 7RH, United Kingdom*

Erko Stackebrandt[10] (485), *Lehrstuhl für Mikrobiologie, Institut Botanik und Mikrobiologie Technischen Universität München, D-8000 Munich 2, Federal Republic of Germany*

J. E. Suárez (179), *Departamento de Microbiología, Universidad de Oviedo, Oviedo, Spain*

[7]Present address: Institute of Medical Microbiology and Immunology, University of Bonn, D-5300 Bonn 1, Federal Republic of Germany.

[8]Present address: Muscle Biology Group, University of Arizona, Tucson, Arizona 85721.

[9]Present address: Department of Biological and Medical Sciences, Brown University, Providence, Rhode Island 02912.

[10]Present address: Institut für Allgemeine Mikrobiologie der Universität, 2300 Kiel, Federal Republic of Germany.

S. M. Sutherland (315, 381), *Departments of Biochemistry and of Microbiology and Immunology, Virginia Commonwealth University, Richmond, Virginia 23298*

G. Szabó (197), *Institute of Biology, University Medical School, Debrecen, H-4012 Hungary*

Mark L. Tamplin (251), *Department of Medical Microbiology and Immunology, College of Medicine, University of South Florida, Tampa, Florida 33620*

Yoshitake Tanaka (367), *School of Pharmaceutical Sciences, Kitasato University, and The Kitasato Institute, Tokyo 108, Japan*

Jill Thompson (303), *Department of Biochemistry, University of Leicester, Leicester LE1 7RH, United Kingdom*

L. Trón (197), *Institute of Biophysics, University Medical School, Debrecen, H-4012 Hungary*

Thomas Troost (325), *School of Medicine, Georgetown University, Washington, D.C. 20057*

G. Valu (197), *Institute of Biology, University Medical School, Debrecen, H-4012 Hungary*

J. C. Vickers (537, 553), *Department of Botany, University of Liverpool, Liverpool L69 3BX, United Kingdom*

S. Vitális (197), *Institute of Biology, University Medical School, Debrecen, H-4012 Hungary*

Gernot Vobis (423), *Fachbereich Biologie-Botanik, Philipps-Universität, D-3550 Marburg, Federal Republic of Germany*

E. M. H. Wellington (537), *Department of Biology, Liverpool Polytechnic, Liverpool, United Kingdom*

R. P. White (597), *Department of Statistics, Rothamsted Experimental Station, Harpenden, United Kingdom*

S. T. Williams (473, 537, 553), *Department of Botany, University of Liverpool, Liverpool L69 3BX, United Kingdom*

PREFACE

The study of actinomycetes is in the midst of a revolution. The introduction of modern scientific technology is reshaping our understanding not only of the biological versatility of actinomycetes but of their impact in the clinical sphere as well. The expansion of this area of microbiology makes it virtually impossible for even the most diligent microbiologist to remain abreast of important developments outside his or her area of particular interest.

In an effort to deal with this problem, several meetings concerned with actinomycetes have been held in different parts of the world. One of the first took place in México in 1970 during the X International Congress of Microbiology. International meetings on actinomycetes have also been held in the German Democratic Republic in 1967, in Venezuela in 1974, in Poland in 1976, and in the Federal Republic of Germany in 1979. The V International Symposium on Actinomycetes Biology was held in Oaxtepec, Morelos, México in mid-August, 1982. So that the major topics would be covered, we invited articles by active and respected scientists in each of the principal areas of actinomycetes research.

The proceedings could not have taken place without the cooperation of the Universidad Autónoma Metropolitana (México), who made this publication possible, and the invaluable support of the Consejo Nacional de Ciencia y Tecnologia (CONACYT). We would like to thank Editora Tipográfica, S.A. Mexico, D. F., México, for the preparation of copy for this work.

PATHOGENESIS OF *ACTINOMYCES ISRAELII* INFECTIONS

G. H. W. Bowden

Department of Oral Biology
Faculty of Dentistry
The University of Manitoba
Winnipeg, Manitoba, Canada

I. INTRODUCTION

The significance of infections which include the pathogen *Actinomyces israelii* diminished considerably with the advent of antibiotics. Cope (1938) and Colebrook (1921) recorded extensive and fatal infections with *A. israelii*; the disease was still considered prevalent in 1950 (Holm, 1950, 1951). However, more recently, extensive infections have become more unusual although 525 fatal cases were recorded in the U. S. A. between 1949 and 1969 (Slack and Gerencser, 1975). Questions on the role of *A. israelii* in the chronic mixed infections typical of actinomycosis have been raised since the earliest examinations of the disease (Naeslund, 1931). Perhaps the most significant of the examinations of the roles of mixed bacteria in actinomycotic lesions were those carried out by Holm (1950, 1951), in which he noted that none of the cultures from over 600 closed actinomycotic lesions was pure (Holm, 1950). The cultures he made from abdominal lesions included coliforms and other Gram-negative rods, whereas thoracic and pulmonary lesions contained *Actinobacillus actinomycetem comitans* and "corroding bacilli". These latter two organisms should be noted as it is likely that they were of oral origin. A more recent source of actinomycotic infections has been associated with the use of intrauterine devices (Barnham *et al.*, 1978).

Over the years, despite many attempts to use different animals to reproduce the disease experimentally, variable results have been obtained (Brown and Von Lichtenberg, 1970; George and Coleman, 1970; Lord and Trevett, 1936; Naeslund, 1931). The most commonly used animal is the mouse (Brown and

ISBN 0-12-528620-1

Lichtenberg, 1970; George and Coleman, 1970), injected intraperitoneally with viable cells of *A. israelii*. Despite the known mixed nature of the infection and although some mixed infections have been studied (Grenier and Mayrand, 1982; Jordan and Kelly, 1982), there is no information on the effects of other bacteria on the pathogenicity of *A. israelii* in the animal models.

II. THE SOURCE OF *A. ISRAELII* IN INFECTION

To date, with the exception of *A. humiferus* strains, *Actinomyces* species have been isolated only from the mouth of man and animals (Bowden and Hardie, 1973; Slack and Gerencser, 1975). This would suggest that those strains causing infections originate in the mouth. It may also be considered significant that *Arachnia propionica*, a pathogen which produces lesions similar to those of *A. israelii* is also a component of the human oral flora (Bowden and Hardie, 1978).

Actinomycotic lesions can occur in any area of the body (Slack and Gerencser, 1975) and only those lesions resulting from human bites (Colebrook, 1920; Cope, 1938) or oral trauma might be thought to have a direct connection with the oral cavity. However, transient bacteremias are known to result from simple oral operations and tooth brushing (Hockett *et al.*, 1977) and the number and extent of the bacteremias increase directly as the oral hygiene of the subject decreases (Silver *et al.*, 1977). *Actinomyces* have been isolated as components of these transient bacteremias (Appelman *et al.*, 1982; Silver *et al.*, 1977). The presence of the other bacteria, *A. actinomycetem comitans* and corroding bacilli (*E. corrodens*) (Kilian and Schiott, 1975), in lesions support the concept of blood-borne infections by oral organisms. However, coliforms are not common inhabitants of the mouth (Hardie and Bowden, 1974) and oral contamination of the blood stream would not explain the presence of coliforms in the abdominal lesions cultured by Holm (1950). It could be that bacteremias or direct contamination from the gut are the source of bacteria from the lower intestinal tract in abdominal actinomycotic lesions. A potential means of infection in pulmonary actinomycosis is inhalation (Slack and Gerencser, 1975).

The conditions in the tissues which predispose the localization and growth of *Actinomyces* and any accompanying microorganisms are not understood. Any information that we have must be drawn from limited animal studies.

III. FACTORS PREDISPOSING ACTINOMYCOTIC LESIONS

Perhaps the most easily understood of actinomycotic lesions are those which develop in the cervico-facial region. These lesions and those associated with oral trauma, such as tooth extraction or periapical infection, can be explained as the result of contamination from the oral cavity. The factors which are responsible for causing localization and extension of actinomycotic lesions elsewhere in the body are not known. Animals studies (Benbehani *et al.*, 1979; Bowden *et al.*, 1981; George and Coleman, 1970) have shown that, in most pure infections, *A. israelii* cannot be considered a virulent pathogen since infected animals seldom die, the lesions resolve, and it is uncommon for the infection to spread from the site of inoculation. In order to establish the disease in animals, workers have injected the organism with mucin (Gale and Waldron, 1955; Meyer and Verges, 1950) or as granules (Bowden *et al.*, 1981; Brown and Von Lichtenberg, 1970). Brown and Von Lichtenberg (1970) considered that the age of the mouse and the injection of whole granules of cells were significant factors in the pathogenesis. However, the large doses of cells given to animals cannot be compared to the small inoculum which would occur as the result of a bacteremia in man. It may be that there must be some irritation or inflammation present (Miller *et al.*, 1978) in order to allow *A. israelii* to establish; the association of actinomycotic lesions with intrauterine devices may support this concept. In the absence of any evidence on the mechanism of establishment of *A. israelii*, one must assume that fragments of dental plaque containing *A. israelii* which enter the blood stream have the potential for causing tissue damage elsewhere in the body. Once the organism becomes established, it produces a lesion typical of actinomycosis. In this respect the observations of Sims (1974) are relevant. Sims suggests that, as the presence of *A. israelli* in a lesion is diagnostic of actinomycosis, any lesion which may become infected with *A. israelii* may be diagnosed as actinomycotic. Thus, some actinomycotic lesions may have been initiated by more virulent pathogens and only subsequently infected with *A. israelii.*

A. The Pathogenic Potential of A. israelii

As stated above, on the basis of animal experiments, *A. israelii* cannot be considered a highly virulent pathogen. Mice infected for up to 122 days will recover and show no distress during the infection despite some loss in weight (Bowden and Pettigrew, 1982; Bowden *et al.*, 1981). No attempts have been made to determine minimal infective doses of *A. israelii* in the animal models. Dosage is often quoted in terms of a standard optical density of a suspension of the organism in saline. More recent studies (Bowden and Pettigrew, 1982; Bowden *et al.*, 1981) have used as many as 22×10^6 viable cells to establish an intraperitoneal infection in mice, although 4.7×10^6 viable cells can produce lesions in the liver. As the death of an experimental animal from intraperitoneal actinomycosis is unusual, calculation of an LD_{50} is difficult. However, Beaman, Gershwin, and Maslam (1979) were able to infect mice by giving *A. israelii* cells intravenously and by inhalation. Beaman *et al.* (1979) quote an LD_{50} of 6×10^8 (CFU) for mice injected intravenously and an LD_{50} (CFU) of 1×10^9 for mice infected intranasally.

TABLE I. Foot Pad Swelling of Mice Injected with Components of A. israelii

Fraction	*Foot pad swelling (mm)*[a]	
	24 h	*48 h*
Whole cell sonicate	*0.64 ± 0.08*	*0.10 ± 0.07*
Washed cell wall	*0.80 ± 0.12*	*0.30 ± 0.23*
Pronase treated wall	*0.60 ± 0.09*	*0.40 ± 0.05*
Purified wall carbohydrate	*0.40 ± 0.10*	*0.30 ± 0.10*

[a]*Mean of six animals ± standard deviation.*

Whether *A. israelii* produces any toxic or aggressive agents is not known and there is no evidence that the extension of actinomycotic lesions depends on the production of toxins. However, sonicates of *A. viscosus* have been shown to stimulate acute inflammation in the mouse footpad model (Engel *et al.*, 1976); similar effects can be produced by either sonicates of or washed cell walls from *A. israelii*. Recent results of footpad swelling tests in control mice showed that *A. israelii* cell walls washed with water and walls treated with Pronase produce inflammatory responses similar to a whole sonicate (Table I). As would be expected with acute inflammation, the reaction decrease at 48 h. This is in contrast to purified cell wall carbohydrate (Bowden and Fillery, 1978) from *A. israelii*, which produces no inflammation. Thus, it is possible that components of the cell wall of *A. israelii* may contribute to an inflammatory response in infected tissues.

B. *The Pathogenicity of Other Organisms Associated with* A. israelii, Actinobacillus, *and* Bacteroides

An organism which has a history of association with *A. israelii* is *A. actinomycetem comitans* (King and Tatum, 1962; Klinger, 1912). Until recently, there were relatively few studies of this species (Kilian and Schiott, 1975; Pulverer and Ko, 1970, 1971). However, studies into the pathogenesis of a specific type of periodontal disease, juvenile periodontis, have shown that *A. actinomycetem comitans* can be a significant pathogen (Slots *et al.*, 1980; Tanner *et al.*, 1979). In particular, the organisms can produce a potent leucotoxin (Taichman and Wilton, 1981; Taichman *et al.*, 1980; Tsai *et al.*, 1979) and have an endotoxic component (Kiley and Holt, 1980). Moreover, there is evidence that strains of *A. actinomycetem comitans* can cause suppression of T- and B-cell responses to mitogens (Shenker and Taichman, 1982). Thus, *A. actinomycetem comitans* may contribute to the extension and persistence of an actinomycotic lesion in which it is a component of the mixed infection.

A second genus which may be found in actinomycotic lesions is *Bacteroides*. The taxonomy of many members of this genus has undergone changes recently (Finegold and Barnes, 1977). The black pigmented strains have been shown to represent a range of different organisms, with the establishment of two asaccharolytic species (*B. asaccharolyticus* and *B. gingivalis*) from man (Coykendall *et al.*, 1980; Van Steenbergen *et al.*, 1981); other pigmented *Bacteroides* can be found in association with animals (Coykendall *et al.*, 1980; Slots and Genco, 1980). The potential pathogenicity of the pigmented *Bacteroides* (Sundqvist *et al.*, 1979; Van Steenbergen *et al.*, 1982) is well known, as they often occur in mixed anaerobic infections without *A. israelii*. Mayrand and McBride (1980) have indicated that succinate could be an important factor in mixed infections. These authors showed that in such infections the requirement of *B. asaccharolyticus* for succinate could be met by including *Klebsiella pneumoniae* in the inoculum. *A. israelii* produces quantities of succinate as an end-product of metabolism and could fulfill this role in mixed infections. In further studies of mixed infections, Grenier and Mayrand (1982) have shown that mixtures of bacteria which include *A. israelii* can produce transmissible infections in the guinea pig. However, *A. israelii* does not seem to be a particularly significant component of the mixtures. Although *A. israelii* with E. *nucleatum* and *B. gingivalis* produced lesions in four of four guinea pigs, the most virulent mixture was *B. gingivalis, C. ochracea, E. saburreum*, and *E. nucleatum*. A mixture of *B. gingivalis, E. saburreum*, and *A. israelii* was noninfective. Despite the apparent lack of an absolute requirement for *A. israelii* in infections in this study, the establishment of *A. israelii* as a component of a mixed infection in experimental animals does provide a basis for further study of *A. israelii* infections.

Other bacteria, such as the coliforms, which are associated with actinomycotic lesions obviously can play a pathogenic role as they are common pathogens in other situations. It is perhaps of more interest to decide whether the pathogens mentioned above augment the pathogenic potential of *A. israelii* or play a major role in the virulent bacterial community in actinomycotic lesions.

IV. POTENTIAL INTERRELATIONSHIPS IN ACTINOMYCOTIC LESIONS

Although there is almost no data on the pathogenesis of mixed infections with *A. israelii* in animals models, certain potential relationships can be considered. A possible role for succinate was mentioned above and it does seem very likely than a symbiotic relationship could exist between *A. israelii* and some *Bacteroides* in lesions.

Perhaps the most interesting of the potential interrelationships is that of *A. israelii* providing "protection" for other bacteria within the lesion. There is some recent data on the relationship of *A. israelii*, *Eikenella corrodens*, and *A. actinomycetem comitans* in the mouse model (Jordan and Kelly, 1982). Mice infected with either *Actinobacillus* or *Eikenella* alone were able to clear the organism rapidly. However, when the mice were infected with a mixture of *A. israelii* and either *Actinobacillus* or *Eikenella*, both organisms persisted, and cells of the *Eikenella* or *Actinobacillus* were present in the granules and lesions together with *A. israelii*. Thus, these initial studies suggest that *A. israelii* may enable other bacteria to survive host responses which would normally eliminate them from the body.

Another aspect of "protection" which may be significant is the possibility that A. *israelii* may cause some modification of the host immune response.

A. Modification of the Immune Response

Relatively early studies on the pathogenesis of actinomycosis considered the possibility of immune hypersensitivity (Slack and Gerencser, 1975). Emmons (1938) and Mathieson *et al.* (1935) could not obtain positive skin reactions in sensitized animals, in infected experimental animals, or interestingly, in patients with actinomycosis. However, Mathieson *et al.* (1935) did detect skin responses in normal subjects. These results suggested that in cases of actinomycosis there may be a reduction of hypersensitivity responses. Moreover, in a note in their book, Slack and Gerencser (1975) state that they had no success with skin tests of a variety of *A. israelii* extracts in infected rabbits and guinea pigs. Despite the reported lack of skin responses in actinomycosis patients, humoral responses often have been demonstrated in patients with extensive lesions (Colebrook, 1920; Georg *et al.*, 1968). The most recent data on the presence of antibodies in actinomycotic patients is that of Holmberg and his co-workers (Holmberg, 1981; Holmberg *et al.* 1975a,b). These workers demonstrated precipitating antibody to several antigens in patients sera, using crossed immunoelectrophoresis with a standard antigen extract. As actinomycosis progresses slowly in man, it seems that antibody alone is not effective in resolving the lesions.

The early studies on patients with actinomycosis, demonstrating negative skin reactions, could suggest that there was some suppression of the immune system. It seems unlikely that antibody responses are suppressed, in view of the results mentioned above. This leaves the possibility of suppression or interference with the cell-mediated response. There is some evidence that extracts of *A. israelii* can suppress the responses of lymphocytes to mitogens (Miller *et al.*, 1978). Another piece of evidence relating to an effect of an *Actinomyces* on cell-mediated responses is provided by the study of Burkhardt *et al.* (1981) on oral colonization of rats by animal strains of *A. viscosus*. *In vitro* lymphocyte transformation tests were carried out by using cells from three groups of germ free rats: 1) rats immunized with *A. viscosus*, 2) rats immunized with *A. viscosus* and colonized orally with *A. viscosus*, 3) non-immunized rats colonized orally with *A. viscosus*. The immunized rats colonized with *A. viscosus* showed a significant reduction in lymphocyte response when compared to the non-colonized, immunized controls. This reduction persisted for up to 28 days after oral colonization, but by 100 days, the responses of the colonized rats were equivalent to those of the immunized rats. One reason suggested for this effect was that sensitized cells were depleted from the spleen to the site of inflammation in the oral cavity. Fitzgerald and Birdsell (1982) have also shown that oral colonization by *A. viscosus* can modify the response of splenocytes to bacterial antigens, LPS, and Con A. A similar situation could occur in patients with extensive actinomycotic lesions. The major component of sensitized cells may be located in association with the lesions or the organisms may modify the responses of host lymphocytes. Some recent results from our laboratory suggest that some reduction in the cell-mediated immune response may occur early in the experimental infection using the mouse model of actinomycosis.

B. A Reduction in the Footpad Swelling Response in the Mouse Model of Actinomycosis

Some initial results suggest that injection of *A. israelii* cells intraperitoneally may reduce the ability of mice to respond in footpad swelling tests early in the infection. Groups of six C57BL/6 mice weighing 15 gm were infected with a mean of 19.0 X 10^6 cells of *A. israelii*, as previously described (Bowden *et al.*, 1981). However, in addition to these mice which were injected in the peritoneum, other groups of mice were injected in the axilla with *A. israelii*. Three antigens were used to test for responses: 1) Pronase treated cell walls (Bowden *et al.*, 1981); 2) purified cell wall carbohydrate (Bowden and Fillery, 1978); and 3) a protein fraction of an *A. israelii* acid extract (Bowden *et al.*, 1976).

In one experiment, mice were infected in the peritoneum or the axilla for 56 and 100 days and tested at each time period by injection of 10 μg of antigen into the footpad. The mean specific increase (footpad swelling test-control) at 24 and 48 h was calculated (Bowden *et al.*, 1981). Table II shows the results of footpad swelling tests with antigens from *A. israelii* at the two time periods. The mean specific increases (swelling test-control) are shown in Table III.

Following the measurement of foot swelling at 48 h, the mice were bled, post-mortem examinations were made, and the footpads taken for histology. In some groups, a mouse was sacrificed at 24 h to determine the nature of the early footpad infiltrate. Antibodies in the sera from infected and control mice were measured by haemagglutination. Passive haemagglutination was used to

TABLE II. Footpad Swelling Induced by Different Antigens in Control Mice and Mice Infected with A. israelii

Injection site	*Time of experiments (days)*	*Footpad swelling (mm) resulting from antigen*		
		Pronase cell wall	*Acid extracted protein*	*Cell wall carbohydrate*
Peritoneum	*56*	*0.57 ± 0.12*	*0.38 ± 0.06*	*0.36 ± 0.07*
	100	*0.34 ± 0.1*	*0.77 ± 0.13*	*0.18 ± 0.08*
Axilla	*56*	*0.73 ± 0.08*	*0.63 ± 0.2*	*0.45 ± 0.09*
	100	*0.37 ± 0.07*	*0.63 ± 0.17*	*0.31 ± 0.06*
Control	*56*	*0.3 ± 0.07*	*0.17 ± 0.08*	*0.30 ± 0.12*
	100	*0.01 ± 0.01*	*0.16 ± 0.2*	*0.02 ± 0.04*

measure antibody to a steroyl derivative of cell wall carbohydrate as previously described (Bowden *et al.*, 1981). Tanned cell haemagglutination (Thorley *et al.*, 1975) was used to detect antibody to the acid extract protein. The antibody levels detected in mouse sera from the various groups is shown in Table IV.

The results of footpad swelling tests for Pronase cell wall and acid extract protein (Table III) showed that the response at 56 days by mice infected in the peritoneum was significantly reduced compared to those infected in the axilla (Pronase-cell wall, $p < 0.05$; acid extracted protein, $p < 0.01$). Histological examination of foot pads from the mice infected in the peritoneum for 56 days showed little infiltrate after 48 h. However, after 100 days of infection, the mice injected in the peritoneum or in the axilla responded equivalently. Cell wall carbohydrate produced little infiltrate in any of the footpads. All the mice produced antibody to acid extract protein at 56 days although the titers were higher after 100 days (Table IV). Cell wall carbohydrate produced low titers,

TABLE III. Mean Specific Increase (△ Footpad Swelling) in Test Mice Infected with A. israelii

Injection site	Time of experiments	△ Footpad swelling resulting from antigen		
		Pronase cell wall	Acid extracted protein	Cell wall carbohydrate
Peritoneum	56	0.27	0.21	0.06
	100	0.33[a]	0.61[a]	0.16
Axilla	56	0.4[b]	0.46[c]	0.15
	100	0.34[a]	0.46[a]	0.28[a]

[a]*Significant difference from controls ($p < 0.01$)*
[b]*Difference significant ($p < 0.05$)*
[c]*Difference significant ($p < 0.01$).*

with some mice giving a negative response. In one sense, these results at 56 days mimic those described for patients with extensive actinomycotic lesions by Mathieson *et al.* (1935). One of the differences between disease in man and experimental infection in mice relates to the period of extension of the lesions. In the mouse model, the intraperitoneal lesions are extensive early in the infection and resolve slowly (Bowden *et al.*, 1981; Slack and Gerencser, 1975). In man, the virulent disease follows the reverse pattern with extension of the lesions with time.

TABLE IV. Antibody Levels to Wall Carbohydrate and Acid-extracted Protein in the Serum from Control Mice and Mice Infected with A. israelii

Injection site	Time of experiment (days)	Antibody titer to antigen	
		Acid extracted protein	Cell wall carbohydrate
Peritoneum	56	30 (10- 80)[a]	12 (0-20)
	100	160 (80-320)	40 (10-80)
Axilla	56	64 (40-160)	20 (0-10)
	100	120 (80-150)[a]	40 (0-80)
Control	56	0	6 (0-10)
	100	0	8 (0-10)

[a]*Reciprocal of the mean (range).*

V. SUMMARY

A. israelii is found in association with a variety of other bacteria in actinomycotic lesions. It seems most likely that *A. israelii* originates from the oral cavity, as this is its natural habitat where it is a component of the normal oral flora. *A. israelii* probably enters the body as one of the components of transient bacteremias which occur after even the simplest oral cleansing, such as tooth brushing. In many cases, the organisms associated with *A. israelii* in the lesions are of oral origin *e. g.*, *A. actinomycetem comitans*, corroding Gram-negative rods, and *Bacteroides* species.

In experimental animal infections, *A. israelii* is sufficiently virulent to produce lesions but fatal infections are rare. Thus, *A. israelii* cannot be considered highly virulent for animals and this probably holds true for man. It is possible that the other bacteria in the lesion might augment the virulence of *A. israelii*. *A. actinomycetem comitans* produces a leucotoxin and a potent endotoxin; the pathogenic potential of members of the genus *Bacteroides* is well known. Some interactions between the organisms in actinomycotic lesions can be proposed; in particular, *A. israelii* may provide succinate which has been shown to be important in mixed infections with *B. gingivalis*, an oral black-pigmented *Bacteroides*. A further significant aspect may involve some modification or suppression of the host response caused by extensive infections with *A. israelii*. Moreover, *A. actinomycetem comitans* has been shown to reduce the responses of lymphocytes to mitogens. Some recent results on pure infections with *A. israelii* in mice have suggested that the ability to mount a cell-mediated response may be depressed by injection of cells of *A. israelii* into the peritoneum. These results could explain the negative skin responses to *Actinomyces* of patients with extensive actinomycosis, which were described almost fifty years ago. Extensive actinomycotic lesions may reduce the ability of the host to mount a cell-mediated response although humoral responses are not so affected. In this regard, the use of autologous or mixed vaccines (Slack and Gerencser, 1975) which promote cellular activity and the production of pus (Buchs, 1963) may involve some stimulation of the cellular immune system. Almost nothing is known of the significance of the role played by *A. israelii* in actinomycosis or of the interrelationships between the organisms in this chronic, mixed infection. Studies are just beginning on the immunology of these infections, stimulated in part by the tremendous increase in interest in those oral bacteria which may be involved in periodontal disease. It is to be hoped that, in future years, we will be better able to explain the significance of the involvement of members of the normal oral flora in actinomycosis.

REFERENCES

Appelman, M. D., Sutter, V. L., and Sims, T. N. *J. Periodontol. 53:*319 (1982).
Barnham, M., Burton, A. C., and Copeland, P., *Br. Med. J. 2:*719 (1978).
Beaman, B. L., Gershwin, M. E., and Maslan, S., *Infect. Immun. 24:*583 (1979).
Benbehani, M. J., Jordan, H. V., and Heeley, J. D., *J. Dent. Res. 58:*A340 (1979).
Bowden, G. H., and Fillery, E. D., *in* "Secretory Immunity and Infection". (J. R. McGhee, J. Mesteky, and J. L. Babb, eds.), p. 685. Plenum Press, New York (1978).
Bowden, G. H., and Hardie, J. M., *in* "Actinomycetes Characteristics and Practical Importance". (G. Sykes, and F. A. Skinner, eds.), p. 277. Academic Press, London (1973).
Bowden, G. H., and Hardie, J. M., *in* "Coryneform Bacteria". (I. J. Bousefield, and A. G. Callery, eds.) p. 235. Academic Press, London (1978).
Bowden, G. H., and Pettigrew, N. M., *J. Dent. Res. 61:*309 (1982).
Bowden, G. H., Hardie, J. M., and Fillery, E. D., *J. Dent. Res. 55:*A192 (1976).
Bowden, G. H., Wilton, J. M. A., and Sutton, R., *in* "Actinomycetes" (K. Schaal, and G. Pulverer, eds.), p. 251. Gustav Fisher Verlag, Stuttgart (1981).
Brown, J. R., and Von Lichtenberg, F., *Arch. Pathol. 90:*391 (1970).
Buchs, H., *Dtsch. Zahnaerztl. Z. 18:*1060 (1963).
Burkhardt, J. J., Gargauf-Zollingen, R. R., Schmid, R., and Guggenheim, B., *Infect. Immun. 31:*971 (1981).
Colebrook, L., *Br. J. Exp. Pathol. 1:*197 (1920).
Colebrook, L., *Lancet 1:*893 (1921).
Cope, V. Z., "Actinomycosis". Oxford University Press (1938).
Coykendall, A. L., Kaczmarek, F. S., and Slots, J., *Int. J. Syst. Bacteriol. 30:*559 (1980).
Engel, D., Van Epps, D., and Clagett, J., *Infect. Immun. 14:*548 (1976).
Emmons, C. W., *Public Health Rep. 53:*1967 (1938).
Finegold, S. M., and Barnes, E. M., *Int. J. Syst. Bacteriol. 27:*388 (1977).
Fitzgerald, J. E., and Birdsell, D. C., *J. Periodontal Res. 17:*237 (1982).
Gale, D., and Waldron, C. A., *J. Infect. Dis. 97:*251 (1955).
Georg, L. K., and Coleman, R. M., *in* "The Actinomycetales" (H. Prauser, ed.), p. 35. Gustav Fischer, Jena (1970).
Georg, L. K., Coleman, R. M., and Brown, J. M., *J. Immunol. 100:*1288 (1968).
Grenier, D., and Mayrand, D., *in:* "32nd Meeting Canadian Society of Microbiologists" Abstract 1N11P, Quebec City, (1982).
Hardie, J. M., and Bowden, G. H., *in*: "The Normal Microbial Flora of Man". (F. A. Skinner, and J. G. Carr, eds.), p. 47. Academic Press, London (1974).
Hockett, R. N., Loesche, W. J., and Sodeman, T. M., *Arch. Oral Biol. 22:*91 (1977).
Holm, P., *Acta Pathol. Microbiol. Scand. 27:*736 (1950).
Holm, P., *Acta Pathol. Microbiol Scand. 28:*391 (1951).
Holmberg, K., *in* "Actinomycetes" (K. Schaal, and G. Pulverer, eds.), p. 259. Gustav Fischer Verlag, Stuttgart (1981).
Holmberg, K. H., Nord, C. E., and Wadstrom, T., *Infect. Immun. 12:*387 (1975a).
Holmberg, K. H., Nord, C. E., and Wadstrom, T., *Infect. Immun. 12:*398 (1975b).
Jordan, H. V., and Kelly, D. M., *J. Dent. Res. 61:*231 (1982).
Kiley, P., and Holt, S. C., *Infect. Immun. 30:*862 (1980).
Kilian, M., and Schiott, C. R., *Arch. Oral Biol. 20:*791 (1975).
King, E. O., and Tatum, H. W., *Infect. Immun. 30:*588 (1962).
Klinger, R., *Zentralbl. Bakteriol. Parasitenkd. Infecktionskr. Hyg. Abt. I. Orig. 62:*191 (1912).
Lord, F. T., and Trevett, D. T., *J. Infect. Dis. 58:*114 (1936).
Mathieson, D. R., Harrison, R., Hammond, C., and Henrici, A. T., Am. J. Hyg. *21:*405 (1935).
Mayrand, D., and McBride, B. C., *Infect. Immun. 27:*44 (1980).
Meyer, E., and Verges, P., *J. Lab. Clin. Med. 36:*667 (1950).
Miller, B. J., Wright, J. L., and Colquhoun, B. P. D., *Surg. Gynecol. Obstet. 146:*412 (1978).
Naeslund, C., *Acta Pathol. Microbiol. Scand. 8* (suppl. 6):1 (1931).
Pulverer, G., and Ko, H. L., *Appl. Microbiol. 20:*693 (1970).
Pulverer, G., and Ko, H. L., *Appl. Microbiol. 23:*207 (1971).
Shenker, B. J., and Taichman, N. S., *J. Dent. Res. 61:*333 (1982).

Silver, J. G., Martin, and McBride, B. C., *J. Clin. Periodontol. 4:*92 (1977).
Sims, W., *Br. J. Oral Surg. 12:*1 (1974).
Slack, J. M., and Gerencser, M. A., "Actinomyces, Filamentous Bacteria". Burgess Publishing Company, Minneapolis (1975).
Slots, J., and Genco, R. J., *Int. J. Syst. Bacteriol. 30:*82 (1980).
Slots, J., Reynolds, H. S., and Genco, R. J., *Infect. Immun. 29:*1013 (1980).
Sundqvist, G. K., Ekerbom, M. I., Larson, A. P., and Sjogren, U. T., *Infect. Immun. 25:*685 (1979).
Taichman, N. S., and Wilton, J. M. A., *Inflammation 5:*1 (1981).
Taichman, N. S., Dean, R. T., and Sanderson, C. J., *Infect. Immun. 28:*258 (1980).
Tanner, A. C. R., Haffer, C., Bratthall, G. T., Visconti, R. A., and Socransky, S. S., *J. Clin. Periodontol. 6:*278 (1979).
Thorley, D. J., Holmes, R. K., and Sanford, J. P., *Am. J. Epidemiol. 101:*438 (1975).
Tsai, C. C., McArthur, W. P., Baehni, P. C., Hammond, B. F., and Taichman, N. S., *Infect. Immun. 25:*427 (1979).
Van Steenbergen, T. J. M., Vlaanderen, C. A., and DeGraaf, J., *Int. J. Syst. Bacteriol. 31:*236 (1981).
Van Steenbergen, T. J. M., Kastelein, P., Touw, J. J. A., and DeGraaf, J., *J. Periodontal Res. 17:*41 (1982).

EPIDEMIOLOGIC, ETIOLOGIC, DIAGNOSTIC, AND THERAPEUTIC ASPECTS OF ENDOGENOUS ACTINOMYCETES INFECTIONS

K. P. Schaal
G. Pulverer

Institute of Hygiene
University of Cologne
Köln, F. R. G.

I. INTRODUCTION

Actinomycetes with a fermentative carbohydrate metabolism, which until recently were all classified in the family *Actinomycetaceae*, can cause a variety of human and animal diseases and impairments. The disorders induced in man include actinomycoses, lacrimal canaliculitis, uterine infections, periodontal disease, and caries. From a practical and social point of view, caries and periodontal disease are undoubtedly the most important impairments in which *Actinomycetaceae* may be etiologically involved. However, these maladies are very complex in both etiology and pathogenesis, and actinomycetes constitute only one link in the long chain of cause and effect which finally results in clinical symptoms. Therefore, their detailed discussion is beyond the scope of this paper.

Among the remaining disease entities mentioned above, actinomycosis is the only one which has a pronounced and potentially malignant invasive power and can easily become dangerous to health or even to life. Its clinical and pathogenetic features have been described elsewhere (Pulverer and Schaal, elsewhere in this volume). This paper deals with current knowledge and new findings on incidence, etiology, diagnosis, and treatment of actinomycotic infections. For comparison, a few data on lacrimal canaliculitis and uterine actinomycete affections have been included although our experience with these conditions is still rather limited.

BIOLOGICAL, BIOCHEMICAL,
AND BIOMEDICAL ASPECTS OF ACTINOMYCETES

ISBN 0-12-528620-1

II. INCIDENCE

Actinomycosis as well as lacrimal canaliculitis and actinomycete infections of the uterine cavity are endogenous infective processes incited by facultatively pathogenic actinomycetes which are commonly present on the human mucosal surfaces as members of the indigenous microflora. As such they are neither transmissible nor geographically limited.

Actinomycoses have been reported in almost every part of the world. However, their incidence appears to vary regionally, especially as far as cervicofacial forms are concerned. For instance, actinomycotic involvement of the face and neck seems to occur less frequently in the U.S.A. than in the European countries. This somewhat surprising fact may reflect differences in dental care and usage of antibiotics, but it may also be due to differing definitions of the disease.

In Europe, several calculations of morbidity rates of humans with actinomycoses have been published. On the basis of histological findings, Hemmes (1963) estimated an incidence of actinomycotic infections in the Netherlands which amounted to one case per 119,000 inhabitants per year. For the Cologne area of West Germany, Lentze (1969) reported a morbidity rate of one case per 83,000 Cologne residents per year between 1950 and 1969. Recently, we re-calculated the incidence of actinomycoses for the same region by using the data from material examined in our laboratory during the past ten years (Schaal, 1979). Inclusion of data from both acute and chronic processes led to an estimate of 1:40,000. The difference between the calculation of Lentze and of ours is probably not due to increasing morbidity but to the fact that Lentze excluded acute manifestations. Nevertheless, these data indicate that actinomycosis is not a rare disease since it is currently twice as frequent in the area of Cologne as are typhoid and paratyphoid fever together.

Lacrimal canaliculitis caused by fermentative actinomycetes is probably also quite common. However, as the removal of the lacrimal concretions which accompany or induce the inflammatory reaction and the local administration of antibiotics usually result in complete recovery, an etiological diagnosis is infrequently established. Therefore, it is very difficult to estimate how common the disease really is.

The same is true for the infection or colonization of the uterine cavity with actinomycetes. Gupta *et al.* (1976) observed that cervicovaginal smears from women wearing intrauterine contraceptive devices (IUD) or vaginal pessaries often contained actinomycetes. More recently, several other workers have confirmed these findings (Bhagavan and Gupta, 1978; Christ and Hajna, 1978), but is remains to be determined whether these actinomycetes represent a state of colonization rather than of infection and whether progressive actinomycoses can develop from such conditions.

Since 1981, we have examined 136 IUD's and cervicovaginal swabs for actinomycetes by using culture techniques. Twenty samples (14.7%) yielded growth of filamentous bacteria which all belonged to the genus *Actinomyces*. Other investigators have reported lower or higher percentages of positive results by using either immunofluorescence or cytological procedures for detecting the actinomycetes. Thus, further studies are needed to determine whether these differences are due to regional variation in the actinomycete colonization, to serological cross-reactions with other genital microbes, or to a relative insensitivity of the culture and cytological methods.

III. ETIOLOGY

The etiology of human actinomycoses is complex in two respects: i) Several actinomycete species which even belong to different genera of the family *Actinomycetaceae* may cause clinically indistinguishable lesions; ii) The microbial flora associated with actinomycotic processes almost always contain one or more bacterial species in addition to the causative actinomycete. The problem of the obligate synergistic mixed infection has been discussed in detail elsewhere (Pulverer and Schaal, elsewhere in this volume).

The spectrum of fermentative actinomycetes which may act as the primary pathogens in these multiple infections is possibly broader than had been previously thought (Table I). Among 1,022 clinical isolates obtained in our laboratory during the past ten years, nine different species were recovered. As one would expect, *Actinomyces israelii* was identified most frequently (82% of the isolates). This organism has been recognized as the predominant human pathogen since the first description of the species by Wolff and Israel (1891).

Other members of the *Actinomycetaceae* are comparatively rarely encountered in clinical specimens, provided that these have not been contaminated with mucosal secretions. In the material used in our study, which was derived chiefly from cervicofacial processes, *A. naeslundii, A. viscosus*, and *Arachnia propionica* were the most common "unusual" species; together they constituted about 10% of the clinical isolates. Pathogenicity has been claimed for all three organisms. We support this view because we have observed several typical cases of actinomycosis from which one of these species could be cultured repeatedly over a period of several weeks, or even months, until the lesion was cured. Aside from *A. bovis* which has never been proven to cause human infections, a few additional species may occasionally be detected in clinical specimens. Their etiological role as far as progressive infective processes are concerned remains to be definitely established although single case reports in the literature would indicate that they do not completely lack pathogenicity.

Of our isolates, 7% were not identifiable under routine conditions. Many of them were included in our recent numerical phenetic study (Schofield and Schaal, 1981) of the *Actinomycetaceae*. In this analysis, about one half of the primarily unidentified strains formed two well-defined clusters (clusters 4 and

TABLE I. Actinomycete Species Isolated from Human Actinomycotic Lesions

Species	Localization					Total
	Cervicofacial	*Thoracic*	*Abdominal*	*Extremities*	*Brain*	
Actinomyces israelii	*838 = 82.0%*	*3*	*3*	*1*	*1*	*846 = 81.9%*
Actinomyces naeslundii	*51 = 5.0%*			*2*		*53 = 5.1%*
Actinomyces viscosus	*19 = 1.9%*					*19 = 1.8%*
Actinomyces bovis	*0 = 0.0%*					*0*
Actinomyces odontolyticus	*4 = 0.4%*					*4 = 0.4%*
Arachnia propionica	*25 = 2.4%*					*25 = 2.4%*
Bifidobacterium eriksonii	*3 = 0.3%*				*1*	*4 = 0.4%*
Bacterionema matruchotii	*8 = 0.8%*					*8 = 0.8%*
Rothia dentocariosa	*2 = 0.2%*					*2 = 0.2%*
Not identified	*72 = 7.0%*					*72 = 7.0%*
Total no. of samples	*1022 = 100.0%*					*1033 = 100.0%*

14) which did not contain any reference strain and might represent new species. Thus, it appears that hitherto unknown pathogenic actinomycetes do exist. The other half of the clinical isolates which could not be identified under routine conditions were aberrant forms which joined the named clusters as satellites at low similarity levels.

A decision on the pathogenic role of some of the "unusual" actinomycete species was complicated by the finding that clinical materials occasionally contained two different fermentative actinomycetes species (Table II). Among 1,018 cases examined, 15 samples (1.5%) yielded growth of two morphologically distinct filamentous organisms. The predominant of the two was identified as *A. israelii*, the second one being either *A. naeslundii*, the second serovar of *A. israelii, A. viscosus* or *Bacterionema matruchotii*. In four cases, combinations of *A. naeslundii* plus *A. viscosus* and *A. naeslundii* plus *Rothia dentocariosa* were found. One may imagine that such a mixed actinomycete in-

TABLE II. Cervicofacial Actinomycotic Lesions from Which Two Different Actinomycetes Had Been Isolated

Combination of species/serovars	*Number of cases*
Actinomyces israelii, *Serovars I + II*	*2*
Actinomyces israelii + Actinomyces naeslundii	*3*
Actinomyces israelii + Actinomyces viscosus	*2*
Actinomyces israelii + Bacterionema matruchotii	*4*
Actinomyces naeslundii + Actinomyces viscosus	*3*
Actinomyces naeslundii + Rothia dentocariosa	*1*
Total number of cases examined/cases with two species	*1018/15 (=1.5%)*

fection is easily overlooked. Thus, it may happen that, in a case with typical clinical symptoms, only the unusual organism is identified, while *A. israelii* is missed. This would lead to the conclusion that *A. naeslundii, A. viscosus, B. matruchotii*, or *R. dentocariosa* are the specific pathogens in these infections, which is probably not the case.

Most tabulations, including sex ratios (Slack and Gerencser, 1975), conform with our findings (see Pulverer and Schaal, elsewhere in this volume) that actinomycoses are disproportionately distributed in males and females. The disease involves males about three times more often than females. The sex distribution of actinomycete species showed some interesting differences (Table III). In our material, only *A. israelii* exhibited this preference for male patients. All of the other members of the family *Actinomycetaceae* including *Ar. propionica* were nearly equally distributed between both sexes. If there should be hormonal influences which favor the development of actinomycoses, these seems to be relevant only to the classical *A. israelii* infection.

Our admittedly small sampling of material on lacrimal canaliculitis seemed to indicate that *Ar. propionica* was the most prevalent organism responsible for this form of disease (Table IV). Six of twelve lacrimal concretions which were examined by culture in our laboratory contained this species. From three samples *A. israelii* was isolated and three other species were represented only by single cultures.

TABLE III. Sex Distribution of Actinomycete Isolates

Species	*Male patients*	*Female patients*	*Ratio female/male*
Actinomyces israelii	596	203	1:2.9
Actinomyces naeslundii	31	25	1:1.2
Actinomyces viscosus	13	16	1:0.8
Actinomyces odontolyticus	1	5	1:0.2
Arachnia propionica	13	15	1:0.9
Bifidobacterium eriksonii	1	1	1:1
Bacterionema matruchotii	4	4	1:1
Rothia dentocariosa	1	1	1:1
Total ratio	660	270	1:2.4
Ratio of non-israelii infections	64	67	~ 1:1 (0.95)

Uterine specimens (IUD's) and cervicovaginal swabs showed a slight prevalence of *A. israelii* but *A. viscosus* was also cultured, but the number of positive samples was too small to assess the species distribution with sufficient certainty (Table V).

IV. DIAGNOSIS

The proper and early diagnosis of actinomycoses is of utmost medical importance because, among the diseases and impairments which may be induced by members of the family *Actinomycetaceae*, this condition is often not cured until an etiological diagnosis has been established and specific therapeutic measures have been employed. The other forms of disease in which these

TABLE IV. Actinomyces Species Isolated from Lacrimal Concretions

Species	*Number of cases positive for actinomycetes*	*Number of cases examined*
Actinomyces israelii	3	
Actinomyces naeslundii	1	
Actinomyces viscosus	1	
Actinomyces odontolyticus	1	
Arachnia propionica	6	
Total	12 = 67%	18 = 100%

TABLE V. Actinomyces *Species Isolated from Cervicovaginal Swabs and IUD's*

Species	*Number of specimens positive for actinomycetes*	*Number of specimens examined*
Actinomyces israelii	*11*	
Actinomyces viscosus	*8*	
Actinomyces israelii + Actinomyces viscosus	*1*	
Total	*20 = 14,7%*	*136 = 100%*

organisms may etiologically be involved are less severe and are predominantly amenable to symptomatic treatment.

In actinomycosis, typical clinical symptoms such as firm, painless swellings, abscess formation, and formation of draining sinus tracts usually develop late in the course of the disease and may mimic tuberculosis, carcinoma, or actinomycetoma. Histology also often fails to establish the diagnosis with sufficient certainty. Therefore, the early and reliable recognition of actinomycoses depends mainly on the results of bacteriological examinations.

Suitable clinical specimens for detecting actinomycetes in the bacteriological laboratory are abscess content, sinus discharge, bronchial secretions, or biopsy material. During sampling, care must be taken to avoid contamination of the specimens by mucosal bacteria. Therefore, liquid pus preferably should be collected by needle aspiration through the outer integument after thorough skin desinfection. Materials which have been collected by puncture or incision through the mucosa are almost always contamined with actinomycetes from the mucosal surface. This can easily be proven by comparing the culture results from samples obtained by extraoral and intraoral puncture or incision, respectively (Table VI). In specimens taken through the oral cavity, the isolation rate of *A. israelii* is much lower than that in those obtained through the outer integument. On the other hand, *A. naeslundii* and *A. viscosus* which prevail in the mouth flora are more often encountered in intraorally collected materials, indicating that they are common contaminants.

In cases of pulmonary actinomycosis, percutaneous transtracheal aspiration, direct lung puncture, or needle biopsy are the most dependable ways of obtaining suitable specimens as they bypass areas with normal actinomycete flora.

As facultative anaerobes, pathogenic *Actinomycetaceae* are usually quite resistant to oxygen contact, but may be inactivated when the samples dries in air, is cooled below 20°C, or suffers changes in pH. Thus, specimens should be kept at room temperature and transported and processed as quickly as possi-

TABLE VI. Influence of the Sampling Technique on the Species Composition of Isolates from Cervicofacial Lesions

Species	*Extraoral incision or puncture*	*Intraoral incision or puncture*
Actinomyces israelii	*838 = 82.0%*	*24 = 27.6%*
Actinomyces naeslundii	*51 = 5.0%*	*33 = 27.9%*
Actinomyces viscosus	*19 = 1.9%*	*15 = 17.2%*
Actinomyces odontolyticus	*4 = 0.4%*	*1 = 1.1%*
Arachnia propionica	*25 = 2.4%*	*0 = 0.0%*
Bifidobacterium eriksonii	*3 = 0.3%*	*0 = 0.0%*
Bacterionema matruchotii	*8 = 0.8%*	*1 = 1.1%*
Rothia dentocariosa	*2 = 0.2%*	*0 = 0.0%*
Not identified	*72 = 7.0%*	*13 = 14.9%*
Total	*1022 = 100.0%*	*87 = 100.0%*
Two species per specimen	*15 = 1.5%*	*6 = 7.4%*

ble. In addition, transport measures should be used which maintain anaerobic conditions.

In the laboratory, pus specimens should first be checked macroscopically for the presence of sulfur granules. These appear as yellowish to brownish or reddish particles which are about 1 mm or less in diameter and which may be so numerous that the discharge may resemble semolina soup (Lentze, 1969). To confirm the identity of suspected particles, it is mandatory to examine them also microscopically, first embedded in methylene blue solution under a cover slip and thereafter in Gram-stained smears. If immunofluorescence methods are directly applied to the crushed granule material or pus smears, it is often possible to identify immediately the causative actinomycete species present. However, antigenically aberrant strains or unusual serovars or species may be missed if immunofluorescence is used as the sole means of searching for actinomycetes.

Culture and identification of the fermentative actinomycetes are the most reliable ways for diagnosing actinomycotic infections, provided that carefully adapted culture techniques are employed. These include anaerobic to semianaerobic growth conditions and suitable media and incubation times. As none of the *Actinomycetaceae* is a strict anaerobe, sophisticated equipment is not needed to reduce the atmospheric oxygen tension. In our hands, Fortner's method which utilizes the reducing and CO_2-producing capacity of *Serratia marcescens* in co-culture with the actinomycetes on tightly sealed agar plates has proven especially useful (Schaal and Pulverer, 1981). Alternatively, anaerobic containers such as the Torbal or the GasPak (BBL) jars may be employed. However, as the more aerophilic strains may be suppressed in their highly anaerobic atmosphere, these jars should be used only in combinations with techniques which allow for a parallel culture in air with increased carbon dioxide.

For primary isolation and subsequent cultive of the *Actinomycetaceae* from clinical specimens, high quality general-purpose media are usually recommended. These include fluid thioglycolate broth supplemented with 0.1% (v/v) sterile rabbit serum, brain heart infusion broth, trypticase soy broth, brain heart infusion agar, trypticase soy agar, and Schaedler broth or agar. In our laboratory, the best results (Schaal and Pulverer, 1981) were obtained with the Tarozzi medium which simultaneously provides reduced growth conditions under a paraffin seal and with an agar medium (CC-medium) developed by Heinrich and Korth (1967) some years ago.

From each specimen, three to four different media, including a fluid medium, should be inoculated. These should be incubated at 36°C for up to 14 days and then checked for growth of filamentous or dense granular macrocolonies which also develop in fluid culture and which appear as small, snowball-like particles affixed to the wall of the test tube. When Fortner's method is applied, the plates can be examined microscopically every two days without disturbing the anaerobic atmosphere. This is especially useful because actinomycete colonies are often more typical in early growth stages (spider-like colonies). Jars should not be opened before 10 to 14 days of incubation because younger colonies are usually too small to be successfully subcultured, and repeated exposure to air of the growing organism often results in loss of the isolate.

After pure cultures have been obtained identification of the organism can be achieved by direct or indirect modifications of the immunofluorescence technique or by a set of physiological and chemotaxonomic tests (see Schaal, elsewhere in this volume).

V. THERAPY

Lacrimal canaliculitis and uterine inflammations usually heal spontaneously when the concretions or the IUD have been removed. Administration of antibiotics may be helpful, but is often not necessary.

The modern standard therapy of human actinomycoses consists of chemotherapeutic and surgical measures. The pathogenic *Actinomycetaceae* are susceptible to a wide variety of antibacterial drugs (Schaal, 1979; Schaal and Pape, 1980; Schaal *et al.*, 1979), especially to penicillins and cephalosporins. Nevertheless, not all of these substances are equally well suited for treating actinomycotic processes because both the actinomycetes and the concomitant bacteria must be considered when choosing the appropriate antibiotic. The latter may be resistant to β-lactam compounds and/or may produce β-lactamases, thereby rendering the therapeutic response insufficient. For this reason benzyl penicillin, although usually recommended as the drug of choice (Slack and Gerencsér, 1975), is not always able to cure the disease or to prevent relapses (Fig. 1). *Actinobacillus actinomycetemcomitans* and certain *Bacteroides* species which are often resistant to penicillin G may sustain the inflammation even after chemotherapeutic elimination of the actinomycetes or may protect the latter from the action of the drug. In our experience, aminopenicillins such as ampicillin or amoxycillin (Fig. 2) provide much better and more consistent therapeutic results. Only in cases in which *Bacteroides fragilis, B. thetaiotaomicron*, or resistant *Enterobacteriaceae* are etiologically involved, other modern antibiotics such as carbenicillins, mezlocillin (Fig. 3), cefoxitin (Fig. 4), or possibly lamoxactam may be more effective.

Alternatively, ampicillin or another β-lactam compound may be combined with metronidazole (Fig. 5) or clindamycin (Fig. 6). Metronidazole is highly active against strict anaerobes but does not inhibit any of the facultatively anaerobic actinomycetes. On the other hand, the activity spectrum of clindamycin includes all of the actinomycetes but only a few *Actinobacillus* strains. For this reason these substances should be administered only in combination with a drug which completes their range of action. Such combinations, possibly extended by an aminoglycoside, may be recommended especially for abdominal and thoracic infections which commonly contain β-lactamase producing *Bacteroidaceae*. Tetracyclines (Fig. 7) which have long been used as an alternative to penicillin G also show a good activity *in vitro* but seem to be less effective *in vivo*. Whenever possible chemotherapy should be complemented by surgery in order to remove infected tissue with insufficient blood supply and pus-filled sinus tracts.

Only in exceptional cases will this detailed therapeutic scheme not work. In these cases, the application of heterovaccines as described first by Colebrook (1921) and developed further by Lentze (1969) should be considered. Lentze's vaccine which contains inactivated cells of *A. israelii*, serovars I and II, *Ar.*

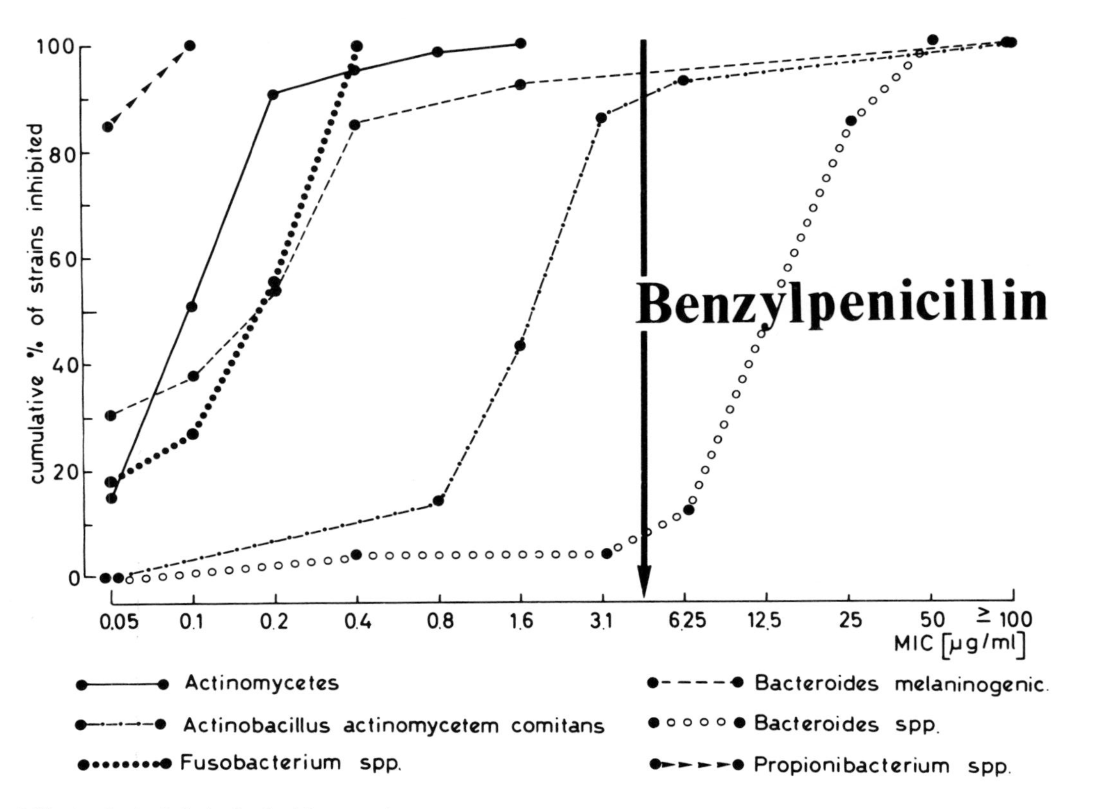

FIGURE 1. Susceptibility in vitro *of clinically significant actinomycetes and selected concomitant bacteria to penicillin G (agar dilution test; DST-medium (Oxoid CM 261); cumulative percentages of inhibited strains).*

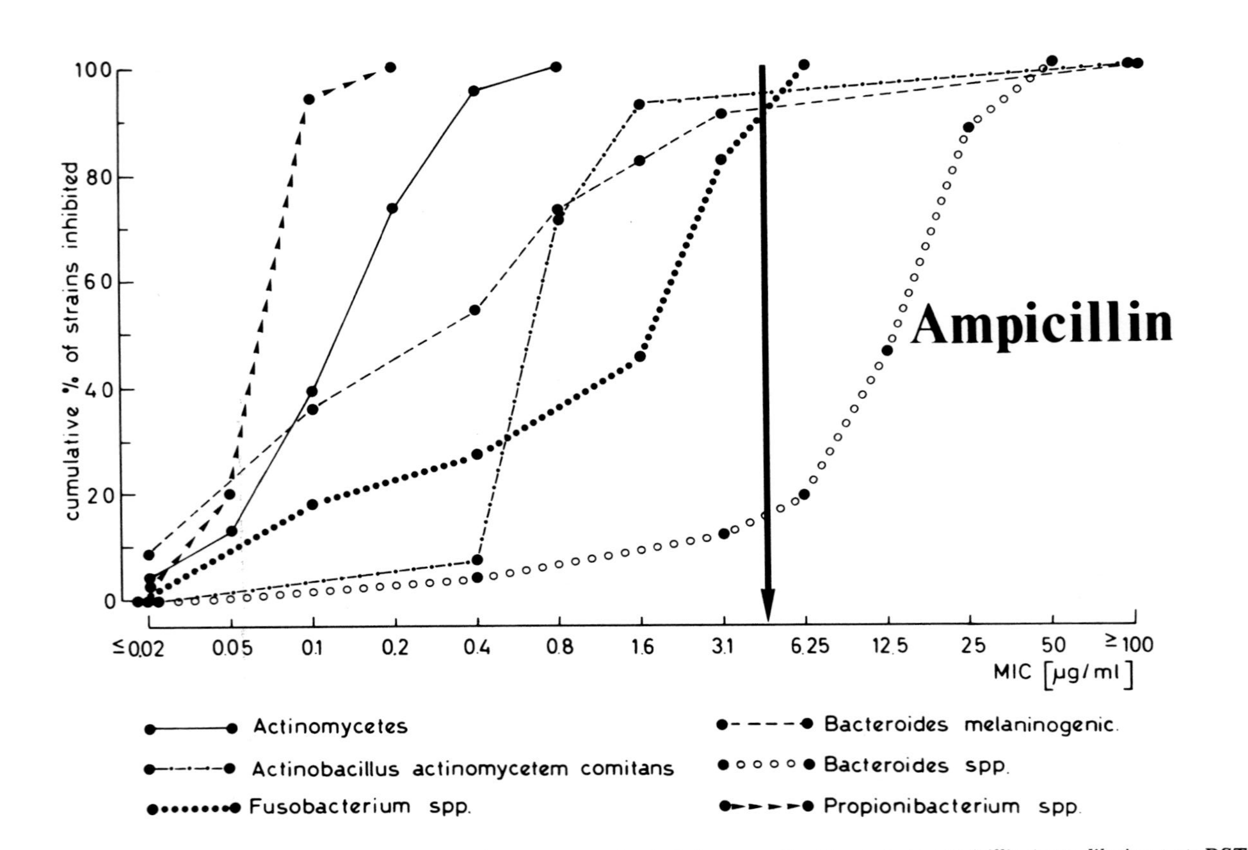

FIGURE 2. Susceptibility in vitro *of clinically significant actinomycetes and selected concomitant bacteria to ampicillin (agar dilution test; DST-medium (Oxoid CM 261); cumulative percentages of inhibited strains).*

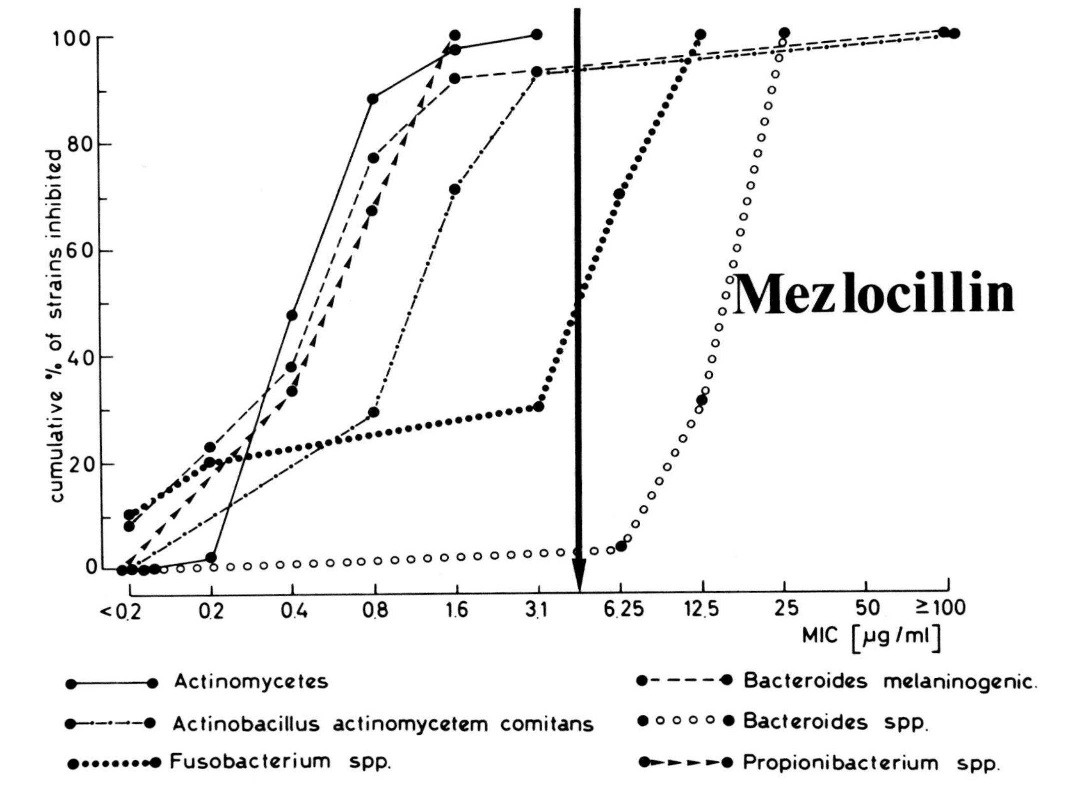

FIGURE 3. Susceptibility in vitro *of clinically significant actinomycetes and selected concomitant bacteria to mezlocillin (agar dilution test; DST-medium (Oxoid CM 261); cumulative percentages of inhibited strains).*

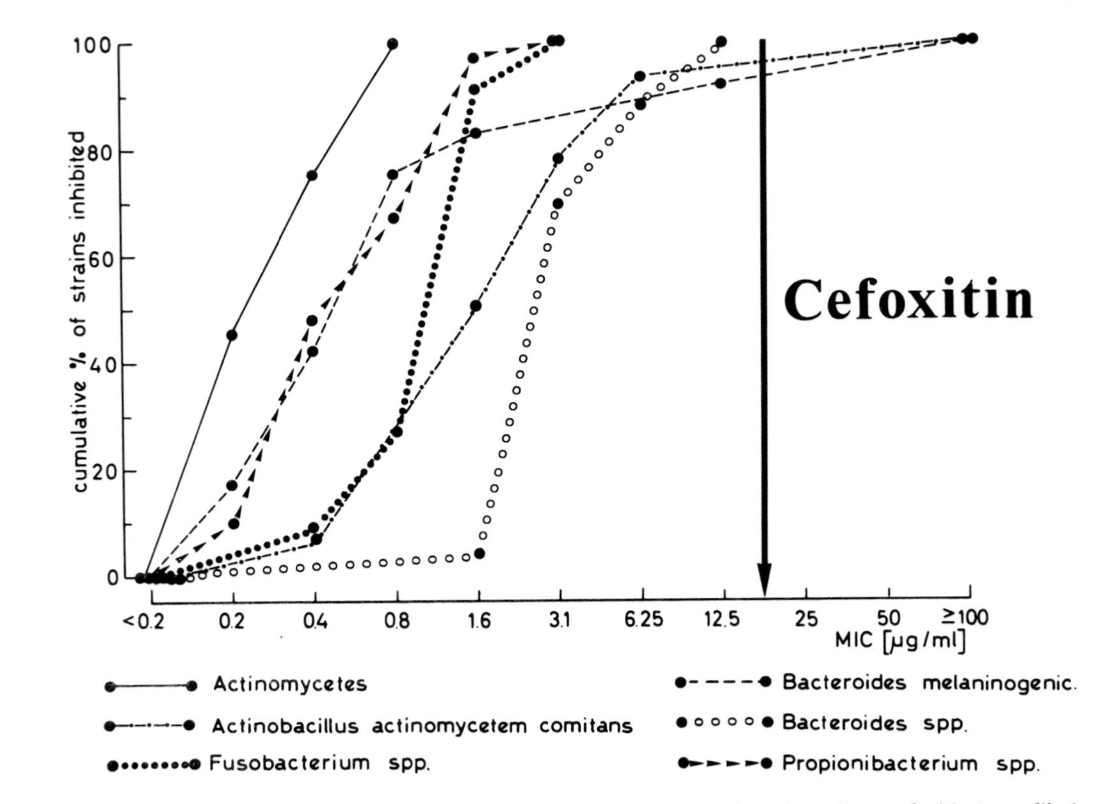

FIGURE 4. Susceptibility in vitro *of clinically significant actinomycetes and selected concomitant bacteria to cefoxitin (agar dilution test; DST-medium (Oxoid CM 261); cumulative percentages of inhibited strains).*

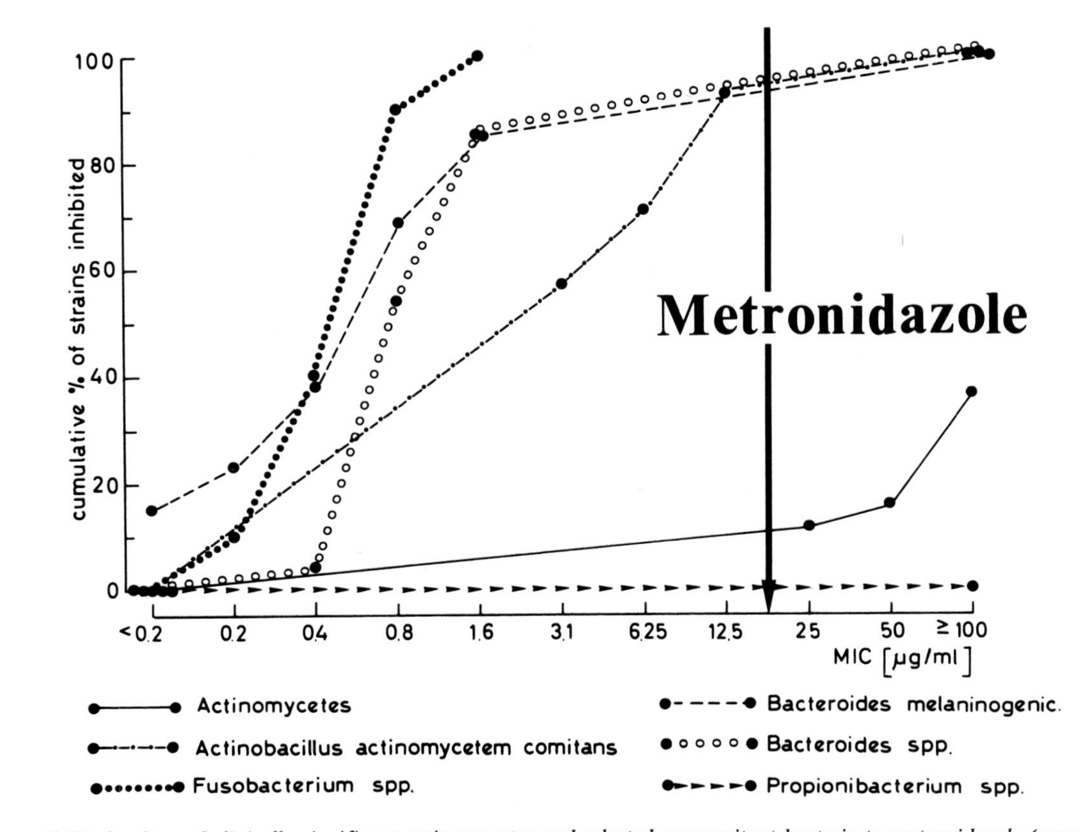

FIGURE 5. Susceptibility in vitro *of clinically significant actinomycetes and selected concomitant bacteria to metronidazole (agar dilution test; DST-medium (Oxoid CM 261); cumulative percentages of inhibited strains).*

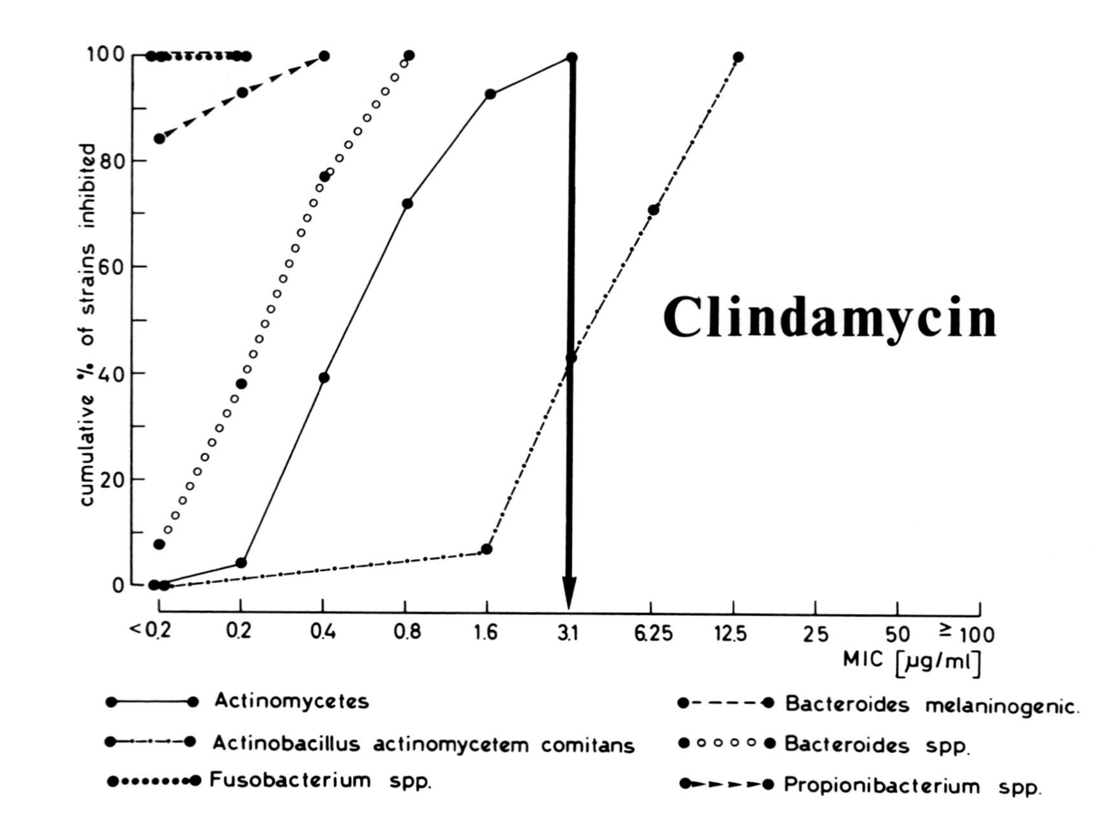

FIGURE 6. Susceptibility in vitro *of clinically significant actinomycetes and selected concomitant bacteria to clindamycin (agar dilution test; DST-medium (Oxoid CM 261); cumulative percentages of inhibited strains).*

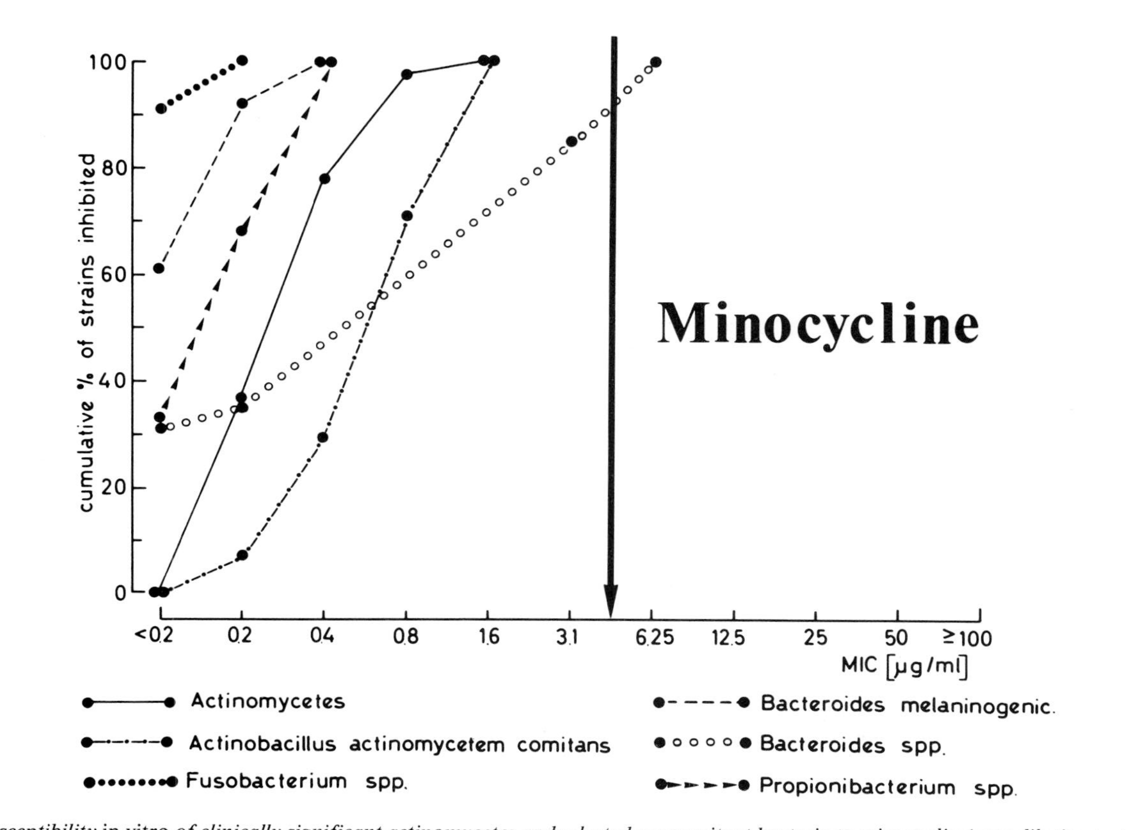

FIGURE 7. Susceptibility in vitro *of clinically significant actinomycetes and selected concomitant bacteria to minocyclin (agar dilution test; DST-medium (Oxoid CM 261); cumulative percentages of inhibited strains).*

propionica, and *Actinobacillus actinomycetem-comitans* has been proven to be an effective tool in treating cervicofacial infections, provided that the causative actinomycete is antigenically similar to the strains used in the vaccine. The typical focal reaction which develops during the first days of application may be used as an immunological confirmation of the diagnosis and also as a way to obtain higher concentrations of antibiotics in the infected tissue.

Finally, some of our clinical observations will illustrate these therapeutic considerations. Until 1965, actinomycoses with long and very long durations were quite common. We have data on one infection which lasted 18 years and on seven others which lasted from two to ten years. In addition, many other cases took several months to be cured. These therapeutic problems obviously were related to the low dosages of penicillin G and of sulfonamide which were used at that time. In Germany since 1970, such therapeutically resistant processes have become very rare which, in our view, is due to the wide application of ampicillin for treating actinomycosis. However, even now, there may be exceptional cases in which actinomycotic infections resist all these therapeutic measures and may become life-threatening. A brief report on one of these cases which we recently observed shall conclude our discussion of the treatment of actinomycoses (Table VII).

In July 1977, a 32 year-old woman developed a cheek abscess which was insufficiently treated by incision and low doses of ampicillin and oxacillin for only seven days. When the second abscess relapse occurred in August, we received a pus specimen and confirmed the diagnosis as actinomycosis by isolating *A. israelii* and typical concomitant bacteria. Therefore, our clinical colleagues decided to administer Lentze's heterovaccine which produced only a weak focal reaction and did not prevent further relapses. As was later determined, the causative actinomycete represented an antigenically aberrant form and, thus, was not susceptible to the induced antibodies. After a series of abscesses in the cervicofacial region, this site healed spontaneously; however towards the end of 1978, the patient developed metastatic abscesses in the left arm. As the patient was thought to have become allergic to penicillins, no antibiotics were applied after October 1978. This resulted in an increasing number of recurring abscesses in the left arm which were all incised; in July 1980, multiple abscesses were formed at other sites including the internal organs. At that time, the general state of health of the patient was deteriorating rapidly. Therefore, we decided to recommend treatment with 40 g carbenicillin per day despite the assumed allergy. Under this therapy which provoked no allergic reaction, the patient's condition improved slowly, the abscesses healed, and no relapses occurred. Apart from the extensive scars which followed the numerous incisions, the patient has remained well to date.

This example shows that treatment of human actinomycoses will often not be successful unless suitable antibiotics in sufficiently high doses and for a sufficiently long period of time are administered. However, it also shows that now even very advanced infections can be cured.

TABLE VII. Systemic Actinomycosis: Case Report Patient M. K., Born September 11, 1944

Date	*Clinical symptoms*	*Treatment*		*Bacteriology*
		Surgical	*Antibiotic*	
27.3.1977	*Painful swelling (left cheek)*	-	-	-
31.3.-7.4.1977	*Abscess (left cheek)*	*Incision*	*Ampicillin + Oxacillin for 7 days*	-
July 1977	*Abscess (left cheek)*	*Incision*	*Penicillin G + cephalexin*	-
August 1977	*Abscess (left cheek)*	*Incision*	*Heterovaccine*	Actinomyces israelii + *typical concomitant bacteria*
September 1977	*Abscess (left cheek)*	*Incision*	-	Actinomyces israelii + *typical concomitant bacteria*
April 1978-October 1978	*Advanced hard swelling (left cheek from the zygomatic arch to the border of the mandible); lockjaw*	*Extirpation of the scarred masseter*	*Cefoxitin + Mezlocillin (medium doses) for 4 weks; heterovaccine*	Actinomyces israelii + Klebsiella pneumoniae, E. coli *and* Pseudomonas aeruginosa
November 1978 -August 1979	*Spontaneous healing of the abscess residues at the left cheek; multiple abscesses left forearm, upper arm, and elbow*	*Incisions*	-	*"Actinomycetes"*
November 1979	*Recurring abscesses left arm*	*Incision*	-	*"Actinomycetes"*, Staphylococcus aureus
May 1980-July 1980	*Recurring abscesses left arm and multiple abscesses in other, possibly also internal organs*	*Incisions as far as possible*	*Carbenicillin 40 g per day for several weeks*	Actinomyces israelii + *various other bacteria*
Autumn-winter 1980	*Steady improvement of the general condition, no recurring abscesses*	-	-	-
Since then	*No relapse*			

ACKNOWLEDGMENTS

The valuable technical assistance of Monika Pinkwart is gratefully acknowledged. We also thank Sigrid Glanschneider for photographic services and Evelyn Heidermann for typing the manuscript. The clinical data for the case report were kindly provided by Prof. Spohn, Karlsruhe, and Prof. Pape, Köln.

REFERENCES

Bhagavan, B. S., and Gupta P. K., *Human Pathol. 9:*567 (1978).

Christ, M. L., and Haja, J., *Acta Cytolog. 22:*146 (1978).

Colebrook, L., *Lancet 1:*893 (1921).

Gupta, P. K., Hollander, D. H., and Frost, J. K., *Acta Cytol. 20:*295 (1976).

Heinrich, S., and Korth, H., *in* "Krankheiten durch Aktinomyceten und verwandte Erreger", (H.-J. Heite, ed.), p. 16. Springer, Berlin (1977).

Hemmes, G. D., *Ned. Tijdschr. Geneeskd. 107:*193 (1963).

Lentze, F., *in* "Die Infektionskrankheiten des Menschen und ihre Erreger" Vol. I (A. Grumbach and O. Bonin, eds.), p. 954. Georg Thieme, Stuttgart (1969).

Schaal, K. P., *Dtsch. Ärztebl. 76:*1997 (1979).

Schaal, K. P., Schütt-Gerowitt, H., and Pape, W., *Infection 7,* Suppl. 1:47 (1979).

Schaal, K. P., and Pape, W., *Infection 8*, Suppl. 2:176 (1980).

Schaal, K. P., and Pulverer, G., *in* "The Prokaryotes: A Handbook on Habitats, Isolation and Identification of Bacteria" (M. P., Starr, H. Stolp, H. G. Trüper, A. Balows and H. G. Schlegel, eds), p. 1923. Springer, New York (1981)

Schofield, G. M., and Schaal, K. P., *J. Gen. Microbiol. 127:*237 (1981).

Slack, J. M., and Gerencser, M. A., "Actinomyces, Filamentous Bacteria". Burgess Publishing Company, Minneapolis (1975).

Wolff, M., and Israel, J., *Arch. Pathol. Anat. 126:*11 (1891).

ADHERENCE OF *ACTINOMYCES VISCOSUS* TO TEETH AND ITS ROLE IN PATHOGENESIS

Dale C. Birdsell

Department of Oral Biology
University of Washington
Seattle, Washington, U. S. A.

Donald R. Callihan

Department of Microbiology
University of Rochester
School of Medicine and Dentistry
Rochester, New York, U. S. A.

James T. Powell
Werner Fischlschweiger

Department of Basic Dental Sciences
College of Dentistry
University of Florida
Gainesville, Florida, U. S. A.

I. INTRODUCTION

Recent investigations have strongly implicated certain members of the oral microbiota as possible etiologic agents in various forms of periodontal diseases. *Actinomyces viscosus*, a Gram-positive, facultative member of the normal oral microbiota, has been implicated as a causative agent in gingivitis (Loesche and Syed, 1978; Syed and Loesche, 1978). This organism was one of the first oral organisms recognized to synthesize potent mitogens (Burckhardt *et*

ISBN 0-12-528620-1

al., 1977; Engel *et al.*, 1977).This aspect of the biological role of *A. viscosus* is dealt with in detail in other papers presented at this symposium. Factors, both associated with the cell structure and released into the culture medium, have been reported to be capable of stimulating bone resorption (Hausman *et al.*, 1977; King *et al.*, 1978; Trummel *et al.*, 1977). Whole cells and cell walls of *A. viscosus* have been shown to promote the selective release of lysosomal enzymes from human peripheral blood polymorphonuclear leukocytes (Taichman *et al.*, 1978). It is clear, therefore, that *A. viscosus* possesses the armamentarium to induce the appropriate host responses to bring about destruction of the hard and soft tissues supporting the teeth.

In order for any of the above responses to occur, however, the organism must attach to (adhere) and colonize (accumulate on) the tooth surfaces. This paper addresses the mechanisms by which *A. viscosus* can adhere to tooth surfaces and focuses primarily on the differential virulence system of strain T14V (virulent) and strain T14AV (avirulent) which was originally described by Hammond, Steel, and Peindl (1976).

II. SURFACE STRUCTURE OF *A. VISCOSUS*

Figure 1 illustrates the ultrastructure of whole cells, cells disrupted by passage through the French Press at 20,000 psi, and cell walls obtained by ballistic disintegration with glass beads (Braun homogenizer) of *A. viscosus* T14V and T14AV. Figures 1A and 1B show the contrast between washed whole cells of strain T14V and of strain T14AV. As has been previously reported (Powell *et al.*, 1978), the surface of cells of strain T14V were covered with an abundance of fibrillar material, whereas the surfaces of similarly prepared cells of T14AV grown in TSBS possessed few fibrils. Transmission electron microscopy of crude lysates following passage through the French Press (Figs. 1C and 1D) revealed that this treatment did not destroy the structural integrity of the cell wall. Most, but not all, of the fibrillar material was removed from strain T14V cell walls; the remaining fibrils were not as numerous as on whole cells and were considerably shortened (Fig. 1C). In contrast to French Press disrupted cells, cell walls of both strains, which had been isolated by differential centrifugation following cell disruption by ballistic disintegration, revealed a definite lack of fibrillar structure and some loss of cell wall integrity.

Powell *et al.* (1978) have also reported the presence of a microcapsule associated with the cell surface of strain T14AV grown on TSBS medium (Fig. 2B). No such structure is present on similarly cultured strain T14V (Fig. 2A). When strain T14AV is grown on a chemically defined medium (CDM), however, no microcapsule is present (Powell *et al.*, 1978). Rather, the cell surface contains numerous fibrils which are capable of adsorbing anti-fibril antibodies. In other words, there is no detectable ultrastructural difference between the two strains when grown on the chemically defined medium.

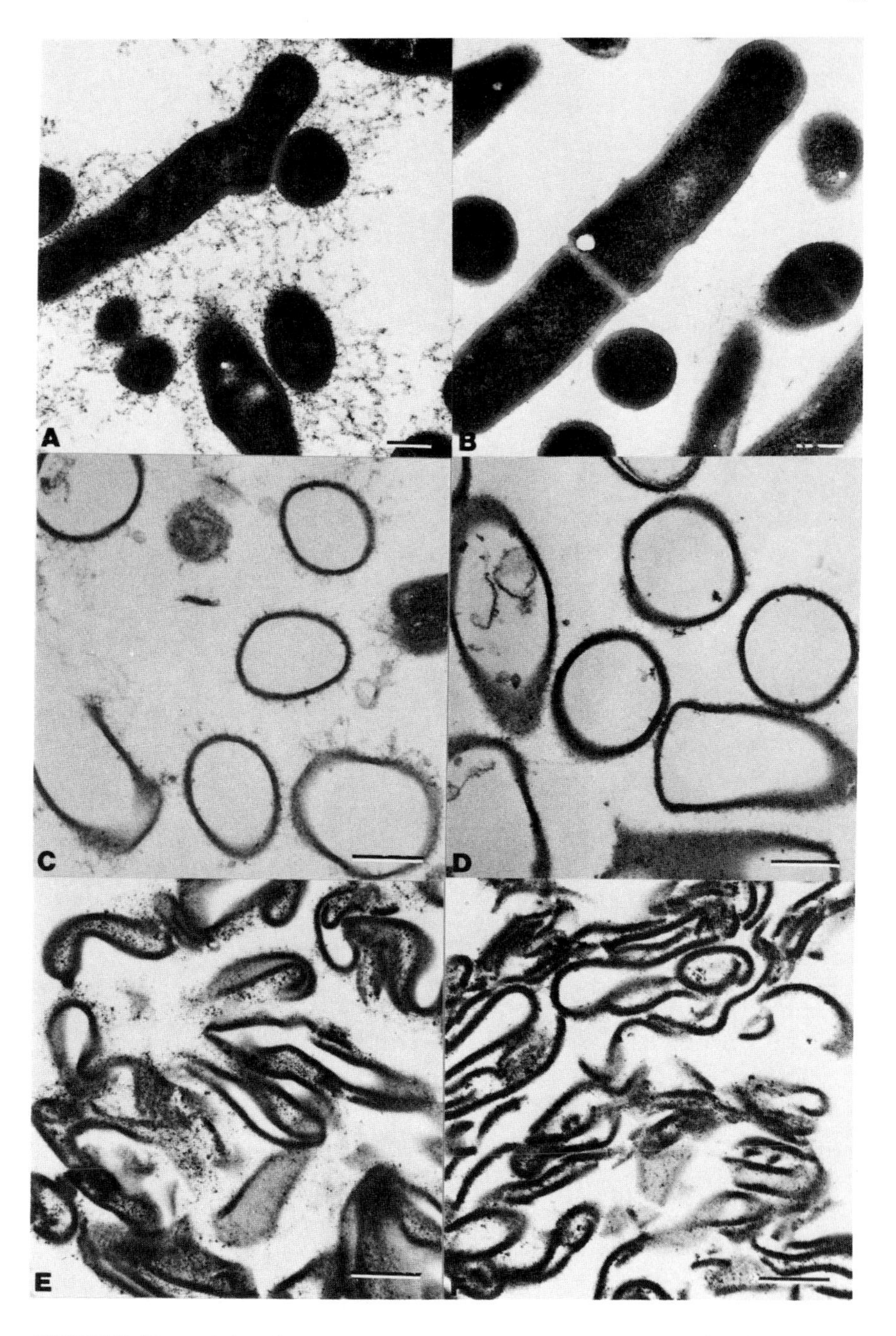

FIGURE 1. Transmission electron microscopy of whole cells and cell walls of A. viscosus *T14V and T14AV.* A, *Whole cells, strain T14V;* B, *Whole cells, strains T14AV; C, French Press treated strain T14V;* D, *French Press treated strain T14AV;* E, *Cell walls of strain T14V;* F, *Cell walls of strain T14AV. The bar represents 100 nm.*

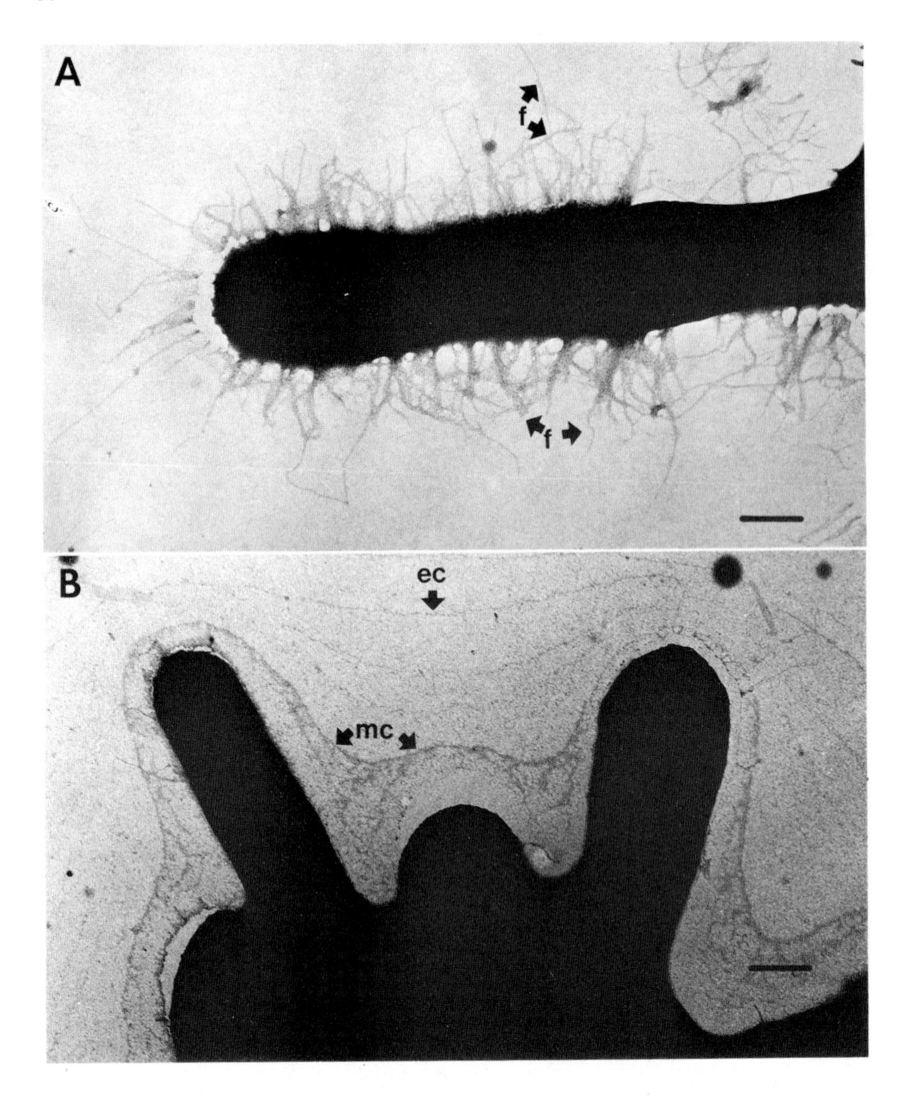

FIGURE 2. Transmission electron micrographs of uranyl acetate-stained TSBS-grown A. viscosus *T14V and T14AV.* A, *strain T14V;* B, *strain T14AV; f, fibrils; mc, microcapsule; ec, extracellular component. The bar represents 100 nm.*

III. ADHERENCE ORGANELLES OF *A. VISCOSUS*

Wheeler, Clark, and Birdsell (1979), using the sphereoidal hydroxyapatite (SHA) assay of Clark, Bammann, and Gibbons (1978), examined the abilities of strains T14V and T14AV to adhere to uncoated and saliva-coated SHA. The fibril-bearing strain T14V adheres much more efficiently to the beads than does the fibril-deficient strain T14AV. In addition, pretreatment with clarified saliva to mimic a salivary pellicle enhances the adherence of strain T14V and decreases the adherence of strain T14AV. These results were confirmed and amplified by introducing antibiotic-resistant strains (T14VJ1 and T14AVT1, both resistant to streptomycin at 200 μg/ml) into the oral cavity of human volunteers. The T14AV strain is much less efficient at the initial interaction with the tooth surface and is cleared more rapidly (Wheeler *et al.*, 1979). From these results one can infer that the interaction between strain T14V and the saliva-coated SHA as well as human teeth is specific. Strain T14AV, grown in chemically defined medium and possessing many surface fibrils, adheres as well as does strain T14V to saliva-coated SHA (Fig. 3). Together, these results strongly suggest that the fibrils are potential mediators of adherence, *i. e.*, they represent the adherence organelle.

A second interaction is a specific, lactose-reversible coaggregation between *A. viscosus* and strains of either *Streptococcus sanguis* (Cisar *et al.*, 1978a) or erythrocytes (Ellen *et al.*, 1978). Utilizing antiserum made monospecific for the "virulence" antigen of strain T14V, Cisar *et al.* (1978a) have demonstrated

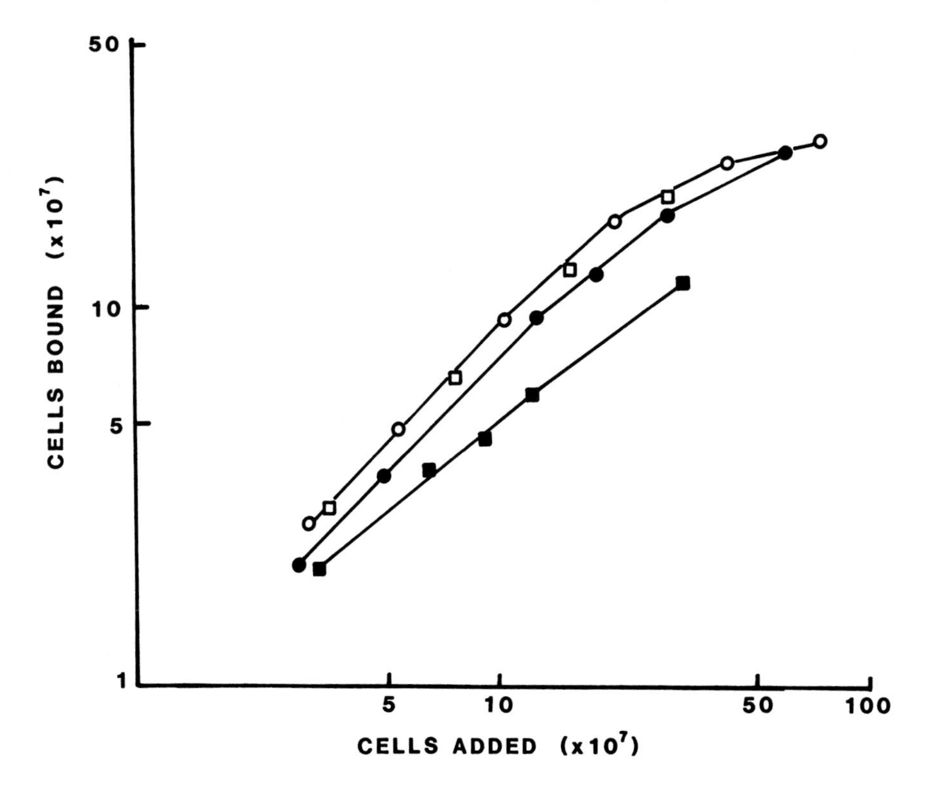

FIGURE 3. Adherence of TSBS- and CDM-grown A. viscosus *T14V and T14AV to saliva-treated hydroxyapatite. CDM-grown T14V(○); CDM-grown T14AV(□); TSBS-grown T14V(●); TSBS-grown T14AV(■).*

by immunoelectron microscopy that the surface fibrils of strain T14V are responsible for the coaggregation with *S. sanguis*. Subsequent studies (Cisar *et al.*, 1979; Costello *et al.*, 1979; McIntire *et al.*, 1978) have confirmed the role of the surface fibrils (frequently termed fimbriae) in the coaggregation between *A. viscosus* and *S. sanguis*. The interaction between *A. viscosus* strain T14V and *S. sanguis* strain 34 has been shown to involve the fibrillar, proteinaceous lectin on *A. viscosus* which interacts with a carbohydrate antigen of *S. sanguis*. The reversibility of the reaction with lactose suggests that the lectin interacts with a galactose-containing cell wall carbohydrate. The complexity of the interactions between *A. viscosus* and *S. sanguis* has been demonstrated by Kolenbrander and Williams (1981). After examining the interactions between a large number of isolates of *A. viscosus* (or *A. naeslundii*) and representative strains of *S. sanguis*, these investigators divided *A. viscosus* into five groups and *S. sanguis* into four groups. Essentially all combinations of interactions have been observed, including those in which the lectin resides on the *S. sanguis* and the carbohydrate on the *A. viscosus* and those that are insensitive to inhibition by lactose.

Figure 4 shows results of studies examining cell wall aggregation of suspen-

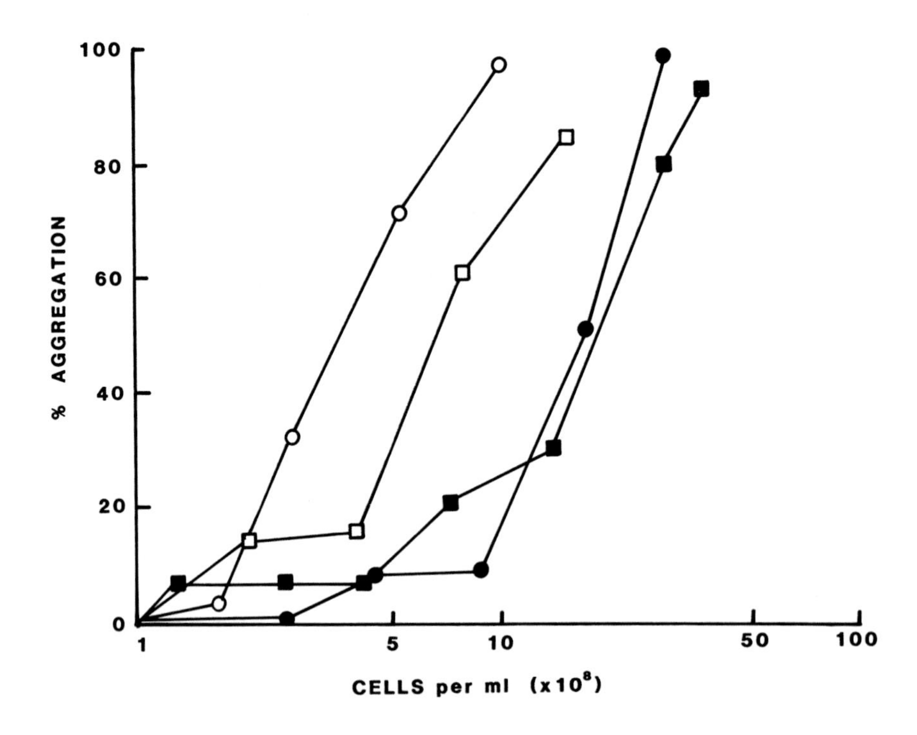

FIGURE 4. Cell to cell aggregation of TSBS- and CDM-grown A. viscosus *T14V and T14AV. CDM-grown T14V(○); CDM-grown T14AV(□); TSBS-grown T14V(●); TSBS-grown T14AV(■).*

sions of *A. viscosus* T14V and T14AV grown either in TSBS (in which strain T14AV lacks surface fibrils) or in chemically defined medium (in which strain T14AV is morphologically indistinguishable from strain T14V (Powell *et al.*, 1979)). Aggregation was measured by allowing suspensions of cells at different cell concentrations to mix at room temperature for 90 min on a rotator (8 rotations/min). Following incubation, the tubes were left undisturbed for 30 min and the turbidity of a sample which had been carefully removed from the top of each suspension was measured at 550 nm. Essentially no difference was noted in the aggregation of strains T14V or T14AV grown in TSBS, although strain T14V possesses fibrils and strain T14AV does not. Approximately 2×10^9 cells per ml are required for a 50% reduction in turbidity. In contrast, there is a significant difference between the aggregations of strains T14V and T14AV grown in chemically defined medium. If this interaction is also mediated by surface fibrils as the data suggest, one must postulate a third function, self-aggregation.

As will be discussed fully below, the question regarding the functions of surface fibrils has not yet been completely answered. Clark *et al.* (1981) have conducted an extensive survey of oral *Actinomyces* and representatives of other filamentous, Gram-positive members of the supra-gingival plaque microbiota. All strains were examined for the presence of surface fibrils by electron microscopy of negatively stained suspensions and for their ability to adhere to saliva-coated SHA. Several representative *Actinomyces* species (both laboratory strains and fresh clinical isolates) were also examined for their ability to adsorb anti-T14V fibril antibodies. As seen in Tables I and II, there is no absolute correlation between the presence of surface fibrils and the ability of the organisms to adhere to saliva-coated SHA or to adsorb anti-fibril antibodies. Also, several strains of *A. naeslundii* contain fibrils which cross-react with *A. viscosus* T14V fibrils (Table II).

IV. ISOLATION AND CHARACTERIZATION OF SURFACE FIBRILS (FIMBRIAE)

Confirmation of biological functions for the surface fibrils requires their isolation, purification, and chemical and biological characterization. Two general methods have been employed to release the fibrils from the cell wall to which they are covalently attached: A) physical treatments to "shear" the fibrils from the wall and B) enzymatic digestion to "release" the fibrils from the cell wall.

A. Physical Methods

Callihan (1977) examined a variety of physical methods to shear the surface fibrils from *A. viscosus* T14V. As shown in Figure 5, use of neither a tissue

TABLE I. Adsorptive and Morphologic Survey of Actinomyces *and Related Species*

Strain	*Serotype*	*Percent adsorption*[a]	*Fimbriae*[b] *present*
A. viscosus			
A780	*1*	*51*	±
T14V	*2*	*100*	+
T14AV	*2*	*23*	±[c]
M100	*2*	*73*	±
UF0086	*2*	*53*	+
UF0318	*2*	*80*	+
UF0372	*2*	*81*	+
A. naeslundii			
W826	*1*	*59*	±
UF0076	*1*	*59*	+
UF0077	*1*	*31*	+
UF0084	*1*	*31*	+
UF0109	*1*	*28*	+
W1544	*2*	*55*	+
N16	*3*	*100*	+
UFO92	*3*	*84*	+
UF0087	*3*	*87*	+
A. israelii			
X523	*1*	*11*	±[c]
UF0111	*1*	*8*	±
UF0512	*1*	*8*	±
W838	*2*	*14*	±[c]
UF0243	*2*	*5*	±[c]
A. odontolyticus			
X363	1	55	±
Rothia dentocariosa			
X599	*1*	*0*	–
X347	*1*	*0*	–
Arachnia propionica			
W857	*1*	*41*	–

[a]*Data is expressed as percent adsorption relative to the adsorption of strain T14V.*
[b]*The presence or absence of surface fimbriae is indicated by + /-; strains with <10 fimbriae per cell width are indicated by ±.*
[c]*These strains possessed a microcapsule surrounding the cell and its associated fibrils.*
(Reproduced from Clark et al., Infect. Immun. 33*:908 (1981) by copyright permission of the American Society for Microbiology.)*

TABLE II. Adsorption Inhibition of A. viscosus *Strain T14V by Anti-strain T14V Sera Preadsorbed with Selected* Actinomyces *Strains.*

Adsorbent strain[a]	*Serotype*	*Relative inhibitory activity removed by preadsorption*[b] *(%)*
A. viscosus		
A 780	*1*	*14*
T14V	*2*	*100*
M100	*2*	*26*
UF0086	*2*	*83*
UF0318	*2*	*84*
UF0372	*2*	*57*
A. naeslundii		
W826	*1*	<5
UF0076	*1*	*20*
UF0077	*1*	*39*
UF0084	*1*	*6*
UF0109	*1*	*10*
W1544	*2*	*39*
N16	*3*	*57*
UF0092	*3*	*61*
UF0087	*3*	*71*
A. israelii		
UF00512	*1*	<5
UF0243	*2*	<5

[a]*Dry weight (8 mg) of each strain was tested as adsorbents. The relative inhibitory activity was not significantly different when 20 mg dry weight of each strain was used as the adsorbent.*
[b]*The relative inhibitory activity of adsorbed antisera was determined as follows:*
[(% inhibition by unadsorbed antiserum - % inhibition after adsorption with Strain X)/% (inhibition by unadsorbed antiserum - % inhibition after adsorption with Strain TI4V)] X 100
(Reproduced from Clark et al., Infect. Immun. 33*:908 (1981) by copyright permission of the American Society for Microbiology.)*

homogenizer nor a Waring Blender removes significant amounts of antigens. In contrast, two passages of cell suspensions through the French Pressure cell at 10,000 psi releases several components which do not sediment at 20,000 *g*. One of these components exhibits a zone of identity with the VA-1 (virulence-associated antigens) component of the Lancefield extract (arrow). As indicated previously, this particular component has subsequently been identified as the surface fibril involved in the adherence of *A. viscosus* to saliva-coated SHA. Figures 6 compares the effect of pressure on the release of physically extractable antigens from *A. viscosus* T14VJ1. The results indicate that maximum release of antigens is obtained at 10,000 psi. Examination of the extent of cell lysis by release of material which absorbs at 260 nm indicates that minimal cell disruption occurs at 10,000 psi. These conditions which optimally remove the VA-1 antigen (adherence fibril) with essentially no cell lysis were subse-

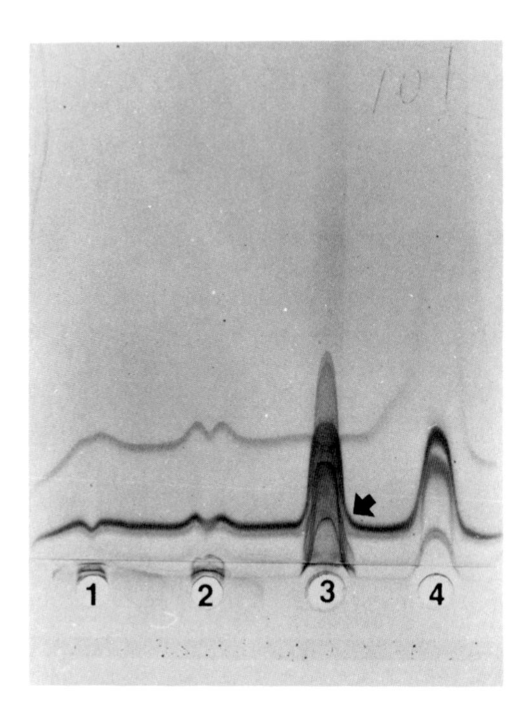

FIGURE 5. Electroimmunoassay showing the effect of extraction method on the ability to release physically extractable antigens from A. viscosus *T14V. Each well contains a 10 μl sample of antigen. Trough contains 50 μl of agarose containing antiserum (30 μl/ml agarose), (1) Treatment with Waring blender for 10 minutes, 20K supernate; (2) treatment with Waring blender for 10 minutes, 20K supernate; (3) two passages through the French Press at 10,000 psi, 20K supernate; and (4)* A. viscosus *T14V standard Lancefield extract, 10 mg/ml.*

quently employed by Wheeler and Clark (1980) to shear the surface fibrils which were then purified by ammonium sulfate precipitation and by differential centrifugation. The purified fibrils were shown to be serologically identical to the VA-1 antigen and to inhibit fibril-mediated adherence of *A. viscosus* T14V to saliva-coated SHA.

More recently, Cisar *et al.* (1981) have employed continuous flow sonication to shear the delicate surface fibrils from *A. viscosus* T14V. Following concentration by ultrafiltration, the material was chromatographed on an Agarose 5 M column and the fibrillar material was found to be eluted in the void volume of the column. The nature of this purified fibril preparation will be discussed in greater detail below.

B. Enzymatic Methods

Early in the development of the study of *A. viscosus* surface fibrils, Cisar and co-workers (Cisar and Vatter, 1979; Cisar *et al.*, 1978b) employed as a

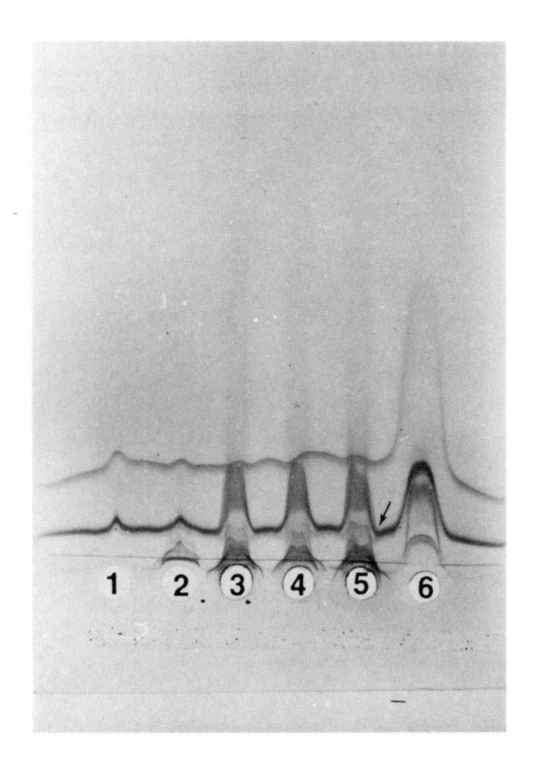

FIGURE 6. Electroimmunoassay showing the effect of pressure on the release of physically extractable antigens of A. viscosus *T14V in the 20K supernate. Each well contains 10 μl sample of antigen; antiserum was added at a concentration of 30 μl per ml agarose. Trough contains 50 μl of* A. viscosus *T14V standard Lancefield extract plus 50 μl of agarose. (1) Untreated control cells; (2) two passages at 5,000 psi; (3) two passages at 10,000 psi; (4) two passages at 15,000 psi; (5) two passages at 20,000 psi; and (6)* A. viscosus *T14V standard Lancefield extract, 25 mg/ml.*

source of fibrils the lysozyme digest of cell walls prepared by ballistic disintegration. As previously described (Birdsell and Fischlschweiger, 1981; Brown *et al.*, 1980), essentially all the surface fibrils are removed by the physical forces necessary to disrupt the integrity of the cell (Fig. 1E). It must be understood that, during the isolation and purification procedures of the cell walls which are to be used as the starting material for isolation of surface components, many surface constituents will have been removed and discarded. The fibrils isolated by Cisar *et al.* (1979) may have represented only a minor component of the total fibrillar material present on the intact cells. Brown, Fischlschweiger, and Birdsell (1982) examined cell wall components released by digestion with the M-1 N-acetylmuramidase of *Streptomyces globisporus*. These authors suggested that there may be a category of fibrils which are present within the matrix of the cell wall. The fibrils isolated in the early work of Cisar *et al.* (1979) may have been such matrix fibrils. The method of choice for isolation of surface fibrils would be the complete solubilization of cell walls of intact organisms under osmotically stabilizing conditions such that cytoplasmic proteins do not contaminate the wall digest preparations. Under these conditions, one would predict that the fibrils would have a fixed orientation, with one end containing that portion involved in the covalent linkage to the cell wall proper. Such preparations should be possible using the M-1 muramidase described above.

V. SURFACE FIBRIL SUB-POPULATIONS

Having recognized that two biological functions, adherence and coaggregation, are attributable to surface fibrils, researchers have wondered whether there were two subpopulations of fibrils or one bifunctional population. Results have recently been published that indicate that the adherence fibrils and the coaggregation fibrils represent two distinct subpopulations (Cisar *et al.*, 1981). The fibril preparation present in the void volume of the Agarose 5 M column gives two distinct peaks in crossed immunoelectrophoresis against rabbit anti-T14V antiserum. Monoclonal antibodies which were selected against fibril-sensitized erythrocytes (and hence against the lactose-sensitive fibrils) form precipitates with only one fibril subpopulation, Ag2 (coaggregation fibrils) (Cisar *et al.*, 1980). Treatment with monoclonal antibodies prior to crossed immunoelectrophoresis against rabbit anti-T14V antiserum yields only one band corresponding to Ag1 (adherence fibrils) of Cisar. Immunolabeling monoclonal antibody-treated cells with peroxidase-labeled goat anti-mouse IgG results in labeling of only a portion of the fibrils. These results strongly indicate that there are at least two subpopulations of fibrils. Further studies will be necessary to define the nature of the second subpopulation, *i. e.*, to determine whether this population represents the fibrils involved in adherence to tooth surfaces. Additional studies will also be necessary to determine whether the self-aggregation function may be assigned either to one of the two populations thus far characterized, to an additional population not yet purified, or to another cell surface polymer.

VI. ROLE OF SURFACE FIBRILS IN PATHOGENESIS

The differential virulence system of *A. viscosus* T14V and of T14AV originally described by Hammond *et al.* (1976) employs monoinfection of germ-free rats. Both strains form plaque; however, only animals infected with strain T14V demonstrate the massive loss of bone which is characteristic of periodontal disease. Whole cells or extracts of both strains have the ability to stimulate lymphocytes, stimulate bone resorption, and cause selective degranulation of polymorphonuclear leukocytes (see Introduction). Purified fibrils do not stimulate blastogenesis of human peripheral blood lymphocytes (Reed, Clark, and Birdsell, unpublished data) nor do they stimulate bone resorption (in Cisar *et al.*, 1978b). Scanning electron microscopic studies of monoinfected rats have suggested the cause of the differential virulence of these two strains (Brecher *et al.*, 1978). Although both strains form dental plaque, the sites of plaque formation are different. Strain T14AV colonizes only pit and fissure sites, whereas strain T14V colonizes supragingivally in addition to pits and fissures. Therefore, although both organisms have the armamentarium to initiate a destructive host response, only strain T14V is in juxtaposition to the gingival tissues. Thus, the surface fibrils which adhere to the tooth surface play an essential role in the pathogenicity of *A. viscosus.*

VII. SUMMARY

A. viscosus has evolved very efficient mechanisms for colonization of the oral environment. If naked or salivary-pellicle-coated tooth surfaces are exposed, one subpopulation of fibrils enables specific adherence to occur. This adherence is lactose insensitive. However, *S. sanguis* represents the initial tooth colonizer due to its presence in high numbers on epithelial surface and in saliva. *A. viscosus* cells possess a second subpopulation of fibrils with a lactose-sensitive specific interaction with carbohydrate antigens of *S. sanguis*, thus enabling adherence (coaggregation) to *S. sanguis* plaque. Finally, if no tooth sites or no *S. sanguis* sites are available, *A. viscosus* is capable of aggregating with other *A. viscosus* cells. Whether the self-aggregation function is associated with the previously purified and characterized fibril subpopulations has yet to be established. The role of fibrils in pathogenesis appears to reside solely in the adherence to and colonization of supragingival plaque. Intimate contact thereby achieved, other cellular constituents or activities bring about activation of host defense parameters with subsequent tissue destruction.

REFERENCES

Birdsell, D. C., and Fischlschweiger, W., *in* "Microbiology in Clinical Dentistry" (F. J. Orland, ed.), p. 33. Publishing Sciences Group, Inc., Littleton (1981).
Brecher, S. M., vanHoute, J., and Hammond, B. F., *Infect. Immun. 22:*603 (1978).
Brown, D. A., Fischlschweiger, W., and Birdsell, D. C., *Arch. Oral Biol. 25*:451 (1980).
Brown, D. A., Fischlschweiger, W., and Birdsell, D. C., *Arch. Oral Biol. 27*:183 (1982).
Burckhardt, J. J., Guggenheim, B., and Hefti, A., *J. Immunol. 118*:1460 (1977).
Callihan, D. R., Master of Science Thesis, University of Florida, Gainesville, Florida (1977).
Cisar, J. O., and Vatter, A. E., *Infect. Immun. 24*:523 (1979).
Cisar, J. O., McIntire, F. C., and Vatter, A. E., *Adv. Exp. Med. Biol. 7*:695 (1978a).
Cisar, J. O., Vatter, A. E., and McIntire, F. C., *Infect. Immun. 19*:312 (1978b).
Cisar, J. O., Kolenbrander, P. E., and McIntire, F. S., *Infect. Immun. 24:*742 (1979).

Cisar, J. O., Barsumian, E. L., Curl, S. H., Vatter, A. E., Sandberg, A. L., and Siraganian, R. P., *J. Reticuloendothel. Soc. 28*:73s (1980).
Cisar, J. O., Barsumian, E. L., Curl, S. H., Vatter, A. E., Sandberg, A. L., and Siraganian, R. P., *J. Immunol. 127:*1318 (1981).
Clark, W. B., Bammann, L. L., and Gibbons, R. J., *Infect. Immun. 19:*846 (1978).
Clark, W. B., Webb, E. L., Wheeler, T. T., Fischlschweiger, W., Birdsell, D. C., and Mansheim, B. J., *Infect. Immun. 33:*908 (1981).
Costello, A. H., Cisar, J. O., Kolenbrander, P. E., and Gabriel, O., *Infect. Immun. 26:*563 (1979).
Ellen, R. P., Walker, D. L., and Chan, R. H., *J. Bacteriol. 134:*1171 (1978).
Engel, D., Clagett, J., Page, R., and Williams, B., *J. Immunol. 118:*1466 (1977).
Hammond, B. F., Steel, C. F., and Peindl, K. S., *J. Dent. Res. 53:*A19 (1976).
Hausan, E., Nair, B. C., and Reed, M. J., *in* "Proceedings Mechanisms of Localized Bone Loss" (J. E., Horton, T. M. Tarpley, and W. F. Davies, eds.), p. 115. Information Retrieval, London (1977).
King, G., Birdsell, D., and Thiems, S., *J. Dent. Res. 57:*144 (1978).
Kolenbrander, P. O., and Williams, B. L., *Infect. Immun. 33:*95 (1981).
Loesche, W. J., and Syed, S. A., *Infect. Immun. 21:*830 (1978).
McIntire, F. C., Vatter, A. E., Boros, J., and Arnold, J., *Infect. Immun. 21:*978 (1978).
Powell, J. T., Fischlschweiger, W., and Birdsell, D. C., *Infect. Immun. 22:*934 (1978).
Powell, J. T., Wheeler, T. T., and Birdsell, D. C., *J. Dent. Res. 58:*A997 (1979).
Syed, S. A. and Loesche, W. J. *Infect. Immun. 21:*821 (1978).
Taichman, N. S., Hammond, B. F., Tsai, C.-C., Baehne, P., and McArthur, W., *Infect. Immun. 21:*594 (1978).
Trummel, C. L., Pabst, M. J., and Cisar, J. O., *J. Dent. Res. 56:*3156 (1977).
Wheeler, T. T., and Clark, W. B., *Infect. Immun. 28:*577 (1980).
Wheeler, T. T., Clark, W. B., and Birdsell, D. C., *Infect. Immun. 25:*1066 (1979).

LYMPHOCYTE INTERACTIONS IN HOST RESPONSES TO ORAL INFECTIONS CAUSED BY *ACTINOMYCES VISCOSUS*

Johann J. Burckhardt[1,2]

Department of Oral Microbiology and General Immunology
Dental Institute
University of Zurich
Zurich, Switzerland

I. INTRODUCTION

Oral microbiota cause gingivitis and periodontitis by triggering an array of defense mechanisms of the host. Several comprehensive reviews have been published in the last few years dealing with the microbiological aspects (Grigsby and Sabiston, 1976; Newman, 1980; Slots, 1979; Socransky, 1977), the histopathological features (Page and Schroeder, 1976), the possible immunopathomechanisms (Attström, 1975; Brandtzaeg, 1973; Horton *et al.*, 1974; Niesengard, 1977; Page and Schroeder, 1981; Seymour *et al.*, 1979a,b), and the resorption of alveolar bone (Hausmann, 1974; Hausmann and Ortman, 1979).

Despite the numerous investigations reviewed in these papers and the considerable advances made towards an unifying hypothesis for the immunopathogenesis of chronic inflammatory periodontal disease, there still remains an urgent need for the new therapeutic approaches (Seymour *et al.*, 1979b). Moreover, assays that would aid in predicting the outcome of the disease in a given patient would certainly be widely welcomed.

[1] *Present address: c/o Research and Development Dept., Diagnostica Division, F. Hoffmann-La Roche and Co. AG, Basel, Switzerland.*
[2] *Leave of absence at the Cellular Immunobiology Unit, University of Birmingham, Alabama, U.S.A. was funded by a fellowship from the Swiss National Science Foundation.*

ISBN 0-12-528620-1

Here, I shall discuss the host response to *Actinomyces viscosus* Ny1 in three parts: (i) by analyzing the animal model of inbred gnotobiotic Sprague Dawley rats, (ii) by dissecting *in vitro* lymphocyte interactions and their modulation by macrophages and other regulatory cells which can result in bone resorption, and (iii) by extrapolating these results to the human situation.

II. ANIMAL MODEL FOR THE ANALYSIS OF THE HOST RESPONSE TO *ACTINOMYCES VISCOSUS* Ny1

A. Description of the Model

Based on the classical experiments of Keyes and Gold (1955), Jordan and Keyes (1964), and Jordan *et al.*, (1965), Guggenheim and Schroeder (1974) assessed the periodontopathic potential of a strain of *Actinomyces viscosus* (van der Hoeven, 1974) by describing the induced lesions on the basis of quantitative cytology. They sensitized young germfree Sprague-Dawley rats, of the inbred RIC strain, by either intravenous or intradermal injections of heat-killed *A. viscosus* Ny1 cells. Three weeks after the start of the weekly injections, the rats were associated orally with viable cultures of the Ny1 strain which rapidly forms extensive dental plaque. When the experiment was terminated six weeks later, blood was collected from the animals; their lower jaws were processed for standardized evaluation of the distance between the alveolar bone crest (ABC) and the cemento-enamel junction (CEJ) and their upper jaws were processed for histometric measurements on random semithin section.

The results showed that the intravenous mode of sensitization induced the highest serum antibody titers and led to the most advanced vertical distances between ABC and CEJ. Also, this group of animals had the lowest number of residual fibroblasts and the most reduced volume of collagen fibers in addition to having the highest number of plasma cells in the cellular infiltrates of the gingival connective tissue. It was concluded from these thorough histometric evaluations that the mode of sensitization influenced the velocity by which plasma cells become superimposed on the picture of the earlier classic delayed hypersensitivity reaction.

Guggenheim and Schroeder (1974) thus postulated that the immune status of the animals was the most significant factor determining the final type of the peripheral infiltrate reaction.

B. Time Course of Sensitization

Based on this study, we began to investigate in parallel experiments time course of both the natural sensitization by oral association with only strain Ny1 and the development of bone loss (Burckhardt *et al.*, 1981a).

The following two parameters revealed *in vitro* the immune status of the rats: (i) serum agglutinins reflected the participation of the humoral or B-lymphocyte axis (including necessary T-helper lymphocyte function) and (ii) the anamnestic response *in vitro* of spleen T lymphocytes to two antigen fractions of strain Ny1 represented the (mainly T) cell-mediated element. The latter test system is discussed further below (IIIA and C); a short comment on the agglutination test should be adequate. The agglutination test was used primarily because whole cells, as used, best represented the full repertoire of bacterial cell surface antigens. The results of the microagglutination system were later confirmed by an enzyme-linked immunosorbent assay (ELISA) in which *A. viscosus* cells were fixed to microtiter plates and the serum antibodies were assessed by using isotype specific second antisera (Burckhardt, unpublished data)

Alveolar bone loss was evaluated by measuring areas and vertical distances on prints enlarged from standardized radiographs of mandibules and maxillae (Gaegauf-Zollinger *et al.*, 1982).

Loss of alveolar bone and the two parameters which measured the immune response *in vitro* were evaluated before application of strain Ny1 to germfree RIC-Sprague Dawley rats at 27 days of age. These same three parameters were determined at 46, 60, and 88 days of age and the results were compared with the values of germfree controls. We found that antibodies of considerable titers, an anamnestic T-cell response, and statistically significant amounts of bone loss were already present 19 days after application. Thereafter, sensitization to *A. viscosus* and total bone loss increased concomitantly until the end of the experiment. However, at some sites, the rate of bone loss decreased with time suggesting that negative feedback mechanisms enabled local bone remodeling.

Thus, the time course of sensitization and of overall bone loss proceeded in parallel. This raised the question as to whether we were observing only an epiphenomenon or a real cause-and-effect relationship.

C. Cause-and-effect Relationship

Combining the two kinds of experiments by following the time course of bone loss and immune reactivity in three groups of germfree rats that had been either sensitized/immunized with *A. viscosus* Ny1 (day 20), orally infected with strain Ny1 (days 38 and 39), or both was considered to conclusively answer the question of a cause-and-effect relationship between sensitization against the microorganism and destructive periodontal disease (Burckhardt *et al.*, 1981b). Indeed, in sensitized animals, the amount of alveolar bone loss was more than double that in monoassociated gnotobiotic rats, thus confirming the allergic nature of the disease. If, however, a similar experiment was performed with conventional rats kept in relative gnotobiosis (König and Guggenheim, 1968), no amplification of bone loss was observed in immunized rats. Also were we unable to detect a difference in serum agglutinins between these two groups. We tentatively explained this finding as being due to suppressor mechanisms, operative in conventional rats, a decrease in which might convert stable periodontal lesions into progressive ones. Surprisingly, we detected soluble suppressor factors in spleen cell cultures from conventional rats but not in those from germfree animals (see below, III C).

If indeed local inflammation, followed eventually by alveolar bone resorption and tooth loss, were the result of a normally limited immune response to oral microorganisms, this would represent the price we pay for the very complex penetration by the dentition of the epithelial barrier. It is this same immune response, on the other hand, that prevents the entry of potential pathogens.

D. Effectiveness of the Animal Model

The above short description of the gnotobiotic rat model and of the main conclusions drawn therefrom is limited and incomplete. Rather than giving a complete bibliography of all relevant animal experiments ever performed in this field, I shall briefly discuss both the advantages and disadvantages of such a model.

The advantages of controlled animal experimentation in periodontal research were well considered by Hulin (1929) whose series of most informative experiments in dogs revealed the allergic nature of periodontal inflammation. He presented radiographs of alveolar bone resorption elicited by local application of egg albumin in sensitized animals. Hulin's work unfortunately lay almost forgotten for more than half a century, although his conclusions are still valid.

The main advantages of the gnotobiotic rat model are that, first, the influence of a single parameter, *e. g.* microorganism, can be investigated under controlled conditions and, second, because the rat strain is inbred, any interference from differences in the animals' histocompatibility complex, which controls the immune response, can be a priori excluded. On the other hand, one cannot overlook the disadvantages that growing rats, not old adult animals are used in the studies and that their dental plaque, consisting of a single microorganism, hardly reflects the human situation. Considering these limitations, the rat model allows a valid insight into the host's response to oral microorganisms in a exaggerated situation, thereby providing results which otherwise would probably take much longer to collect.

III. MECHANISMS OF THE HOST RESPONSE

A. The Role of T Lymphocytes

For increased understanding, the host's allergic responses to antigens, whether of microbial origin or not, need to be dissected, first, into the temporal phases (afferent limb, central processing, efferent limb) and, second, into the main participating cells and molecules. Despite the exponential curve of original articles published, our knowledge is still incomplete.

Recognition of antigen by T lymphocytes requires the help of macrophages (Rosenthal *et al.*, 1978; Shevach and Rosenthal, 1973; Unanue, 1978) or related cells such as epidermal Langerhans cells (Stingl *et al.*, 1978). The amamnestic response *in vitro* of lymphocytes from human peripheral blood was considered, especially in the early seventies, to be a measure of the "cell-mediated immune status" (Lehner, 1972). Several groups have, observed a correlation between the response to antigens derived from oral microorganisms and the severity of gingivitis and periodontitis (Baker *et al.*, 1976; Ivanyi and Lehner, 1971; Lang and Smith, 1977; Lehner *et al.*, 1974; Mackler *et al.*, 1974). We therefore adapted this type of test system to measure the anamnestic response of rat T lymphocytes to antigens from *Actinomyces viscosus* Ny1 (Burckhardt, 1978). Briefly, T cells were purified from spleen cell suspensions by filtration through Ig-anti Ig columns, thereby removing B lymphocytes and most of the adherent cell. However, some macrophage precursors must have passed through the columns because, as was subsequently shown (Burckhardt, 1979), incubation with silica, a well-known toxin for macrophages (Allison *et al.*, 1966), abolished the antigen specific T-cell response. These experiments, in which silica was used, revealed yet another role of macrophages, namely their function as a suppressor cell. This function is discussed further below since it more properly belong to "central processing and modulation" than to the afferent limb of the immune response.

The molecular structures, the elusive "T-cell receptor", which enable T cells to recognize antigens, have been a matter of great controversy in immunology for more than a decade and therefore shall not be discussed here (for a recent review, see Marchalonis, 1982).

What is the role of T cells in the efferent limb, the effector mechanisms of the response? From the accumulated knowledge of T-cell function, this question can be answered briefly as follows: (i) T-cell help is indispensable for oligoclonal B-cell activation that results in the secretion of "antigen-specific" antibodies (Katz and Benacerraf, 1972); (ii) Activated T cells secrete soluble mediators, with the osteoclast-activating factor (OAF) being one of the earliest cytokines described (Horton *et al*,. 1972); (iii) *In vitro*, (T)cell-mediated cytotoxicity for rat fibroblasts requires antigens from strain Ny1 but, surprisingly, no prior sensitization of lymphocytes (Burckhardt *et al*,. 1982); and (iv) Drug-induced suppression by cyclosporin A of the T cell-dependent allergic response hardly affects bone loss in gnotobiotic rats (Guggenheim *et al.*, 1982). Taken together with the histopathological findings which showed that, in progressive and destructive periodontal inflammation, T lymphocytes were outnumbered by B lymphoblasts and plasma cells in rats (Guggenheim and Schroeder, 1974) and in humans (Mackler *et al.*, 1977; Seymour *et al.*, 1979 a,b), it seems that T cells contribute less, and probably more indirectly, to the immunopathomechanisms in the periodontium than do B cells.

B. The Role of B Lymphocytes

Only in the 1940's were immunoglobulins linked with plasmablasts and plasma cells as their sites of synthesis (Bing, 1940; Fagraeus, 1948; Kolouch *et al.*, 1947). Antibodies to oral microorganisms were demonstrated in both human periodontal disease (Ebersole *et al.*, 1982, Genco *et al.*, 1974) and in the animal model (Burckhardt *et al.*, 1981a; Guggenheim and Schroeder, 1974). Almost two decades ago, Brandtzaeg and Kraus (1965) were looking for auto-antibodies to gingival antigens as a possible trigger of periodontal inflammation. Clagett and Page (1978) recently succeeded in isolating auto-antibodies that reacted with double-stranded DNA from inflamed human and canine gingivae.

The unique property of antibodies is their specificity for that antigen which elicited their production through clonal selection and expansion (Burnet, 1959). The discovery of macromolecules capable of the non-antigen-specific activation of B cells was, therefore, quite revolutionary (Andersson *et al.,* 1972). Among these polyclonal B-cell activators (PBA) are the lipopolysaccharides (LPS) from Gram-negative bacteria, lipid A being the active part of the molecule (Andersson *et al.*, 1973), as well as polysaccharides with repeating units, such as dextrans, polymerized flagellin, and pneumococcal polysaccharide (Coutinho and Möller, 1973). For mutan (predominantly $\alpha[1\rightarrow3]$ glucans), we found that mitogenic activity was linked to the few $\alpha\, 1\rightarrow6$ branchings; after oxidation with periodate followed by borohydride reduction, the polymer no longer triggered mouse spleen cells into DNA synthesis and polyclonal antibody synthesis (Burckhardt and Guggenheim, unpublished data).

The growing interest in B-cell differentiation soon led to an increasing number of bacteria with reported PBA property, including *Actinomyces viscosus* (Burckhardt *et al.*, 1977; Engel, *et al.*, 1977; for a review see Clagett and Engel, 1978). We observed the phenomenon when bacterial fractions from strain Ny1 activated different lymphocyte populations from germfree RIC rats assumed to provide low baseline values in lymphocyte cultures supposed to reflect the degree of sensitization to strain Ny1 after oral infection. Later, we compared strain Ny1-PBA and LPS from *E. coli* concerning the isoelectric focussing pattern of Ig synthesized by activated B-cell clones in mouse spleen cell cultures. Among the approximately 60 bands resolved from each PBA, only three discrepancies were discernible (leJeune and Burckhardt, unpublished data). Thus, it seems that strain Nyl-PBA triggered a repertoire of B lymphocytes very similar to that activated by LPS which is known to activate, under optimum conditions, one third of the **B**-cell population (Andersson *et al.*, 1977).

Furthermore, it was interesting to observe the PBA-property of strain Nyl-mitogens for human B cells in culture and to follow their differentiation into B blasts, immature plasmablasts, and finally plasma cells. In contrast to pokeweed mitogen, their induction was independent of the help of Tcells which had been removed from peripheral blood leukocytes drawn from healthy volunteers by prior rosetting with sheep erythrocytes (Burckhardt and Cooper, unpublished data). Although consistent with results reported for water soluble mitogen from *Nocardia* (Gmelig-Meyling and Waldmann, 1981),these findings need further confirmation.

Finally, a series of reports, in which the predominance of auto-antibodies, including rheumatoid factor, secreted after stimulation with PBA (Dresser, 1978, Primi *et al.*, 1977) had been demonstrated, resulted in several serious speculations which considered B-cell activation by PBA not only a primitive form of immunity in relation to oral infection (Clagett and Engel, 1978) but also a major immunopathomechanism in allergic inflammation in general, such as rheumatoid arthritis or systemic lupus erythematosus (Fudenberg, 1980; Möller *et al.*, 1980). The role of Epstein-Barr virus, a well-known human PBA, needs further clarification.

Embedded into such an abundance of stimuli, oligoclonal and polyclonal ones, from oral microorganisms, it is fair to ask:(i) What normally controls these lymphocytes from unrestricted proliferation, differentiation, and antibody-production? and (ii) Once the control system(s) fails which molecular pathomechanisms are involved in the destruction of soft and hard periodontal tissues? These two questions are discussed in the following sections.

C. The Role of Macrophages and Other Regulatory Cells

Macrophages play an essential role in the immune response (reviewed by Unanue, 1978; Nelson, 1981). Since Mosier (1967) first showed the requirement for adherent cells in the generation of antibody-forming cells to sheep erythrocytes *in vitro*, many different systems depending on macrophages have been described, including activation of primed T lymphocytes by antigen *in vitro* (Bergholtz and Thorsby, 1977; Farr *et al.*, 1977; Rosenthal *et al.*, 1977; Shevach and Rosenthal, 1973). However, discordant reports concerned the necessity for macrophages in T-cell activation by mitogens (Burckhardt, 1979; Rosenstreich *et al.*, 1976). The role of macrophages as suppressor cells soon gained increasing attention (Calderon *et al.*, 1974; Gmelig-Meyling and Waldmann, 1981; Mattingly *et al.*, 1979; Metzger *et al.*, 1980; Novogrodsky *et al.*, 1979; Weiss and Fitch, 1977, 1978).

The reports attributing this suppressive regulation of immune responses to prostaglandins are steadily accumulating (for reviews see Goodwin and Webb, 1980; Kuehl and Egan, 1980; Stenson and Parker, 1980), but it might be wise not to exclude other soluble mediators and natural killer (NK) cells modulated by interferons (Herberman and Ortaldo, 1981).

At the beginning of our work on the interactions between *A. viscosus* and cells of the immune system, we soon observed confusing suppressor phenomena. In the presence of commonly used mitogens, spleen cells from conventionally maintained RIC rats showed almost no proliferative response *in vitro*, whereas cells taken from germfree rats responded in the expected way. At that time we were able to attribute the suppressive effect to a soluble, transferable "factor" but did not try to further characterize it (Burckhardt an Guggenheim, unpublished data). Mattingly *et al.* (1979) later showed, in a similar comparison between germfree and conventional rats, that the suppression observed in conventional animals was induced by the microbial stimulation acting via a non-adherent cell population on prostaglandin-secreting macrophages.

Subsequently, we further investigated the role played by macrophages in cultures of rat T-cells which had been purified by filtration of spleen cell suspensions through Degalan Ig-anti Ig columns (Burckhardt, 1978, 1979). By using the well-known macrophage toxin, silica (Allison *et al.*, 1966), the absolute requirement for these cells in antigen-induced, but not in mitogen-triggered, proliferation of T cells became evident (Burckhardt, 1979). During

TABLE I. Effect of Silica in a 24 h Preculture on the Subsequent Uptake of [^{3}H]dTd by T cells Activated with ConA or PHA

Mitogen (μg/ml)	*[^{3}H]dTd uptake by T cells [a] precultured [b] in*	
	Silica (100 μg/ml)	*Medium only*
—	*17.8 ± 3.8* [c]	*1.2 ± 0.3*
ConA		
5	*NT* [d]	*35.2 ± 8.5*
1.25	*305.2 ± 13.6*	*134.0 ± 5.6*
0.6	*200.6 ± 7.4*	*1.1 ± 0.3*
PHA		
8.3	*408.5 ± 9.4*	*12.7 ± 1.6*
2.8	*329.6 ± 10.1*	*7.1 ± 0.7*

[a] *For stimulation by mitogen, 0.5 X 10^6 T cells were used.*
[b] *Preculture contained 50 X 10^6 cells per 15 ml medium.*
[c] *Values are expressed in X (dpm X 10^3) ± SE (N = 4). The uptake of [^{3}H]-dTd present during the final 24 h was measured after three days of culture in microtiter plates.*
[d] *NT: not tested*

these studies, we noticed that a preculture of rat T cells in the presence of silica increased rather than reduced the proliferation in the subsequent cultures (Table I). Interestingly, the spontaneous [^{3}H]-thymidine ([^{3}H]-dTd) uptake by control cultures, without any mitogen, was always higher with silica-pretreatment (Table I and Burckhardt, 1979).

Attempts to resuppress this increased spontaneous proliferation of silica-precultured T cells by adding back macrophages which had been obtained by peritoneal flushing from syngeneic rats were indeed successful (Table II). Taken together, these experiments illustrate the dual role played by macrophages, (i) their requirement for antigen-induced T-cell responses and (ii) their suppressor function by restraining (excessive?) spontaneous proliferation.

In another series of experiments, treatment of adult RIC rats with a single dose of cyclophosphamide (25 mg/kg body weight) resulted in a similar increase in the spontaneous proliferation in T-cell cultures set up five weeks after administration of the drug (Table III). This is well in accord with the results of Mitsuoka, Baba, and Morikawa (1977) in which cyclophosphamide enhanced delayed hypersensitivity (to methylated human serum albumin) in mice. They considered the effect to be due, at least in part, to damage to bone marrow-derived monocytes rather than to that of sensitized effector T cells.

Similarly, Meredith, Kristie, and Walford (1979) showed an important link between aging and autoimmunity in C57Bl mice. By using cyclophosphamide in young animals, they enhanced the auto-antibody production after LPS-stimulation *in vitro* from three- to four-fold up to the levels normally found in 21 to 24 month-old animals. These experiments clearly demonstrate that (i) autoreactive clones of lymphocytes exist in the spleen of young adult mice but normally produce little autoantibody and (ii) aging involves an actual increase in autoreactive B-cell expression thus demonstrating a kind of deregulation which can be mimicked by cyclophosphamide in young adult animals.

Deregulation, in the form of a loss of suppressor activity with aging, also was

TABLE II. Effect of Peritoneal Macrophages on the [^{3}H]-dTd Uptake by T cells Precultured for 24 h in the Presence or Absence of Silica

Macrophages added (%)	*[^{3}H]-dTd uptake by T cells [a] precultured [b] in* Silica (100 μg/ml)	Medium only
0.1	100.9 ± 13.1 [b]	21.6 ± 2.3
1	91.3 ± 15.1	30.4 ± 3.7
10	14.5 ± 5.1	10.9 ± 1.3

[a] *For [^{3}H]-dTd uptake studies, 0.5 X 10^6 T cells were used.*
[b] *Values are expressed in X (dpm X 10^3) ± SE (N = 4). The uptake of [^{3}H]-dTd present during the final 24 h was measured after three days of culture in microtiter plates.*

TABLE III. Effect of Cyclophosphamide Treatment In Vivo *on T-cell Reactivity* In Vitro

Mitogen (μg/ml)	Cyclophosphamide [a]			
	Administered		Not administered	
	Exp. No. 1	Exp. No. 2	Exp. No. 1	Exp. No. 2
—	77.2 ± 3.3 [b]	104.0 ± 4.1	6.1 ± 0.7	12.0 ± 0.9
ConA				
2.5	317.5 ± 52.6	355.9 ± 42.7	132.3 ± 32.3	386.0 ± 29.9
0.6	431.4 ± 28.8	315.8 ± 14.0	268.9 ± 13.9	419.9 ± 19.7
PHA				
2.8	215.0 ± 8.3	183.4 ± 9.2	113.1 ± 12.0	186.3 ± 3.3

[a] *Cyclophosphamide (Endostan ®, Asta) (25 mg/kg bodyweight) was given i.p. five weeks before the cells were tested.*
[b] *See Table II, footnote b.*

recently reported for humans (Becker *et al.*, 1981). They found a statistically significant higher monocyte-mediated suppression of cellular immunity in individuals under 50 years old than in those over 60. It thus appears that aging is accompanied by a decrease in general suppressor activity which allows the increased expression of autoreactive clones.

The molecular mechanisms of macrophage interaction with both lymphocytes and non-lymphoid cells are best seen in the context of the "activated macrophage" (Cohn, 1978; Karnovsky and Lazdins, 1978; North, 1978; Waksman and Namba, 1976), the increasing understanding of interleukin action (Oppenheim and Gery, 1982), and the knowledge of the interaction between interferons and natural killer (NK) cells (Herberman and Orlando, 1981).

The following example may illustrate how rapidly sound and valid data may have to be revised. Suppressor activity in human T lymphocytes had been ascribed to a subpopulation bearing receptors for IgG (T_G cells)(Grossi *et al.*, 1978; Hayward *et al.*, 1978; Moretta *et al.*,1976, 1977). During separation from peripheral blood leukocytes, T_G cells were shown to be non-adherent to plastic, to express receptors for sheep erythrocytes, and to form immune complexes with IgG. These cells even engulfed ox erythrocytes which had been coated with IgG and displayed characteristic azurophilic granules in the cytoplasm when stained with May-Grünwald-Giemsa (Grossi *et al.*, 1978). Therefore, they were also called "large, granular lymphocytes". Functionally, in contrast to the subpopulation bearing receptors for IgM (T_M-cells) which expressed the mandatory helper function (Moretta *et al.*, 1976, 1977), the T_G population was able to suppress the production of immunoglobulin by B cells in a pokeweed mitogen-driven system. However, from studies using monoclonal antibodies specific for human T-cell subsets and monocytes (Reinherz and Schlossman, 1980), it was suggested that T_G-cells might belong to the NK-cell population (Reinherz, *et al.*, 1980). Abo and Balch (1981) have elegantly confirmed this with their monoclonal antibody (HNK-1) by identifying a differentiation antigen of human NK and K cells. Thus, it remains to be seen whether the concept of (human) suppressor T cells needs to be revised in the near future.

The crucial question of how all these cells induce bone resorption shall be discussed in section IV.

IV. MECHANISMS OF BONE RESORPTION

Supernatants from human leukocytes stimulated *in vitro* were first shown to possess bone resorbing activity when cultured with fetal rat bones (Horton *et al.*, 1972). Those authors named this activity osteoclast-activating factor (OAF). It was later shown that OAF-production was elicited not only by stimulated leukocytes but also by the classical and alternative complement pathways and was mediated by prostaglandins (PGE) (Sandberg, 1977). Inhibitors of prostaglandin synthetase effectively blocked the production of OAF (Yoneda and Mundy, 1979); Bockman and Repo (1981) elegantly demonstrated that rapid synthesis and release of PGE preceded and was necessary for the bone resorption caused by OAF. Physiological levels of parathyroid hormone were ineffective and therefore clearly distinct from OAF. Bockman and Repo (1981) discussed several possible mechanisms by which PGE could operate and also considered positive feedback loops on monocytes.

Briefly summarized, bone resorption occurs after a variety of mainly non-(antigen)specific stimuli, such as mitogen-induced lymphokines, immune complexes, or complement activation via the alternative pathway, have acted on macrophages or triggered prostaglandin synthesis and release *in situ*. Reports describing locally active mediators from the family of the arachidonic acid metabolites, other than PGE_2, may soon accumulate.

V. CONCLUSIONS

In this review, I have described only a little piece of a big puzzle, periodontal disease. However, what the animal model and *Actinomyces viscosus* have taught us may well serve one day to link basic research with application *in praxi*.

I deliberately excluded aspects concerning polymorphonuclear leukocytes, their products, and the soluble mediators generated in the complement cascade and in the clotting system. In summary, I can state that (i) the host's humoral and cellular immune response to oral microorganisms develop temporally in parallel with alveolar bone loss and can be viewed as a cause-and-effect relationship, not only as an epiphenomenon; (ii) normally, this allergic response which depends on the local presence of antigens is regulated by several suppressor mechanisms, mainly from macrophages and natural killer cells; (iii) a deregulation, the onset of which during aging is likely to be genetically encoded, permits a broader, polyclonal B-cell activation and immune-complex formation and may be classified as a form of auto-aggressive disease; and (iv) Regardless of how the respective signals are generated, activation of osteoclasts results in an imbalance of bone metabolism.

Consequently, successful therapeutic approaches in allergic periodontitis are unlikely to be based on manipulation of some of the host's antigen-specific reactions, *e. g.* by interfering in the fine-tuned network of idiotype-anti-idiotype reactions (Jerne, 1974; Lindenmann, 1973). Such manipulation is known to be feasible in inducing tolerance to transplantation antigens (Binz and Wigzell, 1976). Rather, they will stem from a more thorough understanding of the immunopharmacological events which form the leukocyte-to-osteoclast axis and from further analyses of the genetics involved in the regulation of the immune response in general and of suppressor mechanisms in particular.

ACKNOWLEDGMENTS

I thank Prof. B. Guggenheim for his continuous support and enthusiasm and M.D. Cooper for his great interest and invaluable encouragement.

REFERENCES

Abo, T., and Balch, C. M., *J. Immunol. 127:*1024 (1981).
Allison, A. C., Harrington, J. S., and Birbeck, M., *J. Exp. Med. 124:*141 (1966).
Andersson, J., Sjöberg, O., and Möller, G., *Eur. J. Immunol. 2:*349 (1972).
Andersson, J., Melchers, F., Galanos, C., and Lüderitz, O., *J. Exp. Med. 137:*943 (1973).
Andersson, J., Coutinho, A., and Melchers, F., *J. Exp. Med. 145:*1511 (1977).
Attström, R., *J. Clin. Periodontol. 2:*25 (1975).
Baker, J. J., Chan, S. P., Socransky, S. S., Oppenheim, J. J., and Mergenhagen, S. E., *Infect. Immun. 13:*1363 (1976).
Becker, M. J., Drucker, I., Farkas, R., Steiner, Z., and Klajman, A., *Clin. Exp. Immunol. 45:*439 (1981).
Bergholtz, B. O., and Thorsby, E., *Scand. J. Immunol. 6:*779 (1977).
Bing, J. *Acta Med. Scand. 103:*547 (1940).
Binz, H., and Wigzell, H., *Nature 262:*294 (1976).
Bockman, R. S., and Repo, M. A., *J. Exp. Med. 154:*529 (1981).
Brandtzaeg, P., *Int. Dent. J. 23:*438 (1973).
Brandtzaeg, P., and Kraus, F. W., *Odontol. Tidskr. 73:*285 (1965).
Burckhardt, J. J., *Scand. J. Immunol. 7:*167 (1978).
Burckhardt, J. J., *Scand. J. Immunol. 10:*229 (1979).

Burckhardt, J. J., Guggenheim, B., and Hefti, A., *J. Immunol. 118:*1460 (1977).
Burckhardt, J. J., Gaegauf-Zollinger, R., and Guggenheim, B., *J. Periodontal Res. 16:*147 (1981a).
Burckhardt, J. J., Gaegauf-Zollinger, R., Schmid, R., and Guggenheim, B., *Infect. Immun. 31:*971 (1981b).
Burckhardt, J. J., Gaegauf-Zollinger, R., Gmür, R., and Guggenheim, B., *Infect. Immun.* (1982). (In press)
Burnet, F. M., "The clonal selection theory of acquired immunity." Cambridge University Press, London (1959).
Calderon, J., Williams, R. T., and Unanue, E. R., *Proc. Nat. Acad. Sci. (U.S.A.) 71:*4273 (1974).
Cohn, Z. A., *J. Immunol. 121:*813 (1978).
Coutinho, A., and Möller, G., *Nature New Biology, 245:*12 (1973).
Clagett, J. A., and Engel, D., *Dev. Comp. Immunol. 2:*235 (1978).
Clagett, J. A., and Page, R. C., *Arch. Oral Biol. 23:*153 (1978).
Dresser, D. W., *Nature 274:*480 (1978).
Ebersole, J. L., Taubman, M. A., Smith, D. J., Genco, R. J., and Frey, D. E., *Clin. Exp. Immunol. 47:*43 (1982).
Engel, D., Clagett, J., Page, R., and Williams, B., *J. Immunol. 118:*1466 (1977).
Fagraeus, A., (1948) Antibody production in relation to the development of plasma cells. Thesis, Stockholm. Cited in *Scand. J. Immunol. 13:*99 (1981).
Farr, A. G., Dorf, M. E., and Unanue, E. R., *Proc. Nat. Acad. Sci. (U.S.A.) 74:*3542 (1977).
Fudenberg, H. H., *Scand. J. Immunol. 12:*459 (1980).
Gaegauf-Zollinger, R., Burckhardt, J. J., and Guggenheim, B., *Arch. Oral Biol.* (1982). (In press)
Genco, R. J., Mashimo, R. A., Krygier, G., and Ellison, S. A., *J. Periodontol. 45:*330 (1974).
Gmelig-Meyling, F., and Waldmann, T. A., *J. Immunol. 126:*529 (1981).
Goodwin, J. S., and Webb, D. R., *Clin. Immunol. Immunopathol. 15:*106 (1980).
Grigsby, W. R., and Sabiston, Jr., C. B., *J. Oral Pathol. 5:*175 (1976).
Grossi, C. E., Webb, S. R., Zicca, A., Lydyard, P. M., Moretta, L., Mingari, C., and Cooper, M. D., *J. Exp. Med. 147:*1405 (1978).
Guggenheim, B., Gaegauf-Zollinger, R., Hefti, A., and Burckhardt, J. J., *J. Periodont. Res. 16:*26 (1981).
Guggenheim, B., and Schroeder, H. E., *Infect. Immun. 10:*565 (1974).
Hausmann, E., J. *Periodontol.* 45:338 (1974).
Hausmann, E., and Ortman, L., *J. Periodontol. 50* (special issue):7 (1979).
Hayward, A. R., Layward, L., Lydyard, P. M., Moretta, L., Dagg, M., and Lawton, A. R., *J. Immunol. 121:*1 (1978).
Herberman, R. B., and Ortaldo, J. R., *Science 214:*24 (1981).
Horton, J. E., Oppenheim, J. J., and Mergenhagen, S. E., *J. Periodontol. 45:*351 (1974).
Horton, J. E., Raisz, L. G., Simmons, H. A., Oppenheim, J. J., and Mergenhagen, S. E., *Science 177:*793 (1972).
Hulin, C., *Odontologie 67:*324 (1929).
Ivanyi, L., and Lehner, T., *Arch. Oral Biol. 16:*1117 (1971).
Jerne, N. K., *Ann. Immunol. 125 C:*373 (1974).
Jordan, H. V., and Keyes, P. H., *Arch. Oral Biol. 9:*401 (1964).
Jordan, H. V., Fitzgerald, R. J., and Stanley, H. R., *Am. J. Pathol. 47:*1157 (1965).
Karnovsky, M. L., and Lazdins, J. K., *J. Immunol. 121:*809 (1978).
Katz, D. H., and Benacerraf, B., *Adv. Immunol. 15:*1 (1972).
Keyes, P. H., and Gold, H. S., *Oral Surg. Oral Med. Oral Pathol. 8:*492 (1955).
König, K. G., and Guggenheim, B., *Adv. Oral Biol. 3:*217 (1968).
Kolouch, F., Good, R. A., and Campell, B., *J. Lab. Clin. Med. 32:*749 (1947).
Kuehl, F. A., and Egan, R. W., *Science 210:*978 (1980).
Lang, N. P., and Smith, F. N., *J. Periodont. Res. 12:*298 (1977).
Lehner, T., *J. Oral Pathol. 1:*39 (1972).
Lehner, T., Wilton, J. M. A., Challacombe, S. J., and Ivanyi, L., *Clin. Exp. Immunol. 16:*481 (1974).
Lindenmann, J., *Ann. Immunol.* (Paris) 124 C:171 (1973)
Mackler, B. F., Altman, L. C., Wahl, S., Rosenstreich, D. L., Oppenheim, J. J., and Mergenhagen, S. E., *Infect. Immun. 10:*844 (1974).

Mackler, B. F., Frostad, K. B., Robertson, P. B., and Levy, B. M., *J. Periodontal Res. 12:*37 (1977).
Marchalonis, J. J., *Immunol. Today 3:*10 (1982).
Mattingly, J. A., Eardley, D. D., Kemp, J. D., and Gershon, R. K., *J. Immunol. 122:*787 (1979).
Meredith, P. J., Kristie, J. A., and Walford, R. L., *J. Immunol. 123:*87 (1979).
Metzger, Z., Hoffeld, J. T., and Oppenheim, J. J., *J. Immunol. 124:*983 (1980).
Mitsuoka, A., Baba, M., and Morikawa, S., *Nature 262:*77 (1976).
Möller, E., Ström, H., and Al-Balaghi, S., *Scand. J. Immunol. 12:*177 (1980).
Moretta, L., Ferrarini, M., Mingari, M. C., Moretta, A., and Webb, S. R., *J. Immunol. 117:*2171 (1976).
Moretta, L., Webb, S. R., Grossi, C. E., Lydyard, P. M., and Cooper, M. D., *J. Exp. Med. 146:*184 (1977).
Mosier, D. E., *Science 158:*1573 (1967).
Nelson, D. S., *Clin. Exp. Immunol. 45:*225 (1981).
Newman, H. N., *J. Clin. Periodontol. 7:*251 (1980).
Niesengard, R. J., *J. Periodontol. 48:*505 (1977).
North, R. J., *J. Immunol. 121:*806 (1978).
Novogrodsky, A., Rubin, A. L., and Stenzel, K. H., *J. Immunol. 122:*1 (1979).
Oppenheim, J. J., and Gery, I., *Immunol. Today 3:*113 (1982).
Page, R. C., and Schroeder, H. E., *Lab. Invest. 33:*235 (1976).
Page, R. C., and Schroeder, H. E., *J. Periodontol. 52:*477 (1981).
Primi, D., Hammarström, L., Smith, C. J. E., and Möller, G., *J. Exp. Med. 145:*21 (1977).
Reinherz, E. L., and Schlossman, S. F., *Cell 19:*821 (1980).
Reinherz, E. L., Moretta, L., Roper, M., Breard, J. M., Mingari, M. C., Cooper, M. D., and Schlossman, S. F., *J. Exp. Med.* 151:969 (1980)
Rosenstreich, D. L., Farrar, J. J., and Dougherty, S., ;*J. Immunol. 116:*131 (1976).
Rosenthal, A. S., Barcinski, M. A., and Blake, J. T., *Nature 267:*156 (1977).
Sandberg, A. L., Raisz, L. G., Goodson, J. M., Simmons, H. A., and Mergenhagen, S. E., *J. Immunol. 119:*1378 (1977).
Seymour, G. J., Powell, R. N., and Davies, W. I. R., *J. Oral Pathol. 8:*249 (1979a).
Seymour, G. J., Powell, R. N., and Davies, W. I. R., *J. Clin. Periodontol. 6:*267 (1979b).
Shevach, E. M., and Rosenthal, A. S., *J. Exp. Med. 138:*1213 (1973).
Slots, J., *J. Clin. Periodontol. 6:*351 (1979).
Socransky, S. S., *J. Periodontol. 48:*497 (1977).
Stenson, W. F., and Parker, C. W., *J. Immunol. 125:*1 (1980).
Stingl, G. S., Katz, S. I., Shevach, E. M., Rosenthal, A. S., and Green, I., *J. Invest. Dermatol. 71:*59 (1978).
Unanue, E. R., *Immunol. Rev. 40:*227 (1978).
van der Hoeven, J. S., *Caries Res. 8:*193 (1974).
Waksman, B. H., and Namba, Y., *Cell. Immunol. 21:*161 (1976).
Weiss, A., and Fitch, F. W., *J. Immunol. 119:*510 (1977).
Weiss, A., and Fitch, F. W., *J. Immunol. 120:*357 (1978).
Yoneda, T., and Mundy, G. R., *J. Exp. Med. 149:*279 (1979).

POLYCLONAL B-CELL ACTIVATION IN RESPONSE TO *ACTINOMYCES VISCOSUS* — ITS NATURE AND GENETICS[1]

James A. Clagett
L. David Engel

Departments of Periodontics, Microbiology-Immunology, and Pathology
School of Dentistry and Medicine
University of Washington
Seattle, Washington, U. S. A.

I. INTRODUCTION

The kinds and effectiveness of immune mechanisms available to mammals by which they may limit the spread and shorten the duration of infections by microorganisms, such as *Actinomyces viscosus*, are central to the survival of the species. Specific immune mechanisms exist in which clones of antigen-sensitized T and B lymphocytes grow and produce immune mediators that directly or indirectly restrict the growth of microorganisms. The development of antigen-specific lymphocytes is probably a relatively recent evolutionary adaptation compared with other mechanisms of host defense, such as phagocytosis. During the past twenty years, evidence has accumulated which clearly demonstrates the existence of a primitive form of immunity, nonspecific for antigen, which is mediated by lymphocytes and is termed polyclonal cell activation (Clagett and Engel, 1978). Polyclonal cell activation differs from antigen-specific immunity in that a large fraction of lymphocytes (5 to 33%) may be recruited, activated, and induced to differentiate. Although antigen-specific receptors are not involved in this process, a cell surface receptor whose expression is also genetically controlled is responsible for cell activation (Watson and Riblet, 1974, 1975). However, the functional effects *in vivo* and *in vitro* of the primitive or polyclonal cell activators mimic the antigen-

[1] *This work was supported by Public Health Service grants DE-02600 and DE-07063.*

ISBN 0-12-528620-1

specific response. Man (Greaves *et al.*, 1974), mice (Clagett and Engel, 1980; Engel *et al.*, 1977), and rats (Burckhardt *et al.*, 1977) have retained the polyclonal cell activation system. *A. viscosus* and nearly all other bacterial species examined thus far possess one or more substances capable of inducing polyclonal cell activation. Bacterial substances which nonspecifically activate B lymphocytes are termed polyclonal B-cell activators (PBA's). *A. viscosus* contains a potent PBA, termed AVIS (Engel *et al.*, 1977), which contains protein and carbohydrates and is associated with the cell wall of the bacteria.

The nature of the proliferative and differentiative responses *in vitro* by B lymphoytes in response to PBA(s) from *A. viscosus* gives clearer insights into the ways the host meets and usually overcomes microbial challenge. A genetic analysis of the capability to respond to PBA's would resolve the question of whether or not every member of a species carries the same genetic potential to resist infection. Understanding the mode of inheritance and determining the number of genes controlling B-lymphocyte responses to PBA's from *A. viscosus* would add another piece to the mosaic of our knowledge of polyclonal B-cell activation. At the same time, such information would raise the possibility that some chronic inflammatory diseases which result in part from polyclonal activation of host B cells by bacterial, viral, fungal, or parasitic B-cell activators have a genetic basis.

II. DNA SYNTHESIS RESPONSES BY ATHYMIC MURINE SPLENOCYTES TO AVIS

One milliliter of splenocyte cultures (5 X 10^5) cells from athymic nude (nu/nu) C57BL/6J mice were established in Falcon 3033 tubes (Clagett *et al.*, 1980). AVIS was prepared as previously described (Clagett *et al.*, 1980); the concentration of AVIS which produced maximal DNA synthesis was 100 to 200 μg/ml. Replicate splenocyte cultures, with and without AVIS, were incubated at 37°C and were assayed daily for six days to determine the level of DNA synthesis by the addition of ^{125}I-UdR (0.5 μCi) to each culture four hours prior to sampling. The radioactivity incorporated into DNA was determined in a gamma counter. In other experiments (Clagett *et al.*, 1980), the numbers of recoverable cells in control (no AVIS added) and in experimental (AVIS added) cultures were counted to determine the level of polyclonal B-cell growth (data not shown). The peak of DNA synthesis reached maximal levels on day 3 of culture and subsequently declined over the remaining three days (Fig. 1). AVIS induced a 4- to 5-fold increase in cell number; maximal numbers were reached four days after initiation of the culture (data not shown). Cultures without AVIS showed little spontaneous DNA synthesis and no increase in total cell numbers (data not shown). The AVIS-induced DNA synthesis response and the growth characteristics of splenic cells of euthymic C57BL/6J mice did not differ from that of athymic littermates.

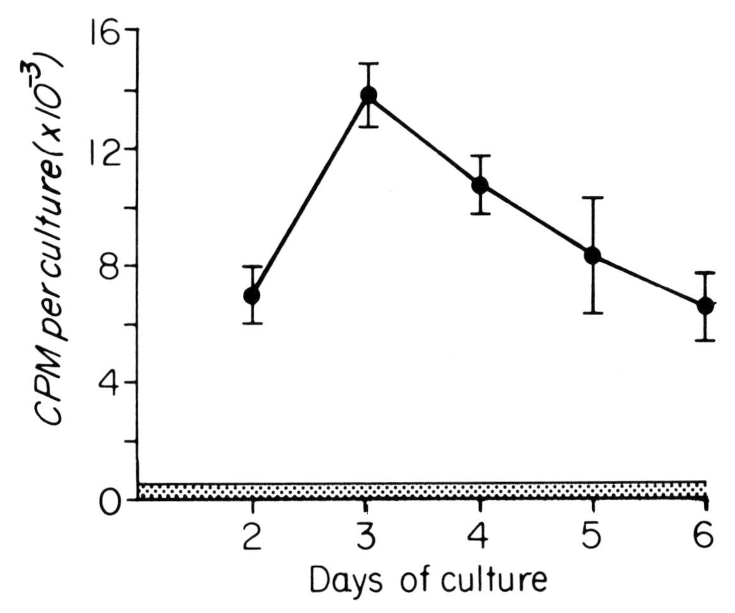

FIGURE 1. Kinetics of [^{125}I] deoxyuridine incorporation in AVIS-stimulated cultures of splenocytes from homozygous nude C57BL/6J mice. Data are presented as mean ± 1 standard deviation. The level of [^{125}I]deoxyuridine incorporation in control cultures is indicated by the shaded area.
(Reproduced from Clagett, J., Engel, D., and Chi, E., Infect. Immun. 29:*234 (1980) by copyright permission of the American Society for Miciobiology.)*

III. IgM AND IgG SUBCLASS PRODUCTION IN SPLENOCYTES FROM ATHYMIC NUDE (nu/nu) MICE

Splenocytes congenitally deficient in mature T cells were cultured as described previously (Clagett *et al.*, 1980) and cultures were harvested daily to determine the frequency of cells synthesizing IgM and/or IgG_1, IgG_2, IgG_3. The presence of cytoplasmic immunoglobulins (c-Ig) in the cultured cells was tested by staining fixed preparations with anti-mouse IgM conjugated with fluorescein (FITC anti-IgM) and counterstaining the cells with anti-mouse conjugated with rhodamine (RITC anti-mouse IgG_1, -IgG_2, -IgG_3), respectively. After five days in culture with AVIS, approximately 20% of the recoverable cells contained cytoplasmic IgM (Fig. 2, panel A). In contrast to cells synthesizing only IgM, an increasing number of cells bearing only IgG_2 or IgG_3 were observed (Fig. 2, panel B). This cohort represented up to 12% of the cells. An additional 5 to 8% of the splenocytes contained both c-IgM and IgG_2 plus IgG_3. AVIS induced no significant increase in the frequency of cells synthesizing c-IgG_1 or c-IgA during these experiments. Without AVIS, the frequency of cells synthesizing c-Ig's remained significantly below 1% of the recoverable

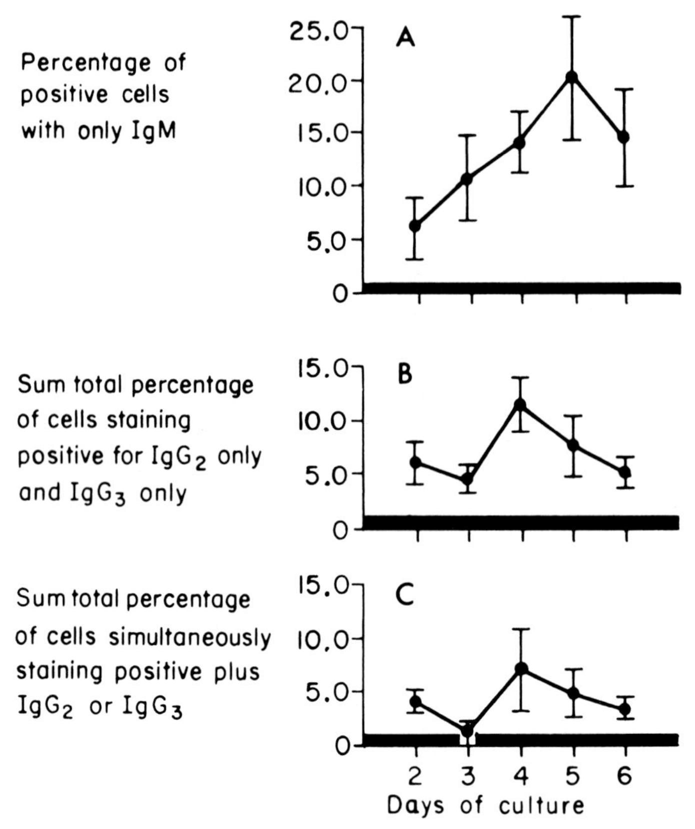

FIGURE 2. (A) Temporal appearance of cells synthesizing only IgM. (B) Appearance of cells synthesizing only IgG_2 or only IgG_3. The frequency of cells synthesizing only IgG_2 and that of cells synthesizing only IgG_3 were added together since these cells were present at approximately equal frequencies. (C) Appearance of cells synthesizing IgM simultaneously with IgG_2 or IgG_3. The frequency of cells synthesizing IgM plus IgG_2 and that of cells synthesizing IgM plus IgG_3 were added together since these cells were present at equal frequencies. The data for (A), (B), and (C) are shown as the mean ± 1 standard deviation, and the solid black area represents the mean background values ± 2 standard deviations.
(Reproduced from Clagett, J., Engel, D., and Chi, e., Infect. Immun. *29:234 (1980) by copyright permission of the American Society for Miciobiology.)*

cells. Early in the culture periods, cytoplasmic-Ig-containing cells were usually large (> 12 μm in diameter) and had a blast cell morphology. During the final days of the experiments, plasma cells and plasmablasts were the predominant c-Ig-containing cell types observed (data not shown).

IV. FUNCTIONAL ANALYSIS

Gronowicz and Coutinho (1975, 1976) have shown that each PBA induces proliferation in a subset of B cells. We determined whether the subset activated by AVIS was identical to, overlaps with, or independent of the subsets activated by two very different PBA's, bacterial lipopolysaccharide (LPS) and dextran sulfate (DxS). Splenocyte cultures containing either AVIS, LPS, or DxS, and 10^{-5} M BUdR were established. After 48 h, the cells were washed three times, incubated with a light-enhancing dye (Hoechst # 33258) and then exposed to white light for 60 min. (All cells that had divided and incorporated the BUdR into their DNA are effectively killed by this procedure (Zoschke and Bach, 1971)). After the remaining cells were washed and cultured an additional 12 to 18 h to permit the death of all cells containing BUdR (Zoschke and Bach, 1971), the cells were counted. The response to a second stimulation of the cells with either the same or one of the other PBA's was measured by determining the amount of added ^{125}I-UdR that was incorporated into the cells. Other cultures were not exposed to BUdR but, during the termination of the second incubation period (see above), were incubated with Hoechst dye and exposed to white light to serve as control for dye and light toxicity. The results of these controls were used as references for the effects of BUdR suicide. T cells should not be affected by the BUdR technique, since the do not proliferate in response to PBA's. Phytohemagglutinin P (PHA-P) was included during the second culture period as a control for toxicity and for nonspecific killing, since it is a mitogen for T cells.

The data presented in Table I show that when cells which had been cultured with AVIS in the presence of BUdR for 48 h and then exposed to Hoechst dye and white light, were further stimulated with AVIS, their response was reduced by 88%. By comparison, removing the AVIS-responsive cells reduced the DxS response by 33%. When the DxS responsive cells were eliminated first, the proliferative response to AVIS (10,915 cpm) was reduced by 66% or 80%, using either 32,095 cpm or 56,509 cpm as the control value, respectively. In all cases, the DNA synthesis response to PHA-P did not decrease, but was marginally, although not significantly, increased by 31%. This increase probably reflected an enrichment of T cells due to the loss of B cells by BUdR suicide. Two points are clear from this data. Some of the DxS-stimulated cells became responsive to AVIS; hence, depleting DxS-responsive cells had a pronounced effect on the proliferative response to AVIS. By contrast, AVIS stimulated cells which were largely unresponsive to DxS, since the removal of AVIS-responsive cells reduced only moderately (33%) the response to DxS.

V. GENETIC CONTROL OF PROLIFERATIVE RESPONSES TO AVIS

How one member of a species handles an infectious agent may depend on the presence and number of genes in its genome which control critical aspects of the cellular defense system. Since we believe that polyclonal responses are a primitive but rapid way to deal with bacterial agents, the genetic control of the B cell response to AVIS becomes an important question. We screened the pro-

TABLE I. Use of BUdR Suicide Technique to Analyze the Functional Heterogeneity of B Cells Responding to AVIS

Condition	Mitogens: First Mitogen (0-2 days)	Mitogens: Second Mitogen [a] (3-5 days)	^{125}I-UdR cpm per culture
Experimental	AVIS + BUdR	DxS	20,816
	"	AVIS	4,124
	"	Medium	3,057
	"	PHA-P	39,197
Experimental	DxS + BUdR	DxS	3,670
	"	AVIS	10,915
	"	Medium	3,022
	"	PHA-P	34,127
Control	Medium + BUdR	DxS	24,721
	"	AVIS	32,095
	"	Medium	1,386
	"	PHA-P	38,314
Control	AVIS + Medium	DxS + dye + light	31,479
	"	AVIS + dye + light	34,870
	"	Medium + dye + light	26,888
	"	PHA-P + dye + light	29,669
Control	DxS + Medium	DxS + dye + light	36,035
	"	AVIS + dye + light	56,509
	"	Medium + dye + light	27,466
	"	PHA-P + dye + light	35,182

[a] *After 48 h of culture in the presence of the first mitogen plus BUdR, the cells were washed, the Hoechst dye was added, and the cells were exposed to light for 1 h. The cells were washed three times and counted. Smears were made for microscopic examination. The cells were allowed to incubate for an additional 24 h (day 3). The cells were recounted, adjusted to the same number of viable cells and cultured as indicated. In all cases, exposure to BUdR and light depleted lymphoblasts by >99% and the level of ^{125}I-UdR incorporation was the same or less than unstimulated cells.*

liferative response of 18 inbred strains of mice which differed at numerous genetic loci including H-2, the major histocompatibility complex. The results of this experiment are shown in Figure 3. All strains responded, although there were minor variations in the magnitude of the proliferative response. The SM/J mice consistently demonstrated a 5- to 8-fold higher proliferative response to AVIS than did the other strains. Highly elevated proliferative responses of SM/J splenocytes to DxS, LPS, and purified protein derivative also showed that SM/J mice were hyperresponsive to several PBA's (data not shown). That the magnitude of the PHA-P proliferative responses of all strains were very similar agrees with Heiniger *et al.* (1975).

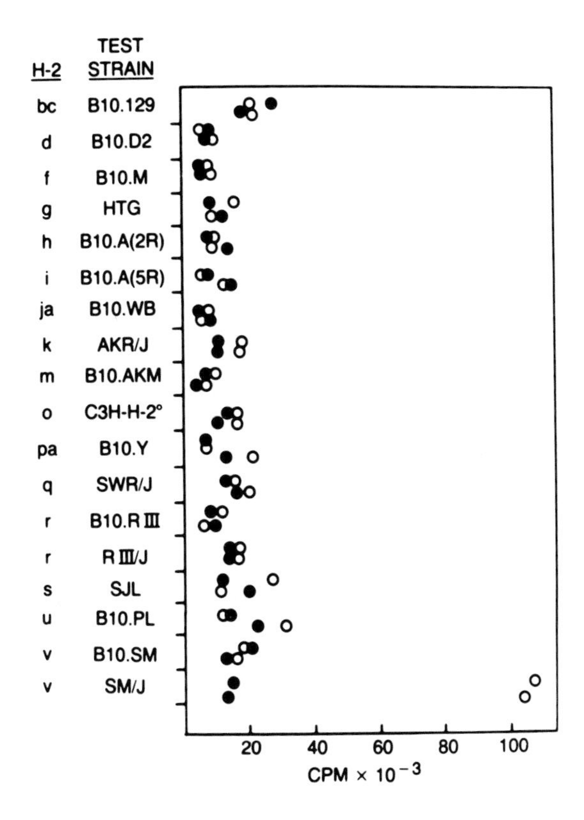

FIGURE 3. The AVIS-induced proliferative responsiveness of splenocytes from a panel of inbred mouse strains. 4 X 10 [5] cells/well were cultured for 48 h in the presence or absence of 100 μg/ml of AVIS mitogen. The cells received a pulse of ^{125}I-deoxyuridine 4 h before harvest. The data represent the mean counts per minute (cpm) of six replicate AVIS-stimulated cultures minus the mean cpm of six replicate nonstimulated cultures. Nonstimulated cultures were generally in the range of 200-1,200 mean cpm. Each point represents the data from one animal. Symbols: O, *test strain;* ●, *C57BL/6J control.*
(Reproduced From Engel, D. et al., J. Exp. Med. 154: *726 (1981) by copyright permission of The Rockefeller University Press.)*

VI. GENETIC ANALYSIS OF PROLIFERATIVE RESPONSES TO AVIS

We were interested in determining the details of the genetic control of proliferative response to AVIS in SM/J mice. The aspects studied included mode of inheritance (dominant, semi-dominant), H-2 linkage, and sex-linkage. When the proliferative responses of F_1 progeny of (B6 X SM) were compared to age- and sex-matched SM/J and C57BL6/J mice, the F_1 mice exhibited intermediate levels of proliferation to AVIS (Fig. 4) and to LPS (data not shown). The proliferative response of (B6 X SM) F_1X B6 backcross progeny was intermediate between F_1 and B6 parental strain. These data show clearly that

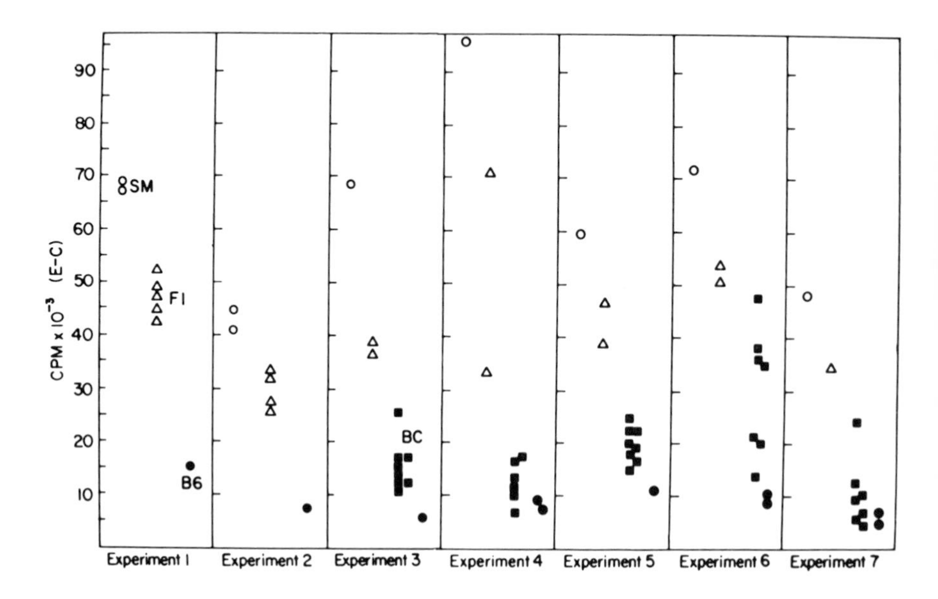

FIGURE 4. Analysis of the AVIS-induced proliferative responsiveness of splenocytes from (B6 X SM) F_1 and (B6 X SM)F_1 X B6 backcross mice. Culture conditions were the same as described in Fig. 3. The data represent the mean cpm of six replicate AVIS-stimulated cultures minus the mean cpm of six replicate nonstimulated cultures. Nonstimulated cultures were generally in the range of 200-1,200 mean cpm. Experiments 1 and 2 compare the (B6 X SM) F_1 responses with the responses of SM/J and B6 parental strains. Experiments 3-7 compare the responses of backcross mice with the responses of SM/J, F_1, and B6 mice. Symbols: O, *SM/J;* △, *(B6 X SM)F_1;* ■, *(B6 X SM)F_1 X B6 backcross;* ●, *B6.*
(Reproduced from Engel, D. et al., J. Exp. Med. 154:*726 (1981) by copyright permission of The Rockefeller University Press.)*

AVIS-induced proliferation was under polygenic control. Thirty-six backcross mice were H-2 typed using antiserum specific for K and D alloantigens. Animals that were H-$2^{v/b}$ and H-$2^{b/b}$ had identical proliferative responses to AVIS and to LPS (Table II). The fact that the B10.SM congenic-resistant strain carries the Sm/J H-2^{v} haplotype, but had a proliferative response similar to that of the B6 suggests that the genes conferring hyper-responsiveness to AVIS are not present on the segment of chromosome 17 which had been transferred to B10.SM.

In other experiments (data not shown), a regulatory role of T cells and of macrophages in this hyperresponsiveness to AVIS was excluded. The removal of T cells from splenocyte preparations by antibody and complement and the removal of macrophages by poisoning with silica had no effect on the magnitude of the DNA synthesis response of SM/J mice to AVIS or to LPS. In contrast, removal of B cells by anti-mouse Ig plus complement or by rosetting with anti-mouse Ig coupled to red cells abrogated mitogenesis. This suggested

TABLE II. Lack of Association between H-2 Type and Mitogenic Responses in (B6 X SM) F_1 X B6 Backcross Mice

H-2 genotype [b]	Number of mice	DNA synthesis [a]	
		AVIS [c]	LPS [c]
b/b	20	18,741 (2,268)	31,189 (3,121)
v/b	16	17,478 (2,438)	30,073 (3,571)

[a] *Splenocytes were cultured as described in Table I, and DNA synthesis was quantitated by incorporation of a pulse of* ^{125}*IUdR. Data are expressed as the mean cpm (SEM).*
[b] *Genotypes were determined by a cytotoxicity assay using specific antisera to K and D alloantigens and rabbit complement.*
[c] *AVIS and LPS mitogens were used at 100* μg/ml.
(Reproduced from Engel, D. et al., J. Exp. Med. 154:*726 (1981) by copyright permission of The Rockefeller University Press.)*

that the genetically controlled hyperresponsiveness to mitogens is expressed solely through the B cell.

VII. DISCUSSION

The data presented in this paper showed that mice have retained a primitive, but complex and genetically regulated cellular response to *A. viscosus*. The response was polyclonal in nature: B cells proliferated and IgM and IgG classes of immunoglobulin were produced. AVIS, the PBA from *A. viscosus*, acted directly on the B lymphocyte without the participation of either T cells or macrophages. Several pieces of evidence supported this notion. AVIS stimulates T-deficient splenocyte cell suspension to proliferate to a degree similar to that of T-sufficient preparations (Engel *et al.*, 1977). Euthymic and athymic C57BL/6J mice responded equally well to AVIS, and the hyperresponsiveness of SM/J splenocytes was not diminished by T-cell depletion. Although macrophages are important in antigen-specific responses as antigen-presenting cells, the polyclonal response to AVIS neither required nor was affected by the removal and killing of splenic macrophages. This does not imply, however, that macrophages do not play a critical role in the clearance of *Actinomyces* species from infected tissues.

That the magnitude and temporal appearance of DNA synthesis and of immunoglobulin response induced by AVIS were similar to that of LPS suggests that the subsets of B cells activated by AVIS and LPS may be overlapping. By BUdR suicide techniques, stimulation with either LPS or AVIS abolished the subsequent response to either PBA. In marked contrast, the selective killing of the cells responsive to DxS altered only moderately the proliferative response to AVIS. The subset of B cells activated by DxS appeared to be only partially overlapping with that of AVIS and LPS (data not shown).

These findings also raise the distinct possibility that the etiologic factor responsible for the histologic and pathologic changes occurring in sites of *Actinomyces* infection is the polyclonal activation of B cells. For examples, in bacteria-laden areas of a host, such as the gingival sulcus and gastrointestinal tract, PBAs from *Actinomyces* and other predominant flora release PBA's which activate B cells to sequester *in situ*, to proliferate, to produce immunoglobulin and other mediators of inflammation, and finally to differentiate into plasma cells (Clagett and Engel, 1978). Equally intriguing is the possibility that the control of the inductive signal to the B-lymphocyte is under complex genetic control. The implication is that some individuals are at risk of immunopathologic tissue injury, yet at the same time, they possess the capacity to a potent immunologic pathway.

VIII. SUMMARY

In addition to antigenetically specific T- and B-lymphocyte responses, man and animals have maintained a primitive but effective system for defense against bacterial infection. B lymphocytes of the organism respond on a polyclonal basis to mitogenic substances released by bacteria colonizing the host. These substances which are called polyclonal B cell activators (PBA's) induce B lymphocytes to manifest many of the properties of an antigen-activated response. Specifically, *A. viscosus* contains a potent PBA which we have termed AVIS. Proliferation and immunoglobulin synthesis in murine splenocytes which were induced *in vitro* by AVIS did not require T cells or macrophages. IgM-producing plasmablasts were observed in significant numbers after two days of culture. In contrast, cells producing IgG_2 or IgG_3 did not appear in significant numbers until three to four days of culture. At that time, 10% of the recoverable cells contained only IgG_2 or IgG_3; 8% stained for both IgM and either IgG_2 or IgG_3.

We examined genetic control of B cell responsiveness to AVIS by testing splenocytes from a large panel of inbred mouse strains. The responsiveness of B cells of one strain, SM/J, was observed to be several-fold higher than that of any other strain. The hyperresponsiveness of SM/J cells to AVIS was not effected by either T cells or macrophages; therefore, genetic control of high responsiveness appears to be expressed solely through the B cell. Examination of (B6 X SM)F_1 X B6 backcross mice revealed that the hyperresponsiveness is under polygenic, autosomal, non-*H-2*-linked gene control. These data raise the possibility that the human polyclonal B-cell response to AVIS and to other PBA's is similarly governed. Furthermore, the propensity to develop diseases which result in part from the polyclonal activation of host B cells by bacterial, viral, fungal, and parasitic B-cell activators may have a genetic basis and, hence, be reliably predicted.

ACKNOWLEDGMENTS

The authors would like to thank Ms. Erna Lund-Gibson for manuscript preparation.

REFERENCES

Burckhardt, J. J., Guggenheim, B., and Hefti, A., *J. Immunol. 118:*1460 (1977).
Clagett, J., and Engel, L. D., *Dev. Comp. Immunol. 2:*235 (1978).
Clagett, J., Engel, L. D., and Chi, E., *Infect. Immun. 29:*234 (1980).
Engel, D., Clagett, J., Page, R., and Williams, B., *J. Immunol. 118:*1466 (1977).
Greaves, M. F., Janossy, G., and Doenhoff, M. J., *J. Exp. Med. 140:*1 (1974).
Gronowicz, E., and Coutinho, A., *Scand. J. Immunol. 4:*429 (1975).
Gronowicz, E., and Coutinho, A., *Scand. J. Immunol. 5:*55 (1976).
Heiniger, H.-J., Taylor, B. A., Hards, E. J., and Meier, H., *Cancer Res. 35:*825 (1975).
Watson, J., and Riblet, R., *J. Exp. Med. 140:*1147 (1974).
Watson, J., and Riblet, R., *J. Immunol. 114:*1462 (1975).
Zoschke, D. C., and Bach, F. H., *J. Immunol. Methods 1:*55 (1971).

MECHANISMS OF PATHOGENESIS AND HOST RESISTANCE TO *NOCARDIA*[1]

Blaine L. Beaman

Department of Medical Microbiology and Immunology
University of California School of Medicine
Davis, California, U. S. A.

I. INTRODUCTION

Of the many species of animals susceptible to naturally acquired infection with *Nocardia* (Ajello *et al.*, 1961; Al-Doory *et al.*, 1969; Bushnell *et al.*, 1979; Gezuelle, 1972; Monga *et al.*, 1978; Orchard, 1979; Pier *et al.*, 1970; Watson, 1977), several have been used to study host-parasite interactions (Beaman, 1973; Folb *et al.*, 1976; Gonzalez-Ochoa and Sandoval-Cuellar, 1976; Kurup *et al.*, 1970; Macotella-Ruiz and Mariat, 1963; Uesaka *et al.*, 1971). Each animal model selected for the study of any infectious disease should be representative of the naturally occurring process. Only recently have good models been established for the study of nocardial infections that appear to be very similar to the natural process which has been described for humans and other animals (Al-Doory *et al.*, 1969; Alteras *et al.*, 1980; Balikian *et al.*, 1978; Curry, 1980; Mason and Hathaway, 1969; Mohapatra and Pine, 1963; Smith and Hayward, 1971). We now have established excellent murine models with *N. asteroides*, *N. brasiliensis*, and *N. caviae* (Beaman, 1980; Beaman and Scates, 1981; Beaman *et al.*, 1978a, 1980a,b, 1982). The diseases produced in these models appear to be almost identical to those clinically described diseases which were produced in the human from whom each nocardial strain was isolated. The laboratory mouse represents a useful animal model for the study of the mechanisms both of pathogenesis and of host resistance because there are dozens of mouse strains that permit

[1]*Most of the information presented in this review was supported by Public Health Service Grants AI-13167 and AI-15114 from the National Institute of Allergy and Infectious Diseases.*

ISBN 0-12-528620-1

dissection *in vivo* of specific single host resistance factors. Further, the immunobiology of mice is more completely defined than is that of any other laboratory animal. Although the murine system differs from the human in many properties, it has been determined that both systems are sufficiently similar to permit analogous studies to be carried out in the field of immunology and host-parasite interactions (Boswell *et al.*, 1980; Fidler *et al.*, 1980; Fletcher *et al.*, 1977; Lizzio and Wargon, 1974; Pritchard and Micklem, 1974).

II. ANIMAL MODELS FOR PULMONARY AND SYSTEMIC NOCARDIOSIS

In humans, one of the most frequently encountered forms of infection caused by *N. asteroides* results from either inhalation or aspiration of nocardial cells into the lungs, thereby leading to pulmonary nocardiosis. The organisms may remain localized within the lungs and lead to pulmonary inflammation, acute pneumonia, abscess formation, or occasionally granulomatous infiltration. The lesions in the lung can vary from a fulminating, aggressive invasion of the lung tissue to a mild or benign self-limited process that may be subclinical and therefore remain undiagnosed. Pulmonary nocardiosis is extremely variable in its clinical manifestations; it mimics a variety of other pulmonary disorders or infections. Frequently, the organisms in the lung enter the circulatory or lymphatic system by which they are spread to other regions of the body. The brain and central nervous system are the most frequent target organs that are infected during dissemination from the lung; however, the kidneys, heart, adrenals, eyes, and other organs may also be affected (Balikian *et al.*, 1978; Beaman *et al.*, 1976; Byrne *et al.*, 1979; Causey and Sieger, 1974; Frazier *et al.*, 1975; Krick *et al.*, 1975; Peterson *et al.*, 1978; Presant *et al.*, 1973; Pulverer and Schaal, 1978; Rosett and Hodges, 1978; Saenz Lope and Gutierrez, 1977; Satterwhite and Wallace, 1979, Sher *et al.* 1977; Stropes *et al.*, 1980; Stuart, 1979; Terezhalmy and Bottomley, 1978).

N. asteroides GUH-2 was isolated from a fatal human infection. The patient had received a kidney transplant prior to developing pulmonary nocardiosis. The organism disseminated from the lungs to the kidneys, the organ for which this strain appeared to have a predilection. It has been demonstrated that this strain of *N. asteroides* is an ideal model for studying host-parasite interactions in mice (Beaman, 1976; Beaman and Maslan, 1977; Beaman *et al.*, 1978a,b, 1980a), since virtually all clinical aspects of the disease described in humans may be reproducibly induced in the normal murine host without the necessity of adding specific adjuvants such as mineral oil or hog gastric mucin. (Previously, this had been a critical problem in establishing *N. asteroides* in an animal model system).

III. ANIMAL MODELS FOR NOCARDIAL MYCETOMAS

A mycetoma is defined as a chronic, progressive, suppurative, and granulomatous process involving the subcutaneous tissues as well as the muscle and bone. This disease process is further characterized by the induction of localized swelling with the production of multiple draining sinuses that discharge a purulent exudate which contains characteristic granules composed of microcolonies of the etiological agent (Barnetson and Milne, 1978; Gonzalez-Ochoa, 1962; Greer, 1974). Mycetomas may be caused either by various fungi or by a variety of actinomycetes; *N. asteroides*, *N. brasiliensis*, and *N. caviae* are important causes of actinomycotic mycetoma in humans (Barnetson *et al.*, 1978; Gonzalez-Ochoa, 1962; Greer, 1974; Sanyal *et al.*, 1978; Talwar and Sehgal, 1979; Thammayya *et al.*, 1972).

N. brasiliensis-17E was isolated from a 40 year-old male construction worker who had developed mycetomatous lesions over 30 percent of his right leg and foot. Several draining sinuses were located in the severely swollen and inflammed foot, and several abscesses were randomly located over the foot and leg. Purulent discharge that contained minute, white granules exuded from the sinuses. When *N. brasiliensis*-17E was injected intravenously into mice, it was observed that the organ tropism and clearance of this organism were quite different from those observed when *N. asteroides* and *N. caviae* were used. *N. brasiliensis*-17E had a predilection for the cooler parts of the body. The bacteria persisted for long periods in the leg muscles, feet, nose, and skull. In addition, from four to six months after injection, typical mycetomas, identical to those described above in the patient from which *N. brasiliensis*-17E was isolated, developed in the tail, feet, legs, and nasal region of the mice. Often the internal organs demonstrated no sign of infection. Histological analysis demonstrated a chronic, progressive, purulent, granulomatous process developing in the mouse. Thus, *N. brasiliensis*-17E induced chronic mycetomas in all mice infected (Beaman, unpublished data).

The murine model utilizing *N. caviae* 112 paralleled the disease process which this organism caused in an otherwise healthy woman (Causey, 1974; Causey *et al.*, 1974). In the human, the infection began in the lungs. There were many episodes of "pleurisy and bronchitis" that persisted for four years. The etiology of this pulmonary infection was not established until dissemination of the organism had occurred. Thereafter, the patient developed massive, chronic subcutaneous lesions as well as multiple abscesses in the lungs and kidneys prior to death. Normal organisms could not be isolated from the patient during the early stages of the disease process; however, after several years, the organisms were isolated from the subcutaneous abscesses. The data suggested the presence of L-forms of *Nocardia* during the first four years of the disease. This strain of *N. caviae* was shown to induce a similar chronic and latent recrudescing disease in mice;

L-forms have been demonstrated to be an integral part of the pathogenesis of this organism (Beaman, 1980; Beaman *et al.*, 1980, 1981).

IV. IMMUNOBIOLOGY OF MICE INFECTED WITH *NOCARDIA*

Genetically immunodeficient murine models have been used to dissect the host response *in vivo* to *Nocardia* (Beaman, 1980; Beaman and Scates, 1981; Beaman *et al.*, 1978a,b, 1982; Deem *et al.*, 1982; Folb *et al.*, 1977). The nude mouse is athymic and severely depleted of T lymphocytes. Even though the nude mouse is not completely devoid of lymphocytes with T-cell characteristics, it is nonetheless totally deficient in those having critical T-cell function (Beaman *et al.*, 1978a,b; Pritchard and Micklem, 1974; Rama-Rao *et al.* 1977). Therefore, these mice serve as an excellent model for studying the role of T lymphocytes in host resistance to *Nocardia* (Beaman *et al.*, 1978a,b).

Mortality curves, 50 percent lethal dose (LD_{50}), organ clearance data, and histological analysis were obtained over a period of three months following several different routes of inoculation with either *N. asteroides* GUH-2, *N. brasiliensis*-17E, or *N. caviae* 112 (Beaman and Scates, 1981; Beaman *et al.*, 1978a,b). The data from these studies clearly established that T lymphocytes are important in host resistance to systematic and chronic nocardial infections. Nude mice are initially more resistant to the early, acute nocardial infection than are their immunologically intact littermates; however, 48 to 72 hours after infection, nude mice begin to demonstrate an increased susceptibility to *Nocardia*. One to two weeks after infection, it became clear that athymic mice are significantly less able to cope with *Nocardia* than are their normal littermates (Beaman and Scates, 1981; Beaman *et al.*, 1978a,b). In addition, it has been observed that athymic mice are not able to develop typical mycetomatous lesions when infected by either *N. brasiliensis*-17E or *N. caviae* 112. Instead, nude mice develop large, purulent abscesses and aggressive systemic disease which results in animal mortality several months earlier than that observed when immunologically intact animal are used (Beaman and Scates, 1981; Beaman *et al.*, 1978a,b).

Hereditarily asplenic mice of a B6.CBA background have been shown to express an age-dependent T-cell deficiency with the development of decreased immunoglobulin production (Fletcher *et al.*, 1977; Lizzio and Wargon, 1974). Unlike the athymic mice, young asplenic mice which have a decreased T-cell function are most susceptible to acute, systemic infections with *Nocardia* (Beaman and Scates, 1981; Beaman *et al.*, 1978a). Further, mice that survive the early stages of infection demonstrate a marked inability to clear the organisms from the body at later stages. Asplenic mice infected with either *N. caviae* 112 or *N. brasiliensis*-17E always die from systemic nocardiosis before progressive mycetomas can be established (Beaman and Scates, 1981; Beaman *et al.*, 1978a; Beaman, unpublished data).

CBA/N mice have an X-linked chromosomal deficiency that results in the functional loss of certain subpopulations of B cells. Thus, these mice have a poor antibody response to certain T-dependent and T-independent antigens, and they produce abnormally low levels of both IgM and IgG_3 (Beaman *et al.*, 1982; Boswell *et al.*, 1980; Fidler *et al.*, 1980). The mating of CBA/N females with DBA/2 males results in $CBD2/F_1$ males which are B-cell deficient, whereas the $CBD2/F_1$ female littermates are immunologically intact and serve as controls (Beaman *et al.*, 1982; Boswell *et al.*, 1980; Fidler *et al.*, 1980). By comparing the immune response and host resistance of $CBD2/F_1$ males with $CBD2/F_1$ females, it is possible to ascertain the role that B cells and antibody play in host-parasite interactions with *Nocardia* (Beaman *et al.*, 1982).

It has been demonstrated that B-cell deficient $CBD2/F_1$ male mice are no more susceptible to infection with *N. asteroides* than are the normal female control animals (Beaman *et al.*, 1982). Preimmunized male mice are unable to produce detectable levels of antibody against *Nocardia*, whereas the normal female littermates produce high titers both of agglutinating antibody and of antibody detectable in a radioimmunoassay (RIA) system (Beaman *et al.*, 1982). Both the immunized males and females demonstrate equal ability to mount a delayed type hypersensitivity response to nocardial antigens. Preimmunized, B-cell deficient male and normal female mice appear to have identical response and resistance to nocardial infection (Beaman *et al.*, 1982). The data obtained from the experiments utilizing $CBD2/F_1$ mice suggest that B cells and antibody formation are not critical to the development of host responsiveness to infection by *Nocardia*. However, these data do not imply that either specific augmentation of B-cell responses against nocardial antigens or passive transfer of high affinity antibody could not significantly modulate disease.

The study of immunologic reactions against a single, specific external factor can best be carried out in germ-free mice. Germ-free mice offer an unique environment in which to study host-*Nocardia* relationships since the host is isolated from the influences of other exogenous microbial interactions as well as that of the normal microflora. *N. asteroides* GUH-2 was administered either intravenously (I.V.) or intranasally (I.N.) into conventionally grown, LPS-treated germ-free, and untreated germ-free NIH:S mice (Beaman *et al.*, 1980a). The number of organisms within the adrenals, blood, brain, kidneys, liver, lungs, and spleen were quantitated at 3, 24, 72, and 168 h after I.V. injection. The immunoresponsiveness of treated and untreated germ-free mice infected with *Nocardia* and histological alterations that occurred following infection were determined. This work showed that germ-free mice respond very differently that do conventionally grown mice following either I.N. or I.V. infection. Germ-free mice are not able to inhibit the rapid growth of *N. asteroides* in the brain, adrenals, kidneys, or lungs. In contrast, *N. asteroides* grows more slowly in the organs of the conventionally grown mice. LPS treatment of germ-free mice when treated with LPS are significantly more resistant to *Nocardia* than are the conventionally grown animals regardless of route of administration. These data establish the importance of both the resident microflora and exposure of the host to exogenous microorganisms in enhancing host resistance to infection with *Nocardia*. In addition, the state of macrophage activation and the development of cell-mediated immunity in the conventionally housed or LPS-treated germ-free mice appear to be critical in protection of the host against acute, systemic nocardiosis (Beaman *et al.*, 1980a).

The data obtained *in vivo* concerning nocardial interactions with nude, B-cell deficient, or germ-free mice strongly suggest that both a functional T cell population and the activation of macrophages are paramount to host resistance to these organisms. In order to study further the interrelationship between T lymphocytes and the state of macrophage activation, athymic nude and control littermate BALB/c mice were treated either with dextran sulfate 500 to suppress macrophage function or with *Corynebacterium parvum (Propionibacterium acnes)* to non-specifically activate macrophage function. Non-lethal doses of *N. asteroides* GUH-2 were administered I.N. into these pretreated mice, and pulmonary clearance responses were quantified and compared to unmanipulated control animals. Dextran sulfate significantly inhibited the ability of the heterozygous (nu/+) mice to clear *N. asteroides* from the lungs. In contrast, dextran sulfate appeared to have no measurable effect on the pulmonary response of athymic (nu/nu) BALB/c mice to I.N. challenge with *N. asteroides*. These data suggest that alveolar macrophage activity in T-cell deficient BALB/c mice is already maximally suppressed toward killing of *N. asteroides*. On the other hand, non-specific activation of macrophage activity by *P. acnes* greatly enhanced pulmonary clearance of *Nocardia* in both athymic and heterozygous (nu/+) littermate mice. Therefore, non-specific activation of alveolar macrophages may occur independent of T-cell function (Beaman, unpublished data). The systemic response to I.V. challenge with *N. asteroides* appeared to be similar to that observed in athymic mice given *Nocardia* I.N. (Beaman, unpublished data).

The ability of T lymphocytes to protect the host from I.V. challenge with *Nocardia* was determined by adoptive transfer of purified T cells obtained from immune, BALB/c heterozygous (nu/+) mice into athymic (nu/nu) BALB/c recipients. Those mice that received T cells from non-immune donors demonstrated only slightly enhanced clearance of *Nocardia*; whereas, the recipient nude mice that received T cells from immunized heterozygous littermates demonstrated good antibody titers against nocardial antigens and greatly enhanced clearance of *Nocardia* (Deem *et al.*, 1982). These data clearly demonstrate that adoptive transfer of T lymphocytes from pre-immunized animals can protect the host from a lethal challenge with *Nocardia* and demonstrate the necessary role of T cells and of cell-mediated immunity in host resistance to systemic nocardiosis.

V. *IN VITRO* MODELS FOR STUDYING HOST-PARASITE INTERACTIONS

The studies done *in vivo* described above demonstrate that host susceptibility to *Nocardia* differs significantly, based entirely upon the route of exposure (Beaman and Scates, 1981; Beaman *et al.*, 1978a,b, 1980b). These observations imply that the defenses of the host and phagocytic cells within different regions of the body have different capacities to destroy cells of *Nocardia*. Organ clearance and cellular distribution following I.V. challenge with different strains of *Nocardia* clearly establish specific organ tropisms as well as differences in clearance within these organs that appear to reflect differences in phagocytic cells and perhaps lymphocyte function (Table I; Beaman *et al.*, 1980b).

In vitro interactions of *Nocardia* with human polymorphonuclear (PMN) neutrophils and monocytes obtained from peripheral blood have been studied (Filice *et al.*, 1980b). It has been shown that PMN neutrophils and monocytes are not able to kill *N. asteroides*; however, more recent studies indicate that PMN neutrophils may inhibit the growth of these bacteria (Filice, personal communication).

It has been demonstrated that rabbit alveolar macrophages as well as mouse peritoneal macrophages obtained from normal animals are unable to kill cells of the more virulent strains of *N. asteroides* during a 3 h incubation period. Further, if these infected macrophages are incubated for longer periods of time, the nocardial cells grow within the macrophages and ultimately kill them (Beaman, 1977, 1979; Beaman and Smathers, 1976; Bourgeois and Beaman, 1974; Deem and Beaman, 1983; Filice *et al.*, 1980b). Glass adherent peritoneal exudate cells (mostly macrophages) that are non-specifically activated with either *P. acnes* or *Toxoplasma gondii* are able to kill some of the cells of the more virulent strains of *N. asteroides* after a 6 h incubation and retard the intracellular growth of the remaining bacteria (Filice, 1980a). In contrast, activated alveolar macrophages appear to be less able to kill *N. asteroides* in the absence of lymphocytes, even though they can retard the intracellular growth of the microorganism (Beaman, 1979). Utilizing rabbit alveolar macrophages, it was found that a combination of host components obtained from pre-immunized animals are necessary for enhancing the alveolar macrophages' ability to maximally kill cells of *N. asteroides* GUH-2. Thus, there appears to be a requirement that macrophages from immunized rabbits be combined with lymphocytes from specifically primed lymph nodes in combination with immune serum and lung lining material rich in surfactant. Deletion of any one of these com-

TABLE 1. Selective Organ Tropisms of N. asteroides *GUH-2,* N. brasiliensis 17E, *and* N. caviae 112 *Three Hours after Intravenous Injection into Swiss Webster Mice*

	Organ [a]							
Organism	*Blood*	*Adrenals*	*Brain*	*Kidneys*	*Lungs*	*Liver*	*Spleen*	*% Total*
N. asteroides	*0.04*	*0.04*	*0.02*	*0.33*	*1.96*	*83.33*	*17.33*	*103.0*
N. brasiliensis	*0.04*	*0.03*	*0.06*	*0.90*	*61.11*	*36.67*	*0.70*	*99.2*
N. caviae	*0.03*	*0.03*	*0.01*	*0.04*	*0.07*	*62.10*	*1.9*	*64.2*

[a] *Percent of total inoculum*

ponents results in decreased killing of *Nocardia* by the alveolar macrophages; however, no killing occurs without the addition of specifically primed lymphocytes (Davis-Scibienski and Beaman, 1980c).

The ability of Kupffer Cells (liver-associated macrophages) to kill *N. asteroides* has been studied. It was found that Kupffer cells obtained from normal mice are not able to either kill or inhibit the growth of *N. asteroides* GUH-2. In fact, the nocardial numbers frequently increase more rapidly within Kupffer cells than in the tissue culture medium alone. This is in sharp contrast to the behavior of Kupffer cells obtained from pre-immunized mice. These "immune" Kupffer cells are able to kill *Nocardia* during a 12 h incubation, and they inhibit the growth of the remaining nocardial cells during prolonged incubation. In addition, the incubation of 1×10^5 normal Kupffer cells obtained from non-immune mice with 7.5×10^5 T cells obtained from the spleens of pre-immunized mice results in a dramatic increase in the ability of the normal Kupffer cells to kill *N. asteroides* GUH-2. The addition of normal lymphocytes to normal Kupffer cells does not enhance the ability of these macrophages to kill *Nocardia* (Deem and Beaman, 1983).

Spleen cells from normal and pre-immunized mice have been studied *in vitro*. It was observed that whole spleen cell preparations obtained from normal mice result in a significant loss in viable cell counts after 12 h incubation. Pre-immunization of the mice either fails to enhance or enhances only slightly the splenic macrophage's ability to kill *Nocardia* (Deem and Beaman, 1983). The spleens obtained from normal and pre-immunized mice were fractionated. The B cells were obtained by a panning technique that employed purified rabbit anti-mouse kappa light chain antibody. T cells were removed by treatment with anti-theta antibody plus complement. The macrophages were removed by separation with iron carbonyl. The cell populations were distinguished by immunofluorescent staining procedures for identification of B cells and T cells; esterase staining was used to quantify the macrophage population (Deem *et al.*, 1983). *N. asteroides* GUH-2 was added to wells that contained only B cells (greater than 99% purity), T cells, null cells (>99.9% macrophages removed and no detectable B cells), or null cells (T cells and B cells removed and >99.9% macrophage free). The B cells from the immunized mice did not kill or inhibit *Nocardia.* On the other hand, T cells obtained from immunized mice produced a greater reduction in the numbers of *Nocardia* than did purified macrophage populations alone. T cells from the spleens of non-immunized mice did not exhibit the ability to kill *Nocardia.* These data appeared to demonstrate a direct lymphocyte-mediated killing of *N. asteroides.* It is very unlikely that the macrophage-depleted lymphocyte preparations contained enough contaminating macrophages to result in the greater than 80% killing that was observed; also, macrophage-rich preparations from the same spleens of immune animals were found to be able to kill only 40 to 50% of the nocardial cells within the same 12 h period of incubation (Deem *et al.*, 1983).

In order to visualize the T lymphocyte-nocardial interactions, spleen cells of immunized mice were treated to remove both B cells and macrophages and the resulting T-cell preparations were incubated with *N. asteroides* GUH-2 for 6 h. After incubation, suspensions were observed under phase-contrast microscopy and prepared for electron microscopy. Light microscopy demonstrated that numerous lymphocytes were attached along the filamentous cells of *Nocardia.* No macrophages were observed in the preparations. Electron microscopy clearly demonstrated a very tight adherence of the outer layer of the nocardial cell wall to the cytoplasmic membrane of the lymphocyte. Further, the outer cell wall layer appeared to be pulled away from the peptidoglycan portion of the wall. Many lymphocytes appeared to have pseudopod-like (or microvilli) extensions that were adherent to the nocardial surface and frequently these extensions appeared to displace the rigidity of the bacterial cell wall in what appeared to be an attempt to penetrate the bacterial cell. These observations support an active bactericidal role for T lymphocytes obtained from immunized mice (Deem *et al.*, 1983). As already indicated above, macrophages from different regions of the body appear to behave differently towards *N. asteroides* GUH-2. Therefore, from normal mice, splenic macrophages demonstrated greater capacity *in vitro* to kill cells of *Nocardia* than did peritoneal macrophages which in turn appeared to be more bactericidal than did alveolar macrophages; Kupffer cells obtained from normal mice appeared to be least capable of killing *N. asteroides.* The order of nocardicidal activity of macrophages obtained from normal mice is splenic macrophages > peritoneal macrophages > alveolar macrophages >Kupffer cells. Macrophages from immunized mice appear to be quite different; thus, Kupffer cells obtained from immune animals are affected most, whereas splenic macrophages appear to be affected least in these bactericidal assays. The order of nocardicidal activity of macrophages from immune mice is

Kupffer cells > alveolar macrophages > peritoneal macrophages > splenic macrophages. These observations probably reflect differences in the interactions of these macrophages with subpopulations of lymphocytes as well as differences in specific enzyme levels that may be modulated by both the bacterial pathogen and lymphocytes or lymphocyte products (Deem and Beaman, 1983).

VI. POSSIBLE MECHANISMS OF NOCARDIAL PATHOGENESIS

It has been clearly established that the components in the cell envelope of *Nocardia* undergo significant shifts or changes in chemical and structural composition during the growth cycle (Beaman, 1975; Beaman and Burnside, 1973; Beaman and Shankel, 1969; Beaman *et al.*, 1971, 1974, 1978c, 1981). These alterations are correlated with dramatic effects in the virulence of the organisms as well as on host-parasite interactions (Beaman, 1973, 1976, 1979; Beaman and Bourgeois, 1981; Beaman and Maslan, 1978; Davis-Scibienski and Beaman, 1980a,b). In every strain of *N. asteroides*, *N. brasiliensis*, and *N. caviae* that have been studied, it was found that organisms in the logarithmic phase of growth were at least ten times more virulent for mice than was the same culture in the coccobacillary, stationary phase (Beaman, unpublished data). Further, in *N. asteroides* GUH-2, it has been shown that the log phase cells are about 1400 times more virulent than are stationary phase cells (Beaman and Maslan, 1978). In addition to the specific phase of growth, it has been shown that the methods used to grow the organisms in broth culture can also have a profound effect on the degree of this difference in virulence between log and stationary phase cells (Beaman and Maslan, 1978).

In order to determine the mechanisms of nocardial virulence and the effects of nocardial cellular development on host-parasite interactions *in vitro*, cells of *N. asteroides* GUH-2 from different phases of growth were incubated with alveolar macrophages obtained from both immunized and non-immunized rabbits (Beaman, 1979; Davis-Scibienski and Beaman, 1980b). It was shown that cells of *N. asteroides* GUH-2 at all stages of development are able to grow within "normal" alveolar macrophages. However, log phase cells are significantly more antiphagocytic and more toxic to macrophages than are cells in the stationary phase (Beaman, 1979). Stationary phase cells, on the other hand, do not grow within normal macrophages as well as do log phase cells. In "activated" macrophages obtained from immunized rabbits, it was shown that the growth of stationary phase nocardial cells is effectively inhibited for at least 12 h. In contrast, activated macrophages only temporarily retarded the growth of log phase organisms. Specific antibody against *N. asteroides* GUH-2 enhances both phagocytosis and intracellular killing of log phase cells at 6 h after infection, whereas there is little effect of antibody on stationary phase cells of the same culture of *N. asteroides*. Once again, however, the log phase cells that survive this initial contact are able to grow within these activated macrophages even in the presence of specific antibody (Beaman, 1979).

The effect of the phase of growth of *N. asteroides* GUH-2 as well as cellular viability on phagosome-lysosome fusion within rabbit alveolar macrophages was determined. In previous studies, it has been shown that the relative virulence of different strains of *N. asteroides* correlates with their ability to inhibit phagosome-lysosome fusion within macrophages by the phagocytized cells of *Nocardia* (Davis-Scibienski and Beaman, 1980a). Study of the same culture of *N. asteroides* GUH-2 at different stages of growth has demonstrated the same direct correlation between the virulence of the organism and its ability to inhibit phagosome-lysosome fusion. Viable, log phase cells of *N. asteroides* GUH-2 inhibited macrophage lysosomes from fusing with more than 80% of the phagocytized bacterial cells. In contrast, only 40% of the late stationary phase cells were able to inhibit phagosome-lysosome fusion (Davis-Scibienski and Beaman, 1980b). Therefore, it appears that the inhibition of phagosome-lysosome fusion in macrophages may be one of the mechanisms of pathogenicity of *Nocardia*.

It seems likely that much of the virulence of *Nocardia* is due to components that either are on the surface of the cell or are an integral part of the cell wall. It has been shown that *Nocardia* have a complex cell envelope composed of several free and bound lipids, peptides or proteins, and polysaccharide compounds which include trehalose dimycolate, glycolipids, nocobactin, peptidolipids, arabinogalactan mycolate, peptidoglycan, perhaps sulfolipid, and a variety of other lipoidal substances (Azuma *et al.*, 1973; Beaman, 1975; Beaman and Burnside, 1973; Beaman *et al.*, 1971, 1974, 1981; Ioneda *et al.*, 1970; Michel and Bordet, 1976; Prabhudesi *et al.*, 1981; Ratledge and Patel, 1976; Yano *et al.*, 1978). Trehalose-mycolate has been shown to play a role in the virulence of *Mycobacterium* and to be toxic for cells. Beaman and Moring (unpublished data) have found that virulent forms of *N. asteroides* contain more trehalose dimycolate than do less virulent strains. Further, log phase cells of *N. asteroides* GUH-2 contain significantly more trehalose dimycolate than do the less virulent stationary phase cells. It was also discovered that the degree of polyunsaturation in both the α- and β-chain of the mycolic acid molecule is more directly correlated with the degree of virulence of the organism than with the absolute amounts of trehalose dimycolate present (Beaman *et al.*, 1981; Beaman, unpublished data). All virulent strains of *Nocardia* (especially in log phase of growth) contained relatively long ($> C_{56}$) mycolic acids that were polyunsaturated in both the α- and β-chain portion of the mycolic acid molecule, while fewer saturated mycolic acids were detected in the α-chain (Beaman, unpublished data). In contrast, the less virulent mutants of the same organism or the less virulent stationary phase cells of the same organism have relatively shorter mycolic acids ($< C_{56}$) that are more highly saturated. These mycolates have no double bonds in the α-chain molecule (Beaman *et al.*, 1981; Beaman, unpublished data).

Whether mycolic acids or trehalose dimycolate plays a role in inhibition of phagosome-lysosome fusion remains to be determined. In *Mycobacterium*, it appears that a surface sulfolipid is responsible for inhibition of phagosome-lysosome fusion. Sulfolipids have been detected on the surface of *N. asteroides*. It is possible that these sulfolipids are important in preventing fusion; however, no experimental data are yet available on the role of sulfolipids in nocardial pathogenesis (Beaman, 1976; Davis-Scibienski and Beaman, 1980a; Prabhudesi *et al.*, 1981).

It has been demonstrated that cells of *N. asteroides* GUH-2 are not killed by either peripheral blood polymorphonuclear leukocytes or blood monocytes even though the nocardiae stimulated an oxidative metabolic burst with the production of the superoxide anion by these phagocytes (Filice *et al.*, 1980b). The probable mechanisms of the resistance of *N. asteroides* GUH-2 to this superoxide anion may reside in both the amount and type of superoxide dismutase (SOD) and catalase produced (Beaman *et al.*, 1983). It was shown that *N. asteroides* GUH-2 produced a unique SOD that was both secreted into the medium and remained on the cell surface. Log phase cells produce more SOD than do stationary phase organisms. Further, the cytoplasm of the virulent, log phase cells of *N. asteroides* GUH-2 have considerably more catalase than do either the stationary phase cells or other less virulent strains of *N. asteroides* (Beaman, unpublished data). Thus, there appears to be a direct correlation between virulence, resistance to oxidative killing and the amounts, type, and location of SOD and catalase produced by *N. asteroides* (Beaman *et al.*, 1983; Beaman, unpublished data).

In order for *Nocardia* to successfully grow within host tissues, it must be able to compete with the host for bound iron. *Nocardia* produces a strong iron chelating compound called nocobactin (Ratledge and Patel, 1976). Preliminary data showed that log phase cells of *N. asteroides* GUH-2 produce more nocobactin than did cells in stationary phase grown under identical conditions (Beaman, unpublished data). Further, *N. asteroides* GUH-2 had more nocobactin than did the avirulent strains *N. asteroides* 10905 under the same growth conditions. Therefore, there appears to be an additional correlation between the amount of nocobactin produced by the organism and its virulence (Beaman, unpublished data).

VII. CONCLUSIONS

The virulence of *Nocardia* appears to be due to the effect of several unique properties which interact in a complex, dynamic system. Each component of this system adds one step to the patogenic spectrum. Thus, the most virulent strain of *N. asteroides* GUH-2 growing in log phase (LD_{50} for I.V. injection into mice is approximately 10^3 CFU) has significantly increased amounts of trehalose dimycolate, increased mycolic acid size ($>C_{56}$), increased polyunsaturation of α- and β-chains of mycolic acid molecules, increased superoxide dismutase that is both cell surface associated and secreted into the environment, increased catalase, and increased nocobactin. In contrast, the avirulent strain of *N. asteroides* 10905 in stationary phase of growth (LD_{50} for I.V. injection into mice is approximately 10^9 CFU) has little trehalose dimycolate, short mycolic acid size ($< C_{56}$), no unsaturation in the α-chain of the mycolic acid molecule, and reduced amounts of SOD, catalase, and nocobactin (Beaman and Bourgeois, 1981; Beaman and Maslan, 1978; Beaman *et al.*, 1981, 1983; Beaman, unpublished data).

Each of the cellular components described above appear to play an integral role in the ability of the organism to establish itself and grow within the host. At the same time, the host has developed a multiple defense complex that is capable of circumventing the organisms invasiveness. This defense against *Nocardia* is necessarily multifaceted, and involves both innate mechanism as well as immunologically specified events. Macrophage activation, the inhibitory effects of PMN phagocytes, cell-mediated immunity, and a T-cell population capable of direct lymphocyte-mediated cytotoxicity for *Nocardia* all appear to work in conjunction. A deficiency in any one of these constituents appears to give the more virulent strains of *Nocardia* an overwhelming advantage over the host (Beaman, 1977, 1979; Beaman *et al.*, 1978a, 1980a; Deem and Beaman, 1983; Deem *et al.*, 1982, 1983; Filice *et al.*, 1980a,b; Folb *et al.*, 1977; Ortiz-Ortiz *et al.*, 1979). Since *Nocardia* can gain entrance into host cells, and survive for long periods of time sequestered from these defense mechanisms (probably as L-forms), the host must continually be on the defensive. The host has little capacity to prevent the organism from initiating chronic, persistent, or latent infections that may remain in a state of dormancy. At some later time, the host's defenses may weaken or become altered and the silent, undetected *Nocardia* may once again emerge to produce a chronic, progressive disease (Beaman, 1976, 1980; Beaman and Scates, 1981; Bourgeois and Beaman 1974, 1976)

ACKNOWLEDGMENTS

I thank Marilyn Wheeler for her expert typing of this manuscript.

REFERENCES

Ajello, L., Walker, W., and Dungwort, D., *J. Am. Vet. Med. Assoc. 138:*370 (1961).
Al-Doory, Y., Pinkerton, M. E., Vice, T. E., and Hutchisson, V., *J. Am. Vet. Med. Assoc.* 155:1179 (1969).
Alteras, I., Feuerman, E. J., and Dayan, I., *Int. Soc. Trop. Dermatol. 19:*260 (1980).
Azuma, I., Kanetsuna, F., Tanaka, Y., Mera, M., Yanagihara, Y., Mifuchi, I., and Yamamura, Y., *Jap. J. Microbiol. 17:*154 (1973).
Balikian, J. P., Herman, P. G., and Kopit, S., *Radiology 126:*569 (1978).
Barnetson, R. S., and Milne, L. J. R., *Brit. J. Dermatol. 99:*227 (1978).
Beaman, B. L., *Infect. Immun. 8:*828 (1973).
Beaman, B. L., *J. Bacteriol. 123:*1235 (1975).
Beaman, B. L., *in* "Biology of the Nocardiae" (M. Goodfellow, G. H. Brownell, and J. A. Serrano, eds.), p. 386. Academic Press, London (1976).
Beaman, B. L., *Infect. Immun. 15:*925 (1977).
Beaman, B. L., *Infect. Immun. 26:*355 (1979).
Beaman, B. L., *Infect. Immun. 29:*244 (1980).
Beaman, B. L., and Bourgeois, A. L., *J. Clin. Microbiol. 14:*574 (1981).
Beaman, B. L., and Burnside, J., *Appl. Microbiol. 26:*426 (1973).
Beaman, B. L., and Maslan, S., *Infect. Immun. 16:*995 (1977).
Beaman, B. L., and Maslan, S., *Infect. Immun. 20:*290 (1978).
Beaman, B. L., and Scates, S. M., *Infect. Immun. 33:*893 (1981).
Beaman, B. L., and Shankel, D. M., *J. Bacteriol. 99:*876 (1969).
Beaman, B. L., and Smathers, M., *Infect. Immun. 13:*1126 (1976).
Beaman, B. L., Kim, K. S., Salton, M. R. J., and Barksdale L., *J. Bacteriol. 108:*941 (1971).
Beaman, B. L., Kim, K. S., Laneelle, M. A., and Barksdale, L., *J. Bacteriol. 117:*1320 (1974).
Beaman B. L., Burnside, J., Edwards, B., and Causey, W., *J. Infect. Dis. 134:*286 (1976).
Beaman, B. L., Gershwin, M. E., and Maslan, S., *Infect. Immun. 20:*381 (1978a).
Beaman, B. L., Goldstein, E., Gershwin, M. E., Maslan, S., and Lippert, W., *Infect. Immun. 22:*867 (1978b).
Beaman, B. L., Serrano, J. A., and Serrano, A. A., *Zentralbl. Bakteriol. 6,* suppl.:201 (1978c).
Beaman B. L., Gershwin, M. E., Scates, S., and Ohsugi, Y., *Infect. Immun. 29:*733 (1980a).
Beaman, B. L., Maslan, S., Scates, S., and Rosen, J., *Infect. Immun. 28:*185 (1980b).
Beaman, B. L., Bourgeois, A. L. and Moring, S. E., *J. Bacteriol. 148:*600 (1981).
Beaman, B. L., Gershwin, M. E., Ahmed, A., Scates, S. M., and Deem, R., *Infect. Immun. 35:*111 (1982).
Beaman, B. L., Scates, S.M., Moring, S.E., Deem, R., and Misra, H. P., *J. Biol. Chem.* 258:91 (1983).
Boswell, H. S., Scher, I., Nerenberg, M. I., and Singer, A., *Fed. Proc. 39:*807 (1980). (Abstract)
Bourgeois, L., and Beaman, B. L., *Infect. Immun. 9:*576 (1974).
Bourgeois, L., and Beaman, B. L., *J. Bacteriol. 127:*584 (1976).
Bushnell, R. B., Pier, A. C., Fichtner, K. E., Beaman, B. L., Boos, H. A., and Salman, M. D., *Am. Assoc. Vet. Lab. Diag. 22:*1 (1979).
Byrne, E., Brophy, B. P., and Perrett, L. V., *J. Neurol. Neurosurg. Psychiatry 42:*1038 (1979).
Causey, W., *Appl. Microbiol. 28:*193 (1974).
Causey, W. A., Arnell, P., and Brinker, J., *Chest, 65:*360 (1974).
Causey, W. A., and Sieger, B., *Am. Rev. Respir. Dis. 109:*134 (1974).
Curry, W. A., *Arch. Int. Med. 140:*818 (1980).
Davis-Scibienski, C., and Beaman, B. L., *Infect. Immun. 28:*610 (1980a).
Davis-Scibienski, C., and Beaman, B. L., *Infect. Immun. 29:*24 (1980b).
Davis-Scibienski, C., and Beaman, B. L., *Infect. Immun. 30:*578 (1980c).
Deem, R., and Beaman, B. L., *J. Reticuloendothel. Soc. (1983). (In preparation)*
Deem, R. L., Beaman, B. L., and Gershwin, M. E., *Infect. Immun. 38:*914 (1982).
Deem, R., Doughty, F., and Beaman, B. L., *J. Immunol. 130:*2401 (1983)
Fidler, J. M., Morgan, E. L., and Weigle, W. O., *J. Immunol. 124:*13 (1980).

Filice, G. A., Beaman, B. L., and Remington, J. S., *Infect. Immun. 27:*643 (1980a).
Filice, G. A., Beaman, B. L., Krick, J. A., and Remington, J. S., *J. Infect. Dis. 142:*432 (1980b).
Fletcher, M. P., Ikeda, R. M., and Gershwin, M. E., *J. Immunol. 119:*110 (1977).
Folb, P. I., Jaffe, R., and Altman, G., *Infect. Immun. 13:*1490 (1976).
Folb, P. I., Timme, A., and Horowitz, A., *Infect. Immun. 18:*459 (1977).
Frazier, A. R., Rosenow, III, E. C., and Roberts, G. D., *Mayo Clin. Proc. 50:*657 (1975).
Gezuelle, E., *Sabouraudia, 10:*63 (1972).
Gonzalez-Ochoa, A., *Lab. Invest. 11:*1118 (1962).
Gonzalez-Ochoa, A., and Sandoval-Cuellar, A., *Sabouraudia, 14:*255 (1976).
Greer, K., *VA. Med. Mon. 101:*193 (1974).
Ioneda, T., Lederer, E., and Rozanis, J., *Chem. Phys. Lipids 4:*375 (1970).
Krick, J. A., Stinson, E. B., and Remington, J., *Ann. Intern. Med. 82:*18 (1975).
Kurup, P. V., Randhawa, H. S., Sandu, R. S., and Abraham, S., *Mycopathol. Mycol. Appl.* 40:113 (1970).
Lizzio, B. B., and Wargon, L. B., *Immunology 27:*167 (1974).
Macotella-Ruiz, E., and Mariat, F., *Bull. Soc. Pathol. Exot. 89:*426 (1963).
Mason, K. N., and Hathaway, B. M., *Arch. Pathol. 87:*389 (1969).
Michel, G., and Bordet, C., *in* "The Biology of the Nocardiae" (M. Goodfellow, G. H. Brownell and J. A. Serrano, eds.), p. 141. Academic Press, London (1976).
Mohapatra, L. N., and Pine, L., *Sabouraudia, 2:*176 (1963).
Monga, D. P., Kapur, M. P., and Dixit, S. N., *Mykosen, 21:*152 (1978).
Orchard, V. A., *N. Z. Vet. J. 27:*159 (1979).
Ortiz-Ortiz, L., Parks, D. E., Lopez, J. S., and Weigle, W. O., *Infect. Immun. 25:*627 (1979).
Peterson, D. L., Hudson, L. D., and Sullivan, K., *Arch. Intern. Med. 138:*1164 (1978).
Pier, A. C., Takayama, A. K., and Miyahara, A. Y., *J. Wildl. Dis. 6:*112 (1970).
Prabhudesi, A. V., Kaur, S., and Khuller, G. K., *Indian J. Med. Res. 73:*181 (1981).
Presant, C. A., Wiernik, P. H., and Serpick, A. A., *Am. Rev. Respir. Dis. 108:*1444 (1973).
Pritchard, H., and Micklem, H. S., *in* "Proceeding First International Workshop Nude Mice", p. 127. Gustav Fischer Verlag, Stuttgart (1974).
Pulverer, G., and Schaal, K. P., *in* "Nocardia and Streptomyces" (M. Modarski, W. Kurylowicz, and J. Jeljaszewicz, eds.), p. 417. Gustav Fischer Verlag, Stuttgart (1978).
Ratledge, C., and Patel, P. V., *J. Gen. Microbiol. 93:*141 (1976).
Rama-Rao, G., Rawls, W. E., Perey, D. Y. E., and Tompkins, W. A. F., *J. Reticuloendothel. Soc. 21:*13 (1977).
Rosett, W., and Hodges, G. R., *Am. J. Med. Sci. 276:*279 (1978).
Saenz Lope, E., and Gutierrez, D. C., *Acta Neurochir. 37:*139 (1977).
Sanyal, M., Thommayya, A., and Basu, N., *Mykosen, 21:*109 (1978).
Satterwhite, T. K., and Wallace, R. J., *J. Am. Med. Assoc. 242:*333 (1979).
Sher, N. A., Hill, C. W., and Eifrig, D. E., *Arch. Ophtalmol. 95:*1415 (1977).
Smith, I. M., and Hayward, A. H. S., *J. Comp. Pathol. 81:*79 (1971).
Stropes, L., Bartlett, M., and White, A., *Am. J. Med. Sci. 280:*119 (1980).
Stuart, M., *Dis. Colon Rectum 22:*183 (1979).
Talwar, P., and Sehgal, S. C., *Sabouraudia, 17:*287 (1979).
Terezhalmy, G. T., and Bottomley, W. K., *Oral Surg. 45:*200 (1978).
Thammayya, A., Basu, N., Sur-Roy-Chowdhury, D., Banerjee, A. K., and Sanyal, M., *Sabouraudia, 10:*19 (1972).
Uesaka, I., Oiwa, K., Yasuhira, K., Kobara, Y., and McClung, N. M., *Jap. J. Exp. Med.* 41:443 (1971).
Watson, W. A., *Res. Vet. Sci. 23:*171 (1977).
Yano, I., Kageyama, K., Ohno, Y., and Masui, M., *Biomed. Mass Spectrom. 5:*14 (1978).

THE CELL WALL AS A DETERMINANT OF PATHOGENICITY IN *NOCARDIA*: THE ROLE OF L-FORMS IN PATHOGENESIS[1]

Blaine L. Beaman

Department of Medical Microbiology and Immunology
School of Medicine
University of California
Davis, California, U. S. A.

I. INTRODUCTION

The cell wall of *Nocardia* is a dynamically changing, complex structure that undergoes chemical and ultrastructural modification during the development of the organism (Beaman, 1975). Nutritional and environmental conditions alter both the synthesis and incorporation of different constituents of the envelope during growth (Beaman and Shankel, 1969). These changes in cell wall structure result in alterations of growth patterns, of cell surface characteristics, and of cell-to-cell interactions. Each of these modifications are translated into specific biological consequences for the organism within a specific environmental niche. Further, a close relationship between certain cell wall substances and the pathogenic mechanisms of *Nocardia* has been demonstrated (Beaman, 1975; Beaman and Maslan, 1978). Therefore, either qualitative or quantitative perturbations of some of these substances in the cell envelope result in dramatic shifts in host-parasite interactions (Beaman, 1975; Beaman *et al.*, 1981).

Some of the major functions of the cell wall are to serve as a physical barrier to extracellular substances, to determine size and shape of the organism, and to protect the cell from osmotic lysis. What are the biological consequences to the nocardial cell of removing or destroying the integrity of

[1] *This work was supported by a Public Health Service Grant from the National Institutes of Health, NIAID, Grant #AI-13167.*

ISBN 0-12-528620-1

the wall? What are the effects on pathogenesis and host-parasite interactions of inducing cells of *Nocardia* to grow in a cell wall-less state?

II. *IN VITRO* REMOVAL OF THE NOCARDIAL CELL WALL

Certain strains of *Nocardia (Rhodococcus) rubra* produce large, involuted cellular forms when grown in brain heart infusion (BHI), but not when grown in nutrient broth (Beaman and Shankel, 1969). Electron microscopy and biochemical analyses have demonstrated that the stability and integrity of the cell wall are greatly affected by as yet undefined substances in BHI. However, similar cellular alterations can be induced during growth in a chemically defined mineral salts medium by adding either 0.5% (w/v) D,L-alanine, 5% (w/v) arabinose, 5% (w/v) galactose, or 5% (w/v) glycine (Beaman, B. L., Ph. D. thesis, University of Kansas, Lawrence, 1968). By limiting either iron, manganese, or zinc but not copper, molybdenum, or boron in the culture medium, *N. opaca* grow as large, bulbous cells and involuted forms (Webley, 1960). Temperatures above 35°C have been shown to induce involuted forms of some strains of *N. corallina* (Webb, *et al.*, 1954); increased CO_2 and certain fatty acids significantly affect the morphology of *Nocardia* (Webley, 1954). Because the modifications of cellular morphology produced by these various conditions appear to be similar, it is reasonable to suggest that the regulation of cell wall biosynthesis and of cell division may be modulated by exogenous nutritional and environmental factors.

When grown in a medium containing 2% (w/v) glycine, *Nocardia* becomes susceptible to lysis by either lysozyme, penicillin, or D-cycloserine (Bourgeois and Beaman, 1976). If an osmotic stabilizer such as 0.11 M mannitol, 0.35 M sucrose, or 5% (w/v) NaCl is added to the medium prior to treatment with lysozyme, spheroplasts are produced. Other methods (*e. g.*, EDTA with lysozyme) that have been used successfully for producing protoplasts or spheroplast of bacteria are generally not effective against *Nocardia*. Cell wall-less forms of *N. rubra* have been recovered after mutagenesis with quinacrine (Prasad and Bradley, 1972). Further, it has been demonstrated that phagocytic cells, such as macrophages, are able to remove the cell wall of several strains of *N. asteroides* (Bourgeois and Beaman, 1974) and *N. caviae* (Beaman and Scates, 1981). The mechanisms by which phagocytic cells induce protoplasts or spheroplasts are not entirely understood, but it is known that the lysosomal bodies within these professional phagocytes are rich in degradative enzymes, including lysozyme. Therefore, the mechanism of induction of the wall-less state of *Nocardia* by macrophages may be similar to that produced *in vitro* by glycine and lysozyme. Additionally, the combination of increased osmolarity within the phagosome and the rigidity of the phagosomal membrane would serve the same protective function in preventing osmotic lysis of the protoplast as does mannitol, sucrose, or NaCl in the broth medium.

III. GROWTH OF NOCARDIAL L-FORMS *IN VITRO*

Protoplasts or spheroplasts of *Nocardia* that are induced either *in vitro* or *in vivo* can be grown as L-forms when transferred into an appropriate medium (Beaman, 1980; Beaman *et al.*, 1978; Bourgeois and Beaman, 1974, 1976; Beaman and Scates, 1981). Most nocardial L-forms can be cultured on Barile, Yaguchi, and Eveland (BYE) agar; however, optimal induction, isolation, and growth are strain specific and are influenced by the stage of cellular development of the *Nocardia* at the time of induction into the wall-less state. Many additional factors were shown to affect the ability of nocardial L-forms to grow *in vitro*. It was observed that different lots of BHI and yeast extract, as well as different batches of serum, greatly influenced the yield of L-forms. Also, some L-form preparations appeared to be unable to grow in BYE in the wall-deficient state beyond the second or third transfer. Often, these cells either reverted to a walled form, or they did not grow at all, as though some necessary metabolite were missing from the culture medium. Temperature and CO_2 levels within the incubator were significant factors affecting growth of L-forms, and optimal growth was obtained at 37°C and 5% CO_2.

Phase contrast microscopy revealed that the protoplast or spheroplasts of *Nocardia*, when transferred to fresh BYE, first become phase dense, enlarged in size, and then became highly refractile (Fig. 1A). Numerous intracellular granules developed within the large, refractile cells, while smaller refractile spheres and dense granules grew and extended from the larger bodies (Fig. 1B). In broth cultures, numerous clumps of the large refractile spheres were encircled by masses of smaller, more refractile spheres, and irregular granules accumulated; on soft BYE agar, the colonies developed a central core that grew into the surface of the agar and is surrounded by granules, spheres, and membranous extensions (Fig. 2A). The colony morphology produced by different preparations of L-forms (even of the same nocardial strain) was quite variable in size and appearance (Fig. 2B,C). L-forms of *Nocardia* that form colonies that have an appearance different from the typical "fried egg" type (Fig. 2C) are referred to as L-form variants (Beaman and Scates, 1981).

IV. ULTRASTRUCTURE OF L-FORMS

Considerable structural variation occurs within L-forms of *Nocardia* (Beaman, 1980; Beaman and Scates, 1981; Beaman *et al.*, 1978; Bourgeois

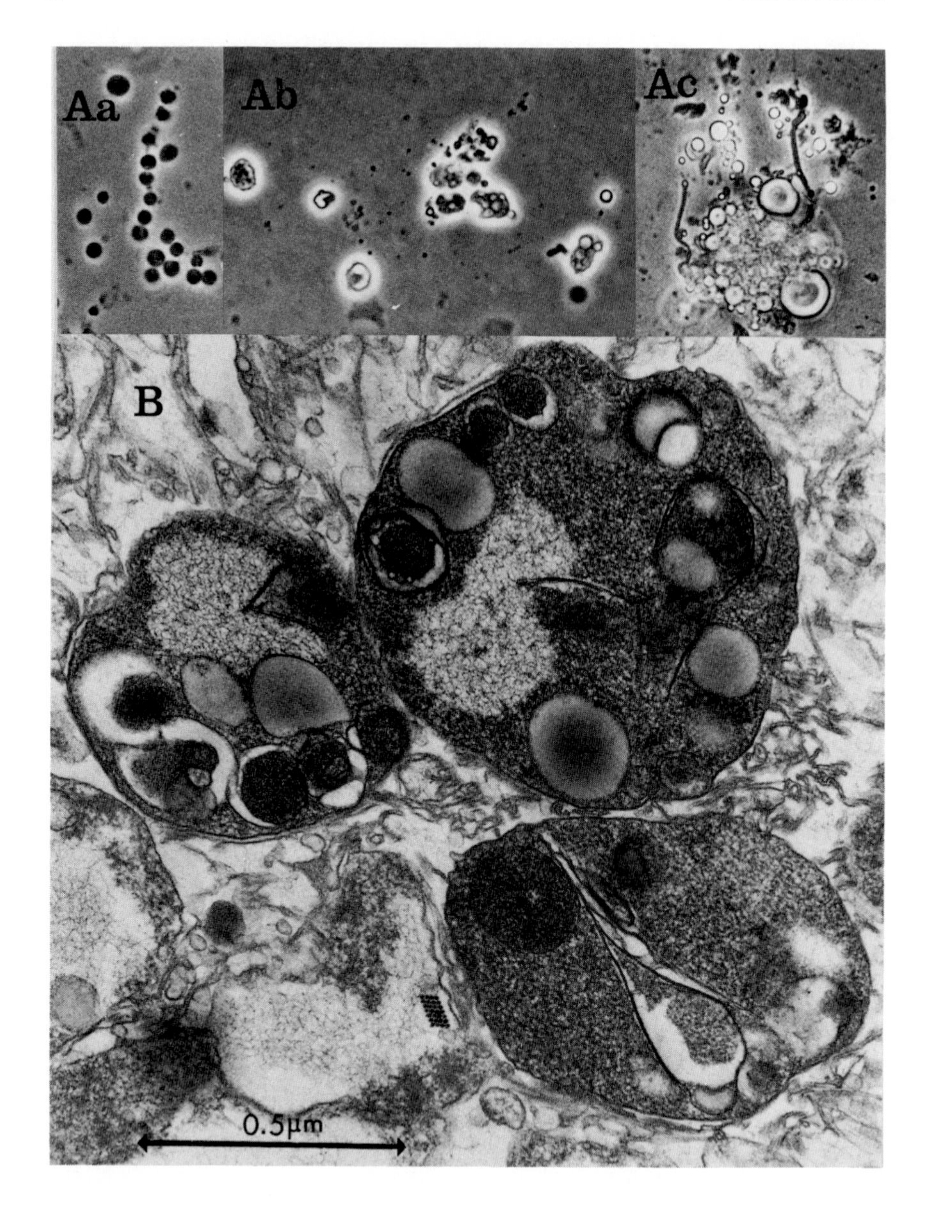

FIGURE 1. Phase contrast microscopy and electron microscopy of developing L-forms of N. asteroides *GUH-2 grown in BYE broth. A. Induction of spheroplasts in BYE supplemented with 1.5% glycine, 20% sucrose, lysozyme and 10% horse serum. a) 3 day incubation; b) 2 week incubation; c) as in b, but transferred to fresh BYE agar for 5 days. B. Electron micrograph of growing L-forms of* N. asteroides *GUH-2 after 4 biweekly transfers into fresh BYE broth.*

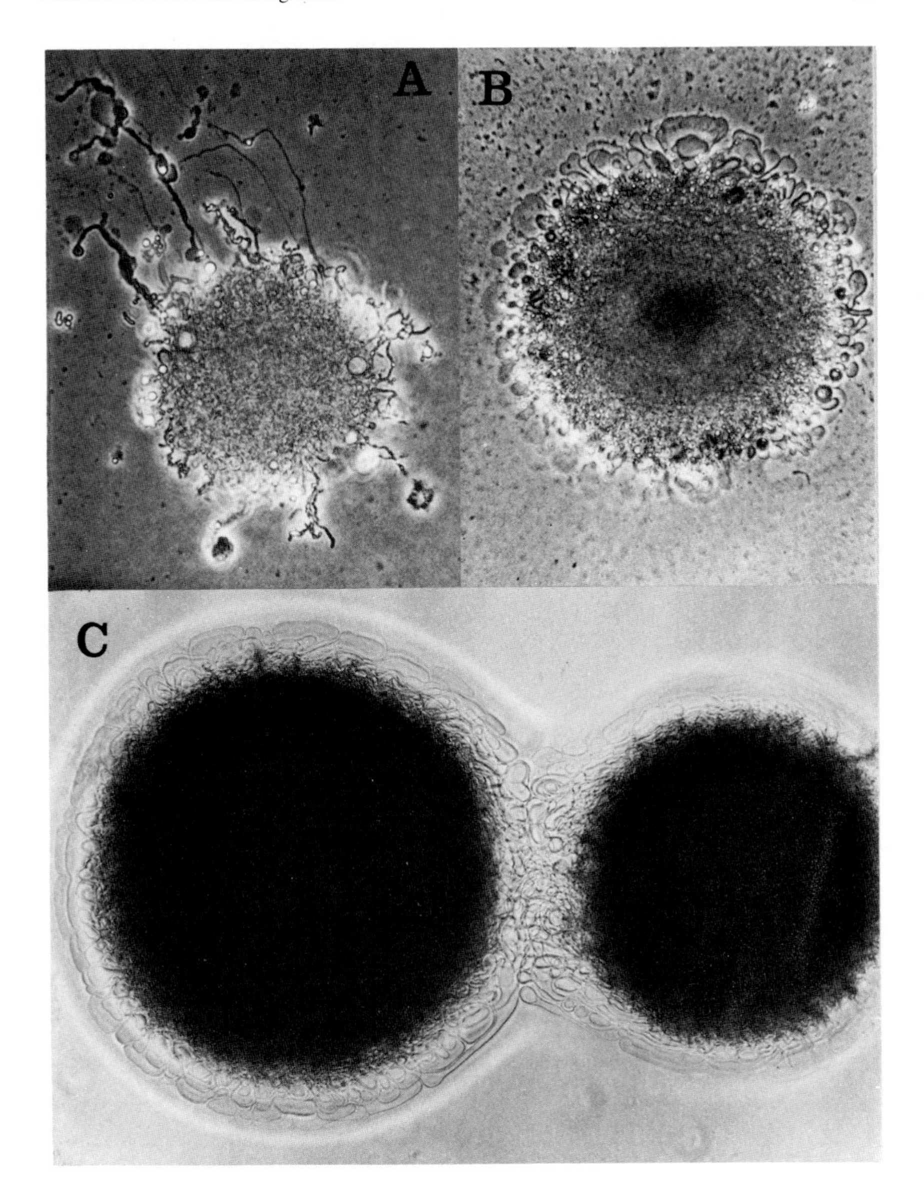

FIGURE 2. L-form colonies of N. caviae *112 grown on BYE agar. A. Developing L-form colony (1 week on BYE agar). B. Two week L-form colony with typical "fried egg" morphology. C. L-form variant colonies showing extensive development of peripheral layers of membrane.*

and Beaman, 1976). Significant quantities of granular and membranous substances (composed of mycolic acids, peptidolipid, arabinogalactan, and other cell wall constituents) are secreted into the culture medium during growth of the L-forms; however, usually no cell wall is discernible in Type A L-forms.

Actively growing L-forms contain large amounts of intracytoplasmic membrane, numerous membrane-bound bodies, large lipoidal inclusion bodies, and phosphate storage granules (Fig. 2B and 3A). These large inclusion bodies frequently have core structures that appear to be microtubular in nature (Fig. 1B). Variants that have altered cellular morphology but that still have a detectable, although defective, cell wall are called as Type B L-forms. The cells of Type B L-forms are Gram-negative and osmotically fragile. Type B L-forms of *N. asteroides* characteristically possess large lipid inclusions that give the cells a beaded appearance (Fig. 3B). These lipid inclusions are acid-fast, Sudan Black B positive for lipids, and Gram-positive.

The chemical composition of L-forms of *Nocardia* has been determined by the use of gas chromatography, thin layer chromatography, gel electrophoresis, and amino acid analysis. Muramic acid has been found to be absent in whole cell preparations of both Type A and Type B L-forms and of L-form variants. Type B L-forms possesses detectable, but significantly decreased, amounts of meso-diaminopimelic acid and glucosamine. Arabinose and galactose polymers have been detected in all L-form preparations, but Type B L-forms contain more arabinogalactan than do Type A cells. All L-forms have 90% less arabinogalactan than do normal organisms. Further, the major sugar detected in L-forms is glucose. Major shifts in the quantitative composition of both fatty acids and mycolic acids occurs in the L-forms. Type A L-forms have less mycolic acid and less tuberculostearic acid than do Type B cells. All types of cell wall defective variants that have been studied have significantly less mycolic acid than do the parental strains grown under the same conditions. The data suggest that the L-forms continue to synthesize and secrete most of the cell wall components during growth. However, since they lack a peptidoglycan, these components become structured in the extracellular milieu to form membranous or granular aggregates no longer associated with the surface of the nocardial cell.

V. REVERSION OF NOCARDIAL L-FORMS

L-forms of bacteria share many basic properties with *Mycoplasma*, including colony morphology. The one definitive property that permits certain differentiation of L-forms from *Mycoplasma* is the ability of the L-form to resynthesize the cell wall and, thereby, to revert back to the

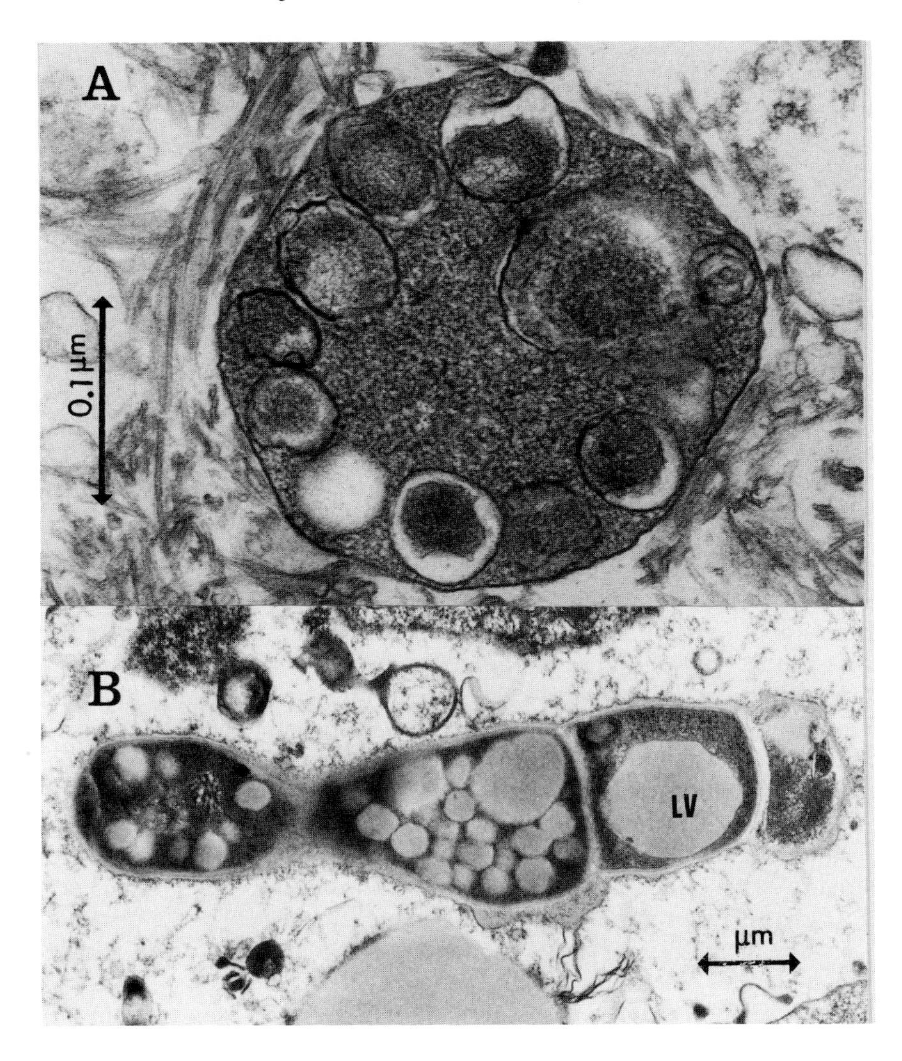

FIGURE 3. A. Electron micrograph of L-form of N. asteroides *GUH-2 grow for 2 weeks in BYE broth. Note the development of numerous membrane-bound bodies (Type A L-form). B. Type B L-form of* N. asteroides. *Note large lipid inclusions (LV).*

parental cell type. However, during the growth and reversion of L-forms, a unique opportunity exists for profound and basic alterations to occur during the reassembly of the cell envelope through one or more of the following mechanisms: indiscriminant uptake of exogenous DNA, protoplast fusion, loss of either chromosomal DNA or plasmids as the result of uncontrolled and randomized cell division, or mutations resulting from membrane-DNA perturbations during DNA replication.

In over 100 L-form revertants of *Nocardia* studied, it was found that each revertant differed from the parental strain in at least one property. Indeed, some revertants exhibit multiple alterations in cellular, physiological, or biochemical traits and in colony morphology (Beaman and Bourgeois, 1981; Beaman *et al.*, 1981). Most of the changes that occur in these revertants are stable, and probably reflect mutational events. Therefore, it appears that the removal of the cell wall from *Nocardia* and the subsequent growth of L-forms either potentiate or select for mutational events that result in altered gene expression. The longer the cells are maintained in the L-form state of growth, the greater the number of changes that are observed. These observations suggest a mechanism for taxonomic heterogeneity among species of *Nocardia*.

VI. POSSIBLE NOCARDIAL L-FORMS WITHIN CLINICAL MATERIAL

During the past several years, there have been many reports in the literature of initial failure to visualize or to isolate *Nocardia* in clinical material that later, usually at autopsy, was shown to contain this pathogen (Causey, 1974; Causey *et al.*, 1974; Greer, 1974; Neu *et al.*, 1967). In addition, recurrences of nocardial infections after apparent clinical cure have been noted (Stropes *et al.*, 1980). There have not been many studies that have attempted to isolate cell wall defective variants of *Nocardia* from clinical material; therefore, it is not known whether these clinical cases, difficult to diagnose and to treat, were due to L-forms. Also, the inability to visualize or to isolate normal organisms from infected material does not prove that L-forms are present in the clinical material. All these observations are compatible with, and supportive of, the presence of an altered cellular state of *Nocardia* within infected tissues.

In one series of studies, L-forms of *N. asteroides* was isolated from the cerebrospinal fluid (CSF) of a middle-aged male that had developed systemic nocardiosis (Beaman, unpublished data). In another investigation, L-forms of *Nocardia* sp. from the CSF of a patient with a chronic central nervous system infection of undetermined etiology were isolated (Beaman, unpublished data). Both of these studies demonstrated that L-forms of *Nocardia* could be isolated from clinical material at a time in which normal organisms could not be recovered. Further, they support the concept that altered cellular forms of *Nocardia* may play an important role in nocardial pathogenesis.

VII. *IN VIVO* MODIFICATION OF NOCARDIAL CELL WALLS

When cells of *Nocardia* growing *in vitro* are transferred from a medium such as BHI broth, trypticase soy broth, or nutrient broth into mice, these cells undergo several tinctorial and ultrastructural changes (Beaman *et al.*, 1980). These changes reflect significant alterations in the nocardial cell envelope as the organisms adapt and grow within the host (Beaman, 1972; Beaman and Scates, 1981; Beaman *et al.*, 1978). In general, with *N. asteroides*, the least virulent strains undergo the greatest ultrastructural modification during growth *in vivo*; whereas, the more virulent organisms tend to be affected less. However, the same correlations have not been noted with strains of *N. caviae* and *N. brasiliensis*. All strains of these latter two species that have been studied appear to be altered significantly during growth *in vivo*. In fact, although *N. caviae* 112 is quite virulent and produces progressive mycetomas in normal mice, the cells of this strain of *Nocardia* undergo the most structural modification *in vivo*. Further, there is a good correlation, with all species of *Nocardia*, between the degree of modification *in vivo* of the outer cell envelope and the organisms ability to produce granules resulting in the development of mycetomatous lesions (Table I).

VIII. L-FORMS IN EXPERIMENTAL INFECTIONS: HOST IMMUNITY AND MYCETOMAS

Intravenous injection of approximately 10^6 cells from a 48 hour culture of *N. caviae* 112 into normal mice was followed by a rapid increase in bacterial numbers in the brain, spinal cord, and kidneys. This acute phase of infection lasted for about two weeks and many of the mice died during this period. The mice that survived the acute form of illness appeared to recover and remained healthy for six months to one year after the initial infection. However, after six months, many of the mice began to develop lesions that tended to originate at the base of the skull and along the spinal column. These lesions often extended into muscle and down the legs. The bone usually became heavily infected and was ultimately destroyed. The spleens of the mice remained uninfected; however, they became greatly enlarged, frequently weighing more than 2 g. The lesions consisted of clusters of bacterial granules surrounded by polymorphonuclear (PMN) leukocytes, macrophages, lymphocytes, and fibrous tissue characteristic of mycetomas (Alteras *et al.* 1980; Barnetson and Milne, 1978). The granules consisted of masses of either spherical or irregularly shaped, Gram-negative cells encircled by a ring of pleomorphic, Gram-positive filaments. Both

TABLE I. The Relationship Between Virulence, In vivo *Modification of the Cell Envelope, and Induction of Mycetomas by* Nocardia

Strain	LD_{50} *during growth phase*[a]			*Visualized degree of modification*[b]	*Mycetomas*[c] *produced*
	Log	*Early stationary*	*Stationary*		
N. asteroides *GUH-2*	3×10^4	3×10^6	5×10^7	±	*No*
N. asteroides *C*	$< 1 \times 10^5$	3×10^7	7×10^7	*N.D.*[d]	*No*
N. asteroides *14759*	3×10^6	8×10^6	2×10^7	+	*No*
N. brasiliensis *17E*	4×10^5	2×10^6	4×10^6	+ +	*Yes*
N. asteroides *Mahvi*	5×10^5	1×10^7	5×10^7	+	*No*
N. asteroides *AI*	9×10^5	8×10^7	2×10^8	*N.D.*	*No*
N. caviae *112*	2×10^6	2×10^7	5×10^7	+ + +	*Yes*
N. caviae *270*	*N.D.*	1×10^7	*N.D.*	+ + +	*Yes*
N. asteroides *R4G*	5×10^6	3×10^7	3×10^8	+ +	*Yes*
N. asteroides *10905*	3×10^7	1×10^8	5×10^8	+ + +	*Yes*
N. asteroides *AniRev*	6×10^7	2×10^8	1×10^9	+ + +	*Yes*

[a] LD_{50} *(calculated by the Reed-Muench method) was based on intravenous inoculation of single cell suspensions of* Nocardia *in saline at different stages of growth into 4-6 week-old female Swiss Webster mice. Values are colony forming units (CFU) per mouse (6 or 10 mice/group/dilution).*

[b] *(+): little visible change, some enhanced acid-fastness, cellular beading, some irregular staining in Gram reaction; (+): moderate cellular alterations observed, enhanced acid-fastness, increased beading, Gram variability, some ultrastructural modification in cell envelope as visualized by electron microscopy; (+ +): significant cellular alterations observed, strong acid-fastness and characteristic beading, Gram variable, increased outer layer of cell wall, and increased lipid inclusions visualized by electron microscopy; and (+ + +): very marked ultrastructural alterations observed in both cell wall and cellular composition, acid-fast beads, Gram variable. L-form-like cells readily visualized.*

[c] *Mycetomas are defined as having bacterial granules surrounded by chronic inflammation and granulomatous infiltration.*

[d] *N.D. = not determined.*

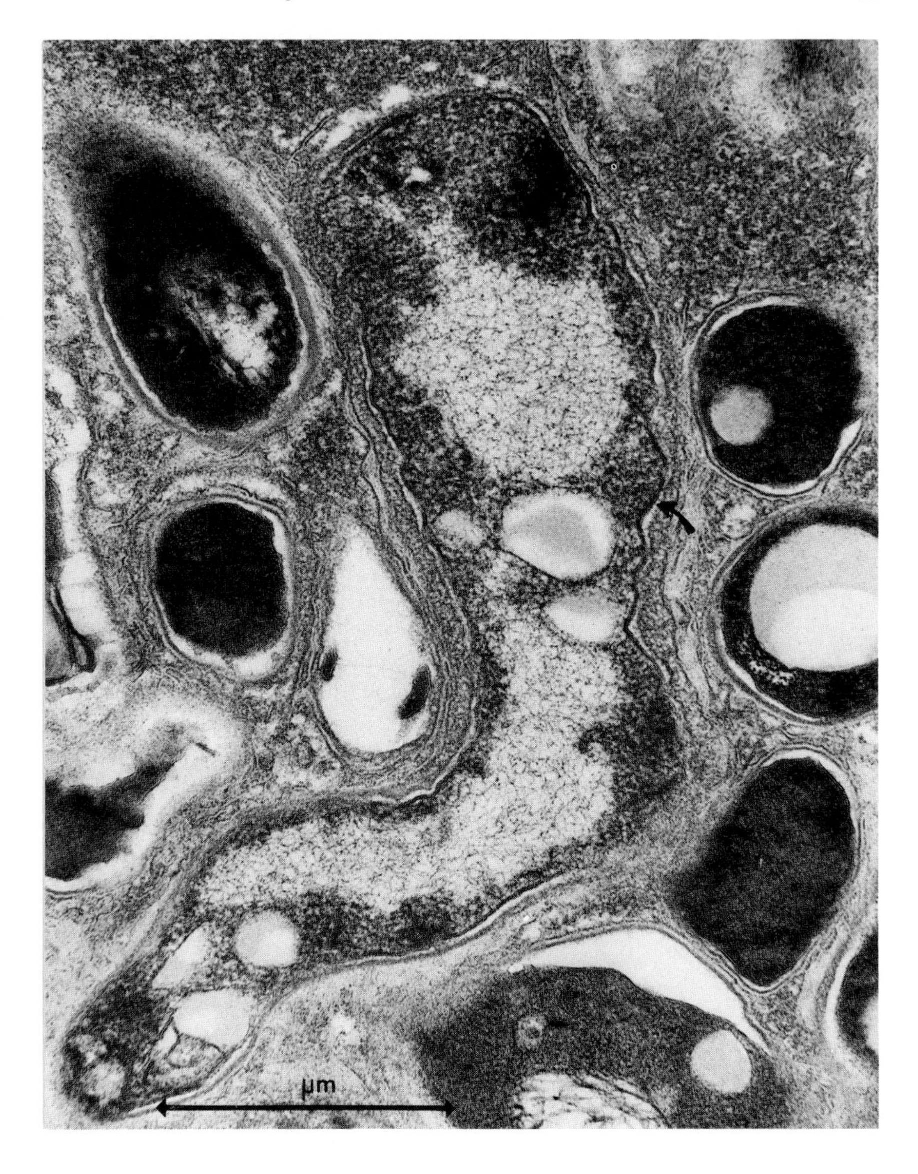

FIGURE 4. Electron micrograph of a bacterial granule of N. caviae *112 within a mycetoma in a mouse, approximately 1 year after intravenous inoculation with normal cells. The large club-shaped cell (arrow) has no cell wall.*

bulbous and club-shaped organisms were prominent within the granules. In addition, some bacterial granules consisted entirely of small granular elements and Gram-negative cells. Electron microscopy revealed the presence of cell wall-less organisms (Fig. 4); when these granules were plated on appropriate culture medium, normal nocardial cells as well as

L-forms could be recovered. Studies of organ distribution and clearance of intravenously administered cells of *N. caviae* demonstrated that L-forms and L-form variants could persist within the brain, spinal cord, and kidneys for more than one year. Neither normal nocardial cells nor L-forms were recovered from the blood, heart, liver, lungs, or spleen of mice one year after infection.

Similar experiments utilizing congenitally immunodeficient murine hosts (athymic or asplenic mice), instead of normal mice, have demonstrated significant differences based entirely upon the immunocompetence of the animals (Beaman and Scates, 1981). The results of studies involving large numbers of different strains of mice which had been infected by a variety of inoculation routes has demonstrated quite clearly that T lymphocyte deficient animals were significantly more susceptible to *N. caviae* than were normal mice. The development of typical, progressive mycetomatous lesions occurred only in immunocompetent mice and could not be demonstrated in immunodeficient animals. Instead, the T cell-deficient mice responded to *N. caviae* with acute, pyogenic lesions and abscesses in which lymphocyte and macrophages were not usually present. L-forms of *N. caviae* could not be isolated from immunodeficient mice, but appeared to be induced only in the immunologically intact host. Asplenic mice were shown to be most susceptible to early acute and fulminating infection and died before chronic mycetomas could be established. L-forms and L-form variants of *N. caviae* have been found to be highly correlated with persistence of the organisms; these L-forms played a major role in granule formation and mycetoma induction (Beaman and Scates, 1981).

Thus, the data from these studies suggest that cells of *N. caviae* grow within the murine host during the acute phase of infection. This results in the induction of a cell-mediated immune response in normal mice that involves T-cell interactions with macrophages and PMN leukocytes. After two or three weeks, this T cell-mediated response inhibits further growth of the bacteria. During this process, the cell walls of the organisms are damaged sufficiently to permit their removal. Cells that are within phagocytic vacuoles in either macrophages or PMN leukocytes can be protected from osmotic lysis during the removal of the cell wall; these organisms with their cell wall stripped away appear to be refractory to the enzymatic milieu within the phagolysosomes. For a period following this process, the host response subsides and the infection becomes quiescent. There is little pathologic evidence of infection; however, the presence of L-forms can be demonstrated. Over a period of time, these L-forms increase in number and, several months after the initial infection, the L-forms increase sufficiently to form microcolonies within the tissues. These L-form colonies increase in size and once again induce a cell-mediated response. Frequently, many of the L-forms within a colony (granule) revert to the walled form. These revertants continue to grow at the periphery of the granule; this results in enhancement of the host response. Thus, a lesion characteristic of a typical mycetoma is produced. It is not known if this same bacterial process occurs with other mycetoma-producing strains of *Nocardia*, but the evidence supports a role for L-forms in persistence, latency, and pathogenesis of mycetomas even in strains of *N. brasiliensis* (Beaman, unpublished data).

IX. PATHOGENICITY OF NOCARDIAL L-FORMS

It appears that L-forms of *Nocardia* may be intimately involved in nocardial pathogenesis; some clearly play a role in nocardial persistence and latency of infections. The ultimate proof of the etiology of L-forms in disease would be the fulfillment of "Koch's Postulates". Therefore, L-forms of *N. caviae* 112 and *N. asteroides* GUH-2 were induced *in vitro* by growth in BYE L-form broth, supplemented with glycine and lysozyme. After repeated serial passages in L-form medium over a period of several months, it was concluded that stable L-forms of these two strains were being maintained. These L-forms were then inoculated, either intravenously or intraperitoneally, into groups of female Swiss Webster mice (4-6 weeks old). The mice were monitored for approximately one year after infection to determine the persistence of the L-forms within the tissues, the latency of infection, and the pathological responses induced by the L-forms. At the same time, both heat- and formalin-killed L-forms were injected into mice to serve as controls of the host response. The mice demonstrated no detectable pathologic change as the result of injection of dead L-forms during the course of these experiments. In contrast, L-forms of both *N. asteroides* GUH-2 and *N. caviae* 112 induced a pathologic response characteristic of mycetomatous lesions. The L-forms of *N. asteroides* induced a self-limited infection that persisted for about one month and then completely resolved. Normal organisms of *N. asteroides* GUH-2 could never be isolated from these lesions; however, L-form colonies were usually, but not always, recovered (Fig. 5).

L-forms of *N. caviae* differed from those of *N. asteroides* in that they persisted within the host for at least one year and they induced a more chronic, progressive development of mycetomas which were indistinguishable from those induced by injection of the normal organism. In addition, in some mice, the L-forms persisted in high numbers without inducing detectable pathologic responses. Most of the animals that received L-forms of *N. caviae* by intravenous injection remained healthy in appearance for 6-8 months and, thereafter, developed large mycetomatous lesions which frequently reached a weight of 50 g (in a 30 g mouse). Usually, normal organisms as well as L-forms were recovered from these lesions. Thus, during the development of the bacterial granule that induced the

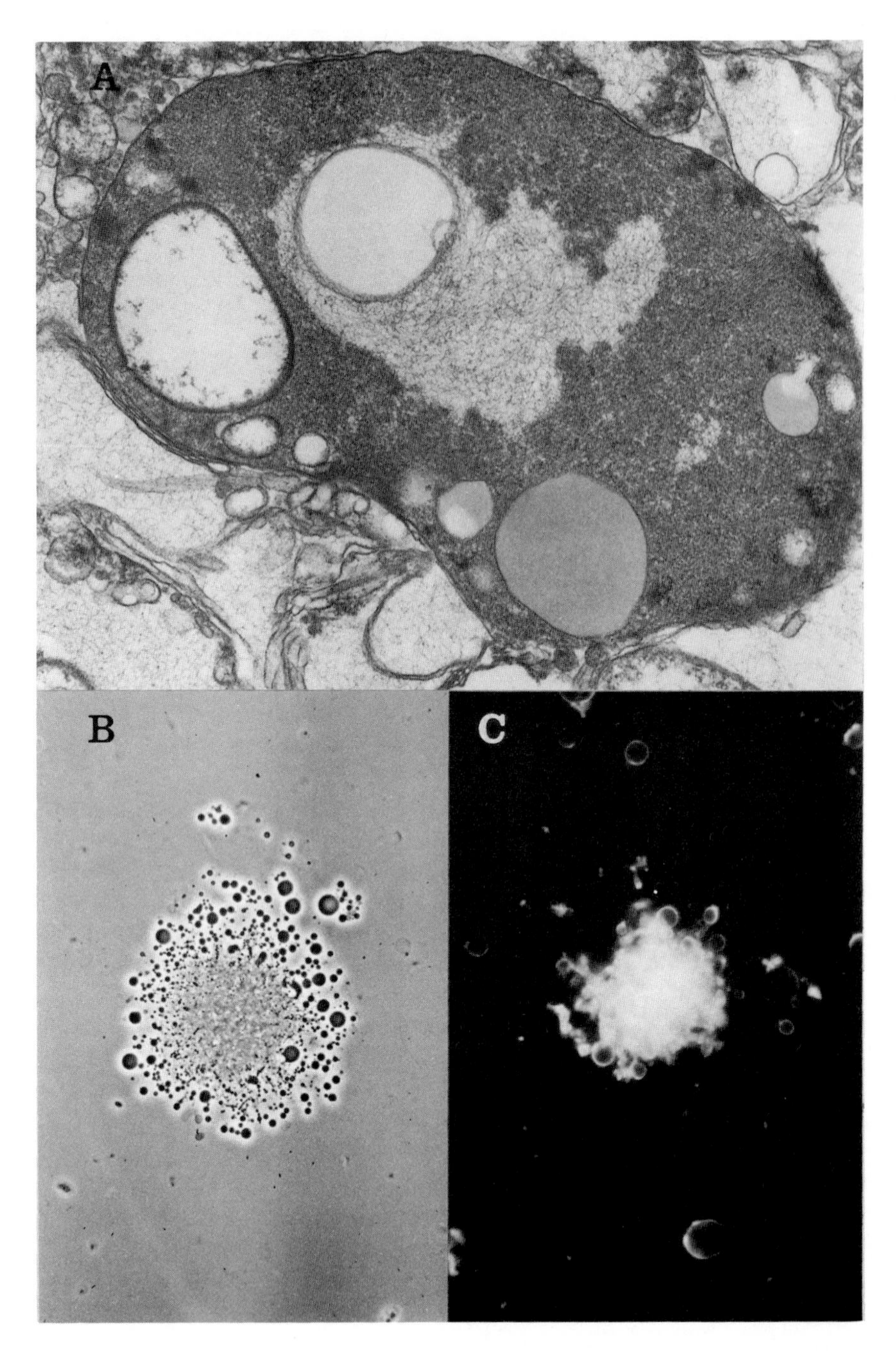
A
B
C

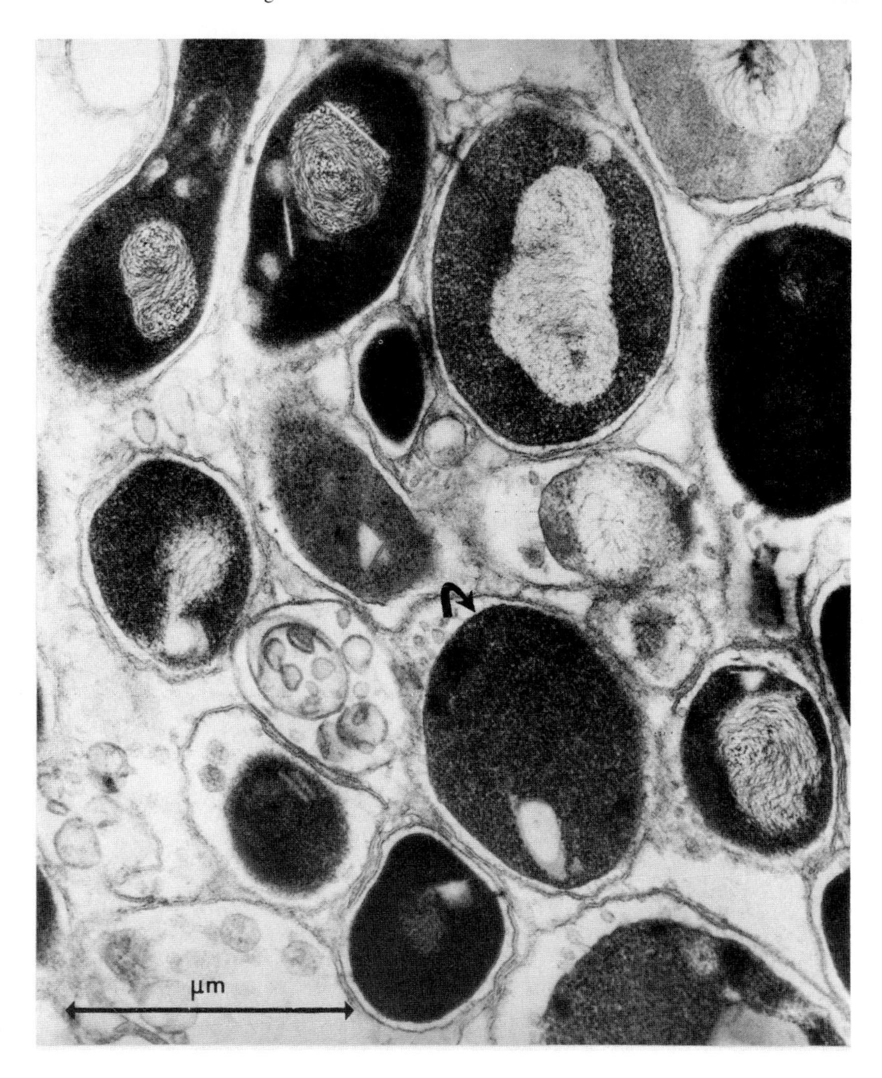

FIGURE 6. Electron micrograph of a granule of L-forms of N. caviae *within a mycetoma approximately 1 year after intravenous inoculation of the mouse with L-forms induced* in vitro. *Arrow points to the cytoplasmic membrane of L-form cell. No cell wall was visualized and only L-forms were recovered from the lesions in this mouse.*

FIGURE 5. L-forms of N. asteroides *GUH-2 from infected mouse. A. Electron micrograph of L-form of* N. asteroides *within the "granule", 2 weeks after intraperitoneal inoculation. B. L-form colony of* N. asteroides *GUH-2 isolated from the lesion. No normal bacteria were isolated. C. Immunofluorescent stain of coverslip impression of colony shown in B. The fluorescence is specific for* N. asteroides.

mycetomatous response, many of the L-forms reverted to the walled organism. However, in at least one mouse, only L-forms of *N. caviae* could be isolated and normal organisms were neither visualized nor recovered from the tissues. Electron microscopy and histology demonstrated that the granules within the mycetomas consisted of Gram-negative spheres, irregularly shaped cells, club-shaped cells, and small granular elements (Fig. 6). Cell wall material could not be visualized. The L-forms that were isolated from this mouse were injected into additional mice. Many of these animals developed mycetomatous lesions six months to one year after infection, and L-forms of *N. caviae* 112 could once again be recovered. Therefore, "Koch's Postulates" appear to have been fulfilled proving that L-forms of *N. caviae* are pathogenic and cause mycetomas.

X. CONCLUSIONS

The cell wall can be removed from most nocardiae without loss of cellular viability by growing them in a hypertonic medium rich in glycine and lysozyme or D-cycloserine. The resultant protoplasts or spheroplasts can grow in the cell wall deficient state when incubated in the appropriate culture medium. Similar results can be obtained in an animal host with some strains of *Nocardia*. Even though the L-forms have no cell wall, they still synthesize and secrete cell wall material. Therefore, the cell wall determinants of nocardial pathogenesis are not lost during growth of the cells in a wall-less state.

That L-forms can be pathogenic was demonstrated by using murine models to study L-forms of both *N. caviae* and *N. asteroides* induced *in vivo* or *in vitro*. Furthermore, L-forms were able to induce mycetomas in experimental animals and appeared to be intimately involved in the pathologic response of the host. L-forms of *N. caviae* were shown to be able to persist within the host's tissues for extended periods and to play a role in the latency of disease. Thus, from these observations, it appears that the structural integrity of the cell wall is not essential for nocardial pathogenicity, but that specific cell wall components may be required in determining the pathogenic capabilities of these organisms.

ACKNOWLEDGMENTS

I thank Marilyn Wheeler for typing this manuscript.

REFERENCES

Alteras, I., Feuerman, E. J., and Dayan, I., *Int. Soc. Trop. Dermatol. 19:*260 (1980).
Barnetson, R. S., and Milne, L. J. R., *Br. J. Dermatol. 99:*227 (1978).
Beaman, B. L., *Infect. Immun. 8:*828 (1972).
Beaman, B. L., *J. Bacteriol. 123:*1235 (1975).
Beaman, B. L., *Infect. Immun. 29:*244 (1980).
Beaman, B. L., and Bourgeois, A. L., *J. Clin. Microbiol. 14:*574 (1981).
Beaman, B. L., and Maslan, S., *Infect. Immun. 20:*290 (1978).
Beaman, B. L., and Scates, S. M., *Infect. Immun. 33:*893 (1981).
Beaman, B. L., and Shankel, D. M., *J. Bacteriol. 99:*876 (1969).
Beaman, B. L., Bourgeois, A. L., and Moring, S. E., *J. Bacteriol. 148:*600 (1981).
Beaman, B. L., Serrano, J. A., and Serrano, A. A., *Zentralbl. Bakteriol. 6* suppl.:201 (1978).
Beaman, B. L., Maslan, S., Scates, S., and Rosen, J., *Infect. Immun. 28:*185 (1980)
Bourgeois, L., and Beaman, B. L., *Infect. Immun. 9:*576 (1974).
Bourgeois, L., and Beaman, B. L., *J. Bacteriol. 127:*584 (1976).
Causey, W., *Appl. Microbiol. 28:*193 (1974).
Causey, W. A., Arnell, P., and Brinker, J., *Chest 65*:360 (1974).
Greer, K. E., *VA. Med. Mon. 101:*193 (1974).
Neu, H. C., Silva, M., Hazen, E., and Rosenheim, S. H., *Ann. Intern. Med. 66:*274 (1967).
Prasad, I., and Bradley, S. G., *J. Gen. Microbiol. 70:*571 (1972).
Stropes, L., Bartlett, M., and White, A., *Am. J. Med. Sci. 280:*119 (1980).
Webb, R. B., Clark, J. B., and Chance, H. L., *J. Bacteriol. 67:*498 (1954).
Webley, D. M., *J. Gen. Microbiol. 11:*420 (1954).
Webley, D. M., *J. Gen. Microbiol. 23:*87 (1960).

DEFENSE AGAINST *NOCARDIA ASTEROIDES* IN MAN

Gregory A. Filice

Infectious Disease Section
Medical Service
Veterans Administration Medical Center
and
Infectious Diseases Section
Department of Medicine
University of Minnesota
Minneapolis, Minnesota

Blaine L. Beaman

Department of Medical Microbiology and Immunology
School of Medicine
University of California
Davis, California

James A. Krick

Department of Medicine
University of California School of Medicine
Davis, California

Jack S. Remington

Department of Immunology and Infectious Diseases
Research Institute
Palo Alto Medical Foundation
Palo Alto, California
and
Division of Infectious Diseases
Department of Medicine
Stanford University School of Medicine
Stanford, California

ISBN 0-12-528620-1

I. INTRODUCTION

In the last three decades *Nocardia asteroides* has been recognized as an important cause of severe disease in man (Beaman *et al.*, 1976; Curry, 1980; Krick *et al.*, 1975; Palmer *et al.*, 1974; Simpson *et al.*, 1981; Weed *et al.*, 1955; Young *et al.*, 1971). *Nocardia* are different in several respects from other actinomycetes and very different from the more common bacterial pathogens of man. The disease caused by *N. asteroides* is distinctive; the course is typically indolent or subacute, and the typical lesions are abscesses which are usually in the lung and often in other organs (Weed *et al.*, 1955). Since *N. asteroides* and nocardiosis are different in many ways from other bacterial pathogens and their associated diseases, host defenses against *N. asteroides* in man may also be different.

Clinical observations of nocardiosis in man have suggested the contributions of certain mechanisms of host defense. Although nocardiosis occurs in apparently healthy people, it seems to be more common in the immunocompromised. Most of the latter have had lymphomas, organ transplants, or steroid therapy—conditions in which cell-mediated immunity is usually markedly abnormal and in which humoral immunity and neutrophil function are usually less abnormal. This suggests that cell-mediated immunity is normally an important defense against *N. asteroides* in man. However, since the histologic hallmark of nocardiosis in man is abscess formation with a marked neutrophil predominance, neutrophils may have a role in defense against *N. asteroides* in man. People with chronic granulomatous disease seem to have increased susceptibility to nocardiosis (Bujak *et al.*, 1973; Idriss *et al.*, 1975); this suggests that oxidative metabolites contribute to inhibition or killing of *N. asteroides* by phagocytes. Chronic lung disease may predispose some people to nocardiosis (Palmer *et al.*, 1974). Abnormalities of immunoglobulins without other abnormalities of host defense probably do not increase risk for nocardiosis.

Extensive work has shown that cell-mediated immunity is of major importance in animal models of *Nocardia* infection (Beaman *et al.*, 1978; Folb *et al.*, 1977; Krick and Remington, 1975). It has been demonstrated that animals (Hiramine *et al.*, 1981) and humans (Ortiz-Ortiz and Bojalil, 1972) can show delayed hypersensitivity to nocardial antigens. Immunoglobulins do not seem to be protective in an animal model (Krick and Remington, 1975), but they may act in concert with other aspects of host defense (Beaman, 1979). Since much of this work will be discussed by others at this symposium, it will not be referred to here.

These clinical and experimental observations suggest that cell-mediated immunity and neutrophils may each contribute to defense against *Nocardia* in man. The studies described herein were done to study the interactions of *N. asteroides* with these two components of host defense.

II. INTERACTION OF *NOCARDIA* AND HUMAN NEUTROPHILS AND MONOCYTES

A. Lack of Killing of Nocardia *by Neutrophils and Monocytes*

Since sites of infection with *N. asteroides* in man are infiltrated predominately by neutrophils, a conventional bactericidal assay was used to determine if *Nocardia* are killed by neutrophils or monocytes *in vitro* (Filice *et al.*, 1980). Five strains of *N. asteroides* isolated from humans were studied. Before use, *Nocardia* were passaged in mice. Infected kidneys were transferred into brain heart infusion (BHI) broth, and subcultures were grown to the early stationary phase of growth. Coccobacillary forms were separated from other forms and aggregates by differential centrifugation and then washed twice in phosphate buffered saline (PBS, pH 7.2) before resuspension in the proper medium. For comparison, *Staphylococcus aureus* strain 502A and a strain of *Listeria monocytogenes*, serotype 4b, were also studied. *Staphylococcus* and *Listeria* were incubated 16 h in trypticase soy broth, washed twice in PBS, and then resuspended in the appropriate medium.

Blood was obtained from 11 healthy adults and separated into granulocyte and mononuclear cell preparations by Ficoll-Hypaque density gradient centrifugation. The cells were washed twice in Hanks' balanced salt solution (HBSS) and counted in a hemacytometer. The proportion of mononuclear cells that were monocytes was determined by allowing mononuclear cells to ingest neutral red. Reaction mixtures which consisted of 4×10^6 phagocytes (neutrophils or monocytes), autologous serum (0.1 ml), 1×10^7 bacteria, and HBSS to give a final volume of 1 ml and control tubes without phagocytes were simultaneously incubated at 37°C with mild agitation. At the beginning of incubation and at various intervals thereafter (up to four hours), 0.05 ml portions of the preparations were added to separate tubes containing 0.45 ml of saline or distilled water. The contents of these tubes were sonicated to disrupt phagocytes, mixed on a vortex mixer, and serially diluted in PBS. Portions (0.1 ml) of the dilutions were plated on BHI agar.

Cytocentrifuge preparations indicated that both neutrophils and monocytes ingested large numbers of the challenge bacteria whether they were *Nocardia, Staphylococcus*, or *Listeria*. The procedures of serial dilution and quantification of colony-forming units which were used in the bacterial assay often show marked variability; therefore, for one to have confidence that killing has occurred and that the reductions were not artifactual, any reductions in number of colony-forming units of challenge bacteria must be greater than that observed for *Nocardia*. Here, both neutrophils and monocytes consistently killed most *Staphylococcus* and *Listeria*. In contrast, the numbers of viable *Nocardia* in preparations with neutrophils and monocytes averaged, respectively, only 10 and 21 percent less than that in control tubes. Results from a representative experiment are shown in Figure 1.

Although certain parameters were altered to reduce the possibility that these minimal reductions in numbers of viable *Nocardia* were artifacts of the experimental conditions, substantial killing of *Nocardia* was not observed.

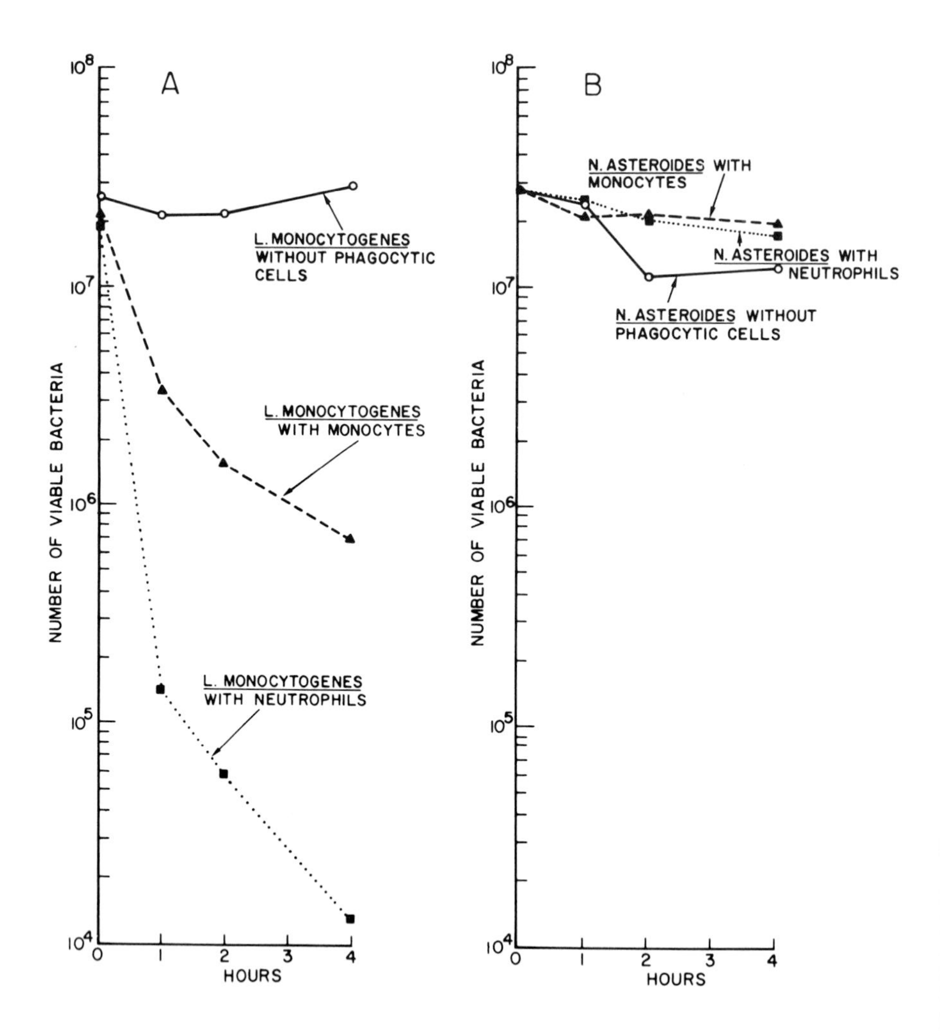

FIGURE 1. Results from one experiment, representative of 28, in which human phagocytes were challenged with L. monocytogenes, S. aureus, *or* N. asteroides *in parallel. The number of viable* L. monocytogenes *(left) or* N. asteroides *strain GUH-2 (right) in preparations of neutrophils or monocytes or in preparations without phagocytes was measured at the time of challenge (zero hour) and at various intervals thereafter.*
(Reproduced from Filice et al., J. Infec. Diseases 142:*432 (1980) by copyright permission of the University of Chicago Press.)*

In addition, opsonization of *Nocardia* with serum from patients convalescent from *Nocardia* infections did not result in substantial killing by phagocytes. From these results, it was concluded that neutrophils and monocytes did not kill substantial numbers of *Nocardia* under experimental conditions in which the comparison bacteria were readily killed.

B. Resistance of Nocardia *to Products of the Oxidative Metabolic Burst*

Resistance to killing by neutrophils in a conventional bactericidal assay is an unusual characteristic among bacteria (Mandell, 1974). *Salmonella typhi* (Miller *et al.*, 1979) and *Toxoplasma gondii* (Wilson *et al.*, 1980) appear to survive within phagocytes because the oxidative metabolic burst does not occur when they are phagocytized. To determine if the oxidative metabolic burst occurred during interaction of *Nocardia* and either neutrophils or monocytes, chemiluminescence was measured (Filice *et al.*, 1980), Neutrophils and monocytes were prepared as described for the bactericidal assays, except that they were washed twice in Krebs-Ringer phosphate buffer containing 5.5 mM glucose (KRPG, pH 7.4), instead of HBSS. Heat-killed *Candida albicans* was used as a positive control. *Candida* and *Nocardia* were pre-incubated in autologous serum and all components were pre-warmed to 37°C. Duplicate 1.5 ml reaction mixtures consisted of 4×10^6 neutrophils or monocytes, autologous serum (0.15 ml), and either 1.6×10^8 *Candida* or $\sim 6 \times 10^8$ *Nocardia* (range 3×10^8 to 9×10^8) in KRPG. Chemiluminescence was measured in a liquid scintillation counter with one photomultiplier tube with its window set at 0 to ∞.

Chemiluminescence consistently occurred when *Nocardia* interacted with neutrophils or monocytes (Fig. 2); this suggested that the oxidative metabolic burst occurred. Since the *Nocardia* were alive, it was not known *a priori* whether chemiluminescence was generated by neutrophils or monocytes or by *Nocardia*. It has been previously reported (Cohen *et al.*, 1978) that neutrophils exposed to 5×10^{-5} M chlorpromazine lack an oxidative metabolic burst when stimulated. In this study, when neutrophils were preincubated in chlorpromazine, washed, and then challenged with *Nocardia*, chemiluminescence did not occur (Fig. 2). This suggested that the chemiluminescence that occurred when untreated neutrophils and *Nocardia* interacted was generated by the neutrophils and was further evidence that the oxidative metabolic burst occurred. Since *Nocardia* were not killed, it appeared as if *Nocardia* were resistant to the injurious effects of metabolites of the oxidative metabolic burst.

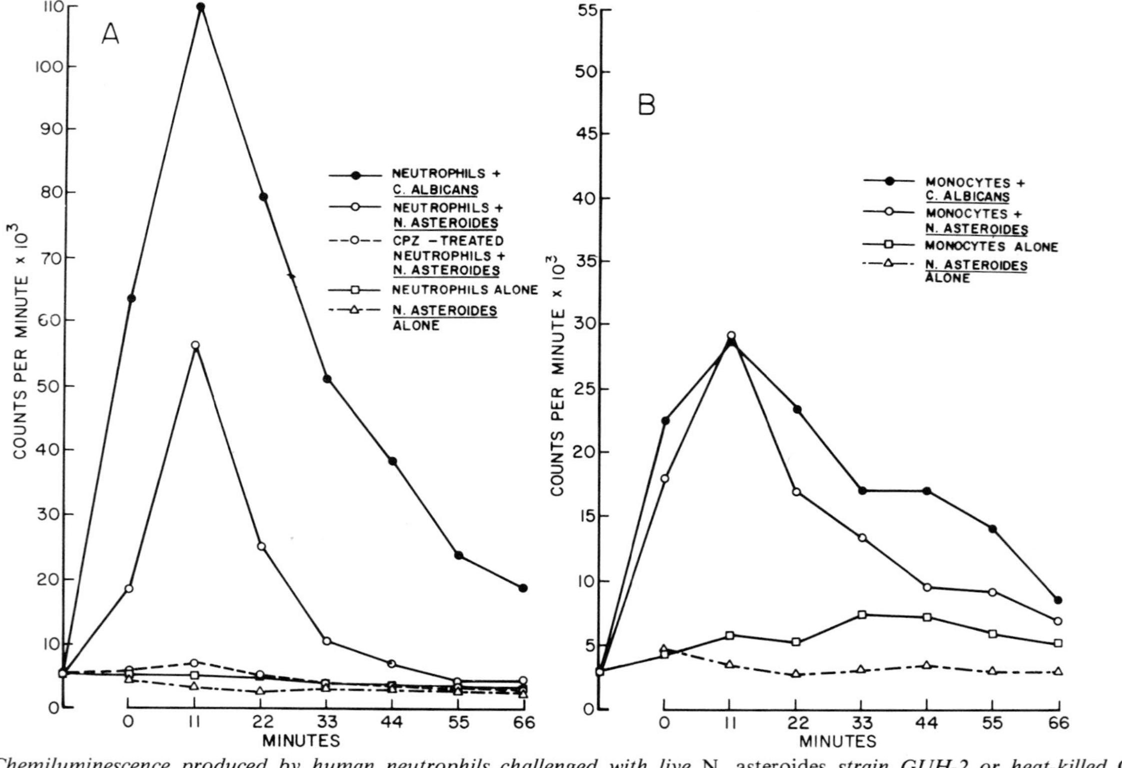

FIGURE 2. A. Chemiluminescence produced by human neutrophils challenged with live N. asteroides *strain GUH-2 or heat-killed* C. albicans. *Chemiluminescence was not produced by neutrophils treated with chlorpromazine (CPZ) and challenged with* N. asteroides, *by resting neutrophils, or* N. asteroides *without neutrophils. B. Chemiluminescence produced by human monocytes challenged with* N. asteroides *or heat-killed* C. albicans. *Chemiluminescence was not produced by resting monocytes or* N. asteroides *without monocytes. For each time period, the mean of three consecutive base-line readings are plotted on the ordinates.*
(Reproduced from Filice et al., J. Infec. Diseases 142*:432 (1980) by copyright permission of the University of Chicago press.)*

III. INTERACTIONS OF *NOCARDIA* WITH ACTIVATED MOUSE PERITONEAL MACROPHAGES

Since defects in cell-mediated immunity seem to result in enhanced susceptibility to nocardiosis, interactions of *Nocardia* with activated macrophages were studied (Filice *et al.*, 1980). Mouse peritoneal macrophages were studied, because they were readily obtainable and because they could be activated *in vivo* before they were harvested. Two strains of *Nocardia*, GUH-2 and ATCC 14759, each originally isolated from a patient with nocardiosis, were studied. Coccobacillary forms were prepared as described in Part II.A. Activated peritoneal macrophages (defined by their ability to inhibit the multiplication of *Toxoplasma gondii*) were harvested from Swiss-Webster female mice that had been injected with *Corynebacterium parvum* or chronically infected with *T. gondii*. In parallel experiments, control macrophages (macrophages that allowed multiplication of *T. gondii*) and activated macrophages were challenged with either *T. gondii* or *N. asteroides*.

Peritoneal exudate cells were harvested from the mice, suspended in medium 199 containing heat-inactivated fetal calf serum, (10%, final concentration) and allowed to adhere to glass for 3 h. Monolayers were washed to remove nonadherent cells. The remaining cells, referred to as macrophages (Filice *et al.*, 1980), were challenged immediately thereafter with *Nocardia*. After one to three hours, monolayers were washed, fresh medium was added, and macrophage preparations were reincubated. At various intervals, viable *Nocardia* either associated or not with macrophages were quantitated, and preparations run in parallel were Gram stained.

Examination of preparations stained at the end of the incubation period revealed that *Nocardia* associated with control macrophages were in the form of short filaments, whereas almost all *Nocardia* associated with activated macrophages were in the coccobacillary form and appeared as they did in the original inoculum. At later time points, progressive elongation and branching of *Nocardia* filaments were observed in control macrophage preparations; it appeared as if *Nocardia* filaments has extended from within the macrophages to the outside; many of these filaments appeared to have extended to and then grown through neighboring macrophages (Fig. 3). In activated macrophage preparations, similar filamentous growth appeared to have originated in a few isolated macrophages, but in general, growth of *Nocardia* was markedly inhibited. By 20 h, almost all macrophages in control monolayers were destroyed, whereas most activated macrophages were still intact.

Quantitation of viable *Nocardia* in preparations of *Nocardia* and macrophages in a typical experiment are shown in Figure 4. The number of *Nocardia* recoverable immediately after the one-hour challenge appeared to be fewer from activated than from control macrophage preparations (Fig. 4a), but the difference was not significant. After the supernatants were removed and the monolayers were washed at the end of the one-hour challenge period, the numbers of macrophage-associated *Nocardia* were similar in both control and activated macrophages (Fig. 4b). Thereafter, the numbers of *Nocardia* in the whole preparations (sum of *Nocardia* associated with macrophages and of *Nocardia* in the supernatant medium) reflected the interactions of macrophages with the *Nocardia* that were associated with macrophages at the end of the one-hour challenge period. Three hours after challenge, the numbers of *Nocardia* in both activated and control macrophage preparations had not changed from those observed at the end of the challenge period. At six hours, the numbers of *Nocardia* in activated macrophage preparations were significantly reduced when compared with the numbers in similar preparations at three hours ($p < 0.001$) or with the

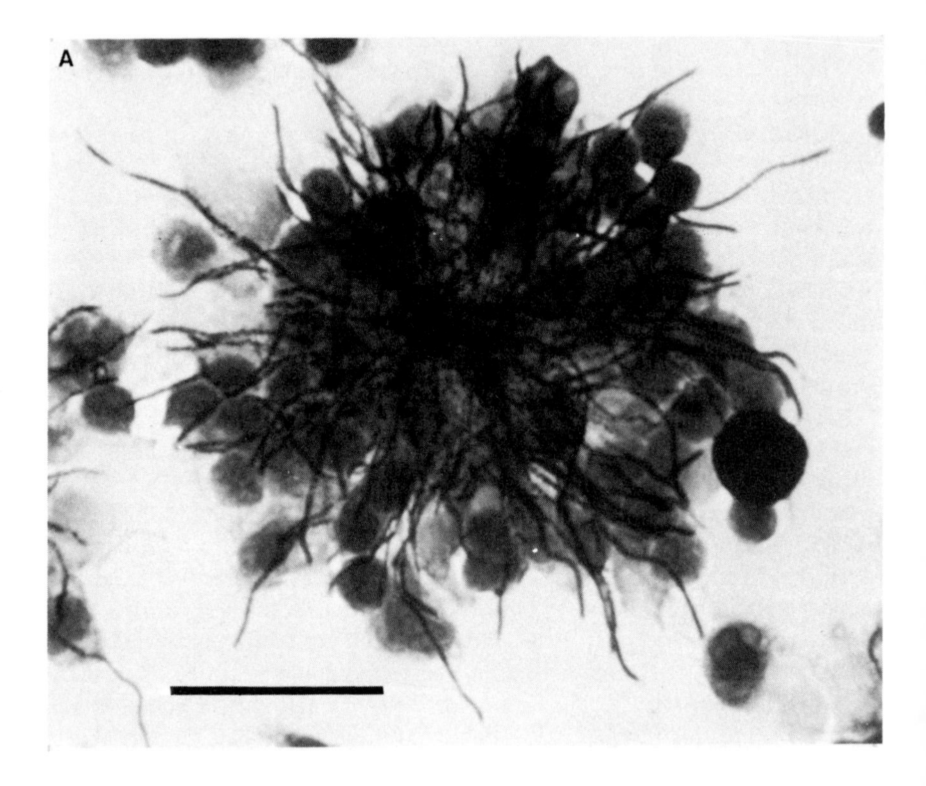

FIGURE 3. Interactions of N. asteroides *GUH-2 with activated macrophages from* Toxoplasma-*infected mice and with macrophages from controls (Gram-stained preparations). Macrophages were challenged with coccobacillary* N. asteroides *for 1 h. The supernatant was removed, fresh medium was added, and the preparations were reincubated. a) Control macrophage preparation 6 h after initiation of challenge. Extensive filament formation by the* Nocardia *is apparent (bar = 25 μm). b) Activated macrophage preparation 6 h after initiation of challenge.* Nocardia *in most activated macrophages remain in the coccobacillary form (arrow) (bar = 25 μm).*
(Reproduced from Filice, G. A., Beaman, B. L., and Remington, J. S., Infect. Immun. 27:*643 (1980) by copyright permission of the American Society for Microbiology.)*

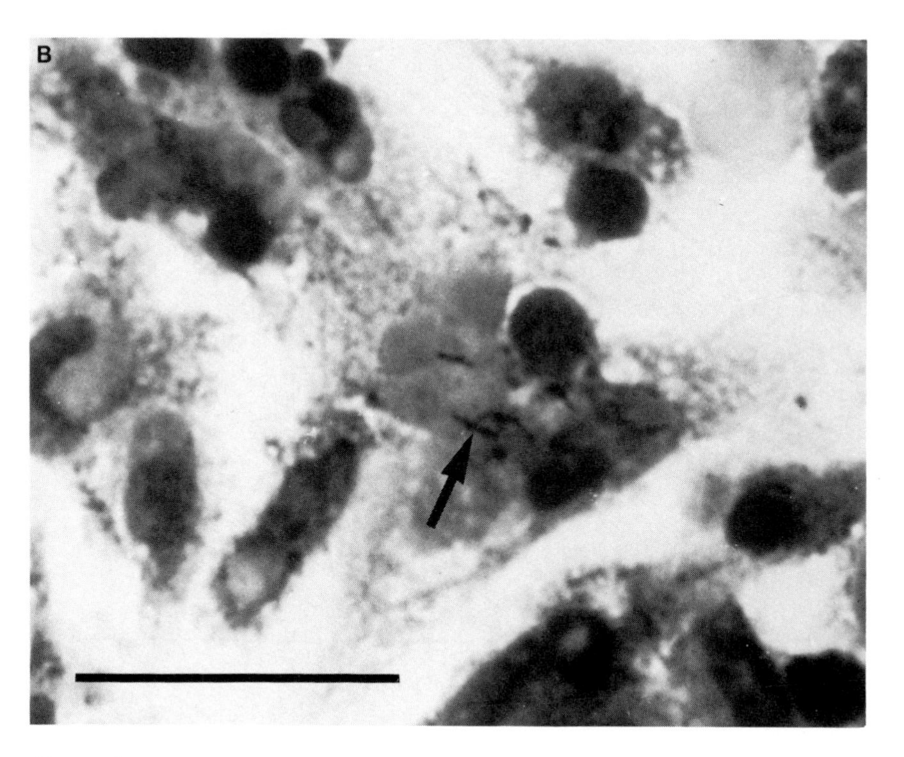

Figure 3b.

numbers in control macrophage preparations at six hours ($p < 0.002$) (Fig. 4b). At 12 hours, the numbers of *Nocardia* increased in both types of macrophage preparations, but the numbers in activated macrophage preparations were still significantly less than those in control macrophage preparations ($p < 0.001$).

It was concluded that activated macrophages killed 40 to 50% of *Nocardia* that were phagocytized and inhibited the growth of the remaining *Nocardia* when compared with control macrophages. Under these conditions, the inhibition of growth of the remaining *Nocardia* was temporary.

IV. CONCLUSIONS AND POSSIBLE IMPLICATIONS FOR HOST DEFENSE IN MAN

These studies showed that activated mouse peritoneal macrophages inhibited the growth of *Nocardia in vitro*. Since *Nocardia* grows by elongation from coccobacillary forms to long filaments, inhibition of the growth of *Nocardia* by activated macrophages could be observed by direct examination of stained preparations. Results of preliminary experiments indicated that the same appeared to be true for human neutrophils. The lack of filament formation could have been misleading, if the phagocytes induced

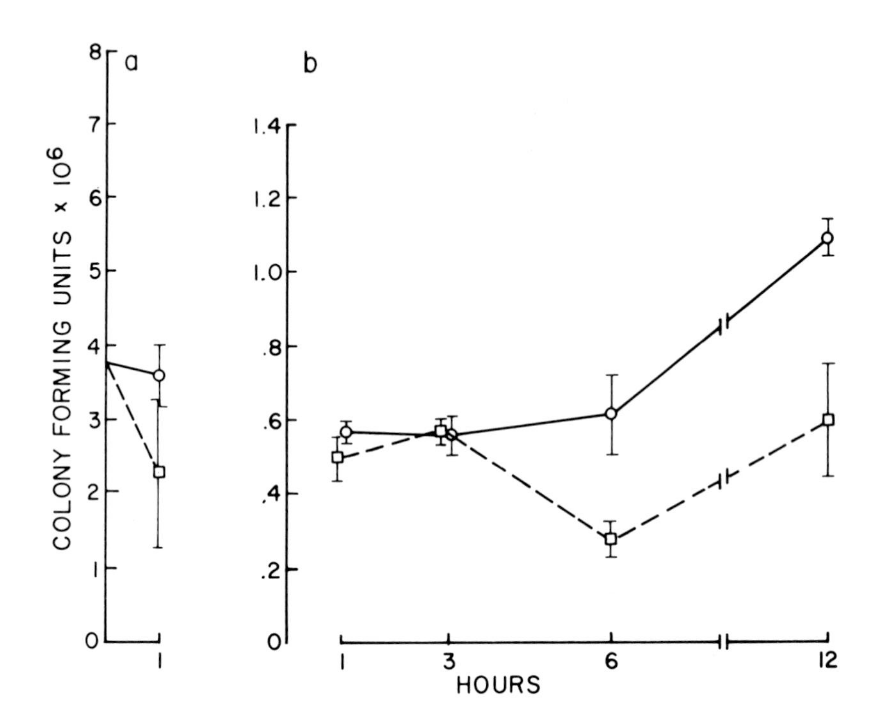

FIGURE 4. Interactions of N. asteroides *GUH-2 with activated macrophages from* Toxoplasma-*infected mice (□) and with macrophages from controls (O). For each time point after challenge, data represent the mean (± standard deviation) of 4 samples. a) Mean colony-forming units of* N. asteroides *in activated and control macrophage preparations 1 h after 3.8 x 10* ^{6} *viable* Nocardia *were added. For each sample, counts of* Nocardia *in supernatant media and* Nocardia *associated with macrophages were added; the means reflect the effect of macrophages on the inoculum. b) Mean colony-forming units of macrophage-associated* Nocardia *in macrophage preparations immediately after washing (1 h) and of* Nocardia *in whole preparations at later time points (see text).*
(Reproduced from Filice, G. A., Beaman, B. L., and Remington, J. S., Infect. Immun. 27:*643 (1980) by copyright permission of the American Society for Microbiology.)*

Nocardia to divide into separated coccobacillary forms instead of forming filaments. However, this would have become apparent by an increase in coccobacillary forms in stained preparations and by an increase in colony-forming units; neither was observed.

Quantitation of colony-forming units during interaction of *Nocardia* and phagocytes indicated that activated macrophages killed substantial portions of *Nocardia*, but that neutrophils did not. Similar observations have been made for alveolar macrophages which were probably activated, since they were harvested from rabbits immunized with repeated injections of *Nocardia* in incomplete Freund adjuvant (Beaman, 1979).

From these and other observations, one might infer how host defenses in healthy people control and eradicate early *Nocardia* infection and how alterations in host defenses allow clinically apparent disease to develop. When *Nocardia* first gain access to the body, neutrophils may be recruited to the site of infection, phagocytose the *Nocardia*, and inhibit the growth of *Nocardia*. Macrophages may also be present in the site, but experimental data suggest that resident peritoneal (Filice *et al.*, 1980) or alveolar (Beaman, 1979) macrophages that are not activated do not inhibit the growth of *Nocardia*. Since neutrophils are short-lived cells, the inhibition by neutrophils would probably be temporary. Meanwhile, the slowly growing *Nocardia* would likely be recognized by macrophages, lymphocytes, or both, the cell-mediated immune system would be stimulated locally, and after five to seven days, macrophages would be activated. Since activated macrophages kill *Nocardia in vitro*, it seems reasonable to surmise that they can do so *in vivo* and that they would eradicate the infecting *Nocardia*. Since *Nocardia* are common soil organisms, it seems likely that *Nocardia* commonly infect man and that the events just described usually occur on a subclinical scale.

Occasionally *Nocardia* will infect a person with a generalized deficiency of cell-mediated immunity or with a specific inability to respond to *Nocardia* infection with a cell-mediated immune response. In such a person, macrophage activation may not occur or may be inadequate to kill and eradicate the *Nocardia*. Neutrophil recruitment would likely continue, and *Nocardia* growth might be slowed down but not stopped. That would result in an indolent or subacute disease characterized histologically by abscess formation, similar to that observed in nocardiosis in man. This sequence of events is speculative and further evidence is required to prove if it is true. However, it is consistent with the clinical and experimental evidence outlined above.

The mechanisms of inhibition and killing of *Nocardia* by phagocytes are not known. The observation that nocardiosis has been recognized in people with chronic granulomatous disease more often than would be expected in healthy people suggests that the oxidative metabolic burst contributes to inhibition or killing by phagocytes in healthy people. On the other hand, the observation that the oxidative metabolic burst occurred in neutrophils that encountered *Nocardia* even though neutrophils do not kill many *N. asteroides in vitro* suggested that *Nocardia* are resistant to metabolites of the oxidative metabolic burst. Recent experiments to resolve these apparently conflicting observations have indicated that *N. asteroides* is relatively, but not completely, resistant to these metabolites. It may be that these metabolites contribute to the inhibition of *Nocardia* by phagocytes or to the killing by activated macrophages, but that the production of these metabolites in neutrophils is insufficient for substantial killing to occur.

ACKNOWLEDGMENTS

This work was supported in part by a research grant from the Veterans Administration and by Grants AI 04717, CA 06341, and AI 13167 from the National Institutes of Health. Dr. Filice was the recipient of an Edith Milo Fellowship Award.

REFERENCES

Beaman, B. L., *Infect. Immun. 26*:355 (1979).
Beaman, B. L., Burnside, J., Edwards, B., and Causey, W., *J. Infect. Dis. 134*:286 (1976).
Beaman, B. L., Gershwin, M. E., and Maslan, S., *Infect. Immun. 20*:381 (1978).
Bujak, J. S., Ottesen, E. A., Dinarello, C. A., and Brenner, V. J., *J. Pediat. 83*:98 (1973).
Cohen, H. J., Chonaviec, M. E., Ellis, S., and Laforet, G, *Pediat. Res. 12*:462 (1978).
Curry, W. A., *Arch. Int. Med. 140*:818 (1980).
Filice, G. A., Beaman, B. L., and Remington, J. S., *Infect. Immun. 27*:643 (1980).
Filice, G. A., Beaman, B. L., Krick, J. A., and Remington, J. S., *J. Infec. Dis. 142*:432 (1980).
Folb, P. I., Timme, A., and Horowitz, A., *Infect. Immun. 18*:459 (1977).
Hiramine, C., Hojo, K., and Yano, I., *Int. Arch. Allergy Appl. Immunol. 65*:220 (1981).
Idriss, Z. H., Cunningham, R. J., and Wilfert, C. M., *Pediatrics 55*:479 (1975).
Krick, J. A., and Remington, J. S., *J. Infec. Dis. 131*:665 (1975).
Krick, J. A., Stinson, E. G., and Remington, J. S., *Ann. Int. Med. 82*:18 (1975).
Mandell, G. L., *Infect. Immun. 9*:337 (1974).
Miller, R. M., Garbus, J., and Hornick, R. B., *Science 23*:737 (1979).
Ortiz-Ortiz, L. and Bojalil, L. F., *Clin. Exp. Immunol. 12*:225 (1972).
Palmer, D. L., Harvey, R. I., and Wheeler, J. K., *Medicine 53*:391 (1974).
Simpson, G. L., Stinson, E. B., Egger, M. L., and Remington, J. S., *Rev. Infect. Dis. 3*:492 (1981).
Weed, L. A., Anderson, H. A., Good, A., and Baggenstoss, A. H., *New England J. Med. 253*:1137 (1955).
Wilson, C. B., Tsai, V., and Remington, J. S., *J. Exp. Med. 151*:328 (1980).
Young, L. S., Armstrong, D., Blevins, A., and Lieberman, P., *Am. J. Med. 50*:356 (1971).

HOST-PARASITE RELATIONSHIP IN INFECTIONS DUE TO *NOCARDIA BRASILIENSIS*

Librado Ortiz-Ortiz

Departamento de Inmunología
Instituto de Investigaciones Biomédicas
Universidad Nacional Autónoma de México
México, D.F., México

Emma I. Melendro

Departamento de Inmunología y Reumatología
Instituto Nacional de la Nutrición
México, D. F., México

Carmen Conde

Centro de Fijación de Nitrógeno
Universidad Nacional Autónoma de México
Cuernavaca, Morelos, México

I. INTRODUCTION

Actinomycotic mycetoma, caused by nocardiae, is usually due to *Nocardia brasiliensis*, although in rare instances *N. caviae* or *N. asteroides* may be responsible. A mycetoma is a chronic, suppurative, purulogranulomatous disease of the subcutaneous tissues and bones, which is characterized by multiple tumefactions from which "granules" are expressed into the pus or are formed in the tissues (Mariat *et al.*, 1977). The majority of mycetomas appear in the foot, particularly on the dorsum of the fore part of the foot, *e. g.*, instep and toe webs. However, many extrapedal cases are seen on other parts of the body that come into contact with soil while working, sitting, or lying, *e. g.*, hand, buttocks, perineum, and back (Mahgoub, 1973). Although no primary visceral localiza-

ISBN 0-12-528620-1

tion of mycetomas has been reported, visceral metastases due to the migration of the parasite from primary subcutaneous foci through the lymphatic vessels have been observed (Destombes *et al.*, 1958; El Hassan and Mahgoub, 1972; Lavalle, 1966).

Mycetomas occur in the tropics and subtropical areas of the world and are frequently encountered in countries near the Tropic of Cancer, but also can be found north of this area. Mycetomas due *N. brasiliensis* are more frequent in the Americas, particularly in Mexico where it has been isolated in 94% of the cases (González-Ochoa, 1975; Mariat, 1963).

It is generally agreed that mycetomas develop following a puncture wound. Penetration of human skin by thorns of various plants, splinters of wood, sharp stones, snake bites, insect bites, farm implements, knives, and machetes, etc. (Abbot, 1954; Bocarro, 1893; Gammel, 1927) facilitates the entry of potentially pathogenic organisms into the body where they may develop. Most of the causal agents have, in fact, been isolated from either soil or plants. González-Ochoa (1962) and González-Ochoa and Sandoval (1960) have reported that the natural "habitat" of *N. brasiliensis* is the soil.

Mycetomas appear only when host conditions are favorable. Although factors which predispose to the disease are not yet well understood, it is known that males are more sensitive to the disease (Mahgoub, 1973). There are no reports on any particular symptoms of disease or malnutrition in the patients studied. It is possible, however, that a predisposing immunological or endocrinological states exists.

II. EXPERIMENTAL INFECTION

Human mycetoma has been simulated in a murine model since the infected animals present all the clinical features of the human disease associated with *Nocardia* (González-Ochoa, 1973). This animal model has permitted the study of some aspects of the host-parasite relationship (Melendro *et al.*, 1978; Ximénez *et al.*, 1980).

The first trials in which experimental animals were inoculated intraperitoneally with *N. brasiliensis* were done by Mackinnon and Artagaveytia-Allende in 1956. Three weeks post-infection, they noticed abscesses and grains around the pancreas and between the liver and diaphragm. In addition, they also managed to produce grains *in situ* in guinea pigs infected intratesticularly. González-Ochoa (1962) and González-Ochoa and Sandoval (1960) later demonstrated the pathogenicity of strains of *N. brasiliensis* taken from the soil when lesions and grains were produced *in situ* 20 to 30 days post intraperitoneal infection in mice.

Destombes *et al.* (1961) reported a detailed histopathological study of lesions experimentally induced by *N. asteroides* and by *N. brasiliensis* in mice which had been inoculated intraperitoneally with 5 to 10 mg of the respective fresh cultures. The use of adjuvants in the production of infection by means of repeated inoculation of the microorganism has suggested that an immunological element of sensitization is required. Thus, Macotela-Ruiz and Mariat (1963) made use of this phenomenon of sensitization to induce infection in mice, hamsters, and guinea pigs by first injecting the animal with killed *N. brasiliensis* or *N. asteroides* and then infecting them with living organisms. These results contrast with those obtained by Ximénez *et al.* (1980) in which previous inoculation of mice with killed *N. brasiliensis* protected the animals against subsequent challenge with living homologous organisms. That the experimental protocols used by both groups were distinct may explain the differences observed. One obvious difference is the high dose used for infection by the former group (60 mg in mice), which may have overwhelmed the immune state.

González-Ochoa and Kumico Hojyo (1967) induced a typical mycetoma, which had sinus formation and no tendency to spontaneous cure, in mice by a

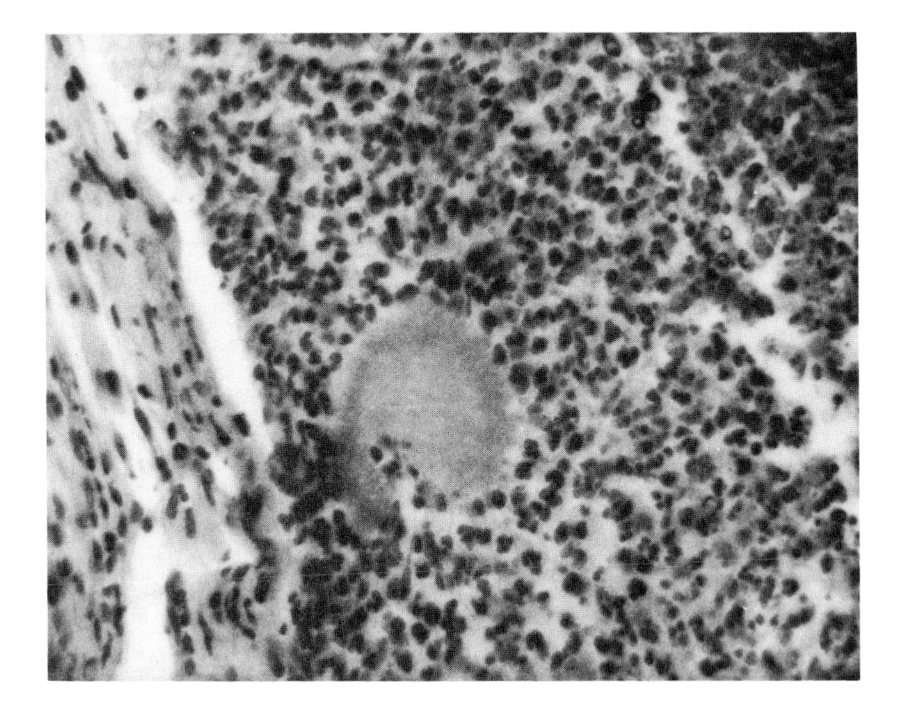

FIGURE 1. Typical actinomycotic granule in a tissue sample obtained from a mouse two weeks after N. brasiliensis *infection.*

single inoculation of *N. brasiliensis*. The doses used in these experiments ranged from 0.2 to 20 mg of the organism. Infections with heavy inoculum (10 or 20 mg) caused spontaneous sloughing off of the infected foot or limb; thereafter, the infection spread to the lungs and liver. Similar infection protocols have been used by others (Melendro *et al.*, 1978; Rico *et al.*, 1981; Ximénez *et al.*, 1980). Recently, reports have appeared in which the mycetoma was induced, either intraperitoneally or subcutaneously in the footpad, by inoculation of a high number of the bacilli suspended in saline (Folb *et al.*, 1976, 1977; Zlotnik and Buckley, 1980). Thus, when Folb (1977) used 10 mg (dry weight) of the organisms suspended in saline, he found the formation of granuloma and large numbers of foam-laden macrophages containing within their cytoplasm *N. brasiliensis* in varying stages of degeneration. In our laboratory, mycetomas with all the characteristics of the human disease have been induced with 2×10^7 organisms suspended in saline. There did not seem to be a difference in the virulence of the different growth phases of *N. brasiliensis*, as opposed to what has been described for *N. asteroides* (Conde *et al.*, 1982). Mice inoculated with *N. brasiliensis* showed a typical mycetoma two weeks post-injection (Fig. 1) (Conde *et al.*, 1982). Interestingly, at that time, while areas of the infected footpad also showed the presence of anti-*Nocardia* globulin (Fig. 2), anti-*Nocardia* antibodies were not detected in the circulation. Anti-*Nocardia* antibody in the blood was found only after 45 days of infection. The presence of serum complement in the infected footpad suggests that immune complexes may contribute to the pathology of the disease (Conde *et al.*, 1983).

III. RESISTANCE TO INFECTION

A. Role of Antibody

The presence of antibody to *Nocardia* in individuals infected with *N. brasiliensis* has been reported. Although such antibodies have served as an aid in diagnosis of chronic infection, their role in resistance to infection is only beginning to be appreciated. Previous studies have indicated that T-cell deficient mice are more susceptible to infection with *N. brasiliensis* than are those with an intact immunological apparatus (Rico *et al.*, 1981). Passive transfer of specific antibody in B mice indicates that, in the absence of cell-mediated immunity (CMI), the antibody seems to favor the course of *Nocardia* infection. The fact that B animals that received specific antibody showed earlier sloughing off of the infected leg than did B mice without antibody treatment suggested that antibody facilitated the growth of the microorganisms rather than destroying them (Table I). Further evidence has been accumulating which supports the facilitating role of antibody in *Nocardia* infection. Thus, studies in which spleen cells that had been depleted of those B lymphocytes which bore receptors for *Nocardia* were transferred into lethally irradiated mice demonstrated that these

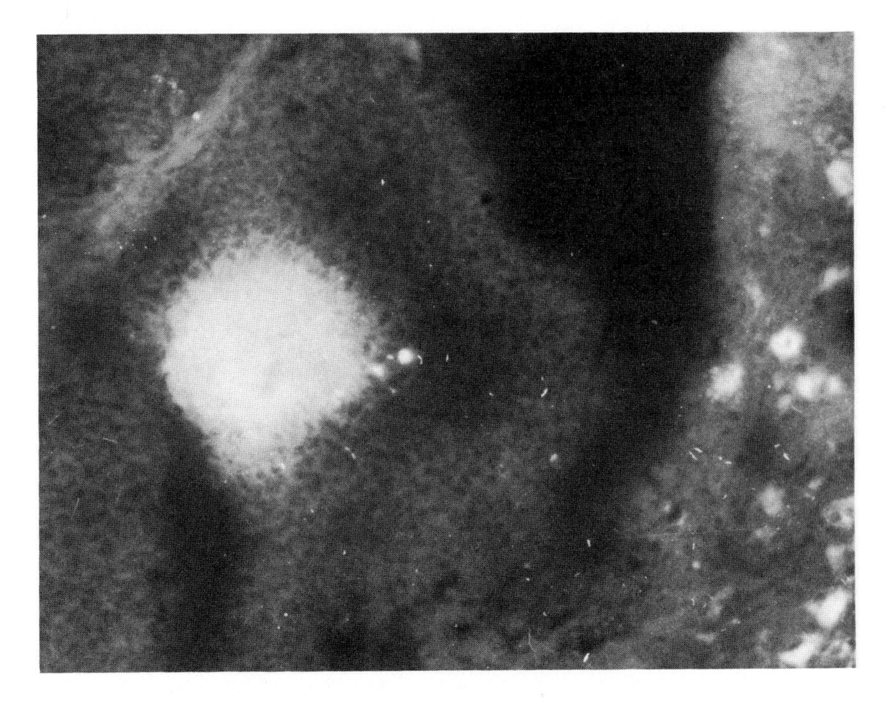

FIGURE 2. Tissue sample obtained from a mouse two weeks after N. brasiliensis *infection, which shows immunofluorescence when treated with fluorescein isothiocyanate-conjugated $F(ab')_2$ fraction of goat anti-mouse Ig's for 60 min at room temperature.*

animals lacked the ability to form antibodies to *Nocardia*. However, as will be discussed later, that these animals were capable of mounting an effective delayed-type hypersensitivity (DTH) reaction and could completely control their *Nocardia* infection established the importance of CMI in halting this disease process (Rico *et al.*, 1982).

B. Role of CMI

Immunity to infection by bacteria which can survive and multiply in host macrophages is effected mainly by CMI mechanisms. The development of this type of immunity is invariably accompanied by a state of specific DTH to antigens of the infecting organism and depends on the acquisition by the host of

TABLE I. Effect of Passively Transferred Antibody on Resistance to Infection by Nocardia *in Mice* [a]

			Mycetoma		
			Spontaneous loss of infected leg		
Mice	*Transferred with antibody*	*Histology (positive/total)*	*Day 12*	*18*	*22*
B	+	*12/12*	*6/12*	*12/12*	-
	-	*14/14*	*9/14*	*10/14*	*10/14*
Normal	+	*10/10*	*0/10*	*2/10*	*8/10*
	-	*10/10*	*0/10*	*0/10*	*2/10*

[a]*Mice were given 250 μg of anti-*Nocardia *antibody intraperitoneally, on days -1, 7, and 14.* Nocardia *infection was elicited on day 0 by footpad inoculation of 0.05 ml of incomplete Freund's adjuvant containing 2 X 10^8 viable* N. brasiliensis. *Histological studies of the infected footpad and the footpad of non-infected control mice were performed to determine the presence of mycetoma.*

macrophages with increased antibacterial mechanisms (Melendro *et al.*, 1978). DTH to *Nocardia* antigens has been previously reported to occur either in experimentally infected animals, in infected individuals, or in healthy individuals who live in areas in which *N. brasiliensis* has been isolated from the soil (Ortiz-Ortiz and Bojalil, 1972; Ortiz-Ortiz *et al.*, 1972a,b,c). However, the role played by DTH in the pathogenesis of the disease is unknown.

Studies on the role of CMI indicate that the microbicidal activity of macrophages is enhanced in mice infected with *Nocardia* (Melendro *et al.*, 1978). Changes in host resistance caused by *N. brasiliensis* was measured using a method based on the cross-resistance that develops between infections caused by different intracellular parasitic bacteria. In this same study, it was also observed that DTH to *Nocardia* antigens was closely correlated to increased resistance to *Listeria monocytogenes*, suggesting that both properties were the consequence of a single immunological event. Studies performed *in vitro* also demonstrated that the microbicidal ability of macrophages, very likely activated by CMI, is enhanced in mice infected with *N. brasiliensis* (Table II) (Melendro *et al.*, 1978).

Resistance to *N. brasiliensis* may be increased by vaccination with either viable or killed *N. brasiliensis*, the degree of resistance correlating with the DTH response in the vaccinated animals, whereas another acid-fast bacilli, BCG, affords only mild protection and low DTH reactivity (Ximénez *et al.*, 1980). However, as already mentioned, the antibody levels to *Nocardia* were similar in either *Nocardia*- or BCG-treated groups, indicating that antibody does not play an important role in resistance to infection by *N. brasiliensis* (Ximénez *et al.*, 1980). This experiment, although ruling out any role for antibody in protection against *Nocardia*, did not directly show T-cell participation in

TABLE II. Effect of Peritoneal Exudate (PE) Cells from Either Normal or Nocardia-*infected Mice on* L. monocitogenes [a]

Group No.	Listeria *added to:*	Nocardia *antigen (100 μg)*	*Number of* L. monocytogenes *per macrophage after:* 0 h	1 h	5 h
1	*Normal PE cells*	–	6.00 ± 1.08[b]	8.00 ± 0.47	22.96 ± 1.52
2	*Normal PE cells*	+	4.06 ± 0.37	8.02 ± 0.62	17.78 ± 1.16
3	*Immune PE cells*	–	6.10 ± 0.49	7.35 ± 0.39	14.16 ± 1.08
4	*Immune PE cells*	+	4.32 ± 0.36	3.06 ± 0.25[c]	0.84 ± 0.24

[a]*Coverslip cultures of PE cells from either normal or* Nocardia-*infected mice were incubated for 24 h in the presence or absence of a* Nocardia cytoplasmic extract (NE; 100 μg/ml), *and then were infected with* L. monocytogenes; *after 1 h, non-attached cells were washed away. At 0, 1, and 5 h intervals after the incubation, attached cells were stained with Giemsa and, for each coverslip, the bacteria in at least 50 macrophages in several randomly chosen fields were counted using an oil immersion objective.*
[b]*Mean ± SE of 50 cells counted for each value.*
[c]*P<0.001 when compared to group 2.*
(Reproduced from Melendro, et al., Int. Arch. Allergy. Appl. Immunol. 57: *74 (1978) by copyright permission of S. Karger AG., Basel.)*

such resistance. To resolve this problem, experiments were done to determine the effect of specific deletion of those B lymphocytes which bore receptors for *Nocardia* on *Nocardia* infection. For this purpose, murine spleen cells bearing Ig's which reacted with a *Nocardia* cytoplasmic extract (NE) were isolated from normal animals; after incubation with NE-coated sheep erythrocytes (Walker *et al.*, 1979), those spleen cells which formed rosettes were separated from non-rosetting spleen cells (T cells, non-*Nocardia* specific B cells, and macrophages) on a density gradient. Lethally irradiated (900 R) BALB/c mice were then reconstituted by i.v. injection of 5 X 10^7 of the non-rosetting spleen cell population. One week later, 1.7 X 10^6 (1 ID_{50}) *Nocardia* organisms were injected into the footpads of these mice. Throughout the experiment, mice which had been given spleen cells depleted of B cells bearing NE receptors had less swelling of the infected foot than did unmanipulated mice similarly exposed to *Nocardia*. Furthermore, on day 75, mice reconstituted with the non-rosetting spleen cells were free from mycetomas, as determined histologically. In contrast, five of the ten control mice which had been reconstituted with undepleted spleen cell populations developed typical mycetomas (Table III). Interestingly, the sera of mice reconstituted with spleen cells depleted of B lymphocytes bearing receptors for NE did not contain hemagglutinating antibodies after 55 days of *Nocardia* infection. However, the production of other Ig's was unaltered (Table III). In comparison, sera of mice reconstituted with normal spleen cells had hemagglutinating titers ranging from 1:16 to 1:64 after

TABLE III. Effect of Specific Deletion of B Lymphocytes Bearing Ig's Which Reacted with NE on the Immune Response of Mice after Nocardia *Infection* [a]

Transfer inoculum	*Mycetoma*[b]	*Antibodies to:* NE	*Antibodies to:* Burro erythrocytes	*DTH to NE*
Nocardia-*depleted B cells*	*0/15*[c]	*0/4*	*5/5*	*3/3*
Whole spleen sells	*5/10*	*4/4*	*5/5*	*3/3*

[a]*Lethally irradiated (900 R) mice which had received 5 X 10^7 of either spleen cells depleted of the NE-reactive B population or whole spleen cells were injected one week later with 1.7 X 10^6 (1 ID_{50})* Nocardia. *Antibody determinations and DTH tests were done 55 days after infection. Antibodies to NE were determined by passive hemagglutination with SRBC coated with NE. Antibodies to burro erythrocytes were assayed by the Jerne and Nordin technique (1963) as modified by Golub* et al. *(1968).*
[b]*Animals were observed for 75 days before sacrifice for histological studies.*
[c]*Values expressed as positive/total.*

55 days of infection; these titers of antibodies lasted throughout the experiment (75 days). However, both groups had positive DTH reactions to NE (Table III), which indicated T-cell reactivity against the NE. Similar results were obtained when mice reconstituted with spleen cells depleted of B lymphocytes bearing receptors for NE were challenged with more than 1 ID_{50} *Nocardia* organisms (Rico *et al.*, 1982). Therefore, the elimination of B lymphocytes bearing receptors for NE, which consequently prevented the formation of antibody to NE in reconstituted mice, did not significantly affect the induction or expression of CMI to *Nocardia.*

IV. B-LYMPHOCYTE ACTIVATION BY *N. BRASILIENSIS*

A. Mitogenic Response to Nocardia *Extract (MNE)**

Spleen cells from mice infected with *N. brasiliensis* showed an increased rate of tritiated thymidine uptake when compared with those from normal, uninoculated mice (Del Bosque, 1978). This observation led to the proposal that this acid-fast microorganism had mitogenic activity. The ability of extracts from nonpathogenic *Nocardia* to promote a mitogenic response in murine spleen cells has been clearly defined (Adam *et al.*, 1973; Bona *et al.*, 1974). Likewise, it

**The method of preparation of the extract (MNE) used for B-lymphocyte activation differed from that followed for obtaining NE.*

has also been shown that an extract from the pathogenic actinomycete *N. brasiliensis* was mitogenic for murine B lymphocytes (Ortiz-Ortiz *et al.*, 1979). This mitogenic activity did not require the presence of T cells since it was generated in T-depleted spleen cells, either those from congenitally athymic nude (nu/nu) mice or those from +/+ mice which had been treated with antithymocyte serum plus complement (Fig. 3).

B. Polyclonal B Cell Activation by MNE

1. Studies in vitro. An extract of pathogenic *N. brasiliensis* can nonspecifically stimulate B lymphocytes. The presence of T cells does not appear to be mandatory for the *in vitro* polyclonal response to MNE, since no appreciable differences were observed between the responses of spleen cells from nude and littermate (nu/+) mice (Table IV) (Ortiz-Ortiz *et al.*, 1979). The ability of MNE to activate antibody-forming cell (PFC) precursors and to stimulate division of B cells, which results in an increased number of cells producing antibodies to many non-cross-reacting antigens, may be relevant to certain autoimmune phenomena. Of special interest is the fact that, in C3H/HeJ mice, MNE induced an antibody response to the protein antigen HGG, whereas butanol-extracted LPS did not (Table IV). This suggests that stimulation by *Nocardia* represents a model for cell activation which results in the secretion of rheumatoid-like factors (Kunkel and Tan, 1964) by B cells in response to infection. As has been suggested elsewhere (Parks and Weigle, 1979), the majority

TABLE IV. Polyclonal B-cell Activation by Nocardia *Extract in Either C3H/HeJ Nude (nu/nu) or Littermate (nu/+) Spleen Cells*[a]

Strain	*Bacterial extracts*[c]	*Direct PFC/culture*[b]					
		SRBC		*TNP*		*HGG*	
		Day		*Day*		*Day*	
		3	*5*	*3*	*5*	*3*	*5*
(nu/nu)	*None*	*10*	*13*	*397*	*44*	*4*	*0*
	Nocardia *extract*	*364*	*176*	*1782*	*1982*	*19*	*8*
	LPS	*ND*	*58*	*2351*	*522*	*ND*	*0*
(nu/+)	*None*	*8*	*44*	*90*	*178*	*0*	*0*
	Nocardia *extract*	*325*	*200*	*1525*	*1276*	*25*	*0*
	LPS	*285*	*307*	*1350*	*784*	*0*	*0*

[a] *Washed, viable spleen cells (10^7) from C3H/HeJ mice were incubated in 1 ml serum-free cultures for 48 h.*
[b] *Mean value of three or four cultures per group. PFC's were determined by the Jerne and Nordin technique (1963) as modified by Golub* et al. *(1968). SRBC, sheep erythrocytes; TNP, trinitrophenol; HGG, human gamma globulin.*
[c] Nocardia *extract (100 μg) or butanol-extracted LPS (lipopolysaccharide) (10 μg).*
(Reproduced from Ortiz-Ortiz et al., Infect. Immun. 25:*627 (1979) by copyright permission of the American Society for Microbiology.)*

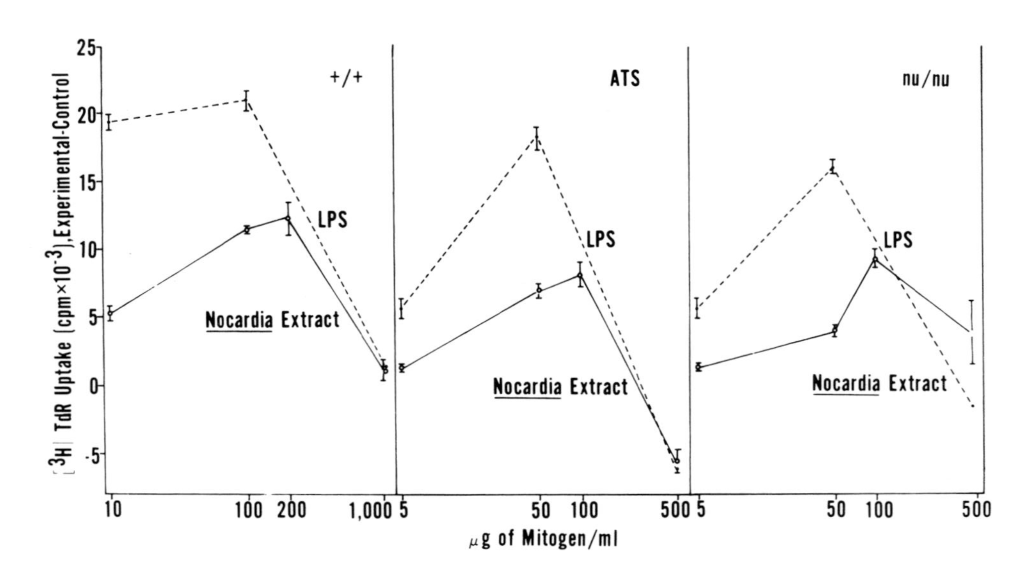

FIGURE 3. Dose response profiles of the mitogenic response to Nocardia *extract. Lymphocytes were cultured in microculture plates at a cell density of 5 X 10^5 viable cells per 0.2 ml RPMI containing 5% fetal calf serum and in the presence of various concentrations of* Nocardia *extract (O) or butanol-extracted LPS (●). Cells were pulsed with 1.0 μ Ci ^{3}H-thymidine during the final 24 h of culture. The mean ± standard error of the counts per minute from quadruplicate cultures is presented.*

(Reproduced from Ortiz-Ortiz et al., Infect. Immun. 25:627 (1979) *by copyright permission of the American Society for Microbiology.)*

of cells stimulated during polyclonal activation may represent antigen-nonreactive cells, the number of antigen receptors of which is below the threshold required for antigen-specific triggering. Therefore, the role of these low affinity antibodies in the etiology of autoimmune diseases may be limited.

2. Studies in vivo. Similar to the effect of *N. brasiliensis in vitro*, infection of mice with *N. brasiliensis* induced polyclonal B cell activation which was optimal three to four days post-infection (Table V). When *Nocardia*-infected animals were immunized with sheep erythrocytes, an adjuvant effect was observed. This effect was more pronounced when the erythrocytes were given three days after infection, *i. e.*, at the time of optimal polyclonal activation (Table VI). These data are interesting since they indicate that *Nocardia* or its products induced *in vivo* an initial polyclonal expansion of B cells which, when antigen was administered, resulted in increased antibody production to said antigen.

C. Bypass of the Requirement for Helper T Cells in an In Vitro *Primary Response*

MNE has also been found to substitute for the T-cell requirement in a T-cell dependent response. Thus, whereas nude spleen cells do not produce antibody when challenged *in vitro* with sheep erythrocytes alone, the addition of MNE to these cultures stimulated specific antibody formation to the erythrocytes in amounts greater than those that were produced against the *Nocardia* extract alone (Table VII). This replacement of T-cell function *in vitro* has previously been described as a by-product of non-specific B-cell activation.

TABLE V. Polyclonal B-cell Activation Induced by N. brasiliensis *Infection in Mice*[a]

Days after infection	*Group*	*PFC/10⁶ spleen cells*[b]	
		SRBC	*TNP*
2	*Normal*	2	17
	Infected	2	22
3	*Normal*	2	12
	Infected	7	40
5	*Normal*	2	14
	Infected	5	27
7	*Normal*	2	11
	Infected	3	8

[a]*Mice were infected with* 5×10^7 *viable* Nocardia. *At different time intervals post-infection, cell suspensions were obtained from excised spleens, and the PFC's were determined.*
[b]*Mean value of five mice per group.*

TABLE VI. Effect of N. brasiliensis *Infection on the Immune Response of Mice to SRBC*[a]

Day of SRBC immunization	*PFC/10⁶ spleen cells*[b]	
	Normal	*Infected*
-3	404	848
0	438	667
+3	496	1194

[a] *Mice were infected as described in Table V. They were immunized with SRBC on days -3, 0, and +3 of* Nocardia *infection.*
[b] *Mean value of three or four mice per group.*

V. EFFECT OF *N. BRASILIENSIS* INFECTION ON THE RESPONSE TO MITOGENS

Mice infected with *N. brasiliensis* showed a diminished response to Con A and LPS than did normal mice. The response of spleen cells from infected animals to LPS was evidenced soon after infection (6 days), whereas the response to Con A was first observed 50 days post-infection (Table VIII). These data indicate that *N. brasiliensis* infection in mice has a suppressive effect on T and B cell activity. Interestingly, when co-cultured *in vitro* with normal spleen cells (NSC), spleen cells taken six days post-infection were unable to suppress the response of NSC to the mitogens (Table IX), whereas spleen cells obtained 50 days post-infection were able to suppress the response (Table X) (Melendro *et*

TABLE VII. Antibody Formation In Vitro *to SRBC in C3H/HeJ Nude (nu/nu) or Littermate (nu/+) Spleen Cells*[a]

Antigen[b]	*Bacterial extract*[c]	*Direct PFC/culture*[d]	
		nu/nu	*nu/+*
None	None	15 ± 2	50 ± 10
SRBC	None	49 ± 6	787 ± 24
SRBC	Nocardia *extract*	314 ± 7	955 ± 27
SRBC	LPS	238 ± 7	1139 ± 44
None	Nocardia *extract*	180 ± 15	153 ± 15
None	LPS	61 ± 15	127 ± 37

[a] *Viable spleen cells (10^7) were cultured for 4 days in 1 ml of RPMI containing 5% fetal calf serum.*
[b] *One drop of a 1% solution per culture.*
[c] Nocardia *extract (100 μg) or butanol-extracted LPS (10 μg).*
[d] *Mean ± standard error of triplicate pools containing three or four cultures each.*
(Reproduced from Ortiz-Ortiz et al., Infect. Immun. 25:*627 (1979) by copyright permission of the American Society for Microbiology.)*

TABLE VIII. Response to Con A and to LPS of Spleen Cells Obtained from Mice Infected with N. brasiliensis[a]

Days after infection	Group	^{3}H-thymidine incorporation	
		Con A	LPS
6	Normal	295,115	86,684
	Infected	177,933	26,841
36	Normal	390,529	111,348
	Infected	243,928	31,224
50	Normal	204,433	79,439
	Infected	74,384	13,211

[a]*Mice were infected with* 5×10^{7} *viable* N. brasiliensis. *At different time intervals mice were killed, spleens were taken and cell suspensions were prepared. Spleen cells at a concentration of* 5×10^{5} *per culture were stimulated with either 0.2 μg of Con A or 20 μg of LPS as described in Fig. 3.*

al., manuscript in preparation). It is possible that *Nocardia* infection initially acts directly on B lymphocytes in some way to suppress their activity. The presence of suppressor cells has been described in infections produced by other acid-fast bacilli and, in infection by mycobacteria, macrophages, as well as T cells, have been found to have suppressor activity (Ellner, 1978; Katz *et al.*, 1979; Wadee *et al.*, 1980). Furthermore, in a study in which lymphoid cells cultured in the presence of killed *Mycobacterium* showed a diminished response to mitogens (Wadee *et al.*, 1980), a factor, synthesized by macrophages which possibly had been stimulated by a lipid component from the cell wall of the bacilli, was reported to activate suppressor T cells.

TABLE IX. Effect of Spleen Cells Obtained from Mice Six Days After N. brasiliensis *Infection*

Composition of culture[a]		^{3}H-thymidine incorporation[b]			
		Con A		LPS	
N	I	Mean cpm	% Unresponsiveness	Mean cpm	% Unresponsiveness
2.5×10^{5}	—	258325	0	74592	0
—	2.5×10^{5}	178907	31	30550	59
5.0×10^{5}	—	261622	0	76527	0
—	5.0×10^{5}	186700	29	32947	57
2.5×10^{5}	2.5×10^{5}	447850	-2	105920	-7

[a]*Cultures consisted of the indicated number of normal (N) and of unfractionated spleen cells from* Nocardia-*infected (I) mice.*
[b]*Results were expressed as the arithmetic mean of four cultures.*

TABLE X. Suppressive Effect of Spleen Cells Obtained from Mice 60 Days after N. brasiliensis *Infection*

Composition of culture[a]		^{3}*H-thymidine incorporation*			
		Con A		*LPS*	
N	*I*	*Mean cpm*	*% Unresponsiveness*	*Mean cpm*	*% Unresponsiveness*
2.5×10^5	—	297,290	0	155,850	0
—	2.5×10^5	170,727	43	46,620	70
5.0×10^5	—	268,177	0	158,463	0
—	5.0×10^5	103,126	62	40,996	74
2.5×10^5	2.5×10^5	154,150	67	83,860	59

[a] *See legends* a *and* b, *Table IX.*

VI. CONCLUSIONS

The objectives of this review were to summarize, interpret, and evaluate the various experimental findings related to the host-parasite relationship in mice infected with *N. brasiliensis*. It is apparent that we are still a long way from a complete knowledge of the host's mechanism of resistance to this microorganism. However, advances have been made in determining the roles played by antibody and by CMI in resistance to *Nocardia* infection. The body of evidence seems to indicate that CMI must be considered to be an essential mechanism of defense to *N. brasiliensis* infection.

N. brasiliensis has many interesting properties. It induces B-cell activation and mitogenesis (Ortiz-Ortiz *et al.*, 1979) and has a deleterious effect on T- and B-lymphocyte activities as evidenced by suppression of the lymphocyte response of the infected animals to Con A and LPS. Furthermore, this suppressive effect caused by *Nocardia* is, in part, mediated by suppressor cells present in the spleen of the infected mice (Melendro *et al.*, manuscript in preparation). These properties of *N. brasiliensis* may contribute to its capacity to evade the immune response of the host.

ACKNOWLEDGMENT

We thank Veronica Yakoleff, Department of Immunology, Instituto de Investigaciones Biomédicas, U.N.A.M., for reviewing this manuscript.

REFERENCES

Abbot, P. H., Dissertation for the Degree of M.D., Cambridge, England (1954).

Adam, A., Ciorbaru, R., Petit, J. F., Lederer, E., Chedid, L., Lemansans, A., Parat, F., Rosselet, J. P., and Berger, F. M., *Infect. Immun. 7*:855(1973).

Bocarro, J. E., *Lancet 2*:797(1893).

Bona, C., Damais, C., and Chedid, L., *Proc. Nat. Acad. Sci. U.S.A. 71*:1602(1974).

Conde, C., Melendro, E. I., and Ortiz-Ortiz, L., *Infect. Immun. 38*:1291(1982).

Conde, C., Mancilla, R., Fresan, M., and Ortiz-Ortiz, L., *Infect. Immun. 40*:1218(1983).

Del Bosque, J., Tesis Licenciatura. Facultad de Ciencias, U.N.A.M., México (1978).

Destombes, P., Camain, R., and Nazimoff, O., *Bull. Soc. Pathol. Exot. 51*:863(1958).

El Hassan, A. M., and Mahgoub, E. S., *Trans. R. Soc. Trop. Med. Hyg. 66*:165(1972).

Ellner, J. J., *J. Immunol. 121*:2573 (1978).

Folb, P. I., Jaffe, R., and Altmann, G., *Infect. Immun. 13*:1490(1976).

Folb, P. I., Timme, A., and Horowitz, A., *Infect. Immun. 18*:459(1977).

Gammel, J. A., *Arch. Dermatol. Syphilol. 15*:241(1927).

Golub, E. S., Mishell, R. I., Weigle, W. O., and Dutton, R.W., *J. Immunol. 100*:133(1968).

González-Ochoa, A., *Rev. Inst. Salubr. Enferm. Trop. Méx. 22*:15(1962).

González-Ochoa, A., *Can. J. Microbiol. 19*:901 (1973).

González-Ochoa, A., *Rev. Salud Pública, Méx. 35*:85(1975).

González-Ochoa, A., and Kumico Hojyo, T., *in* "Fifth International Congress of Chemotherapy", p. 323. Diemer Medizineischen Akademie, Vienna (1967).

González-Ochoa, A., and Sandoval, M. A., *Rev. Inst. Salubr. Enferm. Trop. Méx. 20*:147(1960).

Jerne, N. K., and Nordin, A. A., *Science 140:* 405 (1963).

Katz, P., Goldstein, R. A., and Fauci, A. S., *J. Infect. Dis. 140*:12(1979).

Kunkel, H. G., and Tan, E. M., Adv. Immunol. 4:351(1964).

Lavalle, P., *Gac. Med. Méx. 96*:545(1966).

Mackinnon, J. E., and Artagaveytia-Allende, R. C., *Trans. R. Soc. Trop. Med. Hyg. 50*:31(1956).

Macotela-Ruiz, E., and Mariat, F., *Bull. Soc. Pathol. Exot. 56*:46(1963).

Mahgoub, E. S., *in* "Mycetoma", p. 6. W. Heinemann Medical Book, London (1973).

Mariat, F., *Bull. Soc. Pathol. Exot. 56*:35(1963).

Mariat, F., Destombes, P., and Segretain, G., *Contrib. Microbiol. Immunol. 4*:1(1977).

Melendro, E. I., Contreras, M. F., Ximénez, C., García, A. M., and Ortiz-Ortiz, L., *Int. Arch. Allergy Appl. Immunol. 57*:74(1978).

Ortiz-Ortiz, L., and Bojalil, L. F., *Clin. Exp. Immunol. 12*:225(1972).

Ortiz-Ortiz, L., Bojalil, L. F., and Contreras, M. F., *J. Immunol. 108*:1409(1972a).

Ortiz-Ortiz, L., Contreras, M. F., and Bojalil, L. F., *Infect. Immun. 5*:879(1972b).

Ortiz-Ortiz, L., Contreras, M. F., and Bojalil, L. F., *Sabouraudia 10*:147(1972c).

Ortiz-Ortiz, L., Parks, D. E., López, J. S., and Weigle, W. O., *Infect. Immun. 25*:627(1979).

Parks, D. E., and Weigle, W. O., *Immunol. Rev. 43*:217(1979)

Parks, D. E., Doyle, M. V., and Weigle, W. O., *J. Immunol. 119*:1923(1977).

Rico, G., Ochoa, R., Oliva, A., and Ortiz-Ortiz, L., *in* "Actinomycetes" (K. P. Schaal and G. Pulverer, eds.), p. 229. Gustav Fischer Verlag, Stuttgart (1981).

Rico, G., Ochoa, R., Oliva, A., González-Mendoza, A., Walker, S., and Ortiz-Ortiz, L., *J. Immunol. 129*:1688(1982).

Sjöberg, O., Andersson, J., and Möller, G., *Eur. J. Immunol. 2*:326 (1972).

Wadde, A. A., Sher, R., and Rabson, A. R., *J. Immunol. 125*:1380 (1980).

Walker, S. M., Meinke, G. C., and Weigle, W. O., *Cell. Immunol. 46*:158(1979).

Ximénez, C., Melendro, E. I., González-Mendoza, A., García, A. M., Martínez, A., and Ortiz-Ortiz, L., *Mycopathologia 70*:117(1980).

Zlotnik, H., and Buckley, H. R., *Infect. Immun. 29*:1141(1980).

SEROLOGIC RELATIONSHIPS AMONG AEROBIC AND ANAEROBIC ACTINOMYCETES IN HUMAN AND ANIMAL DISEASE

A. C. Pier

Division of Microbiology and Veterinary Medicine
University of Wyoming
Laramie, Wyoming, U. S. A.

I. INTRODUCTION

Actinomycetes of major importance in human and animal disease are found primarily within the genera *Actinomyces*, *Dermatophilus*, and *Nocardia*. Actinomycosis, as a frequent disease of companion animals, livestock, and man, traditionally occurs as invasive, suppurative granulomatous lesions of subcutaneous tissues, lymph nodes, parenchymatous organs, brain, and bone. In recent years, two rather specialized occurrences of *Actinomyces* species in man have become apparent, *i. e.*, the frequent occurrence of several species in dental plaque and the occurrence of *A. israelii* and *A. viscosus* in cervical lesions associated with intrauterine contraceptive devices. Because of the microaerophilic to anaerobic nature of *Actinomyces* spp. and its relatively slow growth rate, diagnosis by culture is often delayed or precluded by laboratory routine. Accurate early diagnosis is essential for successful therapy and control. Serologic methods have been developed which expedite diagnosis (Pine *et al.*, 1981) and which permit subspecific strain identification (Slack and Gerencser, 1975).

Dermatophilosis occurs as an exudative dermatitis of wild and domestic animals and man. It has fewer diagnostic complications than does actinomycosis or nocardiosis because of the obvious lesions on the skin; however, the fastidiousness and slow growing nature of the causative organism *Dermatophilus congolensis* in culture often impede definitive diagnosis. Serologic methods have been utilized to facilitate diagnosis, particularly in cases of long-standing infection (Pier *et al.*, 1964).

BIOLOGICAL, BIOCHEMICAL,
AND BIOMEDICAL ASPECTS OF ACTINOMYCETES

ISBN 0-12-528620-1

Nocardiosis has the same extensive clinical variety as actinomycosis. The disease occurs as localized mycetoma, as suppurative granulomatous lesions of the cutaneous lymphatics, lungs, pleura lymph nodes, mammary glands, and parenchymatous organs of the abdomen, with extension into brain and bone. Nocardiosis is particularly important as a pathogen of immunocompromised hosts (Beaman *et al.*, 1976). The causative agents of nocardiosis come from soil and other environmental foci, but outbreaks in both cattle and man imply that epidemics occur in confined areas (Pier, 1962; Stevens *et al.*, 1981). Immunologic methods have been applied in the diagnosis and epidemiology of nocardiosis in both animals and man (Pier and Fichtner, 1971; Pier *et al.*, 1968; Salman *et al.*, 1962; Schainhaus *et al.*, 1978; Thurston *et al.*, 1968). It will be the purpose of this paper to summarize the current status of major applications of serologic and immunologic techniques for these agents. No attempt at accumulation of a complete literature review was made.

II. *ACTINOMYCES*

As mentioned above, the diagnosis of actinomycosis is complicated by the deep-seated location of most lesions, the similarity of lesions to those caused by other agents, and the habits of the causative *Actinomyces* species in culture. Serologic methods have been developed which facilitate organism identification in clinical samples (Pine *et al.*, 1981) and which can subspecifically type isolates (Slack and Gerencser, 1975).

As a result of a number of serological procedures, including agglutination of whole cells, immunodiffusion, and fluorescent antibody (FA) staining (both direct and indirect), the genus *Actinomyces* has been divided into four serogroups. These serologic groups coincide with established species designations (Slack and Gerencser, 1975).

Subspecific serotyping of *Actinomyces* sp. has been studied by a number of investigators over the last 12 years (Georg, 1970; Gerencser and Slack, 1976; Slack and Gerencser, 1975). The major developments in this area have been through use of differential staining of *Actinomyces* strains with FA. In each species of *Actinomyces*, two different serotypes have been identified, with the exception of *A. naeslundii* in which four serotypes are recognized. These serotypes and their known cross reactions are presented in Table I. Extensive interspecies cross reactions occur between serotype 2 of *A. viscosus* and both serotypes of *A. israelii* and serotypes 1,2 and 3 of *A. naeslundii*. There is also extensive intraspecies cross reactions among the serotypes of *A. naeslundii*. The existence of these cross reactions makes necessary considerable adsorption of antiserum in order for diagnostic significance or strain typing to be accomplished with accuracy (Gerencser and Slack, 1976; Slack and Gerencser, 1975). Notable among the serotype reactions is the absence of cross reactions between the two serotypes of *A. bovis* and the serotypes of other species. The

TABLE I. Serologic Relationships of Actinomyces *as Determined by the Fluorescent Antibody Technique* [a]

			A. bovis		A. israelii		A. viscosus		A. naeslundii				A. odontol.	
Group	*Species*	*Type*	*1*	*2*	*1*	*2*	*1*	*2*	*1*	*2*	*3*	*4*	*1*	*2*
B	A. bovis													
	(S)	*1*	+ [b]	-	-	-	-	-	-	-	-	-	-	-
	(R)	*2*	-	+	-	-	-	-	-	-	-	-	-	-
D	A. israelii	*1*	-	-	+	±	-	±	±	-	-	±	-	-
		2	-	-	±	+	±	±	-	-	-	-	-	-
F	A. viscosus													
	(animal)	*1*	-	-	-	-	+	-	-	-	-	-	-	±
	(human)	*2*	-	-	±	-	±	+	±	+	+	-	-	-
A	A. naeslundii	*1* [c]	-	-	±	-	±	±	+	±	+	±	-	-
		2	-	-	-	-	-	+	+	+	+	-	-	-
		3	-	-	-	-	-	+	±	+	+	±	-	-
		4	-	-	±	-	-	-	±	±	-	+	-	-
E	A. odontolyticus	*1*	-	-	-	-	-	-	-	-	-	-	+	±
		2	-	-	-	-	±	-	-	-	-	-	±	+

[a] *Gerencser and Slack (1976); Slack and Gerencser (1975).*
[b] *+, positive reaction; ±, reduced titer reaction; -, no reaction*
[c] *evidence for three subtypes.*

smooth colony forms of *A. bovis* regularly occur in serotype 1; rough colony forms are seen generally in serotype 2 (Slack and Gerencser, 1975). Serotype 1 of *A. viscosus* is also antigenically distinct from the other *Actinomyces*; animal strains of *A. viscosus* comprise serotype 1, whereas human strains are found in serotype 2 (Gerencser and Slack, 1969). The reader is referred to Slack and Gerencser (1975) and Gerencser and Slack (1976) for a detailed description of the complicated lattice of cross reactions that occurs among the other serotypes of the genus. Details of methods for production of antigen and antisera for FA application have been reported (Gillis and Thompson, 1978; Lai and Listgarten, 1979; Slack and Gerencser, 1975).

The most practical applications of serology in the diagnosis of actinomycosis appears to be in the subspecific identification of isolates for epidemiologic purposes and in the detection of pathogenic *Actinomyces* in clinical specimens by FA. The latter applications have been used particularly in the study of human dental plaque (Marucha *et al.*, 1978) and of cervical vaginal smears (Pine *et al.*, 1981). Both specific antigenic material which is capable of being stained directly by FA and antiserum which is capable of blocking known reactants can be detected in some patients specimens. A quantitative fluorescent immunoassay system for antibodies and surface antigens has been described (Gillis and Thompson, 1978). Other applications of serology deal with recognition of specific cellular structures by use of absorbed FA sera and of monospecific antibodies (Lai and Listgarten, 1979; Revis *et al.*, 1982). Serologic adjuncts to diagnosis of actinomycosis are a definite advantage when positive results are obtained; however, negative results do not rule out the possibility of the existence of actinomycosis. Positive results also are open to interpretation as in the case of an FA positive cervical-vaginal smear when no disease is demonstrable (Pine *et al.*, 1981). One is constantly reminded that *Actinomyces* are in fact normal residents of the mucosa and that disease is a temporary episode in the course of infection.

III. *DERMATOPHILUS*

The applications of serology to *Dermatophilus congolensis*, the obligate parasite that is the agent of dermatophilosis, are considerably more effective than are those to *Actinomyces* and *Nocardia*. The diagnostic use of FA, prepared against whole cell antigens, has been reported (Pier *et al.*, 1964). This technique remains highly effective in confirming the identity of atypical isolates associated with exudative dermatitis in animals. It is also highly effective in specifically identifying organisms in exudate remnants in recuperative animals in which typical configurations of *D. congolensis* are not readily apparent. However, to date, subspecific serotypes have not been demonstrated within this single species of the genus, *Dermatophilus*.

Animals infected with *D. congolensis* develop detectable antibody if the exposure is of sufficient magnitude and duration (Bida and Kelley, 1976; Lloyd, 1981; Lloyd and Jenkinson, 1981; Makinde, 1980, 1981; Pulliam *et al.*, 1967; Richard *et al.*, 1976). Agglutinins, precipitins, radial immunodiffusion, passive hemagglutination, complement fixation, and ELISA techniques have all been used to demonstrate antibodies in the serum, in the milk, and at the skin surface of infected cattle. However, the appearance of antibody is not constant in primary infections. Precipitins have been demonstrated in the serums of 40% of cattle in geographic areas known to have dermatophilosis. The importance of antibody to the animal is not clear as experimental reinfection of animals with demonstrable antibody has been reported (Richard *et al.*, 1976). Whether such reinfection would have occurred following natural exposure remains unresolved; however, apparent reinfection (or recrudescence) is known to occur in both cattle and sheep, but it is not clear if existing antibody levels influence that reinfection. There is virtually no information on the occurrence of antibodies to *D. congolensis* in two other important hosts, dogs and man; in both species, infection has been reported, but the condition appears to be underdiagnosed.

IV. *NOCARDIA*

Agents causing nocardiosis show many of the taxonomic and diagnostic complications as do those causing actinomycosis. Nomenclature of the agents of nocardiosis and actinomycosis have been substantially confused in the medical literature. Both conditions effect deep-seated, suppurative granulomatous processes that extend by contiguity, erosion, lymphatic and vascular dissemination, as well as effecting localized mycetoma formation. Both genera of organisms have similar morphologic features in tissue and both appear as Gram positive rods and branching filaments, microcolonies, and mineralized tissue granules in histologic preparations. Put simply, they cannot be distinguished reliably by histologic methods alone.

The use of serologic procedures and skin testing for cutaneous hypersensitivity have been applied to the diagnosis of nocardiosis in animals and man for over 20 years (Bojalil and Zamora, 1963; Ortiz-Ortiz *et al.*, 1972; Pier and Enright, 1962; Pier *et al.*, 1968; Shainhaus *et al.*, 1978; Zamora *et al.*, 1963). Two major types of antigenic preparations have been used in these studies. The first are culture filtrate antigens that are used without further absorption or chemical manipulation. They have been used extensively in detecting cutaneous hypersensitivity and in serologic reactions for *N. asteroides* in both animals and man. These reactions include immunodiffusion, complement fixation and passive hemagglutination (Pier *et al.*, 1968; Salman *et al.*, 1982; Shainhaus *et al.*, 1978; Thurston *et al.*, 1968). These culture filtrate antigens also have been the basis of a subspecific typing system in which four serotypes of *N. asteroides* and a species-specific antigen for *N. caviae* were established (Pier and Fichtner, 1971). The second type of antigenic preparation used extensively in studies of nocardiosis has been polysaccharide antigens extracted and chemically purified from nocardial cells. These latter antigens have been used extensively in tests for cutaneous hypersensitivity and immunodiffusion (Ortiz-Ortiz *et al.*, 1972; Zamora *et al.*, 1963). These antigens are both highly specific and sufficiently sensitive to serve as adjuncts to diagnosis of nocardiosis in animals and man. The culture filtrate antigens have shown a high degree of specificity in serological tests of cattle with nocardiosis, tuberculosis (*M. bovis*), and paratuberculosis (Pier *et al.*, 1968). Serologic or cutaneous hypersensitivity cross reactions were not a problem. Some cross reaction was apparent however when these antigens were used in complement fixation serology of human patients with tuberculosis (*M. tuberculosis*) (Shainhaus *et al.*, 1978). Whether this was a difference in host reaction or in parasite antigenicity has yet to be determined, but the specificity of the culture filtrate antigens in sufficiently high to recommend their use in the diagnosis of nocardiosis in both animals and man.

The polysaccharide extract antigens have been shown to have two components, I and II. Polysaccharide I is shared by different species of *Nocardia* as well as by some mycobacteria; polysaccharide II appears to be species specific within *N. asteroides* and *N. brasiliensis* (Ortiz-Ortiz *et al.*, 1972). It is not known whether this second polysaccharide (*e. g.* NaII) interacts with *N. asteroides* of different serotypes (Pier and Fichtner, 1971) on a general or type-specific basis. A study of immunological analysis of *Nocardia* at various stages during morphogenesis has shown that antigenic variations do not occur during growth (Emeruwa, 1980). A number of other investigations have been conducted utilizing serologic methods to diagnose nocardiosis (Blumer and Kaufman, 1979; Humphreys *et al.*, 1975) or to differentiate between infecting strains (Richard *et al.*, 1976; Ridell; 1975; Shigidi *et al.*, 1980). Unfortunately, these studies overlooked existing information that might have been used as a basis for informative comparisons or for a higher degree of diagnostic efficiency. Sufficient data is available to show that there are at least four serotypes of *N. asteroides*; therefore, antigens to be utilized in serologic diagnosis of this infection should be made from each serotype rather than from a single strain, until it is determined whether the antigen is group specific or type specific. The use of group-specific antigens would appear to invite the possibility of cross reactions with the agents causing mycobacteriosis. Similarly, the concept of cross reactivity must be interpreted with care because the patient, whether animal or human, may well have been affected by both nocardiosis and mycobacteriosis. Whereas isolation in culture establishes the presence of an organism in the host, a negative culture does not assure its absence.

Serological typing of *N. asteroides* has proven to be a very useful epidemiological tool. The serological types of 129 isolates of human, animal, and environmental sources have been determined (Pier and Fichtner, 1981) and are summarized in Table II.

Serological types were compared with sensitin types (Magnusson, 1976) and found to have close similarity. One of the most useful applications of serologic typing in *N. asteroides* has been in epidemiologic studies of group infections. Studies in herds of dairy cattle with nocardial mastitis show that the infecting organisms within the herd are usually the same serotype, implying animal-to-animal transfer of infection (Pier and Fichtner, 1981) (Table III). In an interesting study of an outbreak of human nocardiosis in immunosuppressed patients, it was found that isolates from all human cases confirmed by culture,

TABLE II. Serologic Types of N. asteroides

Source	*Number of Isolates*	*Serologic type (%)*							
		I	*II*	*III*	*IV*	*I+IV*	*II+IV*	*III+IV*	*Unclassified*
Human	*44*	*25*	*7*	*14*	*27*	—	*5*	*14*	*9*
Bovine	*40*	*38*	*3*	*20*	*10*	*10*	—	*18*	*3*
Other animal	*11*	*36*	*9*	—	*18*	—	—	*18*	*18*
Environmental	*10*	*20*	—	*10*	*40*	—	*30*	—	—

TABLE III. Application of Serological Typing to Group Infections of Nocardiosis

Bovine nocardiosis [a]			Human nocardiosis [b]	
Herd	*Location*	*Serotypes*	*Ocurrence*	*Serotypes*
			Renal Clinic ICA patients	*8/8: III*
A.	*Iowa*	*5/5: III + IV*	*Renal Clinic ICA Environment*	*12/12: III*
B.	*Iowa*	*6/7: III*	*Renal Clinic Non-ICA Environment*	*1/5: II + IV*
		1/7: IV		*3/5: III*
C.	*California*	*4/4: III*		*1/5: IV*
D.	*Mississippi*	*3/5: I + IV*	*Outside clinic patients*	*1/8: I*
		2/5: I		*2/8: II + IV*
E.	*Kentucky*	*1/2: I + IV*		*2/8: III*
		1/2: I		*1/8: III + IV*
F.	*Kentucky*	*2/2: I + IV*		*2/8: IV*

[a] *Pier and Fichtner (1962).*
[b] *Stevens* et al., *(1981).*

that occurred in an intensive care renal unit, were caused by the same serotype (Stevens *et al.*, 1981). All 12 isolates from the environment of this renal unit also were the same serotype, whereas serotypes of isolates from other areas of the hospital and from nocardiosis patients of other hospitals in the area showed more random distribution of serotypes (Table III). These results would suggest a patient-to-environment-to-patient cycle of infection existed in the renal unit.

Thus it is apparent that there is considerable antigenic complexity within the genera *Actinomyces* and *Nocardia* and apparently less complexity within the genus *Dermatophilus*. However, there is no reported antigenic interaction between these genera. The currently available serologic tools applicable to these genera are recommended as aids of considerable value in the diagnosis and epidemiologic studies of infecting strains of both nocardiosis and actinomycosis and in organisms identification and serologic surveys for dermatophilosis.

V. SUMMARY

The serologic relationships of the major pathogenic actinomycetes, namely *Actinomyces*, *Dermatophilus* and *Nocardia*, are reviewed. *Actinomyces* includes five species, each having its own group antigen. Each species of *Actinomyces* has two serotypes, with the exception of *A. naeslundii* in which four serotypes are recognized. The serotypes of *A. bovis* and serotype 1 of *A. viscosus* are antigenically distinct, but considerable cross-reaction occurs among the serotypes of other species. The use of specifically absorbed fluorescent antibody preparations is used in serotyping *Actinomyces*, in identifying organisms in clinical specimens, and in detecting antibody through FA blocking reactions.

Serologic procedures including immunodiffusion, complement fixation, agglutination, passive hemagglutination, and ELISA techniques have been used to demonstrate antibodies to *D. congolensis* in serum, milk, and skin extracts of cattle with dermatophilosis. A FA procedure has been used diagnostically to identify *D. congolensis* in clinical materials. Antigenic diversity has not been demonstrated among isolates of *D. congolensis*.

Several antigenic preparations have been applied to the diagnosis of nocardiosis and to taxonomic and subspecific identification of *Nocardia* species. Culture filtrate and cell wall extract antigens have been used successfully in demonstrating cutaneous hypersensitivity in cattle and humans infected with *N. asteroides, N. brasiliensis*, and *N. caviae*. Culture filtrate antigens have been used to demonstrate precipitins, complement fixing, and hemagglutinating antibodies in cattle and humans infected with *N. asteroides*. The culture filtrate antigens have established the existence of four serotypes within *N. asteroides*. This serotyping is useful epidemiologically when studying group infections in cattle and humans.

REFERENCES

Beaman, B. L., Burnside, J., Edwards, B., and Causey, W., *J. Infect. Dis. 124:*286 (1976).
Bida, S. A., and Kelley, D. C., *in* "Dermatophilus Infection in Animals and Man" (D. H. Lloyd and K. C. Sellers, eds.), p. 229. Academic Press, New York (1976).
Blumer, S., and Kaufman, L., *J. Clin. Microbiol. 10:*308 (1979).
Bojalil, L. F., and Zamora, A., *Proc. Soc. Exp. Biol. Med. 113:*40 (1963).
Emeruwa, A. C., *Ann. Microbiol. (Paris) 131A:*249 (1980).
Georg, L. K., *Pan Am. Health Organ. Sci. Publ. 205:*71 (1970).
Gerencser, M. A., and Slack, J. M., *Appl. Microbiol. 18:*80 (1969).
Gerencser, M. A., and Slack, J. M., *J. Dent. Res. 55:*A184 (1976).
Gillis, T. P., and Thompson, J. J., *J. Clin. Microbiol. 7:*202 (1978).
Humphreys, D. W., Crowder, J. G., and White, A., *Am. J. Med. Sci. 269:*323 (1975).
Lai, C. H., and Listgarten, M. A., *Infect. Immun. 25:*1016 (1979).
Lloyd, D. H., *Vet. Rec. 7:*426 (1981).
Lloyd, D. H., and Jenkinson, D. M., *Br. Vet. J. 137:*601 (1981).
Magnusson, M., *in* "The Biology of the Nocardiae" (M. Goodfellow, G. H. Brownell, and J. A. Serrano, eds.), p. 236. Academic Press, London (1976).
Makinde, A. A., *Vet. Rec. 26:*383 (1980).
Makinde, A. A., *Res. Vet. Sci. 30:*374 (1981).
Marucha, P. J., Keyes, P. H., Wittenberger, C. L., and London, J., *Infect. Immun. 21:* 786 (1978).
Ortiz-Ortiz, L., Bojalil, L. F., and Contreras, M. F., *J. Immunol. 108:*1409 (1972).
Pier, A. C., *Proc. U.S. Livestock San. Assoc. 66:*409 (1962).
Pier, A. C., and Enright, J. B., *Am. J. Vet. Res. 28:*284 (1962).
Pier, A. C., and Fichtner, R. E., *Am. Rev. Respir. Dis. 103:*698 (1971).
Pier, A. C., and Fichtner, R. E., *J. Clin. Microbiol. 13:*548 (1981).
Pier, A. C., Richard, J. L., and Farrell, E. F., *Am. J. Vet. Res. 25:*1014 (1964).
Pier, A. C., Thurston, J. R., and Larsen, A. B., *Am. J. Vet. Res. 29:*397 (1968).
Pine, L., Malcolm, G. B., Curtis, E. M., and Brown, J. J., *J. Clin. Microbiol. 13:*15 (1981).

Pulliam, J. D., Kelley, D. C., and Coles, E., *Am. J. Vet. Res. 28:*447 (1967).
Revis, G. J., Vatter, A. E., Crowle, A. J., and Cisar, J. O., *Infect. Immun. 36:*1217 (1982).
Richard, J. L., Thurston, J. R., and Pier, A. C., *in "Dermatophilus* Infections in Animals and Man" (D. H. Lloyd and K. C. Sellers, eds.), p. 216. Academic Press, New York (1976).
Ridell, M., *Int. J. Syst. Bacteriol. 25:*124 (1975).
Salman, M. D., Bushnell, R. B., and Pier, A. C., *Am. J. Vet. Res. 43:*332 (1982).
Shainhaus, Z., Pier, A. C., and Stevens, D.A., *J. Clin. Microbiol. 8:*516 (1978).
Shigidi, M. T. A., Mirghani, T., and Musa, M. T., *Res. Vet. Sci. 28:*207 (1980).
Slack, J. M., and Gerencser, M. A., "Actinomyces, Filamentous Bacteria". Burgess Publ. Co., Minneapolis (1975).
Stevens, D. A., Pier, A. C., Beaman, B. L., Morozumi, P. A., Lovett, I. S., and Houang, E. T., *Am. J. Med. 71:*928 (1981).
Thurston, J. R., Phillips, M., and Pier, A. C., *Am. Rev. Respir. Dis. 97:*240 (1968).
Zamora, A., Bojalil, L. F., and Bastarrachea, F., *J. Bacteriol. 85:*549 (1963).

THERMOPHILIC ACTINOMYCETES: THEIR ROLE IN HYPERSENSITIVITY PNEUMONITIS

Viswanath P. Kurup

Department of Medicine
The Medical College of Wisconsin
Milwaukee, Wisconsin, U. S. A.
and
The Research Service
Veterans Administration Medical Center
Wood, Wisconsin, U. S. A.

I. INTRODUCTION

Members belonging to the group of actinomycetes are widely distributed in nature. The majority of the organisms are free-living and are frequently isolated from the soil and other environments. Some of these organisms are capable of producing potent antibiotics and biochemical products of economic importance. Few actinomycetes cause plant diseases, but their devastating effects are phenomenal (Waksman, 1959). Human diseases caused by members belonging to *Actinomyces, Nocardia*, and *Actinomadura* have attracted considerable interest in recent years. The increased interest in these organisms has contributed to a better understanding of the diseases which they cause (Emmons *et al.*, 1977). Of considerable interest is hypersensitivity pneumonitis (HP), an immunological lung disease, which results from the inhalation of thermophilic actinomycetes (Pepys, 1969). Susceptible individuals may develop hypersensitivity lung disease upon inhalation of dust laden with thermophilic actinomycetes found in the working or living environment. Many other organisms, organic materials, and chemicals also have been implicated in HP.

Hypersensitivity pneumonitis, also know as extrinsic allergic alveolitis,

[1] *This work was supported by Specialized Center of Research (SCOR) Grant HL 15389 from the National Heart, Lung, and Blood Institute and by the Veterans Administration.*

ISBN 0-12-528620-1

results from the sensitization of susceptible individuals by inhaled antigens. The disease process usually involves the middle and upper bronchial airways as well as the lung parenchyma and represents an interstitial lymphocytic or granulomatous allergic reaction of the pulmonary tissue (Fink, 1974). Clinically, the disease is characterized by intermittent episodes of chills, fever, cough, and shortness of breath which occur four to eight hours after the inhalation of specific sensitizing agents (Roberts and Moore, 1977). Chest X-ray examination of the patients reveals diffuse nodular infiltrates or interstitial fibrosis depending on the stage of the disease process (Fink, 1974). Pulmonary function abnormalities range from severe diffusion defects to varying degrees of restriction and obstruction (Banaszak and Thiede, 1974). Regardless of the etiologic agent, the similarity of lung changes and symptomatology suggests that the pathogenesis is due to common mechanisms.

II. DISEASES CAUSED BY THERMOPHILIC ACTINOMYCETES

Recent increased interest in the health implications of working and living environments has resulted in the identification of a variety of diseases associated with noxious environmental agents. In the past, HP caused by actinomycetes was inadequately reported or even unknown but is now being increasingly recognized as an important group (Roberts and Moore, 1977) among the occupational lung diseases. The various diseases caused by thermophilic actinomycetes are presented in Table I. Microbial dust smaller than 5 μm can be easily inhaled and deposited on the alveoli. Spores and hyphal fragments of thermophilic actinomycetes are usually much smaller and can cause HP in susceptible individuals. One of the requirements for sensitization is a very high concentration of the antigens in the inhaled air. The major diseases caused by thermophilic actinomycetes are farmer's lung, bagassosis, mushroom worker's lung, and ventilation system-induced pneumonitis (Table I).

TABLE I. Etiologic Agents of Hypersensitivity Pneumonitis

Disease	*Source of environment*	*Antigen*
Farmer's lung	*Moldy hay, moldy corn*	M. faeni, T. candidus T. vulgaris, S. virdis
Bagassosis	*Moldy bagasse*	T. sacchari, T. candidus, T. vulgaris
Mushroom worker's lung	*Moldy vegetable compost*	T. vulgaris, T. candidus, S. viridis, M. faeni
Ventilation system-induced pneumonitis	*Contaminated forced air heating, humidification, and conditioning systems*	T. candidus, T. vulgaris, S. viridis

A. Farmer's Lung

Farmer's lung results from the inhalation of moldy hay dust contaminated with actinomycetes and fungal spores (Pepys, 1969). Moist hay during stacking promotes the growth of bacteria and fungi. This leads to fermentation of substrate and results in the elevation of the temperature, leading the way for luxurient growth of thermophilic actinomycetes. When hay dust containing actinomycetes is inhaled, susceptible individuals develop HP. A farmer working in a moldy atmosphere can inhale as many as 750,000 spores per minute of which 98% are actinomycetes spores (Lacey and Lacey, 1964). Cattle also develop a condition similar to farmer's lung during the winter months when they are confined to the stable area and are exposed to moldy hay dust (Wilkie, 1976).

Although farmer's lung was described by Campbell in 1932, the exact cause of this disease was not clearly established until recently. It was suggested that farmer's lung may result from a hypersensitivity to actinomycetes or to the products of actinomycetes occurring in a wide variety of organic materials (Dickie and Rankin, 1958; Fuller, 1953). The symptoms have been described as shortness of breath, cough, fever, chills, weight loss, and hemoptysis associated with an acute granulomatous interstitial pneumonitis (Dickie and Rankin, 1958). The organisms implicated are *Micropolyspora faeni, Thermoactinomyces candidus, T. vulgaris*, and *Saccharomonospora viridis* (Banaszak *et al.*, 1970, Kurup *et al.*, 1975; Pepys, 1969). Precipitating antibody to the sensitizing antigens usually will be present in the sera of patients; however, about 20% of normal individuals exposed to the antigens also may show similar reactions (Roberts *et al.*, 1976).

B. Bagassosis

Sugar cane fiber when allowed to stand after extracting the juice will promote the growth of thermophilic actinomycetes, particularly *T. sacchari* (Lacey, 1971). Other species such as *T. candidus* and *T. vulgaris* also may be isolated from moldy bagasse. Workers in the sugar industry exposed to the moldy bagasse develop HP. Bagassosis is an acute respiratory illness primarily affecting workers after the inhalation of *T. sacchari* from contaminated bagasse (Nicholson, 1968; Salvaggio *et al.*, 1966). This disease is similar to farmer's lung and has been reported in all parts of the world. Antibodies to *T. sacchari* antigens frequently can be demonstrated in the sera of patients (Salvaggio *et al.*, 1966).

C. Mushroom Worker's Lung

This hypersensitivity lung disease was first reported by Bringhurst, Byrne, and Gershen-Cohen (1959). Since then, more cases have been reported (Sakula, 1967; Stewart, 1974). Mushroom compost, a pasteurized mixture of horse manure and vegetable matter, harbors a variety of thermophilic actinomycetes (Kleyn *et al.*, 1981). Upon inhalation of the spores of thermophilic actinomycetes, susceptible individuals may develop characteristic HP. Published reports indicate that patients and exposed individuals develop antibodies against *M. faeni*, *T. candidus*, and *T. vulgaris*. Occasionally antibodies against compost extract, *Aspergillus fumigatus*, and mushroom spores have also been reported in patients' sera. Bringhurst and his associates (1959) suggested that nitrogen dioxide liberated during compost processing might have also contributed additional damage to the lungs. During spawning of the compost a large number of spores can be liberated into the environment and may be the cause of this lung disease.

D. Ventilation System Pneumonitis

Thermophilic actinomycetes belonging to the species of *M. faeni*, *T. candidus*, *T. vulgaris*, and *S. viridis* have been found in home and industrial environments (Kurup *et al.*, 1976). Forced air heating and humidification systems promote the growth of thermophilic actinomycetes (Fink *et al.*, 1971; Kurup and Fink, 1975). The inhalation of these microorganisms from such contaminated environments can result in the development of HP. Patients frequently have detectable antibodies against thermophilic actinomycetes. Occasionally sera from patients may not contain demonstrable antibodies against standard antigens, but may have antibodies against organisms isolated from that particular environment. Multiple cases of HP developed subsequent to inhalation of thermophilic actinomycetes from contaminated air exchangers of office buildings have been reported (Arnow *et al.*, 1978). Patients challenged with specific antigens or humidifier water may develop symptoms in four to eight hours and this technique has been used in the definitive diagnosis of the disease.

III. CHARACTERISTICS OF HYPERSENSITIVITY PNEUMONITIS

A. Clinical

The clinical features of HP depend upon the immunologic responses of the patients, the antigenicity of the inhaled dust, and the intensity and frequency of the exposure. In general, the characteristics of the illness are similar, regardless of the offending antigen inhaled. Clinical manifestations, however, differ among acute and chronic stages of the disease.

The common and most easily distinguishable form of HP is the acute form, resulting from intermittent exposure to an antigenic source. The patients become sensitized following exposure and subsequently develop symptoms of dyspnea, cough, fever, chills, malaise, and myalgia occurring 4 to 8 h after inhalation of the antigen (Pepys, 1969). In this form of HP, the symptoms may persist up to 12 h and the patients may recover spontaneously. The symptoms usually reappear each time the patient is exposed to the antigens. The patient remains asymptomatic between exposures.

In the chronic form of HP, irreversible damage to the lung may occur. This may be due to the intense immunological inflammatory reaction within the lung caused by heavy and intermittent exposure to antigens or from mild, continuous, and prolonged exposure to small amounts of antigens. Fibrosis with irreversible pulmonary insufficiency can be seen in these patients (Banaszak and Thiede, 1974). Lung biopsies show interstitial fibrosis with granuloma formation and alveolar walls thickened by infiltration with lymphocytes, plasma cells, and some eosinophils. The patients may develop ventilatory impairment and diffusion defects which do not respond to prolonged avoidance or corticosteroid therapy (Fink, 1974).

B. Radiologic

X-ray findings of HP vary from mild interstitial infiltrate to abnormalities similar to those seen in acute pulmonary edema and interstitial pneumonitis. During acute attacks, soft, patchy, ill-defined parenchymal densities that tend to coalesce may be seen in all lung fields. In chronic phases of the disease, the lung may present the appearance of a fibrotic process with diffuse interstitial disease (Meek and Seal, 1974). The X-ray picture is usually nonspecific and the diagnosis thus depends on other features.

C. Physiologic

Pulmonary function abnormalities are variable in HP. In the acute form, a primary restrictive abnormality is characteristic although some degree of small airway disease and airway obstruction may be present. An increase in the residual volume and ratio of the residual volume and lung capacity suggest the presence of small-airway disease and regional air trapping. The diffusion capacity is frequently abnormal in the early part of the disease and diminishes further with advancing severity. In the chronic forms of the disease with severe fibrosis and destruction of lung tissue, findings typical of pulmonary fibrosis or severe pulmonary emphysema with respiratory insufficiency, or both, are seen (Banaszak and Thiede, 1974).

IV. IMMUNOLOGICAL ASPECTS

A. Humoral Immunological Response

The most characteristic immunological feature of HP is the presence of serum precipitins against the offending antigens (Pepys, 1969). In many cases the immunologic reactivity is directed against more than one species of thermophilic actinomycetes or fungi (Wenzel *et al.*, 1974). The complex nature of the offending antigens usually limits precise quantification of the antibody response in patients. In most patients, the antibody concentrations are high enough to be detected by double diffusion in agar gel (Fletcher *et al.*, 1970). Antibodies of immunoglobulin classes IgG, IgM, and IgA against thermophilic actinomycetes have been detected in patient sera by radioimmunoassay methods (Patterson *et al.*, 1976). Although precipitins are present in the serum of patients with HP, approximately 20% of asymptomatic individuals exposed to the same organic dust also have detectable precipitating antibodies (Roberts *et al.*, 1976). The precipitating antibodies against the sensitizing organic dust in symptomatic patients and in exposed individuals without symptoms cannot be differentiated qualitatively or quantitatively although recent studies using antigen-antibody crossed-immunoelectrophoresis have shown some differences in the presence of certain precipitin arcs (Treuhaft *et al.*, 1979). Both asymptomatic and symptomatic individuals have been shown to have antibodies to the etiologic agents; however, it has rarely been reported that individuals with diseases have no antibodies (Edwards *et al.*, 1974). The failure to detect antibody may be due to the use of insensitive methods or inappropriate antigens (Nielsen *et al.*, 1973).

Because of the presence of precipitins in both symptomatic and asymptomatic individuals, the role of these antibodies in the pathogenesis of HP has not been precisely defined. Immune complex disease usually occurs when the ratio of antibody to antigen is in the zone of antigen excess; hence, the absolute quantity of antibody is not as important as inhaling the proper antigen and the subsequent interaction with the antibody in the lung or in the blood. The immune complex formed in the blood is usually in the zone of antibody excess, whereas that on the air side of the lung may be in antigen excess and therefore may be more important in the immunopathogenesis of the disease.

The 4 to 6 h delay in the appearance of symptoms following exposure to the offending antigens seen in HP is comparable to that observed in immune complex diseases. When patients with HP are skin tested with specific antigens, an immediate reaction may sometimes be produced. In addition, an Arthus-like reaction (4 to 8 h) is usually observed (Fink *et al.*, 1971; Pepys, 1969). This late reaction begins with erythema and edema and may progress to central necrosis, but usually subsides within 24 hours. A true delayed response in the absence of an Arthus reaction has not been observed. Antigen, complement, and immunoglobulin have been detected in lung biopsies (Wenzel and Emanuel, 1974). Peripheral leukocytosis is commonly observed during the acute phases of HP. However, neutrophilic pulmonary infiltrates in the lung have not been regularly observed in lung biopsies. This may be due to the fact that most biopsies are not obtained in the early acute stage of the disease and hence this phenomenon may be overlooked.

B. Cell-Mediated Immunological Response

The role of cell-mediated immunological response in patients with HP was not studied until recently although granulomatous immunologic responses in these patients have been known for some time. Sensitized T lymphocytes in the peripheral circulation of patients with HP have been recently reported by several investigators (Allen *et al.*, 1975; Caldwell *et al.*, 1973). These studies have demonstrated antigen-induced migration inhibition factor (MIF) and blast transformation; both are indicators of cell-mediated immunity. Determinations of cell-mediated immunity *in vitro* in farmer's lung were reported to be of use in discriminating symptomatic and asymptomatic individuals (Marx *et al.*, 1973). Migration inhibition factor was produced by peripheral lymphocytes from a large proportion of patients exposed to specific antigens. Antigen-induced lymphocyte transformation is considered a test *in vitro* for activated T-lymphocytes. In HP, lymphocyte transformation was detected in the presence of specific antigens. In most cases there was positive correlation between the cellular immune response *in vitro* and the presence of circulating precipitins. Various animal model studies also strongly support the fact that activated T cells are involved in the pathogenesis of HP (Harris *et al.*, 1976; Salvaggio *et al.*, 1975).

V. DIAGNOSIS

A. Clinical

A diagnosis of HP can usually be made by an appropriate environmental history, laboratory and serological studies, and a trial of avoidance and re-exposure to the suspected environmental antigen. The more insidious form of the disease may be difficult to diagnose and a lung biopsy may be necessary to establish a diagnosis. Radiologic and pulmonary physiologic examinations may provide additional information.

Confirmation of the diagnosis may be made by cautiously exposing the patients to the suspected antigens and carefully observing clinical reaction. This is done by first making the patient relatively asymptomatic by avoiding the offending antigen for a period of time. The patient is then exposed to the environment and observed for up to 18 hours for characteristic signs and symptoms of HP. It may also be necessary to ascertain the specific etiologic agents by using specific purified antigens in challenge tests.

B. Microbiology

The environment of the patient, which is suspected of having the offending antigen, should be studied for the presence of microorganisms. Various dusts, contaminated water, and other materials from the home or work area of the patient should be cultured in appropriate media for the isolation of thermophilic actinomycetes.

1. Isolation of Thermophilic Actinomycetes. Actinomycetes are higher bacteria, resembling fungi in their colonial morphology. Many actinomycetes also produce spores resembling that of fungi. Although several actinomycetes species are implicated in HP, the most important are those belonging to the three thermophilic genera, namely, *Micropolyspora, Thermoactinomyces*, and *Saccharomonospora* (Kurup and Fink, 1975). The most commonly cited sources of antigen associated with HP are the organisms belonging to these three thermophilic actinomycetes genera (Kurup and Fink, 1975; Kurup *et al.*, 1975, 1980; Lacey 1971). These organisms are capable of growing at elevated temperatures such as in moldy hay, mushroom compost, bagasse, and other vegetable matter or in the heating and humidification systems of buildings (Arnow *et al.*, 1978; Banaszak *et al.*, 1970; Dickie and Rankin, 1958; Fink *et al.*, 1971, 1976).

Specimens for culture should be collected in sterile containers. Samples of dust were obtained by a vacuum cleaner. Scrapings, swabs, and liquid samples from suspected contaminated sources should be collected aseptically. Specimens were processed as soon as possible but may be refrigerated until they are cultured. The samples were suspended in sterile saline and appropriate dilutions plated on trypticase soy agar (TSA), TSA with 30 μg of novobiocin per ml of medium for controlling over-growth of bacteria, or TSA supplemented with casein hydrolysate and half-strength nutrient agar. Liquid samples were first centrifuged at 3,000 rpm for 10 min and the deposit suspended in sterile saline and cultured as above. All inoculated plates were incubated at 55 °C for up to one week. Plates should be stored in plastic bags to prevent evaporation of the medium. Incubation under moderate to high humidity is recommended. Plates should be examined daily for one week for evidence of growth.

2. Identification of Thermophilic Actinomycetes. Isolated organisms can be identified by their colonial morphology, microscopic features, physiologic characteristics, chemical composition, and immunological characteristics (Kurup, 1978, 1979, 1981a, b; Kurup and Fink, 1975, Kurup and Heinzen, 1978; Kurup *et al.*, 1975). Various features useful for distinguishing among the organisms are given in Table II. The important characteristics of each organism is considered separately.

a. Micropolyspora faeni. This organism grows slowly in most isolation media except those containing novobiocin to which it is highly sensitive. In the primary isolates, colonies are light yellow, attain a size of 2 to 4 mm in diameter, and frequently appears as a *Nocardia* colony. It is interesting that *M. faeni* although slow-growing, can grow at 37 °C when it is morphologically indistinguishable from *Nocardia*. With prolonged incubation, a white, tuft-like growth will appear on the surface of the colonies. Microscopic examination of the culture shows thin branching hyphae with chains of smooth, round spores, both on aerial and substrate hyphae. Spores are 0.8 to 1.2 μm in diameter and are produced in chains of 5 to 15 spores.

The cell wall composition shows the presence of arabinose, galactose, and meso-diaminopimelic acid, indicating that the cell wall is of type IV, similar to *Nocardia sp.* (Kurup and Fink, 1975). *Micropolyspora faeni* phage specifically lyse strains of this organism (Kurup and Heinzen, 1978). Biochemical characteristics include hydrolysis of tyrosine and xanthine and failure to hydrolyze casein (Table II). Other species included in this genus do not conform to the description of the genus except *M. rectivirgula*. It is apparent that *M. rectivirgula* and *M. faeni* are synonymous (Kurup, 1981a).

b. Saccharomonospora viridis. *S. viridis* has previously been classified under *Thermomonospora* due to the production of spores only on aerial hyphae; however, recent studies indicate that the cell wall type of this organism

TABLE II. Characteristics Used in the Identification of Thermophilic Actinomycetes

Characteristics	M. faeni	S. viridis	T. candidus	T. sacchari	T. vulgaris	T. intermedius	T. dichotomica
Growth	*slow*	*slow*	*fast*	*slow*	*fast*	*fast*	*slow*
Spores	*chain*	*single*	*single*	*single*	*single*	*single*	*single*
Aerial mycelium	+	+	+	-	+	+	+
Decomposition of							
casein	-	+	+	+	+	+	+
tyrosine	+	-	-	-	+	+	-
xanthine	+	-	-	-	-	-	-
starch	-	+	-	+	+	-	+
esculin	-	-	+	+	-	+	-
arbutin	-	-	+	-	-	+	-
Cell wall type	*IV*	*IV*	*III*	*III*	*III*	*III*	*III*

is type IV, whereas *Thermomonospora* possess a type III cell wall (Kurup, 1979; Nonomura and Ohara, 1971). *S. viridis* grows slowly in most media studied. Initially, the colonies appear brownish-white but will turn to bluish-green when spores are produced. Casein hydrolysate medium support the production of spores. The spores are produced on dichotomously branched sporophores of aerial hyphae and never on substrate hyphae. The spores are 1.5 X 1.00 μm and are oval in shape. Microscopic examination of Gram-stained smears show typical branched actinomycetous hyphae. The spores fail to retain malachite green when spore stain was used (Kurup, 1981a).

The ideal temperature for growth was found to be 45 to 50 °C. Casein and starch were hydrolyzed, but tyrosine, xanthine, or hypoxanthine were not decomposed (Table II). Like *M. faeni*, *S. viridis* is highly sensitive to novobiocin. This organism is frequently isolated from moldy hay, corn and other vegetable matter and from heating and humidification systems of buildings (Kurup *et al.*, 1976).

c. Thermoactinomyces. The most commonly encountered species of this genus are *T. candidus*, *T. sacchari*, *T. vulgaris*, *T. intermedius*, and *T. dichotomica*. They are frequently isolated from the environment (Kurup, 1981b; Kurup and Fink, 1975; Nonomura and Ohara, 1971). The first three species are associated with HP, while the role of the latter species in HP is not known although they share several antigens in common with other species of *Thermoactinomyces* (Kurup *et al.*, 1976). All species of *Thermoactinomyces* are resistant to novobiocin.

T. candidus and *T. vulgaris* grow well in all common laboratory media. These species are fast-growing and produce fluffy or granular, white colonies on most of the media. Both species grow at 55 to 60 °C and usually no growth is detected at 37 °C or less. Spores are produced on both aerial and substrate hyphae and are either produced on small sporophores ranging from 1 to 3 μm or directly on the hyphae. The spores are typical endospores, contain dipicolinic acid and are themoresistant (Kurup *et al.*, 1977a). Casein is hydrolyzed by all species of *Thermoactinomyces* (Table II). *T. vulgaris* hydrolyzes tyrosine and starch but not esculin or arbutin, whereas *T. candidus* hydrolyzes arbutin and esculin and is lysed by specific phages which do not lyse any of the other species (Kurup and Fink, 1975; Kurup and Heinzen, 1978). Antigenic differences were also noted between the two species.

T. sacchari grows very slowly on half-strength nutrient agar and on casein hydrolysate agar. The colonies are usually transparent and the aerial hyphae will be lysed in 3 to 4 days. Growth is usually good at 50 °C. Other characteristics are the same as those of *T. vulgaris* and *T. candidus*, except that *T. sacchari* hydrolyzes casein, starch, and esculin but not tyrosine or arbutin. Spores are endospores and always produced on sporophores 2 to 3 μm in length.

T. dichotomica produce slow-growing, yellowish colonies. This species demonstrates all the characteristics of the genus in that *T. dichotomica* hydrolyzes casein, starch, and gelatin, produces endospores, and grows at 55 to 60 °C. *T. intermedius*, morphologically resembles *T. candidus* and *T. vulgaris* but fails to hydrolyze starch although it decomposes casein and tyrosine. Immunologically, this species shares antigens with *T. vulgaris* and *T. candidus*.

T. candidus and *T. vulgaris* have been isolated from a wide range of substrates which include mushroom compost, moldy hay, moldy corn and vegetable matter, and from heating and humidification systems in residential and commercial buildings (Fink, 1974; Kurup *et al.*, 1976). *T. sacchari* has been isolated from moldy bagasse (Lacey, 1971). *T. dichotomica* and *T. intermedius* are less commonly isolated and their roles in HP have not been documented.

C. Serology

Hypersensitivity pneumonitis is characterized by the presence of circulating antibodies against the offending antigen. Extracts of dusts, liquid samples, or the antigens from cultured organisms may be used to detect precipitating antibodies in the patient serum. Several serological methods such as complement fixation, hemagglutination, indirect immunofluorescence, immunoelectrophoresis, and crossed-immunoelectrophoresis have been used for the detection of antibodies (Fletcher *et al.*, 1970; Kurup *et al.*, 1977b; Marx *et al.*, 1975; Nielsen *et al.*, 1973; Wenzel *et al.*, 1972, 1974). Double diffusion in agar gel is the most widely used method because of its simplicity.

1. Antigen Preparations

a. Crude extracts from samples. Dust and other solid samples (500 mg) are extracted overnight in 0.5% phenol-0.85% NaCl. Following centrifugation, the supernatant should be used either directly or after concentration. Liquid samples may also be used as such or after appropriate concentration for serological testing.

b. Thermophilic actinomycetes antigens. Most of the thermophilic actinomycetes grow well on complex media but fail to grow in any of the simple media. The removal of macromolecular components of the media from antigen is extremely difficult. Occasionally, these components may react with the sera of patients and give a false-positive reaction. These antigens are not suitable for skin testing due to the presence of impurities.

A double dialysis method was developed to exclude media components (Edwards, 1972a; Kurup *et al.*, 1976). In this method the complex medium (TSA) is placed inside a dialysis bag with a molecular weight cut-off of 8,000. The dialysis bag containing TSA is placed in a solution of glycine (0.075 M) and sodium chloride (0.1 M) solution. Following autoclaving, the system is allowed to equilibrate for 48 h. The outer fluid is collected aseptically and inoculated with cultures. The cultures are incubated at 55 °C for two weeks at which time 0.5% formalin is added. After standing for 48 h, the culture filtrate is dialyzed extensively against distilled water and is then freeze-dried. *T. vulgaris*, *T. candidus* and *S. viridis* can be grown in this manner. *T. sacchari* can be cultured in trypticase soy broth (TSB) and starch dialysate and incubated at 50 °C for two weeks. Antigens are reconstituted at 10 to 30 mg/ml for the immunodiffusion test.

A synthetic medium was devised to grow *M. faeni* (Kurup and Fink, 1977, 1979b; Kurup *et al.*, 1981). AOAC broth (Difco), supplemented with 1% lactose and 0.01% spermidine, is inoculated with *M. faeni*. The inoculated cultures are incubated on a shaker at 50 °C for 5 to 7 days. After killing the cultures with 0.5% formalin, the culture filtrates are separated and processed as described above. A concentration of 5 mg/ml is used to detect precipitins in the sera of patients by agar gel diffusion.

c. Purification of antigens. Various methods have been used to purify culture filtrate antigens. Edwards (1972b) fractionated trichloracetic acid and alcohol precipitates of culture filtrate antigens of *M. faeni* using Sephadex G-100 and G-200, and DEAE columns. Three major antigenic fractions which reacted with pooled farmer's lung sera have been isolated. These include two proteic and one glycoproteic antigens. However, these antigens were not evaluated in diagnostic methods.

We have investigated the culture filtrate antigens prepared by growing *M. faeni* in synthetic medium (Kurup and Fink, 1977, 1979b). Antigens from several strains of *M. faeni* were compared by tandem-crossed immunoelectrophoresis using patients sera as well as anti-*M. faeni* rabbit serum. The results indicate that many antigens were common to the various strains of *M. faeni* and at least three antigens were shared by all the strains of *M. faeni* (Kurup and Fink, 1979b). By employing gel filtration chromatography and preparative isoelectric focusing, we have obtained a fraction containing two antigens which reacted with all 18 farmer's lung sera studied and showed no reactivity with any of the control sera studied. These antigens have isoelectric points of 3.8 and 4.0 and molecular weights of 29,000 and 51,000 respectively (Kurup *et al.*, 1981).

2. Serological Methods. Diffusion in agar gel is the serological method commonly used for detecting circulating antibodies in the sera of patients suspected of having HP. Other methods used in the detection of antibodies include hemagglutination, complement fixation, indirect immunofluorescence, and radioimmunoassay (Kurup *et al.*, 1977b; Patterson *et al.*, 1976; Wenzel *et al.*, 1972). However, these methods were not altogether satisfactory due to the lack of pure and standardized antigens.

Double diffusion in agar gel. A modification of the micro-immunodiffusion technique of Wadsworth (1962) was used (Kurup and Fink, 1979a). Plastic templates containing holes were placed on microscopic slides layered with 0.8% agarose in 0.1 M barbital buffer, pH 8.6. Antigens were placed in the center wells and sera from patients in the peripheral wells were allowed to diffuse for 48 hours at room temperature in a moist chamber. Known positive and negative sera were included with each set. After diffusion the templates were removed and the slides washed in physiological saline (0.15 M NaCl) for 24 hours to remove the unreacted antigen and sera. After being washed in distilled water, the slides were stained by 0.2% Coomassie blue R-250 stain in methanol:acetic acid:water (4.5:1:4.5). The precipitin arcs developed were evaluated in comparison with the positive and negative controls.

VI. COMMENTS

The immunopathogenesis of HP and the role of thermophilic actinomycetes antigens are quite complex and the presence of a number of factors is probably required to produce disease. Inhalation of antigens which likely possess characteristics such as nondigestibility and/or adjuvant or complement activating properties is important. The qualitative and quantitative requirements of such exposure may be important but have not yet been defined. The exposed individual must also be susceptible to the development of the disease. This may involve a genetic predisposition or an acquired susceptibility in terms of immunologic or target organ sensitivity to inhaled antigens. More studies on the unique characteristics of antigens associated with HP and on the necessary exposure circumstances are desirable. Studies of the predisposing factors and immunological response, with emphasis on quantitative parameters, as well as studies of the combination of reactivity and local pulmonary immunologic factors are needed for better understanding of the disease.

In order to define the mechanisms of HP in patients, more information on the spectrum of pulmonary reaction to inhaled antigen is needed. An increase in knowledge of mechanisms involved in HP will also increase our understanding of the basic immune mechanisms in the lung. The role of antigens in the genesis of the disease and the sequential events of the disease process need to be elucidated for a fuller understanding of the disease. It is also essential to find out how the host responds to the inhaled antigen and how the antigen is broken down and removed from the system. Knowledge of the role of immune complexes, cell-mediated immune responses, and the mediators responsible for the tissue damage would be of considerable value in designing proper preventive measures and more efficient treatment.

ACKNOWLEDGMENTS

The author is grateful to Dr. Jordan N. Fink, Professor of Medicine and Chief of Allergy and Immunology Section for his interest and suggestions.

REFERENCES

Allen, D. H., Basten, A., Williams, G. V., and Woolcock, A. J., *Am. J. Med. 59:*505 (1975).
Arnow, P. M., Fink, J. N., Schlueter, D. P., Barboriak, J. J., Mallison, G., Said, S. I., Martin, S., Unger G. F., Scanlon, G. T., and Kurup, V. P., *Am. J. Med. 64:*236 (1978).
Banaszak, E. F., and Thiede, W. H., *Geriatrics, 29:*65 (1974).
Banaszak, E. F., and Thiede, W. H., and Fink, J. N., *New Engl. J. Med. 283:*271 (1970).
Bringhurst, L. S., Byrne, R. N., and Gershen-Cohen, J., *J. Am. Med. Assoc. 171:*15 (1959).
Caldwell, J. R., Pearce, C. E., Spencer, C., Leder, T., and Waldman, R. H., *J. Allergy Clin. Immunol. 52:*225 (1973).
Campbell, J. M., *Br. Med. J. 2:*1143 (1932).
Dickie, H. A., and Rankin, J., *J. Am. Med. Assoc. 167:*1069 (1958).
Edwards, J. H., *J. Lab. Clin. Med. 79:*683 (1972a).
Edwards, J. H., *Clin. Exp. Immunol. 11:*341 (1972b).
Edwards, J. H., Baker, J. T., and Davies, B. H., *Clin. Allergy 4:*379 (1974).
Emmons, C. W., Binford, C.H., Utz, J.P., and Kwon-Chung, K.J., *in* "Medical Mycology", 3rd ed. p. 76. Lea and Febiger, Philadelphia (1977).
Fink, J. N., *Clin. Notes Respir. Dis. 13:*3 (1974).
Fink, J. N., Banaszak, E. F., Barboriak, J. J., Hensley, G. T., Kurup, V. P., Scanlon, G. T., Schlueter, D. P., Sosman, A. J., Thiede, W. H., and Unger, G. F., *Ann. Intern. Med. 84:*406 (1976).
Fink, J. N., Banaszak, E. F., Thiede, W. H., and Barboriak, J. J., *Ann. Intern. Med. 74:*80 (1971).
Fletcher, S. M., Rondle, C. J. M., and Murray, I. G., *J. Hyg. 68:*401 (1970).
Fuller, C. J., *Thorax 8:*59 (1953).
Harris, J. O., Bice, D., and Salvaggio, J. E., *Am. Rev. Respir. Dis. 114:*29 (1976).
Kleyn, J. G., Johnson, W. M., and Wetzler, T. F., *Appl. Environ. Microbiol. 41:*1454 (1981).
Kurup, V. P., *J. Allergy Clin. Immunol. 61:*232 (1978).
Kurup, V. P., *Curr. Microbiol. 2:*267 (1979).
Kurup, V. P., *Microbiologica 4:*249 (1981a).
Kurup, V. P., *Science-Ciencia 8:*5 (1981b).
Kurup, V. P., and Fink, J. N., *J. Clin. Microbiol. 2:*55 (1975).
Kurup, V. P., and Fink, J. N., *Infect. Immun. 15:*608 (1977).
Kurup, V. P., and Fink, J. N., *Sabouraudia 17:*163 (1979a).
Kurup, V. P., and Fink, J. N., *Int. Arch. Allergy Appl. Immunol. 60:*140 (1979b).
Kurup, V. P., and Heinzen, R. J., *Can. J. Microbiol. 24:*794 (1978).
Kurup, V. P., Barboriak, J. J., and Fink, J. N., *Biol. Actinomycetes 12:* 53 (1977a).
Kurup, V. P., Barboriak, J. J., and Fink, J. N., *J. Lab. Clin. Med. 89:*533 (1977b).
Kurup, V. P., Barboriak, J. J., Fink, J. N., and Lechevalier, M. P., *Int. J. Syst. Bacteriol. 25:*150 (1975).
Kurup, V. P., Barboriak, J. J., Fink, J. N., and Scribner, G., *J. Allergy Clin. Immunol. 57:*417 (1976).
Kurup, V. P., Fink, J. N., and Bauman, D. M., *Mycologia 68:*662 (1976).
Kurup, V. P., Hollick, G. E., and Pagan, E. F., *Science-Ciencia 7:*104 (1980).
Kurup, V. P., Ting, E. Y., Fink, J. N., and Calvanico, N. J., *Infect. Immun. 23:*508 (1981).
Lacey, J., *J. Gen. Microbiol. 66:*327 (1971).
Lacey, J. and Lacey, M. E., *Trans. Br. Mycol. Soc. 47:*547 (1964).
Marx J. J., Jr., Motszko, C., and Wenzel, F. J., *J. Clin. Microbiol. 1:*480 (1975).
Marx, J.J.Jr., Wenzel, F.J., Roberts, R.C., Grany, R.L., and Emanuel, D.A., *Clin. Res. 21:* 852 (1973)

Meek, J. C., and Seal, R. M. E., *in* "Aspergillosis and Farmer's Lung in Man and Animals" (R. de Haller and Suter, eds.), p. 216. Hans Huber, Bern 1974.
Nicholson, D. P., *Am. Rev. Respir. Dis. 97:*546 (1968).
Nielsen, K. H., Parratt, D., and White, R. G., *J. Immunol. Methods 3:*301 (1973).
Nonomura, H., and Ohara, Y., *J. Ferment. Technol. 49:*895 (1971).
Patterson, R., Roberts, M., Roberts, R. C., Emanuel, D. A., and Fink, J. N., *Am. Rev. Respir. Dis. 114:*315 (1976).
Pepys, J., *in* "Monographs in Allergy" (P. Kallos, M. Hasek, T. M. Inderbitzin, P. A. Miescher, and B. H. Waksman, eds.), No. 4, p. 1. Karger Co., Basel (1969).
Roberts, R. C., and Moore, V. L., *Am. Rev. Respir. Dis. 116:*1075 (1977).
Roberts, R. C., Wenzel, F. J., and Emanuel, D. A., *J. Allergy Clin. Immunol. 57:*518 (1976).
Sakula, A. *Br. Med. J. 3:*708 (1967).
Salvaggio, J. E., Buechner, H. A., Seabury, J. H., and Arquembourg, P., *Ann. Int. Med. 64:*748 (1966).
Salvaggio, J., Phanuphak, P., Stanford, R., Bice, D., and Claman, H., *J. Allergy Clin. Immunol. 56:*364 (1975).
Stewart, C. J., *Thorax 29:*252 (1974).
Treuhaft, M. W., Roberts, R. C., Hackbarth, C., Emanuel, D. A., and Marx, J. J., Jr., *Am. Rev. Respir. Dis. 119:*571 (1979).
Wadsworth, C., *Int. Arch. Allergy Appl. Immunol. 21:*131 (1962).
Waksman, S. A., *in* "The Actinomycetes", Vol. I. p. 265. Williams and Wilkins Co., Baltimore (1959).
Wenzel, F. J., and Emanuel, D. A., *in* "Aspergillosis and Farmer's Lung in Man and Animals" (R. de Haller and F. Suter, eds.), p. 255. Hans Huber, Bern (1974).
Wenzel, F. J., Emanuel, D. A., and Gray, R. L., *Am. J. Clin. Pathol. 57:*206 (1972).
Wenzel, F. J., Gray, R. L., Roberts, R. C., and Emanuel, D. A., *Am. Rev. Respir. Dis. 109:*464 (1974).
Wilkie, B. N., *Can. J. Comp. Med. 40:*221 (1976).

MEDICAL AND MICROBIOLOGICAL PROBLEMS IN HUMAN ACTINOMYCOSES

G. Pulverer

K. P. Schaal

Institute of Hygiene
University of Cologne
Köln, Federal Republic of Germany

I. INTRODUCTION

Actinomycosis is presumably one of the oldest diseases existing on earth. However, it has been recognized as a specific disease entity with characteristic etiological and clinical features for only about 100 years. Bollinger (1877) described the bovine actinomycosis and, in the following year, Israel (1878) published his paper on human actinomycosis. Nevertheless, hardly another disease may be found, the history of which has involved so many misunderstandings. For example, for a long time, aerobic actinomycetes such as nocardiae or streptomycetes were mistakenly thought to cause human actinomycosis. This hypothesis led to the incorrect deduction that humans acquire actinomycosis by chewing grass. Even now, this idea may still be found in certain bacteriological or clinical textbooks. Fortunately, many of these misunderstandings have been clarified during the last decades.

Today there is no doubt that human and bovine as well as other animal actinomycoses are different diseases. Not only do the clinical pictures differ to a considerable extent, but also the causative organisms are distinct in both groups of infection. Furthermore, nocardiosis and actinomycosis are unrelated. Human nocardiosis is an exogenous monoinfection in which *Nocardia asteroides* is the predominant pathogen. On the other hand, human actinomycosis is essentially always an endogenous, mixed infection, in which facultatively anaerobic actinomycetes, currently classified in the family *Actinomycetaceae*, are the primary pathogens. *Actinomyces israelii*

ISBN 0-12-528620-1

is the most important specific agent causing this resident and potentially malignant disease. This microorganism must be regarded as a resident member of the physiological microflora of certain human mucous membranes.

Human actinomycoses may start as acute or chronic inflammatory processes depending on the concomitant bacteria present. For example, when β-hemolytic streptococci or *Staphylococcus aureus* join the mixed actinomycotic flora, acute signs of infection such as abscess or phlegmon may develop. In these acute lesions, *Actinomycetaceae* are already present, but do not yet play the predominant pathogenetic role; therefore, such acute infections can be self-limiting. However, self-limitation is no longer possible when the disease has entered the chronic and progressive phase. Acute actinomycotic processes, if not treated properly, may turn into a chronic infection. Signs of chronic infection, such as a slowly emerging and progressing, painless infiltration, may already characterize the early stages of the disease, especially when virulent aerobic bacteria are not involved. Whereas the symptoms of acute actinomycotic processes are not at all specific, the chronic phase is usually highly pathognomonic. Typical symptoms are infiltrations or indurations which are not very painful, multiple abscesses and draining sinus tracts, as well as a tendency to relapse and to form central scars. This chronic stadium of human actinomycosis should be regarded as potentially malignant and requires appropriately adjusted therapeutic measures to ensure a successful cure. Simple surgical excision of the process without supporting chemotherapy is nearly always insufficient.

Formerly, only this typical chronic phase of infection was recognized as actinomycosis *sensu stricto*. Because of typical histological features in this stadium, which are missing during the acute phase, this view is still defended by pathologists. As already mentioned, most acute processes in which facultatively anaerobic actinomycetes are etiologically involved enter the typical chronic stadium, when adequate therapeutic means have not been used. We shall discuss some tragic experiences in another paper (elsewhere in this volume). Therefore, we wish to propose a broader definition of human actinomycoses. The diagnosis of this disease should not be restricted to the chronic form which undoubtedly is most typical from a clinical point of view, but should be primarily an etiologic one that encompasses all the inflammatory lesions which contain pathogenic members of the *Actinomycetaceae*, provided that contamination of the clinical specimen during sampling can be excluded. This implies that the examination of suitable specimens in the bacteriological laboratory will be sufficient to provide reliable and unequivocal information on the nature of the disease and how to treat it properly.

Since 1952, the Cologne Institute has had the status of a reference laboratory for West Germany with respect to the bacteriological diagnosis of actinomycoses (Pulverer, 1974; Pulverer and Schaal, 1978; Schaal,

1979). This reference laboratory was set up by the former director of the Institute, Prof. F. Lentze, who started investigating bacteriological, diagnostic, and therapeutic problems of human actinomycosis more than 40 years ago (Lentze, 1938, 1948, 1969). Each year we receive, especially from German University Clinics of Dental Surgery, some 1000 clinical samples from suspected cases for bacteriological examination. Therefore, our studies are based upon extensive sampling of material of nation-wide origin. Each year more than 100 new cases of human actinomycosis are diagnosed or confirmed in our laboratory. Since Lentze (1938) started his investigations, we have collected information and strains derived from 3027 patients. The 3027 patients with actinomycotic lesions from which the specimens originated usually were living in various areas of West Germany, but some cases were localed in East Germany, Western Poland, and Austria.

II. RETROSPECTIVE STUDIES

In 2872 of the 3027 cases of human actinomycosis, we received information on the exact localization of the infective processes (Table I). As shown in this table, over 98% of all patients had actinomycotic lesions in the cervicofacial region. During the last 30 years, we have seen only 34 cases of thoracic and 16 cases of abdominal actinomycosis. Only one patient showed typical actinomycotic lesions in one hand, which had developed 18 months before the diagnosis was established. These data suggest that, in our region, actinomycoses of cervicofacial sites are rather common, whereas other areas of the human body appear to be much less frequently involved.

This prevalence of human actinomycosis in the cervicofacial region is

TABLE I. Localization of Actinomycotic Lesions

	Cases	
Localization	*Number*	*%*
Cervicofacial region	*2821*	*98.2*
Thoracic region	*34*	*1.2*
Abdominal region	*16*	*0.6*
Hand	*1*	
Total number of cases	*2872*	*100.0*

TABLE II. Sex of Patients with Actinomycosis

Sex of patients	Cases	
	Number	%
Male	2099	73.2
Female	768	26.8
Total number of cases	2867	100.0

understandable for two reasons. A. *israelii* and other pathogenic *Actinomycetaceae* are essentially always present in the microflora of the oral cavity of healthy adults. On other mucosal surfaces such as in the gut and in the urogenital system, these organisms are encountered less commonly or in lower numbers. The second point is the fact that the cervicofacial region provides various possibilities for the anaerobes to invade the interstitial tissue and to find there favorable growth conditions. Very often actinomycotic processes occur after tooth extraction, fracture of the jaw, introduction of foreign bodies and the like.

In 2867 of the 3027 cases investigated, the sex of the patient was reported. As shown in Table II, human actinomycosis is three times more often diagnosed in males than in females. This sex ratio is independent of the localization of the infection. Not only in our material but also in that of other laboratories, this prevalence of male patients has been observed. We do not know the significance of these findings. One possible explanation of this important finding may be that sex hormones have an influence on the development of human actinomycosis.

In 2790 of the 3027 cases investigated, the age of the patient was reported. Table III demonstrates that in males the age bracket from 20 to 40 years is predominant, whereas in women the age bracket is around 20 years of age. Actinomycosis in patients under 10 and over 70 years of age is comparatively rare but does occur. Our youngest patients were one to two years old. On the other hand, four patients were 80 years old or older. It is interesting that the sex incidence of human actinomycosis mentioned above is not observed in the age brackets under 10 years and over 70 years of age. Figure 1 illustrates this sex-age distribution. In children up to 10 years and in individuals over 70 years of age, males and females develop actinomycosis at nearly the same frequency. Of the actinomycosis patients aged 20 to 50 years, only 20 to 23% were females. Thus, it seems that this ratio again supports our hypothesis of a hormonal influence on human actinomycosis.

What is known about the pathogenesis of human actinomycosis? As already mentioned, there can no longer be any doubt that facultatively

TABLE III. Sex and Age of Patients with Actinomycosis

Sex of patients	Age of patients in years								
	$\leqslant 10$	11-20	21-30	31-40	41-50	51-60	61-70	$\geqslant 71$	Total no. of cases
Male	60	291	573	422	318	202	84	38	1988
Female	57	166	171	108	89	72	43	38	744
Not recorded	4	9	20	11	9	4	1	58	
Cases Total no.	121	466	764	541	416	278	128	76	2790
%	4	17	27	19	15	10	5	3	100

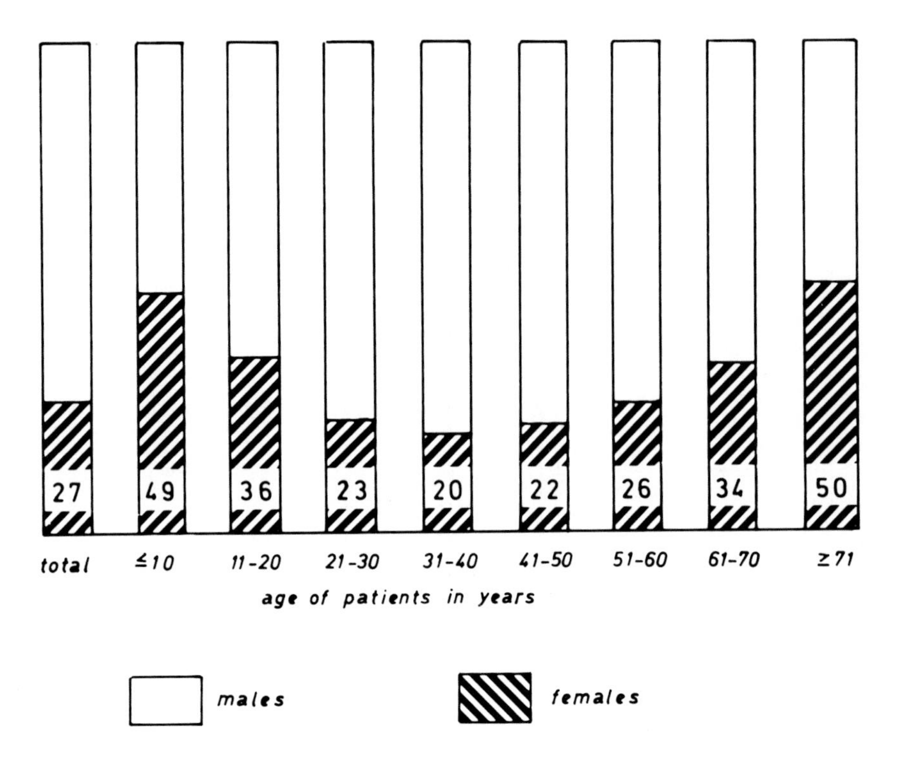

FIGURE 1. Age and sex of patients with actinomycosis

anaerobic actinomycetes are the specific causative agents of actinomycosis. We never examined material from a typical actinomycotic lesion in which we finally failed to demonstrate the actinomycetes. Among the members of the family *Actinomycetaceae*, *A. israelii* is most frequently isolated from actinomycotic lesions. *A. bovis*, the classical agent of bovine actinomycosis has never been identified with certainty in any case of human actinomycosis. As already demonstrated by Lentze (1938, 1948, 1969) and by Holm (1950, 1951), the causative actinomycete is almost always accompanied by other microorganisms. In the material presented in Table IV, the situation is similar. In only 26 of 3027 cases investigated was *A. israelii* isolated in pure culture, without any other bacteria. But in all 26 cases of "monoinfection", we discovered circumstances which could be the reason for not finding other microorganisms. These circumstances include antibiotic therapy before sample collection, prolonged transportation (three to seven days) of the specimen without the use of modern transport medium, and problems with the quality of the culture media used in the laboratory. Therefore, in all 26 cases, there is no reason for postulating with sufficient certainty that monoinfections due to *A. israelii* do occur. In all the other cases, the actinomycete grew together with one or several other aerobic and/or anaerobic bacteria. Predominantly, two or three such companions could be detected, but in a few cases, up to nine concomitant species were encountered. Based upon these observations, it may be concluded that the development of human actinomycosis presupposes two essential conditions:

1) A pathogenic member of the *Actinomycetaceae* such as *A. israelii* must gain access to the interstitial tissue.
2) This organism needs support from a varying collection of concomitant bacteria to establish itself in the tissue and to initiate the typical actinomycotic syndrome.

The specific part in an actinomycotic infection is undoubtedly played by the pathogenic actinomycete species, whereas the various concomitant organisms may be considered nonspecific companions. The synergistic interactions which apparently occur between the actinomycete and its companions await detailed clarification.

In Table V, the aerobic companions of the actinomycoses found in our 3027 cases are listed. In 49% of the samples examined, no aerobic bacteria could be cultured. Staphylococci and streptococci were the most common

TABLE IV. Bacterial Companions of the Actinomycetes in 3027 Cases of Human Actinomycosis

Actinomycetes plus	*Cases*	
	Number	*%*
0 companion	*26*	*0.9*
1 companion	*500*	*16.5*
2 companions	*929*	*30.7*
3 companions	*852*	*28.1*
4 companions	*496*	*16.4*
5 companions	*166*	*5.5*
6 companions	*42*	*1.4*
7 companions	*10*	*0.3*
8 companions	*5*	*0.2*
9 companions	*1*	*0.1*
Total no. of cases	*3027*	*100.0*

aerobic companions, coagulase-negative staphylococci being the most frequently seen. *Enterobacteriaceae*, on the other hand, were only rarely encountered, a fact which may be attributed to the localization of infection in most cases in the cervicofacial region. Of course, in cases of abdominal actinomycosis, *Escherichia coli* was often found. There was no general correlation between the aerobic bacteria present and the clinical features of the disease. However, *Staphylococcus aureus* and β-hemolytic streptococci seemed to be associated with abscesses or phlegmons as observed in cases with an acute early phase.

Table VI summarizes the results of the anaerobic cultures of actinomycotic specimens. In nearly half of the 3027 cases studied, microaerophilic streptococci and/or *Peptococcaceae* were present. Members of the *Bacteroides melaninogenicus* group and fusobacteria were the next most frequent companions, followed by propionibacteria, *Leptotrichia buccalis*, and *Eikenella corrodens*. *B. fragilis, B. thetaiotaomicron*, or other members of the genus *Bacteroides* were relatively rarely encountered. This finding also seems to correlate with the localization of the processes, because these *Bacteroides* species are not regular members of the oral microflora. In general, we may say that the concomitant anaerobic bacteria as well as the facultatively anaerobic actinomycetes involved in human actinomycosis are derived from the same physiological habitat, namely the microflora of human mucous membranes. This corresponds to our statement that human actinomycosis develops as an endogenous mixed infection.

TABLE V. Results of Aerobic Culture in 3027 Cases of Human Actinomycosis

	Cases	
Species/groups	*Number*	*%*
Aerobically sterile	*1482*	*49.0*
Coagulase-negative staphylococci	*844*	*27.9*
Staphylococcus aureus	*394*	*13.0*
α hemolytic streptococci	*258*	*8.5*
β-hemolytic streptococci	*144*	*4.8*
Enterobacteriaceae	*77*	*2.5*
Total number of cases	*3027*	*100.0*

TABLE VI. Results of Anaerobic Culture in 3027 Cases of Human Actinomycosis

Species/groups	*Cases*	
	Number	*%*
Actinobacillus actinomycetem-comitans	*714*	*23.6*
Microaerophilic streptococci/Peptococcaceae	*1402*	*46.3*
Bacteroides melaninogenicus *group*	*1103*	*36.4*
Fusobacteria	*968*	*32.0*
Propionibacteria	*930*	*30.7*
Leptotrichia buccalis	*640*	*21.1*
Eikenella corrodens	*467*	*15.4*
Other Bacteroides *species*	*362*	*12.0*
Total number of cases	*3027*	*100.0*

III. DISCUSSION

A special position may be ascribed to *Actinobacillus actinomycetem-comitans* (Pulverer and Ko, 1970, 1972). The name of this bacterium already indicates its close association with the pathogenic actinomycetes. *A. actinomycetem- comitans* may be found in the oral cavity of healthy adults; the physiological significance of this bacterium is not known and its systematic position also awaits clarification. There are some reports that *A. actinomycetem-comitans* may contribute to the pathogenesis of periodontal disease, but we have some doubts about this. Only rarely is this organism found to be the cause of real monoinfections. Some cases of endocarditis and of brain abscesses have been published. Our experience may be summarized as follows: When *A. actinomycetem-comitans* is present in an infectious process, it is very probable that *A. israelii* or another actinomycete is also there. We have no idea how this obviously existing synergism between *A. actinomycetem-comitans* and *A. israelii* functions. Another observation which should be mentioned here is that the course of actinomycotic infections which contain *A. actinomycetem-comitans* tends to be especially severe and malignant.

The last problem we wish to discuss in this paper is the problem of the so-called sulfur granules or "Drusen". These small and hard granules can be found especially in chronic and advanced actinomycotic lesions. They consist of a conglomerate of filamentous actinomycete microcolonies surrounded by various other bacteria and tissue reaction material and seem to act as a defense-mechanism against the host. These sulfur granules are important not only for diagnosis since this type of granules is only found in actinomycotic processes but also because they can cause chemotherapeutic failures since they prevent the administered antibiotics from penetrating to the centers of the actinomycetes. Problems of diagnosis, incidence, and treatment of human actinomycosis will be discussed in our other paper (elsewhere in this volume).

IV. SUMMARY

In our laboratory during the last 40 years, we have diagnosed 3027 cases of human actinomycosis. In most of these cases, the infection was localized in the cervicofacial region. [The data of our material are carefully analyzed.] The clinical symptomatology, problems of pathogenesis and etiology as well as age and sex correlations in patients are discussed. The results of aerobic and anaerobic bacteriological investigations in these 3027 cases of human actinomycosis are demonstrated.

REFERENCES

Bollinger, O., *Dtsch. Z. Tiermed. 3:*334 (1877).
Holm, P., *Acta Pathol. Microbiol. Scand., 27:*736 (1950).
Holm, P., *Acta Pathol. Microbiol. Scand., 28:*391 (1951).
Israel, J., *Arch. Pathol. Anat. Physiol. Klin. Med. 74:*15 (1878).
Lentze, F., *Muench. Med. Wochenschr. 47:*1826 (1938).
Lentze, F., *Dtsch. Zahnaerztl. Z. 3:*913 (1948).
Lentze, F., *In* "Die Infektionskrankheiten des Menschen und ihre Erreger" (A. Grumbach, and O. Bonin, eds.), p. 954. Georg Thieme-Verlag, Stuttgart (1969).
Pulverer, G., *Postepy Hig. Med. Dosw. 28:*253 (1974).
Pulverer, G., and Ko, H. L., *Appl. Microbiol. 20:*693 (1970).
Pulverer, G., and Ko, H. L., *Appl. Microbiol. 23:*207 (1972).
Pulverer, G., and Schaal, K. P., *in* "Nocardia and Streptomyces" (M. Mordarski, K. Kurylowicz, J. Jeljaszewicz, eds.), p. 417. Gustav Fischer-Verlag, Stuttgart (1978).
Schaal, K. P., *Dtsch. Arztebl. 31:*7997 (1979).

BIOCHEMISTRY OF THE *STREPTOMYCES* SPORE SHEATH[1]

Richard A. Smucker

Chesapeake Biological Laboratory
Center for Environmental and Estuarine Studies
University of Maryland
Solomons, Maryland, U. S. A.

I. INTRODUCTION

Streptomyces aerial mycelia develop a surface fibrous layer just prior to arthrospore development. During hyphal wall involution, the *S. viridochromogenes* fibrous layer develops spines (Rancourt and Lechevalier, 1964), whereas in *S. coelicolor* A 3(2), the fibrous layer remains smooth (Wildermuth *et al.*, 1971). The fibrous layer has been called the "fibrous sheath". These variations in spore surface patterns have been used in taxonomic descriptions of *Streptomyces* (Tresner *et al.*, 1961). The ornamentations have been categorized as smooth, warty, spiny, or hairy.

Depending upon the species studied and the methods of spore preparations, reports of retention of the sheath by free mature spores has varied. In all cases, at least some portion of the sheath has been reported to remain with the mature spore (Rancourt and Lechevalier, 1964; Williams and Sharples, 1970).

Bradley and Ritzi (1968) suggested, on the basis of whole cell carbon replication, that the *S. venezuelae* smooth spore surface fibers were removed by either xylene, benzene, or ethanol. Others have reported that surface ornamentations and fibers are resistant to organic solvent treatments (Smucker and Pfister, 1978; Wildermuth, 1972). The fibrous component of *S. coelicolor* A3(2) has been isolated and shown to have the characteristics of

[1] *This paper is contribution No. 1347, Center for Environmental and Estuarine Studies of the University of Maryland.*

ISBN 0-12-528620-1

chitin fibrils (Smucker and Pfister, 1978). The fibrous component (rodlet mosaic of fibers) in *S. coelicolor* A3(2) smooth spores is not the outermost layer of the spore. What classically has been designated as *Streptomyces* sheath, in fact, has three components, with the fibrous component being the innermost sheath layer. Matselyukh (1978) and Smucker and Simon (1981) independently demonstrated that hairy-spored *Streptomyces (olivaceus* and *bambergiensis*, respectively) also have complex sheath ornaments with at least two distinguishable layers.

The only sheath component that has been isolated and characterized is the rodlet mosaic fibers of *S. coelicolor* (Smucker and Pfister, 1978). This work establishes new information regarding not only the chemistry of the sheath, particularly of the rodlet mosaic fiber portion of the *S. coelicolor* A3(2) sheath, but also the presence of chitin in *S. bambergiensis* (hairy spores).

II. HYDROPHOBIC SPORE BEHAVIOR

Aerial *Streptomyces* spores classically have been harvested by rolling glass beads over the agar lawn of mature aerial spores. Spore chains and single spores harvested by glass beads or rolling drops of water often are collected by centrifugation following treatment with a detergent or solvent. This hydrophobicity has been attributed to the *Streptomyces* spore surfaces, *i. e.*, the sheath. Ruddick and Williams (1972) compared the hydrophobic, sheath-possessing *S. viridochromogenes* with the hydrophilic, non-sheathed *Micromonospora*. Smucker and Pfister (1978) showed that *S. coelicolor* A3(2) rodlet mosaic fibers are resistant to organic solvents and remain intact after conventional epoxy embedding procedures which include the use of propylene oxide (Luft, 1961).

Cultures used in this work were grown on glycerol-asparagine agar (Shirling and Gottlieb, 1966). Liquid nitrogen (Swoager, 1972) was used for the long-term storage of cultures.

When *S. coelicolor* A3(2) spores are harvested from agar lawns with distilled water, dried, and stained with Sudan IV (Erickson, 1947), red patches were visible on an between groups of spores. When similar spore samples were homogenized to break up the spore clumps, subsequent Sudan IV staining failed to show the control staining patterns. These results are similar to those reported by Erickson (1947) for spore chains separated mechanically. Manual agitation of the harvested spores in 2% aqueous phosphotungstic acid (pH 2) for a few minutes separated spore chains with a concomitant loss of the hydrophobicity of the spores. Individual spores did not show hydrophobic behavior. Some loosely associated component of the colony appeared to be responsible for *Streptomyces* spore-aggregate properties. This hydrophobic component has not been isolated and characterized.

III. RODLET MOSAIC

All of the above-mentioned procedures which causes loss of spore hydrophobicity leave the *S. coelicolor* A3(2) fiber mosaic essentially intact (Smucker and Pfister, 1978). Organic solvents, however, induce rearrangements of the fiber interweaving patterns (Fig. 1). These rearrangements suggest that the rodlet mosaic is made of criss-crossed individual fibers. For details of preparation and analysis, see Smucker and Pfister (1978).

Isolated rodlet mosaic fibers of *S. coelicolor* have been characterized as chitin by several techniques, including infrared spectroscopy, thin layer chromatography (TLC) analysis of hydrolysis products, and enzymatic hydrolysis by chitinase (Smucker and Pfister, 1978). The fibers were not isolated from primary mycelia from submerged culture.

If the fibers in *S. coelicolor* rodlet mosaic were chitin, then the fibers in S. *bambergiensis* should also be chitin. To test this hypothesis, *S. bambergiensis* was grown on glycerol-asparagine agar (Shirling and Gottlieb, 1966) overlaid with dialysis tubing. Aerial growth was removed from the dialysis membrane overlay by gentle agitation and scraping of the membrane in 0.05% Triton X-100.

Suspended spores, spore clusters, and fragments of aerial hyphae were homogenized with a tight-fitting Teflon and glass homogenizer. The homogenate was washed with distilled water (dH_2O) and was centrifuged. This process was repeated three times. Fibrils were harvested by using the following extraction protocol: 2N NaOH (1h; 100 °C); wash with dH_2O at room temperature; 2N HCl (1h; 100 °C); dH_2O wash (3 times); acetone wash (3 times); dH_2O wash (3 times); 1% aqueous $KMnO_4$ (1h; 100 °C); ethanol wash; dH_2O wash; and acetone to dry the product. Vegetative growth of *Bacillus megaterium* and of *S. bambergiensis* primary mycelia was used as control for the extraction procedures. No residue from either control was detected after the extraction sequence. The light tan residues from blue crab chitin and from *S. bambergiensis* sporulating growth were subjected to X-ray diffraction analysis. Powder diagrams were recorded in a 114.6 mm circular camera using Cu-KαX-rays. Various exposure times and sample preparations gave slightly different results for the same sample residue. Examples of X-ray powder diagrams are shown in Figure 2. The X-ray data along with the residue extraction pattern (Hackman, 1982) support the hypothesis that *S. bambergiensis* spores contain chitin.

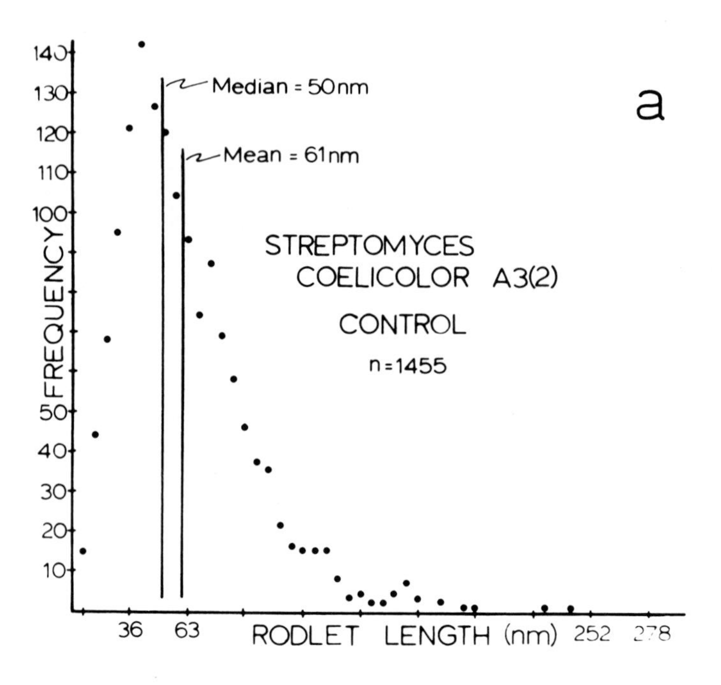

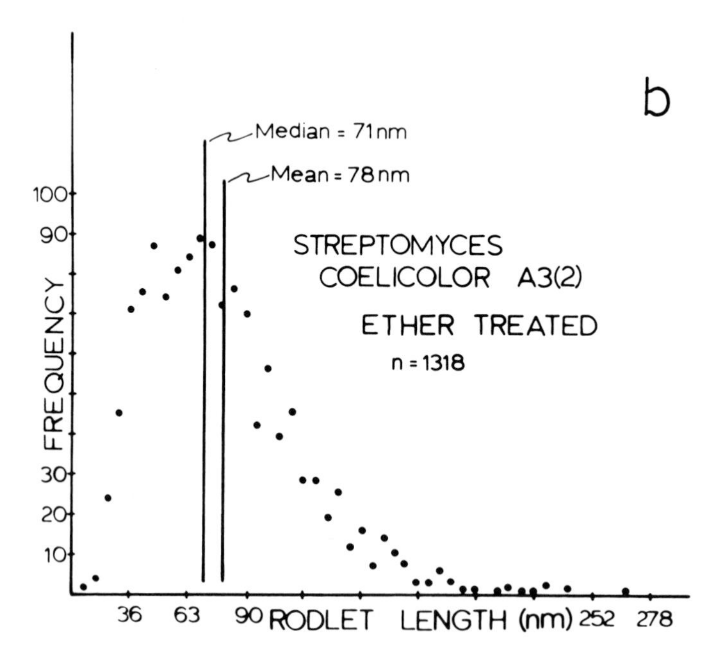

FIGURE 1. Frequency distributions of uninterrupted rodlet fiber lengths. Diethyl ether (b) treatment disrupted the rodlet mosaic pattern of controls (a), thereby exposing longer fiber segments (n = # of measurements).

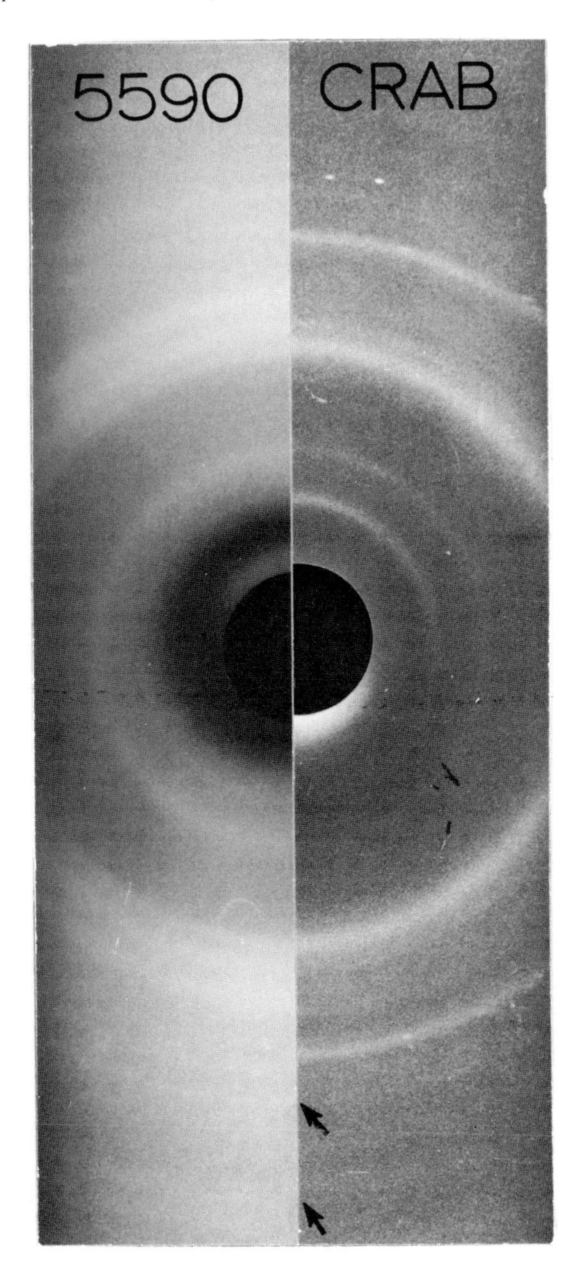

FIGURE 2. X-diffraction powder diagrams of S. bambergiensis *residue and of similarly treated blue crab chitin. Arrows indicate coincidence of faintly visible lines.*

IV. MATERIALS ASSOCIATED WITH THE RODLET MOSAIC FIBRILS

Rodlet fibrils isolated from *S. coelicolor* aerial growth had associated purple pigment (possibly antibiotic) which was removed by treatment with egg white lysozyme. In this case, the fibrils were isolated by treating spores for several hours with lysozyme. After the lysozyme was washed from cells, the cells were disrupted by using the liquid nitrogen cryoimpacting (LNCI) method (Smucker and Pfister, 1978). The resulting material was homogenized several times with a tight-fitting Teflon homogenizer and then treated with trypsin for 12 h. The sample was clarified by centrifugation at 38,000 X *g* for 30 min and then was dialyzed using an Amicon XM300 ultrafilter. The fibrillar product obtained was light purple. Lysozyme treatment released the pigment which was then removed by XM300 ultrafiltration.

S. coelicolor rodlet mosaic fibrils isolated by acetic acid refluxing and Sevag deproteinization (Sevag, 1934) contained 15 to 20% (w/w) glucose (Smucker and Pfister, 1978). After the samples were hydrolyzed with 0.5M H_2SO_4, reduced, and methylated, the nature of the bound glucose was determined by gas chromatography/mass spectroscopy (GC/MS). The data fit a model in which every third glucose residue is a branch point. A trace mannose component was also detected. The occurrence of a glucan associated with the *Streptomyces* chitin is consistent with chitin-glucan complexes commonly found in fungal chitin (Holan *et al.*, 1981).

V. CONCLUSION

Current information supports a *Streptomyces* spore model with a loosely associated, unidentified hydrophobic component. Inside this tenuously associated lipid component is the spore sheath. The outer two sheath layers which were retained after lipid solvent treatment were not characterized chemically. The innermost sheath layer, the rodlet mosaic, was identified as chitin. Tightly associated with the chitin was a highly branched glucan. Sheath components, the chitin fibrils of the rodlet mosaic, and other associated sheath structures were not present in primary mycelia. The production of chitin fiber during sporulation appeared to be physiologically and phylogenetically important in the *Streptomyces*. Production of both peptidoglycan and chitin during the life cycle apparently occurs simultaneously during spore cross-wall development. Formation of these two polymers using the same UDP-N-acetylglucosamine precursor is indicative of the unique position that *Streptomyces* have in procaryotic phylogeny. None of the eubacteria are known to produce chitin, but many fungi do produce chitin in their walls.

ACKNOWLEDGMENTS

Dr. George Gilbert, Chemistry Department, Dennison University, Granville, OH., graciously assisted in the X-ray diffraction analysis. I also wish to thank Dr. Gunther Rodemeyer, Department of Chemistry, Ohio State University, Columbus, OH., for performing the GC/MS analysis.

REFERENCES

Bradley, S. G., and Ritzi, D., *J. Bacteriol. 95:*2358 (1968).
Erickson, D., *J. Gen. Microbiol. 1:*39 (1947).
Hackman, R. H., "Proceedings of the Second International Conference on Chitin and Chitosan" (S. Hirano and S. Tokura, eds.), p. 266. Sapporo, Japan (1982).
Holan, Z., Pokorny, V., Beran, K., Gemperle, A., Tuzar, Z., and Baldrian, J., *Arch. Microbiol. 130:*312 (1981).
Luft, J. H., *J. Biophys. Biochem. Cytol. 9:*409 (1961).
Matselyukh, B. P., *in* "*Nocardia* and *Streptomyces*" (M. Mordarski, W. Kurylowicz, and J. Jeljaszewicz, eds.), p. 440. Gustav Fischer Verlag, New York (1978).
Rancourt, M. W., Lechevalier, H. A., *Can. J. Microbiol. 10:*311 (1964).
Ruddick, S. M. and Williams, S. T., *Soil Biol. Biochem. 4:*93 (1972).
Sevag, M. G., *Biochem. Z. 273:*419 (1934).
Shirling, E. B. and Gottlieb, D., *Int. J. Syst. Bacteriol. 16:*313 (1966).
Smucker, R. A. and Pfister, R. M., *Can. J. Microbiol. 24:* 397 (1978).
Smucker, R. A. and Simon, S. L., *in* "Sporulation and Germination" (H. S. Levinson, A. L. Soneshein, and D. J. Tipper, eds.), p. 317. American Society for Microbiology, Washington (1981).
Swoager, W. C., *Am. Lab. 492:*45 (1972).
Tresner, H. D., Davies, M. C., and Backus, E. J., *J. Bacteriol. 81:*70 (1961).
Wildermuth, H., *Arch. Mikrobiol. 81:*321 (1972).
Wildermuth, H., Wehrli, E., and Horne, R. W., *J. Ultrastruct. Res. 35:*168 (1971).
Williams, S. T. and Sharples, G. P., *Microbios 5:*17 (1970).

EARLY BIOCHEMICAL EVENTS DURING THE GERMINATION OF *STREPTOMYCES* SPORES[1]

C. Hardisson
J. A. Guijarro
J. E. Suárez
J. A. Salas

Departamento de Microbiología
Universidad de Oviedo
Oviedo, Spain

I. INTRODUCTION

Germination of *Streptomyces* spores, considered as the conversion of a resting and dormant cell into a vegetative and metabolically acting organism (Cross and Attwell, 1975), can be divided into three sequential morphological and biochemical stages (Hardisson *et al.*, 1978). During the first stage, the dormant spores became progressively darker in a process requiring both energy and exogenous divalent cations. The second stage is characterized by the swelling of the spores and a notable increase in their metabolic activities and depends upon the supply of an exogenous carbon source. The emergence of germ tubes is the last event and requires both exogenous carbon and nitrogen sources (Table I). For a better understanding of the germination process, it is necessary to establish which are the primary events occurring in the dormant spore to undergo the darkening process. In our laboratory, we are trying to clarify some of these events and the mechanism which induces the appearance of these changes. One of the early biochemical events occurring during spore germination of *Streptomyces antibioticus* is the onset of protein and RNA syntheses. These processes are

[1] *These investigations were supported in part by a grant from Comisión Asesora para el Desarrollo de la Investigación Científica y Técnica, España.*

ISBN 0-12-528620-1

TABLE I. Sequence of Morphological Stages during Germination of Streptomyces antibioticus *Spores in a Minimal Synthetic Medium*

	Incubation time (h)					
	0	1	2	3	4	5
Dormant spores	*100*	*40-60*	*20-40*	*15-25*	*15-25*	*15-25*
Dark spores	—	*35-50*	*60-80*	*55-70*	*15-25*	*5-10*
Swollen spores	—	—	—	*10-20*	*40-60*	*20-35*
Germ tube emission [a]	—	—	—	—	*5-15*	*40-60*

[a] *Germ tube at least as long as the diameter of the swollen spore.*

dependent only on the wetting of the spores and take place even during incubation of the spores in distilled water. In this paper, we report results related to characteristics of the metabolism of RNA and the proteins synthesized during germination of *S. antibioticus* spores and include data on the relative proportion of unstable RNA's, their half-lives, and the degradation pattern of proteins codified by these messengers. Previously, we reported the existence of a functional mRNA fraction in the dormant spores (Hardisson *et al.*, 1980), which is translated early during the germination process. In this report, we also characterize the proteins codified by this stable fraction and the degradation pattern of these polypeptides.

II. MATERIAL AND METHODS

A. Microorganism and Culture Conditions

Streptomyces antibioticus ATCC 11891 spores were obtained, as previously described (Hardisson *et al.*, 1978), after growth in GAE (glucose-asparagine-yeast extract) solid medium for seven to nine days at 28°C. For germination, 20 ml of freshly harvested spore suspensions ($A_{580} = 0.3$; 10^8 spores/ml) were incubated in 100 ml flasks in a buffered minimal synthetic medium (MSM) consisting of the following (g/l): glucose, 10.0; $(NH_4)_2SO_4$, 2.0; asparagine, 2.0; K_2HPO_4, 6.28; KH_2PO_4, 0.81; $MgSO_4 \cdot 7H_2O$, 0.5; $FeSO_4 \cdot 7H_2O$, 0.01. Incubation was carried out at 35 °C in a water bath with shaking at 200 r.p.m. Spore germination was determined by phase-contrast microscopy and by following the changes in absorbance at 580 nm.

B. Measurements of RNA Degradation

Previous experiments showed that the antibiotic rifampicin was a quick and effective inhibitor or RNA synthesis in *S. antibioticus*, blocking the incorporation of radioactive uridine into RNA after 90 seconds of incubation in the MSM (Guijarro *et al.*, submitted for publication), Accordingly, we used this antibiotic to inhibit RNA synthesis in order to study the pattern of degradation of the different RNA species synthesized during germination. Two different approaches were employed.

1. Radioactive Decay of Labeled RNA after Inhibition of Its Synthesis. RNAs synthesized during germination were labeled by incubation of the spores at different times of germination with 1 μCi/ml (37 KBq/ml) of [^{3}H]-uridine (27 Ci/mmol, 0.97 TBq/mmol) for 2 or 5 min, depending on the stage of germination. After the labeling period, rifampicin (15 μg/ml) was added to the cultures to stop RNA synthesis and, at different intervals, samples (0.5 ml) were removed and placed into tubes containing 1.5 ml of ice-cold 20% (w/v) trichloroacetic acid (TCA) (15% final concentration). After 30 min in an ice bath, the samples were filtered through Whatman GF/A glass-fibre filters (2.5 cm diameter) and the precipitates were washed with ice-cold distilled water. The filters were then dried and placed in a toluene-based scintillation fluid; the radioactivity was measured by an Isocap 300 Nuclear Chicago liquid scintillation spectrophotometer.

2. Residual Protein Synthesis after Inhibition of RNA Synthesis. At different times of germination, 1 μCi/ml (37 KBq/ml) of [^{3}H]-leucine (50 Ci/mmol, 1.8 TBq/mmol) and rifampicin (15 μg/ml) were simultaneously added to the cultures. To ensure an adequate supply of the amino acid during the experiment, unlabeled leucine was also added at a concentration of 0.2 μg/ml at 1 and 2 h of incubation or of 1 μg/ml at 3, 4, and 5 h of incubation. From this moment, samples (0.5 ml) were withdrawn and added to tubes containing 1.5 ml of ice-cold TCA. After 30 min at 4 °C, the samples were heated at 90 °C for 15 min, and the radioactivity was counted after treatment of the samples as described above.

C. Measurements of Protein Degradation

Proteins synthesized during spore germination were labeled with [^{3}H]-leucine both for short and long time periods. For short time periods, 1 μCi/ml (37 KBq/ml) of radioactive leucine was added to the spore suspensions at different stages of germination (dormant spores: 0 h; dark spores: 1 and 2 h; swollen spores: 3 and 4 h; germ tube emission: 5 h). After 5 to 15 min (depending on the stage of germination) unlabeled leucine was added to the germination medium (10 mM final concentration) to stop further radioactive incorporation and to decrease the intracellular specific activity of the amino acid. This consequently blocked the reincorporation of labeled leucine released from degraded proteins into newly synthesized polypeptides. Upon incubation, samples were taken as described above for DNA degradation, including heating at 90 °C for 15 min before filtering. When long time periods of labeling were used, spores were incubated in the presence of radioactive leucine for intervals ranging from 1 to 4 h and then, after the addition of unlabeled leucine, the processing of the samples was similar to that described for short intervals of labeling. Proteins coded by the stable mRNA fraction of dormant spores were labeled for 15 min, but rifampicin (15 μg/ml) was included during the labeling period. Spores were then centrifuged at 12,000 X *g* for 5 min, washed twice in distilled water, and finally suspended in distilled water or in the MSM, both supplemented with 10 mM unlabeled leucine. Incubation then continued, and samples were removed at different times and treated as described above.

D. Electrophoresis and Fluorography

Pulse-labeling of the proteins coded by the stable mRNA was done with L-[^{35}S]-methionine (1 μCi/ml; 37 KBq/ml; specific activity 535 mCi/mmol, 19.8 KBq/mmol) for 15 min in distilled water or in MSM in the presence or absence of rifampicin (15 μg/ml). After the labeling period, the samples were cooled in an ice water bath and cell-free extracts for polyacrylamide gel electrophoresis were obtained as follows: spores were suspended in 0.06 M Tris-HCl buffer (pH 6.8) containing 2 mM EDTA and broken with glass beads (0.10-0.11 mm diameter) at 4 °C for 20 min in a Vibrogen homogenizer with continuous ice-water refrigeration (Hetofrig refrigeration unit). After this treatment, spore breakage was greater than 95%. The, 2% SDS and 6% 2-mercaptoethanol (w/v; final concentration) were added and the samples boiled for 2 min. The glass beads were removed by centrifugation at 2,000 X *g* for 2 min, and the extract was again centrifuged at 12,000

TABLE II. Some Features of Streptomyces antibioticus *Spores*

Endogenous respiration [a]	QO_2: *10.69 ± 0.35 μl O_2/h/mg dry weight*
Adenine nucleotide levels [b]	
AMP	*0.20 - 0.30 nmoles/mg dry weight*
ADP	*0.10 - 0.15 nmoles/mg dry weight*
ATP	*0.20 - 0.25 nmoles/mg dry weight*
Adenylate energy charge [b]	*0.47*
Sugar uptake activity [a]	*Glucose, constitutive; galactose and fructose, inducible*
Amino acid pool [b]	*20 Amino acids*
Nucleic acid precursors pool [b]	*5 Nucleosides*
Trehalose content [b]	*About 10% of the dry weight*
Dipicolinic acid	*Absent*
Heat resistance [a]	*Slightly superior to vegetative cells*

[a] *Adapted fom Salas, J.A. and Hardisson, C.,* J. Gen. Microbiol. 125:*25 (1981) by permission of the Society for General Microbiology.*

[b] *Unpublished results*

X *g* for 15 min in an Eppendorf microfuge. The supernatant was then dialyzed overnight in two changes of the same buffer. Electrophoresis was carried out in a 12% slab gel with a 5% stacking gel, following the procedure described by Laemmli (1970), with a 45 mA current and a constant voltage. Gels were fixed in a 20% TCA (w/v) and stained and destained for proteins by the method of Fairbanks *et al.*, (1971). For fluorography, the method of Chamberlain (1979) was used. Gels were immersed in an equimolar solutions of 2 M sodium salicylate (pH 6.0 to 6.5) for 30 min and then dried under vacuum in a Whatman 3MM filter paper. Kodak X-Omat S film was exposed to the gel for seven days at -70 °C.

III. RESULTS

A. Some Features of Streptomyces antibioticus *Dormant Spores*

Table II shows some important biochemical and physiological features of dormant spores *S. antibioticus*. In contrast to bacterial endospores, *Streptomyces* spores show detectable endogenous respiration ($QO_2 = 10.69 + 0.35$ $\mu l O_2$/h/mg dry weight). Adenine nucleotide levels are much higher than those found in bacterial endospores, with a low adenylate energy charge compared to the values described for exponentially growing bacteria. Sugar uptake systems for some hexoses seem to differ depending on the sugar: glucose uptake was constitutive in the dormant spore, whereas galactose and fructose uptake systems were inducible during germination after a 2 hour lag period. This pattern remained whether sporulation took place in the presence of galactose or fructose as the sole carbon and energy source (Salas and Hardisson, 1981). Spores have a complete amino acid pool and five nucleosides, precursors of the nucleic acids. Large quantities of trehalose and calcium are present, *i. e.*, about 10% and 2% of the dry weight, respectively. Some of these typical features of *S. antibioticus* spores, such as the existence of endogenous respiration and ATP levels, have also been reported for *S. viridochromogenes* spores (Ensign, 1978).

B. RNA Degradation during Spore Germination

To study the degradation patterns and half-lives of the different RNA species synthesized during spore germination, it was necessary to develop a methodology that would switch off RNA synthesis at different times of germination. We were able to do this with the antibiotic rifampicin, a selective inhibitor of the RNA polymerase. Spores were prelabeled briefly with [^{3}H]-uridine, as stated in Material and Methods, after which rifampicin was added to the germinating cultures to stop further incorporation of radioactivity into RNA. Under these conditions, the addition of antibiotic to spores germinating in the MSM resulted in an initial transient increase of radioactivity incorporated into RNA (Fig. 1A to F), probably because rifampicin does not inhibit the biosynthesis of RNA once it has started (Hartmann *et al.*, 1967). When the experiment was done with spores after 15 min of ger-

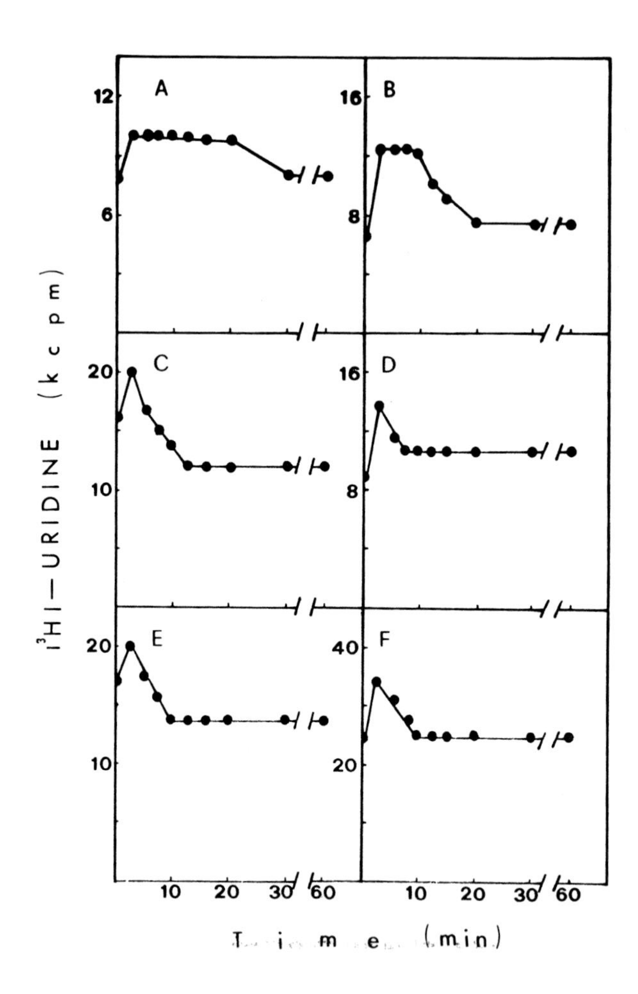

FIGURE 1. Radioactive decay of pulse-labeled RNA synthesized during germination of Streptomyces antibioticus *spores. Spores were germinated in MSM and, at different times of incubation, pulse-labeled with [^{3}H]-uridine (1 μCi/ ml; 37 KBq/ml) for 5 min (15 min of germination) or 2 min (all the other germination times). Rifampicin (15 μg/ml) was then added, and samples (0.5 ml) were removed at intervals and treated as described in Material and Methods. Germination time: (A) 15 min; (B) 1 h; (C) 2 h; (D) 3 h; (E) 4 h; (F) 5 h.*
(Reproduced from Guijarro, J. A. et al., FEMS Microbiol. Lett. 14:*205 (1982) by copyright permission of Elsevier Biomedical Press.)*

mination (Fig. 1A), after this increase, the incorporation leveled off, and then decreased approximately 20%. The remaining 80% indicates the existence of a very stable RNA fraction synthesized during the first minute of germination. Spores labeled after one hour of germination (Fig. 1B) showed a similar pattern, but the plateau lasted a shorter period and the percentage of the RNA degradation was higher (about 40%). This result suggests the existence of a higher fraction of unstable mRNA in spores germinated for one hour. Labeling of RNA synthesized in later stages of germination (2 to 5 h) did not show this initial stable RNA fraction represented by the plateau, and the decrease in labeled RNA ranged from 22 to 40% of the radioactive uridine incorporated (Fig. 1C to F).

An interesting factor in these experiments is the half-life of the mRNA's synthesized during spore germination, *i. e.*, the time necessary for the radioactive counts to reach half of their total decay. Spores germinated for 15 min showed the longest half-life (about 22.5 min), with this value decreasing as the germination process was more advanced (12 min at 1 h and 6 min at 2 h of germination) (Table III).

Another experimental approach was also used to evaluate the half-life of the mRNA's. After inhibition of RNA synthesis by the addition of rifampicin, residual protein synthesis due to the remaining preformed mRNA's was measured by radioactive leucine incorporation. In these experiments, [^{3}H]-leucine and rifampicin were added simultaneously. The results obtained during germination in the MSM are shown in Figure 2 (A to F). In all cases, there was a period of rapid leucine incorporation into protein, followed by stabilization of the incorporated counts, indicating that degradation of the mRNA's took place. The initial period became shorter as the germination process advanced; 60 min was required for spores ger-

TABLE III. Half-life of RNA's Synthesized during Germination of Streptomyces antibioticus Spores

Time of incubation (h)	*Fraction of unstable RNA (%)*	*Half-life of unstable [a] mRNA (min)*	*Half-life of mRNA[b] (%)*
0.25	20	22.5	20
1	40	12	11
2	39	6	7.5
3	22	4	5
4	32	5	5
5	27	6	6

[a] *The time necessary to reach half the decay in [^{3}H]-uridine incorporation into TCA insoluble material after the spores had been pulse-labeled and then rifampicin (15 μg/ml) was added.*
[b] *The time necessary to reach half the total incorporation of [^{3}H]-leucine into TCA insoluble material in the presence of rifampicin (15 μg/ml).*
Reproduced from Guijarro, J. A. et al., FEMS Microbiol. Lett. 14:*205 (1982) by copyright permission of Elsevier Biomedical Press.*

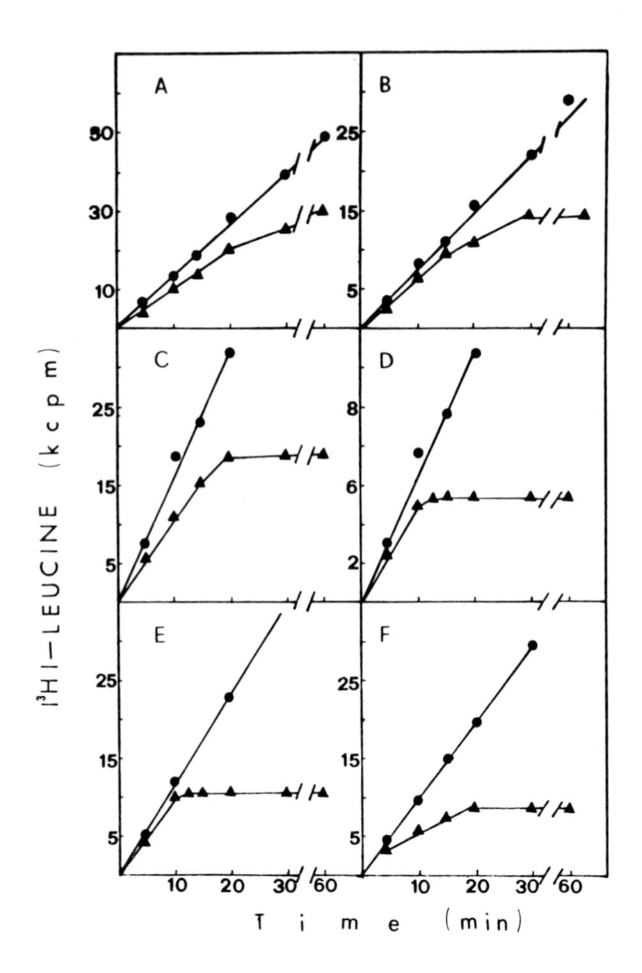

FIGURE 2. Residual synthesis of protein during germination of Streptomyces antibioticus *spores after inhibition of RNA synthesis by rifampicin. Spores were germinated in the MSM and, after different intervals, [^{3}H]-leucine (1 μCi/ml, 37 KBq/ml) (●) or [^{3}H]-leucine plus rifampicin (15 μg/ml) (▲) were added. Samples (0.5 ml) were removed and treated as described in Material and Methods after (A) 15 min; (B) 1 h; (C) 2 h; (D) 3 h; (E) 4 h; (F) 5 h of germination.*
(Reproduced from Guijarro, J. A. et al., FEMS Microbiol. Lett. 14:*205 (1982) by copyright permission of Elsevier Biomedical press.)*

minated for 15 min but only 10 min for spores germinated for 3 h. From the duration of protein synthesis, the half-life of the mRNA's could also be calculated. As seen in Table III, the values obtained by this method were close to those obtained by measuring the radioactive decay of [^{3}H]-uridine labeled RNA.

C. Protein Degradation during Spore Germination

Protein degradation during spore germination of *S. antibioticus* showed a typical pattern when labeled for a short time. Within the first hour after the labeling period of dormant spores, *i. e.*, during the darkening process, about 30% of the protein into which [^{3}H]-leucine had been incorporated was degraded (Fig. 3A). This degradation began within 5 min of the addition of unlabeled leucine within a lag phase and proceeded continuously during the next 30 min (Fig. 3B). Later, for the next 6 h of incubation, protein degradation took place at a lower and constant rate (4.5 to 5.5%/h). When protein was labeled after 1 or 2 h of germination, a similar pattern of degradation was found. However, the percentage of initial and rapid protein degradation decreased to about 25%/h and 18%/h for labeling at 1 and 2 h, respectively (Fig. 3A). Then, the degradation rate decreased to values close to those found with dormant spores (3.3 to 6.0%/h). In contrast, the degradation pattern of proteins synthesized after 3 or 4 h (swelling process) or after 5 h (germ tube emission) was constant at the beginning, a rate ranging from 4.9%/h to 5.5%/h, without the great and rapid initial degradation observed in the experiments carried out during the first 2 h of germination.

The effect of different protease inhibitors on the protein degradation of spores labeled at the onset of incubation was assayed (Fig. 4). None completely blocked the initial protein degradation. *p*-Chloromercuribenzoic acid (pCMB) and *o*-phenanthroline caused about 60% reduction in the protein degradation. Phenylmethylsulfonylfluoride (PMSF), although it did not affect the degradation rate in the first 30 min of incubation, clearly blocked further degradation. Ethylenediaminetetraacetic acid (EDTA) had no effect on the degradation pattern.

Because labeling for short time periods preferentially labels proteins with the highest turnover rate, *i. e.* more unstable proteins, we also decided to study the protein degradation rates during spore germination after longer periods of labeling (1 to 4 h) than those described above for brief labeling (5 to 15 min). The results obtained (data not shown) clearly indicated that the length of the labeling period had no influence on the rate of protein degradation during the germination process. Obviously, the length of longer labeling periods masked the initial and rapid degradation, otherwise observed in the first hour after labeling.

We previously reported the existence of a stable mRNA fraction in the dormant spore of *S. antibioticus* (Hardisson *et al.*, 1980). The degradation pattern of the translation products of this stable fraction was also studied. It must be emphasized that this mRNA fraction was translated upon incubation of dormant spores either in the MSM or in distilled water. Proteins were labeled by incubation of the spores in distilled water or in MSM in the presence of rifampicin (15 μg/ml) and a radioactive precursor, [^{3}H]-leucine. After removal of the rifampicin by centrifugation and

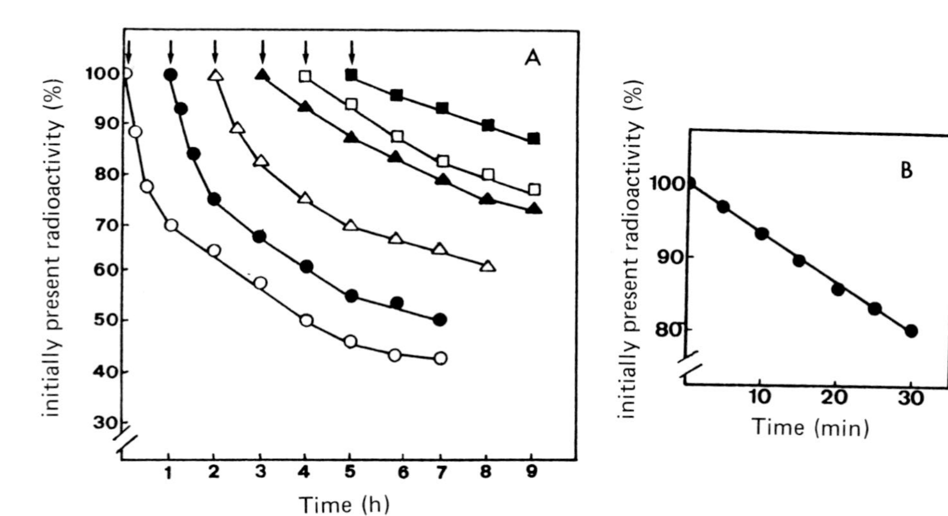

FIGURE 3. Pattern of protein degradation during germination of Streptomyces antibioticus *spores. Spores were incubated in the MSM and, at different time intervals, labeled for 15 min (O h) or 5 min (1, 2, 3, 4 or 5 h) with [^{3}H]-leucine (1 μCi/ml, 37 KBq/ml). Then, unlabeled leucine (10 mM final concentration) was added and samples (0.5 ml) removed at intervals and treated as described in Material and Methods. 1A. Degradation of proteins synthesized at different times of germination. Pulse-labelling at (○) 0 hours, 10,000 c.p.m.; (●) 1 h, 25,000 c.p.m.; (△) 2 h, 38,000 c.p.m.; (▲)* 3 h, 42,000 c.p.m.; *(□) 4 h, 54,000 c.p.m.; (■) 5 h, 60,000 c.p.m. The c.p.m. data indicate the radioactive counts of leucine incorporated into protein. 1B. Initial and rapid protein degradation when the labelling period was done during the first 15 min of germination.*

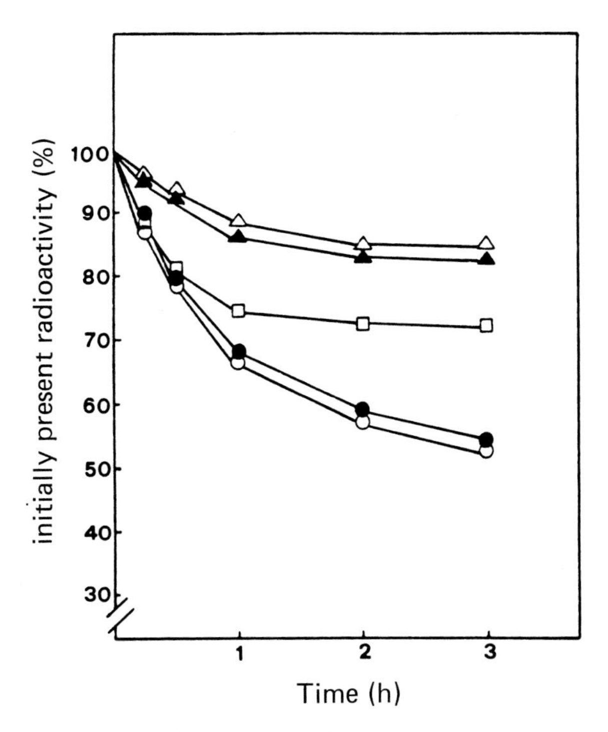

FIGURE 4. Effect of protease inhibitors on the initial degradation of proteins synthesized when germination of Streptomyces antibioticus *spores began. Spores were incubated in the MSM and labeled for 15 min with [^{3}H]-leucine (1 μCi/ml, 37 KBq/ml). Afterward, unlabeled leucine (10 mM final concentration) was simultaneously added with individual protease inhibitors. From this moment, samples were removed and treated as described in Material and Methods. (○) Control; (●) 1 mM EDTA; (△) 1 mM pCMB; (▲) 1 m M o-phenanthroline; (□) 1mM PMSF.*
*(Reproduced from Guijarro, J.*A. et al., Can. J. Microbiol. 29 *(May 1983) by copyright permission of the National Research Council of Canada.*

washings, spores were suspended in distilled water or in MSM containing 10 mM unlabeled leucine, and the fate of the labeled proteins was followed during incubation. Under these conditions, although the mRNA fraction was translated even upon incubation in distilled water, no significant degradation of its polypeptide products occurred, unless spores were incubated in the MSM (Fig 5). During incubation of the spores in distilled water, protein degradation was extremely slow compared to the much greater degradation rates during incubation in the MSM. The degradation pattern of proteins synthesized by dormant spores was unaffected by the presence or absence of rifampicin (Fig. 5). However, in the MSM, more proteins were labeled in its absence (13,000 c.p.m.) than in its presence (8,000 c.p.m.). Consequently, the great similarity found in the two degradation patterns suggests that the whole protein population must be degraded at very similar rates. Alter-

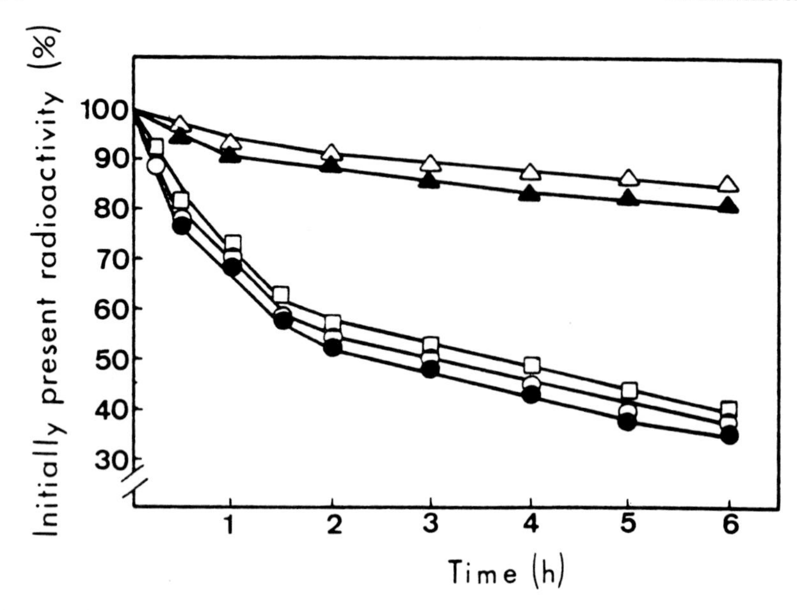

FIGURE 5. Degradation of proteins coded by the stable mRNA fraction of dormant spores of Streptomyces antibioticus. *Spores were labeled for 15 min with [^{3}H]-leucine (1 μCi/ml, 37 KBq/ml) in the MSM (○, ●) or in distilled water (△, ▲, □) with (closed symbols) and without (open symbols) rifampicin (15 μg/ml). Thereafter, the spores were centrifuged and washed with distilled water and finally suspended in the MSM or in distilled water. Incubation continued as samples were withdrawn at intervals and treated as described in Material and Methods. (○) Labeling and germination in the MSM; (●) labeling in MSM plus rifampicin and germination in MSM; (△) labeling and incubation in distilled water; (▲) labeling in distilled water plus rifampicin and incubation in distilled water; (□) labeling in distilled water plus rifampicin and germinaton in MSM.*
*(Reproduced from Guijarro, J.*A. et al., Can. J. Microbiol. 29 *(May 1983) by copyright permission of the National Research Council of Canada.*

natively, the newly synthesized mRNA's (in the absence of rifampicin) might represent only a small fraction of that population; therefore, the products they encode would not significantly change the overall degradation rate.

D. Characterization of the Products of the Stable mRNA Fraction of the Dormant Spore

By using polyacrylamide gel electrophoresis and fluorography, it was possible to characterize the proteins that this stable messenger fraction encoded. Evaluation of the exact number of protein bands in the fluorograms was difficult, but about 20 bands were detected with molecular weights ranging from 18 to 80K (Fig. 6). The pattern of radioactive proteins was similar when different batches of spores were used. There was no detectable qualitative difference between the products of translation of the stable

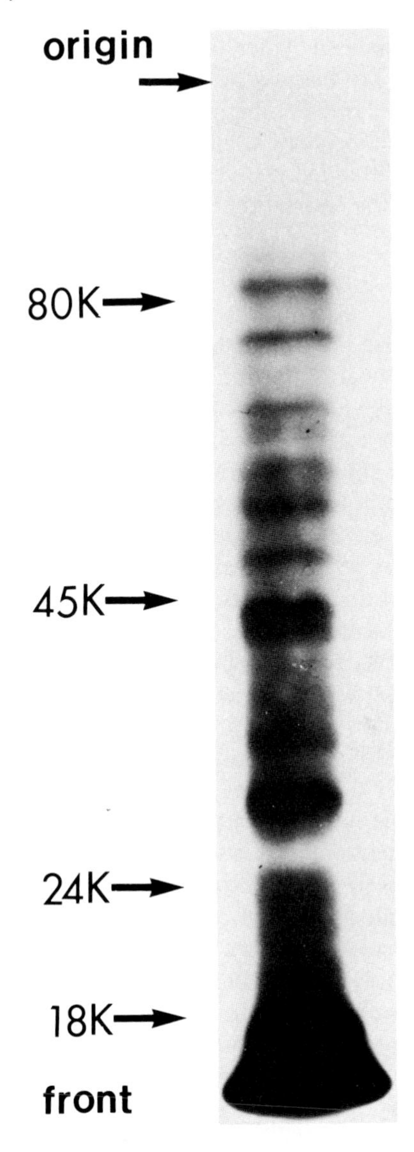

FIGURE 6. Fluorogram of SDS-polyacrylamide gel electrophoresis of the proteins encoded by the stable mRNA fraction of dormant spores of Streptomyces antibioticus. *Spores were incubated in the MSM for 15 min in the presence of rifampicin (15μg/ml) and L-[^{35}S]-methionine (0.2 μCi/ml, 7.4 KBq/ml). Then, cell-free extracts were prepared for electrophoresis as described in Material and Methods, and 10^5 c.p.m. labeled protein were applied in a 12% polyacrylamide gel with a 5% stacking gel. The gel was run at 45 mA for 7 h, fixed in 20% (w/v) TCA and immersed for fluorography in an equimolar solution of 2 M sodium salicylate (pH 6.0-6.5) for 30 min. After the gel was dried in a vacuum, it was exposed to Kodak X-Omat S film for 7 days at -70 °C. Molecular weights of the standards are given at the side.*
*(Reproduced from Guijarro, J.*A. et al., Can. J. Microbiol. 29 *(May 1983) by copyright permission of the National Research Council of Canada.*

mRNA fraction in distilled water or in the MSM, but quantitative differences were found. The activity of spore proteases could not account for the existence of such a large number of protein bands because a similar pattern was found when protease inhibitors (PMSF, pCMB, EDTA, and *o*-phenanthroline) were added just after the labeling period and maintained in the extract during the entire processing of the samples.

IV. DISCUSSION

Previous work in our laboratory has shown that RNA and protein synthesis start immediately after the addition of *S. antibioticus* spores to MSM (Hardisson *et al.*, 1980). These biosynthetic processes may therefore be considered as some of the early biochemical events occurring during germination of *Streptomyces* spores. Surprisingly, although at a lower rate, synthesis of both macromolecules took place even upon the incubation of spores in distilled water. The functioning in distilled water of this biosynthetic machinery must require an endogenous source of amino acid and nucleic acid precursors and an energy donor able to support the biosynthetic requirements. These demands are fulfilled by the existence of a high trehalose content in the dormant spore (about 10% of the dry weight), a complete amino acid pool, five nucleosides, and detectable levels of adenine nucleotides (AMP, ADP, and ATP). Another interesting observation is the presence of a stable mRNA fraction in the dormant spore (Hardisson *et al.*, 1980), which is quickly translated as an early event in the germination process. However, no role has been assigned either to the stable mRNA fraction or to the rapid macromolecule synthesis.

The mRNA's synthesized in the transition from dormant to dark spores showed a very long half-life (about 20 min). These values decreased during germination to reach constant values (5 to 6 min) similar to those found for mycelial growth. This decrease in half-life of the mRNA's, beginning with the start of spore swelling, can be related to activation of the spores' metabolism at this stage of germination. The half-life values we have obtained are quite high when compared to the 2 min values reported for germinating spore of *Bacillus subtilis* (Balassa and Contesse, 1965), vegetative and sporulating cells of *Bacillus thuringiensis* (Glatron and Rapoport, 1972), and exponentially growing *Escherichia coli* (Pato and Von Meyenburg, 1970). This difference probably reflects the considerably slower growth of *Streptomyces*.

Another interesting feature of the RNA's in *Streptomyces* spores is the existence of a higher proportion of unstable RNA with respect to that reported for vegetative bacteria such as *Escherichia coli* (Leive, 1965) and *Bacillus megaterium* (Zimmerman and Levinthal, 1967), 6% and 9% respectively. In this cases, long labeling periods of at least one generation were used with the consequence of a strong selection for more stable

RNA's. In fact, when periods of labeling were short, 35% of unstable RNA was found in germinating spore of *Bacillus subtilis* (Balassa and Contesse, 1965), and values as high as 50 to 60% were also reported for vegetative and sporulating cells of *Bacillus thuringiensis* (Glatron and Rapoport, 1972), and *Escherichia coli* (Pato and Von Meyenburg, 1970).

In contrast to the long half-life of the mRNA species, proteins synthesized early during germination, *i. e.*, before swelling, underwent a very high turnover rate; about 30% of the proteins synthesized in the first 15 min of germination were degraded in the 30 min following the labeling period. Several types of proteolytic enzyme activities have been found in cell-free extracts of *S. antibioticus* spores. Based on the inhibition of this initial protein degradation by specific protease inhibitors, one can assume that different types of proteases participate in the protein degradation (Guijarro *et al.*, 1982b). It must be emphasized that the degradation pattern of proteins encoded by the stable mRNA fraction (*i. e.*, in the presence of rifampicin) was almost coincident with that of the proteins encoded by both this fraction and the newly synthesized mRNR's (*i. e.*, in the absence of rifampicin). This result suggests either that the quickly degraded proteins represent only stable mRNA coded proteins or that the degradation rates of proteins coded by the entire mRNA population have an average half-life strongly resembling that of the polypeptides coded by the pre-existent mRNA fraction. Setlow (1975) has also reported a high turnover rate for proteins synthesized in the first minutes of germination of *Bacillus megaterium* spores.

The products of the stable mRNA fraction were characterized by polyacrylamide gel electrophoresis and fluorography. The mRNA fraction encoded a large number of proteins with a broad range in molecular weight (18 K to 80 K). Although their existence has been reported (Hardisson *et al.*, 1980), the stable mRNA fraction of the dormant spores has not been assigned a function. Spores do not initiate germination upon incubation in distilled water, but in these conditions, the stable mRNA fraction is translated. Taken together, these considerations make it difficult to assign a role for proteins coded by that fraction. Two main hypotheses may be formed. (i) Some of the last messengers formed during sporulation might remain "trapped" in the spore cytoplasm at the end of spore maturation. These mRNA's would have no function in germination; they could be quickly translated after placing the spores in water, and their products would consequently become degraded. In this sense, mRNA-like species have been detected in bacterial endospores that are remnants possibly of sporulation (Setlow, 1981). (ii) Another potential role for the stable mRNA fraction could be repression of the initiation of germination by wet conditions. *Streptomyces* spores form only on the surface of solid media, and dehydration may play an important role in maintaining their dormancy. Incubation of spores in water starts macromolecule synthesis (Hardisson *et al.*, 1980), and water may be necessary, but not sufficient, for the initiation of germination. In this sense, it must be pointed out that only divalent cations initiate *Streptomyces* spore germination (Eaton and Ensign, 1980; Hardisson *et al.*, 1978). Consequently, a mechanism inhibiting the formation of specific germination products could be useful to avoid starting the process under conditions (water incubation) in which germination will not be triggered. However, incubation of spores in water could allow translation of the stable mRNA fraction whose products would repress the synthesis of specific germination products. After divalent cations trigger germination, some metabolite synthesized in the first minutes of germination would produce a conformational change in the receptor(s), with the subsequent release of repressor(s). Then, specific germination products could accumulate and degradation of the protein repressor(s) coded by the stable mRNA fraction could begin. Obviously, these two hypothetical mechanisms could not act simultaneously.

Whatever the answer may be, we must not exclude the possibility that some of the quickly degraded proteins could include abnormal proteins synthesized by an error in gene expression or by a defective protein synthesizing system. There is now strong evidence that bacterial cells can selectively hydrolyze abnormal proteins (Goldberg and Dice, 1974). Such a mechanism for eliminating protein waste or the potentially harmful consequences of imprecise synthesis would be of great utility for the development of germination. Degradation of abnormal proteins appears to be especially important in slowly growing cells which, unlike rapidly growing bacteria, cannot dilute out such proteins by rapid growth.

Studies of bacterial spore germination have been carried out mainly in *Bacillus* endospores. In recent years, it has been proposed that a biophysical or pure hydrolytic event might trigger germination of *Bacillus* spores (Rossignol and Vary, 1979; Scott and Ellar, 1978a,b; Wax and Freese, 1968). Although a great variety of chemical and physical agents have been reported to trigger germination (Foerster and Foster, 1966), remarkably, all of them seem to trigger the same sequence of biochemical events.

However, to date, the only germinants cited as capable of initiating the darkening process of *Streptomyces* spores are divalent cations (Eaton and Ensign, 1980; Hardisson *et al.*, 1978). These metal ions induce the transition from dormant to dark spore, which is accompanied by a loss of spore refractility and consequently by a fall in absorbance. In spite of the importance of *Streptomyces* in cell differentiation studies, only in the last few years has important work been done to identify the triggering mechanism and the subsequent events. However, no clear picture of the initial events of spore germination has yet emerged. Many questions remain to be answered, What are the nature and location of the receptor for divalent cations? What kind of interaction occurs and what are the first events that this mechanism triggers? How does this early event end dormancy? By using several experimental approaches, we are attempting to answer these questions. Recent studies in our laboratory (Salas, Guijarro, and Hardisson, unpublished results) indicated the existence, in dormant spores, of several species of *Streptomyces* of a very high calcium content (about 2% of the dry weight). This calcium was quickly mobilized in the first minutes of germination and excreted to the medium. Preliminary experiments suggest that calcium could be located outside the spore membrane, as has been reported for *S. viridochromogenes* spores (Eaton and Ensign, 1980). Another notable observation made recently in our laboratory was that the triggering of spore darkening by cations induced calcium release from spore integuments, suggesting that the earliest event in *Streptomyces* spore germination might take place outside the spore's cytoplasm (Salas, Guijarro, and Hardisson, unpublished results). How this event occurring outside the membrane is related to the subsequent changes within the spore's cytoplasm remains to be clarified.

REFERENCES

Balassa, G., and Contesse, G., *Ann. Inst. Pasteur 109:*684 (1965).

Chamberlain, J. P., *Anal. Biochem. 98:*132 (1979).

Cross, T., and Attwell, R. W., *in* "Spores VI" (P. Gerhardt., R. N. Costilow, and H. L. Sadoff, eds.), p. 3. American Society for Microbiology. Washington (1975).

Eaton, D., and Ensign, J. C., *J. Bacteriol. 143:*377 (1980).

Ensign, J. C., *Annu. Rev. Microbiol. 32:*185 (1978).

Fairbanks, G., Steck T. L., and Wallach, D. F. H., *Biochemistry 10:*2206 (1971).

Foerster, H. F., and Foster, J. W., *J. Bacteriol. 91:*1168 (1966).

Glatron, M. F., and Rapoport, G., *Biochimie 54:*1291 (1972).

Goldberg, A. L., and Dice, J. F., *Annu. Rev. Biochem. 43:*835 (1975).

Guijarro, J. A., Suárez J. E., Salas, J. A., and Hardisson, C., *FEMS Microbiol. Lett. 14:*205 (1982).

Guijarro, J. A., Suarez, J.E., Salas, J.A., and Hardisson, C., *Can. J. Microbiol.:(1983).*

Hardisson, C., Manzanal, M. B., Salas J. A., and Suárez, J. E., *J. Gen. Microbiol. 105:*203 (1978).

Hardisson, C., Salas J. A., Guijarro, J. A., and Suárez, J. E., *FEMS Microbiol. Lett. 7:*233 (1980).

Hartmann, G., Honikel, K. O., Knusel, F., and Nuesch, J., *Biochim. Biophys. Acta 145:*843 (1967).

Laemmli, U. K., *Nature (Lond). 227:*680 (1970).

Leive, L., *J. Mol. Biol. 13:*862 (1965).

Pato, M. L., and Von Meyenburg, K. L., *Cold Spring Harbor, Symp. Quant. Biol. 35:*497 (1970).

Rossignol, D. P., and Vary, J. C., *J. Bacteriol. 138:*431 (1979).

Salas, J. A., and Hardisson, C., *J. Gen. Microbiol. 125:*25 (1981).

Scott, I. R., and Ellar, D. J., *Biochem. J. 174:*627 (1978a).

Scott, I. R., and Ellar, D. J., *Biochem. J. 174:*635 (1978b).

Setlow, P., *J. Biol. Chem. 250:*631 (1975).

Setlow, P., *in* "Sporulation and Germination" (H. S. Levinson, A. L. Sonenshein, and D. J. Tipper, eds.), p. 13. American Society for Microbiology, Washington (1981).

Wax, R., and Freese, E., *J. Bacteriol. 95:*433 (1968).

Zimmerman, R. A., and Levinthal, C., *J. Mol. Biol. 30:*349 (1967).

MODE OF ACTION OF FACTOR C UPON THE DIFFERENTIATION PROCESS OF *STREPTOMYCES GRISEUS*

G. Szabó
S. Biró
L. Trón[1]
G. Valu
S. Vitális

Institutes of Biology and Biophysics [1]
University Medical School
Debrecen, Hungary

I. INTRODUCTION

By differentiation one generally means the appearance of a new marker — a protein or a biochemically more complex morphological trait. Regulation of the production of a new protein has been described by the Jacob-Monod operon model; however, since no such clear-cut idea has been formed so far about the mechanism for development of complex morphological traits, the following questions remain open: How is the order of their appearance regulated; how are they interrelated; are there any endogenous regulatory substances and, if so, what is their chemical nature and do they affect specific genes or larger genome segments; etc.?

To study these problems, differentiation has been investigated with *Streptomyces griseus* as a model system. First, we have described the development of *S. griseus* strain No. 45-H in detail. The vegetative and reproductive phases of growth and conidium production are clearly distinguishable in this strain. We have established an orderly appearance of the cytomorphological and biochemical characteristics that follow each other consecutively: the various types of nucleoids and cytoplasm; changes in the distribution of periodic acid Schiff(PAS)positive material; changes in the cell-wall structure and composition; changes in the composition and rate of synthesis of macromolecules in the

ISBN 0-12-528620-1

mycelia; changes in the content of DNA, RNA, and protein; and changes in the production and disappearance of several enzymes (Biró *et al.*, 1979a,b; Valu and Szabó, 1970; Valu *et al.*, 1975; Vitális *et al.*, 1963, 1981).

The orderly appearance of the various markers during development reflects an association between those markers. If the marker, *e.g.*, the nucleoid, is of the reproductive type during sporulation, then the same hyphal part is judged in most cases to be reproductive by the appearance of cytoplasm. Having established the normal, regular features of development, we also examined the extent and pattern of appearance of the same markers in a mutant strain (No. 52-1) which does not develop conidia in submerged culture. The individual markers, described in the natural, conidium-producing strain, were found in the mutant as well, but the time and the degree of their appearance differed and the development of the different markers did not correlate with each other within a given hyphal segment. In contrast, in the well-sporulating strain, as stated above, there was a close association between the markers, *e.g.*, between a certain nucleoid type and a given type of cytoplasm. They regularly appeared together in the same hyphal segment, whereas the same cytomorphological traits combined somewhat randomly in the hyphae of the nonsporulating mutant strain. The courses of the different genetic programs and subprograms no longer coordinated so strictly. If, however, the fermentation liquid of the well-sporulating strain was added to the cultivation medium of the nonconidiating mutant, the normal pattern of development was partially restored; some conidia were also produced. The active compound in the fermentation liquid is called factor C (Szabó *et al.*, 1967; Vitális and Szabó, 1969). Our view of this development and the way in which factor C influences differentiation is summarized in the following short statement: By the differentiation of *S. griseus*, we mean orderly sequential and coordinated appearance of the markers that are characteristic of the vegetative and reproductive phases, a process ending in conidium production. When this regular development is disturbed in the mutant strain, factor C can partially restore the normal pattern of differentiation.

II. RESULTS

A. Detection, Quantitative Determination and Isolation of Factor C

Factor C was detected and determined on the basis of its ability to induce cytomorphological changes in the submerged culture of the conidium-nonproducing strain, *S. griseus* No. 52-1. The test was regarded positive if formation of reproductive branches could be detected in the mycelia of the test strain, which otherwise did not produce reproductive branches. Serial dilutions were made and the amount of factor C found in 1 ml of the highest dilution still definitely positive was taken as one unit.

TABLE I. Amino Acid Composition of Factor C

Amino acid	*Amount mol/100 mol*
Aspartic acid	*7.11*
Threonine	*5.93*
Serine	*5.93*
Glutamic acid	*2.77*
Proline	*6.72*
Glycine	*15.42*
Alanine	*7.91*
Valine	*10.28*
Methionine	*5.15*
Isoleucine	*4.74*
Leucine	*8.30*
Tyrosine	*2.77*
Phenylalanine	*4.74*
Lysine	*3.95*
Histidine	*2.77*
Arginine	*3.55*
Glucosamine	*1.98*
Tryptophan	*n.d.*[a]

[a]*n. d., not detected*
(Reproduced from Biró, S. et al., Eur. J. Biochem. 103:*359 (1980) by copyright permission of Springer Verlag.)*

The biologically active substance factor C, was isolated (Biró *et al.*, 1980). Sodium dodecyl sulphate-acrylamide gel electrophoresis of purified factor C yielded a single band with a molecular weight of 34,500. Factor C also migrated as a single band in nondenaturing polyacrylamide gel electrophoresis and was successfully extracted from the gel in a biologically active form. The amino acid composition of factor C is shown in Table I. About 60% of the amino acids were found to be hydrophobic. The analysis revealed that factor C also contained 1.98% glucosamine.

B. Production of Factor C

To understand its natural role, investigation of the conditions for the factor C production was necessary. Factor C was discovered in the fermentation broth of the streptomycin-nonproducing and well-sporulating *S. griseus* No. 45-H strain. It could be detected after 24 h of cultivation; its concentration increased and reached the highest level at 72 h, after which time it leveled off (Fig. 1). Endocellularly, it was already detectable in the 16 hour-old mycelium. The hyphae that produced factor C were probably of the late vegetative type. We failed to locate the site of production of factor C by cell-fractionation, although this factor appeared in the cell membrane fraction earlier than in the

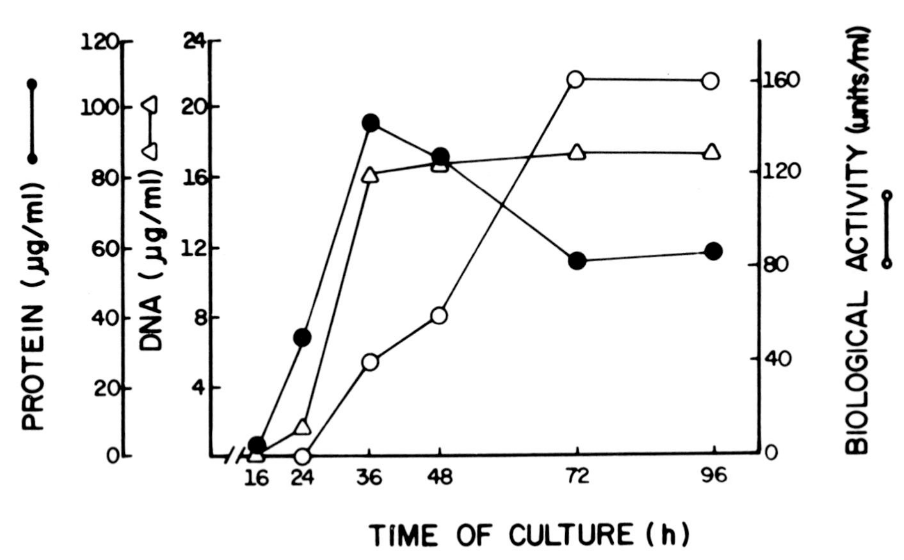

FIGURE 1. Production of factor C by Streptomyces griseus *No. 45-H in submerged culture. The activity of factor C was determined from the supernatant as described in the text. The methods for DNA and protein determination were described by Vargha, Vitális and Szabó (1978).*

cytosol. When ten other *Streptomyces* strains which sporulated well in submerged culture were investigated for factor C production (Table II), another strain of *S. griseus* and two strains of different *Streptomyces* species also produced factor C-like substance(s). We designated these biologically active substances factor C-like materials, because, although they were specific in

TABLE II. Strains Producing a Factor C-like Substance [a]

Strains	Appearance of factor C in:		Sporulation in:		*Production of NPT in submerged culture*	*Production of antibiotic*
	culture liquid	*cell*	*submerged culture*	*surface culture*		
S. griseus *45-H*	+ +	+	+	+	+	-
S. griseus *FBUA 821*	-	+ [b]	+	+	+	-
S. intermedius *FBUA 1071*	+	+	+	+	+	-
Act. levoris *FBUA 1114*	+	+	+	+	+	+

[a]*Strains were cultivated in phosphate buffered soya-bean liquid medium. Factor C and NPT were determined as described by Szabó* et al. *(1967) and Biró* et al. *(1979a). Antibiotic production was screened by applying agar plugs from 10 day-old cultures on* B. subtilis *containing solid media.*
[b]*Only at a short period of the life cycle.*

TABLE III. Sensitivity of Streptomyces *Species (24 Strains) to Factor C*

Strains		*Sensitivity*	*Sporulation in* submerged culture	*Sporulation in* surface culture	*Production of NPT in submerged culture*	*Production of antibiotic*
S. griseus *No. 52-1*		+	-	+	-	+
S. griseus *FBUA*[a] *831*		+	-	+	-	+
S. griseus *FBUA 871*		+	-	+	-	+
S. griseus *FBUA 961*		+	-	+	-	-
S. griseus *FBUA 971*		+	-	+	-	+
Act. cyaneofuscatus	*FBUA 523*	-	-	+	-	+
S. griseobrunneus	*FBUA 752*	-	-	+	-	-
S. annulatus	*FBUA 151*	-	-	-	-	-
S. fimicarius	*FBUA 577*	-	-	+	-	-
S. lipmanii	*FBUA 1121*	-	-	+	-	-
Act. microflavus	*FBUA 1144*	-	-	+	-	-
Act. parvus	*FBUA 1219*	-	-	+	-	+
S. praecox	*FBUA 1241*	-	-	+	-	-
Act. rubiginosohelvolus	*FBUA 1312*	-	-	+	-	-
S. rutgersensis	*FBUA 1341*	-	-	-	-	-
Act. setonii	*FBUA 1375*	-	-	-	-	-
S. griseus *(8 different strains from*	*FBUA)*	-	-	+	-	+

[a] *FBUA, Forschungsinstitut für Bodenkunde der Ung. Akad. Wiss., Budapest, Hungary*

bringing about reproductive forms, they were not identified chemically. Immune serum which had been produced against purified factor C in rabbits precipitated factor C and inhibited its biological activity.

Although these extracts, which had factor C-like activity, immunoprecipitated with the antibodies prepared against purified factor C, this was not enough to prove their identity, because they may have contained common antigenic determinants responsible for the positive immunoprecipitation. It is worth mentioning that, as shown in Table II, the factor C-like substance in three strains was secreted into the medium, whereas in one (strain No. 821) of the four producers, it remained inside the cell and could be detected only after the disintegration of the hyphae. These substances may be present in the fermentation liquid for a few days of cultivation, but they may also appear only for a shorter time. In strain No. 821, the mycelia were disintegrated and tested at every two hours of culture, but biological activity was detected only in the 26 h and 28 h samples. No factor C-like activity was found before or after this short period. Accordingly, it cannot be excluded that the occurrence of this (these) regulatory molecule(s) is more general since we might not have taken samples at the proper time and/or the method used for its (their) detection was not good enough. All the strains that produced factor C-like substances also sporulated well in submerged culture; only one of them produced antibiotics. Data on the production of nucleotide pyrophospho-transferase (NPT) by these strains were included in Tables II and III because in earlier experiments

(Biró *et al.*, 1979c) sporulation was observed in submerged culture for 19 of 80 *Streptomyces* strains, with NPT activity detected in 12 strains. These findings showed a statistically significant positive correlation between sporulation and NPT synthesis. Moreover, NPT activity, which was not detected in submerged cultures of *S. griseus* No. 52-1, appeared in the mycelium on solid medium that favored sporulation (Biró *et al.*, 1979b). Although data obtained for four strains may be inadequate to permit statistical analysis, still a good correlation might show positivity.

C. Mode of Action of Factor C

The mode of action of the molecule that modulates the differentiation process was studied first by collecting data about the gross physiological effects of factor C. It did not inhibit the germination of *S. griseus* spores, but it had a regulatory influence upon the length distribution of growth by germinating tubes (Figs. 2 and 3). That is, fewer spores had very long germinating tubes in the presence of factor C, although the average length did not differ from the control value.

The effect of factor C was phase dependent. Adding the factor to the test

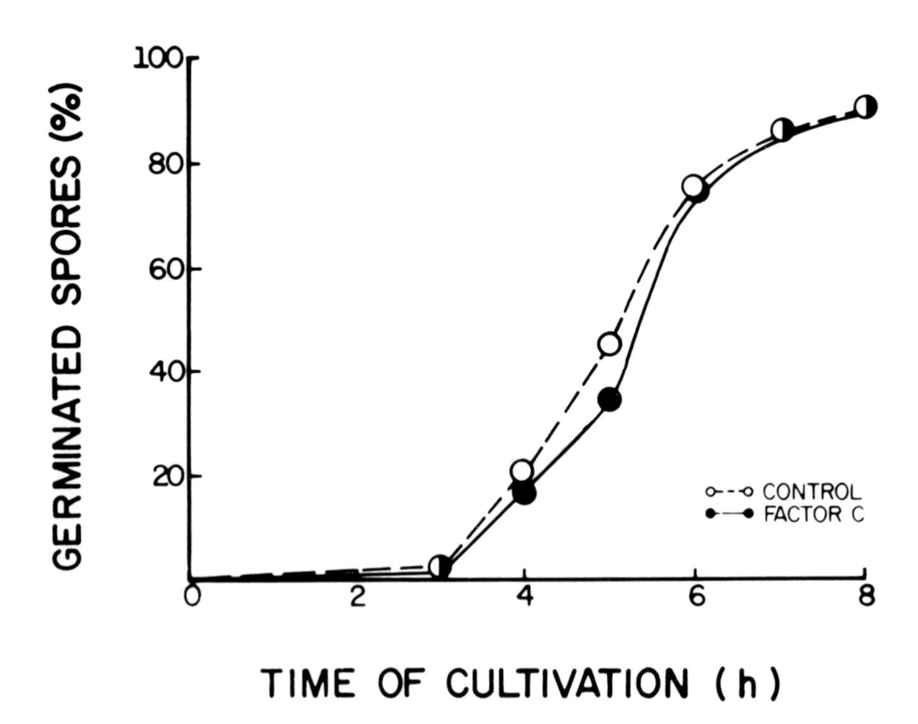

FIGURE 2. Effect of factor C upon germination of Streptomyces griseus *No. 52-1 spores.*

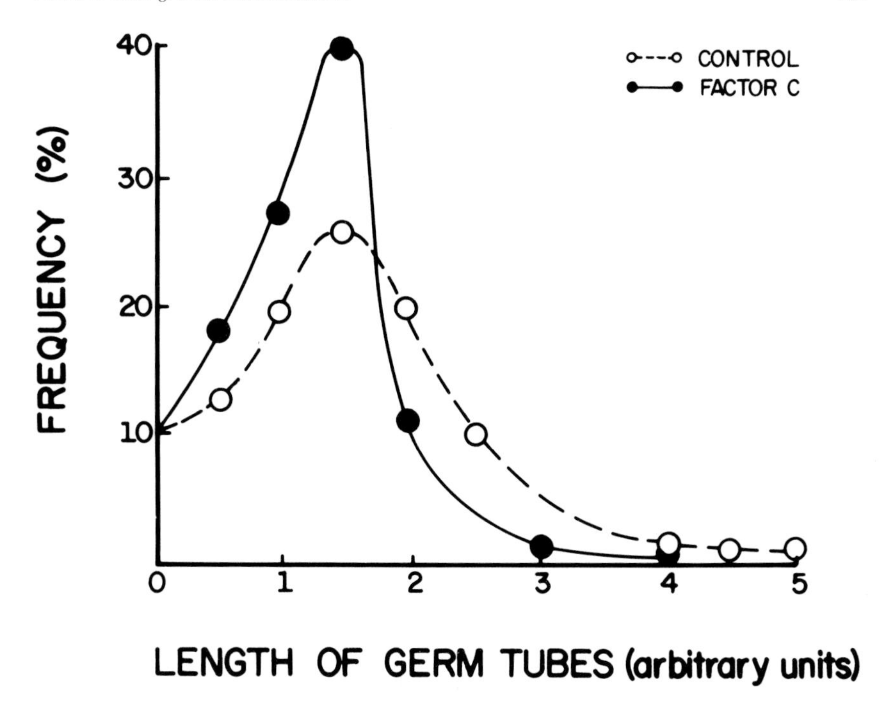

FIGURE 3. Effect of factor C upon the length of germ tubes grown from Streptomyces griseus *No. 52-1 spores during 8 h of germination.*

culture before the end of the vegetative phase (before the culture was 24 h old) was necessary to induce reproductive development. Factor C seemed to be species specific. Of 24 *Streptomyces* strains (Table III) belonging to 12 different species, five proved to be sensitive to factor C. All these sensitive strains belong to the *S. griseus* species. The sensitive strains produced antibiotics with one exception.

Factor C was adsorbed very quickly by the germinating conidia. The disappearance and reappearance of factor C added to the culture medium was followed by determining its biological activity in samples taken between 0 and 72 h. Biological activity almost disappeared 50 min after mixing the spores with factor C (Fig. 4). The mechanism of reappearance of this biological activity after 12 h of incubation in the fermentation liquid of the test strain was not clear. When the uptake of labeled factor C was followed, similar results were obtained. ^{125}I-labeled factor C also disappeared very quickly from the test medium.

1. Effect of Factor C upon in vivo *and* in vitro *Protein and RNA Synthesis.* Protein and RNA syntheses were followed for 15 h in the mycelia of the conidium-nonproducing *S. griseus* No. 52-1 strain in submerged culture with

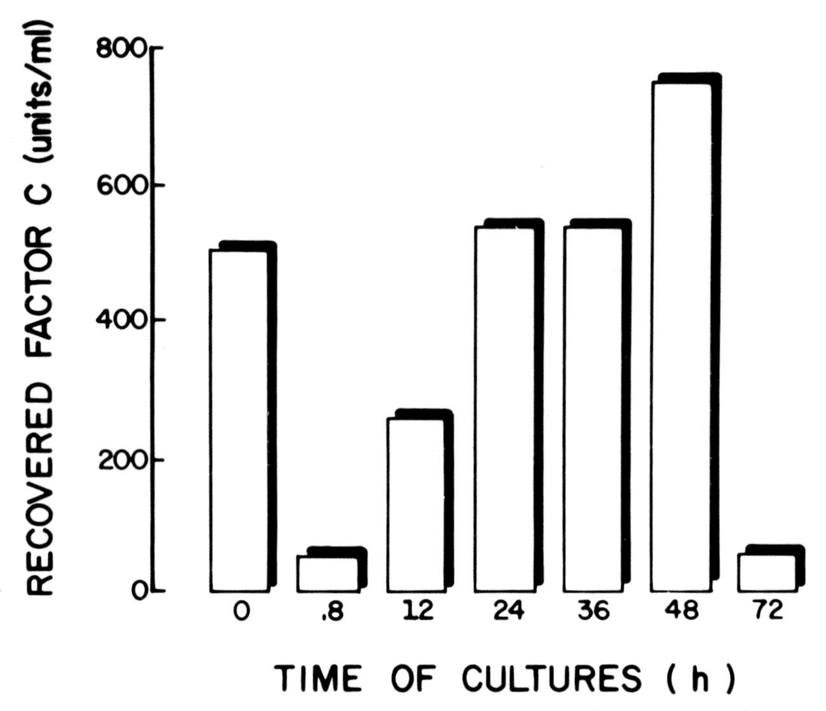

FIGURE 4. Recovery of added factor C from culture medium of Streptomyces griseus *No. 52-1. Incubation took place in phosphate buffered soya-bean liquid medium containing 1.8 x 10 ^{6} spores/ml.*

and without factor C. The protein and RNA content of the germinating spores and, later, that of the mycelia were determined as described by Vargha, Vitális, and Szabó (1978). Figure 5 shows that factor C slightly inhibited both protein and RNA synthesis *in vivo.*

To understand the mechanism of this inhibition, we studied the mode of action of factor C in polypeptide synthesis *in vitro.* The conditions and composition of the incorporation system are shown in the footnote to Table IV. The polyuridylate (poly-U)-directed incorporation of phenylalanine was definitely inhibited by 50 μg factor C/ml. Heat denaturation for 30 min at 75 °C eliminated biological activity and reduced this inhibitory effect. In comparison, the inhibitory effect of 200 μg chloramphenicol/ml is also shown. Factor C proved to be a more potent inhibitor of polypeptide synthesis. Inhibition depended on the concentration of factor C (Fig. 6). The dose-response curve (Fig. 7), plotted on logarithmic scale, showed that inhibition changed linearly when the dose of factor C was elevated exponentially.

To find where in the incorporation system factor C exerted its inhibitory effect, decreasing concentrations of the template (poly-U) and of the ribosomes were examined. As seen in Figure 8, changing the poly-U concentration from 500 to 1500 μg/ml decreased the inhibitory effect of factor C from 35% to

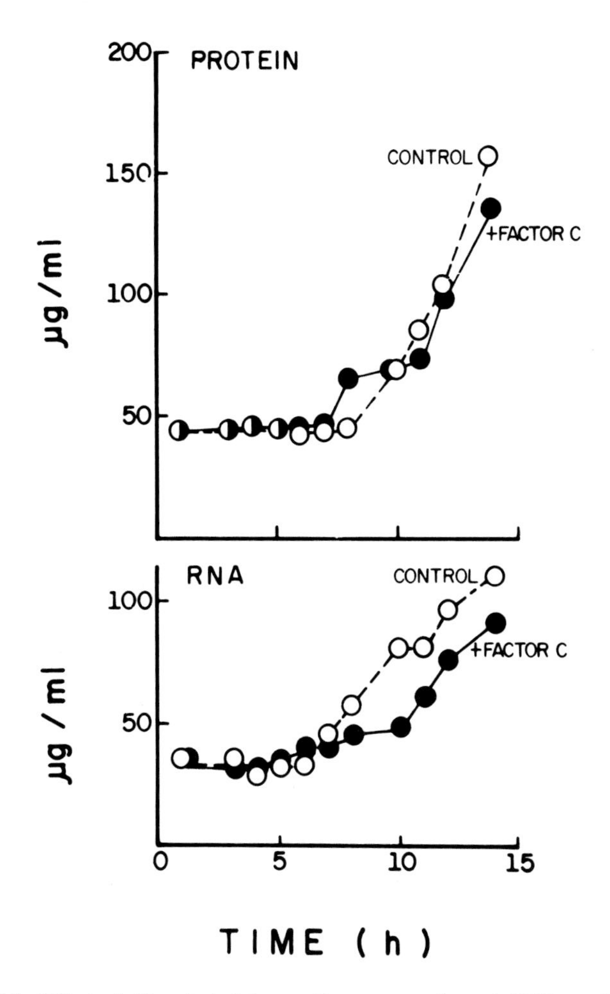

FIGURE 5. Effect of 40 units/ml factor C upon protein and RNA synthesis during the germination and early growth phase of Streptomyces griseus *No. 52-1. The cultures were inoculated at 0 h with 2 x 10 ^{6} spores/ml. Protein and RNA were determined as described by Vargha et al. (1978).*

22%. If the concentration of the ribosomes was decreased four-fold in the incubation mixture, no decrease in the effectiveness of factor C could be detected (Fig. 9). We inferred from these experiments that factor C bound to the template and not to the ribosomes inhibited poly-U-directed incorporation.

To determine which stage of the polypeptide synthesis was inhibited by factor C, we compared the kinetic curves of the inhibitory action of aurintricarboxylate (AT) and factor C. The experiment presented in Figure 10 was conducted with AT, a known inhibitor of translation initiation (Grollman and Stewart, 1968). The incorporation mixture was preincubated for 16 min at 27 °C

TABLE IV. *Inhibition of* in vitro *Polypeptide Synthesis by Factor C, Heat-denatured Factor C and Chloramphenicol with S-30 fractions of* S. griseus *No. 45-H (16 hr)*[a].

Specific activity with poly-U template						
Control	+ Factor C (50 μg/ml)		+ Factor C (50 μg/ml) heat-denaturated		+ Chloramphenicol (250 μg/ml)	
pmole/A_{260}/60'	pmole/A_{260}/60'	% inhibition	pmole/A_{260}/60'	% inhibition	pmole/A_{260}/60'	% inhibition
228.36	22.73	90.05	-	-	101.5	55.6
243.16	24.04	90.12	164.57	32.33	-	-
181.00	21.90	88.00	126.49	30.12	-	-

[a] *Composition of the reaction mixture for* in vitro *polypeptide synthesis. One ml contained: 50 μM HEPES, pH 7.8; 50 μM KCl; 12.5 μM $MgCl_2$; 12.5 μM DTT (dithiothreitol); 2.5 μM ATP; 0.1 μM GTP; 12.5 μM phosphocreatine; 70.0 μg creatine phosphokinase; 25 nM nonradioactive amino acids; 10 A_{260} units E. coli tRNA; 800.0 μg poly-U; 25 A_{260} units S-30 extract; and 25.0 nM [^{14}C]-L-phenylalanine (Phe) (317 mCi/mM). Specific activity: pmole [^{14}C]-Phe incorporated per 1.0 O.D.$_{260}$ unit of cell free extract.*

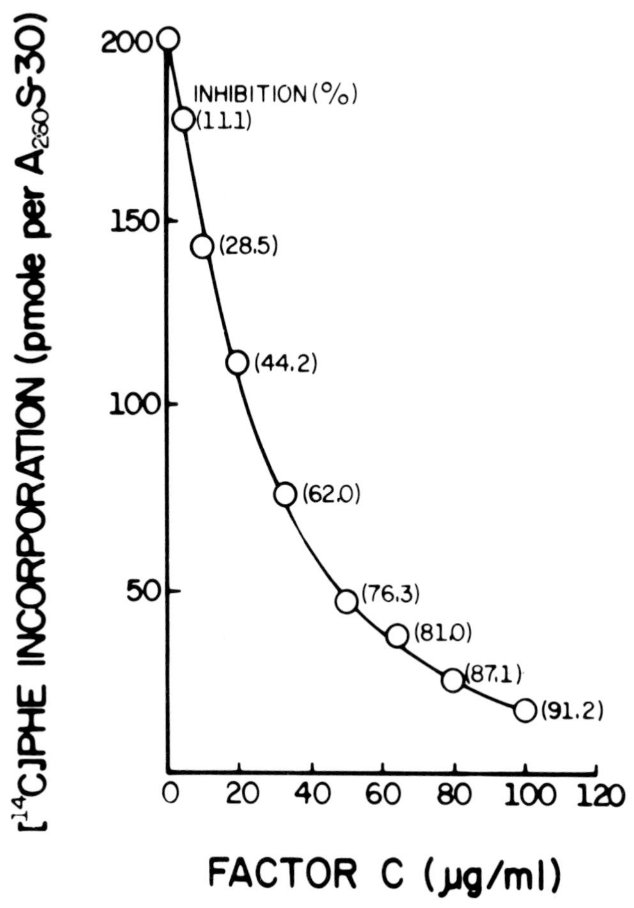

FIGURE 6. Dose dependent inhibition of factor C on [^{14}C]-Phe incorporation. (Experimental conditions are the same as in Table IV.)

with or without AT and factor C. After preincubation, incorporation started at 0 min with the addition of phenylalanine, phenylalanine plus AT, phenylalanine plus factor C, or phenylalanine with both compounds. If the incorporation mixture was preincubated with AT or with factor C, phenylalanine incorporation was strongly inhibited (Fig. 10). If, however, no preincubation with these substances took place, then both compounds were inhibitory but to a much lower degree. If AT and factor C were applied together, no additive effect was observed. If factor C was combined with phenylalanine and added at 0 min (poly-U was present during preincubation), factor C inhibited far less than if given together with or preceding the addition of poly-U. From these experiments, we concluded that the effect of factor C was similar to that of AT. No synergism was found between them; therefore, we thought that

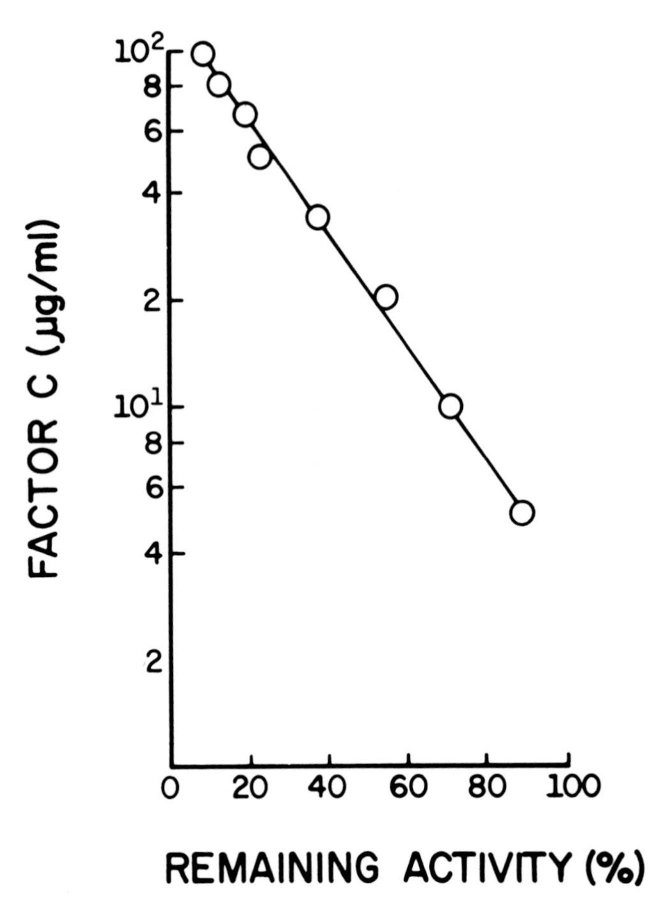

FIGURE 7. Dose dependent inhibition of factor C on [^{14}C]-Phe incorporation. (Experimental conditions are the same as in Table IV.)

factor C inhibited the formation of the initiation complex between the template (poly-U) and the ribosomes and did not influence the elongation stage of polypeptide synthesis.

As mentioned, factor C inhibited RNA synthesis *in vivo*. Therefore, we examined its effect upon RNA polymerase activity, *i. e.*, upon the transcription process. In these experiments, we used purified DNA which had been isolated from various organisms as a template and RNA polymerase from E. *coli*. RNA polymerase activity was followed by determining the amount of labeled ^{14}C-UTP incorporated. Factor C inhibited UTP incorporation (Fig. 11). If the addition of the inhibitory substance preceded that of the RNA polymerase, its effect was more pronounced. We compared DNA preparations from the various sources (Table V). Irrespective of the origin of the template (*S. griseus*,

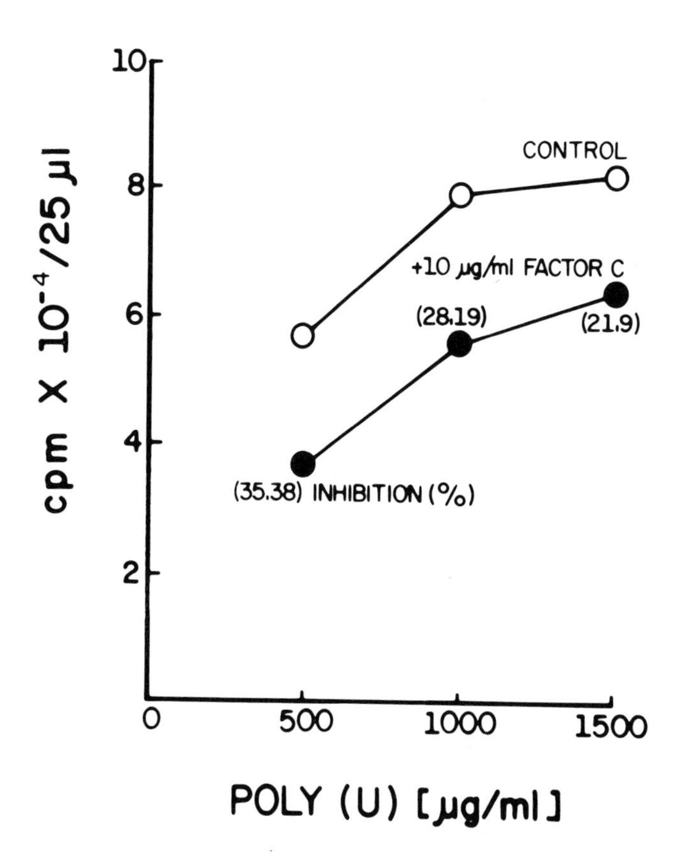

FIGURE 8. Effect of poly(U) concentration on the inhibition of factor C. (Experimental conditions are the same as in Table IV.)

Escherichia coli, Neurospora crassa, calf thymus, and poly[d(AT)]), factor C strongly inhibited UTP incorporation. If the order of addition was also considered we came to the same conclusion. This is presented in Table V as the differential value calculated from the inhibition values obtained in experiments with reverse order of the additions. If factor C was added before RNA polymerase, inhibition was greater than when the order of additions was reversed. These results indicated that factor C binds to DNA. This conclusion was supported by determining the effect of factor C upon the melting profile, the Tm value of DNA. Factor C elevated the Tm point and altered the melting profile of DNA (Fig. 12).

Factor C also inhibited the initiation of transcription followed by the incorporation of γ-labeled ATP (Fig. 13). The degree of inhibition did not depend upon the order in which RNA polymerase and factor C were added.

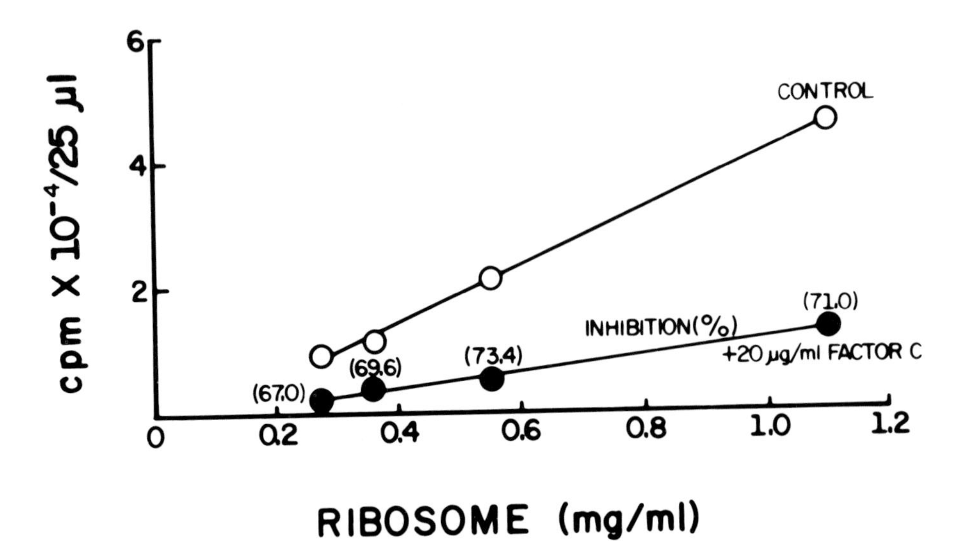

FIGURE 9. Effect of ribosome concentration on the inhibition of factor C. (Experimental conditions are the same as in Table IV.)

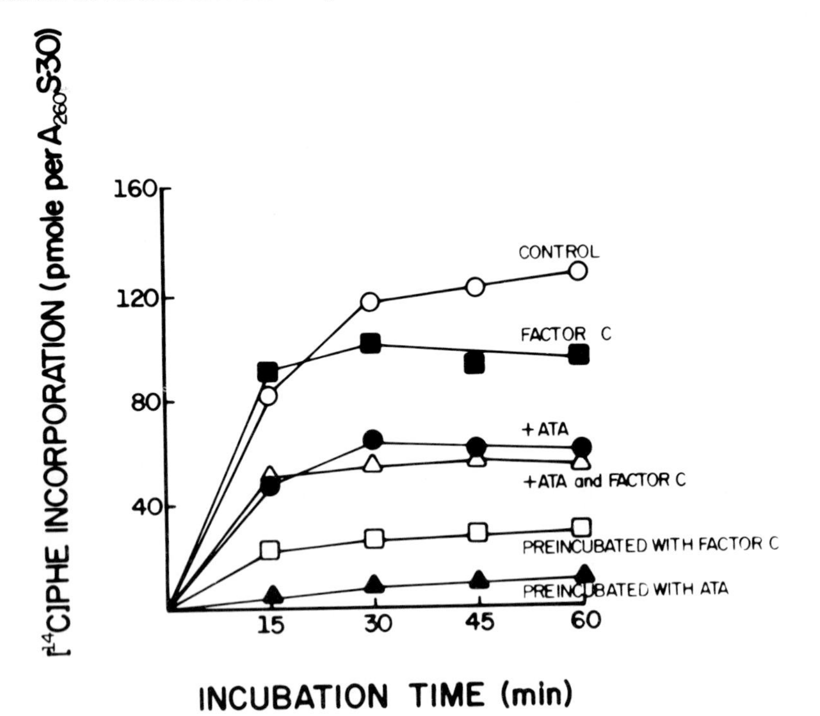

FIGURE 10. Effect of aurintricarboxylic acid (ATA) and factor C on [^{14}C]-Phe incorporation. Comparison of kinetic curves. Samples were preincubated with or without additives at 27 °C for 16 min. Factor C was applied in 1.2×10^{-6} M (42.5 µg/ml) and ATA in 6×10^{-5} M concentrations. (Experimental conditions are otherwise given in Table IV and also in the text.)

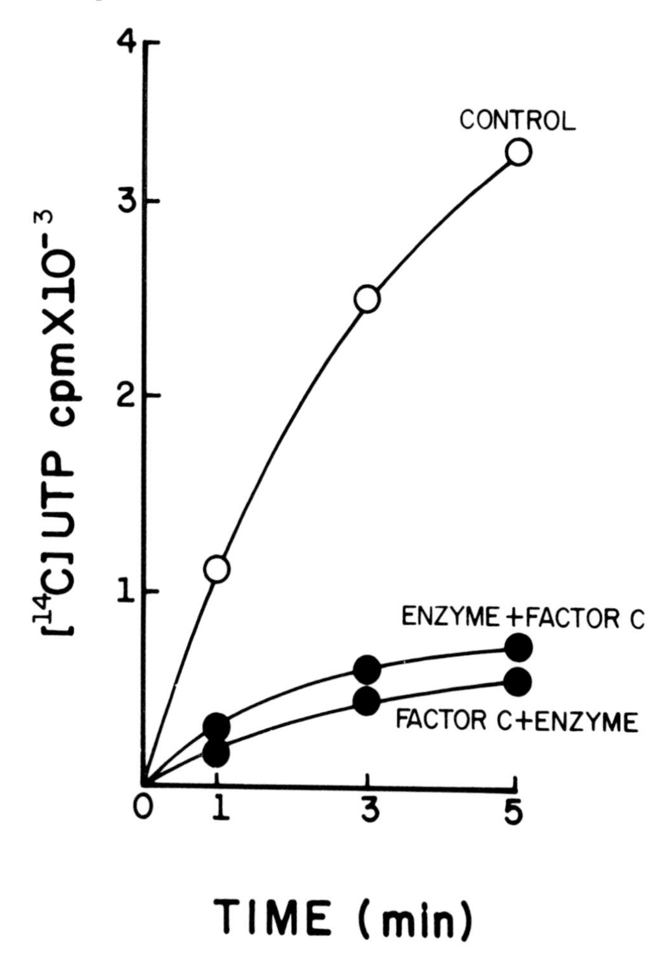

FIGURE 11. Effect of factor C upon [^{14}C]-UTP incorporation using Streptomyces griseus *No. 52-1 DNA as template. Assay mixture: 200 mM Tris-HCl, pH 8.0; 50 mM $MgCl_2$; 1 mM EDTA; 50 mM 2-mercaptoethanol; 100 μg/ml DNA; 0.4 mM ATP, GTP, and CTP; 0.4 mM [^{14}C]-UTP; 50 μg/ml factor C; RNA polymerase in the total volume of 110 μl. Aliquots are adsorbed at given time-intervals to DEAE-cellulose, washed with 0.3 M ammonium formate and put into scintillation fluid for measuring radioactivity.*

III. DISCUSSION AND CONCLUSION

From the above results, we may conclude that, to study differentiation as a complex example of gene expression, the experimental and conceptual model chosen is of crucial importance.

TABLE V. Inhibition Depends on the Order of Adding of RNA Polymerase and Factor C

Order of addition	*Inhibition (%) after 5 min incubation*					
	S. griseus *45-H*	S. griseus *52-1*	E. coli	N. crassa	*Calf thymus*	*Poly [d(AT)]*
1. RNA polymerase + factor C	*71.37*	*77.00*	*85.59*	*77.18*	*74.60*	*66.59*
2. Factor C + RNA polymerase	*77.88*	*82.54*	*82.70*	*81.21*	*83.59*	*79.00*
Differential value[a]	*22.72*	*24.10*	-	*17.69*	*35.42*	*37.41*

[a] *This value was calculated by using the formula [(A-B)/A]·100, were A = [^{14}C]-UTP incorporation (cpm) in case 1; and B = [^{14}C]-UTP incorporation (cpm) in case 2.*

Mutation of *Streptomyces* strains may disturb differentiation. The clear-cut distinction between the vegetative and reproductive forms (phases) of the strains isolated from natural sources disappears. The markers established as

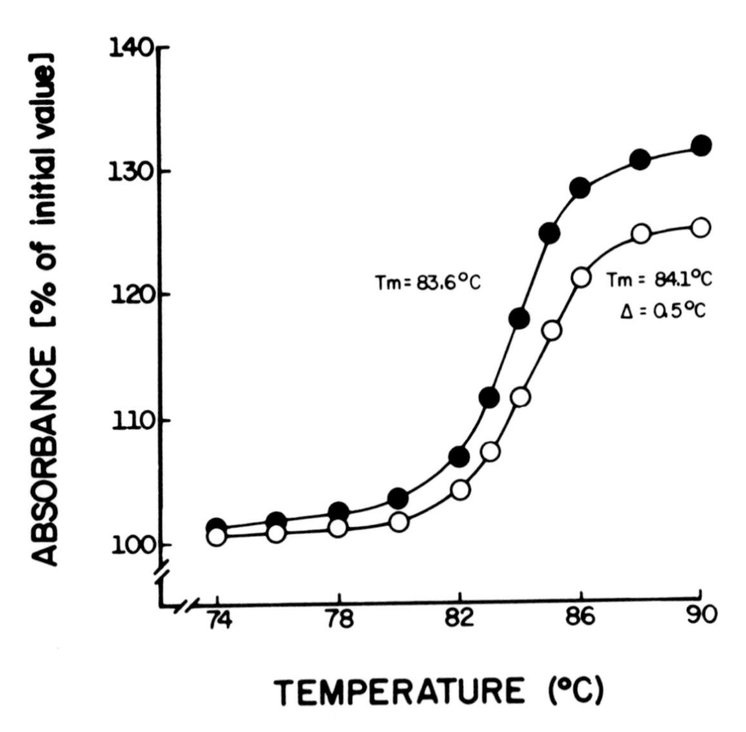

FIGURE 12. Melting profile of Streptomyces griseus *No. 52-1 DNA with and without factor C. Tm point was determined in 0.1 x SSC (sodium chloride-sodium citrate) with Beckman Acta V spectrophotometer. The DNA and factor C concentrations were 250 μg/ml and 6.0 μg/ml, respectively.*

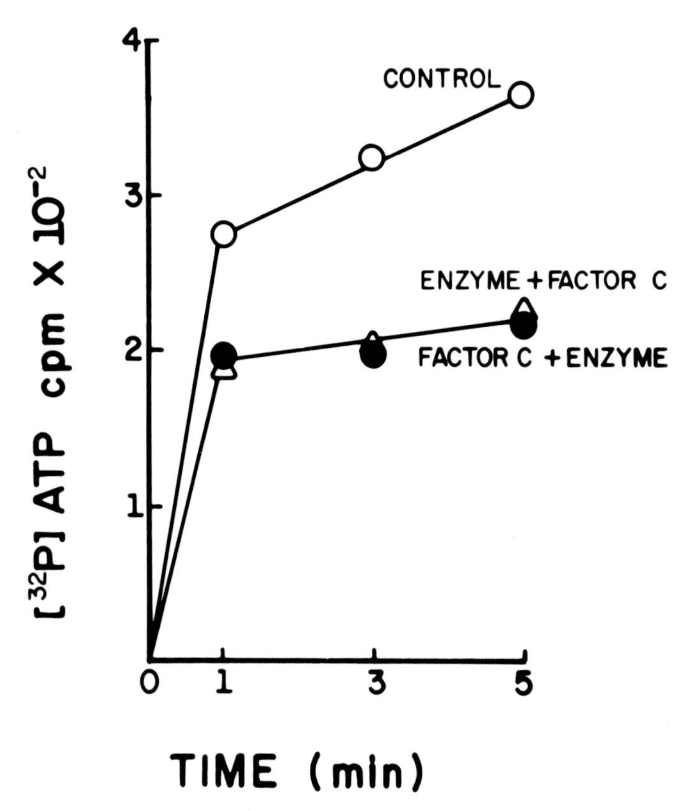

FIGURE 13. Effect of factor C upon $\gamma[^{32}P]$-ATP incorporation using Streptomyces griseus *No. 52-1 DNA template. Assay mixture is the same as in Fig. 11 except the use of $\gamma[^{32}P]$-ATP with a specific activity of 10.13 Ci/mM.*

characteristic of the vegetative and reproductive hyphae appear disorderly and uncoordinated, often in mixed patterns; conidiation is sparse or blocked. Factor C modulates differentiation, the developmental pattern of the wild type is partially restored, and some conidiation may ensue.

We used a mutant that might have all the structural genes needed for spore production, *i. e.* although it sporulates on solid medium, it produces only a small number of spores. However, it seems to be a regulatory mutant with disturbed, disorderly development in submerged culture.

Although it is not possible to understand the mode of action of factor C on the basis of the above experiments, the model is informative in that factor C is a natural constituent in *Streptomyces* cells, *i. e.*, a polypeptide that partially restores the damaged, disorganized development of several nonconidiating *S. griseus* strains in submerged culture. Factor C strongly inhibits poly-U-directed polypeptide synthesis *in vitro* and binds both to RNA and DNA. Therefore, the possibility exists that factor C affects either the transcription of one or more specific regulatory genes or their translational processes.

Factor C may exert its effect by activating or repressing key steps of the developmental process, although by itself, it can only partially rechannel the disturbed differentiation. We may speculate that factor C is an inhibitor of transcription and/or translation; it is produced at the end of the vegetative phase of growth when there is a switch to the genetic program of a new stage, the beginning of the reproductive phase of differentiation. Factor C may be a potent inhibitor of larger DNA segments, those parts of the genome that are inactivated during the reproductive phase. To discern the relationship between the biological specificity of factor C and the results obtained in the *in vitro* translational and transcriptional systems, further investigation is needed.

These speculations may be far from reality, but the prevailing ideas about the genetic regulation of the differentiation process do not offer any better working hypothesis.

REFERENCES

Biró, S., Vitális, S. and Szabó, G., *Acta Biol. Acad. Sci. Hung. 30:*41 (1979a).
Biró, S., Vitális, S. and Szabó, G., *Acta Biol. Acad. Sci. Hung. 30:*381 (1979b).
Biró, S., Vitális, S. and Szabó, G., *Actinomyc. Rel. Org. 14:*48 (1979c).
Biró, S., Békési, I., Vitális, S. and Szabó, G., *Eur. J. Biochem. 103:*359 (1980).
Grollman, A. P. and Stewart, M. L., *Proc. Nat. Acad. Sci. U.S.A. 61:*719 (1968).
Szabó, G., Vályi-Nagy, T. and Vitális, S., *Acta Biol. Acad. Sci. Hung. 18:*237 (1967).
Valu, G. and Szabó, G., *Acta Biol. Acad. Sci. Hung. 21:*99 (1970).
Valu, G., Békési, I. and Szabó, G., *Acta Biol. Acad. Sci. Hung. 26:*151 (1975).
Vargha, Gy., Vitális, S. and Szabó, G., *Acta Microbiol. Acad. Sci. Hung. 25:*299 (1978).
Vitális, S. and Szabó, G., *Acta Biol. Acad. Sci. Hung. 20:*85 (1969).
Vitális, S., Szabó, G. and Vályi-Nagy, T., *Acta Biol. Acad. Sci. Hung. 14:*1 (1963).
Vitális, S., Biró, S., Vargha, Gy., Békési, I. and Szabó, G., *in* "Actinomycetes" (K. P. Schaal and G. Pulverer, eds.), p. 153. Gustav Fischer Verlag, Stuttgart (1981).

BIOCHEMICAL STUDIES ON β-LACTAMASES AND PENICILLIN-BINDING PROTEINS IN *STREPTOMYCES*

Hiroshi Ogawara

Department of Biochemistry
Meiji College of Pharmacy
Tokyo, Japan

I. INTRODUCTION

Streptomyces species are known to produce various kinds of antibiotics including β-lactams. As manufacturers of antibiotics, we insist that they must be produced in the largest possible amounts without having harmful effects on the producing organisms. For this purpose, the producer organisms must protect themselves from the attack of β-lactams, their own metabolites, at least during the producing period. They do so by using three mechanisms: β-lactamases, changes of targets (penicillin-binding proteins) and a permeability barrier (compartmentalization between penicillin-binding proteins and β-lactams) (Ogawara, 1981a). Such β-lactamases and penicillin-binding proteins have attracted our attention because of their relationship to those in pathogenic bacteria. In this work, the properties of these mechanisms in *Streptomyces* are discussed, and new vector systems that are useful for the biochemical studies of β-lactamases, penicillin-binding proteins and other secondary metabolites in *Streptomyces* are presented.

II. β-LACTAMASES

As reported previously (Ogawara, 1981b), most *Streptomyces* strain produce large amounts of β-lactamases constitutively and extracellularly, irrespective of their resistance to β-lactams; these strains can be classified into five groups on the basis of their substrate specificity and physicochemical prop-

BIOLOGICAL, BIOCHEMICAL,
AND BIOMEDICAL ASPECTS OF ACTINOMYCETES

ISBN 0-12-528620-1

erties. A β-lactamase of *S. cacaoi*, which belongs to class 5 by these criteria and which can hydrolyze cloxacillin and methicillin quite well, was purified and its properties were examined (Ogawara *et al.*, 1981). The molecular weight calculated from the mobility in sodium dodecyl sulfate polyacrylamide gel electrophoresis was 34,000, slightly greater than that of other *Streptomyces* β-lactamases. The isoelectric point was 4.7 and the optimal pH was 6.5; the optimal temperature was between 40 °C and 45 °C, with the enzyme losing activity above 50°C. Among the reagents tested, N-bromosuccinimide was the strongest inhibitor, followed by iodine. *p*-Chloromercuribenzoate showed a weak inhibitory effect. These properties are often found among β-lactamases in *Streptomyces* and in other bacteria. Moreover, this enzyme has a high turnover rate (2.84×10^4 mole benzylpenicillin hydrolyzed per mole of enzyme per minute) which is comparable to that for *Staphylococcus aureus* and TEM-type enzymes but which is more than ten-fold that of cloxacillin- and methicillin-hydrolyzing enzymes from Gram-negative bacteria. This β-lactamase of *S. cacaoi* has three other noteworthy properties. First, its enzyme activity is not inhibited by sodium chloride; β-lactamases of Gram-negative bacteria, which can hydrolyze cloxacillin quite well, are inactivated by high concentrations of sodium chloride. Furthermore, the extent to which these two enzymes hydrolyze various penicillins and cephalosporins differs somewhat. This indicates that β-lactamases in Gram-negative bacteria and in *Streptomyces* have different evolutional origins and that their cloxacillin-hydrolyzing ability coincides only by chance. In any case, these types of enzymes are relatively rare among R-factor mediated- and *Streptomyces* β-lactamases; Sykes and Smith (1981) explained this as being due to a lower substrate turnover rate for these enzymes compared with that of other β-lactamases such as TEM-type enzymes. However, this explanation cannot be applied to the *Streptomyces* β-lactamase although the cloxacillin-hydrolyzing β-lactamase is also rare in *Streptomyces*. Rather, it is thought that the enzyme is a result of neutral evolution and that one of the mutated proteins happened to have the ability to hydrolyze cloxacillin. High substrate turnover rate (the relative hydrolysis rates of the enzyme are benzylpenicillin, 100; ampicillin, 38; cloxacillin, 38; methicillin, 73; and carbenicillin, 66) and broad substrate specificity can hardly explain its rare occurrence. These facts support the concept that β-lactamases have envolved convergently from many types of proteins and have one common property, the ability to hydrolyze the β-lactam ring of penicillins, cephalosporins, and other β-lactams such as nocardicin, clavulanic acid, thienamycin, and sulfazecin (Ogawara, 1975; Ogawara and Horikawa, 1979).

The second notable characteristic of this enzyme is its ability to catalyze the hydrolysis of benzylpenicillin and cloxacillin by quite different mechanisms. K_M values for benzylpenicillin increase sharply with decreasing pH. From this curve, involvement of an amino acid residue with pK value of 6.5 to 7.0 was suggested in the binding or catalytic reaction of the enzyme. In contrast, the pattern of K_M versus pH curve for cloxacillin is completely different from that for benzylpenicillin, showing a minimum at pH 7.5. The relationship between V_{max} and pH is significant in that the V_{max} values, when benzylpenicillin is the substrate, change with pH in a bell-shaped curve, whereas those for cloxacillin change only within a small range. Moreover, K_M values for benzylpenicillin change slightly with temperature, whereas those for cloxacillin change far more. Finally, the activation energy required to catalyze the hydrolysis of benzylpenicillin and cloxacillin differs, as inferred from different temperature coefficients ($V_{30°C}/V_{20°C}$) of the hydrolysis rates of both substrates. These facts suggest that the β-lactamase of *S. cacaoi* catalyzes the hydrolysis of benzylpenicillin and cloxacillin by different mechanisms. In this sense, benzylpenicillin and cloxacillin may be said to be different groups of penicillins although they have a common structure of β-lactam. They may act on different primary targets in bacteria (Noguchi *et al.*, 1979).

The third characteristic of this enzyme is that, during the life cycle of *S. cacaoi*, almost no β-lactamase activity is detectable in the membrane or in the cytoplasmic soluble fraction compared to that in the culture supernatant. This contrasts strikingly with such activity in *Bacillus licheniformis* β-lactamase and indicates that in *S. cacaoi* β-lactamase is easily released into the culture medium immediately after its biosynthesis instead of remaining in the membrane or in the cytoplasmic fraction. Despite these facts, the enzyme has many properties in common with other β-lactamases (penicillinases): both can barely hydrolyze cephalosporins and cephamycins; neither is inhibited by diisopropylfluorophosphate; and both are inactivated by low concentrations of iodine and N-bromosuccinimide.

In some *Streptomyces* strains, the trait of β-lactamase biosynthesis or its control is lost at high frequency either spontaneously or after treatment with mutagenic agents such as ultraviolet light, acriflavine, rifampicin, or ethidium bromide (Ogawara and Nozaki, 1977). In *S. lavendulae*, which has no detectable plasmid, many of the mutants that produce very low levels of β-lactamase require arginine or argininosuccinate for growth (Matsubara-Nakano *et al.*, 1980); conversely, all the arginine auxotrophic mutants examined are poor producers of β-lactamase and have the following pleiotropic effects on secondary and primary metabolism: (1) repression of β-lactamase production; (2) loss of the ability for aerial mycelium and spore formation; (3) development of acidic pH and low saturation density of growth in liquid culture; (4) decrease in antibiotic production (growth inhibitory activity against *Bacillus subtilis*) and increase in sensitivity to benzylpenicillin; (5) decrease in production of pigment; and (6), concomitant with reversion of the mutants to prototroph (arginine nonrequiring), restoration of the ability to form aerial mycelia and spores, but no recovery of β-lactamase activity (Nakano and Ogawara, 1980). However, repression of β-lactamase production in mutants with multiple mutations can be recovered by changing the nitrogen source of the culture medium. When polypeptone, polypeptone S, or casamino acid was used as nitrogen source, β-lactamase biosynthesis was repressed. On the other hand, when casein, soybean meal, or serum albumin was used as nitrogen source, β-lactamase activity and the pH of the culture medium increased to the level of the parental strain (over 3 units β-lactamase/ml; alkaline pH). In other words, retarded utilization of a nitrogen source limits the expression of secondary metabolism such as β-lactamase production in the mutant strain, whereas the parental strain remains in a derepressed state of control (Nakano and Ogawara, 1981). Retarded utilization of nitrogen and carbon source is usually essential for high production of antibiotic. In this sense, the above mentioned mutants fall within such a category and the parent strain is a "mutant".

Related to the nitrogen catabolite repression, consumption of a carbon source such as glycerol is also dependent on the nitrogen source associated with the β-lactamase biosynthesis. When casein or soybean meal is the nitrogen source, glycerol as a carbon source is used readily; the concentration of free NH_4^+ decreases; and the activity of β-lactamase increases. Then, in parallel with the increased NH_4^+ concentration in the medium, the level of β-lactamase decreases. Glutamine synthetase, glutamate dehydrogenase, glutamate synthetase, alanine dehydrogenase, and β-lactamase reached high levels of enzyme activity in the parental strain, whereas these enzymes in mutants with rapid utilization of the nitrogen source (polypeptone or polypeptone S) attained only low levels of activity. However, with a slowly utilized nitrogen source (soybean meal or casein), the activity of alanine dehydrogenase in the mutant strains reached the parental level and that of glutamine synthetase recovered partially (M. M. Nakano and H. Ogawara unpublished results). One possible explanation for these facts is that the transposable element is frequently inserted into a common regulatory gene, and such an element is related to the argininosuccinate synthetase gene on the chromosome as well as to nitrogen and carbon catabolite repression. In Gram-positive bacteria such as *Streptococcus*, transposable elements have been found to insert into other location without help from a plasmid or phage (Clewell, 1981). A similar phenomenon may occur in *S. lavendulae*. Another type of β-lactamase non-producing mutant that does not cause a pleiotropic effect is also found frequently. This type of mutant cannot produce β-lactamase even when the nitrogen source is changed. Thus, the structural gene for β-lactamase as well as its control element is unstable in *S. lavendulae*.

In plasmid-containing *S. kasugaensis*, however, no clear correlation is observed between arginine auxotrophy and the repression of β-lactamase biosynthesis although mutants requiring arginine are also obtained at high frequency (Nakano and Ogawara, 1981; Nakano *et al.*, 1980). In this case, arginine auxotrophy is closely correlated with the loss of pigment and spore formation. In addition, once the plasmid is integrated into the chromosome, a part of it or some other element transposes at high frequency to or near either the arginine gene or its control element on the chromosome, although the plasmid cannot transpose at high frequency. Furthermore, arginine nonrequiring revertants can be isolated after ethidium bromide treatment at a frequency of 10^{-9} to 10^{-12}; all the revertants examined regain the original 6.7 megadalton plasmid. In any case, the nature of the genetic element involved in the biosynthesis of β-lactamases and other secondary metabolites in *Streptomyces* should be studied in more detail at a molecular level (Fig. 1).

As described above, *Streptomyces* produce various kinds of β-lactamases constitutively and their properties are thought to be species specific, *e. g.*, cephalosporinases in Gram-negative bacteria. One of the possible reasons for this complexity is that even though unstable genetic elements control the biosynthesis of β-lactamase or its regulation, the structural genes for many β-lactamases may be on the chromosome. In addition, because penicillin-binding proteins should have a major role in defense against β-lactams, for the reasons described above and below (Ogawara, 1981a), and because β-lactamases in *Streptomyces* produce only a limited resistance to β-lactams, the genes of β-lactamases would not be affected by any evolutional and selectable pressure if they have no other physiological functions. On the other hand,

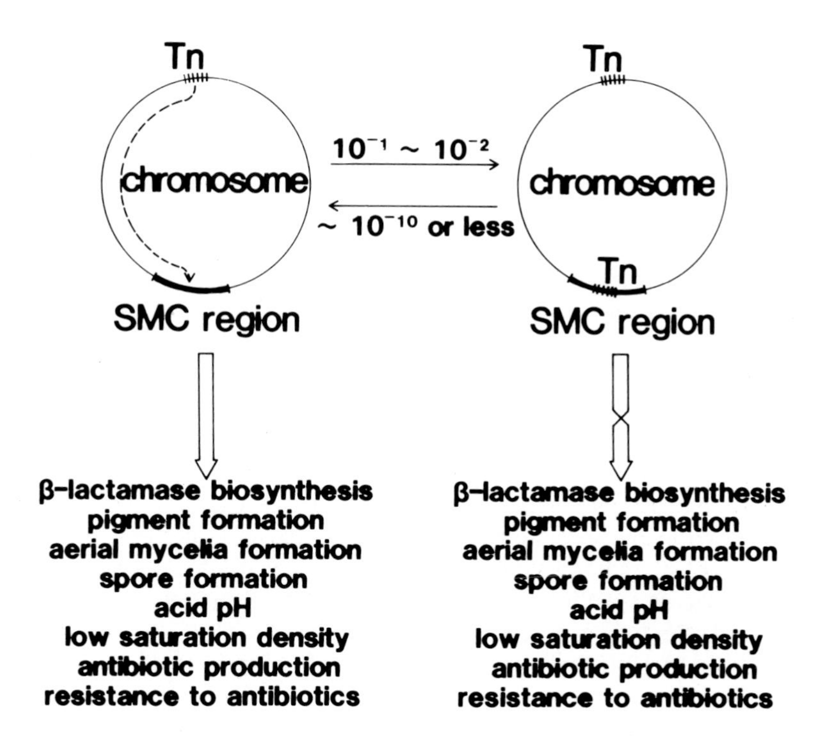

FIGURE 1. Possible control of secondary metabolism by transposable element (Tn). Gene(s) on the secondary metabolism control region (SMC region) may code for a regulatory protein or a regulatory element such as cAMP and ppGpp and is near or on the argininosuccinate synthetase gene.

the properties of β-lactamases in *Streptomyces* vary. What, then, is the physiological role of β-lactamases in *Streptomyces*? What, if any, (original) function, other than defense, could β-lactamases have in pathogenic bacteria? At the present time, there are no answers to these questions. Weinberg (1981) grouped the proposed activities of secondary metabolism into three categories: (1) relief for postulated imbalances of primary metabolites that might occur within a cell during the time when its rate of multiplication slows; (2) facilitation by physiologically active secondary metabolites of diferentiation or mediation of germination in cells; and (3) ecological roles of physiologically and pharmacologically active secondary metabolites. β-Lactamases, proteinaceous secondary metabolites like exotoxins, may fall into the second category. In this respect, it is interesting that introduction of a β-lactamase gene from *Staphylococcus aureus* into *Bacillus subtilis* impaired the biosynthesis of peptidoglycan to some extent (T. Yamakawa, personal communication).

III. PENICILLIN-BINDING PROTEINS

One of the most distinguishing characteristics of penicillin-binding proteins (PBP's) in *Streptomyces* is that their number and their affinity for penicillins differ markedly in β-lactam producing strains as opposed to β-lactam non-producing strains (Ogawara and Horikawa, 1980). When one examines PBP's by autoradiography and fluorography, at least five are detectable in the cytoplasmic membrane fractions of β-lactam non-producing strains of *Streptomyces* such as *S. cacaoi, S. felleus, S. lavendulae*, and *Streptomyces* E750-3 In addition, the fluorographic pattern of the PBP's in *S. cacaoi* is quite similar to that in *Bacillus subtilis* and in other bacteria (Nakazawa *et al.*, 1981). By comparing the affinities of many β-lactam compounds for PBP's in *S. cacaoi* with the minimum inhibitory concentrations, it was found that PBP-2 (molecular weight: 91,000) is the most probable lethal target (Ogawara and Horikawa, 1980); this is also true for *B. subtilis* (Horikawa and Ogawara, 1980). In contrast, two to five PBP's, at most, are detectable in the membrane fraction of β-lactam-producing strains of *Streptomyces*, such as *S. olivaceus, S. clavuligerus*, and *Streptomyces* strain 7371. The most probable reason is that the β-lactams produced or the genes related to production of β-lactams or its control affect the properties or production of the PBP's in *Streptomyces*. This possibility is strongly supported by the following facts. First, β-lactam-producing strains of *Streptomyces, S. olivaceus,* and *S. clavuligerus* are more resistant to benzylpenicillin than are non-producing strains, even though one of the latter group, *S. cacaoi*, produces β-lactamase constitutively. In addition, the former strains become more resistant not only to their own β-lactam compounds, but also to all the other β-lactams or unrelated metabolites. Second, under the same conditions, fluorographic bands of PBP's in β-lactam producers are generally fainter than those in β-lactam non-producing strains such as *S. cacaoi*. Third, clavulanic acid (a natural β-lactam), mecillinam, and methicillin do not bind at all to PBP's of the β-lactam producers, *S. olivaceus* and *S. clavuligerus*. In contrast, in *S. cacaoi*, mecillinam which is known to bind to PBP-2 (molecular weight: 66,000) in *Escherichia coli* was shown to bind to PBP-2 (molecular weight: 91,000) selectively and preferentially. Moreover, methicillin bound at very low concentrations to three PBP's (PBP-1, -3, and -4; molecular weights: 105,000; 64,000; and 55,000 respectively) in *S. cacaoi*. Even clavulanic acid bound to PBP-1 (molecular weight: 105,000) although at very high concentrations. Fourth, PBP's in the β-lactam nonproducing strains, *S. cacaoi, S. felleus, S. lavendulae*, and *Streptomyces* E750-3 produce quite similar fluorographic patterns, whereas, the fluorographic patterns of PBP's in the membrane fraction of β-lactam producers are not so similar. In addition, the PBP's in the culture supernatants and in the cytoplasmic supernatant fractions from two β-lactam producers, *S. olivaceus* and *S. clavuligerus*, gave extremely different patterns. *S. clavuligerus* produced more than ten PBP's some of which appeared in both the culture filtrate and in the cytoplasmic fraction, whereas *S. olivaceus* showed few or no PBP's in these fractions or in the membrane. Fifth, the duration of *Streptomyces* in culture in liquid medium affected the patterns of PBP's, especially in β-lactam-producing strains. Throughout the growth cycle, the β-lactam non-producers, *S. cacaoi, S. felleus*, and *S. lavendulae* showed many PBP's, with those of molecular weights 45,000 to 50,000 always detected. The high molecular weight PBP's in *S. felleus* (molecular weights: 140,000, 127,000 and 90,000) and *S. lavendulae* (molecular weights: 145,000, 140,000, and 91,000) showed little change during the growth phase. In contrast, β-lactam producers, *S. flavogriseus* and *Streptomyces* 7371, showed few PBP's and only in the early logarithmic phase. Furthermore, in *S. clavuligerus*, the patterns of PBP's in early logarithmic phase were clearly different from those in other phases. PBP's of molecular weight 120,000 could be detected in the early logarithmic phase, but PBP-1 (molecular weight: 83,000), PBP-2, (molecular weight: 79,000) and PBP-3 (molecular weight: 47,000) appeared only slightly. The change in the patterns of the PBP's seems to be related to the biosynthesis of cephamycin and clavulanic acid. It can be concluded, therefore, that production of β-lactam compounds does not affect all the PBP's to the same extent; rather it affects each PBP in a different manner and to a different degree, depending on the strain, the growth condition, the growth phase, and so on. In addition, all the PBP's in *Streptomyces* generally have very low affinity for benzylpenicillin. Thus, even though PBP-3 (molecular weight: 64,000) has the highest affinity for benzylpenicillin among the PBP's in *S. cacaoi*, 10 μg/ml is needed to saturate the PBP-3. PBP-2 (molecular weight: 91,000) in S. *cacaoi* and all the PBP's in *S. clavuligerus* could not be saturated with benzylpenicillin at a concentration of 50 μg/ml. PBP-2 in *S. cacaoi* has the lowest affinity and confirms the suggestion that this

PBP may be a lethal target of many β-lactam compounds in this strain, as deduced from the comparison of their antibacterial activities with their affinities for PBP's. Furthermore, it should be noted that none of the PBP's in *S. clavuligerus* are anywhere near being saturated even at 50 μg benzylpenicillin/ml. These results reflect the high minimum inhibitory concentrations of benzylpenicillin (10 to 50 μg/ml, or more) in *Streptomyces* in general, although they are Gram-positive bacteria. The values are more than 30-fold higher in *S. cacaoi* and *S. clavuligerus* than in *B. subtilis* (Ogawara and Horikawa, 1980).

That few PBP's could be detected in the membrane fraction of β-lactam producers of *Streptomyces* does not seem to result from the disappearance of the corresponding PBP's by binding of β-lactams, because the patterns of these proteins are completely different in the two groups of *Streptomyces*. Also, although some PBP's have similar molecular weights, their properties are quite different. For example, the molecular weights of PBP-6 in *S. cacaoi* and PBP-3 in *S. clavuligerus* are similar (47,000), yet the former protein is a minor component (7% of the total PBP's), but the latter is a major band (89% of the total). Furthermore, the binding capacity of PBP-6 for benzylpenicillin in *S. cacaoi* is highly sensitive to heat and is lost after treatment for 10 minutes at 50 °C, whereas that of PBP-3 in *S. clavuligerus* retains about 50% of this capacity after incubation for 10 minutes at 60 °C (Horikawa *et al.*, 1980). Release of [^{14}C]-benzylpenicillin from PBP-1 (molecular weight: 83,000) or from PBP-2 (molecular weight: 79,000) was not observed in *S. clavuligerus* even after incubation for 80 min with a large excess of non-radioactive benzylpenicillin, whereas that from PBP-1 (molecular weight: 105,000) and PBP-2 (molecular weight: 91,000) of *S. cacaoi* was relatively rapid. Conversely, the release from complexes with lower molecular weight components, *i. e.*, PBP-4 (molecular weight: 55,000) and PBP-5 (molecular weight: 50,000), was extremely slow. In any case, one of the distinctive characteristics of PBP's in *Streptomyces* species is that complete dissociation of [^{14}C]-benzylpenicillin does not occur from all the PBP's.

Such differences are quite significant considering that these *Streptomyces* species belong to the same genus. Although *Escherichia coli, Salmonella typhimurium, Pseudomonas aeruginosa*, and *Proteus* belong to different genera, the patterns of PBP's in these bacteria are very much alike. This is also the case with *Bacillus subtilis, B. licheniformis*, and *B. megaterium*. It is concluded that *Streptomyces* species protect themselves from attack by their own metabolites, β-lactams, through production of PBP's having lower affinity and by altering their biosynthesis and other properties to various degrees during the cell cycle (Fig. 2).

To pursue further the roles of PBP's in resistance to β-lactams, β-lactam-resistant strains were isolated by treatment with N-methyl-N'-nitro-N-nitrosoguanidine (H. Nakazawa and H. Ogawara unpublished results). The minimum inhibitory concentrations of benzylpenicillin for these mutants were

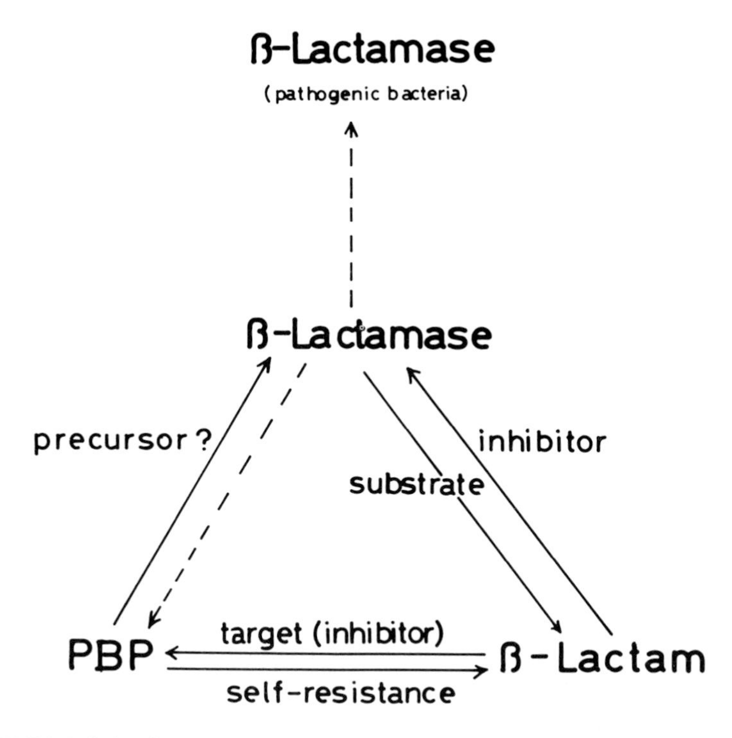

FIGURE 2. Relationships among β-lactamase, penicillin-binding protein (PBP), and β-lactam.

1,000 to 4,000 μg/ml, compared with 250 μg/ml in the parental strain. The minimum inhibitory concentrations of mecillinam, cephamycin C, and methicillin also increased. In these mutants, two particular changes were observed in the fluorographic patterns of PBP's. First, the amount of PBP-1* (molecular weight: 130,000) increased considerably. Second, the affinity of benzylpenicillin, mecillinam, clavulanic acid, and cephamycin C for PBP's, especially PBP-2, a presumed lethal target of many β-lactams, was lowered in these resistant mutants. In addition, the amounts of [^{14}C]-benzylpenicillin bound to whole cells or to the membrane fraction were also lowered. Consequently, it is suggested that these mutants protect themselves from the attack of exogenous β-lactams by changing PBP's and permeability (compartmentalization between PBP's and β-lactams). When these mutants were treated with Triton X-100, EDTA, or toluene, their sensitivity to benzylpenicillin and other β-lactams and the amount of [^{14}C]-benzylpenicillin bound to whole cells or to the membrane fraction almost reached the parental level. The fluorographic patterns of PBP's in the mutants treated with the above reagents also changed resembling those in parental strains similarly treated. Interestingly, the minimum inhibitory concentrations of β-lactam compounds for the mutants were not far below that the parental strain even after treatment with the above reagents. This confirms the previous finding that the major factor involved in resistance to β-lactams in the parent strain is the intrinsic changes of the PBP's, even though this strains does not produce a β-lactam compound and is a Gram-positive bacterium.

IV. CLONING VECTOR

There are no detailed reports on *Streptomyces* at the molecular level, essentially because the production of antibiotics by *Streptomyces* strains has been improved through a combination of random mutation and screening. Consequently, the genetic background of *Streptomyces* is poorly understood compared to that of *Escherichia coli, Bacillus subtilis*, etc. Recently, however, recombinant DNA technology has been applied to *Streptomyces*. As a first steep in using this technique to clarify the molecular mechanism that controls biosynthesis of β-lactamase, PBP's, and other secondary metabolites in this species, we have tried to develop new vector systems in *S. lavendulae* by using a

TABLE I. Numbers of Cleavage Sites of pSL1 with Restriction Endonucleases

Number of sites	*Restriction endonuclease*	*Sequence* [a]
0	Bam *HI*	*G/GATCC*
	Hind *III*	*A/AGCTT*
	Xho *I*	*C/TCGAG*
	Pst *I*	*CTGCA/G*
1	Eco *RI*	*G/AATTC*
	Sal *I*	*G/TCGAC*
	Bgl *II*	*A/GATCT*
	Sma *I*	*CCC|GGG*
	Hinc *II*	*GTPy|PuAC*
	Sst *I*	*GAGCT/C*
	Acc *I*	$GT/\binom{A}{C}\binom{G}{T}AC$
2	Pvu *II*	*CAG CTG*
	Kpn *I*	*GGTAC/C*
	Mlu *I*	*A/CGCGT*
	Sst *II*	*CCGC/GG*
⩾*6*	Alu *I*	*AG CT*
many	Hha *I*	*GCG/C*
	Hae *III*	*GG CC*

[a]*/: cohesive end;* | *: blunt end.*

plasmid and a bacteriophage. Screening a plasmid in some strains of *S. lavendulae* gave us a new, small plasmid named pSL1 from *S. lavendulae* S985 (Nakano *et al.*, 1980). The molecular weight as determined by electrophoretic mobilities of the fragments produced by digestion with various restriction endonucleases and by electron microscopy was 2.6 megadaltons; the copy number was over 100. Table I shows the number of cleavage sites of pSL1 for the various restriction enzymes and their base sequences. There are no cleavage sites for *Bam* HI, *Hind* III, *Xho* I, or *Pst* I, whereas there is one cleavage site each for *Eco* RI, *Sal* I, *Bgl* II, *Sma* I, *Hinc* II, *Sst* I, and *Acc* I. Base sequences recognized by *Sal* I and *Sma* I are useful for the cloning of genes in *Streptomyces. Sma* I and *Hinc*II produce blunt ends of the sequence, so that these sites are available for the cloning of many genes in *Streptomyces* and other organisms. It is interesting that, although pSL1 is a small plasmid (among physically characterized plasmids in *Streptomyces*, pSL1 is the smallest), it contains one or two cleavage sites for many of the endonucleases examined. It is also thought to be quite common.

Despite these advantageous properties, pSL1 has no known physiological function or selectable markers as do many other plasmids in *Streptomyces.* Therefore, it is necessary to provide a suitable selectable marker for it. One method is to make a composite plasmid having either a drug resistance marker or other suitable marker from other bacteria. Another method is to clone a suitable marker of *Streptomyces* into pSL1. The former method was first tried by using a kanamycin-resistant marker on pCR1 (8.7 megadaltons) of *Escherichia coli* since *Streptomyces* strains tested in this study were more sensitive to kanamycin than to other antibiotics. The construction mode of chimera plasmids is shown in Figure 3. pSL1 and pCR1 were ligated through the cohesive ends generated by *Sal* I cleavage. After transformation of *E. coli* with the ligated mixture, transformants were selected by kanamycin resistance, followed by colony hybridization with [^{32}P] pSL1. The ligation of the two plasmids resulted in pMNO having two cleavage sites for *Eco* RI and for *Sal* I. Then, pMN1 was constructed by cleavage with *Eco* RI and ligation with T4 ligase. This composite plasmid has many advantages: (I) The plasmid carries the kanamycin-resistance gene suitable for selection and cloning DNA; (II) it contains most of the DNA sequence of pSL1; (III) it has one cleavage site for *Eco* RI and for *Sal* I so that other DNA sequences may be introduced. (The *Sal* I site is especially convenient because *Sal* I enzyme is known to digest chromosomal DNA in *Streptomyces* quite efficiently; and (IV) the plasmid DNA can be amplified in *E. coli* in the presence of chloramphenicol. With these advantages, pMN1 coupled with pMNO is useful for cloning the particular *Streptomyces* gene DNA in *E. coli*; after enrichment of the gene in *E. coli*, it can be transferred to *Streptomyces*. However, one problem with this hibrid plasmid is that the kanamycin resistance marker from *E. coli* is expressed in *S. lavendulae* only to a limited extent. A similar species barrier for gene expression is known to exist between *B. subtilis* and *E. coli* (Ehrlich, 1978).

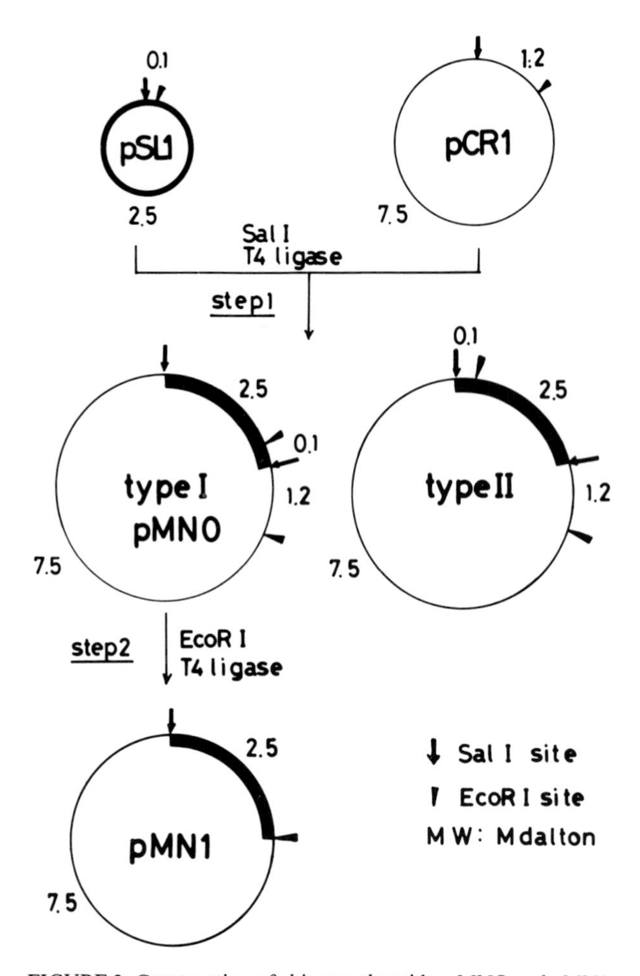

FIGURE 3. Construction of chimera plasmids, pMNO and pMN1.

As described above, most of the plasmids in *Streptomyces* have no known physiological function and are "cryptic". In addition, some plasmids have only a limited host range. Therefore, the phage vector system is quite convenient, if it is not restricted in a wide range of *Streptomyces* species. Actinophage Bα isolated from *S. lavendulae* S283 has a wide host range (Nakano *et al.*, 1981) in that, of eleven strains studied, eight (*S. flavus* IFO3359, *S. microflavus* ETH24414, *S. somaliensis* ATCC19017, *S. kasugaensis, S. parvulus* ATCC12434, *S. lipmanii* ISP5070, *S. lavendulae*, and *S. cattleya* NRRL8057) were sensitive; the rest (*S. californicus* ATCC3312, *S. felleus* NRRL2251, and *S. coelicolor* A3(2)) were resistant to Bα. Bα produced turbid plaques characteristics of a temperate phage on *S. lavendulae* and on other *Strep-*

tomyces strains. This phage has a cleavage site for *Eco* RI, *Pst* I, and *Pvu* II and a molecular weight of about 34 megadaltons. Consequently, actinophage Bα is useful as a cloning vector in many *Streptomyces* strains. To this end, we attempted to isolate deletion mutants by selecting resistant mutants to pyrophosphate and heat (H. Ishihara, M. M. Nakano, and H. Ogawara, unpublished results). One mutant, designated A10, has a deletion of about 2.5 megadaltons near one end of the genome, including a single *Eco* RI site. Actinophage Bα and its mutant, A10, were transfected to the protoplast of *S. lavendulae* S985 at an efficiency of 2×10^4 per μg Bα DNA. It is concluded that, together with the presence of single cleavage site for *Pst* I and *Pvu* II, the deletion mutant is useful for cloning genes in various species of *Streptomyces*.

REFERENCES

Clewell, D. B., *Microbiol. Rev. 45:*409 (1981).

Ehrlich S. D., *Proc. Nat. Acad. Sci. U. S. A. 75:*1433 (1978).

Horikawa, S., and Ogawara, H., *J. Antibiot. 33:*614 (1980).

Horikawa, S., Nakazawa, H., and Ogawara, H., *J. Antibiot. 33:*1363 (1980).

Matsubara-Nakano, M., Kataoka, Y., and Ogawara, H., *Antimicrob. Agents Chemother. 17:*124 (1980).

Nakano, M. M., Ishihara, H., and Ogawara H., *J. Gen. Microbiol. 122:*289 (1981).

Nakano, M. M., and Ogawara, H., *J. Antibiot. 33:*420 (1980).

Nakano, M. M., and Ogawara, H., *in* "Advances in Biotechnology", Vol. 3. (C. Vezina and K. Singh, eds.), p. 31. Pergamon Press, Toronto (1981).

Nakano, M. M., Ozawa K., and Ogawara H., *FEMS Microbiol. Lett. 9:*111 (1980).

Nakano, M. M., Ozawa, K., and Ogawara, H., *J. Bacteriol. 143:*1501 (1980).

Nakazawa, H., Horikawa, S., and Ogawara, H., *J. Antibiot. 34:*1070 (1981).

Noguchi, H., Matsuhashi, M., and Mitsuhashi, S., *Eur. J. Biochem. 100:*41 (1979).

Ogawara, H., *Tanpakushitsu Kakusan Koso 20:*1214 (1975). (in Japanese)

Ogawara, H., *Microbiol. Rev.* 45:591 (1981a).

Ogawara, H., *in* "Actinomycetes" (K. P. Schaal and G. Pulverer eds.), p. 291. Gustav Fischer Verlag, Stuttgart (1981b).

Ogawara, H., Mantoku, A., and Shimada, S., *J. Biol. Chem. 256:*2649 (1981).

Ogawara, H. and Horikawa, S., *J. Antibiot. 32:*1328 (1979).

Ogawara, H., and Horikawa, S., *J. Antibiot. 33:*620 (1980).

Ogawara, H., and Horikawa, S., *Antimicrob. Agents Chemother. 17:*1 (1980).

Ogawara, H. and Nozaki, S., *J. Antibiot. 30:*337 (1977).

Sykes, R. B. and Smith, J. T., *in* "Beta-lactamases". (J. M. T. Hamilton-Miller and J. T. Smith, eds.), p. 369. Academic Press, London (1981).

Weinberg, E. D., *in* "Microbiology—1981". (D. Schlessinger, ed.), p. 356. American Society for Microbiology, Washington (1981).

POLYMERS OF THE CELL WALLS OF *STREPTOMYCES ROSEOFLAVUS* VAR. *ROSEOFUNGINI* 1128 AND ITS UNDIFFERENTIATED VARIANT 1-68

I. B. Naumova

Biology Department
M. V. Lomonosov State University
Moscow, U. S. S. R.

N. S. Agre
N. K. Skoblilova

Institute of Biochemistry and Physiology of Microorganisms
Academy of Sciences of the U.S.S.R.
Pustchino, U. S. S. R.

A.S. Shashkov

N. D. Zelinsky Institute of Organic Chemistry
Academy of Sciences of the U.S.S.R.
Moscow, U. S. S. R.

I. INTRODUCTION

What changes, is any, have occurred in cell wall polymers of streptomycetes due to impairment of cellular differentiation? No information that would allow one to answer this question with a high degree of certainty is available. To get at least one step nearer to the answer to this question, we studied the cell walls of *Streptomyces roseoflavus* var. *roseofungini* 1128 and its variant 1-68 which differs from the parent culture in that it lacks aerial mycelium and spores and its substrate mycelium tend to fragment.

Perhaps in the future, studies on changes in the composition of the cell wall

ISBN 0-12-528620-1

of streptomycete mutants in which individual steps of cellular differentiation are blocked may be of use for understanding the functions of cell wall polymers.

In our studies, we placed emphasis on teichoic acids which have been shown to be active regulators of biochemical processes occurring in the surface regions of bacterial cells. With bacterial mutants, it was demonstrated that impairment of cell division and formation of filamentous forms may be due to (or may be associated with) changes in the structure of teichoic and teichuronic acids of the cell wall mutants (Robson and Baddiley, 1977a,b). Indirect data obtained with "bald" mutants of *S. coelicolor* (Chater and Merrick, 1980) indicate that these changes may also take place in streptomycete mutants which have anomalities in cellular differentiation.

The present work sums up the results of the first step in our study of the cell wall polymers of *S. roseoflavus* var. *roseofungini* 1128 and its "nocardioform" variant 1-68.

II. MATERIALS AND METHODS

The cultures of *S. roseoflavus* var. *roseofungini* 1128 and of its variant 1-68 have been studied previously, and their peculiarities were described in detail (Kalakoutskii and Nikitina, 1977). The conditions for the growth of the cultures and the composition of the peptone-yeast medium (PYM) have been described elsewhere (Skoblilova *et al.*, 1981).

The dissociation of the studied cultures was monitored by periodically inoculating Chapeck's agar with suitable dilutions of PYM-grown cultures. To reveal septa, mycelia were fixed with the Carnoy liquid and was stained with crystalline violet according to Gutstein (Prokofieva-Belgovskaya, 1963). The preparations were examined with a MBI-6 microscope. The number of septa

TABLE I. Frequency of Septation of Hyphae of Strain 1128 and Variant 1-68 (Submerged Cultures on PYM)

Mycelium age (h)	Average distance between septa (μm)	
	Parent culture 1128	"Nocardioform" variant 1-68
9	11.75	5.47
18	6.13	3.40
22	7.14	2.59
31	6.39	2.69
48	6.39	2.57

was counted in hyphae not less than 300 μm long. The average distance between the septa was determined by dividing the length of a hypha by the number of its septa. The ultrastructure of the septa in the sections of hyphae was studied by a standard method (Suyetin *et al.*, 1980) with a JEM-100 B microscope. The cell walls and teichoic acids preparations were obtained as described by Skoblilova *et al.* (1981). Acidic, alkaline, and enzymatic hydrolyses, ammoniolysis and hydroxylaminolysis of teichoic acids and cell walls, determination of the length of teichoic acid chains, as well as analytical procedures, paper chromatography and electrophoresis, were carried out as described by Skoblilova *et al.* (1981). ^{13}C-NMR spectroscopic study was performed under the conditions described in Naumova *et al.* (1982).

III. RESULTS AND DISCUSSION

As the parent strain grows in submerged culture, a number of variants which no longer are capable of forming aerial mycelium and spores emerge (Salekh *et al.*, 1979). It was necessary to obtain a homogeneous biomass devoid of the variant admixtures. By plating the biomass of the studied cultures growing in PYM onto Chapeck medium containing agar, we determined that during the logarithmic growth phase no dissociation was observed in the cultures. Biomass yield of both cultures reached a maximum after 24 h and was followed by a short steady-state phase; after 30 h, the mycelium biomass in the parent culture decreased noticeably, and a great number of variants appeared in the mycelium platings. The mycelial mass of the cultures was thus quite homogeneous in the logarithmic phase. It was this biomass that was used for subsequent biochemical analyses.

Comparing the morphology and the ultrastructure of the mycelium of parent culture 1128 with that of variant 1-68, we paid attention to two details. First, the hyphae of the variant contained much more (two to three fold) septa than did those of the parent culture (Table I and Fig. 1a,b). Second, the septa in the variant hyphae were not monolayered, as in the parent mycelium, but rather were trilayered, *i. e.*, divided into two parts by an intermediate light layer (Fig. 2 a,b).

The analysis of the walls of the cultures under study showed that peptidoglycan comprised 50% of their dry weight and had a structure typical of streptomycetes, *i. e.*, A3γ (Tipper and Wright, 1979). No differences were shown in the structure or in the quantitative content of the peptidoglycan of the two cultures (Table II).

However, the cell walls of the streptomycetes differed greatly in the quantitative content and the structure of polysaccharide polymers. Both organisms had teichoic acid in their cell walls (Table III and IV).

The structure of teichoic acids was studied by chemical methods and ^{13}C-NMR spectroscopy. This study consisted in the degradation of polymers in

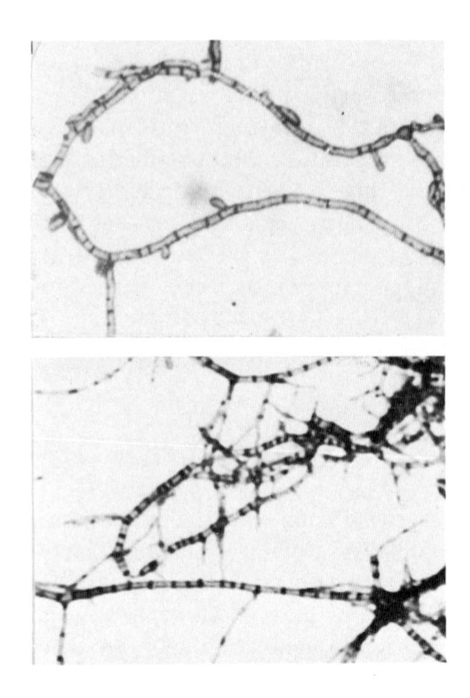

FIGURE 1. A) *Hyphae of the parent strain 1128;* B) *Hyphae of variant 1-68.*

acid or alkaline medium and the subsequent determination of the degradation products. The compounds resulting from acid hydrolysis provided information on the qualitative composition of the polymer. Strong acid hydrolysis (2 N HCl, 3 h, 100 °C) of the teichoic acid of the parent strain yielded the following products: glycerol, its mono- and di-phosphates, glucosamine, lysine, inorganic phosphate, and acetic acid. This hydrolysis indicated that the teichoic acid is a glycerol phosphate polymer containing lysine and N-acetylglucosamine.

The question of the monomeric units of the chain and the type of phosphodiester linkage between them was resolved when the products of alkaline hydrolysis were analyzed. The alkaline hydrolysis of teichoic acid of the parent strain gave glycerol, its mono- and di-phosphates, lysine, 2-acetamido-2-deoxy-α-D-glucopyranosyl-(1-2)-glycerol, and two glycerol phosphodiesters, N-acetylglucosaminyldiglycerol diphosphate (1) and N-acetylglucosaminyltriglycerol tetraphosphate (2). The structure of glycoside as well as of esters (1) and (2) was determined by acid and enzymatic degradation followed by an electrophoretic and chromatographic analyses of the products formed (Naumova *et al.*, 1982; Skoblilova *et al.*, 1982).

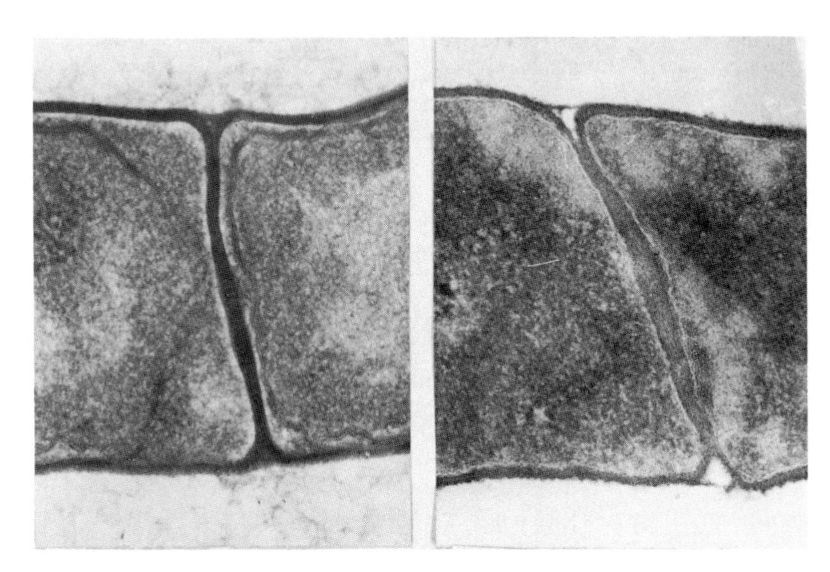

FIGURE 2. A) *Septa of the parent strain 1128;* B) *Septa of variant 1-68. Scale: 0.2* μM.

Formation of great amounts of glycerol mono- and di-phosphates as a result of splitting of the polymer with alkali indicated, first, that the initial chain was poly(glycerol phosphate) in which the phosphodiester bond linked glycerol units (sugar components were not involved in chain formation), and second, that the polymer contained free glycerol residues. The presence of N-acetylglucosaminylglycerol in the hydrolysis products provided evidence in favor of the presence of substituted glycerol units in the chain. The type of phosphodiester bond was determined as follows. Among the products of hydrolysis, phosphodiesters of glycerol (1) and (2) were found, which were indicative of the presence of the 1,3-poly(glycerol phosphates), the substitution of the glycerol unit at position 2 by a sugar component results in formation of alkali-stable glycerol phosphodiesters of various complexity (Ellwood *et al.*, 1963).

TABLE II. Molar Ratio of Cell Wall Peptidoglycan Components

Culture	*Preparation*	*Mur*	*GlcN*	*Ala*	*DAP*	*Glu*	*Gly*
1128	*Peptidoglycan*	*0.9*	*1.2*	*1.9*	*1.0*	*0.8*	*0.8*
	Cell wall	*0.8*	*2.2*	*1.8*	*1.0*	*0.9*	*1.0*
Variant 1-68	*Peptidoglycan*	*0.8*	*1.2*	*1.9*	*1.0*	*0.9*	*0.9*
	Cell wall	*0.9*	*1.9*	*1.8*	*1.0*	*0.9*	*1.0*

TABLE III. Characteristic of Teichoic Acid Preparations

Culture	P total[a]	P NA[b]	P lab[c]	P TA[d]	Lys	GlcNAc	P NA:GlcNAc	PTA:Lys
	(% of dry weight)						(mole:mole)	
1128	7.35	0.01	0.00	7.34	7.36	12.84	3.3:1	4.7:1
1-68	12.00	0.15	0.00	11.85	1.30	5.99	11.4:1	42.9:1

[a]P_{total}: total content of phosphorus
[b]P_{NA}: phosphorus in nucleic acids
[c]P_{lab}: phosphorus mineralized in 1 N HCl for 7 min at 100°C
[d]P_{TA}: phosphorus of teichoic acids.

It is unlikely that there are two neighboring N-acetyl-glucosaminylglycerol units in the chain since in this case the products of alkaline hydrolysis would contain phosphodiester with two amino sugar residues. The absence of glycosyl glycerol phosphomonoester in the hydrolysis products demonstrated that the terminal glycerol residue carrying a phosphomonoester group was not substituted by amino sugar.

The presence of heterogeneous chains with various degrees of substitution by a sugar component was possible. The presence in the mixture of free and totally substituted poly (glycerol phosphate) chains was excluded since the latter cannot be hydrolyzed by alkali and the products of degradation therefore would not contain glycerol phosphodiesters with a sugar substituent.

Two independent methods were used for determining the chain length of the teichoic acid: the amount of formaldehyde resulting from periodate oxidation of the polymer, and the amount of inorganic phosphate formed after the treatment of the polymer with phosphomonoesterase. Both methods yielded com-

TABLE IV. Content of Teichoic Acid Phosphorus, Lysine, and Galactose in the Cell Walls of Cultures of Various Ages

Culture	Age (h)	P[a]TA	Lys	Gal	PTA: Lys (mole: mole)
		(% of dry weight of preparation)			
	9	3.5	3.3	no	5.0:1
1128	18	4.4	4.4	no	4.7:1
	22	4.2	4.3	no	4.6:1
Variant 1-68	9	1.5	0.7	2.1	10.1:1
	18	1.2	0.6	5.6	9.4:1
	22	1.3	0.6	3.1	10.2:1

[a]P_{TA}: phosphorus of teichoic acids.

patible results which allow one to suggest that a chain contains 11 to 13 glycerol phosphate units.

As stated above, lysine was found among the products of hydrolysis of teichoic acid; by the method of circular dichroism, its configuration was determined to be L-lysine. By ammoniolysis and hydroxylaminolysis, it was established that lysine was linked with the polymer by an ester bond.

It should be noted that the content of glucosamine varied in the teichoic acid preparations which had been isolated from the cell walls of various batches of mycelium, with values of the molar ratio phosphorus/glucosamine (P/GlcNAc) ranging from 3.3:1.0 to 5.7:1.0; that of the lysine content, with values of the molar ratio P/Lys ranging from 3.3:1.0 to 4.7:1.0.

The study of the teichoic acid of variant 1-68, which was carried out according to the same scheme, showed that the polymer was also, 1,3-poly (glycerol phosphate) with a chain length of 14 to 15 glycerol phosphate units. However, this polymer contained a smaller amount of substituents. Values of the P/GlcNAc and P/Lys molar ratios in the preparations of teichoic acid from the cell walls of various batches of mycelium ranged from 11.4:1.0 to 47.0:1.0 and from 13.4:1.0 to 42.9:1.0, respectively. The ^{13}C-NMR spectroscopic study of both teichoic acids supported the results of the chemical analysis (Fig. 3 and Table V).

Thus, comparison of the data of the analysis of teichoic acids in the parent strain and its undifferentiated variant 1-68 showed that both acids had 1,3-poly(glycerol phosphate) chains of a similar length. But, unlike the teichoic acid of the parent strain polymer, that of the variant contained a smaller amount of L-lysyl and N-acetylglucosaminyl substituents. In the teichoic acid of the parent strain, 25 to 30% of glycerol units were linked by an α-glucoside bond with an amino sugar residue and 20 to 30% by an esteric bond with L-lysine; in the teichoic acid of the variant, the number of substituted glycerol units was 2 to 10% and 5 to 7%, respectively. The cell walls of the cultures also differed in the content of teichoic acids, which was 50% of the dry weight in the parent strain and 9% in the variant. This regularity could be traced during the analysis of the walls prepared from mycelium of various ages (Table IV).

The cell wall of the variant was found to contain, besides teichoic acid, another polysaccharide which was composed of galactose and glucosamine in the molar ratio 5:1. This polymer was present in the walls of the variant mycelium of various ages (Table IV). The cell wall of the parent strain contained no monosaccharides other than the N-acetylglucosamine of the teichoic acid.

In conclusion, we shall dwell on the peculiarity which we discovered in the structure of the cell walls of parent strain 1128 and its variant 1-68, namely, the presence of lysyl residues in the cell wall (up to 4.4% in that of the parent strain, Table IV). As stated above, the lysine was linked to the teichoic acid by an esteric bond; it was not identified in peptidoglycan. It is of interest that the cell

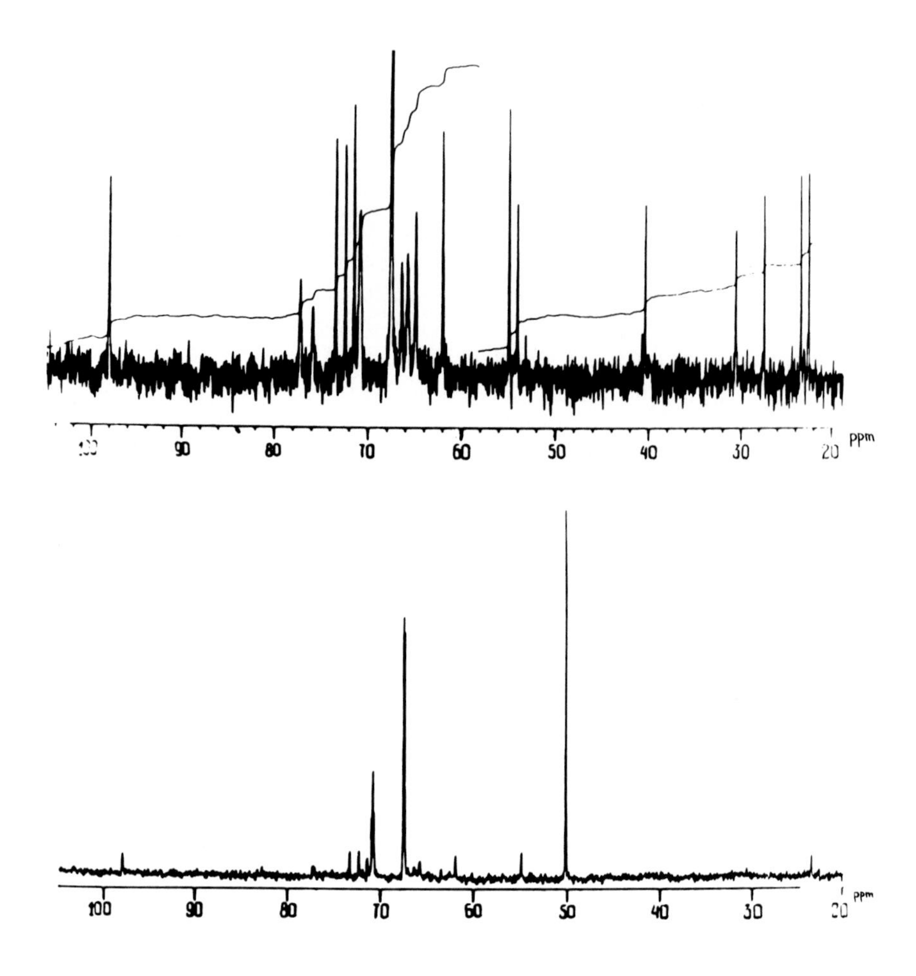

FIGURE 3. ^{13}C-NMR spectrum of teichoic acid of A) Streptomyces roseoflavus *var.* roseofungini *1128;* B) *variant 1-68.*

walls of the parent strain had five to seven times as many residues as did the walls of the variant strain. This was due not only to a lower content of teichoic acid in the cell walls of the variant, but also to the fact that the teichoic acid of the variant contained fewer lysyl groups.

The difference in the content of labile amino acid ester groups in the cell wall of the two morphological streptomycete variants makes us wonder about the role of these compounds in the functioning of teichoic acids. In some cases, polysaccharide polymers of bacterial cell walls are known to be regulators of biochemical processes occurring in the surface regions of the bacterial cell. For example, it has been shown that, in pneumococci, the work of autolytic enzymes is regulated by wall teichoic and lipoteichoic acids (Tomasz *et al.*, 1975).

TABLE V. Chemical Shifts of the Lysyl Teichoic Acid from Streptomyces roseoflavus *var.* roseofungini *1128 and Some Model Compounds*[a]

	C1	C2	C3	C4	C5	C6	OCH_3	CH_3CO
Nonsubstituted glycerol residues in teichoic acid	67.5	70.8	67.5					
2-acetamido-2-deoxy-α-D-glucopyranose-substituted glycerol residues in teichoic acid	66.3[b]	77.1	65.7[b]					
L-lysine-substituted glycerol residues in teichoic acid	64.8	75.8	64.8					
Residues of 2-acetamido-2-deoxy α-D-glycopyranose in teichoic acid	98.0	54.9	72.3	71.4	73.3	61.9		23.4 175.5
Methyl-2-acetamido-2-deoxy-α-D-glycopyranoside	99.3	54.85	72.4	71.3	72.9	61.9	56.4	23.15 175.6
L-lysine residues in teichoic acid	170.3	54.0	27.45	22.5	30.5	40.3		
Methylate of L-lysine (dichlorohydrate)	171.5	54.7	27.3	22.5	30.3	40.0	53.7	

[a] *All the signals of glycerol residues in the spectrum were broadened compared with the signals of the residues of lysine or 2-acetamido-2-deoxy-α-D-glucopyranose.*
[b] *Reference of these signals to C1 or C3 was arbitrary. Difference in chemical shifts was caused by the presence of chiral carbon.*

More recent studies have shown that the presence in teichoic acid of O-D-alanyl units, which partly neutralize the negative charge of the chain, make autolysine incapable of interacting with a lipoteichoic acid chain (Fischer *et al.*, 1981).

It is likely that the teichoic acids under study may be involved in the regulation of the work of autolysines and that the leading role in this process is played by O-L-lysyl residues which, since they have two positively charged amino groups, are more suitable for neutralizing the charge of the phosphate groups of the chain than are O-D-alanyl residues. Apparently, the cell wall of variant 1-68 which appears devoid of lysyl groups at some sites might be more efficient in interaction with an autolytic enzyme which, if localized in these regions, may carry out partial hydrolysis. The above-mentioned differences in the fine structure of the septa of the parent culture and of the variant may speak in favor of this idea. The presence of the intermediate layer, possibly the autolytic zone, in the septa of the variant was probably due to the well-regulated activity of autolytic enzymes, which resulted in mycelium fragmentation.

IV. CONCLUSIONS

A study has been made of the cell walls of *Streptomyces roseoflavus* var. *roseofungini* and of its undifferentiated variant 1-68 which differs from the parent culture by its lack of aerial mycelium and spore formation, by higher frequency of septation and fragmentation of the hyphae of substrate mycelium. Differences were found in the quantitative content and structure of wall polysaccharide polymers. The cell wall of the parent strain contained primarily peptidoglycan and teichoic acid. The cell wall of the variant contained, in addition to these polymers, another polysaccharide comprised of galactose and N-acetylglucosamine. The types of teichoic acid structure in the parent culture and in the variant were similar: a 1,3-poly(glycerol phosphate) chain with lysyl and N-acetylglucosaminyl substituents. However, in the teichoic acid of the parent culture, 25 to 30% of glycerol units had an N-acetylglucosaminyl substituent and 20 to 30% of glycerol units had an O-L-lysyl one; the teichoic acid of the variant had 2 to 10% and 5 to 7% of substituted glycerol units, respectively. Also, the amount of this polymer in the wall of the parent culture was four to seven times higher as compared with that in the wall of the variant.

Peptidoglycans of the cultures studied were identical in structure and belong to the A3γ type that is characteristic of streptomycetes. The content of peptidoglycan in the cell walls of both cultures was equal.

REFERENCES

Chater, K. F., and Merrick, M. J., *in* "Biology of Development in Prokaryotes" (Parish, ed.), p. 330. Academic Press, London (1980).

Ellwood, D. C., Kelemen, M. V. and Baddiley, J., *Biochem. J. 86:*213 (1963).

Fischer, W., Rösel, P., and Koch, H. U., *J. Bacteriol 146:*464 (1981).

Kalakoutskii, L. V., and Nikitina, E. T., *Postepy Hig. Med. Dosw. 31:*313 (1977).

Naumova, I. B., Shashkov, A. S., Skoblilova, N. K., Agre, N. S., and Romanov, V. V., *Bioorg. Khim. 8* (1982). (In press)

Prokofieva-Belgovskaya, A. A.,"Structure and Development of Streptomycetes". Academy of Sciences of the USSR, Moscow (1963).

Robson, R. L., and Baddiley, J., *J. Bacteriol. 129:*1045 (1977a).

Robson, R. L., and Baddiley, J., *J. Bacteriol. 129:*1051 (1977b).

Salekh, S., Agre, N. S., and Kalakoutskii, L. V., *Mikrobiologiya 48:*705 (1979).

Skoblilova, N. K., Agre, N. S., and Naumova, I. B., *Mikrobiologiya 50:*1037 (1981).

Skoblilova, N. K., Agre, N. S., Shashkov, A. S., and Naumova, I. B., *Bioorg Khim. 8* (1982). (In press)

Suyetin, S. O., Parijskaya, A. N., and Kalakoutskii, L. V., *Mikrobiologiya 49:*604 (1980).

Tipper, D. J., and Wright, A., *in* "The Bacteria" (J. R. Sokatch and L. N. Ornston, eds.), vol. 7, p. 291. Academic Press, London (1979).

Tomasz, A., Westphal, M., Briles, E. B., and Fletcher, P., *J. Supramol. Struct. 3:*1 (1975).

LIPIDS OF ACTINOMYCETES: THEIR STRUCTURES AND BIOSYNTHESES[1,2]

Thuioshi Ioneda

Instituto de Química
Departamento de Bioquímica
Universidade de São Paulo
São Paulo, Brazil

I. INTRODUCTION

Excellent reviews dealing with the lipids of actinomycetes have been published, many of which focus on the relationship between composition and classification of actinomycetes (Ioneda and Silva, 1978; Kroppenstedt and Kutzner, 1978; Lechevalier *et al.*, 1981; Minnikin and Goodfellow, 1976, 1978, 1981; Minnikin *et al.*, 1978). Studies regarding the complex lipids of the cell wall (Asselineau, 1981) as well as investigations on the trehalose and arabinose mycolates (Asselineau and Asselineau, 1978; Ioneda *et al.*, 1981) were recently reviewed (Batrakov and Bergelson, 1978). In this article lipids of mycobacteria, nocardiae, and rhodococci will be preferentially discussed.

II. FATTY ACIDS

In mycobacteria, nocardiae, and rhodococci, there are characteristically two types of fatty acids: non-hydroxylated fatty acids and high molecular weight α-alkyl-β-hydroxylated fatty acids (R_1-CHOH-CHR_2-COOH, mycolic acids) (Asselineau, 1966; Bordet and Michel, 1969; Etémadi, 1967).

[1] *This work was supported by grants from Fundação de Amparo à Pesquisa do Estado de São Paulo (FAPESP) and from Financiadora de Estudos e Projetos (FINEP).*
[2] *Part of this work was carried out in the Divisão de Hansenologia (Director: C.E.R. Rudge), Instituto da Saúde, São Paulo.*

ISBN 0-12-528620-1

A. Non-hydroxylated Fatty Acids

Mycobacterium smegmatis ATCC 356, misclassified previously as *M. phlei* ATCC356, possesses a tightly organized multienzyme complex (M. W. ~ 1.39 X 10^6) which catalyzes the synthesis of (C_{16}-C_{18})-saturated fatty acids by using preferentially acetyl-CoA as a primer and malonyl-CoA as the C_2-unit donor. Both NADH and NADPH function as hydrogen donors (Vance *et al.*, 1973). Similar enzyme complexes found in *Nocardia farcinica, N. rhodochrous, Rhodococcus lentifragmentus*, and *R. bronchialis* have estimated molecular weights of about 1.5 X 10^6 (Gesztesi *et al.*, 1980). These preparations incorporate 2-^{14}C-malonyl CoA into the (C_{12}-C_{16})-saturated fatty acids by using acetyl CoA as the primer. Although other acyl CoA homologs are able to prime the fatty acid synthesis, the efficacy drops form acetyl CoA to the next homologs but rises with octanoyl and decanoyl CoA. In addition to C_{16}- and C_{18}-saturated fatty acids, the corresponding unsaturated ($C_{16:1}$ and $C_{18:1}$) fatty acids are produced in large amounts through oxidative desaturation (Campbell and Naworal, 1969; Kashiwabara *et al.*, 1975). Oleic acid ($C_{18:1}\triangle^9$) may be further C-methylated by S-adenosylmethionine to give an intermediary compound, 10-methylene octadecanoic acid (Jauréguiberry *et al.*, 1966), which in turn undergoes enzymatic reduction by NADPH, producing 10-methyl octadecanoic acid (tuberculostearic acid) (Akamatsu and Law, 1970). This fatty acid is a common constituent of mycobacteria (Lennarz *et al.*, 1962), nocardiae (Ioneda *et al.*, 1970; Lanéelle *et al.*, 1965), and rhodococci (Silva and Ioneda, 1981). On the other hand, the fatty acids synthesized by the multienzyme complex may be further elongated by the action of other enzyme systems (Odriozola *et al.*, 1977). Studies carried out in *M. tuberculosis* $H_{37}R_a$ suggest a pivotal role for long unsaturated fatty acids such as cis-5-tetracosenoic acid ($C_{24:1}\triangle^5$) (Asselineau *et al.*, 1970) in the production of precursors, such as cis,cis-3,4 and 15,16-dimethylenetetratriacontanoic acid and biscyclopropane rings containing (C_{43} - C_{56}) fatty acids (Qureshi *et al.*, 1980), for the synthesis of mycolic acids. A homolog series of polyunsaturated acids (phleic acids), in which 4,8,12,16,20-hexatriacontapentenoic acid is the prominent member, represents a special group of fatty acids found in *M. phlei*; in their biosynthesis, long chain fatty acid such as palmitic acid function as the primer to which tetracarbon units are added (Asselineau and Montrozier, 1976).

B. Mycolic Acids

Since the fact that isoniazid (isonicotinic acid hydrazide, INH) strongly inhibits the cellular synthesis of mycolic acids (Takayama *et al.*, 1972; Winder and Collins, 1970) has interested investigators; efforts to elucidate the exact structural nature of these components have been renewed. Individual mycolic acids from mycobacteria have been successfully separated by the technique of high performance liquid chromatography of their *p*-bromophenacyl esters; the corresponding structures have been elucidated by using nuclear magnetic resonance and mass spectrometric techniques (Gray *et al.*, 1982; Qureshi *et al.*, 1978; Steck *et al.*, 1978).

Variation in the composition of the mycolic acid molecular species is observed according to changes in the growth temperature so that in *M. phlei*, for example, the amount of the $C_{78:2}$-mycolic acid is increased with increment of growth temperature (2.7% monocarboxylic mycolic acid at 23°C; 51.5% at 51°C 23°C), whereas that of the $C_{74:2}$-mycolic acid is diminished (46.9% to 5.1%). A similar behavior is observed with the dicarboxylic mycolic acids. Thus, the $C_{62:1}$-mycolic acid content is increased from 1.5% to 40.4% and that of the $C_{58:1}$-mycolic acid is reduced from 35.7% to 9.5% (Toriyama *et al.*, 1978, 1980). In *N. rubra* grown at 15°C, a high level of dienoic acids ($C_{46:2}$, $C_{48:2}$); *i. e.*, 44.2% of the extractable mycolic acids, is obtained compared to 20.4% when grown at 40°C. In turn, at the higher temperature, the amount of monoenoic acids ($C_{46:1}$, $C_{48:1}$) is 38.4%, whereas at the lower temperature, it falls to 19.7% (Tomiyasu *et al.*, 1981). Therefore, the change in mycolic acid composition according to the growth temperature may be regarded as an adaptive response of the enzymatic machinery of the cells in order to maintain proper fluidity.

Based on the type of fatty acid present, two lipid groups will be considered: a) non-hydroxylated fatty acid containing lipids and b) hydroxylated fatty acids (mycolic acids) containing lipids.

III. NON-HYDROXYLATED FATTY ACID CONTAINING LIPIDS

Non-hydroxylated fatty acids may be found in a free form or esterified either to the glycerol moiety of glycerolipids or to the long chain β-glycol moiety of mycosides (Barksdale and Kim, 1977). In some species of mycobacteria, these fatty acids as well as 4,8,12,16,20-hexatriacontapentenoic acid and their homologs esterify the trehalose molecule.

A. Monoacylglycerols

A 1-monoacylglycerol fraction isolated from *M. tuberculosis* Brévannes has palmitic and stearic acids as fatty acid constituents (Noll and Jackim, 1958). In addition to these acids, in the 1-monoacylglycerol fractions of *R. rubropertinctus* (Silva *et al.*, 1980) and of *N. asteroides* (Silva and Ioneda, 1981), hexadecenoic, octadecenoic, and 10-methyl octadecanoic acids have been found.

B. Glycerophospholipids, Triacylglycerols and Glycerophosphoglycolipids

The fatty acid moiety of phosphatidylglycerol, bisphosphatidylglycerol (cardiolipin), phosphatidylinositol, phosphatidylethanolamine, as well as of phosphatidylinositomannosides is represented by non-hydroxylated fatty acids; to date, mycolic acids have not been reported among these lipids. Studies carried out with triacylglycerols and phospholipids of *M. smegmatis* ATCC 19420 and *M. bovis* BCG (Glaxo strain) have demonstrated a preferential location of $C_{18:0}$, $C_{18:1}$, and tuberculostearic acid at C_1 of the glycerol moiety, whereas the position C_2 is occupied by C_{16} fatty acids. In the triacylglycerol, a predominance of (C_{24}-C_{26}) fatty acids at position C_3 has been observed (Walker *et al.*, 1970).

Lipid composition is subject to temperature control so that *M. smegmatis* 607 synthesizes a greater quantity (47.1%) of phosphatidylinositomannosides at 37°C than at 27°C (32.7%), whereas at this temperature, the level of phosphatidylethanolamine is higher (23.4%) compared to that at 37°C (12.3%) (Taneja *et al.*, 1979).

In the cell wall of *N. caviae*, an unusual normal fatty acid-containing glycolipid fraction is found (Pommier and Michel, 1976, 1981). The water soluble moiety consists of αD-glucopyranosyl-(1→2)-D-glyceric acid; the lipid moiety contains myristic, palmitic, stearic, and a small amount of oleic acid. These fatty acids esterify the hydroxyl groups of the glucose unit at C'_2 and C'_3.

C. Esters of Trehalose

The majority of esters of trehalose from various sources contains specific mycolic acids as their fatty acid constituents (Ioneda *et al.*, 1981). However, a fraction of esters of trehalose (Fig. 1) of *M. fortuitum* has palmitic acid attached to trehalose through the C_6 and C_6' hydroxyl groups: 6,6'-dipalmitoyltrehalose (Ia).

Besides this symmetric compound, an asymmetric 4,6-diester of trehalose (Ib) occurs; the fatty acid moiety contains palmitic, stearic, and tuberculostearic acids (Vilkas *et al.*, 1968). *M. chelonei* also possesses several esters of trehalose having distinct behavior on thin layer chromatography as well as different physical characteristics, their constituent fatty acids being non-hydroxylated (Donoso-Fernández and Ioneda, unpublished results). In a glycolipid fraction of *M. phlei*, all the hydroxyl groups of the trehalose molecule are esterified with the 4,8,12,16,20-hexatriacontapentenoic acid (phleic acid) and their homologs, giving the octaphleate of trehalose-lipid P (Ic) (Asselineau *et al.*, 1972). In *M. tuberculosis*, a main sulfolipid fraction identified as 2,3,6,6'-tetracyl-2'-sulfate trehalose (Id) (sulfolipid I) occurs (Goren, 1972). Among a series of trehalose lipids elaborated by *M. paraffinicum*, there exists

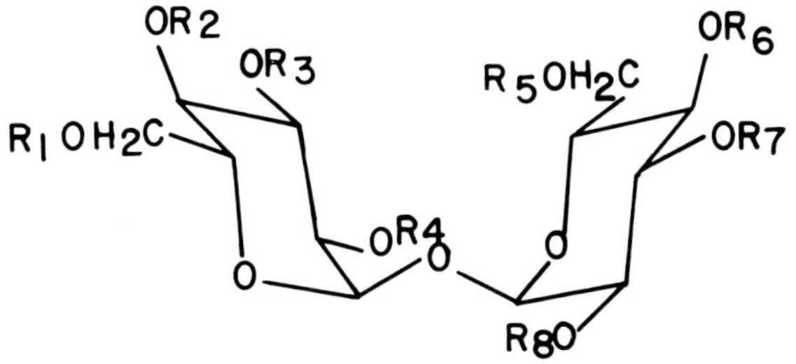

R_1OH_2C — OR_2 — OR_3 — OR_4 — O — R_5OH_2C — OR_6 — OR_7 — R_8O

a = R_1 = R_5 = palmitoyl; R_2 to R_4 and R_6 to R_8: H

b = R_1, R_2: palmitoyl, stearoyl, tuberculostearoyl; R_3 to R_8: H

c = R_1 to R_8: 4,8,12,16,20-hexatriacontapentenoyl (phleyl)

d = R_1, R_3, R_4, R_5: acyl — $CH_3-(CH_2)_{14}-CO-$; $CH_3-(CH_2)_{16}-CO-$

R_8 = sulfyl — $CH_3-(CH_2)_{15}-(CH(CH_3)-CH_2)_n-CH(CH_3)-CO-$

R_6 = R_7 = H — n = 4 to 9 (phthioceranoyl)

R_2 = H — $CH_3-(CH_2)_{14}-CH(OH)-(CH(CH_3)-CH_2)_n-CH(CH_3)-CO-$

n = 2,4 to 9 (hydroxyphthioceranoyl)

e = R_1 = $-OOC-CH_2-CH_2-COOH$; R_2 = R_5 = R_6 = R_7 = H

R_3 = R_8 = $CH_3-(CH_2)_8-CO-$; R_4 = $CH_3-(CH_2)_6-CO-$

f = R_1 = $(C_{34}-C_{44})$-mycoloyl ; R_2 = R_3 = R_4 = R_6 = R_7 = R_8 = H

R_5 = $(C_{12}-C_{16})$-acyl

FIGURE I. Structure of various esters of trehalose.

a singular one 2-octanoyl-3,2'-didecanoyl-6-succinoyltrehalose (Ie), which has not been found so far in any other organism (Batrakov *et al.*, 1981). Besides 6,6'-di(C_{34}-C_{44}) mycoloyl trehalose, 6-mono(C_{34}-C_{44}) mycoloyl trehalose, 6,6'-di(C_{12}-C_{16}) acyltrehalose, 6-(C_{34}-C_{44}) mycoloyl, the (C_{12}-C_{14}) acyl trehalose (If) is also synthesized by this organism (Batrakov *et al.*, 1981). In

Micromonospora sp F3, about 5% of its lipid extract is a glycolipid fraction identified as 6,6'-diacyl trehalose in which (C_{15}-C_{18}) fatty acids are the constituents of the lipid moiety (Tabaud *et al.*, 1971).

IV. MYCOLIC ACID CONTAINING LIPIDS

The discovery (Bloch, 1950) and the elucidation of the structure (Noll *et al.*, 1956) of a toxic glycolipid from *M. tuberculosis* $H_{37}R_v$ represent a remarkable step in the development and progress of investigations regarding the esters of mycolic acids.

A. Monomycoloylglycerols

The same type of mycoloylglycerol first described in *M. tuberculosis* (Bloch *et al.*, 1957; Noll, 1957) and in *M. paratuberculosis* (Lanéelle and Lanéelle, 1970) is found in nocardiae (Silva and Ioneda, 1977, 1980) and in rhodococci (Silva *et al.*, 1980) (Table I). *Corynebacterium pseudotuberculosis*, a C_{32}-mycolic acid producer, also synthesizes 1-mono-(C_{30}-C_{36})-mycoloylglycerol (αD = + 11°; ^{n}D = 1.572) (Ioneda and Silva, 1979).

B. Mycoloyltrehaloses and Mycoloylarabinoses

Since the identification among the lipids of *M. tuberculosis* of a chemically defined entity endowed with toxic properties (Noll *et al.*, 1956) 6,6'-dimycoloyl

TABLE I. 1-Monomycoloylglycerols from Mycobacteria, Nocardiae, and Rhodococci

Organism	$(\alpha)_D$	*Melting point (mp) or refractive index* (n_D)	*Mycolic acids*	*References*
M. tuberculosis	+ 6.4°	mp: 38-39°C	~C_{90}	Noll, 1957
M. tuberculosis	+ 10.0°	mp:45-46°	~C_{90}	*Bloch* et al., *1957*
M. paratuberculosis	—		mycolic-wax[a]	Lanèlle and Lanèlle, 1970
N. rhodochrous	+ 10.8°	mp:41°	C_{40}-C_{44}	Silva and Ioneda, 1977
R. lentifragmentus	+ 11.0°	n_D:1.468	C_{42}-C_{50}	Silva and Ioneda, 1980
N. asteroides	+ 9.8°	n_D:1.476	C_{50}-C_{56}	Silva and Ioneda, 1980
R. bronchialis	+ 9.8°	n_D:1.476	C_{60}-C_{66}	*Silva* et al., *1980*

[a] *mycolic-wax or γ-mycolic acid* CH_3 - $(CH_2)_{17}$ - CH_3 - OCO - (C_nH_{2n-2}) - CHOH - CH($C_{22}H_{45}$) - COOH; n = *32 to 39.*

TABLE II. Various Trehalose Dimycolates of Mycobacteria, Nocardiae, and Rhodococci

Organisms	$(\alpha)_D$	*Mycolic acids*	*References*
N. asteroides *10905*	$+62^o$	$\sim C_{32}$	*Ioneda, Lederer and Rozanis, 1970*
R. erythropolis	$+78.6^o$	$\sim C_{34}$	*Rapp* et al., *1979*
M. paraffinicum	—	$\sim C_{36}$	*Batrakov* et al., *1981*
N. rhodochrous	$+50^o$	$\sim C_{44}$	*Ioneda, Lederer and Rozanis, 1970*
R. lentifragmentus	$+58^o$	$\sim C_{46}$	*Silva, Gesztesi and Ioneda, 1979*
N. asteroides	$+51^o$	$\sim C_{50}$	*Silva, Gesztesi and Ioneda, 1979*
N. caviae	—	$\sim C_{56}$	*Pommier and Michel, 1979*
R. bronchialis	$+47^o$	$\sim C_{66}$	*Silva, Gesztesi and Ioneda, 1979*
M. smegmatis	$+33^o$	$\sim C_{77}{}^a$	*Mompon* et al., *1978*
M. phlei	$+36.5^o$	$\simeq C_{80}{}^b$	*Promè* et al., *1976*
	$+41^o$	$\sim C_{80}{}^a + \sim C_{80}{}^c$	
	$+38^o$	$\sim C_{80}{}^b$	
M. tuberculosis	$+40^o$	$\sim C_{90}$	*Noll* et al., *1956*

[a] *mixture of mycolic acids:* $CH_3-(CH_2)_x-CH-CH(CH_2)-(CH_2)_y-CH=CH-CH(CH_3)-(CH_2)_z-CHOH-CH(C_{22}H_{45})-COOH$ *and* $CH_3-(CH_2)_x-CH=CH-(CH_2)_y-CH=CH-CH(CH_3)-(CH_2)_z CHOH-CH(C_{22}H_{45})-COOH$; $x + y + z = \sim 44$

[b] *mycolic acid,* $C_nH_{2n-4}-CHOH-CH(C_{22}H_{45})COOH$; $n = \sim 56$.

[c] γ*-mycolic acid or mycolic-wax,* $CH_3-(CH_2)_{17}-CH(CH_2)-CH(CH_3)-OCO-C_n-H_{2n-2}-CHOH-CH(C_{22}H_{45})-COOH$, $n = \sim 37$

trehalose (the so-called cord factor), a series of similar compounds whose constituent mycolic acids are characteristic of the species has been found in nocardiae and rhodococci (Ioneda *et al.*, 1970, 1981; Silva *et al.*, 1979). Therefore, on the basis of the molecular weight of the constituent fatty acids, trehalose mycolates found in some mycobacteria, nocardiae, and rhodococci may be arranged as shown in Table II.

Small amounts of monomycoloyltrehalose are also found in the majority of the trehalose mycolate-producing organisms (Batrakov *et al.*, 1981; Ioneda *et al.*, 1981; Pommier and Michel, 1979).

On the other hand, mycolic acids bind to the arabinose unit of the arabinogalactan of the cell wall, giving rise to the fraction not extractable with neutral solvents (Azuma and Yamamura, 1962, 1963; Bruneteau and Michel, 1968; Ioneda *et al.*, 1981; Lanéelle and Asselineau, 1970). Many mycolic acid-producing organisms examined so far contain arabinose mycolates. Therefore, this component may be of paramount importance for the architecture of the cells as well as for their physiology.

C. Mycoloylsucroses, Mycoloylfructoses and Mycoloylglucoses

In many mycolic acid-producing organisms, addition of either sucrose, fructose, or glucose to the culture medium induces the synthesis of the corresponding sugar mycolates: sucrose mycolate (6-mycoloylglucosyl-β-fructoside) (Suzuki *et al.*, 1974), fructose mycolate (6-mycoloyl fructose) (Itoh and Suzuki, 1974), and glucose mycolate (6-mycoloylglucose) (Brennan *et al.*, 1970, Ioneda and Silva, 1978). However, some organisms such as *M. avium* and *M. tuberculosis* do not have the capability to synthesize sucrose lipids (Itoh and Suzuki, 1974), although they do have the ability to produce fructose lipid. Investigations carried out with *N. rhodochrus* show that supplementation of the medium with 1% glucose led to the production of glucose lipids in an amount equal to approximately 45% of the ethanol-diethyl ether extractable material and to reduction of the quantity of trehalose esters to a trace amount (Teixeira and Ioneda, unpublished results). The carbon-13 nuclear magnetic resonance spectrum shows that absorption corresponding to C_6 of the sugar unit is displaced downfield from 61.5 ppm (C_6 with free hydroxyl group) toward 64.1 ppm, indicating that the hydroxyl group at C_6 of the sugar unit is occupied by an ester group as had been observed in studies carried out with other glycolipids (Polonsky *et al.*, 1977). Thus, the glucose mycolate is 6-mycoloylglucose (~ C_{44}-mycolic acids) and both α-and β-anomeric forms occur.

V. BIOSYNTHESIS: MYCOLIC ACIDS AND THEIR ESTER DERIVATIVES

When the carbon source (glucose for glycerol) in the culture medium of *N. rhodochrous* is changed, the trehalose content is considerably reduced and a new glycolipid (6-mycoloylglucose)is synthesized. Studies on the incorporation *in vivo* of 1-^{14}C-acetate or 1-^{14}C-glucose show a continuous synthesis of glucose lipid throughout all growth phases, whereas only a negligible amount of radioactivity is detected in the trehalose mycolate fraction. However, in contrast to these findings, *N. asteroides* grown in glucose elaborates a considerable amount of trehalose dimycolate (20.1% of the lipid extract) compared to that produced when it is grown on glycerol (1.2% of the lipid extract) (Yano *et al.*, 1971). The increased production of mycoloylglucose by *N. rhodochrous* upon glucose stimulation suggest an elevated activity of an enzyme system specific for the transfer of the mycoloyl group from a carrier to the glucosyl unit. In *N. asteroides*, glucose possibly stimulates the transfer of the mycoloyl group to the trehalose molecule.

On the other hand, *N. rhodochrous* grown in galactose synthesizes a rather large amount of trehalose dimycolate (40% of the lipid extract) instead of mycoloylgalactose as might be expected (Breda and Ioneda, unpublished results). Here, galactose or its metabolites may promote the production of trehalose which in turn undergoes acylation. In a similar manner, when lactose is added to the culture medium, a high level of trehalose mycolate is also produced (Breda and Ioneda, unpublished results).

Therefore, it seems reasonable to expect that in *N. rhodochrous* a steric arrangement of the hydroxyl group of the sugar unit at C_4 may play a paramount role in the biological esterification of the hydroxyl group of the hexose unit. Studies carried out with *R. lentifragmentus* cultured in a medium supplemented with glycerol as carbon source shows that addition of INH causes a drastic reduction in the trehalose mycolate content and in that of the glycerol monomycolate as well. Alkaline hydrolysis of the cell residue after it had been extracted with chloroform-methanol liberates an increased quantity of the non-hydroxylated fatty acids (this fraction is almost negligible in the control cells (Gesztesi and Ioneda, unpublished results)). It is known that, in mycobacteria, INH inhibits the synthesis of mycolic acids (Winder and Collins, 1970). Moreover, this drug blocks the elongation of C_{26}-fatty acids and, consequently, the synthesis of very long chain fatty acids (C_{27}-C_{40} and C_{39}-C_{56}) is affected (Takayama *et al.*, 1975). INH also inhibits the desaturation of $C_{24:0}$-fatty acids, thereby blocking the production of cis-5-tetracosenoic acid (Asselineau *et al.*, 1970), a precursor of long chain fatty acids necessary for the synthesis of $\sim C_{80}$-mycolic acids (Davidson and Takayama, 1979). Since *R. lentifragmentus* possesses $\sim C_{46}$-mycolic acids, it may be similarly affected by this drug provided that this organism needs long chain $\sim C_{30}$-fatty acids for condensation with C_{16}-fatty acids in the synthesis of mycolic acid.

The fact that *R. lentifragmentus* survives at a high INH concentration (300 μg/ml constitutes a strong indication that the cells are able to overcome the restrictive growth condition imposed by the drug. Under the new physiological conditions, the cells survive and change their lipid composition so that a fraction of mycolic acids of the bound lipids of the cell wall (mycoloylarabinogalactan) may be replaced by non-hydroxylated fatty acids. Under these conditions, there is a recovery mechanism distinct from that proposed for mycobacteria for the mycolic acid transfer to the cell wall (Takayama and Armstrong, 1977; Takayama *et al.*, 1979).

VI. CONCLUDING REMARKS

The dual occurrence of the fatty acids, *i. e.*, non-hydroxylated and mycolic type fatty acids as well as the presence of many varieties of mycolic acids differing either in molecular size ($\sim C_{32}$ up to $\sim C_{90}$) or in peculiarities of the carbon chain (presence of the double bond, oxygenated function other than carboxyl group, methyl- and α-alkyl branch, cyclopropane ring), affords a very exciting challenge for investigators. Since the discovery of mycolic acids in mycobacteria, a considerable body of data on the lipids of mycobacteria and related organisms has been accumulated; however, many relevant questions remain open to investigation. Some of those fundamental problems are a) the biosynthesis of the different types of mycolic acids, b) the biosynthesis of mycoloyl esters (monomycoloylglycerol, trehalose mycolates, other sugar mycolates), c) the acylation of the arabinose unit of arabinogalactan, d) the physicochemical interaction of mycoloyl esters with artificial and biological membranes, e) the involvement of the trehalose esters in the mechanism of pathogenicity, and f) the mycoloyl esters and immunostimulation.

ACKNOWLEDGMENTS

I wish to thank Dr. F. G. da Nóbrega for reading the manuscript and Mrs. L. Y. Ioneda for drawing the figure.

REFERENCES

Akamatsu, Y. and Law, J. H., *J. Biol. Chem. 245:*709 (1970).
Asselineau, J., "The Bacterial Lipids" Hermann, Paris (1966).
Asselineau, J., *in* "Actinomycetes" (K. P. Schaal, and G. Pulverer, eds), p. 391. Gustav Fischer Verlag, Stuttgart (1981).
Asselineau, C., and Asselineau, J., *Prog. Chem. Fats Other Lipids 16:*59 (1978).
Asselineau, C. P., and Montrozier, H. L., *Eur. J. Biochem. 63:*509 (1976).
Asselineau, C., Lacave, C., Montrozier, H. and Promé, J. C., *Eur. J. Biochem. 14:*406 (1970).
Asselineau, C. P., Montrozier, H. L., Promé, J. C., Savagnac, A. M., and Welby, M., *Eur. J. Biochem. 28:*102 (1972).
Azuma, I., and Yamamura, Y., *J. Biochem. 52:*200 (1962).
Azuma, I., and Yamamura, Y., *J. Biochem. 53:*275 (1963).
Barksdale, L., and Kim, K S., *Bacteriol. Rev. 41:*217 (1977).
Batrakov, S. G., and Bergelson, L. D., *Chem. Phys. Lipids 21:*1 (1978).
Batrakov, S. G., Rozynov, B. V., Koronelli, T. V., and Bergelson, L. D., *Chem. Phys. Lipids 29:*241 (1981).
Bloch, H., *J. Exp. Med. 91:*197 (1950).
Bloch, H., Defaye, J., Lederer, E., and Noll, H., *Biochim. Biophys. Acta 23:*312 (1957).
Bordet, C., and Michel, G., *Bull. Soc. Chim. Biol. 51:*527 (1969).
Brennan, J. P., Lehane, D. P., and Thomas, D. W., *Eur. J. Biochem. 13:*117 (1970).
Bruneteau, M., and Michel, G., *Chem. Phys. Lipids 2:*229 (1968).
Campbell, I. M., and Naworal, J., *J. Lipid Res. 10:*593 (1969).
Davidson, L. A., and Takayama, K., *Antimicrob. Agents Chemother. 16:*104 (1979).
Etémadi, A. H., *Expo. Annu. Biochim. Med. 28:*77 (1967).
Gesztesi, J. L., Breda, M., Silva, C. L., and Ioneda, T., *Biol. Actinomycetes Relat. Org. 16:*17 (1980).
Goren, M. B., *Bacteriol. Rev. 36:*33 (1972).
Gray, G. R., Wong, M. Y. H., and Danielson, S. J., *Prog. Lipid Res. 21:*91 (1982).
Ioneda, T., and Silva, C.L., *in* "Nocardia and Streptomyces" (M. Mordaski, W., Kurylowicz, and J. Jeljaszewicz, eds.) p. 67. Gustav Fischer Verlag, Stuttgart (1978).
Ioneda, T., and Silva, C. L., *Chem. Phys. Lipids 25:*85 (1979).
Ioneda, T., Lederer, E., and Rozanis, J., *Chem. Phys. Lipids 4:*375 (1970).
Ioneda, T., Silva, C. L., and Gesztesi, J. L., *in* "Actinomycetes" (K. P. Schaal, and G. Pulverer, eds.), p. 401. Gustav Fischer Verlag, Stuttgart (1981).
Itoh, S., and Suzuki, T., *Agric. Biol. Chem. 38:*1443 (1974).
Jauréguiberry, G., Lenfant, M., Das, B. C., and Lederer, E., *Tetrahedron 22*, suppl. 8:27 (1966).
Kashiwabara, Y., Nakagawa, H., Matsuki, G., and Sato, R., *J. Biochem. 78:*803 (1975).

Kroppenstedt, R. M., and Kutzner, H. J. *in* "Nocardia and Streptomyces" (M. Mordaski, W. Kurylowicz, and J. Jeljaszewicz, eds.), p. 125. Gustav Fischer Verlag, Stuttgart (1978).
Lanéelle, M. A., and Asselineau, J., *FEBS Lett. 7:*64 (1970).
Lanéelle, M. A., and Lanéelle G., *Eur. J. Biochem. 12:*296 (1970).
Lanéelle, M. A., Asselineau, J., and Castelmuovo, G., *Ann. Inst. Pasteur (Paris) 108:*69 (1965).
Lechevalier, M. P., Stern, A. E., and Lechevalier, H. A., *in* "Actinomycetes" (K. P. Schaal, and G. Pulverer, eds.), p. 111. Gustav Fischer Verlag, Stuttgart (1981).
Lennarz, W. J., Scheuerbrandt, G., and Bloch, K., *J. Biol. Chem. 237:*664 (1962).
Minnikin, D. E., and Goodfellow, M., *in* "The Biology of the Nocardiae" (M. Goodfellow, G. H. Brownell, and J. A. Serrano, eds.), p. 160. Academic Press, London (1976).
Minnikin, D. E., and Goodfellow, M., *in* "Nocardia and Streptomyces" (M. Mordarski, W. Kurylowicz, and J. Jeljaszewicz, eds.), p. 75. Gustav Fischer Verlag, Stuttgart (1978).
Minnikin, D. E., and Goodfellow, M., *in* "Actinomycetes" (K. P. Schaal, and G. Pulverer, eds.), p. 99. Gustav Fischer Verlag, Stuttgart (1981).
Minnikin, D.E., Collins, M.D. and Goodfellow, M., *in* "Nocardia and Streptomyces" (M. Mordarski, W. Kuylowicz, and J. Jeljaszewicz, eds.), p. 85. Gustav Fischer Verlag, Stuttgart (1978).
Mompon, B., Frederici, C., Toubiana, R., and Lederer, E., *Chem. Phys. Lipids 21:*97 (1978).
Noll, H., *J. Biol. Chem. 224:*149 (1957).
Noll, H., and Jackim, E., *J. Biol. Chem. 232:*903 (1958).
Noll, H., Bloch, H., Asselineau, J., and Lederer, E., *Biochim. Biophys. Acta 20:*299 (1956).
Odriozola, J. M., Ramos, J. A., and Bloch, K., *Biochim. Biophys. Acta 488:*207 (1977).
Polonsky, J., Soler, E., Toubiana, R., Takayama, K., Raju, M. S., and Wenkert, E., *Nouv. J. Chim. 2:*317 (1977).
Pommier, M. T., and Michel, G., *Biochim. Biophys. Acta 441:*327 (1976).
Pommier, M. T., and Michel, G., *Chem. Phys. Lipids 24:*149 (1979).
Pommier, M. T., and Michel, G., *Eur. J. Biochem. 118:*329 (1981).
Prome, J. C., Lacave, C., Ahibo-Coffy, A., and Savagnac, A., *Eur. J. Biochem. 63:*543 (1976).
Qureshi, N., Takayama, K., Jordi, H. C., and Schnoes, H. K., *J. Biol. Chem. 253:*5411 (1978).
Qureshi, N., Takayama, K., and Schnoes, H. C., *J. Biol. Chem. 255:*182 (1980).
Rapp, P., Bock, H., Wray, V., and Wagner, F., *J. Gen. Microbiol. 115:*491 (1979).
Silva, C. L., and Ioneda, T., *Chem. Phys. Lipids 20:*217 (1977).
Silva, C. L., and Ioneda, T., *Chem. Phys. Lipids 27:*43 (1980).
Silva, C. L., and Ioneda, T., *Biol. Actinomycetes Relat. Org. 16:*49 (1981).
Silva, C. L., Gesztesi, J. L., and Ioneda, T., *Chem. Phys. Lipids 24:*17 (1979).
Silva, C. L., Gesztesi, J. L., Zupo, M. C., Breda, M., and Ioneda, T., *Chem. Phys. Lipids 26:*197 (1980).
Steck, P. A., Schwartz, B. A., Rosendahl, and Gray, G. R., *J. Biol. Chem. 253:*5625 (1978).
Suzuki, T., Tanaka, H., and Itoh, S., *Agric. Biol. Chem. 38:*557 (1974).
Tabaud, H., Tisnovska, H., and Vilkas, E., *Eur. J. Biochem. 53:*55 (1971).
Takayama, K., and Armstrong, E. L., *J. Bacteriol. 130:*569 (1977).
Takayama, K., Armstrong, E. L., Kunugi, K. A., and Kilburn, J. O., *Antimicrob. Agents Chemother. 16:*240 (1979).
Takayama, K., Schnoes, H. K., Armstrong, E. L., and Boyle, R. W., *J. Lipid Res. 16:*308 (1975).
Takayama, K., Wang, L., and David. H. L., *Antimicrob. Agents Chemother. 2:*29 (1972).
Taneja, R., Malik, U., and Kuhller, G. K., *J. Gen. Microbiol. 113:*413 (1979).
Tomiyasu, I., Toriyama, S., Yano, I., and Masui, M., *Chem. Phys. Lipids 28:*41 (1981).
Toriyama, S., Yano, I., Masui, M., Kusunose, M., and Kusunose, E., *FEBS Lett. 95:*111 (1978).
Toriyama, S., Yano, I., Masui, M., Kusunose, E., Kusunose, M., and Akimori, N., *J. Biochem. 88:*211 (1980).
Walker, R. W., Barakat, H., and Hung, J. G. C., *Lipids 5:*684 (1970).
Winder, F. G., and Collins, P. B., *J. Gen. Microbiol. 63:*41 (1970).
Vance, D. E., Mitsuhashi, O., and Bloch, H., *J. Biol. Chem. 248:*2303 (1973).
Vilkas, E., Adam, A., and Senn, M., *Chem. Phys. Lipids 2:*11 (1968).
Yano, I., Furukawa, Y., and Kusunose, M., *J. Gen. Appl. Microbiol. 17:*329 (1971).

QUANTITATIVE STUDIES OF THE RELATIONSHIP BETWEEN TREHALOSE LIPIDS AND VIRULENCE OF *NOCARDIA ASTEROIDES* ISOLATES

Mark L. Tamplin

Department of Medical Microbiology and Immunology
University of South Florida
Tampa, Florida, U. S. A.

Norvel M. McClung

Department of Biology
University of South Florida
Tampa, Florida, U. S. A.

I. INTRODUCTION

Cord factor is a proposed virulence factor for *Mycobacterium tuberculosis* (Bekierkunst and Artman, 1966; Bloch, 1950; Ioneda *et al.*, 1963, 1970). It has also been isolated from *Corynebacterium* sp. and *Nocardia* sp. (Ioneda *et al.*, 1963, 1970). Cord factor is glycolipid in which mycolic acids are esterified to the 6 and 6' ends of a trehalose molecule (Noll *et al.*, 1956). Extensive characterization of this lipid has been made primarily with that extracted from *Mycobacterium* sp. (Asselineau and Portelance, 1960; Bloch, 1950; Gangadharam *et al.*, 1963; Goren, 1970; Kato *et al.*, 1958; Mompon *et al.*, 1976; Noll *et al.*, 1956). Studies by Block (1950) showed that cord factor preparations were toxic at low doses in mice. Kato *et al.* (1958) reported that purified cord factor causes mitochondrial membrane disruption and uncoupling of oxidative phosphorylation in mouse liver cells. In addition, cord factor stimulates NADase activity in Ehrlich ascites cells (Bekierkunst and Artman, 1966). Other investigators have proposed that cord factor is involved in strain virulence, since there appears to be a direct correlation

BIOLOGICAL, BIOCHEMICAL,
AND BIOMEDICAL ASPECTS OF ACTINOMYCETES

ISBN 0-12-528620-1

between quantity and virulence in strains of *M. tuberculosis* (Asselineau and Portelance *et al.*, 1960; Bekierkunst and Artman, 1966; Gangadharam *et al.*, 1963; Kato *et al.*, 1958). Cord factor was isolated from *Nocardia* sp. by Ioneda *et al.* in 1970. However, there are no published reports dealing with the quantity of cord factor and virulence of *Nocardia* sp. strains for mice.

Since trehalose molecules are primarily associated with cord factor derivatives found in cell wall envelopes, we conducted studies to quantify trehalose lipids in three strains of *Nocardia asteroides*. The relative virulence of each strain for mice was also determined.

II. MATERIALS AND METHODS

Three strains of *N. asteroides* designated 84, 171, and 186 were chosen for study. Strain 84 was isolated from a brain abscess, 171 from a pulmonary infection, and 186 from an ulcer of the skin. All strains were maintained on nutrient agar (Difco, Detroit, MI) supplemented with 2% glucose, pH 7.2. For lipid extraction, each strain was grown in 250 ml of medium containing 0.5% yeast extract, 0.8% nutrient broth and 2.0% glucose. Cultures were incubated at 37 °C in 1000 ml Erlenmeyer flasks and were shaken at 100 rpm until early stationary phase was reached. Triplicate aliquots (1 ml) were removed from the cultures at selected time intervals and filtered through a 0.22 μm filter (Millipore). The filtrate was dried at 60°C until constant weight was reached.

At early stationary phase, the cells were harvested and washed thrice with 250 ml of phosphate buffered saline, pH 7.2. The cells were then extracted for a crude trehalose lipid preparation by a modification of the method of Goren (1970). Briefly, the cells were extracted three times with 100 ml of 95% ethanol at room temperature. The cells were then dried, and were extracted three times with 100 ml of a mixture of chloroform:methanol (1:1) at room temperature. The extract was then aspirated to dryness. The residue was extracted three times with 100 ml of methanol at room temperature. The methanol-insoluble, ethanol-soluble, and methanol-soluble extracts were stored for future trehalose analyses. The methanol insoluble residue was transferred to a 16 x 150 mm test tube for refluxing. This preweighed material was assumed to be a trehalose lipid and was mixed with a 5-fold excess of 5% KOH in methanol. This mixture was refluxed for 5 h in a 70°C water bath and aspirated to dryness. One ml of distilled water was added to the residue and the pH was adjusted to 7.0 with a weak HCl solution. The mixture was then extracted with an equal volume of diethyl ether. This procedure was repeated twice with an equal volume of diethyl ether for adequate extraction of contaminating lipid. This same procedure was used to analyze the ethanol and methanol extracts for trehalose lipid contamination. Each strain was extracted on three separate occasions for trehalose lipids.

Each water extract was tested for the presence of trehalose by using silica gel plates (Merck Silica gel 60 F254) and a l-butanol:acetic acid:water (130:30:50) solvent system. After the solvent front had run 8 cm, the plates were air-dried, sprayed with 50% sulfuric acid, and heated at 110°C for 15 min. Trehalose, arabinose, galactose, and glucose standards were used as controls.

The water extract following refluxing was aspirated to dryness under negative pressure. After drying, 0.4 ml of trimethylchlorosilane, 1.2 ml of hexamethyldisilazane, and 0.2 ml of anhydrous pyridine were added to the residue. The mixture was heated at 60°C for 45 min, centrifuged at 100 X *g*, and 5 μl of the supernatant was injected into a Hewlett-Packard 5840A gas-liquid chromatograph. The parameters for sugar analyses were as follows: column packing, Chromosorb W, 80-100 mesh (AW-DMCS) with 3% OV-1; carrier gas, nitrogen (30 ml/min); FID, 350°C; injection temperature, 210°C; and oven temperature, 150 to 250°C at 5°C per min. Using doubling dilutions of pure derivatized trehalose (Sigma, St. Louis, MO), trehalose peaks were quantified. The 21.7 min peak was found to be quantitative. Standard curves were plotted during each analysis for quantification of unknowns. All samples were assayed in duplicate.

Virulence was determined in each strain grown to early stationary phase as described above. A 25 ml sample of the cultures was centrifuged at 3000 X *g* for 15 min and washed three times with phosphate buffered saline. After determining the dry weight per ml of sample, the concentration of washed cells was adjusted to 0.005 g dry cell weight per ml. An aliquot (1 ml) was centrifuged and resuspended in 1 ml of sterile hog gastric mucin, pH 7.2. A 0.5 ml suspension of each strain was injected intraperitoneally into five mice. Both BALB/c and an Swiss outbred strain of mice were used. BALB/c mice were 30 to 35 g and 10 to 12 weeks of age. The Swiss outbred strain were 20 to 25 g and four to six weeks of age. Following injections, the mice were observed for 21 days for mortality.

III. RESULTS

After differing proportions of nutrient broth, yeast extract, plus a 2% glucose supplement had been used, it was found that 0.8% nutrient broth, 0.5% yeast extract, and 2% glucose provided excellent dispersed growth of *N. asteroides*. This greatly aided in the quantification and handling of harvested cells. The weights of the crude trehalose lipid extracts following methanol extraction are reported per 12 g dry cell weight, which was a common quantity obtained per batch culture (Table I).

Silica gel plate chromatography of each water extract following refluxing

TABLE I. Weights of Crude Trehalose Lipid Extracts following Methanol Extraction of the Chloroform: Methanol Lipid Extract

Strain No.	*Weight of crude trehalose lipid extract (g per 12 g dry cell weight)*	*Average weight (g) ± S.D.*
84	*0.9789*	*0.9610 ± 0.0543*
	1.0042	
	0.9000	
171	*0.0369*	*0.0370 ± 0.0019*
	0.0389	
	0.0352	
186	*0.0351*	*0.0346 ± 0.0038*
	0.0350	
	0.0381	

demonstrated a single spot that migrated identically with that of pure trehalose (R_f = 0.18). Glucose, arabinose, and galactose all migrated above trehalose (R_f = 0.33). Water extracts of the ethanol and methanol extracts following refluxing did not demonstrate spots similar to trehalose. A total of five peaks were usually observed in a typical gas-liquid chromatogram of pure trehalose (Fig. 1). The 16.5 min peak represented the solvent. Standard curves showed the 21.7 min peak to be quantitative.

Chromatography of the other potential cell wall sugars (*i. e.*, glucose, arabinose, and galactose) showed that they were eluted from the column much earlier than was trehalose, thus greatly facilitating the analysis. Chromatography of cell lipid hydrolysates normally only displayed a trehalose peak.

In order to assess our accuracy in detecting trehalose lipids, three different quantities of purified "Peurois" cord factor were refluxed. (The cord factor was kindly provided by Dr. Mayer Goren of the National Jewish Hospital and Research Center, Denver, CO.). We found that we were able to detect only a mean value of 65.8 ± 2.8% of the trehalose in the sample. We felt that this may have been due to degradation of the sugar, since refluxing pure trehalose alone gave a 67.5 ± 2.8% mean value of detection. Therefore, when we estimated the amount of trehalose in our bacterial extracts, we multiplied our values by 1.5 to approximate more closely the quantity present.

Quantification of trehalose lipids in *N. asteroides* strains No. 84, No. 171, and No. 186, which is expressed as moles of trehalose detected per

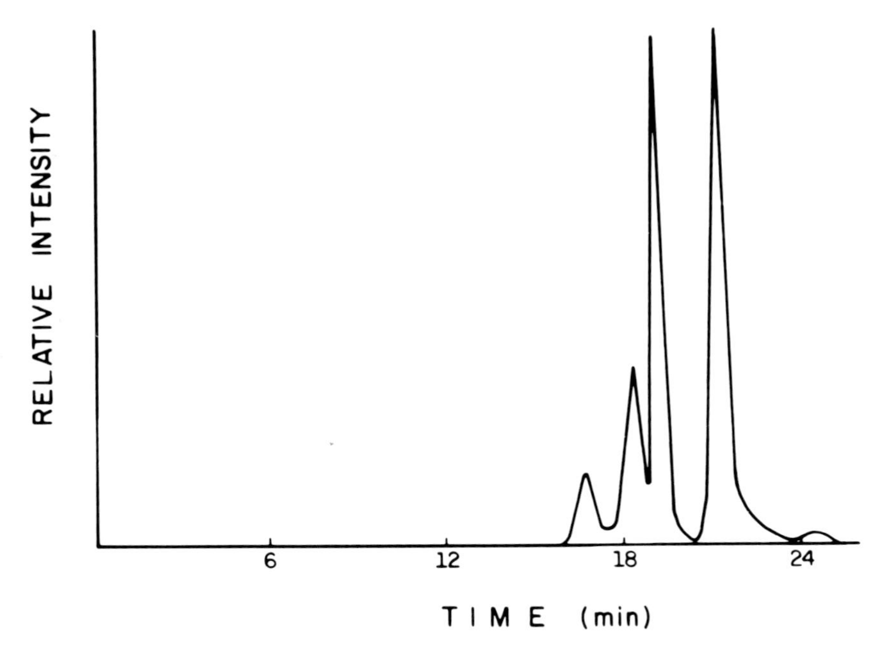

FIGURE 1. Gas chromatogram of a TMS-trehalose derivative.

gram of bacterial dry cell weight, is shown in Table II. The values represent data after three separate analyses of batch cultures. As shown, strain No. 84 had approximately 24-fold more trehalose than did strain No. 171 ($P < 0.05$), and 10-fold more than did strain No. 186 ($P < 0.05$). Strain No. 186 did not produce a significantly different amount of trehalose lipid than strain No. 171.

TABLE II. Analysis of the Quantity of Trehalose in Three Strains of N. asteroides.

Strain	*Moles of trehalose detected/g dry cell weight* [a]
84	$7.2 \times 10^{-6} \pm 4.2 \times 10^{-7}$ [b]
171	$3.0 \times 10^{-7} \pm 9.5 \times 10^{-8}$
186	$6.9 \times 10^{-7} \pm 2.6 \times 10^{-7}$

[a] *Value has been multiplied by 1.5 as stated in Results.*
[b] *Mean ± standard deviation*

TABLE III. Analysis of the Virulence of Three Strains of N. asteroides *in BALB/c and in Outbred Swiss mice.*

Strain No.	Daily mortality			
	Swiss		BALB/c	
	Day 1, 2, 3, 4, 5, 6, 7, 8, 9, 10...21		Day 1, 2, 3, 4, 5...21	
84	1, 1, 1, 1, 0, 0, 0, 0, 1, 0... 0	(5) [a]	2, 0, 0, 1, 0...0	(3)
171	3, 0, 0, 0, 0, 0, 0, 0, 0, 0... 0	(3)	1, 0, 0, 0, 0...0	(1)
186	1, 1, 0, 0, 0, 0, 0, 0, 0, 0... 0	(2)	1, 0, 0, 0, 0...0	(1)

[a] *Total number of mice dead.*

Virulence of *N. asteroides* strains No. 84, No. 171, and No. 186 in BALB/c and an outbred strain of Swiss mice is shown in Table III, in which the results of the 21 day virulence studies are presented. With the Swiss strain, we found that all five mice inoculated with strain No. 84 died; with strain No. 171, only three died; and with strain No. 186, only two died. This same trend was observed with BALB/c mice. However, three mice died when strain No. 84 was used, whereas only one died with both strains No. 171 and No. 186. In all studies, all control mice receiving only gastric mucin survived for the duration of the experiment.

IV. DISCUSSION

These studies showed that the most virulent strain of *N. asteroides* produced the most trehalose lipid of the three strain studied. These detected quantities of trehalose lipid were not grossly different from those reported by Asselineau and Portelance (1960) for BCG strains. They found that concentrations of cord factor varied from 0.6% to only trace amounts per gram dry cell weight. If it is assumed that a mole of cord factor of *N. asteroides* weighs approximately 1,500 g (this value is within the estimate reported to date by Ioneda *et al.*, 1970), then with strain No. 84, cord factor represented 1.10% of the dry cell weight. Strains No. 171 and No. 186 would then contain approximately 0.05% and 0.10%, respectively.

Virulence studies demonstrated that, at the concentrations used, No. 84 was more virulent than were No. 171 and No. 186, and that No. 171 and No. 186 did not significantly differ from each other. The age and weight of the two strains of mice probably affected the degree of bacterial virulence;

however, a general trend of virulence was observed with each strain used (*i. e.*, No. 84 appeared more virulent).

We feel that it is premature to suggest that trehalose lipids (*e. g.*, cord factor) are the only bacterial compounds involved in *N. asteroides* virulence. Other virulence factors, such as iron chelators which may aid in bacterial sequestering of host iron (Beaman, 1976), have been proposed.

Additional studies would aid in defining the role of trehalose lipids in strain virulence. An extensive analysis of numerous *N. asteroides* strains would present a more statistically valid comparison of virulence and quantity of trehalose lipid. Also, there may be a certain type of trehalose lipid (*e. g.*, sulfoglycolipids, saturated and unsaturated glycolipids, branched glycolipids, or acylated glycolipids) that predominates in a virulent strain. Of course, one must determine the experimental conditions to be used to study lipids, since it has been shown that the lipid content in *N. asteroides* varies considerably over their growth cycle *in vitro* (Beaman, 1975).

The results of these studies agree with reports of others studying cord factor in *Mycobacterium* sp. (Asselineau and Portelance, 1960; Goren, 1974; Ioneda *et al.*, 1970) that trehalose lipids seem to play an important role in virulence of both *Mycobacterium* sp. and *Nocardia* sp.

V. SUMMARY

Trehalose lipids, derivatives of "cord factor" (trehalose 6, 6'-dimycolate), have been reported to play a role in the virulence of several pathogenic bacteria. Our studies of three isolates of *N. asteroides* showed that, at the concentrations used, one strain (No. 84) was more virulent than were two other strains (No. 171, 186) for BALB/c and an outbred strain of Swiss mice. The test lipids were extracted and hydrolyzed with base. The hydrolyzate contained a trehalose moiety as determined by thin-layer chromatography. Quantification of these lipid fractions by gas chromatography showed that strain No. 84 contained 24-fold more trehalose lipids than did strain No. 171, and 10-fold more than did No. 186. In mice virulence studies, approximately twice as many mice died in nine days when injected with strain No. 84 than with equivalent amounts of either of the two other strains.

REFERENCES

Asselineau, J. and Portelance, V., *Recent Results Cancer Res. 47:*214 (1960).
Bekierkunst, A. and Artman, M., *Am. Rev. Respir. Dis. 86:*832 (1966).
Beaman, B., *J. Bacteriol. 123:*1235 (1975).
Beaman, B., *in* "The Biology of the Nocardiae" (M. Goodfellow, G. H. Brownell, and J. A. Serrano, eds.), p. 402. Academic Press, New York (1976).
Bloch, H., *J. Exp. Med. 91:*197 (1950).

Gangadharam, P. R. J., Cohn, M. L., and Middlebrook, G., *Tubercle 44:*452 (1963).
Goren, M. B., *Biochim. Biophys. Acta 210:*116 (1970).
Goren, M. B., and Brokl, O., *Recent Results Cancer Res. 47:*251 (1974).
Ioneda T., Lenz, M., and Pudles, J., *Biochem. Biophys. Res. Commun. 13:*110 (1963).
Ioneda, T., Lederer, E., and Rozanis, J., *Chem. Phys. Lipids 4:*375 (1970).
Kato, M., Miki, K., Matsunga, K., and Yamamura, Y., *Am. Rev. Tuberc. Pulm. Dis. 77:*482 (1958).
Mompon, B., Toubiana, R., and Lederer, E., *Chem. Phys. Lipids 12:*213 (1976).
Noll, H., Bloch, H., Asselineau, J., and Lederer, E., *Biochim. Biophys. Acta 20:*299 (1956).

PLASMID INVOLVEMENT IN THE GENETIC DETERMINATION OF ANTIBIOTIC BIOSYNTHESIS IN ACTINOMYCETES[1]

Alfredo Aguilar [2]

Departamento de Microbiología
Facultad de Biología
Universidad de León, Spain and
John Innes Institute
Norwich, England

I. INTRODUCTION

Actinomycetes have a striking genetic capacity to synthesize a large number of antibiotics with very different chemical structures (Hopwood and Merrick, 1977) and plasmids are possibly involved in the synthesis of some antibiotics (Hopwood, 1978; Okanishi, 1979). In the last few years, a number of papers have appeared suggesting that plasmids may control the biosynthesis of several antibiotics. However, in general, the evidence is rather weak and indirect and the nature of this control is not clear. In most cases, the evidence for plasmid involvement or even for the existence of plasmids in actinomycetes is still very tentative.

A recent review on this topic points to the need for critical evidence before concluding that production of a particular antibiotic is controlled by plasmid-

[1] *This paper is dedicated to the memory of Professor Carlos Asensio who, on August 17, 1982, died in an accident at the age of 58. Carlos Asensio was at the pinnacle of his scientific career when he died and had much that he wished to accomplish. He was a wonderful human being. His death came as a shock to those of us who knew, loved, and counted on him for counsel and support. His death leaves a void which will not be easily filled.*

[2] *The author was recipient of a long-term EMBO fellowship during the performance of the experimental part of this work.*

ISBN 0-12-528620-1

borne genes (Hopwood, 1983). Such evidence for plasmid involvement, as mentioned elsewhere (Hopwood, 1978), may be of several kinds. Segregational evidence is obtained when, in a cross, the production of an antibiotic (or any other secondary metabolite) segregates independently of chromosomal markers. This evidence is in total agreement with the genetic definition of a plasmid as originally proposed by Lederberg (1952). For a segregation test, it is necessary to have at least a rudimentary linkage map, a condition that exists for only a few antibiotic-producing actinomycetes. The construction of such maps can be very time-consuming and may be avoided when the plasmid is transferable, either by conjugation or transformation, to strains with well-defined chromosomal maps.

Infectious transfer of antibiotic production concomitant with plasmid transfer can occur by conjugation or transformation or, passively, by means of other conjugative plasmids. Only positive results are relevant because of the possible involvement of non-conjugative plasmids, inadequate transformation systems, and incompatibility with a plasmid already present in the recipient strains. When positive, other tests, such as isolation of plasmid DNA from an antibiotic-producing strain, instability and curing of plasmid DNA with simultaneous loss of antibiotic production, indicate that a plasmid may be involved in antibiotic biosynthesis. Since the rationale of these tests has been discussed elsewhere (Hopwood, 1978), they will not be mentioned here, except to point out the unreliability of so-called "curing agents" such as acridine orange, acriflavine, ethidium bromide, etc. as evidence of plasmid involvement in antibiotic biosynthesis. It has been documented (Schrempf and Goebel, 1979; Komatsu *et al.*, 1981) that these substances often provoke DNA rearrangements, including deletions and insertions, instead of plasmid loss. As an alternative, protoplast formation and regeneration has been suggested (Hopwood, 1981a) as a "mild" test for curing plasmid in the diagnosis of plasmid involvement in antibiotic biosynthesis in *Streptomyces*.

The basic techniques for protoplast formation and regeneration, as well as for gene cloning in *Streptomyces*, are readily available. (A review on the subject (Chater *et al.*, 1982) may be consulted by anyone wishing to become familiar with the practical manipulation of the organisms.)) One interesting property of some *Streptomyces* plasmids is the appearance of "pocks" when spores containing conjugative plasmid DNA are plated on a lawn of plasmidless spores. This phenomenon has been called "lethal zygosis" (Bibb *et al.*, 1977, 1981) by analogy to the phenomenon of this name in *Escherichia coli* (Skurray and Reeves, 1973). In addition, one of the most informative tests is to detect the presence or absence of homologous DNA sequences in the antibiotic-producing strain by using a probe containing a gene or genes coding for antibiotic production or resistance, according to the Southern hybridization technique (Southern, 1975).

In the genus *Streptomyces*, there are several examples of antibiotics, the biosyntheses of which are determined by clusters of chromosomal structural genes, but in which plasmids exert a control, by an as yet undefined mechanism, over the chromosomal genes in order to achieve significant antibiotic production. Finally, there are a few cases in which the structural genes coding for antibiotic production and resistance are plasmid-borne (Hopwood, 1979).

There is little information on the genetic determination of the biosynthesis of even the most important antibiotics and it has not been increased very greatly since the last review on this topic (Hopwood, 1979). However, in the last few years there has been a dramatic increase in the versatility and scope of molecular genetic techniques in the analysis of antibiotic-producing microorganisms (Hopwood, 1981). Among these, recent advances in DNA technology are powerful tools to study these problems (Chater *et al.*, 1982).

The aim of this paper is to summarize the present knowledge about the genetic determination of antibiotics coded by plasmids, either when the plasmid harbors the biosynthetic genes, such as in methylenomycin which will be described in detail, or when the plasmid only carries a gene that serves in some kind of "regulatory" role, the structural genes being located on the chromosome.

II. GENETIC CONTROL OF ANTIBIOTIC BIOSYNTHESIS

A. Methylenomycin A: A Plasmid-Coded Antibiotic

Methylenomycin A in *S. coelicolor* A3(2) is to date the only antibiotic whose biosynthetic pathway is known to be coded by plasmid-borne genes. The evidence is substantial and has been summarized elsewhere (Hopwood, 1979; Hornemann and Hopwood, 1981). Different fertility states can be recognized amongst derivatives of *S. coelicolor* A3(2) with the properties expected if a sex factor is absent ($SCP1^-$), autonomous ($SCP1^+$ or SCP1'), or chromosomally integrated (NF)(Hopwood *et al.*, 1973). These different plasmid situations present close similarities to the different status of F plasmid in *E. coli*. The SCP1 plasmid can be transferred by mating to *S. lividans* and *S. parvulus* (the latter has little chromosomal base sequence homology with *S. coelicolor*) (Westpheling, 1980). Thus, *S. parvulus* produces methylenomycin A when SCP1 is mated into it (Kirby and Hopwood, 1977). Another line of evidence comes from the isolation of apparent point mutations (*mmy*) which lead to loss of methylenomycin production in *S. coelicolor* A3(2) strains carrying SCP1. All the *mmy* mutants isolated are linked to SCP1 rather than to the chromosome (Hornemann and Hopwood, 1981; Kirby and Hopwood, 1977). Physical data on the nature of SCP1 have been difficult to obtain since it cannot be isolated by standard methods which readily recover most eubacterial and some other *Streptomycete* plasmids. Westpheling (1980) obtained some physical information on SCP1 although it was not possible to isolate enough DNA for physical studies and for DNA cloning. SCP1-specific sequences were detected either by positive selection for DNA sequences homologous to SCP1 in different backgrounds with little chromosomal homology, as well as by negative selection for non-homologous sequences between *S. coelicolor* A3(2) SCP1$^+$ and SCP1$^-$ strains (Hopwood *et al.*, 1979).

Recently, more evidence on the genetic determination of methylenomycin has been obtained with another plasmid, pSV1 (Aguilar and Hopwood, 1982). This plasmid was detected by Okanishi, Manome, and Umezawa (1980), by electron microscopy only, in *S. violaceus-ruber* SANK 95570, the original methylenomycin-producing strain (Haneishi *et al.*, 1974). The pSV1 plasmid was detected on agarose gels using a slightly modified version of Westpheling's (1980) lysis method described by Charter *et al.* (1982); its size was estimated as about 110 Md using several plasmids as standards, which is in quite good agreement with the estimate of 100 Md made by Okanishi *et al.* (1980).

pSV1, like SCP1, codes for methylenomycin production and resistance. Some of the experiments supporting this hypothesis are summarized here. pSV1 hybridized with a probe containing a 2.55 Kb *S. coelicolor* fragment, generated by *Pst* I, which codes for methylenomycin A resistance (Fig. 1). This indicated that the plasmid carries at least the resistance genes and suggested that pSV1 might also be responsible for methylenomycin production. This possibility is due to the selective problems that might arise in an antibiotic-producing strain in which the resistance genes coding for production were on the chromosome; spontaneous plasmid loss might lead to the suicide of the strain.

The hybridization was specific for pSV1 and for SCP1$^+$ strains (Fig. 2). In the latter case, hybridization occurred with the chromosomal fractions, indicating that the SCP1 plasmid had been linearized during the isolation. Alternatively, SCP1 could exist as ccc DNA, because it is known (Currier and Nester, 1976; Palchaudhuri and Chakrabarty, 1976), that large ccc DNA's often precipitate with chromosomal and membrane fractions during a clearing centrifugation in the presence of SDS.

The hypothesis that pSV1 codes for methylenomycin production was confirmed by the observations that about 15% of the regenerated protoplasts from *S. violaceus-ruber* had lost the capacity to produce methylenomycin and were sensitive to it and that no plasmid could be detected in any of the samples examined. Protoplast formation and regeneration was used as a "mild" method of plasmid curing, according to Norvick *et al.* (1980) and Hopwood (1981a). Another indication of the involvement of pSV1 in methylenomycin production was obtained by transferring the plasmid, either by mating or transformation, from the original background, *S. violaceus-ruber* SANK 95570, to *S. lividans* 66 or to *S. coelicolor* A3(2), whereupon the recipient strains became methylenomycin producers, and a plasmid, identical to pSV1 in size and in other properties, could be isolated from the recipient strains.

There are close similarities between pSV1 and SCP1, in transmissibility,

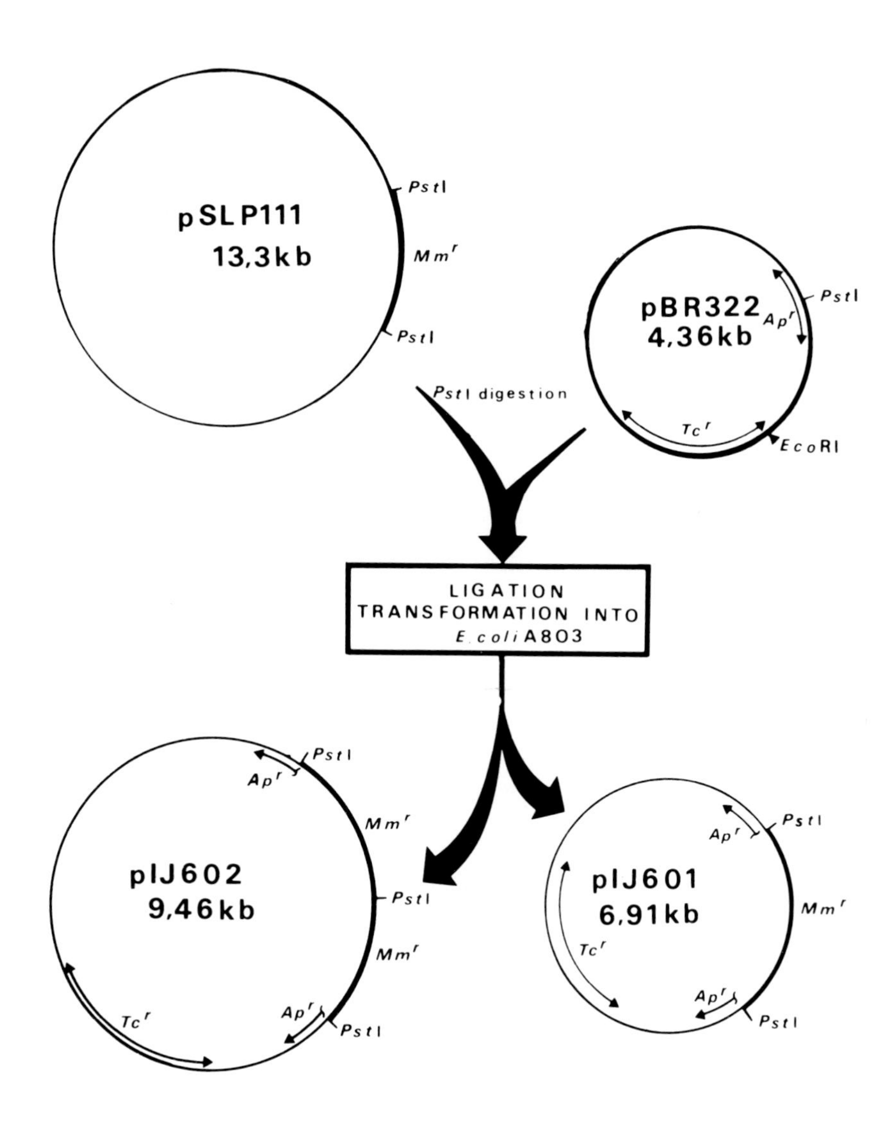

FIGURE 1. Construction of a probe consisting of a 2.55 Kb fragment coding for Mmy^r into pBR322. This fragment had previously been cloned by a "shotgun" experiment from an SCP1-containing strain of S. coelicolor *into the SLP1.2 plasmid by Bibb, Schottel, and Cohen (1980).*

large size, and especially methylenomycin production and resistance. Nevertheless, there are some differences, as yet unknown, which make SCP1 extremely difficult to isolate (Westpheling, 1980). One possibility is that most SCP1 molecules are integrated into the chromosome; only during transfer

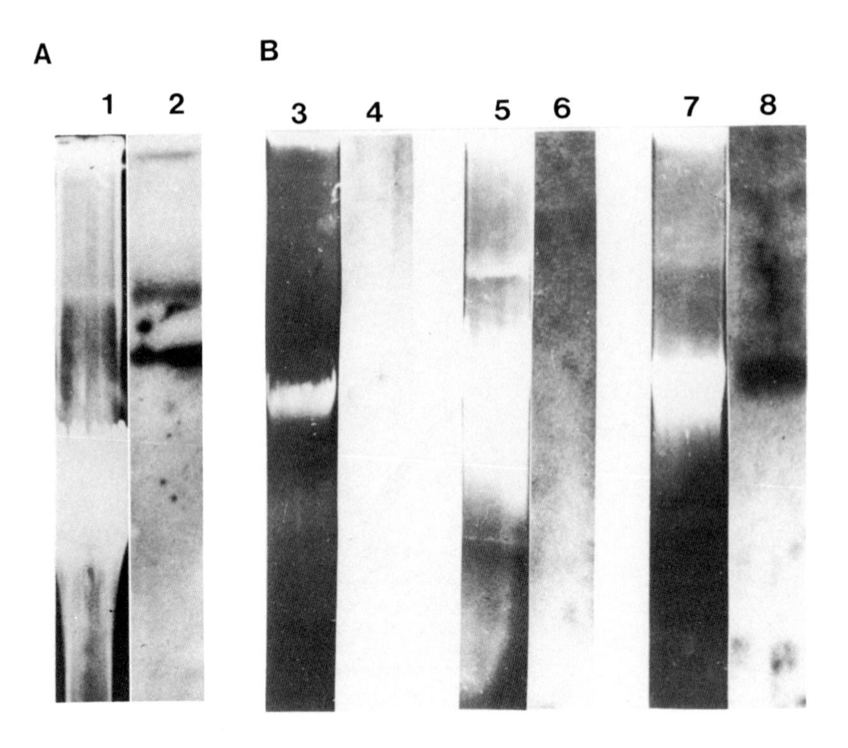

FIGURE 2. Plasmid patterns and corresponding autoradiograms after hybridization with ^{32}P-labelled pIJ601 DNA probe. A) S. violaceus-ruber *SANK95570 (tracks 1 and 2) and B)* S. coelicolor *M130 (SCP1⁻, SCP2⁻), tracks 3 and 4;* S. coelicolor *1190 (SCP1⁻, SCP2⁺), tracks 5 and 6;* S. coelicolor *M146 (SCP1⁺, SCP2⁻), tracks 7 and 8. The DNA patterns were analyzed on 0.7% agarose tris-borate gels. Note hybridization of plasmid pSV1 in track 2 and SCP1 in track 8. (Reproduced in part from Aguilar, A., and Hopwood, D.A.,* J. Gen. Microbiol. 128: *1893 (1982) by copyright permission of the Society for General Microbiology.)*

would SCP1 be located extrachromosomally and therefore have all the characteristics of a plasmid. This hypothesis is compatible with the data obtained by Westpheling (1980). In this context we may consider an analogy between SCP1 and the IncJ plasmid R391 from *E. coli*. This plasmid lacks the genes for autonomous replication, and all attempts to isolate it have been unsuccessful. The reason seems to be that R391 normally exists in a chromosomally integrated form and transfer of resistance markers is concomitant with the transfer of chromosomal markers (Nugent, 1981). However, the case of SCP1 seems to be more complex because there are situations in which SCP1 is integrated at various sites on the chromosome and promotes chromosome transfer either unidirectionally (Hopwood and Wright, 1976a; Vivian and Hopwood 1973) or bidirectionally (Hopwood *et al.*, 1969) as well as situations in which the plasmid exists autonomously (see above). The opposite also occurs: the integration of chromosome regions into the plasmid SCP1, to give SCP1-primes, analogous to the F-primes from *E. coli* (Hopwood and Wright, 1976b).

In order to isolate enough pSV1 plasmid DNA for gene cloning, CsCl-EthBr centrifugation was used to isolate pSV1; however, a DNA plasmid band was not detected by this method. Nevertheless, a small amount of plasmid DNA was obtained after a cleared lysate had been prepared according to Westpheling's method (1980) and run in a preparative agarose gel. Only a small part of the gel was irradiated to locate the plasmid band. Plasmid DNA from the unirradiated section of the gel was then electrophoresed onto filter papers backed with dialysis membranes according to Girvitz *et al.* (1980), from which it was subsequently eluted. The resulting pSV1 plasmid usually had acquired nicks and therefore would not enter agarose gels unless it was broken into linear fragments with restriction endonucleases (Aguilar and Hopwood, unpublished results).

pSV1 plasmid was totally digested with *Bam* HI and the DNA fragments were ligated to the pIJ41 plasmid linearized with *Bam* HI and transformed into *S. lividans* 66 protoplasts as described by Thompson, Ward, and Hopwood (1982). The cleavage of pIJ41 with *Bam* HI inactivates the neomycin resistance gene (Thompson *et al.*, 1982; see Chater *et al.*, 1982), allowing recombinant plasmids to be detected by insertional inactivation. Among thiostrepton-resistant, neomycin-sensitive colonies, it was possible to detect 40 apparently different clones (Fig. 3) having inserts between 1.5 and 6 Kb (Aguilar and Hopwood, unpublished results). Perhaps because of the small size of the cloned fragments, none of the recombinants acquired the ability to produce methylenomycin. However, an alternative explanation is that clones producing methylenomycin did not contain the resistance gene and so perished. Much better recipient strains for transformation would be those that harbor SCP1-linked *mmy* mutations (strains carrying SCP1 in the autonomous state which have lost the ability to make methylenomycin but are still resistant to the antibiotic). Such strains are formed by mating different representatives of the various classes of mutants from *S. coelicolor*, in which they were isolated (Kirby and Hopwood, 1977), into *S. lividans* 66, since so far the pIJ41 vector can be used only in the latter species. Since these matings were successful, the appropriate *S. lividans mmy* strains are now available for use as recipients in the cloning experiments.

B. Other Examples of Plasmid Involvement

Two recent reviews (Chater and Hopwood, 1982; Hopwood, 1979) deal with this topic and include examples in which the evidence for plasmid involvement in antibiotic biosynthesis is minimal, *e. g.*, loss of antibiotic production after treatment with curing agents. Clearly, these examples are in need of further study to confirm the involvement of plasmids. Perhaps *S. reticuli*, producer of leucomycin, is the best example because plasmid deletions are correlated with

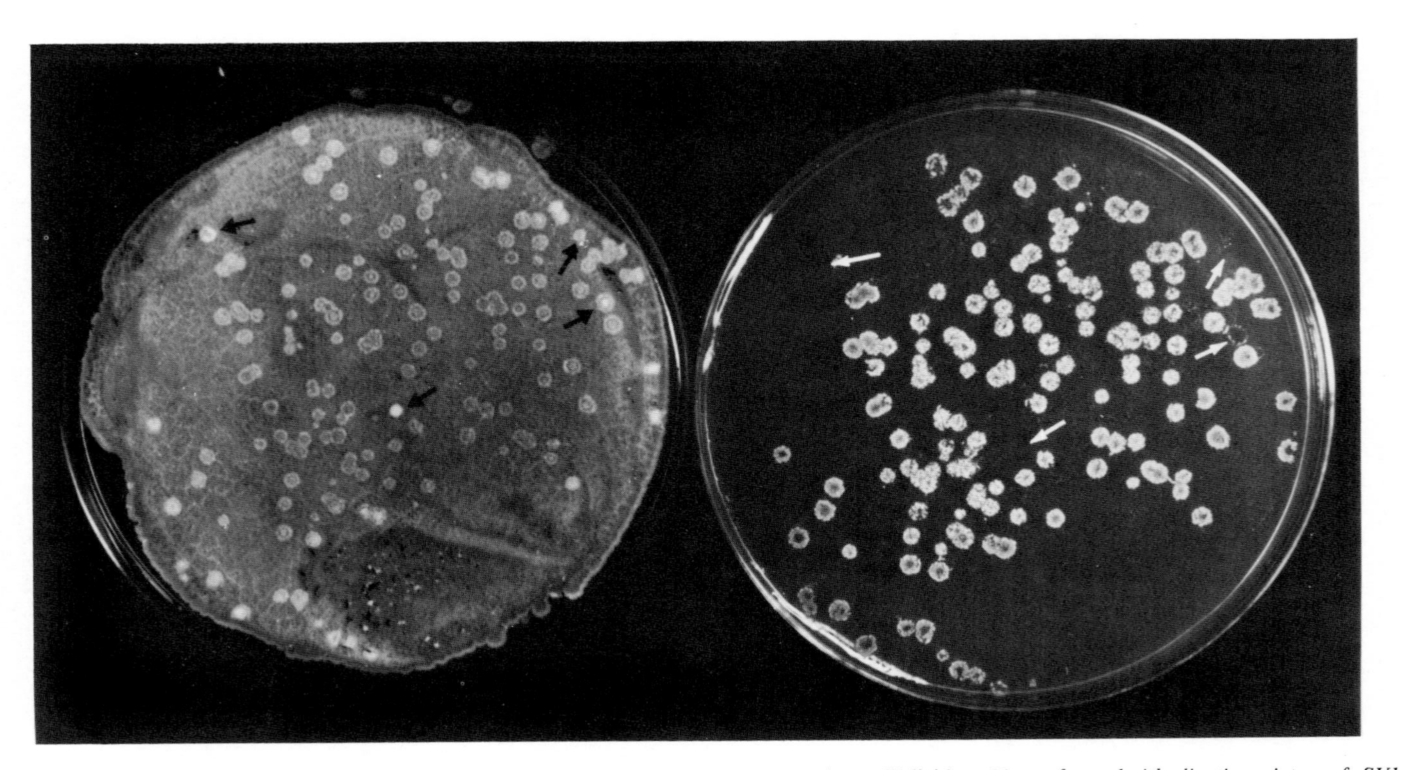

FIGURE 3. Cloning of pSV1 plasmid fragments. Pocks originated by regenerated protoplasts of S. lividans *66 transformed with a ligation mixture of pSV1 and p1J41,* BamH1 *generated fragments. Recombinant clones (arrows) can be recognized in replica plating to a neomycin containing medium (1μg/ml) by their sensivity to the antibiotic.*

loss of antibiotic production. Therefore, the hypothesis of plasmid involvement does not rely simply on a failure to detect plasmid DNA in non-producing variants (Schrempf, 1982; Schrempf and Goebel, 1979). In most cases described, firm evidence on the nature of the genetic control of antibiotic biosynthesis by plasmids is lacking; therefore, very little is known about whether this control is due to plasmid-linked structural biosynthetic genes, to plasmid-linked regulatory genes, or to an indirect pleiotropic effect of plasmid loss, perhaps reflecting an interference with an important metabolic control (Martín and Demain, 1980). A case similar to the production of leucomycin by *S. reticuli* is that of erythromycin by *S. erythreus* NRRL 2328 (Wang *et al.*, 1982). Treatment of the wild type strain with ethidium bromide or acridine orange gave two variants: one, a non-producer, that was blocked at the step of attachment of mycarose to C-3 of erythronolide and the other that produced two- to three-fold more erythromycin than did the parent strain. Both variants showed different plasmid profiles from the parental one. The striking variations in plasmid content and in antibiotic production phenotype must be considered circumstantial evidence at the present level of understanding of this system. In addition to the example described, some other cases have recently come to light; a brief comment will be made on some significant examples.

The β-lactam antibiotic cephamycin C is produced by *S. lactamdurans*. When this strain is plated in solid medium, it spontaneously forms two different kinds of clones: about 95% of the clones form aerial mycelium, but 5% do not (*Amy*$^-$) (Castro *et al.*, unpublished data) and are unable to synthesize cephamycin C. A plasmid, pSC1, has been isolated from the *Amy*$^+$ strains, but not from the *Amy*$^-$. A possible role for pSL1 may be deamination of lysine. *Amy*$^-$ clones cannot grow on lysine as the only carbon and nitrogen source, which suggests that *Amy*$^-$ clones are unable to convert lysine to α-aminoadipic acid. It has been suggested (Castro *et al.*, unpublished data) that this plasmid may originate through insertions, rearrangements, or deletions in the chromosome, thereby inactivating some of the genes controlling the above-mentioned characters. A more detailed study of this interesting case is described by Martín *et al.*, (elsewhere in this volume).

Vyskocil and Zeleny (1982) reported that, after acriflavine treatment of *S. aureofaciens*, a high proportion of non-producer mutants, including pleiotropic mutants with a loss of aerial mycelium and spores as well as with a loss of pigment synthesis, were produced. These pleiotropic effects may be explained if it is assumed that acriflavine induces DNA rearrangements (see above). About 50% of the non-producer mutants showed only loss of antibiotic production and were normal in respect to the other characters. Yashima and Eguchi (1982) have reported a similar case in which pleiotropic phenomena are observed when *Streptomyces* sp., producer of formycin, is treated with acridines.

SPQ1 is a conjugative plasmid originally isolated from *S. gingfengmyceticus*, and it has been postulated that the structural genes encoding the enzymes of gingfengmycin biosynthesis are located on SPQ1 DNA plasmid (Zheng, *et al.*, 1982). The evidence in this case is the presence of SPQ1 plasmid in the gingfengmycin-producing strain and the concomitant transmissibility of SPQ1 and of antibiotic production from *S. gingfengmyceticus* to *S. hygroscopicus* var. *jinggangensis*. Additional evidence is provided by the fact that up to 10% of the transconjugants lost antibacterial activity after incubation at high temperature.

Streptomycin synthesis in *S. griseus* No. 45 also seems to be mediated by a plasmid (Zhuang *et al.*, 1982): cured variants simultaneously had lost this plasmid and the ability to produce streptomycin. Strong evidence for plasmid control of streptomycin synthesis was provided by the finding that streptomycin production was regained after transformation of the variants with the plasmid found in the original strain.

In some cases, plasmid involvement seems to be rather complex. Thus, it has been postulated (Danilenko, personal communication) that *S. antibioticus*, oleandomycin producer, contains a genetic element, eSA1 in different states. In the wild type, eSA1 is located on the chromosome, whereas in other strains, isolated from the wild type after repeated selection for oleandomycin production, eSA1 is contained not only extrachromosomally but also in the chromosome as a multiple tandem repeats. It has been suggested that the increase in oleandomycin production is a consequence of tandem duplication of eSA1 DNA which contains genes for antibiotic production (Danilenko, personal communication).

Omura, Ikeda, and Kitao (1979) reported the presence of a plasmid, pSA1, in the spiramycin producer, *S. ambofaciens*. They postulated that this plasmid was involved in the synthesis of the antibiotic because they failed to detect plasmid DNA in the cleared-lysate fractions of the non-producer variants which had been obtained with acriflavine. However, more recently Ikeda, Tanaka, and Omura (1982a,b) found that the spiramycin-non-producer strains harbored a chromosomal and/or membrane-associated plasmid which, as mentioned above, could not be detected in the cleared lysate fractions. This example highlights the need for extreme caution in concluding that a plasmid controls antibiotic production on evidence of its apparent absence from non-producing variants. Such variants may be pleiotropically altered such that the ease of plasmid isolation is affected. (Hopwood, 1983).

III. CONCLUSION

Perhaps the obvious question in studying the genetic expression by plasmids of secondary metabolites such as antibiotics is: How general is the situation in which the genes coding for antibiotic production and/or resistance are plasmid-borne? With present knowledge it is difficult to give an answer. However, there are several well-studied examples that allow some considerations to be drawn. Thus, it seems to be the exception rather than the rule for a plasmid to carry the structural genes for production and resistance. Methylenomycin coded by SCP1 and pSV1 (see above) is the best example; tylosin coded by a self-transmissible plasmid from *S. fradiae*, still physically undetected, may be another (Baltz *et al.*, 1981). There are many other examples in which plasmids play only a regulatory role and this role is still entirely undetermined. Nonetheless, even though we are far from a complete understanding of these mechanisms, it is clear that a continued study of the regulatory mechanisms of plasmids in antibiotic production is a worthwhile goal in basic research and that the manipulation of the regulatory mechanisms of antibiotic production in industrial microorganisms by gene cloning can have great medical and economic importance.

Two main mechanisms have been suggested for plasmid evolution (Cohen *et al.*, 1978). "Macroevolution" occurs through insertion, deletion, or rearrangements of relatively large segments of plasmid DNA and involves mostly site-specific, illegitimate genetic recombination; "microevolution" appears to be associated with the insertion, deletion, or substitution of very short sequences of DNA (Garaev *et al.*, 1982; Santamaría, personal communication). Natural plasmids may have originated by excision of chromosomal segments harboring antibiotic biosynthetic genes and by their attachment to a sequence of deoxyribonucleotides capable of autonomous replication. These molecules may have originated by "natural genetic engineering" and they may have evolved, by any of the mechanisms mentioned above, to more stable molecules. Mechanisms such as these may explain the wide spread of antibiotic resistance genes from *Streptomyces* to plasmid-carrying eubacteria (Benveniste and Davies, 1973). The fact that most eubacteria only carry resistance genes may be due to microevolution phenomena which lead to rearrangements that would inactivate the production genes. Gene flow between chromosomal and plasmid DNA, such as phage λ, F' plasmids as well as other elements such as bacteriophage Mu and other transposable DNA, seems to be a normal event. Although these processes have been studied almost exclusively in *E. coli* (Campbell, 1981), comparable information is beginning to emerge in *Streptomyces*. For example, SCP1' variants of SCP1 are known to carry chromosomal genes (Hopwood and Wright, 1976) and several phenomena in *Streptomyces* may involve transposon-like elements. Most of these have been reviewed by Chater and Hopwood (1982), but no rigorous demonstration of their presence is yet available.

One interesting question that may be solved in the next few years concerns the regulatory mechanisms that control the expression of plasmid genes coding for antibiotic production exclusively during the idiophase (or stationary phase). It is well known that antibiotics, as most secondary metabolites, fail to appear during exponential growth because the enzymes responsible for their formation are repressed during the tropophase (exponential phase) (Drew and Demain, 1977; Martín and Demain, 1980). DNA cloning techniques will doubtless bring us closer to answering these important questions.

ACKNOWLEDGMENTS

I thank Professors J. F. Martín and D. A. Hopwood for valuable advice and critical reading of the manuscript.

REFERENCES

Aguilar, A., and Hopwood, D. A., *J. Gen. Microbiol. 128:*1893 (1982).

Baltz, R. H., Seno, E. T., Stonesifer, J., Matsushima, P., and Wild, G. M., *in* "Microbiology-81" (D. Schlessinger, ed.), p. 371. American Society for Microbiology, Washington (1981).

Benveniste, R., and Davies, J., *Proc. Nat. Acad. Sci. U. S. A. 70:*2276 (1973).

Bibb, M. J., Freeman, R. F., and Hopwood, D. A., *Mol. Gen. Genet. 154:*155 (1977).

Bibb, M. J., Schottel, L., and Cohen, S. N., *Nature (Lond.) 284:*526 (1980).

Bibb, M. J., Ward, J. M., Kieser, T., Cohen, S. N., and Hopwood, D. A., *Mol. Gen. Gent. 184:*230 (1981).

Campbell, A., *Annu. Rev. Microbiol. 35:*55 (1981).

Chater, K. F., Hopwood, D. A., Kieser, T., and Thompson, C. J., *Curr. Top. Microbiol. Immunol. 96:*69 (1982).

Chater, K. F., and Hopwood, D. A., *in* "The Actinomycetes" (M. Goodfellow, M. Modarski and S.T. Williams, eds.). Academic Press, London (1983). (In press)

Cohen, S. N., Brevet, J., Cabello, F., Chang, A. C. Y., Chou, J., Kopecko, D. F., Kretschmer, P. J., Nisen, P., and Timmis, K., *in* "Microbiology-1981" (D. Schlessinger, ed.), p. 217. American Society for Microbiology, Washington (1978).

Currier, T. C., and Nester, E. W., *Anal. Biochem. 76:*431 (1976).

Drew, S. W., and Demain, A. C., *Annu. Rev. Microbiol. 33:*343 (1977).

Garaev, M. M., Bobkov, A. F., Bobkova, A. F., Kalinin, V. N., Smirnof, V. D., Khudakov, Yu, E., and Tikconenko, T. I., *Gene 18:*21 (1982).

Girvitz, S. C., Bacchetti, S., Rainbow, A. J., and Graham, F. L., *Anal. Biochem. 106:*492 (1980).

Haneishi, T., Kerahara, A., Arai, M., Hata, T., and Tamura, C., *J. Antibiot. 27:*393 (1974).

Hopwood, D. A., *Annu. Rev. Microbiol. 32:*373 (1978).

Hopwood, D. A., *J. Nat. Prod. 42:*596 (1979).

Hopwood, D. A., *in* "Actinomycetes" (K. P. Schaal and G. Pulverer eds.), p. 523. Gustav Fischer, Stuttgart (1981a).

Hopwood, D. A., *in* "Genetics as a Tool in Microbiology" (S. W. Glover and D. A. Hopwood, eds.), Society for General Microbiology Symposium 31, p. 187. Cambridge University Press, Cambridge (1981b).

Hopwood, D. A., Harold, R. J., Vivian, A., and Ferguson, H. M., *Genetics 62:*461 (1969).

Hopwood, D. A., and Merrick, M. J., *Bacteriol. Rev. 41:*595 (1977).

Hopwood, D. A., and Wright, H. M., *in* "Second International Symposium on the Genetics of Industrial Microorganisms" (K. D. Macdonald, ed.), p. 607. Academic Press, London (1976a).

Hopwood, D. A., *in* "Biochemistry and Genetic Regulation of Commercially Important Antibiotics" (L. C. Vining, ed.), p.1. Addison-Wesley, Reading (1983).

Hopwood, D. A., Bibb, M. J., Ward, J. M., and Westpheling, J., *in* "Plasmids of Medical, Environmental and Commercial Importance" (K. N. Timmins and A. Pühler, eds.), p. 245.Elsevier/North Holland, Amsterdam (1979).

Hopwood, D. A., Chater, K.F., Dowding, J.E., and Vivian, A., Advances in *Streptomyces coelicolor* genetics. *Bacteriol. Rev. 37:*371 (1973).

Hopwood, D. A. and Wright, H. M., *J. Gen. Microbiol. 95:*107 (1976b).

Hornemann, U., and Hopwood, D. A., *in* Antibiotics IV. Biosynthesis" (J. W. Corcoran, ed.), p. 123. Springer-Verlag, Berlin (1981).

Ikeda, H., Tanaka, H., and Omura, S., *J. Antibiot. 35:497 (1982a).*

Ikeda, H., Tanaka, H., and Omura, S., *J. Antibiot. 35:*507 (1982b).

Kirby, R., and Hopwood, D. A., *J. Gen. Microbiol. 98:*239 (1977).

Komatsu, K., Leboul, J., Harford, S., and Davies, J., *in* "Microbiology 1981" (D. Schlessinger, ed.), p. 384. American Society for Microbiology, Washington (1981).

Lederberg, J., *Physiol. Rev. 32:*403 (1952).
Martin, J. F., and Demain, A. L., *Microbiol. Rev. 44:*230 (1980).
Norvick. R., Sánchez-Rivas, C., Grass, A., and Edelman, I., *Plasmid 3:*348 (1980).
Nugent, M. E., *J. Gen. Microbiol. 126:*305 (1981).
Omura, S., Ikeda, H., and Kitao, C., *J. Antibiot. 32:*1058 (1979).
Okanishi, M., *in* "Genetics of Industrial Microorganisms" (O. K. Sebek and A. I. Laskin, eds.), p. 134. American Society for Microbiology, Washington (1979).
Okanishi, M., Manome, T., and Umezawa, H., *J. Antibiot. 33:*88 (1980).
Palchaudhuri, S., and Chakrabarty, A., *J. Bacteriol. 126:*410 (1976).
Schrempf, H., and Goebel, W., *in* "Plasmids of Medical Environmental and Commercial Importance" (K. N. Timmius. A. Pühler, eds.), p. 259. Elsevier/Nort Holland, Amsterdam (1979).
Schrempf, H., *J. Chem. Technol. Biotechnol. 32:*292 (1982).
Skurray, R. A., and Reeves, P., *J. Bacteriol. 113:*58 (1973).
Southern, E. M., *J. Mol. Biol. 98:*503 (1975).
Thompson, C. J., Ward, J. M., and Hopwood, D. A., *J. Bacteriol. 151:*668 (1982).
Vivian, A., and Hopwood, D. A., *J. Gen. Microbiol. 76:*147 (1973).
Vyskocil, P., and Zeleny, K., *Biotechnol. Lett. 4:*277 (1982).
Wang, Y-G., Davies, J. E., and Hutchinson, C. R., *J. Antibiot. 35:*335 (1982).
Westpheling, J., Ph. D. Thesis University of East Anglia, Norwich, England (1980).
Yashima, S., and Eguchi, Y., *in* "Proceedings of the Fourth International Symposium on Genetics of Industrial Microorganisms", p. 90. Kyoto (1982).
Zheng, Y. X., Xu, X. X., and Zhang, T. L., *in* "Proceedings of the Fourth International Symposium on Genetics of Industrial Microorganisms", p. 35. Kyoto (1982).
Zhuang, Z., Zhu, Y., Tan, H., and Xue, Y., *in* "Proceedings of the Fourth International Symposium on Genetics of Industrial Microorganisms", p. 76. Kyoto (1982).

EXTRACHROMOSOMAL GENETIC ELEMENTS THAT CONTROL SPECIFIC ENZYMES INVOLVED IN ANTIBIOTIC BIOSYNTHESIS: POSSIBLE INVOLVEMENT OF AN INTRACELLULAR PLEIOTROPIC EFFECTOR

Juan F. Martín
Maria T. Alegre
Jose M. Castro
Paloma Liras

Departamento de Microbiología
Facultad de Biología
Universidad de Leon
Leon, Spain

I. CHARACTERISTICS OF SECONDARY METABOLITES

Antibiotics and other secondary metabolites (also termed special metabolites) (Martin, 1978) are defined as compounds which have great chemical diversity and unusual structures and which, although not essential for growth of the organisms naturally producing them, probably confer a survival advantage to these microorganisms (Martin and Demain, 1980). Production of such secondary metabolites is restricted to only some species of a genus or even to only certain strains of a particular species.

Moreover, secondary metabolites are usually produced as families of closely related compounds, whereas in primary metabolism, only a final product is formed in each biosynthetic pathway. The proportion of each component in a mixture depends on both genetic and environmental factors. This raises an interesting question: Do specific genes encode all the enzymes which are necessary for the transformations in the biosynthesis of these different members of the family of antibiotics, or are there genes that the cell uses primarily for other purposes and only subsidiarily for antibiotic biosynthesis?

ISBN 0-12-528620-1

As we will show later, there is no doubt that much, but probably not all, of the genetic information used for biosynthesis of antibiotics is specific for such purposes and does not seem to be used in primary metabolism.

A second interesting question is how much of this genetic information is located in the chromosome and how much is encoded in extrachromosomal genetic elements. This subject has been reviewed by Hopwood (1983) and Aguilar (elsewhere in this volume) and therefore will not be discussed here.

Our purpose in this article is to review briefly the present state of the research concerning the first question and to discuss in some detail our recent results on the possible location of genetic information coding for specific enzymes involved in the biosynthesis of the β-lactam antibiotic cephamycin C and the polyene macrolide antibiotic candicidin.

II. GENETIC INFORMATION REQUIRED FOR BIOSYNTHESIS OF ANTIBIOTICS

In spite of the chemical diversity of secondary metabolites and the variety of biosynthetic pathways involved, most secondary metabolites are assembled from a few key intermediates (Martin and Liras, 1982) that include amino acids, sugar-phosphate derivatives, acetyl-CoA, malonyl-CoA, a variety of CoA derivatives of mono- and dicarboxylic acids, aromatic intermediates, isoprenoid units, purine and pyrimidine bases, etc.

A. Primary Biosynthetic Pathways Providing Precursors for Antibiotics

The biosyntheses of these key intermediates follow the same pathways and the information for such biosyntheses is coded by the same genes as for primary metabolites. For example, lysine auxotrophs of *Penicillium chrysogenum* blocked in the first part of the pathway are unable to form α-aminoadipic acid, a lysine precursor, and also an intermediate in the biosynthesis of α-aminoadipyl-cysteinyl-valine (ACV) tripeptide that is later converted into penicillin (Demain, 1981; Martin *et al.*, 1979). Similarly, mutants blocked in the common stem of the aromatic amino acid pathway are unable to synthesize rifamycin (Ghisalba *et al.*, 1978). Many other examples of auxotrophic mutants blocked in the formation of different precursors (or inducers) of antibiotics have been reported in the literature (Drew and Demain, 1977). In these cases, the utilization of the precursor (itself a primary metabolite required for growth) occurs only when the intermediate is in excess of the level required for growth (see review by Martin and Liras, 1982). In summary, in many cases, biosynthesis of secondary metabolites may be understood in terms of the existence of an overflow of precursor of these secondary metabolites which are coded by the same genetic information rather than as intermediates of primary metabolism.

Clusters of genes required for the biosynthesis of one specific precursor (*e.g*,valine, a precursor of β-lactams) will certainly be involved in the biosynthesis of antibiotics. Since antibiotics are synthesized from several moieties arising from different biosynthetic pathways (Martin and Liras, 1982), different clusters of genes containing information for the biosynthesis of diverse precursors of antibiotics are likely to be scattered around the chromosome.

These primary precursors of secondary metabolites may be provided by any microorganism. Therefore, a rational approach for constructing hybrid antibiotics is to put together genes that code for primary precursors from overproducing microorganisms with the converting genes of an antibiotic-producing strain. For example, genes from a valine (or α-aminoadipic acid or cysteine)-overproducing strain may be cloned into another strain which efficiently converts these amino acids into β-lactams.

B. Specific Pathways Involved in Antibiotic Biosynthesis: "Silent" Genes

A second part of the biosynthetic pathways of antibiotics is carried out by specific enzymes that convert the precursors into the final products. These pathways are highly specific in each producing strain (see below for the PABA synthetase enzyme). Therefore, the genetic information coding for these enzymes appears to be present only in certain strains, although there is an increasing number of reports of the existence of "silent" genes which are not normally expressed even though they are always present in the cell. In one of the best examples reported (Schlegel and Fleck, 1980), recombinant strains obtained by crosses between *Streptomyces hygroscopicus*, blocked in the production of the non-polyene macrolide turimycin, and mutants of *S. violaceus*, blocked in the biosynthesis of violamycin (an anthracycline antibiotic), formed new antibiotics having biological activity. One of the recombinant strains produced a new anthracycline named iremycin. This antibiotic has a new aglycone (γ-rhodomycinone), not found in either of the parental strains, that is attached to the sugar L-rhodosamine which is formed by the parental *S. violaceus* strain. Genetic information for the biosynthesis of the sugar in the hybrid strain comes from the *S. violaceus* parent, but the origin of the genes for biosynthesis of γ-rhodomycinone is unclear. It seems, therefore, that "silent" genes exist in some strains coding for compounds (or for compounds which can modify other compounds) which are not usually expressed either because they are subject to strong negative regulatory mechanisms (repression) or because they need to be turned on either by insertion of extrachromosomal genetic elements or by chromosomal rearrangement.

Actinomycetes contain an amount of genetic information much larger than do other eubacteria; therefore, it is likely that they might contain "silent" genes which are only expressed under certain culture conditions. *S. clavuligerus* is capable of producing, under different nutritional conditions, two cephalosporins, penicillin N, clavulanic acid (Higgens and Kastner, 1971; Howarth *et al.*, 1976), three different clavams (Brown *et al.*, 1979), efrotomycin (Wax *et al.*, 1976), holomycin, and a tunicamycin-related antibiotic (Kening and Reading, 1979). There is a good chance that many new compounds might be produced by antibiotic-producing strains in media that favor a wider expression of "silent" genes.

Another important aspect is the recent discovery that an ever-increasing number of taxonomically unrelated microorganisms produce the same type of antibiotic. For example, the aminocyclitol antibiotic butirosin is produced both by streptomycetes and by *Bacillus circulans* (Claridge *et al.*, 1974). The β-lactam ring previously thought to be synthesized only by filamentous fungi is now known to be made by a large variety of actinomycetes (*Streptomyces* and *Nocardia*) (Aoki *et al.*, 1977; Martin, 1981) and, strikingly, by a large number of eubacteria, including *Pseudomonas mesoacidophila* (Kintaka *et al.*, 1981), *Gluconobacter* and *Acetobacter* (Sykes *et al.*, 1981), *Chromobacterium violaceum* (Wells *et al.*, 1982a), and *Agrobacterium radiobacter* (Wells *et al.*, 1982b). This suggests that there is a wider distribution of genes for β-lactam biosynthesis than was previously thought and that the possibility exists that "silent" genes for these compounds may be present in a large variety of other microorganisms.

III. GENETIC INFORMATION CODING FOR ANTIBIOTIC BIOSYNTHESIS IS DISPENSABLE

Secondary metabolites are not essential for the producer cell although they appear to confer a survival advantage in nature to the producer strains (Gottlieb, 1976; Martin and Demain, 1980). The loss of the antibiotic-producing capacity, the so-called strain degeneration phenomenon, is well known. However, in many cases, strain degeneration does not involve complete loss of the genetic information coding for antibiotic biosynthesis but simply a decrease (sometimes drastic) in the production level. Strain degeneration is easily explained in terms of either a reduction in the level of an inducer of the expression of biosynthetic genes or a decrease in the pool of adequate antibiotic precursors. Both may result from the loss of extrachromosomal elements, transposition of jumping genes, or chromosomal rearrangement. In a few cases, however, spontaneous or induced loss of the capacity to produce a particular antibiotic occurs at high frequency and is usually accompanied by a pleiotropic loss of several other functions (see below).

IV. SPONTANEOUS LOSS OF THE ANTIBIOTIC-PRODUCING ABILITY AND OTHER CELLULAR FUNCTIONS

A number of cellular functions are spontaneously and simultaneously lost at high frequency in several *Streptomyces*. In *S. lavendulae*, arginine-requiring mutants (*arg*) arise at high frequency. All the *arg* mutants grow on minimal medium supplemented with argininosuccinate but not on minimal medium supplemented with ornithine or citrulline (Nakano and Ogawara, 1980). The *arg* mutants also show no or very low extracellular β-lactamase activity and do not form aerial mycelium (Matsubara-Nakano *et al.*, 1980). Other effects associated with this phenomenon are lower antibiotic production, decrease in formation of pigment, and increase in sensitivity to penicillin. Nakano and coworkers (1980) found a similar phenomenon in *S. kasugaensis*. They observed a coupled loss and reappearance of plasmid DNA with arginine auxotrophy and suggested that auxotrophy was caused by insertion of a plasmid into chromosomal DNA.

A similar situation exists in *S. coelicolor* A3(2) in relation to chloramphenicol resistance (Sermonti *et al.*, 1978). Chloramphenicol resistance (clm^R) in *S. coelicolor* is mediated by a factor which is transferred at very high rates even in strains that lack the SCP1 fertility factor (Freeman et al., 1977; Sermonti *et al.*, 1977). Chloramphenicol-sensitive mutants (clm^S) have been obtained which are sensitive to chloramphenicol (1μg/ml). Some clm^S mutants present a requirement for arginine (*arg* G). As occurs in *S. lavendulae arg* auxotrophs, those of *S. coelicolor* are able to grow on argininosuccinate but not on ornithine or citrulline (Sermonti *et al.*, 1978). One of these mutants lacks argininosuccinate synthetase (Redshaw *et al.*, 1979). One clm^s mutants gives rise spontaneously to *arg* G derivatives at high frequency. Further evidence indicating that fresh isolates from clm^R cultures show several unequivocal map locations for clm^R has led Sermonti *et al.* (1980) to propose that the gene for chloramphenicol resistance in addition to a gene for aerial mycelium formation and a gene for argininosuccinate synthetase (*arg* G) is located in a transposon. However, further evidence is required to prove unequivocally the existence of such a transposon.

It is interesting that the occurrence of *amy* and *arg* G variants in several other *Streptomyces* has also been reported. In *S. alboniger, S. scabies*, and *S. violaceus-ruber*, mutants lacking aerial mycelium (*amy*) were formed in the presence of intercalatin dyes at high frequency (2 to 20%). All *S. alboniger amy* isolates, 27% of *S. scabies* isolates, and 39% of the *S. violaceus-ruber* isolates *were arg* G auxotrophs, lacking argininosuccinate synthetase (Redshaw *et al.*,1979). In addition, the lack of reversion of *amy* or *arg* G isolates to amy^+ or arg^+ may be the result of a deletion of genetic material.

Finally, formation of aerial mycelium was linked to streptomycin production and streptomycin resistance in *S. griseus* and *S. bikiniensis* (Kirby and

Lewis, 1981). *Amy* clones which were also melanine-negative have been isolated from *S. reticuli* by Schrempf (1982) and the phenomenon has been shown to be due to chromosomal rearrangement.

In conclusion, it seems that unstable genes control aerial mycelium formation, arginine biosynthesis, pigmentation, and several other features in several *Streptomyces*. Do such genes also control antibiotic biosynthesis, and if so, at which enzymatic level?

V. SPONTANEOUS LOSS OF UNSTABLE GENES CODING FOR CEPHAMYCIN BIOSYNTHESIS AND AERIAL MYCELIUM IN *STREPTOMYCES LACTAMDURANS*

A. Aerial Mycelium Negative Spontaneous Variants

S. lactamdurans NRRL 3802 has now been reclassified as *Nocardia lactamdurans* since it appears that its cell wall components are characteristic of the genus *Nocardia* (Wesseling and Lago, 1981) and it produces the β-lactam antibiotic cephamycin C in both complex and defined medium. Single clone isolates of this strain spontaneously segregate clones (about 5%) which lack aerial mycelium (*amy*). These spontaneous *amy* variants are unable to produce cephamycin C as measured both by bioassay against *Escherichia coli* Ess 22-31 (supersensitive to β-lactam antibiotics) and by high performance liquid chromatography (HPLC) (Table I), and do not seem to form extracellular serine proteases (Ginther, 1979). No production of cephamycin C by *amy* variants takes place on either solid or liquid fermentation medium. *Amy* clones do not revert to amy^+ with any significant frequency when grown in either complex or minimal solid medium. However, when growing in minimal medium containing lysine as the only carbon and nitrogen source, all the *amy* population acquire a partial degree of aerial mycelium formation after six days (the wild type forms aerial mycelium in two to three days. However, all these clones behave as conventional *amy* mutants when plated once again in either complex or minimal medium with carbon and nitrogen sources other that lysine. There is, therefore, a clear-cut *amy* phenotype which does not behave as such in lysine minimal medium, suggesting that aerial mycelium formation is controlled at the level of an intracellular precursor or effector which is modulated by the lysine concentration in the medium and probably by other cellular parameters as well.

The precursor of cell wall formation lacking in *amy* mutants may be 2,6-diaminopimelic acid (DAP), a compound which exists in the cell walls of *Streptomyces*. Lysine retroregulates DAP-decarboxylase (an enzyme converting DAP into lysine) (Kirpatrick *et al.* 1973; Mendelovitz *et al.*, elsewhere in this volume) channelling DAP into aerial mycelium formation (Martin and Aharonowitz, 1983).

The existence of a difusible positive effector, *i. e.*, a precursor or inducer in *S. lactamdurans* is supported by the finding that *amy* clones growing in the vicinity of wild type clones started to form aerial mycelium at the edges closest to the wild type colony (Castro, Liras, and Martin, unpublished results).

B. Amy *Variants Lack α-Aminoadipyl-Cysteinyl-Valine Tripeptide*

All *amy* variants are unable to synthesize cephamycin C (*cep*) in either complex or defined medium (Table I). *Amy, cep* clones which are induced to form aerial mycelium in lysine minimal medium after six days do not produce cephamycin.

Cephamycin C is synthesized as other β-lactam antibiotics by condensation of the component amino acids α-aminoadipic acid, cysteine, and valine to form the tripeptide δ (α-aminoadipyl)cysteinyl-valine (ACV) by the action of the enzyme tripeptide synthetase. The ACV tripeptide is later cyclized by the tripeptide cyclase to form the isopenicillin N nucleus which is isomerized to penicillin N. Penicillin N is converted by ring expansion into deacetoxycephalosporin C (DACPC) which is finally transformed into cephamycin C by enzymes which introduce a carbamoyl moiety at the C-3 methyl group or a methoxy group at the C-7 position of the cephem nucleus.

Amy variants are unable to form the ACV tripeptide and therefore cannot synthesize cephamycin, in contrast to the wild type in which peptide is found although not accumulated in large amounts (Liras and Martin, unpublished results). Neither biosynthesis of ACV tripeptide nor antibiotic production is restored in *amy cep* strains by supplementation of the medium with 5 mM α-aminoadipic acid, cysteine, or valine (Castro, Liras, and Martin, unpublished data). Uptake of α-aminoadipic acid in *amy cep* variants is similar to the uptake of the wild type, but no cephamycin biosynthesis occurs in complex medium supplemented with 5 or 10 mM α-aminoadipic acid under conditions in which the wild type produced cephamycin. These results suggest that these variants lack tripeptide synthetase.

TABLE I. Production of Cephamycin C in Solid Medium[a] *by* Amy$^+$ *and* Amy$^-$ *Variants of* Streptomyces lactamdurans *3802*

	Number of clones tested	*Average diameter of the Inhibition zone (mm)*
Amy$^+$	*40*	*18.5 ± 4.7*
Amy$^-$	*61*	*7.3 ± 1.0*

[a]*Agar plugs (7 mm diameter) with colonies grown on solid cotton medium (cotton flour, 10 g; glucose, 25 g; agar, 20 g; distilled water, 1000 ml) were incubated at 28°C for four days and assayed on plates seeded with* E. coli *Ess 22-31 (supersensitive to β-lactam antibiotics).*

C. Amy *Variants Lack Several Aminotransferases*

Amy cep were unable to grow in L-lysine (5 mM) as the only carbon and nitrogen source in contrast to the wild type (only a very slow growth and aerial mycelium occurs after six days in minimal medium containing lysine as the only carbon and nitrogen source), suggesting that these variants lack the enzyme system required to metabolize lysine. Lysine may be catabolized by several pathways which are summarized in Figure 2. Pathway B which exists in *S. lactamdurans* (Kern *et al.*, 1980) and in *Flavobacterium*, converts lysine into piperideine-6-carboxylic acid by the action of the ϵ-deaminating enzyme L-lysine ϵ-aminotransferase (E.C. 2.6.1.36, L-lysine : 2-oxoglutarate 6-aminotransferase). A modification of this pathway is found in *Pseudomonas* sp. (Fig. 1, pathway A) (Fothergill and Guest, 1977). Piperideine 6-carboxylate is later transformed into α-aminoadipic acid which serves as precursor of cephamycin C (Kern *et al.*, 1980). L-Lysine is also metabolized in *Pseudomonas aeruginosa* by an oxygenase pathway (route C) in which L-lysine is converted to 5-amino-valerimide and then to 5-amino-valerate. Alternatively, L-lysine may be decarboxylated to cadaverine (pathway D) which is later deaminated by cadaverine aminotransferase to piperidine, which, in turn, is finally converted into 5-amino-valerate. Further metabolism of either 5-amino-valerate or α-aminoadipic acid to acetyl-CoA involves additional deaminations by aminotransferases (Fig. 1).

A very interesting result is that α-aminoadipate, cadaverine, or 5-amino-valerate could not be used as the only carbon and nitrogen source by *amy* variants, whereas they were clearly metabolized by the wild type strain of *S. lactamdurans*, as occurred with lysine. Since catabolism of all these compounds takes place by the action of different 2-oxoglutarate-dependent aminotransferases, it is suggested that *amy* strains have spontaneously lost the expression of the gene for aminotransferases. Assays "in vitro" show that, in effect, the level of aminotransferases in *amy cep* is reduced compared with that of the wild type. Plating in lysine as the only carbon and nitrogen source induces a slow growth rate, transaminases, and aerial mycelium formation, suggesting that lysine produces a regulatory effect partially neutralizing the defects attributed to the pleiotropic mutation.

In summary, about 5% of *S. lactamdurans* clones spontaneously lose the ability 1) to form aerial mycelium, 2) to synthesize the tripeptide ACV precursor of cephamycin, and 3) to produce extracellular serine proteases and several aminotransferase activities. There is partial evidence suggesting that this is due to the loss (or the switch-off) of genetic information coding for the formation of an intracellular effector which, in turn, controls the expression of several genes.

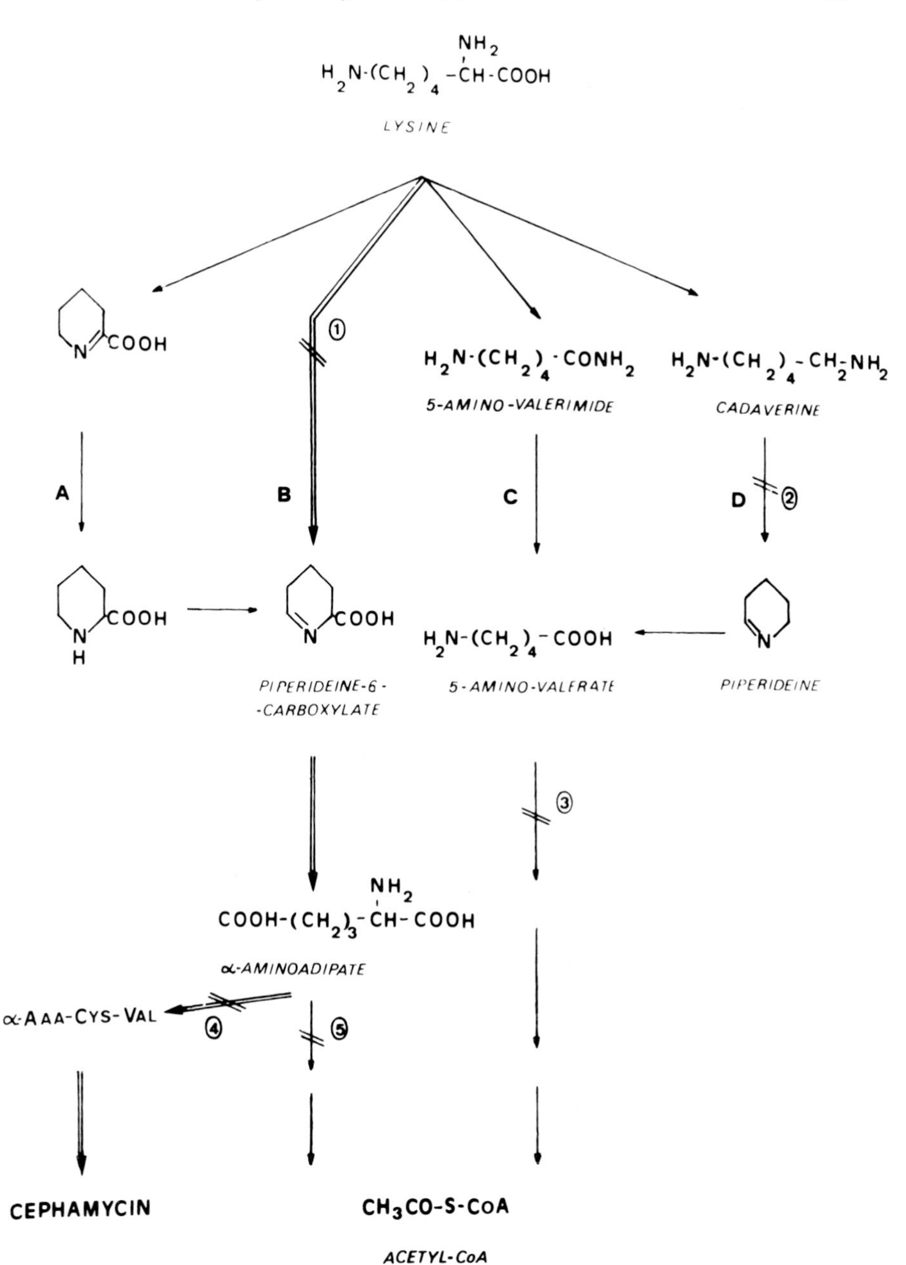

Figure 1. Catabolic pathways of lysine in different microorganisms. The biosynthetic pathway of cephamycin is indicated by double-line arrows. Note the conversion of lysine into α-aminoadipate and the incorporation of α-aminoadipate into ACV tripeptide and cephamycin. The following aminotransferases appear to be deficient in amy mutants of S. lactamdurans*: 1) Lysine ε-aminotransferase, 2) cadaverine aminotransferase, 3) 5-amino-valerate aminotransferase, 4) tripeptide synthetase and 5) α-aminoadipate aminotransferase.*

VI. HIGH-FREQUENCY LOSS OF PABA-SYNTHETASE IN *STREPTOMYCES GRISEUS*.

Candicidin is a polyene macrolide antibiotic produced by *S. griseus* IMRU 3570. It is synthesized by the condensation of three different moieties: 1)*p*-aminoacetophenone starter unit which derives from *p*-aminobenzoic acid, 2) a polyketide chain formed by polymerization of two- and three-carbon units (in the form of malonyl-CoA and methylmalonyl-CoA)that finally forms the aglycone by a head to tail lactone bond, and 3) the rare aminosugar mycosamine (3-amino-3,6-dideoxy-D-mannopyranose). *p*-Aminobenzoic acid (PABA) is a key intermediary metabolite in candicidin biosynthesis and is formed from chorismic acid (the branching point intermediate of the aromatic amino acid biosynthetic pathway) by the action of PABA synthetase. This enzyme exists only in strains producing polyene macrolide antibiotics with an aromatic moiety. It is found in *S. griseus* IMRU 3570 (producer of candicidin), in *S. levoris* (producer of levorin), and in *S. coelicolor* var. *aminophilus* (producer of perimycin) but has not been found in *S. nododus* that synthesizes the non-aromatic polyene macrolide amphotericin B or in other strains of *S. griseus* like the cephamycin-producing strain NRRL 3851 (Gil and Martin, unpublished results).

The biosynthesis of candicidin is under control of several regulatory mechanisms. Some of the control mechanisms are highly specific, *e. g.*., feedback regulation of candicidin biosynthesis by aromatic amino acids at the DAHP synthetase and PABA synthetase levels. Other regulatory mechanisms such as phosphate regulation are general phenomena affecting many biosynthetic pathways (Martin, 1977; Martin and Demain, 1980)

A. PABA Synthetase Negative Mutants

Treatment of *S. griseus* with the so-called curing agents, which are also known to produce chromosomal rearrangements, led to the formation of a high percentage (up to 25%) of clones which have completely lost PABA synthetase activity (*pab*) and therefore candicidin biosynthesis or which are able to form only a basal level of this enzyme. The same result was obtained by treatment with UV light or by protoplasting and regeneration of the protoplasts. Protoplasting is a gentle method of removing plasmids from both *Streptomyces* (Aguilar and Hopwood, 1982) and in *Staphylococcus aureus* (Novick *et al.*, 1980). In addition, these *pab* mutants had lost the ability to form extracellular proteases (*prt*) when incubated at high (non-permissive) temperature (38 °C)(Alegre and Martin, unpublished data). *Pab* mutants that are completely deficient in PABA- synthetase failed to revert (less than 10^{-8} reversion rate). Candicidin biosynthesis by these mutants could not be restored by addition of *p*-aminobenzoic acid (Gil and Martin, unpublished results).

PABA synthetase formation and candicidin biosynthesis by the wild type is extremely sensitive to phosphate. All PABA synthetase activity is repressed by 10 mM phosphate. However, PABA synthetase formation in all *pab* mutants having a residual enzyme activity is insensitive to phosphate (*pho* R) (Gil and Martin, unpublished data).

These results indicate that an unstable genetic element which is involved in the formation of PABA synthetase, and in the temperature sensitive formation of extracellular proteases, exists in *S. griseus* and is removed or rearranged by curing agents or protoplasting and regeneration of protoplasts. Such unstable genetic element controls the expression of the structural gene for protease (*prt*) but does not code for the protease itself because, after curing (or chromosomal rearrangement), a normal expression of proteases at low (permissive) temperature still occurs, suggesting that the structural gene itself has not been altered by the curing (rearrangement) treatment.

B. Extrachromosomal Genetic Element pSG1

An extrachromosomal genetic element pSG1 has been found in clear lysates of the candicidin producer *S. griseus* IMRU 3570 (Fig. 2). It has an approximate molecular weight of 22 Kb by comparison with plasmids of *E. coli* V517 and plasmid SCP2 of *S. coelicolor*. This extrachromosomal element presents some unusual features. It is better observed when clear lysates are run in 0.8% agarose gel at low voltage (0.6 volts/cm). It could not be isolated from the cesium chloride/ethidium bromide equilibrium density centrifugation. However it was clearly observed on total DNA lysates (Fig. 3).

The abnormal behavior of plasmid pSG1 might be related to association with chromosomal DNA, protein, or membrane fragments (Ikeda *et al.* 1982, Marahiel *et al.* 1981). Protein association with the ends of linear plasmid DNA has been found in *Streptomyces* sp. producing lankacidin (Hirochika and Sakaguchi, 1982).

An extrachromosomal band similar to pSG1 has been found in *S. levoris* producer of levorin, an antibiotic almost identical to candicidin, but not in strains producing other polyene macrolide antibiotics. *Pab* mutants lacking the enzyme or those which have a low level of phosphate-derepressed enzyme, still show a faint band of plasmid (Fig. 3). Therefore plasmid has not been completely removed by the curing treatment. Deletions or rearrangements may have occurred in the extrachromosomal element (or in the chromosome) and they may be the responsible of the loss of PABA synthetase as is the case in *S. reticuli* (Schrempf, 1982).

One intriguing question is how the pleiotropic mutation (*pab,prt*) affects phosphate regulation of PABA synthetase residual activity which becomes completely derepressed to phosphate. Our results suggest that phosphate regulation is exerted by the same intracellular effector which controls expression of both *pab* and *prt* genes.

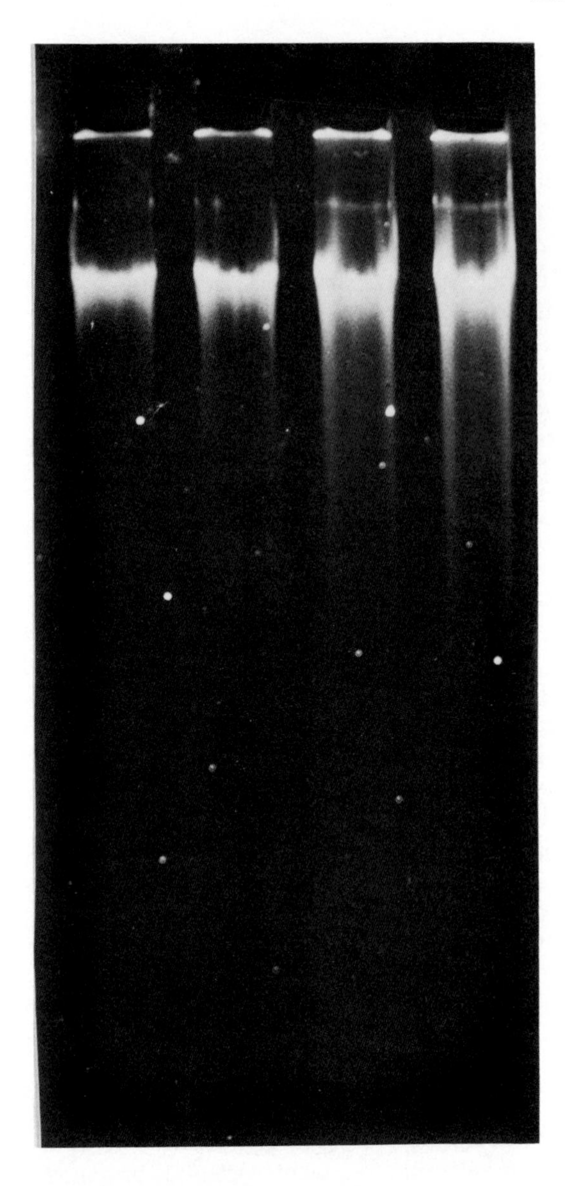

Figure 2. Extrachromosomal genetic element pSG1 in clear lysates of S. griseus *IMRU 3570. Increasing concentrations of DNA from left to right.*

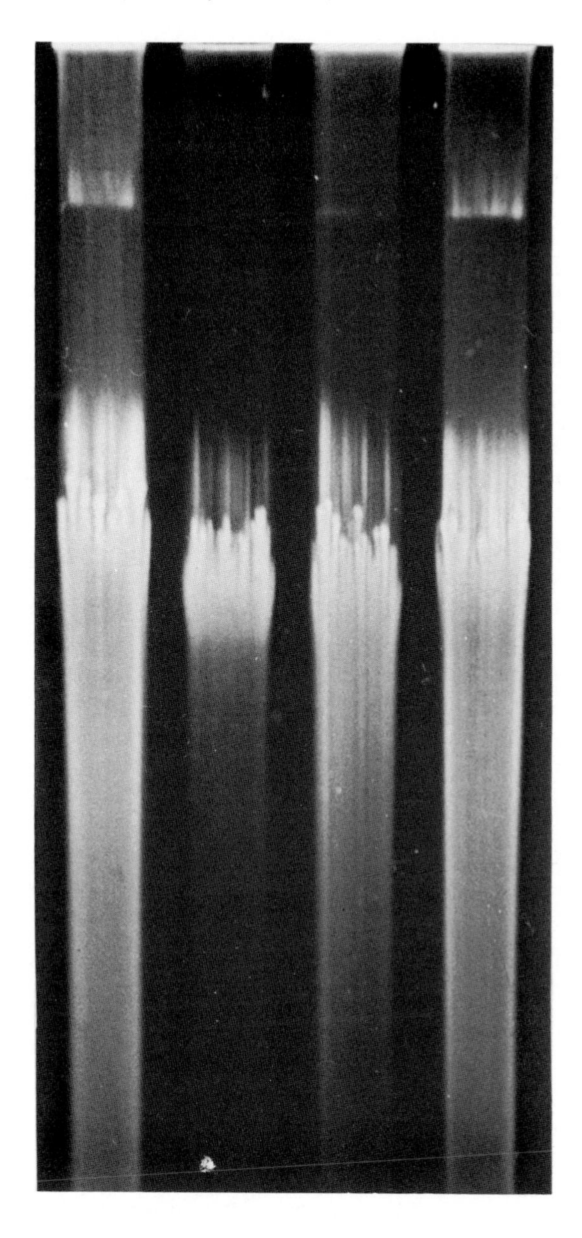

Figure 3. Total DNA lysates of S. griseus *IMRU 3570. Band 1 and 4, wild type; band 2,* S. lactamdurans, *a nonproducer of polyenes; band 3,* pab *mutant.*

VII. DISCUSSION

The existence of intracellular effectors controlling several cellular processes, including differentiation (aerial mycelium formation and sporulation) and biosynthesis of antibiotics or other secondary metabolites, has been reported in some cases and probably is widespread in *Streptomyces*.

One of the best known examples is the control of differentiation of *S. griseus* and strepomycin biosynthesis by factor A. Factor A (2,5-isocapryloyl-3-R-hydroxymethyl-γ-butyrolactone) (Kleiner *et al.*, 1976) is formed by all streptomycin-producing wild type and industrial strains of *S. griseus* and *S. bikiniensis* (Khokhlov *et al.*, 1973; Khokhlov and Tovarova, 1979). It has also been found in 26 of 175 *Streptomyces*, in three of 14 *Actinomyces* and in one of 11 *Nocardia* tested (Hara and Beppu, 1982). It has been reported that 1 μg of pure factor A added at the time of inoculation to a mutant of *S. griseus* blocked in streptomycin biosynthesis induces the production of 1 g of streptomycin, *i. e.*, the induction coefficient is 10^6 (Khokhlov and Tovarova, 1979). The stimulatory effect of factor A is intense when the compound is added at the time of inoculation but does not occur when the compound is added after 48 h of growth. Mutants unable to synthesize the inducer lack the ability to form streptomycin. Factor A restores the ability to produce streptomycin in 114 of 119 non-producers. Such mutants are not capable of synthesizing the streptidine moiety of streptomycin unless supplemented with factor A.

Factor A is also involved in differentiation of *S. griseus*. Antibiotic-producing strains (factor A^+) differentiate normally and form aerial spores in solid medium. Non-producing mutants are not able to differentiate unless supplemented with factor A. The addition of factor A to the mutants also causes formation of well-developed intracellular membranes similar to those existing in the producing strain.

In their study of 95 mutants blocked in streptomycin biosynthesis, Hara and Beppu (1982) reported that 49 lacked factor A and therefore were unable to synthesize streptomycin. Mutants unable to synthesize streptomycin because they lacked factor A arise at high frequency by incubation at high temperature or by treatment with acriflavine or acridine orange. All the factor A deficient mutants fail to revert (less than 10^{-8} to 10^{-9}) as also occurred in our results both in *S. lactamdurans*(*amy, cep*) and in *S. griseus* (*pab, prt*).

Production of streptomycin by Factor A-deficient mutants cannot be detected after the addition of intermediates of streptomycin biosynthesis, *e.g.*, streptidine and O-α-L-dihydrostreptose-(1$\rightarrow$4)-streptidine. Similarly, our results indicate that *pab* mutants of *S. griseus* obtained by curing and protoplasting do not recover candicidin biosynthesis by supplementation with PABA or the sugar moiety mycosamine. These results suggest that more than one enzyme involved in streptomycin (Hara and Beppu, 1982) or candicidin biosynthesis are deficient in these mutants.

Factor A may not be an universal effector controlling antibiotic production and differentiation in actinomycetes. Other substances producing similar effects have been isolated from different actinomycetes. Pamamycin, an antibiotic produced by *S. alboniger* stimulated aerial mycelium in this strain (McCann and Pogell, 1979). Liras and Martin (unpublished data) have found an aerial mycelium stimulating factor in the cephamycin producer *S. griseus* NRRL 3851(Liras and Martin, 1981). Other factors have been reported in *S. venezuelae* by Scribner *et al.* (1973) and *S. griseus* by Biro *et al.* (1980). A factor isolated from *S. virginiae* induces staphylomycin production by the same strain (Yanagimoto and Terui, 1971; Yanagimoto *et al.,* 1979).

Whether there is an universal effector controlling differentiation and secondary metabolism in all actinomycetes, or there are several ones in different species, is not as important as the phenomenon itself. Genetic manipulation of such effector is of great biological and industrial interest.

An intriguing point is how such an unstable genetic element carries the control of expression of genes coding for antibiotic biosynthesis and differentiation. If the finding that no enzymes involved in antibiotic biosynthesis (*e. g.*, PABA synthetase or aminotransferases) are formed in pleiotropic mutants holds true for all reported cases, this would point to a transcriptional (or translational) control. This control may be carried out 1) by an effector molecule coded by a gene (*eff*) located in an extrachromosomal element which is lost or 2) by altering the expression of such a gene by chromosomal rearrangement if the gene is located in the chromosome.

The occurrence of several chromosomal rearrangements affecting directly (without the need of a pleiotropic effector) the expression of several structural genes (*pab, prt, amy, cep, arg*) cannot be ruled out, but the existence of a common intracellular effector produced by a regulatory gen (*eff*) that exerts a pleiotropic effect on the expression of several other structural genes seems more likely on the basis of current evidence.

REFERENCES

Aguilar, A., and Hopwood, D. A. *J. Gen. Microbiol.* (1982). To be published.
Aoki, H., Kunugita, K., Hosoda, J., and Imanaka, H., *J. Antibiot. 30,* suppl.:s207 (1977).
Biro, S., Bekesi, I., Vitalis, S., and Szabo, G., *Eur. J. Biochem. 103:*359 (1980).
Brown, D., Evans, J. R., and Fletton, R. A., *J. Chem. Soc. Chem. Commun.* 282 (1979).
Claridge, C. A., Bush, J. A., Defuria, M. D., and Price, K. E., *Dev. Ind. Microbiol. 15:*102 (1974).
Demain, A. L., *in* "β-lactam Antibiotics, Mode of Action, New Developments and Future Prospects" (M. R. Salton and G. D. Shockman, eds.), p. 567. Academic Press, New York (1981).
Drew, S. W., and Demain, A. L., *Annu. Rev. Microbiol. 31:*343 (1977).
Fothergill, J. D., and Guest, J. R., *J. Gen. Microbiol. 99:*139 (1977).
Freeman, R. F., Bibb, M. J., and Hopwood, D. A., *J. Gen. Microbiol. 98:*453 (1977).
Ghisalba, O., Traxler, P., and Nuesch, J., *J. Antibiot. 31:*1124 (1978).
Ginther, C. L., *Antimicrob. Agents Chemother. 15:*522 (1979).
Gottlieb, S., *J. Antibiot. 29:*987 (1976).
Hara, O., and Beppu, T., *J. Antibiot. 35:*349 (1982).
Higgens, C. E., and Kastner, R. F., *Int. J. Syst. Bacteriol. 21:*326 (1971).

Hirochika, H., and Sakaguchi, K., *Plasmid 7:*59 (1982).
Hopwood, D., *in* "Biochemistry and Genetic Regulation of Commercially Important Antibiotics" (L. C. Vining, ed.). Addisson-Wesley, Reading (1983). (In press)
Howarth, T. T., Brown, A., and King, T. J., *J. Chem. Soc. Chem. Commun. 1976:*226 (1976).
Ikeda, H., Tanaka, H., and Omura, S., *J. Antibiot. 32:*497 (1982).
Kening, M., and Reading, C., *J. Antibiot. 32:*549 (1979).
Kern, B. A., Hendlin, D., and Inamine, E., *Antimicrob. Agents Chemother. 17:*679 (1980).
Khokhlov, A. S., and Tovarova, I. I., *in* "Regulation of Secondary Products and Plant Hormone Metabolism" (M. Luckner, and K. Scheiber, eds.), p. 133. Pergamon Press, New York (1979).
Khokhlov, A. S., Anisova, L. N., Tovarova, I. I., Kleiner, E. M., Kovalenko, I. V., Krasilnikova, O. I., Kornitskaja, E. Y., and Pliner, S. A., *Z. Allg. Mikrobiol. 13:*647 (1973).
Kintaka, K., Aibara, K., Asai, M., and Imada, A., *J. Antibiot. 34:*1081 (1981).
Kirby, R., and Lewis, E., *J. Gen. Microbiol. 122:*351 (1981).
Kirpatrick, J. R., Doolin, J. L., and Godfrey, O. W., *Antimicrob. Agents Chemother. 4:*542 (1973).
Kleiner, E. M., Pliner, S. A., Soifer, V. S., Onoprienko, V. V., Balashova, T. A., Rozynov, B. V., and Khokhlov, A. S., *Bioorg. Khim. 2:*1142 (1976).
Liras, P., and Martin, J. F., *in* "Actinomycetes" (K. P. Schaal, and G. Pulverer, eds.), p. 539. Gustav Fischer Verlag, Stuttgart (1981).
Marahiel, M. A., Lurz, R., and Kleinhauf, H., *J. Antibiot. 34:*323 (1981).
Martin, J. F., *Adv. Biochem. Eng. 6:*105 (1977).
Martin, J. F., *in* "Antibiotics and Other Secondary Metabolites: Biosynthesis and Production" (R. Hutter, T. Leisinger, J. Nuesch, and W. Wehrli, eds.), p. 19. Academic Press, New York (1978).
Martin, J. F., *in* "Actinomycetes" (K. P. Schaal, and G. Pulverer, eds.), p. 417. Fischer-Verlag, Stuttgart (1981).
Martin, J. F., and Aharonowitz, Y., *in* "Antibiotics Containing β-Lactam Structures" (A. L. Demain, and N. Solomon, eds.). Springer-Verlag, Berlin (1983). To be published.
Martin, J. F., and Demain, A. L., *Microbiol. Rev. 44:*230 (1980).
Martin, J. F., and Liras, P., *in* "Biotechnology" *Vol. 1* (H. Rehm, ed.). Springer-Verlag, Berlin (1982).
Martin, J. F., Luengo, J. M., Revilla, G., and Villanueva, J. R., *in* "Genetics of Industrial Microorganisms" (O. K. Sebek, and A. I. Laskin, eds.), p. 83. American Society for Microbiology, Washington (1979).
Matsubara-Nakano, M., Kataoka, Y., and Ogawara, H., *Antimicrob. Agents Chemother. 17:*124 (1980).
McCann, P. A., and Pogell, B. H., *J. Antibiot. 32:*673 (1979).
Nakano, M. M., and Ogawara, H., *J. Antibiot. 33:*420 (1980).
Nakano, M. M., Ozawa, K., and Ogawara, H., *J. Bacteriol. 143:*1501 (1980).
Novick, R., Sanchez-Rivas, C., Gruss, A., and Edelman, I., *Plasmid 3:*348 (1980).
Redshaw, P. A., McCann, P. A., Pentella, M. A., and Pogell, B. M., *J. Bacteriol. 137:*891 (1979).
Schlegel, B., and Fleck, W. F., *Z. Allg. Mikrobiol. 20:*527 (1980).
Schrempf, H., *J. Chem. Technol. Biotechnol. 32:*292 (1982).
Scribner, H. E., Tang, T., and Bradley, S. G., *Appl. Microbiol. 25:*873 (1973).
Sermonti, G., Petris, A., Micheli, M., and Lanfaloni, L., *J. Gen. Microbiol. 100:*347 (1977).
Sermonti, G., Petris, A., Micheli, M., and Lanfaloni, L., *Mol. Gen. Genet. 164:*99 (1978).
Sermonti, G., Lanfaloni, L., and Micheli, M., *Mol. Gen. Genet. 177:*453 (1980).
Sykes, R. B., Cimarusti, C. M., Bonner, D. P., Bush, K., Floyd, D. M., Georgopapadokou, N. H., Koster, W. H., Liu, W. C., Parker, W. L., Principe, P. A., Rathnum, M. L., Slusarchyk, W.A. , Trejo, W. H., and Wells, J. S., *Nature (Lond.) 291:*489 (1981).
Wax, R., William, M., Weston, R., and Birnbaum, J., *J. Antibiot. 29:*670 (1976).
Wells, J. S., Trejo, W. H., Principe, P. A., Bush, K., Georgopapadokou, N., Bonner, D. P., and Sykes, R. B., *J. Antibiot. 35:*184 (1982a).
Wells, J. S., Trejo, W. H., Principe, P. A., Bush, K., Georgopapadokou, N., Bonner, D. P., and Sykes, R. B., *J. Antibiot. 35:*295 (1982b).
Wesseling, A. C., and Lago, B. D., *Dev. Ind. Microbiol. 22:*641 (1981).
Yanagimoto, M., and Terui, G., *J. Ferment. Technol. 49:*611 (1971).
Yanagimoto, M., Yamada, Y., and Terui, G., *Hakkokogaku Kaishi 57:*6 (1979).

CONTROL OF DEVELOPMENT AND SECONDARY METABOLITE PRODUCTION IN STREPTOMYCES[1]

Burton M. Pogell

Department of Medicinal Chemistry/Pharmacognosy
School of Pharmacy
University of Maryland
Baltimore, Maryland, U. S. A.

I. INTRODUCTION

In our laboratory, we have been interested in the role of extrachromosomal DNA (plasmids) and specific endogenous effectors in the regulation of development and of other secondary metabolite production in streptomycetes. This paper presents a brief summary of our research activities in this area.

The growth cycle of a typical streptomycete proceeds as follows on solid media. Spores germinate to form substrate mycelium which appear smooth, waxy, and translucent. As the colonies mature, aerial mycelium develop and the colonies become powdery and opaque. To complete the growth cycle, spores are again produced, predominantly in the aerial and to a lesser extent in the substrate mycelium (Redshaw *et al.*, 1976).

In our laboratory, the primary evidence that extrachromosomal DNA has a role in the control of aerial mycelium formation and the expression of many other specialized functions in streptomycetes came from curing studies on spores with intercalating dyes. Germination and outgrowth of spores in the presence of several of these dyes resulted in a high frequency of occurrence of permanent aerial mycelium-negative (Amy^-) isolates. A frequency as high as 20% was obtained when *Streptomyces alboniger* spores were grown in the presence of ethidium bromide (Redshaw *et al.*, 1976).

[1] *This work was supported by grants from the National Science Foundation, National Institutes of Health and Merck and Co.*

ISBN 0-12-528620-1

Three unusual observations were associated with these mutants (Redshaw *et al.*, 1979): (i) The pleiotrophic nature of the curing was revealed in the simultaneous loss of several secondary metabolite functions, including the ability to produce the characteristic "earthy" odor (geosmin formation) of streptomycetes, pigments found in wild-type strains, and a specific antibiotic which stimulated aerial mycelium formation. (ii) Arginine auxotrophs ocurred at a high frequency. In fact, in *S. alboniger*, all stable Amy^- isolates were Arg^-. The missing enzyme step was specifically identified as argininosuccinate synthetase in all of these Amy^- strains. The gene coding for this enzyme has been designated as *arg* G in streptomycetes (Sermonti *et al.*, 1980). (iii) Loss of these functions correlated with the occurrence of a genetic deletion since no revertants were detected in any isolate. With three different streptomycetes, we tested as many as 2×10^9 organisms from seventeen different Amy^- isolates and in no instance was a single Amy^+ or Arg^+ revertant seen. The characteristics of *S. alboniger* Amy^- isolates are summarized in Table I. When we could measure the parameters quantitatively, secondary metabolite formation appeared to be completely repressed, *i. e.*, there was no detectable pamamycin or geosmin in any Amy^- isolate. Also, no measurable leakiness of the *arg* G gene function was seen. Bentley and Meganathan (1981) have obtained further quantitative substan-

TABLE I. Properties of S. alboniger *Amy^- Isolates*

	Function	*Frequency*
i.	*Cannot form aerial mycelia but can sporulate*	*84/84*
ii.	*Do not have characteristic "earthy" odor*	*84/84*
iii.	*Do not produce "niger" pigment*	*48/48*
iv.	*a. No measurable formation of aerial mycelium-stimulating factor*	*15/15*
	b. Remain Amy^- in presence of stimulating factor	*10/10*
	c. Produce inhibitors of aerial mycelium formation	*10/12*
v.	*Continue to excrete antibiotic activity against* S. lutea	*84/84*
vi.	*Require arginine for growth on minimal medium and remain Amy^-*	
vii.	*No detectable reversion to Amy^+ or Arg^+ (six isolates, 2×10^6 to 8×10^6 organisms plated; four isolates 1.4×10^8 to 1.8×10^9 organisms plated)*	*10/10*

tiation for the loss of geosmin and 2-methylisoborneol formation in Amy⁻ isolates of *S. sulfureus* and *antibioticus*. Neither of these compounds was detected using a sensitive radiogas chromatography assay. With this assay, as little as 2 μg geosmin/l (0.1% of the level formed in Amy^+ cultures) would have been detected.

II. MOLECULAR MODEL

The generality of the loss of several secondary metabolite functions and the association with loss of the *arg* G gene has recently been confirmed in several other streptomycetes (McCann-McCormick *et al.*, 1980; Nakano and Ogawara, 1980; Redshaw *et al.*, 1979; Sermonti *et al.*, 1980). The molecular model shown in Figure 1 was proposed to explain these results (Pogell, 1979). In wild-type *S. alboniger*, an extrachromosomal element codes for either these differentiation functions or a gene involved in the regulation of expression of specialized functions. This plasmid can exist free or in episomal integration in the chromosome near the *arg* G gene. Curing of the plasmid also presumably results in the loss of part of this Arg gene. It is conceivable that loss or inversion of a transposable element could pro-

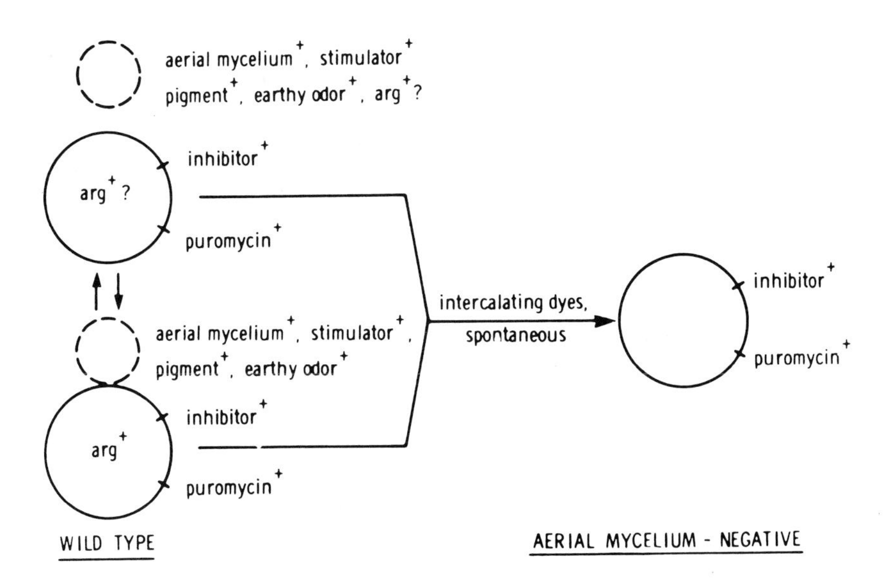

FIGURE 1. Proposed model for the formation of mutants unable to form aerial mycelium.

duce the same results. Recently, Harry Meade of Biogen, Inc. (private communication) has obtained experimental evidence to support part of this model. Using a cloned *arg* G gene from *Streptomyces cattleya*, he observed the complete absence of this gene in Amy$^-$ *arg* G$^-$ mutants using Southern blot analysis (Southern, 1975).

Although we have not yet succeeded in isolating any plasmid from *S. alboniger*, studies using *S. coelicolor* have provided not only experimental evidence to support this model, but also a system in which to attempt cloning of Amy$^+$ genes. A plasmid (SCP2) can be reproducibly isolated from the wild-type in CsCl-ethidium bromide gradients (Fig. 2) (Akagawa *et al.*, 1980). No evidence for the presence of cccDNA in two Amy$^-$ *arg* G$^-$ strains (one spontaneous and the other obtained after treatment with acridine orange) was seen either in gradients (Fig. 2) or on agarose gels. That the Amy$^-$ strain in an SCP2$^-$ strain showed very low recombination frequency, whereas a high recombination frequency was seen with the wild-type Amy$^+$, and that no "lethal zygosis" appeared when the Amy$^-$ isolates were crossed with an SCP2$^-$ strain were genetic evidence which also supported the idea that the SCP2 plasmid is absent in Amy$^-$ isolates. Furthermore, after transformation and regeneration of protoplasts of the Amy$^-$ strain 1098/4 (Amy$^-$ *phe*$^-$ *arg* G$^-$) with SCP2* plasmid, these colonies then produced "lethal zygosis" with an SCP2$^-$ strain (M130, Amy$^+$ *his*$^-$ *ura*$^-$ SCP2$^-$). We were able to recover Amy$^+$ *arg* G$^+$ *phe*$^-$ and Amy$^+$ prototrophs from such crosses from which plasmid could again be isolated. Restriction enzyme analysis revealed that these plasmids were identical with the original SCP2*.

These studies have been further extended by DNA-DNA hybridization studies using SCP2 plasmid labeled with ^{32}P by nick translation. Both liquid phase hybridization and Southern blot techniques were adapted to quantitate the amount and organization of plasmid DNA sequences in these strains. These results confirmed the presence of plasmid in Amy$^+$ *S. coelicolor* (about 10 copies per chromosome) and its absence in Amy$^-$ strains and, more interestingly, showed the presence of small amounts of homologous sequences in *S. coelicolor* Amy$^-$ DNA as well as in *S. alboniger* Amy$^+$ DNA, which hybridize with the SCP2 plasmid (Table II), but which are probably not identical with plasmid sequences (Fig. 3) (Suter *et al.*, 1980). The nature of this DNA material is yet to be determined.

III. PRESENCE OF SPECIFIC EFFECTORS WHICH CONTROL DEVELOPMENT IN STREPTOMYCETES

There are several known compounds produced by streptomycetes which affect aerial mycelium formation. A-factor (2-S-isocapryloyl-3R-hydroxymethyl-γ-butyrolactone), isolated from *S. bikiniensis* and *griseus*,

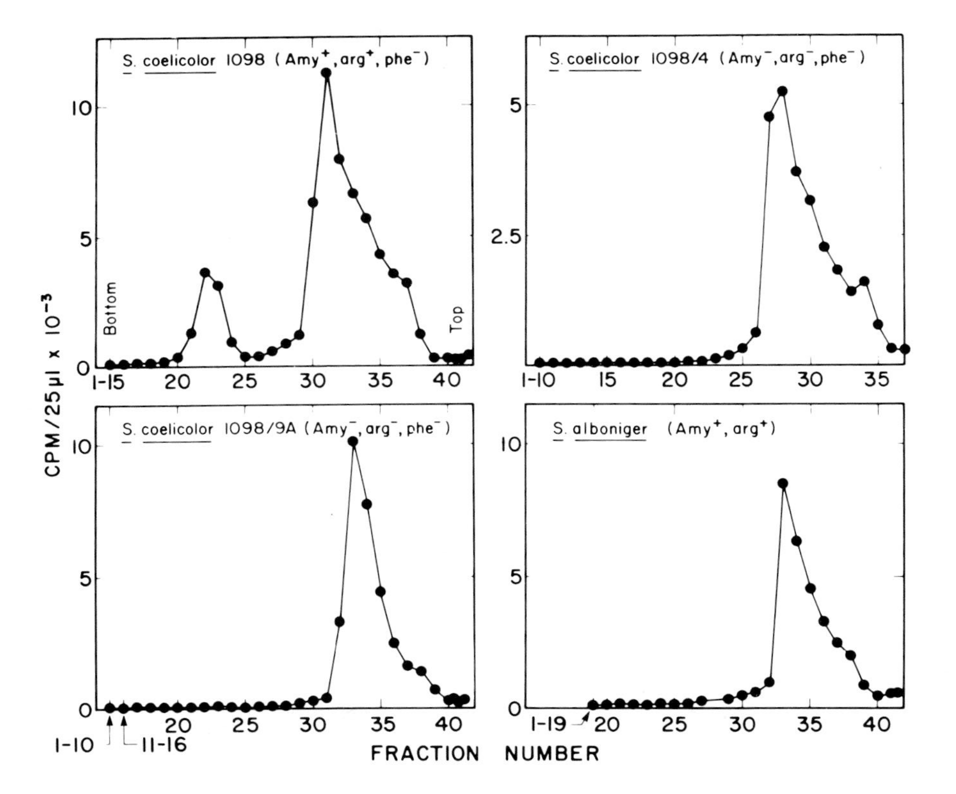

FIGURE 2. Plasmid detection in developmental variants of streptomycetes. In CsCl-ethidium bromide density gradients, ccc DNA (plasmid) sediments to the lower part of the gradient.

TABLE II. Hybridization of Streptomyces coelicolor *Total DNA with ^{32}P-SCP2 Plasmid*[a]

Exp.	*Conditions*	*DNA*	*Concentration (μg/ml)*	*% ds*
A	74 °C; 41 h	S. coelicolor *Amy*$^+$	10	83,76
		S. coelicolor *Amy*$^-$	150	30,28
		0.3 copy SCP2	0.14	72
		Calf thymus		10
B	74 °C; 41 h	S. coelicolor *Amy*$^-$	150	11
		0.007 copy SCP2	0.0028	7
		0.033 copy SCP2	0.014	14
		Calf thymus		4
C	68 °C; 60 h	S. coelicolor *Amy*$^+$	2000	72
		S. coelicolor *Amy*$^-$	2000	40
		1.1 copies SCP2	6.3	59
		Calf thymus		10

[a]Liquid phase DNA-DNA hybridizations were carried out with 20 μl aliquots containing 10^3 cpm of plasmid probe and the DNA concentrations indicated. Double and single stranded DNA were separated by hydroxyapatite and the results are presented as per cent double stranded DNA (% ds).

restores formation of aerial mycelium and streptomycin in mutants which lost these functions (Khokhlov *et al.*, 1973; Kleiner *et al.*, 1976). A short peptide isolated from *S. griseus*, Factor C, was found to restore conidia-producing hyphae in mutants defective in this process (Biro *et al.*, 1980). Methylenomycin A, an epoxide ring antibiotic isolated from *S. violaceoruber* and *S. coelicolor*, inhibits aerial mycelium formation in certain strains of *S. coelicolor* (Kirby *et al.*, 1975). Lincomycin, an antibiotic which inhibits protein synthesis, exerts concentration dependent effects on development in *S. alboniger*. Very low concentrations cause a marked enhancement of aerial mycelium appearance, and somewhat higher levels completely repress this effect without obvious inhibition of vegetative growth (McCann and Pogell, 1979).

We have been able to isolate specific endogenous effectors from *S. alboniger* which regulate development in this organism. Pamamycin, a specific stimulator of aerial mycelium formation, which also has potent antibiotic activity, has been purified to homogeneity (McCann and Pogell, 1979). In addition, two fractions which inhibit aerial mycelium development without affecting growth have been separated from pamamycin by silicic acid column chromatography. Figure 4 illustrates the assay for these effectors on agar plates and Table III summarizes the properties of pamamycin. No evidence for pamamycin formation could be found in fifteen different Amy isolates of *S. alboniger* by assaying for either aerial mycelium stimulation activity or antibiotic activity. The complete absence of pamamy-

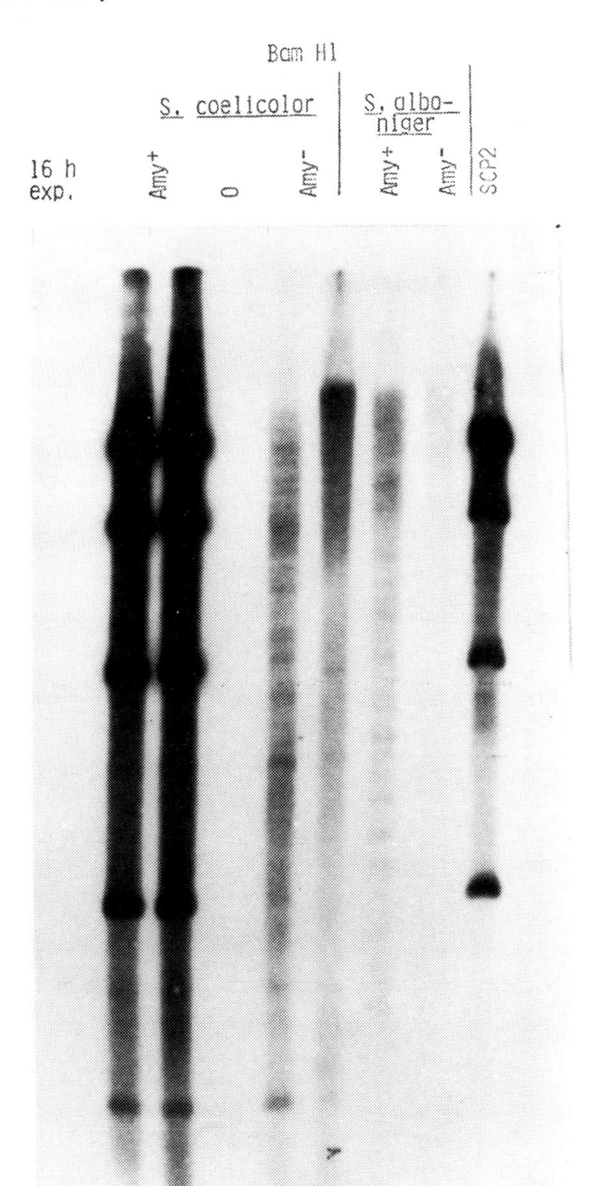

FIGURE 3. Hybridization pattern by the Southern blotting technique of Bam *HI restriction digests of total DNA from various streptomycetes with* ^{32}*P-labeled SCP2 plasmid.* Bam *HI cleaves SCP2 into four fragments. In this overexposure, note the presence of faint but significant hybridization of SCP2 with the* S. coelicolor *Amy⁻ and wild-type* S. alboniger *DNA's, compared to the absence of hybridization with the control of calf thymus DNA (O). The absence of bands at the position of plasmid fragments agrees with our inability to isolate free plasmid from the Amy⁻* S. coelicolor *strains.*

FIGURE 4. Assays of factors which differentially stimulate or inhibit aerial mycelium formation in S. alboniger: *(a) Stimulation by hexane-soluble material from methanol extract of 4 mg of dried wild-type* S. alboniger; *(b) Inhibition by a similar extract from 3 mg of liquid-grown cells.*

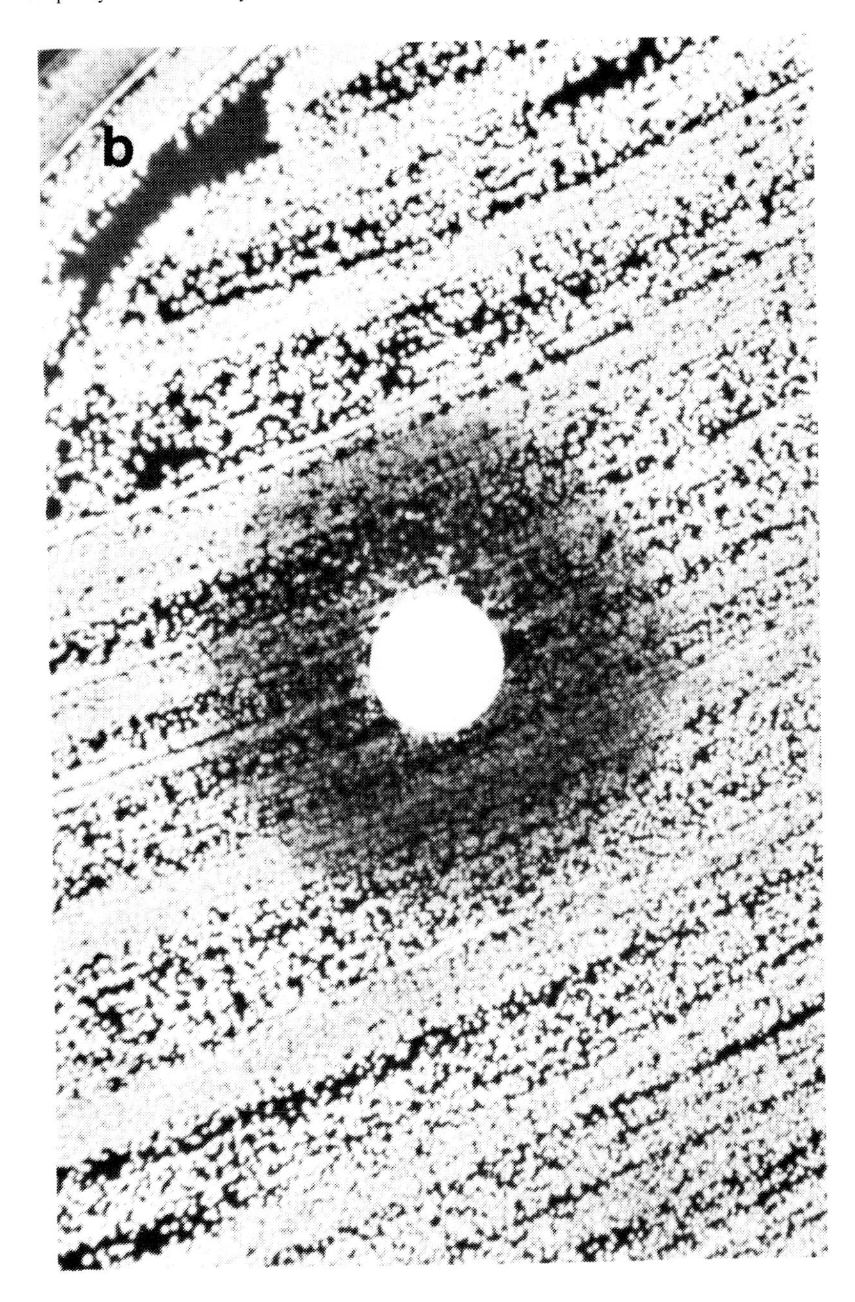
b

TABLE III. Properties of Pamamycin, a Stimulator of Aerial Mycelium Formation

1. *Molecuar weight: 621*
2. *Elemental composition:* $C_{36}H_{63}NO_7$
3. *Insoluble in water; soluble in wide range of organic solvents*
4. *No aromatic, -OH, -NH, amide, or ester groups*
5. *Probably 1 or 2 carbonyl groups per molecule; N is in tertiary linkage*
6. *Highly saturated alicyclic compound*
7. ID_{50} *for* S. aureus*: 0.1 to 0.2 U/ml brain heart infusion broth; 1U = 1 μg*

cin was confirmed in one isolate after a large-scale purification through to the silica gel column chromatography step. Also, at least one other factor required for aerial mycelium formation was lost in the Amy⁻ isolates, since no aerial mycelium was formed on agar in the presence of added pamamycin. Preparations with inhibit formation of aerial mycelium can still be extracted from Amy⁻ isolates.

The mode of action of pamamycin has turned out to be extremely interesting. Recent studies have shown that the effector binds tightly to bacterial membranes and results in inhibition of the uptake of nucleosides and inorganic phosphate in both *Staphylococcus aureus* and *S. alboniger* (Fig. 5) (Chou and Pogell, 1981). Under the same conditions, no other cellular functions were affected. Studies done *in vitro* with isolated membrane vesicles from *S. aureus* confirmed that respiration-dependent nucleoside transport was inhibited. However, inhibition of nucleoside uptake alone cannot be the mechanism for growth inhibition of *S. aureus* since this organism does not require exogenous nucleosides to grow. A more plausible site for the action of pamamycin may be related to the inhibition of phosphate uptake. Thus, limitation of inorganic phosphate is known to inhibit the growth of microorganisms and also to stimulate secondary metabolism in streptomycetes, including the production of antibiotics (Martin, 1977).

Very little is known about the nature of the inhibitors of aerial mycelium formation, but the two fractions have the unusual property of competitively reversing the antibiotic activity of pamamycin against *Sarcina lutea* and *S. aureus* when assayed on agar (Pogell, 1979; Chou, W.-G., unpublished observations).

IV. TRANSFER OF AMY⁺ CHARACTER BY PROTOPLAST FUSION

Before attempting to clone genes required for the expression of the Amy⁺ character into our Amy⁻ strains, we wanted to see if stable Amy⁺ recombinants could be obtained by protoplast fusion. Crosses of *S. coelicolor* M110 (Amy⁺

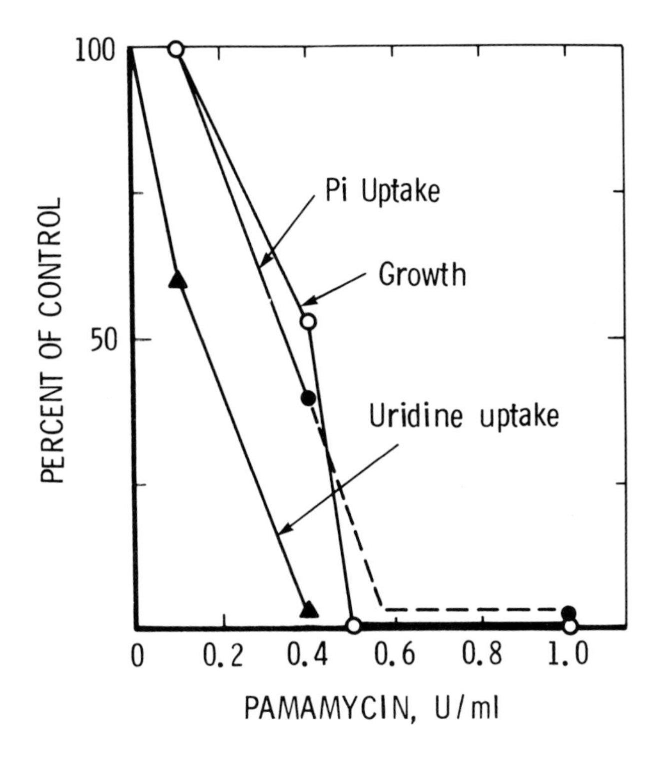

FIGURE 5. Effect of pamamycin on growth and uptake of uridine and inorganic phosphate in S. alboniger.

ura⁻ his⁻ SCP2*⁺) with 1098/4 (Amy⁻ *phe⁻ arg* G⁻ SCP2⁻) yielded a variety of stable Amy⁺ isolates. In all cases, these colonies were now *arg* G⁺. In addition to obtaining several *phe⁻* isolates, stable Amy⁺ prototrophs were also found.

Two interesting new observations concerning aerial mycelium formation were revealed in these crosses. (1) *S. coelicolor* strain M110 behaved unusually when grown on Hickey-Tresner agar in that no aerial mycelium is formed (Fig. 6a). Abundant formation of aerial mycelium is found with this strain on several other media. Strain 1098/4 remains Amy⁻ under all conditions tested. However, in the above crosses, we were able to obtain stable isolates which appeared fully Amy⁺ on Hickey-Tresner agar (Fig. 6a). (2) The other unusual observation in these crosses was the finding that cultures of 1098/4 secreted a substance into agar which was diffusable and able to convert M110 phenotypically into Amy⁺ on Hickey-Tresner agar (Fig. 6b). As seen in Figure 6c, a culture of 1098/4 grown at a greater distance from M110 did not affect the morphology of the latter strain. Strain M110 again appeared Amy⁻ when regrown in the absence of 1098/4. The nature of this phenomenon has yet to be determined.

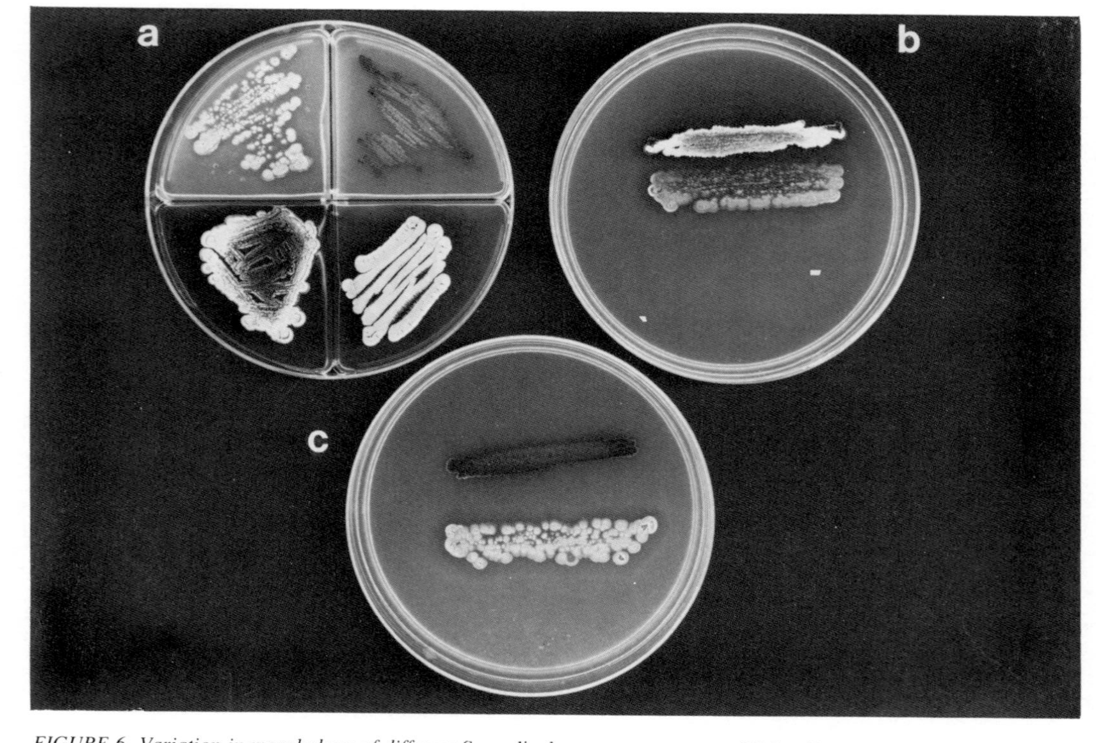

FIGURE 6. Variation in morphology of different S. coelicolor *mutants grown on Hickey-Tresner agar. (a) Upper left:* S. coelicolor *1098/4 (Amy$^-$ phe$^-$ arg G$^-$ SCP2); Upper right:* S. coelicolor *M110 (Amy$^+$ his$^-$ ura$^-$ SCP2*$^+$); Lower left:* S. coelicolor *recombinant (Amy$^+$ phe$^-$); Lower right:* S. coelicolor *recombinant (Amy$^+$ prototroph); (b) and (c): Upper culture,* S. coelicolor *M110 and Lower culture,* S. coelicolor *1098/4.*

ACKNOWLEDGMENTS

Unpublished research described in this article was carried out by Nand K. Gaur, German C. Naharro, and Ci-Jun Liu.

REFERENCES

Akagawa, H., Suter, M., and Pogell, B. M., *Abstr. Annu. Meet Am. Soc. Microbiol.* 187 (1980).

Bentley, R., and Meganathan, R., *F.E.B.S. Lett. 125:*220 (1981).

Biro, S., Bekesi, I., Vitalis, S., and Szabo, G., *Eur. J. Biochem. 103:*359 (1980).

Chou, W.-G., and Pogell B. M., *Antimicrob. Agents Chemother. 20:*443 (1981).

Khokhlov, A. S., Anisova, L. N., Tovarova, I. I., Kovalenko, E. M., Krasilnikova, O. I., Kornitskaya, E. Ya., and Pliner, S. A., *Z. Allg. Mikrobiol. 13:*647 (1973).

Kirby, R., Wright, L. F., and Hopwood, D. A., *Nature (Lond.) 254:*265 (1975).

Kleiner, E., Pliner, S. A., Soifer, V. S., Onoprienko, V. V., Balashova, T. A., Rozynov, B. V., and Khokhlov, A. S., *Bioorg. Khim. 2:*1142 (1976).

Martin, J. F., *Adv. Biochem. Eng. 6:*105 (1977).

McCann, P. A., and Pogell, B. M., *J. Antibiot. 32:*673 (1979).

McCann-McCormick, P. A., Monaghan, R. L., Baker, E. E., Goegelman, R. T., and Stapley, E. O., *in* "Abstracts of the VI International Fermentation Symposium", p. 31. London, Ontario (1980).

Nakano, M. M., and Ogawara, H., *J. Antibiot. 33:*420 (1980).

Pogell, B. M., *in* "Third International Symposium on the Genetics of Industrial Microorganisms" (O.K. Sebek and A.I. Laskin, eds), p. 218. American Society for Microbiology, Washington (1979).

Redshaw, P. A., McCann, P. A., Sankaran, L., and Pogell, B. M., *J. Bacteriol. 125:*698 (1976).

Redshaw, P. A., McCann, P. A., Pentella, M. A., and Pogell, B. M., *J. Bacteriol. 137:*891 (1979).

Sermonti, G., Lanfaloni, L., and Micheli, M. R., *Molec. Gen. Genet. 177:*453 (1980).

Southern, E. M., *J. Mol. Biol. 98:*503 (1975).

Suter, M., Akagawa, H., and Pogell, B. M., *Abstr. Annu. Meet. Amer. Soc. Microbiol.* 187 (1980).

RESISTANCE TO ANTIBIOTICS IN ANTIBIOTIC-PRODUCING ORGANISMS[1]

Eric Cundliffe
Richard H. Skinner
Jill Thompson

Department of Biochemistry
University of Leicester
Leicester, England

I. INTRODUCTION

Many antibiotic-producing organisms are faced with an acute problem of self-defense against their own products. In principle, there are at least three mechanisms by which this might be achieved: (i) modification of the normal intracellular target site, thereby rendering it unavailable to the antibiotic; (ii) inactivation of intracellular antibiotic; and (iii) exclusion of the antibiotic from the cytoplasm. Mechanism (iii), if totally efficient, would obviously enable an organism to tolerate high extracellular levels of a drug. Even a partially efficient exclusion system would allow self-protection, provided that any drug reaching the cytoplasm could be inactivated. However, in high-level producers, one might not expect antibiotic-inactivation systems to operate in the absence of a reasonably effective permeability barrier. Otherwise, such organisms would need to inactivate virtually all the antibiotic produced. In addition to precluding their status as high-level producers, this might be energetically infeasible. Presumably, therefore, producing-organisms lacking a drug-exclusion system must resort to target-site modification.

In addition to withstanding the effects of extracellular drug, producing-organisms must also protect themselves against intracellular antibiotic dur-

[1]*Research in the authors' laboratory is supported by a Project Grant from the Medical Research Council (to E. C.) and by a Research Fellowship and Project Grant from the Science and Engineering Research Council (to R.H.S.).*

ISBN 0-12-528620-1

ing biosynthesis and export. Those which indulge in target-site modification face no such problem. However, others must either produce the antibiotic in a separate cellular compartment or must produce inert antibiotic-precursors which are only activated during or following secretion. The latter procedure is probably quite common and this raises certain problems when "antibiotic-inactivating" enzymes are identified in extracts of producing-organisms. Such enzymes might normally act at an earlier stage in the biosynthetic pathway and might not, *in vivo*, contribute to the mechanism of resistance to mature antibiotic. Given such considerations, the physiological roles of "antibiotic-inactivating" enzymes should not readily be assumed, particularly when, as is usually the case, the relevant permeability properties of the cell surface are not known.

The most direct way to resolve such issues is to use recombinant DNA technology in order to examine the properties of putative resistance determinants in a "clean" background. Recently, such experiments have been carried out. A plasmid carrying DNA fragments from various organisms, including *S. fradiae* (the neomycin-producer), was introduced into the host *Streptomyces lividans*. Clones resistant to neomycin contained either aminoglycoside acetyltransferase (AAC) activity or aminoglycoside phosphotransferase (APH) activity, both of which previously had been characterized in extracts of *S. fradiae* (Thompson *et al.*, 1982c). Interestingly, however, such clones displayed only low levels of resistance to neomycin. High-level resistance (as in *S. fradiae*) was obtained only when clones of either type were crossed and was associated with the presence of both AAC and APH activities. The AAC and APH enzymes each inactivated neomycin; however, it is apparent that neither the acetyltransferase gene nor the phosphotransferase gene alone was an effective resistance determinant. Although all the implications of these findings are not obvious, these data reinforce the need for caution in assessing the contributions of antibiotic-inactivating enzymes to resistance phenotypes.

Other cloning experiments confirmed that resistance to thiostrepton in *S. azureus* involves methylation of ribosomal RNA (rRNA) and that a similar mechanism of target-site modification (but involving a different methylase) mediates resistance to erythromycin in *S. erythreus* (Thompson *et al.*, 1982c). In contrast, resistance to viomycin in *S. vinaceus* and to capreomycin in *S. capreolus* involves antibiotic-modification. These resistance mechanisms are discussed in detail below.

II. MODIFICATION OF ANTIBIOTIC-TARGET SITES

Various antibiotic producers are known to possess resistant drug-target sites although further details are not available. These include resistance to rifamycin, involving RNA polymerase, in *S.* (now *Nocardia*) *meditteranei*

(Watanabe and Tanaka, 1976) and pseudomonic acid resistance, involving isoleucyl-tRNA synthetase, in *Pseudomonas fluorescens* (Hughes *et al.*, 1980). Also, antibiotic-resistant ribosomes are present in the fungus *Myrothecium verrucaria* which produces T-2 toxin (Hobden and Cundliffe, 1980) and in *S. tenjimariensis*, the producer of istamycin (Yamamoto *et al.*, 1981). However, two mechanisms of self-protection in producing-organisms, in which modification of antibiotic target sites is involved, have been characterized. Both relate to inhibitors of protein synthesis (namely, thiostrepton and erythromycin) and both involve modification of the ribosome.

Thiostrepton, a modified polypeptide antibiotic, is produced by *S. azureus*. Normally, the drug binds tightly to a single site on the larger (50S) subunit of the bacterial ribosome and inhibits protein synthesis. Eucaryotic ribosomes are not affected by the drug. The mode of action of thiostrepton has been studied in detail, but is discussed here only briefly (for a review, see Gale *et al.*, 1981). The most convenient assay for the action of this drug involves inhibition of GTP hydrolysis catalyzed by the protein known as elongation factor G (factor EF-G) in conjunction with the ribosome. Normally, during the "translocation reaction" of protein synthesis, factor EF-G associates with the ribosome and promotes a strictly controlled hydrolysis of GTP. However, in the absence of mRNA and other components required for protein synthesis, the ribosome and factor EF-G catalyze "uncoupled" hydrolysis of GTP, and this is greatly inhibited by thiostrepton. This assay was used in the critical experiment concerning the mechanism of resistance to thiostrepton in *S. azureus* (Table I).

Although *Streptomyces*, in general, are quite sensitive to thiostrepton, *S. azureus* is totally unaffected by the drug and this character is reflected in the properties of its ribosomes *in vitro*. *S. azureus* possesses a methylase enzyme which acts upon 23S rRNA (Cundliffe, 1978) by introducing a single methyl group into residue adenosine 1067 (Thompson *et al.*, 1982b). The product of this methylation is 2'-O-methyladenosine; ribosomes containing such modified RNA do not bind thiostrepton (Cundliffe and Thompson, 1979). The "thiostrepton-resistance methylase" does not act upon intact ribosomes nor upon their 50S subunits; the preferred substrate is naked 23S rRNA (Thompson and Cundliffe, 1981). The action of this enzyme is illustrated in Table I. Ribosomes reconstituted *in vitro* and containing specifically methylated RNA are insensitive to the action of thiostrepton.

Presumably, in *S. azureus*, thiostrepton-resistance methylation occurs at an early stage either during or following transcription, but prior to ribosomal assembly. The methylase is apparently produced constitutively. This might have been predicted, given that the enzyme could not protect pre-existing ribosomes were its synthesis to be induced immediately prior to the onset of drug production.

TABLE I. Properties of Reconstituted 50S Ribosomal Subunits Containing 23S RNA Subjected to the Action of the Thiostrepton-Resistance Methylase[a]

50S particles	*Thiostrepton*[b] *input*	*GTP hydrolysis*[c]
Reconstituted with unmethylated RNA	*0*	*170*
	3	*24*
Reconstituted with methylated RNA	*0*	*185*
	30	*175*

[a]*Total ribosomal RNA from* E. coli *was incubated with purified thiostrepton-resistance methylase from* S. azureus *in either the presence or absence of S-adenosyl-methionine (cofactor for methylation). Proteins from* E. coli *50S ribosomal subunits were then added. The reconstituted 50S particles were assayed, in the presence of a 4-fold excess of* E. coli *30S ribosomal subunits, for their ability to support GTP hydrolysis in the presence of factor EF-G. The 30S particles alone were inactive in supporting GTP hydrolysis.*
[b]*Molar/min/pmol 50S particles.*
[c]*pmol γ-[^{32}P]GTP hydrolyzed(min/pmol 50S particles.*

Various antibiotic-producing organisms, other than *S. azureus*, also possess enzymes indistinguishable in activity from the thiostrepton-resistance methylase of *S. azureus* (Table II). The presence of the methylase in *S. laurentii* (which produces thiostrepton) was predictable as was its occurrence in producers of siomycin and sporangiomycin (Thompson and Cundliffe, 1980). These latter antibiotics are closely similar in structure to thiostrepton and share with it a common mode of action. Although nosiheptide (synonym multhiomycin) partially resembles thiostrepton in

TABLE II. Organisms Known to Possess Thiostrepton-Resistance Methylase Activity

Organism	*Antibiotic produced*
Streptomyces azureus *ATCC 14921*	*thiostrepton*
Streptomyces laurentii *ATCC 31255*	*thiostrepton*
Streptomyces sioyaensis *NRRL B-5408*	*siomycin*
Planomonospora parontospora *subsp.* antibiotica *ATCC 23864*	*sporangiomycin*
Streptomyces actuosus *NRRL 2954*	*nosiheptide (multhiomycin)*
Streptomyces bernensis *UC 5144*	*berninamycin*

structure and competes with it for ribosomal binding (Cundliffe and Thompson, 1981), the discovery of the enzyme in *S. actuosus* was surprising. More surprising was the finding of this methylase in *S. bernensis* because berninamycin is quite unlike thiostrepton; however, again, it appears that these two drugs act upon ribosomes in similar fashion (Thompson *et al.*, 1982a). Each of the organisms listed in Table II possesses ribosomes which are resistant to both thiostrepton and to the endogenous antibiotic. Moreover, ribosomes from *S. azureus* (or *Escherichia coli* ribosomes containing 23S RNA subjected to the action of the thiostrepton-resistance methylase from *S. azureus*) are resistant to each of the antibiotics listed in Table II.

The erythromycin-producer, *S. erythreus*, also demonstrates target-site modification. Erythromycin inhibits protein synthesis by binding to the bacterial 50S ribosomal subunit, and clinical isolates of staphylococci have long been known to demonstrate resistance to the drug. More specifically, resistance is expressed against macrolides (such as erythromycin and spiramycin), lincosamides (including lincomycin) and streptogramin B-type antibiotics. This is the so-called MLS-resistance phenotype which, in staphylococci, is inducible and involves the action of a plasmid-encoded methylase on 23S rRNA (Lai and Weisblum, 1971; Lai *et al.*, 1973). The product of methylation is N^6,N^6-dimethyladenine; hence, when this residue was detected in 23S rRNA from *S. erythreus*, it was inferred that a similar mechanism of resistance might operate in the erythromycin-producer (Graham and Weisblum, 1979). The validity of that hypothesis has recently been established (Skinner and Cundliffe, 1982). A methylase (apparently produced constitutively in *S. erythreus*) acted upon 23S rRNA from *Bacillus stearothermophilus* so that reconstituted ribosomes containing such modified RNA were resistant to MLS antibiotics (Table III). Erythromycin was not suitable for use in these experiments, because it is a poor inhibitor of cell-free protein synthesis directed by polyuridylic acid (polyU). Again, the product of methylation was N^6,N^6-dimethyladenine and the stoichiometry of methylation was less than two. Since monomethyladenine is apparently not present in 23S rRNA from *B. stearothermophilus* (Tanaka and Weisblum, 1975), this implies that the *S. erythreus* enzyme caused dimethylation of a single adenine residue within 23S RNA. In other experiments (Skinner, unpublished data), it has been shown that the staphylococcal MLS-resistance methylase and the corresponding enzyme from *S. erythreus* act upon 23S RNA in a mutually exclusive fashion. The conclusion that the two enzymes act at the same site seems inescapable and it will be interesting to determine whether they are homologous proteins.

The two cases of ribosomal methylation discussed above represent the only examples of target-site modification as yet characterized in antibiotic-producing organisms. It therefore remains to be seen how prevalent this stratagem may be. In particular, resistance mechanisms in streptomycetes

TABLE III. Properties of Reconstituted 50S Ribosomal Subunits Containing RNA Subjected to the Action of the Erythromycin-Resistance Methylase[a]

50S particles	*Activity in protein synthesis (normalized)*		
	Control	*Spiramycin added*	*Lincomycin added*
Reconstituted with unmethylated RNA	*100*	*8*	*5*
Reconstituted with methylated RNA	*100*	*64*	*44*

[a] *Total rRNA from* B. stearothermophilus *was incubated with erythromycin-resistance methylase from* S. erythreus *in either the presence or absence of S-adenosyl-methionine. Then, 50S subunit proteins from* B. stearothermophilus *ribosomes were added. The reconstituted 50S particles were supplemented with 30S ribosomal subunits, and their activity in polyphenylalanine synthesis, directed by polyU, was determined. Antibiotics were used at 20 μM.*

which produce MLS antibiotics other than erythromycin deserve closer scrutiny. In some such organisms, induction of resistance to MLS antibiotics has been linked circumstantially with the appearance of either mono- or di-methyladenine (or both) within 23S rRNA (Fujisawa and Weisblum, 1981). Further studies will be necessary to determine whether any casual connection exists between these events.

III. INACTIVATION OF ANTIBIOTICS

The second major type of resistance mechanism, employed by producing-organisms, involves inactivation of antibiotics by enzymic modification. This is illustrated here by the resistance of *S. vinaceus* and *S. capreolus* to their respective products, viomycin and the capreomycins. These antibiotics are structurally similar cyclic peptides (Fig. 1) and are active against Gram-positive bacteria, especially mycobacteria and streptomycetes. Viomycin and capreomycin (especially the latter) are among the "second choice" options available for the treatment of tuberculosis. Both drugs inhibit bacterial protein synthesis and have binding sites on both the 50S and 30S ribosomal subunits.

In common with many other antibiotic-producing organisms, *S. vinaceus* and *S. capreolus* are resistant to their products *in vivo*. Moreover, *S. capreolus* is also resistant to viomycin, whereas *S. vinaceus* is partially resistant to capreomycin IA, but sensitive to capreomycin IB. In crude extracts of these two organisms, cell-free protein synthesis directed by polyU was partially resistant to inhibition by viomycin and capreomycin (Skinner and

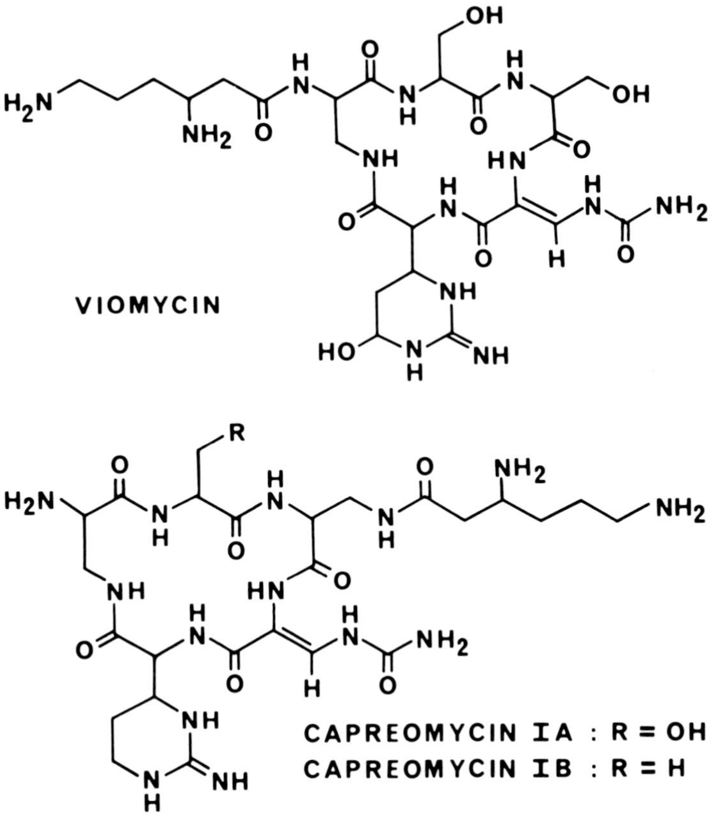

VIOMYCIN

CAPREOMYCIN IA : R = OH
CAPREOMYCIN IB : R = H

FIGURE 1. Structures of viomycin and the major capreomycins. Capreomycins IIA and IIB differ from IA and IB, respectively, in lacking the β-lysine moiety; IIA and IIB are minor components of the capreomycin complex and, together, usually represent less than 10% of the total. (Adapted from Skinner, R. H., and Cundliffe, E., J. Gen Microbiol. 120:*95 (1980) by copyright permission of the Society for General Microbiology).*

Cundliffe, 1980). However, when such extracts were further fractionated into ribosomes and postribosomal supernatant, which were used in reciprocal combination with the corresponding preparations from *S. coelicolor*, the ribosomes of *S. vinaceus* and *S. capreolus* proved to be fully sensitive to the action of the two drugs. Evidently, enzymes in the post-ribosomal supernatants of *S. vinaceus* and *S. capreolus* were capable of inactivating viomycin and capreomycin. As shown in Table IV, each organism possessed an ATP-dependent ability to inactivate both viomycin and capreomycin IA (but not capreomycin IB). Subsequent experiments, involving the use of γ-[^{32}P]ATP as cofactor, characterized the antibiotic-inactivating enzymes as phosphotransferases, each of which were active upon viomycin and capreomycins IA and IIA (Table V). However,

TABLE IV. Inactivation of Viomycin and Capreomycin IA by Extracts of the Producing-Organisms[a]

Source of extract	*Cofactor*	*Antibiotic*	*Inhibition of Protein Synthesis* in vitro *(%)*
S. coelicolor	-	*viomycin*	*90*
S. coelicolor	*ATP*	*viomycin*	*93*
S. vinaceus	-	*viomycin*	*91*
S. vinaceus	*ATP*	*viomycin*	*14*
S. vinaceus	-	*capreomycin IA*	*72*
S. vinaceus	*ATP*	*capreomycin IA*	*4*
S. capreolus	-	*capreomycin IA*	*92*
S. capreolus	*ATP*	*capreomycin IA*	*20*
S. capreolus	-	*viomycin*	*94*
S. capreolus	*ATP*	*viomycin*	*23*

[a] *Viomycin and capreomycin IA were incubated with post-ribosomal supernatant from the organisms indicated, with or without ATP. Then, the ability of the antibiotics to inhibit cell-free protein synthesis, directed by polyU, was determined. Such assays employed ribosomes and supernatant from* S. coelicolor.
(Adapted from Skinner, R. H., and Cundliffe, E., J. Gen Microbiol. 120:*95 (1980) by copyright permission of the Society for General Microbiology.)*

capreomycins IB and IIB were not substrates, since they lack the hydroxyl group which is the site of phosphorylation in the former molecules (Skinner and Cundliffe, 1980). The presence of such a phosphotransferase in *S. capreolus* could account for resistance *in vivo* both to viomycin and to capreomycin IA, but not to capreomycin IB. Therefore, the presence of another enzyme able to inactivate the latter was suspected. This proved to be an acetyltransferase which was active upon all the capreomycins (Table VI). Purification of acetylcapreomycins IA and IB revealed that neither compound inhibited cell-free protein synthesis.

As discussed earlier, the most unequivocal way of testing the hypothesis that a particular resistance-phenotype is due to the possession of a particular gene (and, therefore, a particular enzyme) is to clone the resistance-determinant in another organism and to characterize the resistance-mechanism. This has been done in the case of the viomycin-resistance determinant of *S. vinaceus* (Thompson *et al.*, 1982c). The recipient was *S. lividans*, into which a plasmid carrying fragments of *S. vinaceus* DNA was introduced. Several viomycin-resistant transformants were examined. Most exhibited high-level resistance to viomycin (similar to that with *S. vinaceus*)

TABLE V. Substrate Specificities of Phosphotransferases in Extracts of S. vinaceus *and* S. capreolus[a]

Substrate	*[^{32}P] phosphoryl groups transferred (pmol)*	
	S. vinaceus *extract*	S. capreolus *extract*
Viomycin	*76*	*455*
Capreomycin IA	*293*	*604*
Capreomycin IB	*1*	*2*
Capreomycin IIA	*287*	*488*
Capreomycin IIB	*0*	*8*

[a]*Antibiotics (600 pmol) were incubated with post-ribosomal supernatant from* S. vinaceus *or* S. capreolus *together with* γ-*[^{32}P]ATP.*
(Adapted from Skinner, R. H., and Cundliffe, E. J. Gen Microbiol. 120:*95 (1980) by copyright permission of the Society for General Microbiology.)*

and were cross-resistant to capreomycin IA. Such clones contained viomycin phosphotransferase activity as previously characterized in extracts of *S. vinaceus* (Table V: note that "viomycin phosphotransferase" of *S. vinaceus* is also active upon capreomycin IA). The remainder were resistant only to low levels of viomycin, were sensitive to capreomycin IA, and lacked viomycin phosphotransferase. Thus, it would appear that resistance of *S. vinaceus* to viomycin and capreomycin IA is determined by possession of the gene encoding viomycin phosphotransferase. There is, however, one slight anomaly in this context in that *S. vinaceus* was substantially more sensitive *in vivo* to capreomycin IA than to viomycin, even though both antibiotics were excellent substrates for the phosphotransferase *in vitro* (Table

TABLE VI. Substrate Specificity of Acetyltransferase Activity in Extracts of S. caprelous[a]

Substrate	*[^{14}C]acetyl groups transferred (pmol)*
Viomycin	*7*
Capreomycin IA	*29*
Capreomycin IB	*49*
Capreomycin IIA	*27*
Capreomycin IIB	*61*

[a]*Antibiotics (600 pmol) were incubated with post-ribosomal supernatant from* S. capreolus *together with 1-[^{14}C]acetyl-COA.*
(Adapted from Skinner, R. H., and Cundliffe, E. J. Gen Microbiol. 120:*95 (1980) by copyright permission of the Society for General Microbiology.)*

V). This could be explained if the cell membrane of *S. vinaceus* were less permeable to viomycin than to capreomycin IA, thus contributing to the observed resistance.

As yet, the determinants of resistance to capreomycin in *S. capreolus* have not been cloned. Hence, no definitive comments can be made concerning the relative contributions of the phosphotransferase and the acetyltransferase to the resistance phenotype of that organism.

IV. DISCUSSION

Two fundamentally different strategies of self-protection, employed by antibiotic-producers, have been described. It might be asked which is "better" or preferable. Certainly, based on the limited evidence so far available, it seems that target-site modification confers higher levels of resistance than does antibiotic-modification. Viewed in that light, the former mechanism may be regarded as more efficient. However, such considerations ignore the reason(s) why antibiotics are produced. Some may play a regulatory role during cell growth or differentiation so that their producers may need to be sensitive to their action (even if only partially so) at some stage(s) of the growth cycle. Under such circumstances, total resistance (as might be mediated by target-site modification) would obviously be undesirable. There may also be more direct constraints on the "choice" of resistance-mechanisms facing producers. Some antibiotics may not have vital groups which are susceptible to modification. Conversely, it is possible that not all drugs could be prevented from binding to their targets by modifications of the latter consistent with maintenance of their function.

Other unanswered questions concern the possibility that enzymes involved in resistance-mechanisms may fulfill additional functions *in vivo*. Thus, antibiotic-modifying enzymes may participate in the synthesis and/or export of antibiotics produced as inactive precursors, *e. g.*, streptomycin phosphotransferase is involved in the biosynthesis of streptomycin (Walker and Skorvaga, 1973). In such cases, other enzymes might be needed to "activate" the product *e. g.*, phosphostreptomycin phosphatase in *S. griseus* (Miller and Walker, 1970) and acetylkanamycin hydrolase in *S. kanamyceticus* (Satoh *et al.*, 1976). Conceivably, viomycin is produced by *S. vinaceus* in phosphorylated form, in which case one would predict the existence of a phosphoviomycin phosphatase. In this context, it may be significant that inorganic phosphate inhibits viomycin biosynthesis (Pass and Raczynska-Bojanowska, 1968), possibly by inhibiting the putative phosphatase. Certainly, it would be interesting to know whether, phosphoviomycin accumulates in the presence of inorganic phosphate, as does phosphostreptomycin under similar circumstances (Miller and Walker, 1970).

Alternative functions for target-site modification enzymes may be less obvious, even if such functions exist. For example, the thiostrepton-resistance methylase is unlikely to be involved in drug production and there is no reason to suppose that it recognizes thiostrepton. Nor is it at all apparent that, in the absence of thiostrepton, *S. azureus* benefits from the constitutive expression of the methylase gene. It therefore remains unclear how the genes for thiostrepton production and resistance came to congregate in the same organism and whether the levels of drug production and resistance developed immediately thereafter.

It would also be interesting to know whether enzymes of similar activity (such as the viomycin and capreomycin phosphotransferases of *S. vinaceus* and *S. capreolus*, or the thiostrepton-resistance methylase of *S. azureus* and the similar enzyme in *S. bernensis*) are homologous proteins. That is, whether their genes are descended (or derived) from common ancestors. This raises the broader question of the origin(s) of antibiotic-resistance determinants in general. The attractive hypothesis, that producing-organisms may represent the source (at least of some) of the resistance-determinants encountered in clinical isolates, was first proposed by Walker and Walker (1970) following their observation that streptomycin was phosphorylated by extracts of *S. bikiniensis*. Subsequently, actinomycetes which produce aminoglycoside-aminocyclitol antibiotics were shown to possess aminoglycoside-inactivating enzymes similar to those encoded on R plasmids (Benveniste and Davies, 1973; for a recent review, see Davies and Smith, 1978). Although the occurrence of such enzymes is not restricted to those organisms producing drugs which are suitable substrates, the above hypothesis still has far-reaching implications. At the very least, it suggests that studies of antibiotic-resistance in producing-organisms may have predictive potential. Novel resistance mechanisms may be characterized (and possibly circumvented by strategic modification of drug molecules) prior to their emergence in clinical situations.

The potential value of target-site modification enzymes in aiding studies of the targets themselves is also obvious. The thiostrepton-resistance methylase has been used to pinpoint the thiostrepton binding site within the ribosome (Thompson *et al.*, 1982b), and the erythromycin-resistance methylase can be expected to be equally useful. Given a knowledge of the modes of action of those antibiotics, such studies will aid in the elucidation of structure-function relationships among ribosomal components. Enzymes affecting other drug-target sites can be expected to be equally useful.

ACKNOWLEDGMENTS

We are grateful to Dino and Teddy Bedlington for their enthusiastic support.

REFERENCES

Benveniste, R., and Davies, J., *Proc. Nat. Acad. Sci. U.S.A. 70:*2276 (1973).
Cundliffe, E., *Nature (Lond.) 272:*792 (1978).
Cundliffe, E., and Thompson, J., *Nature (Lond.) 278:*859 (1979).
Cundliffe, E., and Thompson, J., *J. Gen. Microbiol. 126:*185 (1981).
Davies, J., and Smith, D. I., *Annu. Rev. Microbiol. 32:*469 (1978).
Fujisawa, Y., and Weisblum, B., *J. Bacteriol. 146:*621 (1981).
Gale, E. F., Cundliffe, E., Reynolds, P. E., Richmond, M. H., and Waring, M. J., "The Molecular Basis of Antibiotic Action" (2nd ed.). John Wiley and Sons, Chichester (1981).
Graham, M. Y., and Weisblum, B., *J. Bacteriol. 137:*1464 (1979).
Hobden, A. N., and Cundliffe, E., *Biochem. J. 190:*765 (1980).
Hughes, J., Mellows, G., and Soughton, S., *FEBS Fed. Eur. Biochem. Soc. Lett. 122:*322 (1980).
Lai, C. J., and Weisblum, B., *Proc. Nat. Acad. Sci. U.S.A. 68:*856 (1971).
Lai, C. J., Weisblum, B., Fahnestock, S. R., and Nomura, M., *J. Mol. Biol. 74:*67 (1973).
Miller, A. L., and Walker, J. B., *J. Bacteriol. 104:*8 (1970).
Pass, L., and Raczynska-Bojanowska, K., *Acta Biochim. Pol. 15:*355 (1968).
Satoh, A., Ogawa, H., and Satomura, Y., *Agric. Biol. Chem. 40:*191 (1976).
Skinner, R. H., and Cundliffe, E., *J. Gen. Microbiol. 120:*95 (1980).
Skinner, R. H., and Cundliffe, E., *J. Gen. Microbiol. 128:*2411 (1982).
Tanaka, T., and Weisblum, B., *J. Bacteriol. 123:*771 (1975).
Thompson, J. and Cundliffe, E., *J. Bacteriol. 142:*455 (1980).
Thompson, J. and Cundliffe, E., *J. Gen. Microbiol. 124:*291 (1981).
Thompson, J., Cundliffe, E., and Stark, M. J. R., *J. Gen. Microbiol. 128:*875 (1982a).
Thompson, J., Schmidt, F. J., and Cundliffe, E., *J. Biol. Chem. 257:*7915 (1982b).
Thompson, C. J., Skinner, R. H., Thompson, J., Ward, J. M., Hopwood, D. A., and Cundliffe, E., *J. Bacteriol. 151:*678 (1982c).
Walker, J. B., and Skorvaga, M., *J. Biol. Chem. 248:*2435 (1973).
Walker, M. S., and Walker, J. B., *J. Biol. Chem. 245:*6683 (1970).
Watanabe, S., and Tanaka, K., *Biochem. Biophys. Res. Commun. 72:*522 (1976).
Yamamoto, H., Hotta, K., Okami, Y., and Umezawa, H., *Biochem. Biophys. Res. Commun. 100:*1396 (1981).

REGULATION AND PROPERTIES OF INTRACELLULAR PROTEINS IN ACTINOMYCETES[1]

S. G. Bradley
J. S. Bond[2]
S. M. Sutherland

Department of Microbiology and Immunology and
Department of Biochemistry
Virginia Commonwealth University
Richmond, Virginia, U. S. A.

I. REGULATION OF METABOLIC PROCESSES

A. Enzyme Activity and Concentration

Many metabolites can participate in several alternative reactions; therefore, the rates of various reactions must be regulated in order that delicately adjusted supplies of all the intermediates which are needed to meet the immediate requirements of an organism are available. All enzymes possess attributes than allow their activities to be regulated. The environmental factors that affect enzyme activity include pH of the milieu, concentrations of the substrate and product, and availability of required cofactors (Bradley, 1978). Several different regulatory processes are known, in which the enzyme activity of a cell is specifically adjusted. One of the control systems involves the related phenomena of enzyme induction and repression, in which deoxyribonucleic acid is transcribed selectively. As a result, the amount of a given enzyme synthesized is altered according to the needs of the organism. β-Lactamase production by a mutant strain of *Streptomyces lavendulae*, for example, is subject to nitrogen catabolite repression. When casamino acids are supplied as a nitrogen source,

[1] *This work was supported by a grant (PCM-7807868) from the National Science Foundation.*
[2] *J. S. B. is a recipient of a Research Career Development Award from the National Institutes of Arthritis, Diabetes, Digestive and Kidney Diseases.*

ISBN 0-12-528620-1

β-lactamase biosynthesis is repressed. Conversely, when casein is supplied as a nitrogen source, β-lactamase biosynthesis is enhanced (Ogawara, elsewhere in this volume). Similarly, the glycosyl residue of mycolic acids is influenced by the carbon source supplied in the growth medium. Some nocardiae grown in sucrose synthesize 6-mycoloylglucosyl-β-fructoside; when grown in fructose, they synthesize 6-mycoloyl fructose; and when grown in glucose, 6-mycoloyl glucose. The increased production of mycolyl glucose during growth in glucose medium indicates, but does not prove, that there is an induced enzyme system in *Rhodococcus* (formerly *Nocardia*) *rhodochrous* for the specific transfer of the mycoloyl residue from a carrier to the glucosyl unit (Ioneda, elsewhere in this volume).

B. Noncovalent and Covalent Modification

Other specific regulatory processes modify the level of the catalytic activity of enzyme molecules that already exist. The catalytic activity of an enzyme may be modified by metabolic effectors that bind specifically, but noncovalently, to sites on the molecule. Presumably, the conformation of the enzyme is altered by the binding or removal of the effector in response to its concentration within the cell. These shifts in the conformation of the enzyme alter the capability of the active site to catalyze a reaction. The activity of *Rhodococcus* trypsin, for example, is stimulated fourfold by $MgCl_2$ (Shannon *et al.*, 1982). Conversely, the activity of RNA polymerase in *Streptomyces griseus* is inhibited by factor C, a polypeptide that restores the capability of a particular variant to conidiate in submerged cultures. It is not known whether factor C acts at the level of the enzyme or of the nucleic acid substrate (Szabo *et al.*, elsewhere in this volume).

As another means to control the rate of enzyme-mediated reactions, enzymes are interconverted between active and inactive forms through their covalent modification by the action of other enzymes. Once a polypeptide has been made by a ribosome, it undergoes covalent modification to yield its normal, biologically active form. The N-formyl group of N-formyl methionine is removed by a deformylase, or the methioninyl residue is eliminated by a methionine aminopeptidase. Intrachain disulfide cross-linkages are formed enzymatically after the polypeptide chain has been assembled. A variety of methylated amino acids are formed by post-translational events; for example, ε-N-dimethyllysine is present at 0.6 to 2.2 μmoles per gram protein in some streptomycetes (Paik and Kim, 1975). Dicarboxylic amino acids and histidine residues of proteins are also methylated. It is possible that methylation plays a role in storage and secretion of protein. Phosphorylation of serine residues of proteins by phosphokinase and cleavage of phosphate residues by phosphatase are important means for regulating the activities of some enzymes. ADP-ribosylation of proteins may constitute a general type of covalent modification that is important in the metabolism of proteins (Hayaishi and Ueda, 1977). Deamidation of glutaminyl and asparaginyl residues of proteins can occur spontaneously. Deamidation may have a determinative role as a molecular timer of biological events (Robinson and Rudd, 1974). The change from uncharged residues to acidic, negatively charged residues as a consequence of deamidation may increase the susceptibility of a protein molecule to proteolytic degradation. Post-translational proteolytic modification is an important process in the metabolism of many proteins. Limited proteolysis is an obligatory step in the activation of some enzymes (Dreyfus *et al.*, 1978).

C. Selective Degradation of Intracellular Proteins

Intracellular protein degradation is important in determining both the steady-state level of an enzyme and the rate of change in concentration from one steady-state level to another; it appears to be an important process in the regulation of enzyme and protein levels in actinomycetes (Bradley, 1981). In nature, actinomycetes are exposed to different nutrients as a result of translocation from one site to another or influx of new nutrients. Synthesis of new enzymes may be necessary to adapt to new environmental conditions. Filamentous and/or inherently slowly growing actinomycetes and actinomycetes in the stationary phase of growth must be able to degrade some of their proteins in order to synthesize others if exogenous amino acids or energy or both are limiting. Protein degradation has been demonstrated in *R. erythropolis* by use of a variety of experimental approaches. The amino acids released by protein degradation may be reutilized for new protein synthesis. Mannitol dehydrogenase activity disappears rapidly after its substrate is depleted from the medium. It is a relatively unstable enzyme *in vitro*, but cofactors or substrate with cofactor afford protection against inactivation *in vitro* (Bradley *et al.*, 1980).

II. PROPERTIES OF INTRACELLULAR PROTEINS

A. Properties Related to Protein Degradation

A substantial body of information now exists concerning the processes of protein degradation. Proteins that are degraded slowly possess a number of characteristics that distinguish them from rapidly degraded proteins, indicating that the information determining the stability of an enzyme resides in its structure (Bond *et al.*, elsewhere in this volume). Several workers have reported that proteins that have 40 to 80 kilodalton subunits with isoelectric points below pI 6.0 possess a number of properties which differ from those of proteins that have 20 to 40 kilodalton subunits with isoelectric points above pI 6.0 (Goldberg and St. John, 1975). Acidic proteins, for example, appear to be synthesized more rapidly than are basic proteins (Duncan *et al.*, 1980). It has been reported that acidic proteins in the soluble fraction are generally larger than are neutral or basic proteins (Woodworth *et al.*, 1982a). Other investigators have failed to demonstrate a significant correlation between subunit size and charge (Dice and Goldberg, 1975; Dice *et al.*, 1979); the latter workers have concluded that subunit size and charge are independent variables. It is important that this controversy be resolved (Woodworth *et al.*, 1982b). In addition, many workers have observed that the majority of bacterial proteins are acidic, whereas proteins of eucaryotic cells include many basic polypeptide species as well (O'Farrell *et al.*, 1977). To determine whether there is a correlation between subunit size and charge and, should one exist, whether it is found in both eucaryotic and procaryotic cells, we have examined the soluble proteins from several bacteria, eucaryotic protists, and mouse tissues.

B. Correlations between Subunit Size and Charge

Proteins in the soluble fractions of extracts of *R. erythropolis* were separated by chromatography on diethylaminoethyl cellulose (DEAE-cellulose); the elution buffer was 20 mM Tris (hydroxylmethyl)-aminomethane containing varied concentrations of NaCl (20 mM Tris-NaCl). The resulting profile was compared with those of selected procaryotes and eucaryotes. The median protein of *Rhodococcus, Bacillus*, and *Escherichia* was eluted from DEAE-cellulose by 0.3 M NaCl; the median protein of mouse kidney or liver was eluted by 0.1 M NaCl; and that of *Rhodotorula* and *Naegleria gruberi*, by 0.2 M NaCl. Accordingly, the median protein of procaryotes was more anionic (acidic) than was that of eucaryotes. Moreover, the median protein of mouse liver or kidney was less anionic than was that of the eucaryotic microorganisms (Table I). Procaryotes had little protein that was eluted by 0 or 0.1 M NaCl, whereas eucaryotes had little protein that was eluted by 1.0 M NaCl. Little protein of the yeasts *Rhodotorula* and *Saccharomyces* was eluted by 0 M NaCl, whereas nearly 20% of the protein of the amebae *N. fowleri* and *N. gruberi* was eluted with 0 M NaCl. Accordingly, the soluble fraction of procaryotes contained relatively little "weakly anionic" (neutral) protein, whereas that of eucaryotes contained relatively little "strongly anionic" (acidic) protein (Table I).

The median molecular sizes of proteins separated by DEAE-cellulose chromatography of soluble fractions from the procaryotes *Rhodococcus, Bacillus, Escherichia*, and *Pseudomonas* were 45 kilodaltons or less, whereas the median molecular sizes of several fractions from the eucaryotes *Saccharomyces, Naegleria*, and the laboratory mouse were 45 kilodaltons or

TABLE I. Separation of Proteins in the Soluble Fraction of Extracts of Selected Procaryotes and Eucaryotes by Chromatography on a DEAE-cellulose Column

Source of soluble protein	*% Protein eluted by 20 mM Tris containing NaCl at*				
	0 M	*0.1 M*	*0.2 M*	*0.3 M*	*1.0 M*
Rhodococcus erythropolis *305*	0	8	21	50	21
Bacillus licheniformis *A5*	0.5	1	31	43	24
Escherichia coli *K-10*	6	15	22	22	35
Pseudomonas aeruginosa *PAO*	2	9	42	29	18
Rhodotorula *sp.*	5	31	30	29	7
Saccharomyces cerevisiae	7	46	22	14	11
Naegleria fowleri *nN68*	23	27	25	18	6
Naegleria gruberi *EGB*	19	30	26	17	7
Mouse (BALB/c) liver	39	30	13	13	4
Mouse (BALB/c) kidney	25	28	24	15	6

more. Accordingly, there were relatively more large proteins in the soluble fractions of eucaryotes than in those of procaryotes (Table II). Conversely, there were relatively more small proteins in the soluble fractions of procaryotes than in those of eucaryotes (Figs. 1 and 2). The median sizes of proteins in the fractions eluted by 0.3 M NaCl were larger than those eluted by 0.1 M NaCl for soluble fractions from both procaryotes and eucaryotes. Thus, the more anionic (acidic) proteins in the soluble fractions were relatively larger than the weakly anionic (neutral) proteins (Table II).

TABLE II. Median Molecular Sizes of Proteins Separated by Chromatography of Soluble Fractions from Selected Procaryotic and Eucaryotic Cells on DEAE-cellulose

Source of soluble proteins	*Median molecular size in kilodaltons of proteins eluted by 20 mM Tris containing NaCl at*				
	0 M	*0.1 M*	*0.2 M*	*0.3 M*	*1.0 M*
R. erythropolis	-[a]	17	33	41	33
B. lichenformis	35	27	35	35	45
E. coli	34	26	44	44	26
P. aeruginosa	25	32	41	41	32
Rhodotorula *sp.*	25	28	42	42	34
S. cerevisiae	48	40	40	48	40
N. fowleri	57	57	57	69	104
N. gruberi	38	30	38	100	48
Mouse liver	35	35	61	61	46
Mouse kidney	36	47	47	47	64

[a]*No proteins were eluted.*

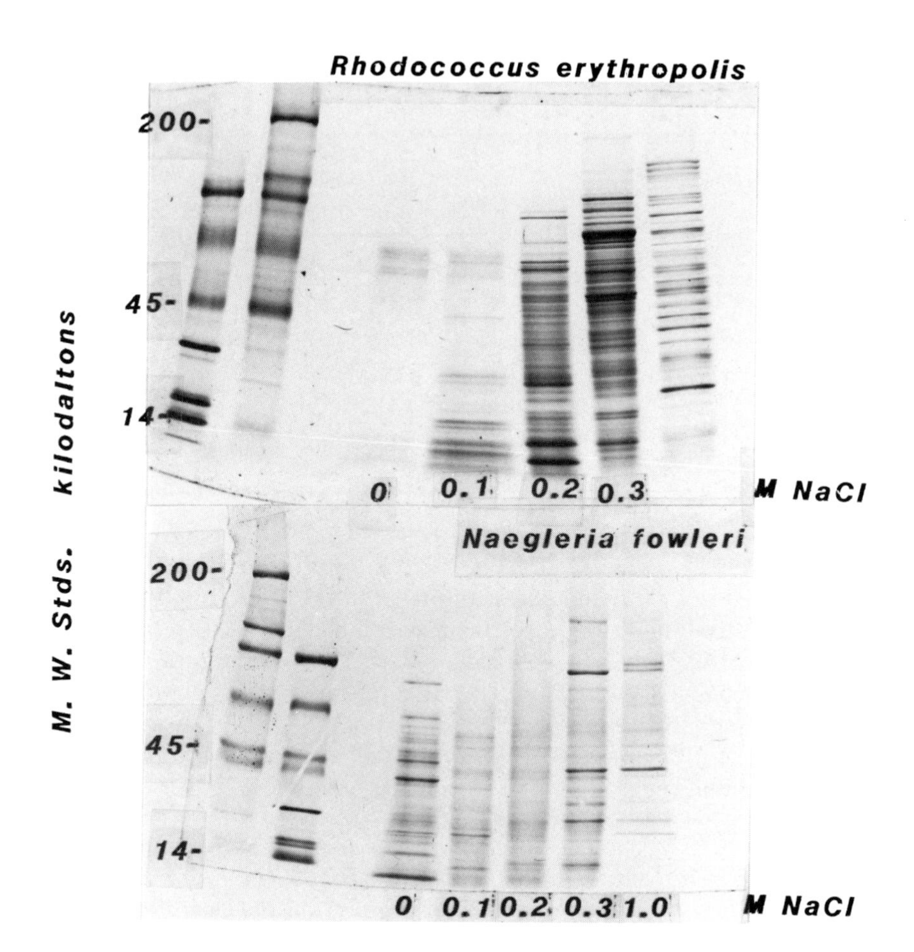

FIGURE 1. Molecular sizes of soluble proteins separated by chromatography of samples from Rh. erythropolis *305 and* N. fowleri *nN68 on DEAE-cellulose. The proteins were denatured with sodium dodecyl sulfate and subjected to electrophoresis in a 5 to 20% exponential gradient of polyacrylamide gel. Two lanes are molecular weight standards: 14, 21, 30, 45, 68, and 94 kilodaltons in one lane and 45, 68, 94, 130, and 200 kilodaltons in the other lane. Five lanes are DEAE-cellulose fractions eluted by 20 mM Tris containing 0, 0.1, 0.2, 0.3, or 1.0 M NaCl.*

The median isoelectric points of protein fractions separated by DEAE-cellulose chromatography of soluble fractions from the procaryotes were pI 5.9 to 7.7, whereas those from eucaryotes were pI 6.9 to 7.2. The median isoelectric points of proteins in the fractions eluted by 0 M or 0.1 M NaCl were higher than were those of fractions eluted by 0.3 M or 1 M NaCl, with the exception of the two yeasts, *Rhodotorula* and *Saccharomyces* (Table III). Subjective examination of the isoelectric focusing gels gives the impression that there are

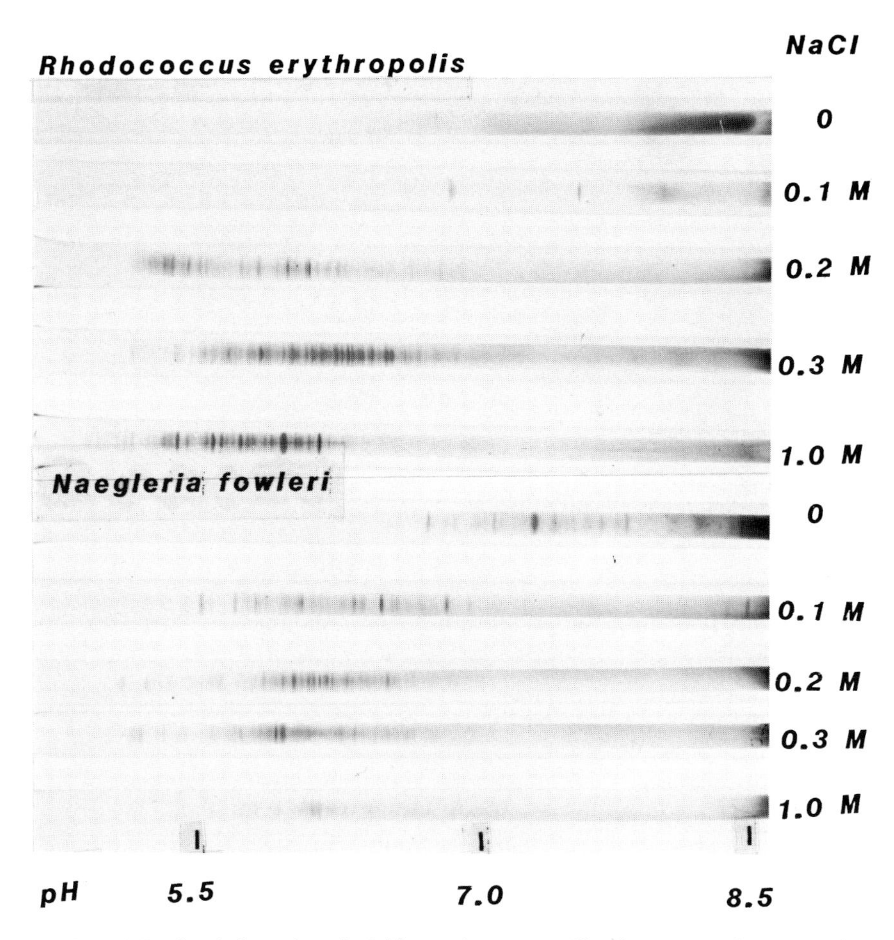

FIGURE 2. Isoelectric focussing of soluble proteins separated by chromatography on DEAE-cellulose of samples from Rh. erythropolis *305 and* N. fowleri *nN68. The five cylindrical gels are DEAE-cellulose fractions eluted by 20 mM Tris containing 0, 0.1, 0.2, 0.3, or 1.0 M NaCl.*

more proteins with lower isoelectric points in procaryotes (*e. g., Rhodococcus*) than in eucaryotes (*e. g. Naegleria*) and that there are more proteins with higher isoelectric points in the eucaryotes (*e. g. Naegleria*) than in the procaryotes (*e. g. Rhodococcus*). Compilations based upon median values (Table III) obscure the differences between the proteins of procaryotes and eucaryotes. (Figs. 1 and 2).

Our data indicate that molecular size and charge are not entirely independent variables, and that this correlation (acidic proteins tend to be large and basic proteins tend to be small) is found in both eucaryotic and procaryotic organisms. The basis of the correlation between size and charge is unknown at this time. In *Rhodococcus*, the more acidic proteins appear to be synthesized

TABLE III. Median Isoelectric Points of Proteins Separated by Chromatography of Soluble Fractions from Selected Procaryotic and Eucaryotic cells on DEAE-cellulose

Source of soluble protein	*Median pH of proteins eluted by 20 mM Tris containing NaCl at*				
	0 M	*0.1 M*	*0.2 M*	*0.3 M*	*1.0 M*
R. erythropolis	-[a]	7.7	6.9	6.9	6.8
B. lichenformis	-	7.5	6.5	5.9	6.8
E. coli	-	7.5	6.8	6.9	6.9
P. aeruginosa	-	7.4	6.9	6.8	6.8
Rhodotorula *sp.*	-	7.0	6.8	7.2	-
S. cerevisiae	-	6.8	6.8	6.8	-
N. fowleri	7.2	7.0	6.9	7.0	-
N. gruberi	7.2	7.0	6.9	6.9	-
Mouse liver	7.0	6.8	6.7	6.7	-
Mouse kidney	7.2	7.0	6.9	6.7	-

[a] *Dash denotes insufficient proteins for analysis.*

more rapidly than are basic proteins, but there is no marked correlation between subunit charge and the rate of protein degradation (Bradley *et al.*, 1980).

III. CONCLUDING REMARKS

Several mechanisms for the regulation of enzyme activities in actinomycetes have been studied in depth: (a) transcriptional control of the rate of enzyme synthesis (Goldberger *et al.*, 1976); (b) control of enzyme activity by noncovalently bound ligands (Switzer, 1977); and (c) intracellular protein degradation (Bradley, 1981). Earlier views that bacterial proteins are completely stable in exponentially growing cells can no longer be sustained. A substantial body of data has accumulated documenting that selective protein degradation occurs in actinomycetes (Bond *et al.*, elsewhere in this volume).

The steady-state level of an enzyme is dependent upon both its rate of synthesis and its rate of degradation. If two enzymes are synthesized at the same rate, the one with the longer half-life will have a higher steady-state level. The half-life of an enzyme is also a factor in determining how rapidly the enzyme concentration will change in response to environmental changes. Enzymes with short half-lives can be induced rapidly as a result of increased biosynthesis and will disappear rapidly when specific biosynthesis stops. Enzymes with longer half-lives respond more slowly to changes in the rate of protein synthesis. Accordingly, intracellular protein degradation is important in determining both the rate of change of enzyme concentration and the steady-state level per se. It should be noted that many enzymes and regulatory proteins, particularly those critical to differentiation, are likely to be present at very low concentration within the cell and therefore are not detectable by current methods.

REFERENCES

Bradley, S. G., *In* "Nocardia and Streptomyces" (M. Mordarski, W. Kurylowicz and J. Jeljaszewicz, eds.), p. 287. Gustav Fischer Verlag, Stuttgart (1978).
Bradley, S. G., *in* "Actinomycetes" (K. Schaal and G. Pulverer, eds.), p. 367. Gustav Fischer Verlag, Stuttgart (1981).
Bradley, S. G., Bond, J.S., and Shannon, J. D., in "Genetics and Physiology of Actinomycetes" (S. G. Bradley, ed.), p. 1. National Technical Information Service, Springfield (1980).
Dice, J. F., and Goldberg, A. L., *Proc. Nat. Acad. Sci. U.S.A. 72:*3893 (1975).
Dice, J. F., Hess, E. J. and Goldberg, A. L., *Biochem. J. 178:*305 (1979).
Dreyfus, J. C., Kahn, A., and Scharpira, F., *Curr. Top. Cell. Regul. 14:*243 (1978).
Duncan, W. E., Offermann, M. K. and Bond, J. S., *Arch. Biochem. Biophys. 199:*331 (1980).
Goldberg, A. L., and St. John, A. C., *Ann. Rev. Biochem. 45:*747 (1975).
Goldberger, R. F., Deeley, R. G., and Mullinix, K. P., *Adv. Genet. 18:*1 (1976).
Hayaishi, O., and Ueda, K., *Ann. Rev. Biochem. 46:*95 (1977).
O'Farrell, P. Z., Goodman, H. M., and O'Farrell, P. H., *Cell 12:*1133 (1977).
Paik, W. K., and Kim, S., *Adv. Enzymol. 42:*227(1975).
Robinson, A. B., and Rudd, C. J., *Curr. Top. Cell. Regul. 8:*247 (1974).
Shannon, J. D., Bond, J. S., and Bradley, S. G., *Arch. Biochem. Biophys. 219:*80 (1982).
Switzer, R. L., *Ann. Rev. Microbiol. 31:*135 (1977).
Woodworth, T. W., Keefe, W. E., and Bradley, S. G., *J. Protozool. 29:*246 (1982a).
Woodworth, T. W., Keefe, W. E., and Bradley, S. G., *J. Bacteriol. 150:*1366 (1982b).

REGULATION OF TRYPTOPHAN METABOLISM AND ITS RELATIONSHIP TO ACTINOMYCIN D SYNTHESIS

Edward Katz
Dale Brown[1]
Michael J. M. Hitchcock[2]
Thomas Troost[3]
John Foster [4]

Department of Microbiology
Georgetown University
Schools of Medicine and Dentistry
Washington, D. C., U. S. A.

I. INTRODUCTION

The metabolic sequence from L-tryptophan via kynurenine (Fig. 1) is utilized in the biosynthesis of NAD (Chaykin, 1967) and of certain insect pigments (Katz and Weissbach, 1967) and is most likely employed in the formation of certain microbial pigments and antibiotics such as actinomycin, triostin, anthramycin, etc. (Katz and Weissbach, 1967; Sivak *et al.* 1962; Troost and Katz, 1979). We have been investigating the biosynthesis of actinomycin, a chromopeptide antibiotic produced by a number of actinomycetes (*e. g., Streptomyces, Micromonospora, Streptosporangium*). Isotope labeling experiments have shown that benzene-ring labeled tryptophan, kynurenine, 3-hydroxykynurenine, and 4-methyl-3-hydroxyanthranilic acid (MHA) are precursors of the actinomycin chromophore, actinocin, providing evidence that this pathway functions in its synthesis (Herbert, 1974; Sivak *et al.*, 1962;

[1] *Laboratory of the Biology of Viruses, National Institute of Allergies and Infectious Diseases, National Institutes of Health, Bethesda, Maryland, U.S.A.*
[2] *Bristol Laboratories, Syracuse, New York, U.S.A.*
[3] *School of Medicine, Georgetown University, Washington, D.C., U.S.A.*
[4] *Department of Microbiology, Marshall University School of Medicine, Huntington, West Virginia, U.S.A.*

ISBN 0-12-528620-1

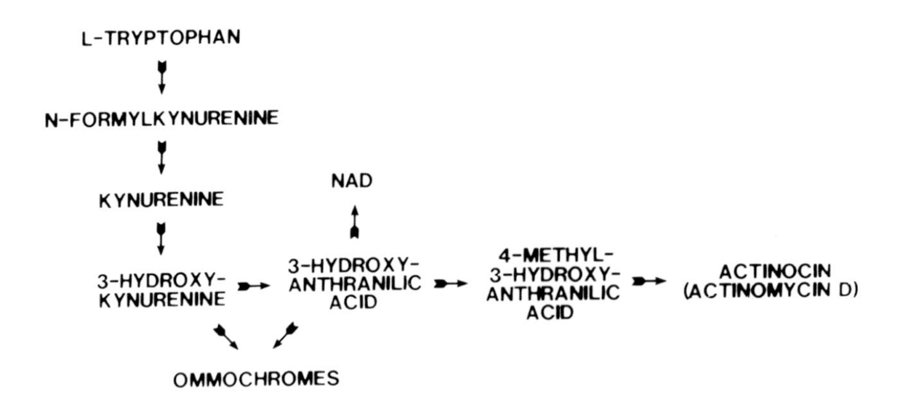

FIGURE 1. Metabolic pathway from L*-tryptophan to actinocin, NAD and the ommochromes.*

Troost and Katz, 1979). Further, actinomycin-negative and low-producing mutants accumulate MHA (Troost and Katz, 1979), and this metabolite is also present in the culture medium when antibiotic synthesis is inhibited or following isotope dilution experiments with ^{14}C-labeled tryptophan and unlabeled MHA (Beavan *et al.*, 1967; Weissbach *et al.*, 1965). In addition, the enzyme, phenoxazinone synthase, purified from extracts of *Streptomyces antibioticus*, catalyzes the formation of actinocin from two molecules of MHA (Katz and Weissbach, 1962, 1967). Our laboratory has been studying certain additional enzymes (tryptophan pyrrolase, kynurenine formamidase, hydroxykynureninase) in the metabolic sequence and their regulation with respect to the biosynthesis of actinocin by *S. parvulus*.

II. RESULTS

A. Growth and Actinomycin Synthesis

For the experimental studies *S. parvulus* was first grown in NZ-Amine medium at 30°C for 48 h to generate an abundant vegetative mycelium; actinomycin synthesis was then effected in a chemically defined production medium (Williams and Katz, 1977). Zero hour for some experiments refers to the time of inoculation of *S. parvulus* into this medium. The organism grew readily, utilizing primarily glutamic acid and histidine as carbon and nitrogen sources (Williams and Katz, 1977). By 21 h, the amino acids were almost completely depleted from the medium (Fig. 2) and within one to three hours later, synthesis of the antibiotic was initiated. D-Fructose was consumed slowly and only about 30% of the sugar was utilized by the end of the fermentation.

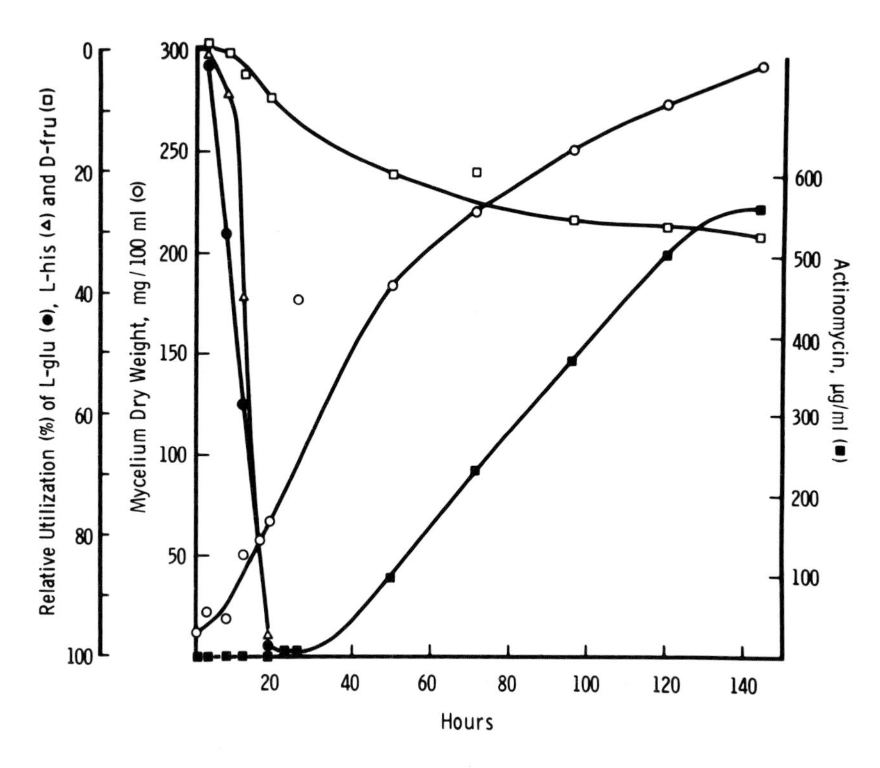

FIGURE 2. Time course of substrate utilization, growth, and actinomycin D formation by S. parvulus.

B. Relationship between Enzyme Synthesis and Actinomycin Formation

An earlier investigation revealed that there is a temporal relationship between the appearance of phenoxazinone synthase activity and actinomycin production (Gallo and Katz, 1972). This enzyme activity was low in young, actively-growing cultures, but it increased some 20-fold just prior to, and during the onset of, antibiotic synthesis. Tryptophan pyrrolase, kynurenine formamidase and hydroxykynureninase activities were also low initially (Fig. 3); but, by 21 to 24 h, there was an increased synthesis of these enzymes and by 48 h the specific activity of tryptophan pyrrolase had increased 40-fold (Katz, 1980; Hitchcock and Katz, unpublished results); kynurenine formamidase, 4- to 7-fold (Brown *et al.*, 1980); and hydroxykynureninase, 30- to 60-fold (Troost *et al.*, 1980).

Unlike tryptophan pyrrolase and hydroxykynureninase, kynurenine formamidase activity is readily detected in non-producing cultures and is never completely repressed (Brown *et al.*, 1980). These results suggested that there are two kynurenine formamidase activities present—a constitutive one (of

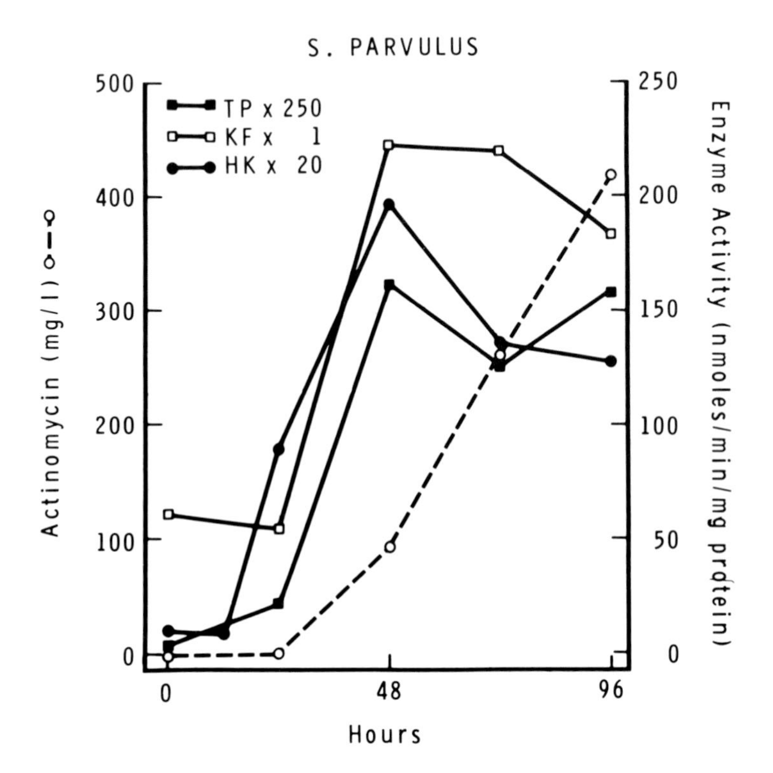

FIGURE 3. Temporal relationship of synthesis of tryptophan pyrrolase, kynurenine formamidase, and hydroxykynureninase activities to actinomycin production by S. parvulus.

unknown function) and an inducible form associated with actinomycin synthesis. To establish this point, extracts prepared from 0 h- and 48-h-old mycelium were examined by gel filtration on Sephadex G-75 columns (Brown *et al.*, 1980). The elution profile with extracts prepared from 48 h mycelium (*i. e.*, during actinomycin synthesis) revealed two distinct peaks of formamidase activity. About 10% of the activity was present in a higher molecular weight enzyme (42,000, formamidase I) and about 90% of the activity was found in a lower molecular form (24,000, formamidase II). By contrast, similar treatment of extracts from 0 h mycelium (*i. e.*, prior to antibiotic production) showed that formamidase II activity was virtually absent and that almost all of the activity was associated with formamidase I. In a more extensive experiment, gel filtration of extracts prepared from mycelium at various times prior to and during the onset of actinomycin formation established that, between 0 and 48 h, the actual increase in formamidase II activity was 21-fold (Brown *et al.*, 1980). Thus, the temporal expression of kynurenine formamidase II activity and its extent are similar to that observed with phenoxazinone synthase, tryptophan pyrrolase, and hydroxykynureninase.

C. *Effect of Transcriptional and Translational Inhibitors*

The effect of actinomycin D, rifampicin, chloramphenicol, and puromycin added at 21 h upon enzyme induction (derepression) was also examined (Brown *et al.*, 1980; Foster and Katz, 1981; Katz, 1980). Kynurenine formamidase II activity increased in the control; however, no further synthesis and, in fact, some degradation of the enzyme activity previously synthesized (formamidase I) was seen in the antibiotic-treated cultures (Fig. 4). Moreover, rifampicin or chloramphenicol inhibited formamidase II synthesis when supplied at different times during the induction period (Fig. 5). Similar findings were obtained when tryptophan pyrrolase synthesis was studied under similar conditions (Foster and Katz, 1981). Thus, it appears that both transcriptional and translational events are required for the appearance of tryptophan pyrrolase and formamidase II activities during this time.

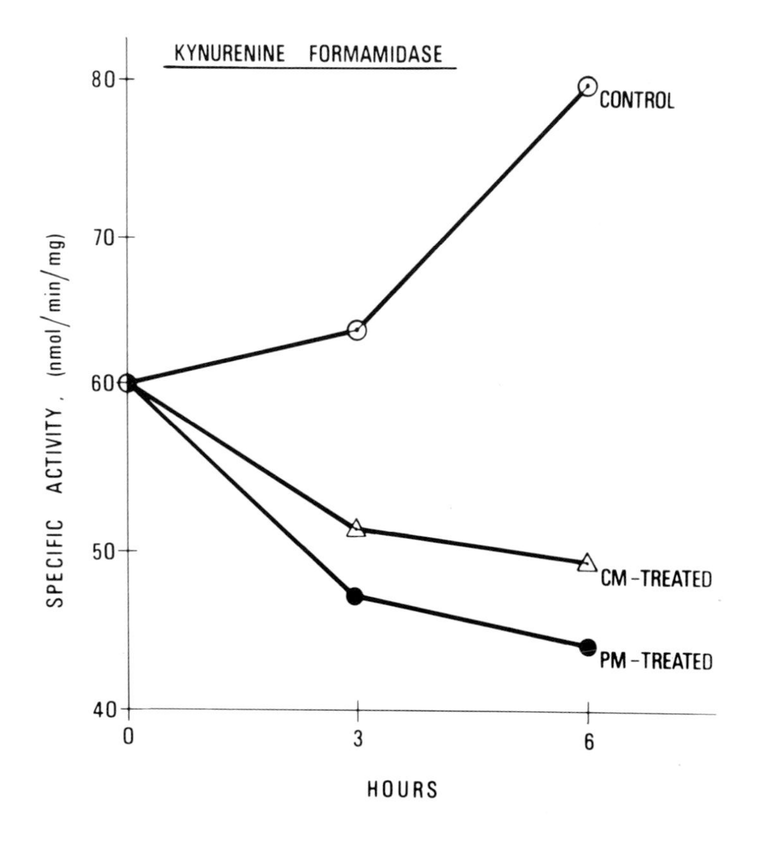

FIGURE 4. Influence of chloramphenicol (CM) and puromycin (PM) upon induction of kynurenine formamidase II activity by S. parvulus. *Time of addition of antibiotics was at 21 h; mycelium was harvested at 0, 3, and 6 h post-treatment.*

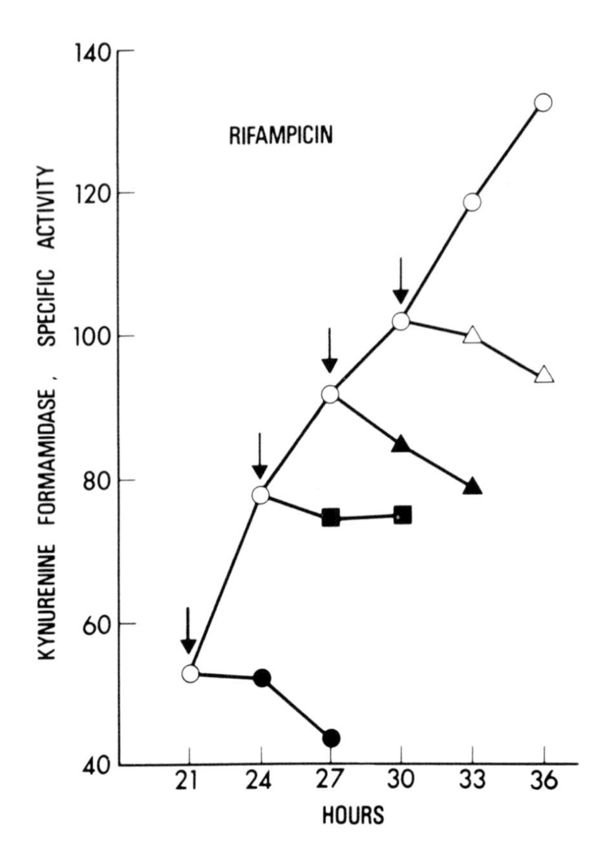

FIGURE 5. Effect of rifampicin upon induction of kynurenine formamidase II activity by S. parvulus. *Arrow (↓) denotes the time of addition of rifampicin.*

D. Regulation of Expression

1. *Carbon Catabolite Regulation.* Glucose, generally an excellent source of carbon for growth is not satisfactory for the biosynthesis of many antibiotics and other secondary metabolites. Complex, slowly utilized substances, such as polysaccharides, oligosaccharides, and oils, are often better carbon sources than is glucose. In a medium containing both a rapidly metabolizable carbon source and a more slowly utilized one, the former is generally consumed first without significant antibiotic production. After its depletion, the alternate carbon source is consumed for antibiotic formation.

The effect of glucose (or glycerol) when used as carbon source was investigated with *S. parvulus* to establish whether enzymes of the tryptophan→actinocin pathway were under carbon catabolite regulation (Brown *et al.*, 1980; Foster and Katz, 1981; Katz, 1980; Troost *et al.*, 1980). This was indeed the case with respect to tryptophan pyrrolase, kynurenine formamidase

TABLE I. Carbon Catabolite Repression Control of Enzyme Synthesis[a]

Medium	Specific activity, nmoles/min/mg		
	Tryptophan pyrrolase	Kynurenine formamidase	Hydroxy-kynureninase
Fructose	0.646	223	14.50
Fructose plus glucose	0.001	35	0.10
Glycerol	0.018	33	0.48

[a] *Measurement of enzyme activities was carried out using extracts prepared from 48-h cultures of* S. parvulus.

II, and hydroxykynureninase (Table I). More extensive experiments in which the synthesis of tryptophan pyrrolase (Fig. 6) and kynurenine formamidase II (Brown *et al.*, 1980) activities were examined confirmed that glucose (or glycerol) completely repressed enzyme and actinomycin formation. It appears unlikely that c-AMP mediates carbon catabolite regulation during actinomycin biosynthesis. High levels of c-AMP did not reverse glucose repression of tryptophan pyrrolase or kynurenine formamidase II synthesis in contrast to the effect seen with inducible enzymes in enteric bacteria.

2. *Nitrogen Metabolite Regulation.* A regulatory mechanism that controls the use of nitrogen sources has been reported for bacteria, yeast, and fungi (Dubois *et al.*, 1974; Magasanik *et al.*, 1974; Marzluf, 1981). Ammonia (or some other readily utilizable nitrogen source) represses synthesis of enzymes that are involved in the metabolism of alternate nitrogen sources; such enzymes include arginase, nitrite reductase, nitrate reductase, glutamate dehydrogenase, etc. In enterobacteria, nitrogen metabolite regulation appears to involve glutamine synthetase; in fungi, an NADP-specific glutamate dehydrogenase has been implicated in the repression of enzymes of nitrogen metabolism.

The temporal relationship of growth, glutamate, and histidine utilization, and antibiotic synthesis by *S. parvulus* was noted earlier (Fig. 2). It was observed that within 1 to 3 h after depletion of the amino acids from the medium, there was enzyme induction (derepression) followed by antibiotic production. Conceivably, the addition of certain amino acids to the medium at this time in the fermentation could maintain the repressed state by preventing expression of genes coding for synthesis of enzymes essential for antibiotic production. To test this hypothesis, we examined the effect of L-glutamate addition (10 mM) at 21 to 23 h upon the synthesis of tryptophan pyrrolase and actinomycin production (Foster and Katz, 1981). Both tryptophan pyrrolase and actinomycin synthesis were repressed for an additional 19 to 20 h under conditions of glutamate supplementation; as noted previously, the initiation of enzyme and actinomycin synthesis occurred only after depletion of the amino acid in the

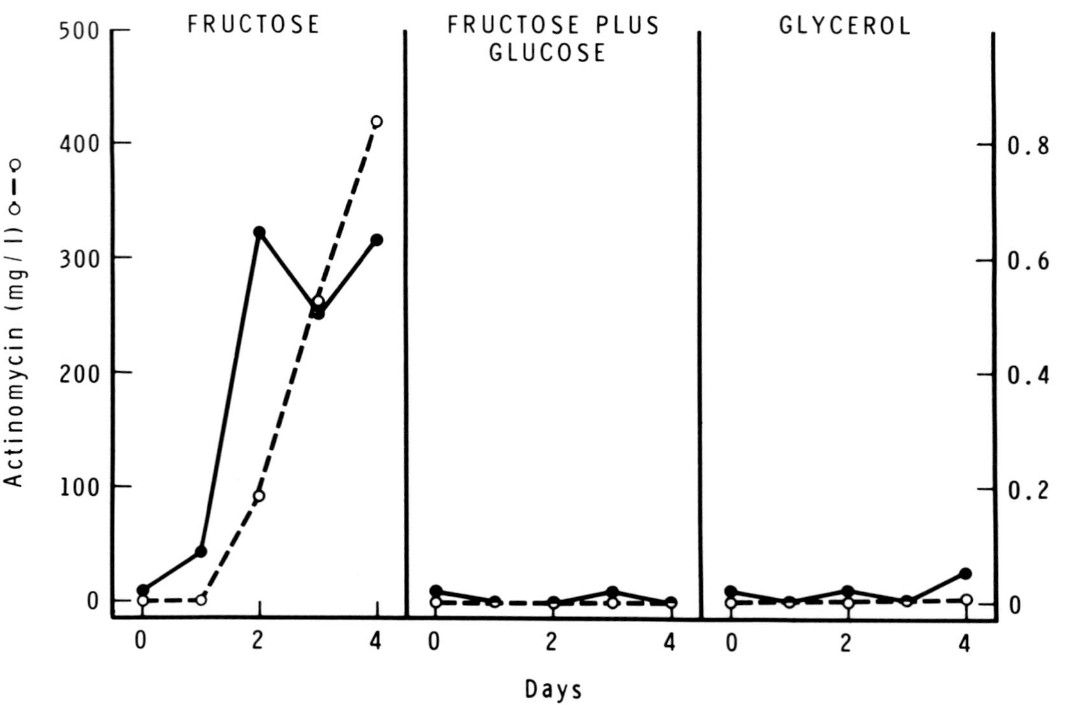

FIGURE 6. Repression of synthesis of tryptophan pyrrolase activity during growth in the presence of D-glucose or glycerol.

medium. Similar results were obtained in studies of the synthesis of formamidase II activity and actinomycin production during short-term incubations (Katz, 1980). We also observed that the extent of repression by L-glutamate was concentration dependent and that addition of glutamate at different times during the period of enzyme induction (derepression) severely repressed further enzyme synthesis (Fig. 7). It was also found that a number of amino acids repress synthesis of tryptophan pyrrolase and formamidase II activities (Table II). In most cases the repression appeared to be transient in nature for after the supply of an amino acid was depleted from the medium, derepression of enzyme synthesis and antibiotic production occurred. A more striking effect was noted with D-valine; addition of this amino acid to the fermentation repressed synthesis of formamidase II activity and actinomycin for several days (Table III). The type of repression seen in this study may be due to the slow rate of metabolism of the amino acid by the organism (Katz, unpublished results). Conceivably, repression of enzyme synthesis may provide the basis for the marked inhibition of actinomycin production by D-amino acids which we described some years ago (Katz, 1960).

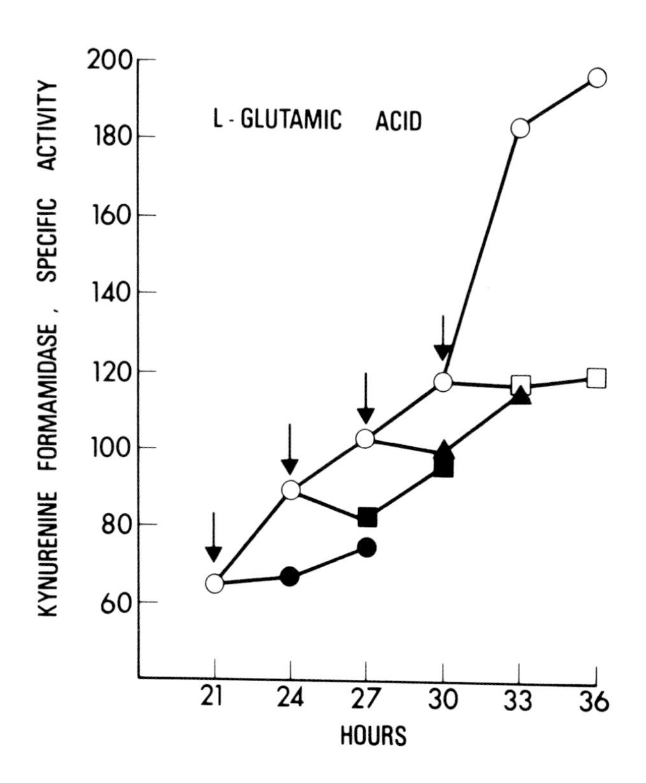

FIGURE 7. Effect of L-glutamate (5 mM) upon induction of kynurenine formamidase II activity. Arrow (↓) denotes the time of addition of L-glutamate.

TABLE II. Repression of Tryptophan Pyrrolase (TP) and Kynurenine Formamidase (KFase) Synthesis by Amino Acids

Compound	*Repression*		*Compound*	*Repression*	
	TP %	*KFase %*		*TP %*	*KFase %*
L-*Alanine*	72	100	L-*Histidine*	18	45
L-*Aspartic Acid*	79	59	L-*Valine*	64	52
D-*Aspartic Acid*	77	83	L-*Isoleucine*	15	0
L-*Asparagine*	-	3	L-*Leucine*	-	6
L-*Threonine*	30	77	*Glycine*	-	0
L-*Lysine*	2	43	L-*Serine*	-	67
L-*Methionine*	61	44	L-*Arginine*	11	52
L-*Glutamic Acid*	92	98	L-*Phenylalanine*	-	85
D-*Glutamic Acid*	87	60	L-*Tyrosine*	28	79
L-*Glutamine*	-	0	L-*Tryptophan*	-	12
L-*Proline*	55	68	NH_4Cl	-	16

From the studies with amino acids it was concluded that 1) the most pronounced repression was observed with glutamate, aspartate, and alanine; in contrast, their amides (glutamine, asparagine), and α-keto acid derivatives exerted almost no effect; 2) the repression did not appear to be due to ammonium ion, and a combination of NH_4^+ and an α-keto acid was equally ineffective; 3) D- as well as L-amino acids repress; 4) not all readily metabolizable compounds repress (*e. g.*, glycine); 5) not all members of a family of amino acids repress (*e. g.*, phenylalanine, tyrosine vs tryptophan); and 6) precursors of actinomycin may be strong or weak repressors. Whether the repression is due to the amino

TABLE III. Effect of D-*Valine on Induction of Kynurenine Formamidase and Actinomycin Synthesis by* Streptomyces parvulus

D-*Valine* [a] *mM*	*Kynurenine formamidase* [b] *Δ nmol/min/mg*	*Repression %*	*Actinomycin μg/ml*	*Percent inhibition %*
None	81.6	0.0	121	100
0.1	38.3	56.7	85	30
0.25	21.2	74.0	52	57
0.5	4.5	94.0	6	95
1	-7.2	100.0	2	98
2	-40.8	100.0	1	99

[a] D-*Valine was added at 21 h.*
[b] *Kynurenine formamidase (control) specific activity at 21 h was 56.1 nmol/min/mg; at 48 h, 137.7 nmol/min/mg.*

acid *per se* or a metabolite derived from it remains uncertain at the present time. Thus, neither analogs of L-glutamate and L-aspartate nor their α-keto acid derivatives repressed tryptophan pyrrolase synthesis appreciably.

We also observed that a number of amino acids exhibited a quantitatively similar repression of synthesis of an inducible β-galactosidase activity in *S. parvulus* (Foster and Katz, 1981). By contrast, no repression of a constitutive enzyme, glucose-6-phosphate dehydrogenase, was seen, suggesting that regulation by amino acids may be specific for certain inducible enzyme systems in *Streptomyces*.

E. Growth Experiments

One possible reason for the repressive effects of amino acids on inducible enzyme synthesis is that their metabolism may stimulate the growth rate of *S. parvulus*. Addition of glutamate at 24 and 48 h did enhance both the rate and extend of growth of the organisms, and these results are consistent with the hypothesis cited above. However, isoleucine, an amino acid which does not repress synthesis of tryptophan pyrrolase or formamidase II activities to any significant extent, provided an even greater increase in the rate and extent of growth (Foster and Katz, 1981). Hence, it is concluded that repression of enzyme synthesis does not appear to be related directly to the growth rate.

F. Early Synthesis of Tryptophan Pyrrolase

The synthesis of enzymes in the pathway tryptophan $\rightarrow$ actinocin and subsequent formation of actinomycin may be due to a release from amino acid repression rather than to a response to some metabolic intermediate or factor. If this be the case, mycelium incubated in a medium without glutamate should be able to synthesize the biosynthetic enzymes and actinomycin. To test this hypothesis initially, we inoculated washed cells, which had been previously grown for 15 h in GHF medium, either into filtered, aged GHF medium depleted of amino acids by prior incubation with *S. parvulus* for 23 h or into fresh GHF medium lacking glutamate and histidine. Under both sets of conditions, synthesis of tryptophan pyrrolase (and actinomycin D) began at, or shortly after, 16.5 h (Fig. 8) which is some 4.5 h before its occurrence in the control culture (Foster and Katz, 1981). In the second experiment we used washed mycelium previously grown for 5, 12, or 15 h in GHF medium. *S. parvulus*, incubated *in fresh medium prepared without glutamate and histidine* rapidly initiated enzyme synthesis (Fig. 9) and actinomycin production (Foster and Katz, 1981). Thus, it is evident that enzyme and antibiotic synthesis can take place at any time in the absence of a repressible compound such as glutamate, although growth in glutamate-containing media does result in the production of much higher antibiotic titers.

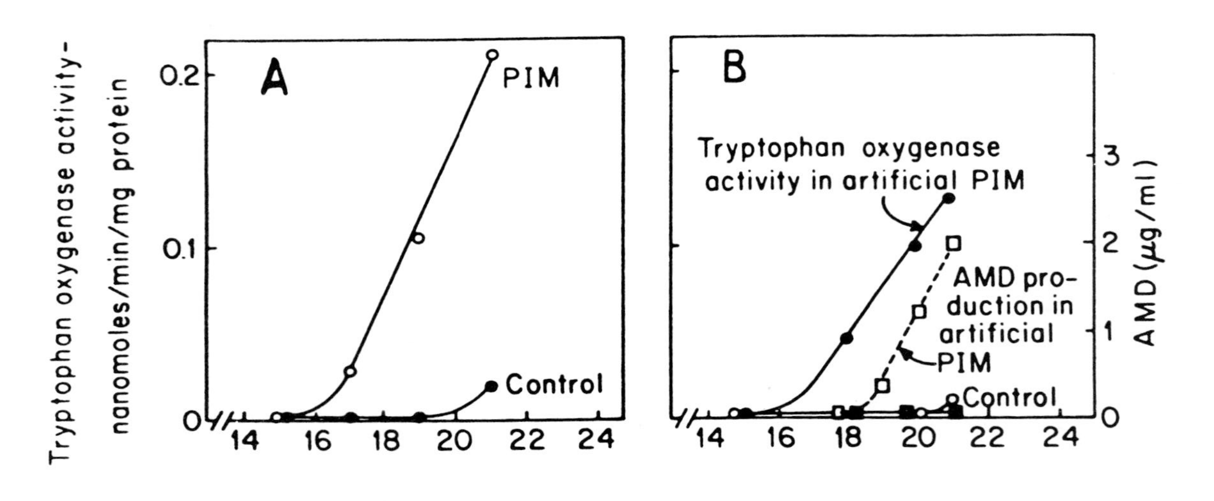

FIGURE 8. Tryptophan pyrrolase (oxygenase) induction and actinomycin production in (A) amino acid-depleted medium (PIM), or in (B) fresh GHF medium minus L*-glutamate and* L*-histidine (artificial PIM).* S. parvulus *was grown for 15 h in GHF medium, washed, and used as inoculum. (A) tryptophan pyrrolase in PIM* (O) *or in GHF medium (□); (B) tryptophan pyrrolase (●) and actinomycin synthesis (□) in artificial PIM, or tryptophan pyrrolase (O) and actinomycin synthesis (■) in GHF medium.*

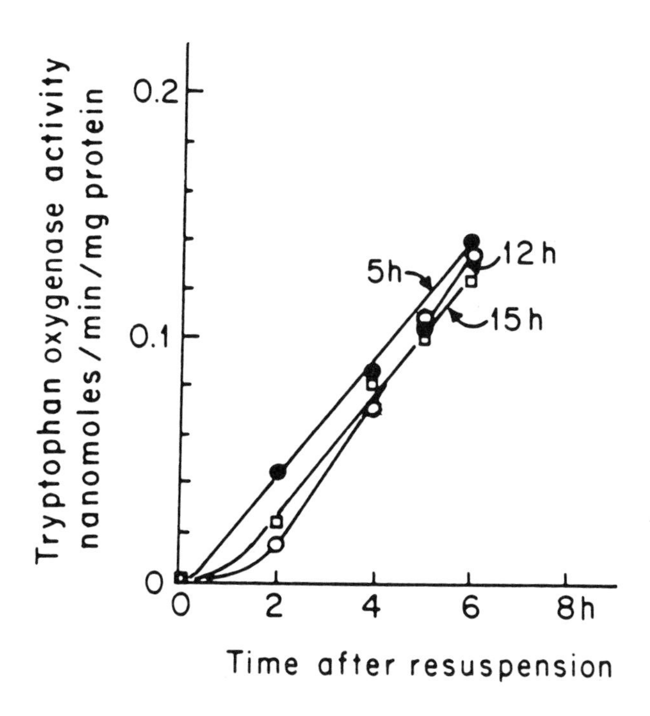

FIGURE 9. Tryptophan pyrrolase (oxygenase) induction in fresh GHF medium minus L-glutamate and L-histidine (artificial PIM). Mycelium grown in GHF medium was harvested, washed, and inoculated into artificial PIM. Tryptophan pyrrolase activity using 5 h (●), 12 h (O) and 15 h (□) mycelium. Actinomycin synthesis (data not shown) was initiated soon after enzyme induction in each case.

G. Enzyme Activity in Mutants

Several classes of mutants of *S. parvulus* that fail to produce actinomycin or that synthesize low levels of the antibiotic have been isolated. The mutants obtained thus far that affect actinomycin synthesis appear to be blocked in the synthesis of a primary metabolite (tryptophan) or to be defective at a regulatory site on the genome.

One strain, AM8, was obtained after 8-methoxypsoralen plus near ultraviolet treatment (Troost *et al.*, 1980). The inability of this organism to produce actinomycin was found to correlate with its failure to synthesize nor mal amounts of the enzymes in tryptophan→ actinocin pathway (Fig. 10). Only slightly elevated levels of pyrrolase and hydroxykynureninase activities were noted after 48 h incubation. However, the enzyme activities displayed in the mutant declined subsequently, presumably, as a consequence of degradation or inactivation of previously synthesized enzyme. Formamidase activity was also present, but at markedly reduced levels in the AM8 strain; there appeared to be virtually no synthesis of formamidase II activity. Although the nature of

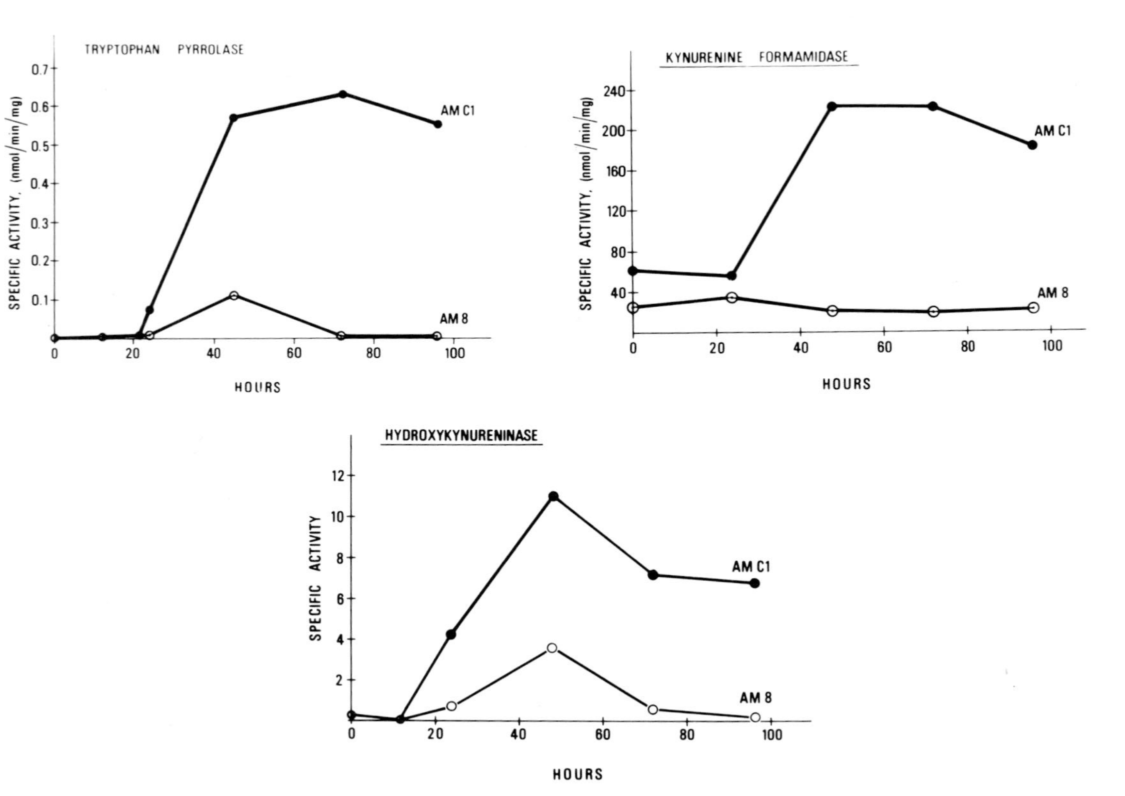

FIGURE 10. Enzyme activities (tryptophan pyrrolase, kynurenine formamidase, and hydroxykynureninase) in wild type and mutant strains (AM 8) of S. parvulus.

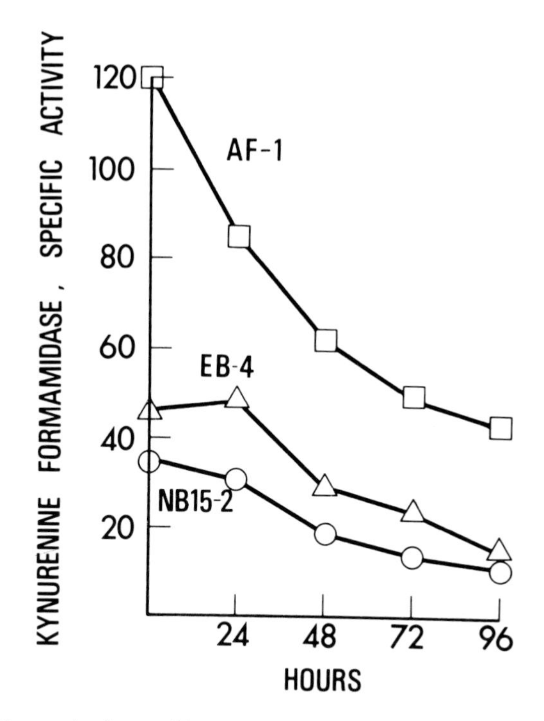

FIGURE 11. Kynurenine formamidase activity in extracts of acriflavin (AF-1), ethidium bromide (EB-4), and novobiocin (NB-15-2)-"cured" strains of S. parvulus. *For comparison see Fig. 2.*

the mutation in AM8 that is responsible for the absence of antibiotic synthesis is not understood, it seems unlikely from these results that the mutation is in a structural gene coding for one of the enzymes involved in tryptophan metabolism. It is possible that the mutation is at a regulatory site on the genome. The expression of these gene(s) must be tightly coupled to that of functional actinomycin synthesis since, in its absence, expression is transient and is not maintained.

S. parvulus mutants that fail to synthesize actinomycin D have also been isolated after acriflavin, ethidium bromide, and novobiocin treatment (Ochi and Katz, 1978). Studies thus far have been limited to examination of kynurenine formamidase activities (Katz, 1980). In contrast to that in the parent strain, formamidase activity in the mutant strains was highest at the time of transfer from complex medium to the chemically-defined production medium (Fig. 11). Thereafter, the activity declined significantly. Gel filtration of a partially purified preparation derived from extracts of an acriflavin-treated mutant established that formamidase I activity was present; however, we were unable to detect any formamidase II activity. Similar results were obtained with extracts prepared from novobiocin- or ethidium bromide-treated strains.

A third type of mutant has been isolated which appears to have normal levels of enzymes in the tryptophan→actinocin pathway, but is partially blocked in the synthesis of the primary metabolite, tryptophan. Although growth of the bradytroph is comparable to that seen with the parent, synthesis of actinomycin is delayed for some 36-48 h as compared to the wild type strain. The mutant initially accumulates appreciable amounts of anthranilic acid (250-300 μg per ml) which disappears during actinomycin synthesis. The metabolic block in the mutant, B356, may involve a defect in synthesis of PR transferase, the second enzyme in the pathway from chorismic acid→tryptophan. The specific activity of the enzyme in extracts derived from the mutant is one-sixth to one-tenth that found in the parent strain. On the other hand, anthranilate synthetase, the first enzyme in the pathway is derepressed in the mutant; the specific activity of the enzyme in extracts is some 5- to 15-fold greater than in the parent which undoubtedly accounts for the accumulation of anthranilic acid in the medium. If tryptophan is supplied to the mutant, the time and the extent of synthesis of actinomycin compares favorably to that found in the parent strain. Moreover, the addition of tryptophan to growing cultures of B356 represses synthesis of anthranilate synthetase to levels normally found in the parent. Some of the phenotypic properties of the wild type and mutant strains are presented in Table IV.

TABLE IV. Phenotypic Properties of S. parvulus *(Parent) and B356 (Mutant)*

Enzyme or compound		*Parent*	*B356*
Anthranilate synthetase	*Relative specific activity*	*1*	*5-12*
		Tryptophan - no effect, maximally repressed	*Tryptophan - repressed*
		Feed-back sensitive, 50% inhibition by 3×10^{-5} M	*Feed-back insensitive at 1×10^{-3} M*
Phosphoribosyl transferase	*Relative specific activity*	*1*	*0.05-0.2*
Anthranilic acid		*0*	*250-350 μg*/ml
Actinomycin		*550-600 μg/ml*	*200-300 μg/ml (delayed)*

III. DISCUSSION

In conclusion, we investigated the formation of certain enzyme activities (tryptophan pyrrolase, kynurenine formamidase, and hydroxykynureninase) in *S. parvulus* that catalyze specific reactions in the metabolic sequence from tryptophan to the actinomycin chromophore, actinocin. We suggest that this pathway is regulated specifically for the specialized function of actinomycin biosynthesis. The time and extent of induction (derepression) of these enzymes, their repression by glucose, glycerol, and amino acids, and their low level of expression or absence from certain actinomycin non-producing mutants indicate that they have a functional role in this pathway during antibiotic formation. Inhibitor studies provided evidence that both transcriptional and translational events are required for the appearance of these enzymes during this time. Severe nitrogen metabolite regulation may account for the inhibition of actinomycin synthesis observed with D-valine. Studies with "curing" agents suggest (but do not establish) that extrachromosomal determinants may participate in the synthesis and/or regulation of kynurenine formamidase II activity.

From the results it is concluded that there are two distinct kynurenine formamidase activities in *S. parvulus*. The higher molecular weight form (formamidase I) appeared to be constitutive and was present at low levels in both producing and non-producing cultures, whereas the synthesis of the lower molecular enzyme activity (formamidase II) was induced (derepressed) just prior to and during the onset of antibiotic synthesis. The role of formamidase I is still unknown. Possibly, it may participate in NAD biosynthesis, or it may be the vestigial element of a pathway of tryptophan catabolism that is no longer operative *per se*. By contrast, formamidase II is envisaged as catalyzing the second step in the pathway from tryptophan → actinocin. The relationship between formamidase I and II in *S. parvulus* is not understood at the present time. Formamidase isozymes have not been found previously in a bacterium, but have been observed in certain eukaryotic forms, *e. g.*, chick, Drosophila (Bailey and Wagner, 1974; Moore and Sullivan, 1975). It has been suggested that during evolution gene duplication and subsequent divergence of one gene resulted in the ability to synthesize isozymes that both participate in kynurenine synthesis (Moore and Sullivan, 1975). Evidence for a relationship between the two forms of the enzyme in *S. parvulus* will require the purification and subsequent characterization of the two proteins. Also, the assignment of definitive roles for these two enzymes will necessitate a more detailed study of strains in which mutations in the structural gene(s) for one or the other isozyme prevent expression.

IV. SUMMARY

The formation of certain enzymes (tryptophan pyrrolase, kynurenine formamidase, hydroxykynureninase) that catalyze specific reactions in the metabolic sequence from L-tryptophan to the actinomycin chromophore, actinocin, was examined using *Streptomyces parvulus*. Two forms of kynurenine formamidase were found. Formamidase I (M.W. 42,000) was synthesized constitutively, whereas formamidase II (M.W. 24,000) was present only just prior to and during AMD synthesis. It is envisaged that this pathway of tryptophan metabolism is regulated specifically for the specialized function of AMD biosynthesis. Evidence in support of this hypothesis is based upon the time and extent of induction (derepression) of these enzymes, their repression by glucose, glycerol and amino acids and their absence from certain AMD$^-$ mutants. Inhibitor studies indicate that both transcriptional and translational events are required for the appearance of these enzymes during this time. Severe nitrogen metabolite regulation may account for the inhibition of AMD synthesis observed with D-valine. Several classes of mutants that affect AMD synthesis were isolated. Studies with "curing" agents suggest that extrachromosomal determinants may participate in the synthesis and/or regulation of enzymes in the tryptophan actinocin pathway.

REFERENCES

Bailey, C. B., and Wagner, C., *J. Biol. Chem. 249:*4439 (1974).
Beavan, V., Barchas, J., Katz, E., and Weissbach, H., *J. Biol. Chem. 242:*657 (1967).
Brown, D., Hitchcock, M. J. M., and Katz, E., *Arch. Biochem. Biophys. 202:*18 (1980).
Chaykin, S., *Annu. Rev. Biochem. 36:*149 (1967).
Dubois, E., Grenson, M., and Wiame, J. M., *Eur. J. Biochem. 48:*603 (1974).
Foster, J. W., and Katz, E., *J. Bacteriol. 148:*670 (1981).
Gallo, M., and Katz, E., *J. Bacteriol. 109:*659 (1972).
Herbert, R., *Tetrahedron Lett. 51:*4525 (1974).
Katz, E., *J. Biol. Chem. 235:*1090 (1960).
Katz, E., *in* "Biochemical and Medical Aspects of Tryptophan Metabolism" (O. Hayaishi, Y. Ishimura and R. Kido, eds.), p. 159. Elsevier/ North-Holland Biomedical Press, Amsterdam (1980).
Katz, E., and Weissbach, H., *J. Biol. Chem. 237:*882 (1962).
Katz, E., and Weissbach, H., *Dev. Ind. Microbiol. 8:*67 (1967).
Magasanik, M., Prival, M. J., Brenchley, J. E., Tyler, B. M., DeLeo, A. B., Streicher, S. L., Bender, R. A., and Paris, C. G., *Curr. Top. Cell. Regul. 8:*118 (1974).
Marzluf, G. A., *Microbiol. Rev. 45:*437 (1981).
Moore, G. P., and Sullivan, D. T., *Biochim. Biophys.* Acta *397:*468 (1975).
Ochi, K., and Katz, E., *J. Antibiot. 31:*1143 (1978).
Sivak, A., Meloni, M. L., Nobili, F., and Katz, E., *Biochim. Biophys.* Acta *57:*283 (1962).
Troost, T., and Katz, E., *J. Gen. Microbiol. 111:*121 (1979).
Troost, T., Hitchcock, M. J. M., and Katz, E., *Biochim. Biophys. Acta 612:*97 (1980).
Weissbach, H., Redfield, B. G., Beavan, V., and Katz, E., *Biochem. Biophys. Res. Commun. 19:*524 (1965).
Williams, W. K., and Katz, E., *Antimicrob. Agents Chemother. 11:*281 (1977).

REGULATION OF ERYTHROMYCIN FORMATION IN *STREPTOMYCES ERYTHREUS*

Sergio Sanchez
Rosa del Carmen Mateos
Laura Escalante
Julieta Rubio
Hector López
Amelia Farres

Instituto de Investigaciones Biomédicas
Universidad Nacional Autónoma de México
México, D. F., México

and

María Elena Flores

Facultad de Química
Universidad Autónoma del Estado de México
Toluca, México

I. INTRODUCTION

The antibiotic erythromycin is a secondary metabolite formed by strains of *Streptomyces erythreus* at the end of the trophophase (Smith *et al.*, 1962). Since its discovery (McGuire *et al.*, 1952), five structurally related types of erythromycin have been characterized. Of these, erythromycin A is the form preferentially synthesized by the streptomycete (Ostrowska-Krysiak, 1974) and represents the type used in clinical practice.

Since it is a broad-spectrum antibiotic and has low toxicity for humans, erythromycin is of great medical and economic importance (Oleinick, 1975). In spite of the practical interest of erythromycin, very little has been published concerning the regulatory aspects of its biosynthesis (Corcoran, 1975; Raczynska-Bojanowska *et al.*, 1970a; Spizek *et al.*, 1965). The aim of this work

BIOLOGICAL, BIOCHEMICAL,
AND BIOMEDICAL ASPECTS OF ACTINOMYCETES

ISBN 0-12-528620-1

is to present the most recent advances in the synthesis and regulation of erythromycin formation. For this purpose, several regulatory studies are reviewed and integrated with our own observations.

Erythromycin antibiotics of which erythromycin A is perhaps the best known example possess a 14-membered polyfunctional macrocyclic aglycone ring (erythronolide), substituted with both a basic (desosamine) and a neutral sugar moieties. The neutral sugars are cladinose (found in the A, B, and E types of erythromycin) and mycarose (in types D and C).

Through the detailed examination of erythromycin biosynthesis in several laboratories, the propionate origin of the macrolide ring (Corcoran, 1973) and the glucose origin of the sugars (Grisebach, 1967) have been established. The participation of S-adenosyl-methionine as a donor of the methyl group for the neutral and amino sugars (Grisebach *et al.*, 1967) has also been determined. In addition, with the use of radioactive precursors, it has been shown that initiation of erythronolide biosynthesis begins with a primer of activated propionate (propionyl-CoA) (Vanek *et al.*, 1961). The remaining portion of the lactone ring is fashioned from the linear addition of six units of methylmalonyl-CoA to the activated propionate (Friedman *et al.*, 1964). Furthermore, the analysis of intermediates accumulated by mutants of *S. erythreus* which were blocked at different positions of the antibiotic pathway (Martin and Egan, 1970) has also played an important role in the elucidation of the erythromycin pathway, especially when taken together with results of feeding experiments (Ostrowska-Krysiak, 1974). These studies have indicated, as shown in Figure 1, that the major pathway in erythromycin A formation involves the conversion of erythromycin D into erythromycin C which is then converted into erythromycin A, rather than D to B to A

Enzymatic studies have shown that propionate kinase (phosphotransferase adenosine triphosphate (ATP): acetate, EC 2.7.2.1.), first enzyme of the erythromycin formation pathway, is a limiting step in the antibiotic formation. The enzyme catalyzes the synthesis of propionyl phosphate from propionate and ATP and shows high substrate specificity (Raczynska-Bojanowska *et al.*, 1970a). Enzyme activity increases in parallel with erythromycin biosynthesis and reaches a maximum during the stationary phase of growth.

Another important step in erythromycin formation is carried out by propionyl-CoA carboxylase (propionyl-CoA, carbodioxide ligase, EC 6.4.1.3.). This enzyme catalyzes the formation of methylmalonyl-CoA units from propionyl-CoA, carbon dioxide and ATP. As with propionate kinase, propionyl-CoA carboxylase also reaches maximum activity at the post-exponential growth phase (Raczynska-Bojanowska *et al.*, 1970a).

For a long time, erythromycin A was regarded as the final product in the antibiotic pathway. However, erythromycin A is slowly metabolized to erythromycin E (Martin *et al.*, 1975), a compound having only 10 to 15% the activity of erythromycin A.

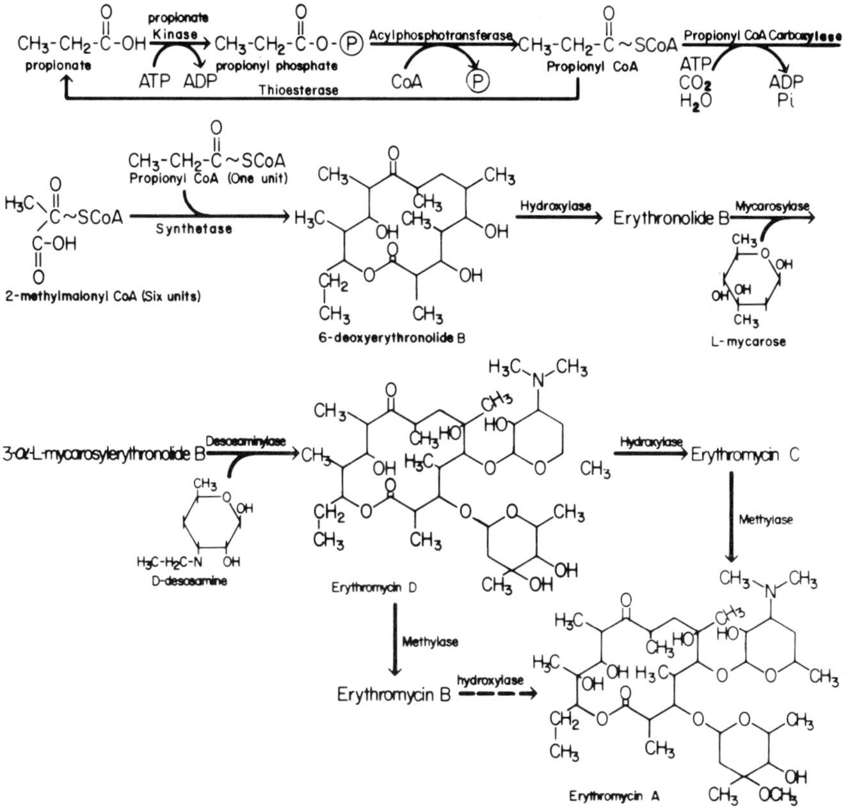

FIGURE 1. Biosynthetic pathway of the erythromycins.

The course of antibiotic fermentation has been described by several authors. In complex media, erythromycin biosynthesis starts at the end of the exponential growth phase and proceeds linearly during a defined period of time, depending upon the medium being used. In other words, little, if any, synthesis of antibiotic occurs while the cells are still growing; its formation follows a fairly well-defined course with respect to fermentation time. This behavior can be correlated either to the growth phase or to changes in the fermentation medium such as the carbohydrate and nitrogen concentration, variation in pH, aeration rate, and so forth. For instance, Smith *et al.* (1962) reported that, in a complex medium, erythromycin is formed at the end of the logarithmic growth phase. Under these conditions, biosynthesis continues linearly as long as sugar is present in the fermentation broth. Furthermore, addition of sucrose towards the end of the fermentation cycle reinitiates erythromycin biosynthesis. This effect is also obtained with other sugars such as glucose, maltose, and fructose. However, addition of the carbohydrate in the presence of a suitable nitrogen source prevents further stimulation of erythromycin biosynthesis. This effect is probably due to the channelling of intermediates to fill growth requirements rather than due to a secondary metabolite formation. In other words, under appropriate conditions, *i. e.*, in the presence of a suitable carbon and nitrogen source, the cell growth can be reinitiated.

Stark and Smith (1961) monitored the pH variation during erythromycin fermentation in a complex medium. The initial pH of the medium was slightly above neutrality and ranged from 6.0 to 6.5 until the end of the process, when an increase in the pH value due to autolysis was observed.

Brinberg (1959) studied the effect of oxygen transfer on erythromycin formation in a complex medium. He observed a close relationship between the presence of nutrients and the demand for oxygen. Thus, an increase in aeration up to 29.6 mg O_2 $L^{-1}min^{-1}$ was effective in raising the amount of antibiotic produced. Similar conclusions were obtained by Ettler, Paca, and Cechner (1976) in a pilot plant fermentor. They observed high erythromycin production after elimination of oxygen limitation in the mycelium growth by increasing the impeller speed of the fermentation tank.

II. CARBON REGULATION

A. Induction of Biosynthesis

In batch cultures containing a nutritionally complete medium, erythromycin formation by *S. erythreus* takes place only after most of the cellular growth has already occurred; in other words, the microorganism appears to be programmed to produce antibiotic only when the specific growth rate decreases below a certain level (Bu'Lock, 1975).

Several nutritional factors are probably involved in the onset of antibiotic biosynthesis. Changes in the intracellular concentration of certain primary metabolites seem to stimulate secondary metabolism through mechanisms which are only beginning to be understood (Lara *et al.*, 1982).

Erythromycin biosynthesis is stimulated by propanol and to a lesser extent by propionate. Brinberg (1959) reported that addition of 0.2 to 0.5% propanol to the culture medium increases the yields of antibiotic up to 80% without affecting growth. Such stimulation is exerted not only by the precursor properties of propanol (Raczynska-Bojanowska *et al.*, 1970b), but also through an inductive action on propionyl-CoA carboxylase, one of the first enzymes of the antibiotic biosynthetic pathway (Raczynska-Bojanowska *et al.*, 1970a). In addition to propanol, citrate also increases the activity of propionyl-CoA carboxylase and, as in the case of propanol, the effect is prevented by actinomycin K (Raczynska-Bojanowska *et al.*, 1970a).

B. Glucose Transient Repression

The synthesis of many antibiotics is strongly suppressed by rapidly used carbon sources (Martin and Demain, 1980). Among such carbon sources, glucose seems to be one of the most effective in exerting a negative influence on antibiotic formation (Hu and Demain, 1979). Similarly, in *S. erythreus*, D-glucose caused transient repression on erythromycin formation (Escalante *et al.*, 1982). As pointed out by these authors, the use of D-glucose as major carbon and energy source results in changes in maximum growth and in production of erythromycin, *i. e.*, this carbohydrate brings about a strong but temporary repression of antibiotic formation and this effect is proportional to the sugar concentration (Fig. 2).

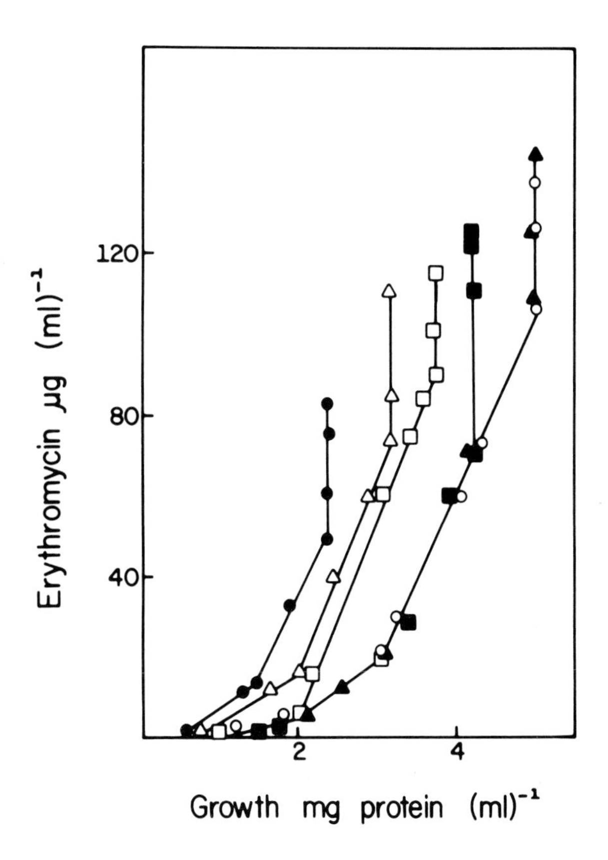

FIGURE 2. Effect of different D*-glucose concentrations on erythromycin formation. Cultures were grown in CM medium (Escalante* et al., *1982) with (△) 1; (□) 2; (■) 5; (O) 10; and 6▲) 20 mg ml^{-1}* D*-glucose. Control without sugar (●).*

C. Effect of Other Carbon Sources

When *S. erythreus* was provided with different carbohydrates as major carbon and energy source, changes in maximum growth and in production of erythromycin resulted. Among the carbon sources tested (Fig. 3), glycerol, sucrose, and D-mannose increased the extent of growth and delayed antibiotic production. Under these conditions, the cells seemed to use the carbon source for growth rather than for antibiotic production. Lactose and sorbose did not influence this relation; D-fructose and D-galactose exerted an intermediate action. Similar effects by glycerol have been observed on cephalosporin formation in *S. clavuligerus* (Aharonowitz and Demain, 1978).

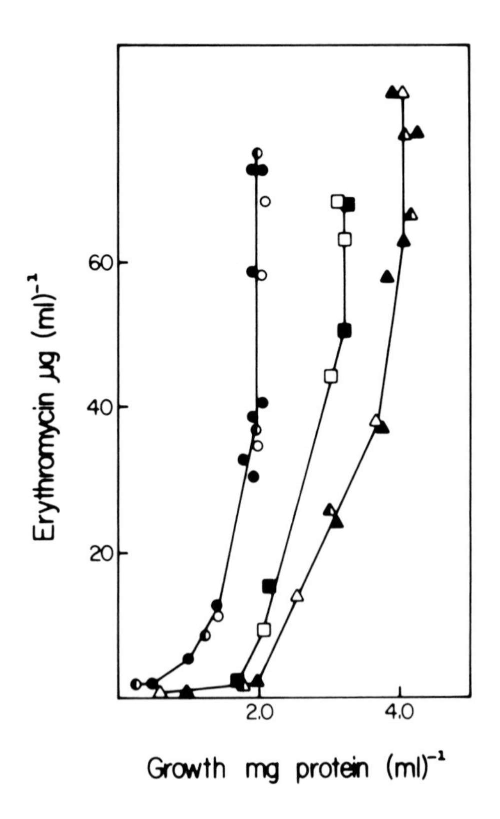

FIGURE 3. Effect of different carbon sources on erythromycin formation. Cultures were grown in CM medium either supplemented or not with the carbohydrates (10 mg ml^{-1}). Lactose (●), sorbose (◐), D-fructose (■), D-galactose (□), glycerol (△), sucrose (◭), and D-mannose (▲). Control without sugar (O).

III. NITROGEN REGULATION

A. Ammonium Regulation

Nitrogen regulation plays an important role in the formation of many antibiotics and other secondary metabolites (Aharonowitz and Demain, 1979; Gräfe *et al.*, 1977; Sanchez *et al.*, 1981). We found that ammonium strongly suppressed erythromycin biosynthesis in *S. erythreus*. Suppression was proportional to the ion concentration, whereas growth and pH values were not significatively affected (Fig. 4). As shown in the same figure, maximum effect was obtained with 100 mM ammonium. The degree of erythromycin suppression was correlated to the type of ammonium salt used. Thus, ammonium sulfate caused 92% suppression, whereas the chloride form affected only by 50% antibiotic formation (Table I). Preliminary evidence was obtained suggesting that the ion repressed rather than inhibited antibiotic formation (data not shown).

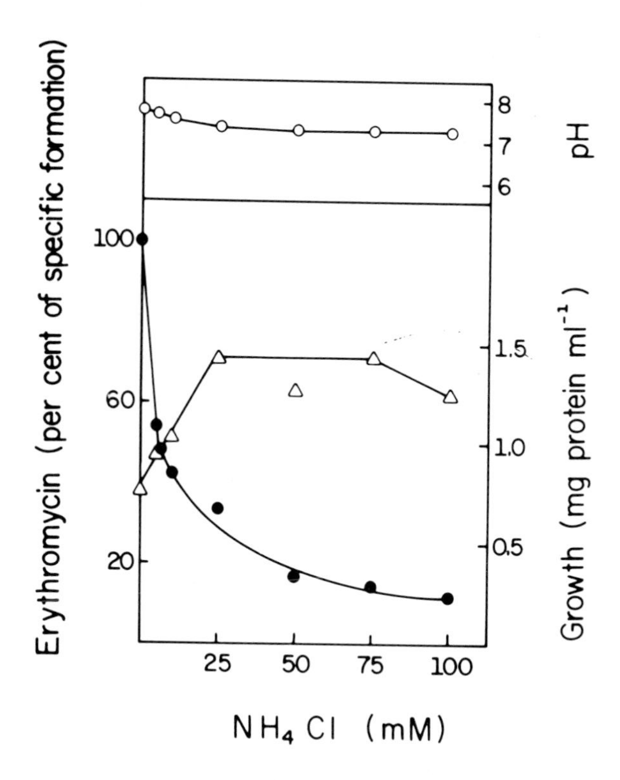

FIGURE 4. Effect of different ammonium concentrations on the maximum growth (△), specific erythromycin formation (●), and final pH of the medium (O).

TABLE I. Effect of Different Ammonium Salts on Erythromycin Formation

Condition[a]	*Specific erythromycin formation (μg/mg protein)*
Control	*9.37*
$(NH_4)_2SO_4$	*0.75*
NH_4Cl	*4.75*
NH_4NO_3	*3.20*

[a] *Cultures were grown in a complex medium either supplemented or not with different ammonium salts (50 mM each) on a rotary shaker (160 rev. min^{-1}) at 29°C as previously reported (Escalante* et al., *1982).*

B. Effect of Amino Acids

Considering that some amino acids are required for the synthesis of the methyl and amino groups of the sugar desosamine and that some other amino acids have the ability to produce the propionate and methylmalonate units of erythromycin, an important regulatory relationship may be expected to exist between the metabolism of these amino acids and antibiotic formation. However, in the example under study, these aspects have not been extensively explored and only a few observations, merely describing their nutritional properties in relation to antibiotic production, have been reported (Stark and Smith, 1961).

A high intracellular concentration of glutamate and alanine preceding antibiotic formation was reported by Roszkowski *et al.* (1969). In addition, they found that *S. erythreus* when fed with alanine results in better growth and erythromycin formation than when fed with other aminoacids, including glutamate. Furthermore, alanine utilization has been found to be associated with a high activity of alanine dehydrogenase (L-alanine: NAD oxidoreductase EC 1.4.1.1.). However, the regulatory importance of these observations, although quite suggestive, has not been clarified.

IV. REGULATION BY ERYTHROMYCIN

A. Feedback Regulation

Several authors have suggested that erythromycin concentration may be subject to feedback regulation. Spizek *et al.* (1965) reported that erythronolide B inhibits its own biosynthesis when added to the fermentation medium. Nevertheless, the metabolic relevance of this mechanism is not completely understood since the intracellular accumulation of erythronolide B must be preceded by an additional regulatory mechanism to prevent its further metabolism. In addition, the same authors have demonstrated that this intermediate is efficiently converted into erythromycin A (Spizek *et al.*, 1965).

Experiments *in vitro* with a partially purified sample of transmethylase (S-adenosyl methionine: erythromycin C O-methyl transferase), the last step in the antibiotic formation pathway, have shown that the enzyme is inhibited by the products of the activity, *i. e.*, erythromycin A and S-adenosyl L-homocysteine inhibited 43 and 76%, respectively (Corcoran, 1975). However, location of the transmethylase turned out to be of great importance since most of the antibiotic is excreted into the culture medium.

B. Activation

That erythromycin is able to stimulate its own formation was initially observed by Smith *et al.* (1962). As these authors pointed out, addition of a concentrated erythromycin solution to the fermentation medium enhances the biosynthetic rate of antibiotic production by 50%. This effect was observed with addition on the fourth day of fermentation; however, when added on the second day, no stimulation was observed.

Looking for a feedback regulatory action of erythromycin on its own

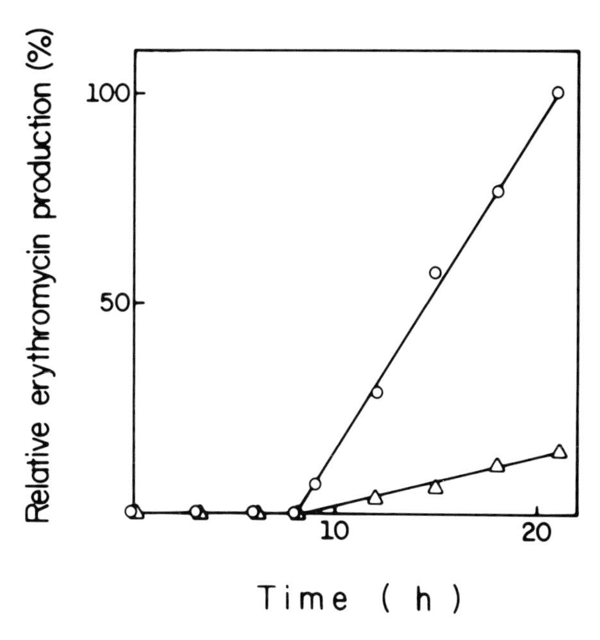

FIGURE 5. Effect of erythromycin on its own formation. Cells were grown in CM medium for 36 h and then transferred to a resting cell system containing 2.4 mM $MgSO_4 \cdot 7H_2O$, 20 mM K_2HPO_4 and 100 mM 3-(N-morpholino) propanesulfonic acid buffer (Sigma), pH 7.0, either with (O) or without (△) erythromycin (10 μg ml^{-1}).

biosynthesis, we obtained results which are in agreement with the stimulatory effect. In fact, concentrations of erythromycin, which caused no growth inhibition, stimulated erythromycin formation in a resting cell system of *S. erythreus* (Fig. 5). Further evidence of this action was obtained by growing cells in the presence of 2-[^{14}C]-sodium propionate and then measuring the radioactivity incorporated into the antibiotic produced (Smith *et al.*, 1962). Labelled erythromycin was first detected at 44 h incubation and its formation was increased fivefold in the presence of non-labelled erythromycin (Fig. 6). In addition, most of the labelled antibiotic was erythromycin A (Fig. 7), as visualized by thin layer chromatography (Martin *et al.*, 1975). To investigate this problem further, the effect of chloramphenicol (100 μg ml^{-1}) on erythromycin stimulation was studied. The addition of this antibiotic did not prevent the stimulatory action of erythromycin, indicating that erythromycin exerted an activating effect on its own formation (Fig. 8).

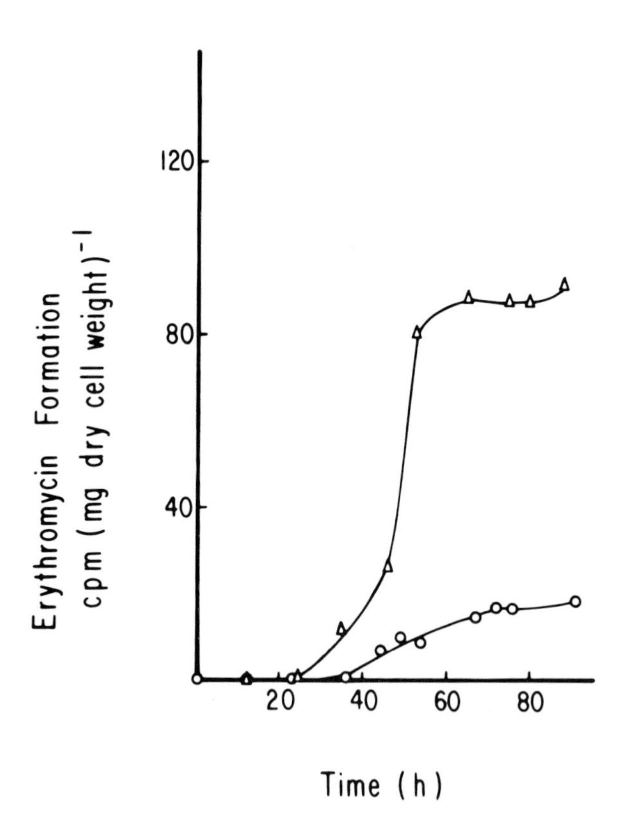

FIGURE 6. Effect of erythromycin on its own formation. Cells were grown in CM medium containing 0.8 μM 2-[^{14}C] sodium propionate (48.2 μCi μmol^{-1}) either with (△) or without (O) erythromycin (200 μg ml^{-1}).

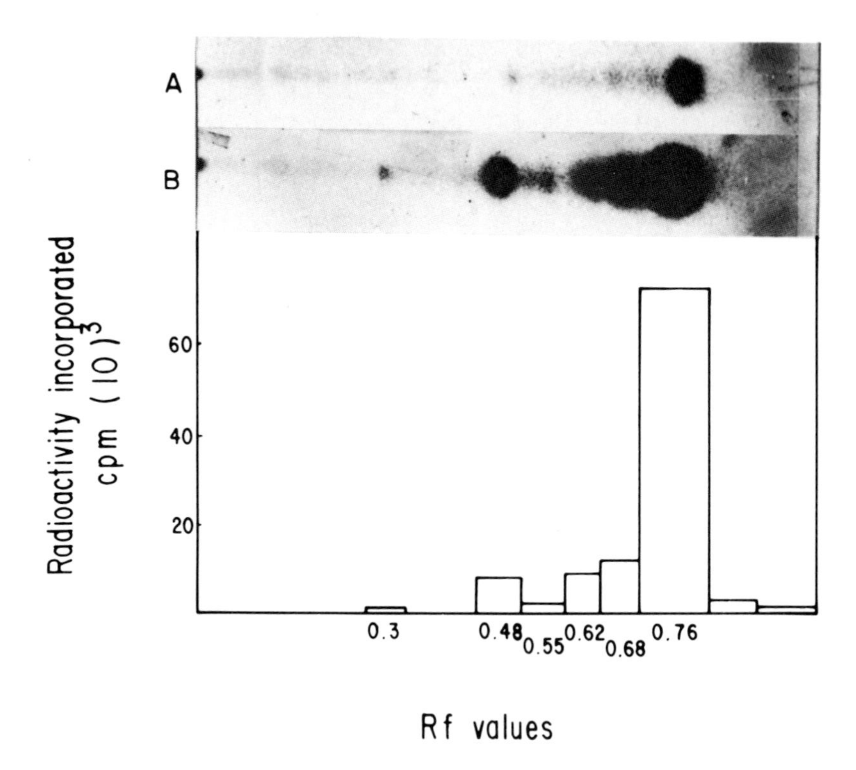

FIGURE 7. Thin layer chromatography (Martin and Goldstein, 1970) of the erythromycins produced by cells of S. erythreus *(plate B) and radioactivity incorporated into the antibiotics. Cells were grown in CM medium supplemented with erythromycin (50 μg ml^{-1}) and 0.8 μM 2-[^{14}C] sodium propionate (48.2 μCi μmol^{-1}). In the plate A, a standard of erythromycin A was chromatographed.*

V. CONCLUDING REMARKS

Erythromycin, first reported in 1952 by McGuire and associates, is a useful and extremely safe antibiotic with activity against a spectrum of microbes which include many of the infectious agents commonly encountered in clinical practice. It is a macrolide composed of a 14-membered lactone ring with neutral and amino-deoxy sugars bound to the ring through a glycoside linkage. It is synthesized by *S. erythreus* at the end of the trophophase from the lactone precursors (propionyl-CoA and methylmalonyl-CoA) with subsequent attachment of the sugars.

Sophisticated large-scale fermentation processes for the commercial production of erythromycin have been perfected. However, three decades after the discovery of erythromycin, the underlying biosynthetic reactions and the regulatory mechanisms controlling its production have not been fully elucidated.

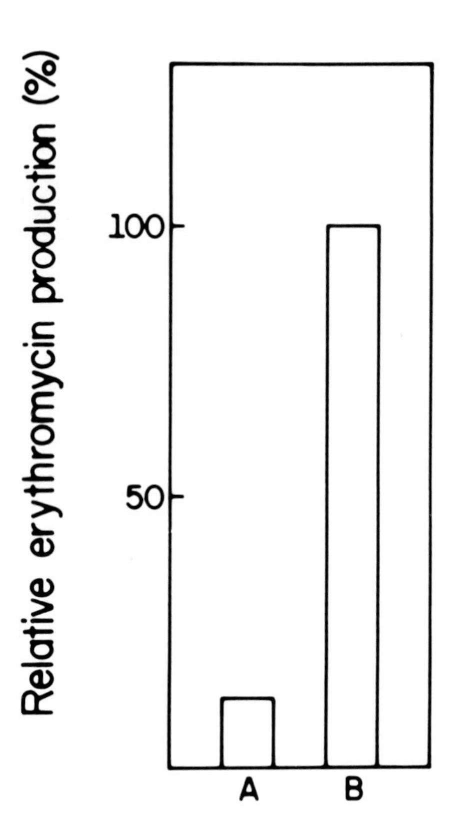

FIGURE 8. Effect of chloramphenicol on erythromycin stimulation. Cells were grown in CM medium and then transferred to a resting cell system containing cloramphenicol (100 μg ml^{-1}) either in the presence (column B) or not (column A) of erythromycin (10 μg ml^{-1}).

Several mechanisms appear to be involved in controlling erythromycin formation. Of these, transient glucose repression, ammonium regulation, and induction of propionyl CoA-carboxylase seem to play important roles in the onset of formation of the antibiotic. On the other hand, feedback repression may have a role in the cessation of erythromycin biosynthesis. In addition, it is known that erythromycin is able to activate its own formation; however, the regulatory significance of this action is not completely understood.

The mechanisms which control erythromycin formation are only beginning to be understood. A better understanding would have great economic impact since the yield of antibiotic may then be rationally increased.

REFERENCES

Aharonowitz, Y., and Demain, A. L., *Antimicrob. Agents Chemother. 14:*159 (1978).
Aharonowitz, Y., and Demain, A. L., *Can. J. Microbiol. 25:*61 (1979).
Brinberg, S. L., *Antibiotiki 4:*15 (1959).
Bu'Lock, J. D., *in* "The Filamentous Fungi" (J. E. Smith and D. R. Berry, eds.), Vol. 1, p. 33. John Wiley and Sons Inc., New York (1975).
Corcoran, J. W., *in* "Genetics of Industrial Microorganisms" (Z. Vánek, Z. Hostálek, and J. Cudlin, eds.), p. 339. Streptomycetes Academia, Prague (1973).
Corcoran, J. W., *Methods Enzymol. 43:*487 (1975).
Escalante, L., López, H., Mateos, R. C., Lara, F., and Sanchez, S., *J. Gen. Microbiol. 128:*2011 (1982).
Ettler, P., Páca, J., and Cechner, V., *in "Proceedings of the Fifth International Fermentation Symposium",* p. 46. Berlín (1976).
Friedman, S. M., Kaneda, T., and Corcoran, J. W., *J. Biol. Chem. 239:*2386 (1964).
Gräfe, U., Bocker, H., and Thrum, H., *Z. Allg. Mikrobiol. 17:* 201 (1977).
Grisebach, H., "Biosynthetic Patterns in Microorganisms and Higher Plants", p. 32. Wiley, New York (1967).
Grisebach, H., Achenbach, H., and Hofheins, W., *Tetrahedron Lett. 234:*274 (1961).
Hu, W. S., and Demain, A. L., *Process Biochem. 14:*2 (1979).
Lara, F., Mateos R. C., Vázquez, G., and Sánchez, S., *Biochem. Biophys. Res. Commun. 105:*172 (1982).
Martín, J. F., and Demain A. L., *Microbiol. Rev. 44:*230 (1980).
Martin, J. R., and Egan, R. S., *Biochemistry 9:*3439 (1970).
Martin, J. R., Egan, r. S., Goldstein A. W., and Collum. P., *Tetrahedron 31: 1985 (1975).*
McGuire, J. M., Bunch, R. L., Anderson, R. C., Boaz, H. E., Flynn, E. H., Powell, H. M., and Smith J. W., *Antibiot. Chemother. 2:*281 (1952).
Oleinick, N. L., *in* "Antibiotics" (J. W. Corcoran and F. E. Hahn, eds.), Vol. 3, p. 396. Springer-Verlag, New York (19757.
Ostrowska-Krysiak, B., *Postepy Hig. Med. Dosw. 28:*515 (1974).
Raczynska-Bojanowska, K., Rafalski, A., and Ostrowska-Krysiak, B., *Acta Biochim. Pol. 17:*332 (1970a).
Raczynska-Bojanowska, K., Ruczaj, Z., Ostrowska-Krysiak, B., Roszkowski, J., Gaworowska-Michalik, J., and Sawnor-Korszynska, D., *Acta Microbiol. Pol. 2(19):*103 (1970b).
Roszkowski, J., Rafalski, A., and Raczynska-Bojanowska, K., *Acta Microbiol. Pol. 1(18):*59 (1969).
Sánchez, S., Paniagua, L., Mateos, R. C., Lara, F., and Mora, J., *in* "Advances in Biotechnology" (M. Moo-Young and C. Vezina, eds.), Vol. 3, p. 147. Pergamon Press, Toronto (1981).
Smith, R. L., Bungay, H. R., and Pittenger, R. C., *Appl. Microbiol. 10:*293 (1962).
Spizek, J., Chick, M., and Corcoran, J. W., *Antimicrob. Agents Chemother. 5:*138 (1965).
Stark, W. M., and Smith, R. L., *Prog. Ind. Microbiol. 3:*210 (1961).
Vánek, Z., Puza, M., Majer, J., and Dóleziolová, L., *Folia Microbiol. 6:*408 (1961).

FEEDBACK REGULATORY STEPS INVOLVED IN DIAMINOPIMELIC ACID AND LYSINE BIOSYNTHESIS IN *STREPTOMYCES CLAVULIGERUS*[1]

S. Mendelovitz
F. T. Kirnberg
N. M. Magal
Y. Aharonowitz

Tel Aviv University
George S. Wise Faculty of Life Sciences
Department of Microbiology
Tel Aviv, Israel

I. INTRODUCTION

Streptomyces clavuligerus was first described by Higgens and Kastner (1971). This strain produces several β-lactam antibiotics (Howarth *et al.*, 1976; Nagarajan, 1972) among which penicillin N, clavulanic acid, and cephamycin C are major products. The penicillin N and cephalosporin type of antibiotics are produced by both fungi and streptomycetes and possess a common β-lactam structure built up mainly from three primary metabolites, L-cysteine, L-valine, and L-α-aminoadipic acid (Demain, 1974). Whitney *et al.* (1972) showed that the α-aminoadipic acid moiety of the cephalosporins produced by *S. clavuligerus* is derived from L-lysine. However, in fungi such as *Penicillum, Cephalosporium*, and *Paecilomyces*, α-aminoadipic acid is an intermediate in the lysine biosynthetic pathway rather than a catabolic product (Umbarger, 1978). Physiological studies on the control of β-lactam biosynthesis have shown that lysine causes a reduction in antibiotic titers when added, for exam-

[1] *This research was supported by a grant from the United States — Israel Binational Science Foundation (BSF), Jerusalem, Israel.*

ISBN 0-12-528620-1

ple, to the fermentation medium of a penicillin-producing culture. This phenomenon has been attributed to feedback regulatory effects of lysine on homocitrate synthetase, the first step in the fungal lysine biosynthetic pathway (Drew and Demain, 1977; Luengo *et al.*, 1980). Thus, by inhibition and/or repression of homocitrate synthetase, lysine reduces the availability of its own precursor, α-aminoadipic acid, which is required for penicillin production. In contrast, when amino acids of the aspartate family are added to a chemically defined medium of *S. clavuligerus*, L-lysine and DL-*meso*-diaminopimelic acid (DAP) clearly stimulates antibiotic production (Mendelovitz and Aharonowitz, 1982); other amino acids of the aspartate family have no such effect. These results suggest that the carbon flow from aspartate via diaminopimelate and lysine to α-aminoadipic acid in *S. clavuligerus* may be rate-limiting for antibiotic production. In order to obtain a better understanding of the regulatory role that lysine plays in the biosynthesis of the cephalosporin type of antibiotics in *S. clavuligerus*, we have begun to characterize the regulatory properties of the key enzymes of the aspartate family pathway of *S. clavuligerus* that are involved in lysine biosynthesis (Mendelovitz and Aharonowitz, 1982). Some properties of aspartokinase, dihydrodipicolinate synthetase, and homoserine dehydrogenase have been described. In the present report, we extend these studies to other enzymes of the lysine pathway, including diaminopimelate decarboxylase and diaminopimelate epimerase. In addition, we present results for a regulatory mutant which accumulated diaminopimelate in the intracellular free amino acid pool, as well as results for the distribution of the DAP isomers in the cell wall peptidoglycan and pools of free amino acids.

II. MATERIALS AND METHODS

Streptomyces clavuligerus NRRL 3585 was used throughout this study. Culture conditions for the biochemical studies were the same as described previously (Mendelovitz and Aharonowitz, 1982), except for those alterations described in the legends of Table II.

Aspartokinase (AK) activity was measured in a 30 to 65% saturated ammonium sulfate fraction by the formation of aspartylhydroxamate. Our assay (Mendelovitz and Aharonowitz, 1982) was a modification of that described by Stadtman *et al.* (1961). Dihydrodipicolinic acid synthetase (DDPS) activity was measured as described by Yugari and Gilvarg (1965). For further details, see Mendelovitz and Aharonowitz (1982). Homoserine dehydrogenase (HSD) was measured as described by Mendelovitz and Aharonowitz, (1982). Diaminopimelic acid decarboxylase (DAP-Dcase) was measured by the conversion of [1,7-^{14}C]-diaminopimelate to radioactive CO_2. The following components were added to a single sidearm manometer vessel with a center well to give a final volume of 1 ml: 200 mM Tris HCl, pH 8.0; 0.2 mM pyridoxal phosphate; 1 mM 2,3-dimercaptopropanol (BAL); 20 mM [^{14}C]-diaminopimelate; 5000 cpm/μmol of commercial mixed isomers; and appropriate amounts of enzyme, as required. The sealed flasks were then incubated for 1 h at 30°C after which 1 ml of 1 M H_2SO_4 was added. $^{14}CO_2$ was fixed on the folded strips of filter paper placed in the center well of the vessel to which 0.2 ml of dimethyl benzyl ammonium hydrochloride was added. After two hours of equilibration at 4°C, the filter paper strips were transferred to vials containing 3 ml of a toluene-based scintillation fluid. One unit of DAP-Dcase is defined as the amount of enzyme that will catalyze the formation of 1 μmole CO_2 under the assay conditions outlined above. Diaminopimelic acid epimerase activity was assayed under conditions similar to those reported by Misono and Soda (1980). Analysis of the reaction products was carried out by high performance liquid chromatography (HPLC) according to the method described by Tisdall and Anhalt (1979).

Preparations of peptidoglycan from *S. clavuligerus* were made according to Leyh-Bouille *et al.* (1970). The final fraction was hydrolyzed in 6N HCl overnight at 105°C, lyophilized, and assayed by HPLC.

Free amino acid pools were extracted according to a method described by Grandgenett and Stahly (1971). Standard volumes of amino acid extracts were applied to a Dowex AG-50W column of a Durrum-550 automatic amino acid analyzer along with a known amount of norleucine as the internal standard.

III. RESULTS

A. Feedback Regulatory Effects on Enzymes of the Aspartic Acid Pathway

When *S. clavuligerus* was grown in a chemically defined medium, aspartokinase activity in cell extracts was strongly feedback inhibited only when threonine and lysine were simultaneously added (Table I). This concerted feedback inhibition was very specific; combination of lysine with other amino acids of the aspartic acid family were not inhibitory. Methionine and isoleucine did not reverse the concerted feedback effect exerted by lysine plus threonine. DDPS activity in these cell extracts was partially feedback inhibited only by DL-*meso*-DAP or AAA (Table I). In order to achieve 50% inhibition of DDPS activity, 20 mM of DL-*meso*-DAP or about 40 mM of AAA was required. HSD activity was significantly inhibited by threonine and homoserine, whereas isoleucine, an end product of the homoserine branch, had no effect on HSD activity (Table I). DAP-Dcase was feedback inhibited by L-lysine (Table I). However, despite the relatively high concentrations of lysine used, only partial inhibition was obtained; thus, 50 mM of L-lysine was required in order to reduce the activity by 70%.

In order to determine whether any of the amino acids of the aspartic acid family or their intermediates have an effect on the biosynthesis of these four

TABLE I: Effect of Aspartic Acid Family of Amino Acids on the Activity of Their Biosynthetic Enzymes

Amino acid added (10 mM)	*Relative enzyme activity (%)*			
	AK [a]	*DDPS* [b]	*HSD* [c]	*DAP-Dcase* [d]
None	100	100	100	100
L-threonine	120	103	26	91
L-lysine	150	96	98	61
L-methionine	112	102	82	104
L-isoleucine	96	99	105	-
L-homoserine	105	81	48	100
DL-meso-*DAP*	84	72	100	-
α-aminoadipate	113	66	-	101
L-lysine + L-threonine	4	96	20	-
L-lysine + DL-meso-*DAP*	-	75	-	-
L-lysine + L-methionine	127	93	-	-
L-lysine + L-isoleucine	146	90	-	-
L-threonine + L-methionine	86	94	23	-
L-threonine + L-isoleucine	104	94	23	-
L-isoleucine + L-methionine	-	92	-	-

[a] *For AK activity the standard hydroxamate assay was used. The protein content in each assay was 0.94 mg. In the reaction with no added amino acid, 21.6 nmol of aspartohydroxamate per min was formed which was taken as 100% activity.*

[b] *DDPS activity was measured as previously described; 100% activity represents a specific activity of 0.063 U/mg protein.*

[c] *HSD activity was measured by determining the initial rate of decrease of absorbance at 340 nm; 100% activity represent a decrease of 0.05 A_{340} per min in a reaction mixture containing 0.64 mg protein.*

[d] *DAP-Dcase activity was measured by determining the release of radioactive CO_2 from a mixture of ^{14}C-labeled* DL-meso-*DAP as substrate; 100% activity is equivalent to the release of 0.3 μ*moles of CO_2 per h per mg protein.

enzymes, cells were grown separately in the presence of 10 mM of each amino acid and the specific enzyme activity in cell extracts was then measured. Table II shows that the presence of either methionine or isoleucine at this concentration during growth reduced aspartokinase activity by some 50 and 70%, respectively. Isoleucine had an inhibitory effect on growth which could be explained by such a decrease in a key biosynthetic enzyme. DDPS was not significantly repressed by any of the amino acid tested; only isoleucine and DL-*meso*-DAP showed some repressive effect (15 and 22%, respectively). Isoleucine also repressed the biosynthesis of HSD. This enzyme was only partially repressed by threonine and DL-*meso*-DAP. DAP-Dcase biosynthesis was not affected by the amino acids tested.

TABLE II: Effect of Aspartic Acid Family of Amino Acids on Enzyme Biosynthesis [a]

Amino acid added to culture medium (10 mM)	*Relative enzyme specific activity (%)*[b]			
	AK	*DDPS*	*HSD*	*DAP-Dcase*
None	*100*	*100*	*100*	*100*
L-threonine	*120*	*109*	*87*	-
L-lysine	*121*	*100*	*124*	*139*
L-methionine	*47*	*103*	*104*	-
L-isoleucine	*29*	*84*	*54*	-
L-homoserine	*88*	*100*	-	-
DL-meso-*DAP*	*95*	*77*	*84*	*92*
α-aminoadipate	*84*	*108*	*125*	*112*
L-lysine + DL-meso-*DAP*	*120*	*126*	*111*	-
L-lysine + L-methionine	-	*100*	-	-
L-lysine + L-isoleucine	*63*	*94*	-	-
L-threonine + L-methionine	-	*101*	*69*	-
L-threonine + L-isoleucine	-	*95*	*136*	-

[a]*Cultures were grown for 24 h in the chemically defined medium. Amino acids were added to the growth medium and cultures grown for an additional 24 h. Specific activities were determined for AK, DDPS, and HSD in an ammonium sulfate fraction of each cell extract; for DAP-Dcase the specific activity was measured after passing the crude extract through a Pharmacia G-10 column.*

B. Isolation of Mutants Impaired in Regulation of Their Aspartokinase Activity

The data presented in the previous section indicate that the carbon flow from aspartate to lysine is strictly controlled by a highly sensitive, concerted feedback inhibition mechanism acting on the first enzyme of the biosynthetic pathway, aspartokinase. The strong feedback inhibition of homoserine dehydrogenase by threonine and homoserine and the lack of significant inhibition of DDPS suggest the possibility that an impaired regulation of aspartokinase activity would cause an increase flux of carbon through the lysine specific pathway. To tests this hypothesis, *S. clavuligerus* was treated with several lysine analogs, among which only S-(2-aminoethyl)-L-cysteine (AEC), at concentrations higher than 1 mg/ml, was able to inhibit growth on solid media. This inhibition was reversed by the addition of lysine to the medium. Mutants resistant to AEC were isolated after mutagenesis with N-methyl-N'-nitro-N-nitrosoguanidine (NTG).

Cell-free extracts were made for a number of independently isolated AEC-resistant mutants grown on a chemically defined medium and aspartokinase activity was assayed. Assays were made with or without additions of threonine and lysine in order to observe any change in the concerted feedback mentioned earlier. Table III compares results for aspartokinase activi-

TABLE III: Feedback Inhibition of Aspartokinase in AEC-*resistant Mutants of* S. clavuligerus[a]

Strain	Relative (%) aspartokinase activity in presence of			
	no addition	L-lysine (8mM)	L-threonine (8mM)	L-lysine + L-threonine (0.4 mM)
wild type	100	173	115	35
aec*(R)11*	100	139	104	100
aec*(R)20*	100	151	87	29

[a]*For experimental details see legends to Tables I and II.*

ty for two AEC-resistant mutants with that of the wild type strain. Strain *aec*(R)20, although resistant to AEC, did not show a regulatory change in its aspartokinase activity; on the other hand, the resistant strain *aec*(R)11 possessed an impaired aspartokinase phenotype. To date, 70% of the AEC-resistant mutants that we have isolated appeared to have a similar impairment in their aspartokinase activity.

C. *Accumulation of DAP in the Intracellular Free Amino Acid Pool of* AEC-*Resistant Mutants*

Analysis of the amino acid pool was carried out to determine if the lack of feedback control of aspartokinase might increase the concentration of lysine or its biosynthetic intermediate, DAP, in the intracellular free amino acid pool. The data in Table IV show that both wild type and *aec*(R)11 strains possessed almost the same amount of lysine during 72 h of growth. However, surprisingly, cells of the mutant strain accumulated 12-fold more DAP compared with the maximum values obtained by the wild type strain. Thus, the DAP concentration in the free pool of the wild type represented, at most, 1.3% of the total amino acid concentration, whereas in the mutant strain, it was 10 to 14%.

D. *DAP Isomers in* S. clavuligerus

Since DAP has been reported to be a cell wall component of *S. clavuligerus* (Higgens and Kastner, 1971), an analysis was carried out to determine which of the two DAP isomers is present in the cell wall. The data presented in Figure 1B identify LL-DAP as the cell wall component in *S. clavuligerus*. HPLC analysis of the distribution of the DAP isomers in the free pool showed that both isomers, LL-DAP and *meso*-DAP, were present (Fig. 1C and D). Furthermore, the difference in DAP concentrations in the wild type and mutant strains,

TABLE IV. L-Lysine and Diaminopimelic Acid (DAP) Levels in the Free Amino Acid Pools of S. clavuligerus *Wild Type and AEC-Resistant Mutant Strains*

		Intracellular concentration μmoles/g DCW[a]				
Strain	*Amino acid*	*0 h*	*6 h*	*24 h*	*48 h*	*72 h*
wild type	L-*lysine*	7.80	5.90	10.20	4.60	3.60
aec(*R*)*11*	L-*lysine*	4.30	7.40	4.50	6.30	5.20
wild type	*DAP*	0.87	3.50	1.20	0.79	0.38
aec(*R*)*11*	*DAP*	33.30	43.30	18.1	13.40	11.50

[a]*DCW: dry cell weight*

reported in the previous section, was well documented in the analysis shown in Figure 1.

Incubation of *meso*-DAP and a preparation of the optical active isomers of DAP with *S. clavuligerus* cell extracts revealed an active epimerase which was able to convert *meso*-DAP to LL-DAP, and vice versa (Kirnberg and Aharonowitz, manuscript in preparation). This observation may explain the distribution of DAP isomers shown in Figure 1.

IV. DISCUSSION

In the experiments reported here, we examined the idea that genetically introduced alterations in the regulatory mechanism of key enzymes of the synthetic pathway of the aspartic acid family of *S. clavuligerus* could affect the accumulated amounts of end products or intermediates of the lysine pathway in the free amino acid pool.

Our results indicate that feedback regulatory mechanisms may operate at three distinct steps to control the flow of carbon from aspartic acid to lysine. Thus, aspartokinase, the first enzyme in the biosynthetic pathway, was regulated by a concerted feedback inhibition of threonine and lysine. Such regulation may affect the availability of aspartate-β-semialdehyde which is a common intermediate for both the lysine specific branch and the homoserine branch leading to threonine, isoleucine, and methionine. However, since threonine and homoserine could each feedback inhibit the homoserine dehydrogenase, whereas DDPS was not significantly feedback regulated by end products of the pathway, a preferential flow of carbon towards the lysine branch would probably occur upon excess of aspartate-β-semialdehyde formation. Diaminopimelate in *S. clavuligerus* occurs both as an intermediate in the biosynthesis of L-lysine and as a cell wall constituent in its LL form. This fact may explain the lack of strict regulation on the proximal specific steps in

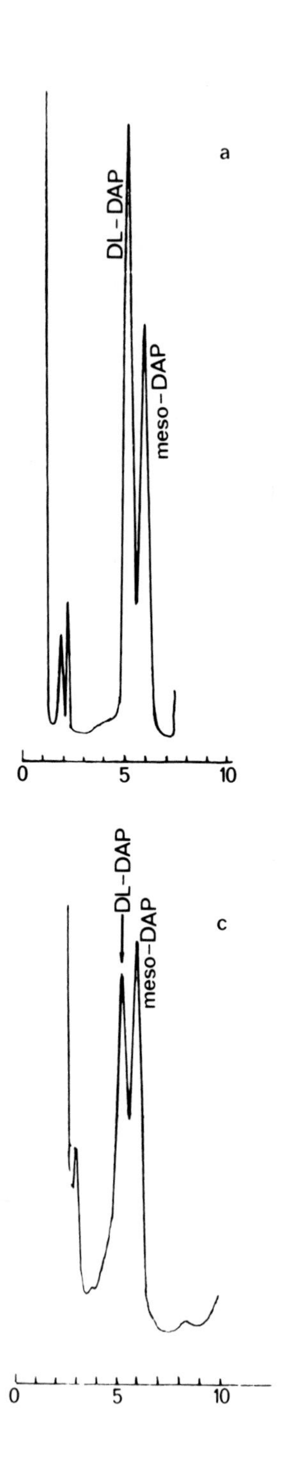
a
DL-DAP
meso-DAP
0
5
10
c
DL-DAP
meso-DAP
0
5
10

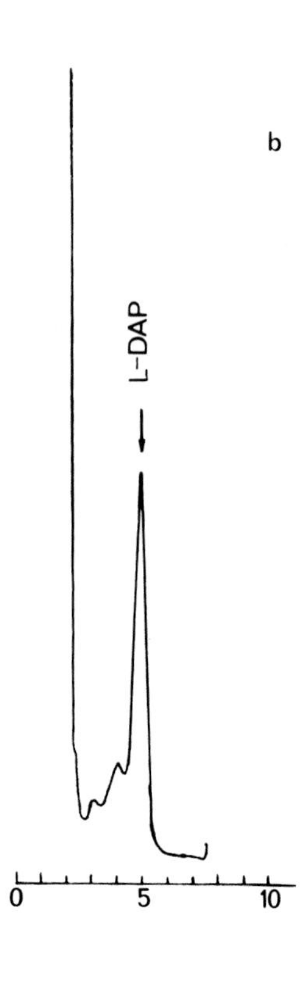
b
L-DAP
0
5
10

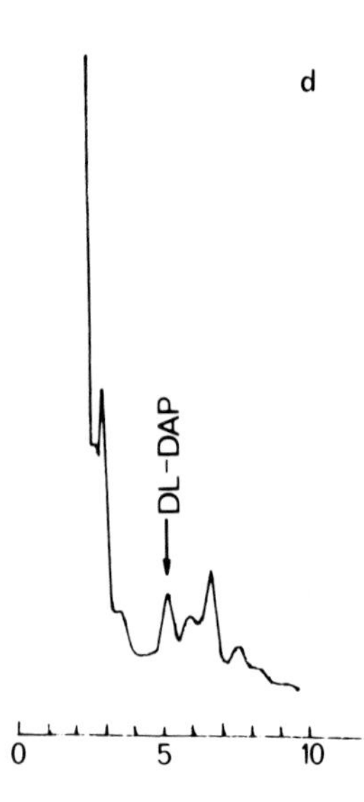
d
DL-DAP
0
5
10

the lysine branch. The data in Table I demonstrate that DAP-Dcase, which converts *meso*-DAP into L-lysine, was susceptible to feedback inhibition by L-lysine although relatively high concentrations of L-lysine were required. We propose, therefore, that feedback inhibition of lysine biosynthesis in *S. clavuligerus* occurs when (a) an excess of lysine and threonine concertedly inhibit the first common step in the pathway and (b) when only lysine is accumulated, in which case, DAP-Dcase may be affected and then, sequentially, DAP may exert some inhibition on its own biosynthesis by reducing DDPS activity.

Our studies indicate that feedback inhibition mechanisms operating in the aspartic acid pathway of *S. clavuligerus* may be less complicated than are those of other bacteria (Umbarger, 1978). Extracts of *Escherichia coli* and *Salmonella typhimurium* possess three aspartokinases and two homoserine dehydrogenases which are subject to feedback controls by single and different combinations of the amino acids of the aspartic acid family. In this respect, the situation in *S. clavuligerus* resembles that in *Brevibacterium flavum* (Shiio, 1973) which was found to be lacking isoenzymes for these two activities. In addition to the observation that inhibition profiles of aspartokinase did not change in extracts of culture of different ages and degrees of purification (Mendelovitz and Aharonowitz, 1982), the fact that lysine and threonine did not separately inhibit *S. clavuligerus* aspartokinase suggests that this activity is not of the isoenzymatic type.

It was expected that toxic analogs of lysine could be used to select for resistant strains, some of which would possess an alteration in the feedback inhibited enzyme aspartokinase. Indeed, Table III shows that such mutants were obtained and many were found to have an aspartokinase activity that was no longer inhibited by lysine plus threonine. Similar mutants were isolated from *B. flavum* (Sano and Shiio, 1970) and were used for lysine production.

Following the same operative approach exploited in *B. flavum*, we demonstrated in this work that the accumulated product of an AEC-resistant mutant of *S. clavuligerus* was diaminopimelic acid rather than lysine. The accumulation of DAP was probably due to feedback inhibition exerted on DAP-Dcase activity by L-lysine, although only at very high concentrations of lysine. When DAP was analyzed by HPLC, both the *meso* and LL isomers were found to have accumulated. That cell wall peptidoglycan contains LL-DAP and that a

FIGURE 1. Distribution of DAP isomers in cell wall hydrolyzate and free amino acid pools of S. clavuligerus. *a) A standard mixture of* DD, LL, *and* meso *isomers. (*DL*-DAP represents the mixture of the two optical isomers which were well separated from the* meso *isomers); b) A chromatogram of the cell wall acid hydrolyzate; c) A chromatogram of the free amino acid pool extracted from cells of mutant* aec*(R)11; and d) A chromatogram of the free amino acid pool extracted from cells of the wild type strain. The samples were run in a Waters Assoc. HPLC apparatus (Models 6000A, U6K, 440 and, RCM-100 were used). The column conditions were as follows: Radial-PAK C18 cartridge; mobile phase: H_2O:ACN::6:4 containing 6mM tetrabutylammonium phosphate, pH 6.5; pressure: 2000 psi; flow: 2 ml/min; detector: 254 nm; and sample size 5μl.*

DAP-epimerase activity is present in cell extracts of *S. clavuligerus* may explain the distribution of both isomers in the free amino acid pool.

We hope that the results represented will provide a basis for encouraging further work directed to the elucidation of other steps in the regulation of the carbon flow from primary substrates toward the final desired secondary metabolites.

ACKNOWLEDGMENTS

We wish to thank G. Cohen for critical reading of the manuscript, V. Kuper and S. Weisblum for technical assistance, and S. Linder for the typing of the manuscript.

REFERENCES

Demain, A. L., *Lloydia 32:*147 (1974).
Drew, S. W. and Demain, A. L., *Annu. Rev. Microbiol. 31:*343 (1977).
Grandgenett, D. P. and Stahly, D. P., *J. Bacteriol. 106:*551 (1971).
Higgens, C. E. and Kastner, R. E. *Int. J. Syst. Bacteriol. 21:*326 (1971).
Howarth, T. T., Brown, A., and King. T. J., *J. Chem. Soc. Chem. Commun. 1976:*266 (1976).
Leyh-Bouille, M., Bonaly, R., Ghuysen, J. M., Tinelli, R., and Tipper, D. J., Biochemistry *9:*2944

Luengo, J. M., Revilla, G., Lopez, M. J., Villanueva, J. R., and Martin, J. F., *J. Bacteriol. 144:*869 (1980).
Mendelovitz, S., and Aharonowitz Y., *Antimicrob. Agents Chemother. 21:*74 (1982).
Misono, H., and Soda, K., *Agric. Biol. Chem. 44:*2125 (1980).
Nagarajan, R., *in* "Cephalosporins and Penicillins: Chemistry and Biology" (E. H. Flynn, ed.), p. 636. Academic Press, New York (1972).
Sano, K., and Shiio, I., *J. Gen. Appl. Microbiol. 16:*373 (1970).
Shiio, I., *in* "Genetics of Industrial Microorganisms (Bacteria)" (Z. Vanek, Z. Hostalek, and J. Cudlin, eds.), p. 249. Elsevier Publication Co., Amsterdam (1973).
Stadtman, E. R., Cohen, G. N., Le-Bras, G., and de Robichon-Szulmajster, H., *J. Biol. Chem. 236:*2033 (1961).
Tisdall, P. A., and Anhalt, J. P., *J. Clin. Microbiol. 10:*503 (1979).
Umbarger, H. E., *Annu. Rev. Biochem. 47:*533 (1978).
Whitney, J. G., Brannon, D. R., Mabe, J. A., and Wicker, K. J., *Antimicrob. Agents Chemother. 1:*247 (1972).
Yugari, Y., and Gilvarg, C., *J. Biol. Chem. 240:*4710 (1965).

CONTROL OF AMMONIUM ION LEVEL IN ANTIBIOTIC FERMENTATION

Satoshi Omura
Yoshitake Tanaka

School of Pharmaceutical Sciences
Kitasato University and The Kitasato Institute
Minato-ku, Tokyo, Japan

I. INTRODUCTION

Antibiotics are biosynthesized from precursors related to primary metabolism. The synthesis is often subject to carbon, nitrogen, or phosphate regulation and is usually suppressed when glucose, ammonium ion, inorganic phosphate, or related substances are present in media at high concentrations (Drew and Demain, 1977). Nitrogen regulation was found in the production of several antibiotics (Table I). In most cases, however, the biochemical and physiological bases of such regulation are unknown.

Today, high level production of several antibiotics (penicillin G, for example, at 30 mg/ml or more) has been achieved by genetic and environmental alterations which release the producing organisms from regulation. Preferred genetic techniques are the traditional "random mutation and random selection" of improved mutants. Directed mutations are successful in a few cases (Chang and Elander, 1979). Accordingly, environmental techniques that are still valuable include the use of media containing slowly utilizable carbon and/or nitrogen sources and feeding the media with limited amounts of glucose, ammonium ion, or other suppressing nutrients by chemostat and fed-batch cultures (Gray and Bhuwapathanapun, 1980). In this paper, the authors present a new and simple culture method by which ammonium ion in the medium is controlled at a level which favors the high production of antibiotics.

Previously, the authors (Tanaka *et al.*, 1981a,c) reported the stimulation of microbial conversion of glycine to L-serine by magnesium phosphate. The

ISBN 0-12-528620-1

TABLE I. Nitrogen Catabolite Regulation in Antibiotic Biosynthesis

Antibiotic		*Producing organism*
Cephamycin	*β-Lactam*	Streptomyces clavuligerus
Penicillin N and Cephalosporin	,,	Cephalosporium acremonium
Thienamycin	,,	S. cattleya
Streptomycin	*Aminoglycoside*	S. griseus
Erythromycin	*Macrolide*	S. erythraeus
Oleandomycin	,,	S. antibioticus
Novobiocin	-	S. niveus
Candihexin	*Polyene*	S. viridoflavum
Patulin	-	Penicillium urticae[a]
Fusidic acid	*Steroid*	Fusarium coccineus
Nourseothricin	*Streptothricin*	S. noursei
Lincomycin	-	S. lincolnensis

[a] *Repression of* m-*hydroxylbenzyl alcohol dehydrogenase*

compound acts as an ammonium ion-trapping agent, thereby decreasing the ammonium level in the medium, and stimulates the enzymatic cleavage of glycine, an ammonium ion-generating reaction involved in L-serine biosynthesis from glycine. In view of the resemblance of the general technical requisites such as low carbon and ammonium levels for high antibiotic production and those for glycine cleavage which is stimulated by an ammonium ion-trapping agent, the authors were interested in the effect of magnesium phosphate and other ammonium ion-trapping agents on antibiotic production.

Several-fold increases in the production of macrolides, their biosynthetic intermediate, and other antibiotics were subsequently found when an ammonium ion-trapping agent was added to the medium. The results are discussed in relation to the nitrogen regulation of the biosynthesis of these antibiotics.

II. EFFECT OF MAGNESIUM PHOSPHATE ON LEUCOMYCIN PRODUCTION

Leucomycin, a 16-membered macrolide antibiotic which is active against penicillin-resistant bacteria and mycoplasma is produced by *Streptoverticillium kitasatoensis* (syn. *Streptomyces kitasatoensis*). Among the 16 components, leucomycins A_1 and A_5 (C-3 hydroxyl) are the most active, followed by the corresponding C-3 acetyl counterparts leucomycin A_3 and A_4. Kitao *et al.* (1979) reported the induction of C-3 hydroxylacylase synthesis by glucose and the inhibition of the induction by butyrate. An important application of this regulation is the selective production of the most active component (A_1 or

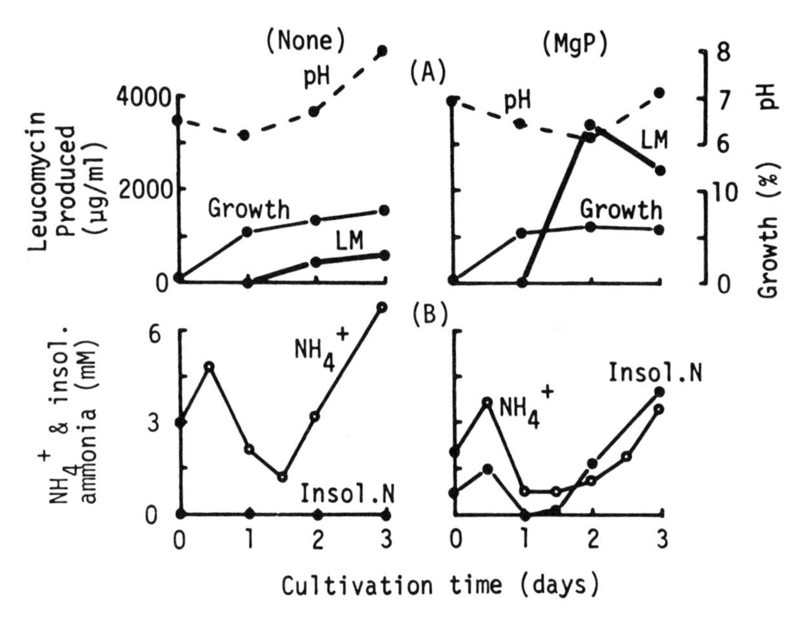

FIGURE 1. Leucomycin production by Streptoverticillium kitasatoensis *KA-429 in a complex medium in the presence or absence of magnesium phosphate (MgP). Medium: 2% glucose, 0.5% peptone, 0.3% dried yeast cells, 0.5% meat extract, 0.5%NaCl, 0.3% $CaCO_3$, and 0 or 1% magnesium phosphate (MgP: $Mg_3(PO_4)_2 \cdot 8H_2O$).*

A_5) in the presence of butyrate. On the other hand, a significant increase in the total amount of the leucomycins has been brought about by lowering the ammonium ion level in the media as follows (Omura *et al.*, 1980c,d). The addition of magnesium phosphate to a complex medium resulted in a five-fold increase in the leucomycin titer; the titer reached a maximum of 3800 μg/ml; whereas it was only 700 μg/ml in the control medium (Fig. 1). Ammonium ion remained at a low level, while active leucomycin production continued. In both media, leucomycin A_3 was the major product. The mycelial growth was almost unaffected, whereas the rise of pH value was delayed in the presence of magnesium phosphate.

In a chemically defined medium containing glycerol and ammonium lactate as carbon and nitrogen sources, respectively, the leucomycin titer similarly increased five- to eight-fold with concomitant decreases in the ammonium ion concentration and pH value and an increase (approximately two-fold) in mycelial growth. Separate pH control experiments showed that the enhancement of leucomycin titers by magnesium phosphate was not due to a pH effect.

In both complex and defined media, the decreases in ammonium ion concentrations were accompanied by increases in insoluble nitrogen content (Fig. 1). Observation by microscopy revealed the formation of insoluble materials that were different in size or shape from magnesium phosphate (Fig. 2). This in-

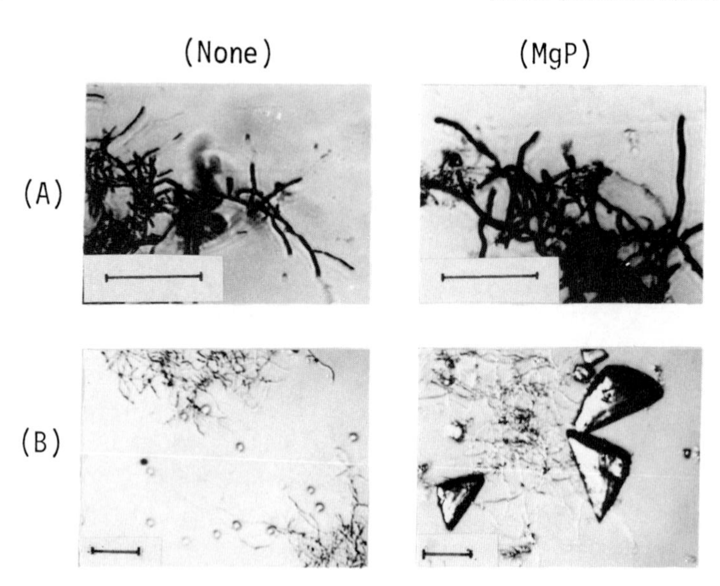

FIGURE 2. Micrographs of Stv. kitasatoensis *mycelia (A) and crystalline materials formed in the medium (B) in the presence or absence of MgP (bar: 30 μm.)*

dicated that magnesium phosphate trapped ammonium ion and transformed it into insoluble complexes thereby maintaining ammonium ion in the medium at a low level.

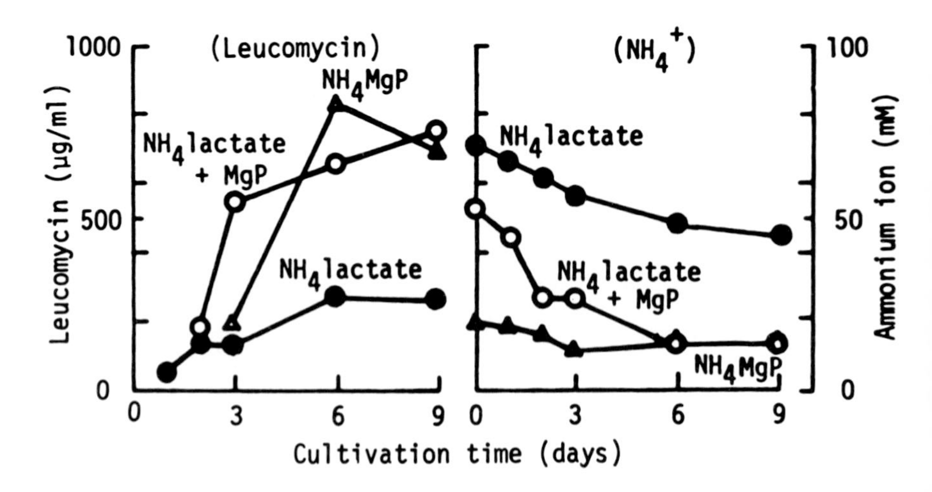

FIGURE 3. Effects of soluble and insoluble nitrogen sources on leucomycin production and ammonium ion level in defined media. Medium: 3% glycerol, 0.5% glucose, 1% nitrogen source as indicated, 0.05% K_2HPO_4, 0.05% $MgSO_4 \cdot 7H_2O$, 3 mg of trace metal salts (Fe, Mn, Co, Zn, and Cu) per liter, and 0 or 1% MgP as indicated, pH 7.5. NH_4MgP: ammonium magnesium phosphate ($NH_4MgPO_4 \cdot 6H_2O$).

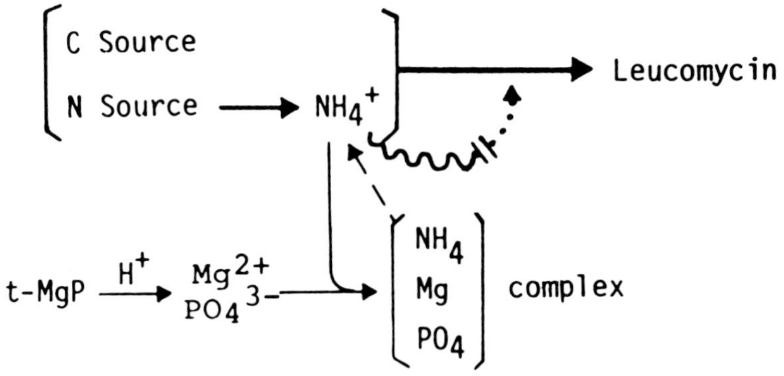

C Source
N Source → NH_4^+ → Leucomycin

t-MgP $\xrightarrow{H^+}$ Mg^{2+} PO_4^{3-} → NH_4 Mg PO_4 complex

FIGURE 4. Proposed function of MgP in stimulating leucomycin production by Stv. kitasatoensis *∿∿∿···> regulation and its relief; - - - -> slow supply of ammonium ion.*

Insoluble nitrogen sources such as sodium urate and ammonium magnesium phosphate supported higher leucomycin production and lower ammonium ion levels in the medium than did soluble nitrogen sources such as ammonium lactate (Fig. 3). Thus, a reverse correlation was demonstrated between the ammonium ion level and the leucomycin titer in a defined medium. A small amount (about 3 mM) of ammonium sulfate inhibited leucomycin production, indicating that leucomycin biosynthesis in *Stv. kitasatoensis* is regulated by nitrogen.

The authors conclude from these results that magnesium phosphate traps ammonium ion in the medium to create an ammonium ion-restrained condition which favors the efficient production of leucomycin, as illustrated in Figure 4.

Magnesium phosphate is nearly insoluble under neutral and alkaline conditions but is soluble under acidic conditions. In the defined medium described above, a substantial portion of magnesium phosphate was solubilized in six days thereby releasing magnesium and inorganic phosphate ions into the medium. Fortunately, leucomycin biosynthesis in *Stv. kitasatoensis* was much less susceptible to phosphate regulation. The amount of potassium phosphate required for 50% inhibition of leucomycin production was 125 mM (Tanaka *et al.*, 1981b), which is far lower than that required for phosphate-sensitive synthesis of antibiotics, *i. e.*, 1 to 10 mM (Weinberg, 1973).

III. ENHANCED PRODUCTION OF VARIOUS ANTIBIOTICS BY AMMONIUM ION-TRAPPING AGENTS

The usefulness of magnesium phosphate in leucomycin production is based on its ability to trap ammonium ion under physiological conditions and its weak toxic effect on microbial growth. There may be other such agents which promote antibiotic production. Several compounds that are presumably nontoxic to microorganisms were tested for their ability to react with ammonium ion under neutral conditions to yield an insoluble complex. As expected, magnesium phosphate and natural and synthetic zeolites, known to be ammonium ion-trapping agents, formed insoluble complexes when their acidic solutions were neutralized with ammonium hydroxide but did not when neutralized with sodium hydroxide. After calcium phosphate, sodium phosphotungstate, and sodium urate had been selected in the same way, antibiotic production was carried out in media supplemented with these substances. The maximum titers produced in the presence or absence of a trapping agent are summarized in Table II.

The production of various classes of antibiotics such as macrolides, aminoglycosides, β-lactams, and others was promoted by a factor of two to seven by one or more of the selected ammonium ion-trapping agents. In a set of fermentations in which magnesium phosphate or the natural zeolite had been used, production of about one-third of the antibiotics tested was enhanced more than two-fold; one-third was not affected, and one-third was suppressed by more than 50%. Much, if not all, of the enhancement may result from

TABLE II. Enhancement of Production of Antibiotics by Ammonium Ion-trapping Agent

Organism	*Antibiotic*	*NH_4^+-trapping agent*	*Antibiotic titers (μg/ml)*	
			No addition	*Addition*
Streptoverticillium kitasatoensis	*Leucomycin*	*Magnesium phosphate (MgP)*	700.0	3800.0
Streptomyces ambofaciens	*Spiramycin*	*Sodium phosphotungstate*	150.0	450.0
S. fradiae	*Tylosin*	*Natural zeolite (ZL)*	59.0	149.0
S. clavuligerus	*Cephamycin*	*ZL*	20.1[a]	22.5[a]
S. cattleya	*Thienamycin*	*ZL*	27.4[a]	29.5[a]
S. *sp.*	*Dihydrostreptomycin*	*MgP*	55.0	145.0
S. *sp.*	*Chloramphenicol*	*MgP*	7.0	32.0
S. aureofaciens	*Tetracycline*	*MgP*	28.0	45.0
S. rosa *subsp.* notoensis	*Nanaomycin*	*NH_4^+-saturated ZL*	85.0	750.0
Cephalosporium caerulens	*Cerulenin*	*ZL*	40.0	280.0

[a] *Diameter (mm) of inhibition zone.*

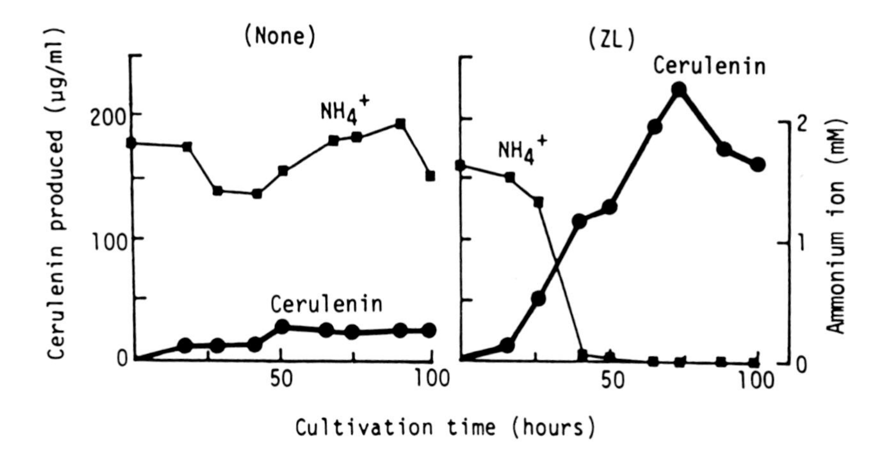

FIGURE 5. Effects of a natural zeolite on cerulenin production by Cephalosporium caerulens *KF-140 and ammonium ion concentration in the medium. Medium: 3% glycerol, 1% glucose, 0.5% peptone, 0.2% NaCl, and 0 or 1% a natural zeolite (ZL) (Fusseki Kako, Tokyo).*

release from regulation by ammonium ion. However, secondary effects such as growth and pH control may contribute to some extent. The results obviously suggest that nitrogen regulation affects antibiotic biosynthesis more significantly than was expected. Two examples presented below will explain this in more detail.

A. Cerulenin Production by Cephalosporium caerulens

Cerulenin is an antifungal antibiotic produced by the fungus *C. caerulens*. It specifically inhibits the condensing enzyme involved in polyketide and fatty acid biosyntheses (Omura, 1981). The production of cerulenin was inhibited by magnesium phosphate and calcium phosphate but was enhanced about five-fold by a natural zeolite. The enhancement was accompanied by a decrease in ammonium ion in the medium and with an increase (about two-fold) in the mycelial growth, as shown in Figure 5. The maximum titer reached 250 μg/ml in a 30 liter fermentor jar. Heating the zeolite or washing it with water increased the extent of the enhancement but washing it with an ammonium sulfate solution abolished the enhancing effect of zeolite (Masuma *et al.*, 1982).

Cerulenin production was inhibited by a small amount of an ammonium salt which had been added initially (Fig. 6). The inhibition was reversed at a later period of culture, presumably because the ammonium ion was consumed thus producing a low ammonium level not inhibitory to cerulenin biosynthesis.

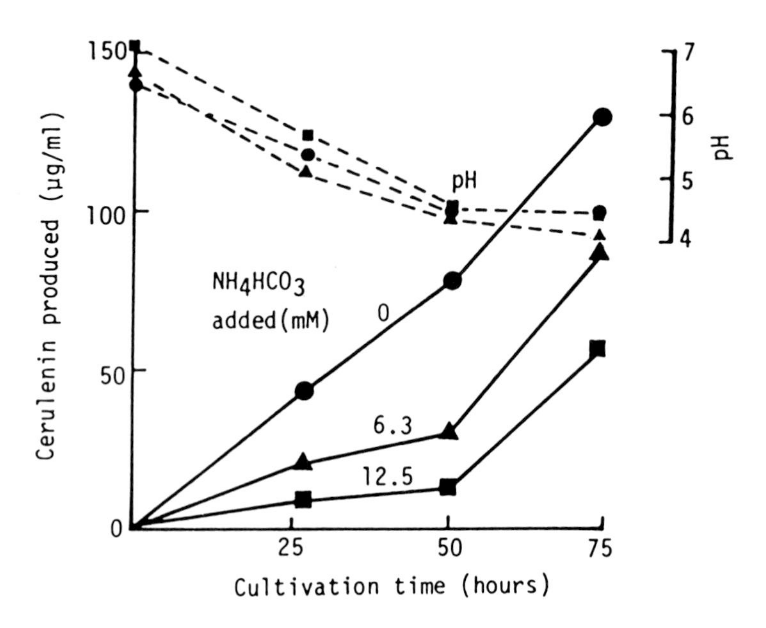

FIGURE 6. Effect of ammonium bicarbonate on cerulenin production by C. caerulens.

B. Nanaomycin Production by Streptomyces rosa *subsp.* notoensis

Nanaomycin, a quinone antibiotic active against fungal dermatophytes and mycoplasma, was discovered by Omura *et al.*. (1974). (An application is currently on file for approval of nanaomycin A as a veterinary drug.) The production of nanaomycin was studied in three different media A, B, and C supplemented with various ammonium ion-trapping agents. As shown in Table III, production was promoted by magnesium phosphate, calcium phosphate, and sodium urate in medium A. In medium B, improved for higher nanaomycin titer, nanaomycin production was inhibited by magnesium phosphate but was promoted by sodium urate. Also, in medium C which is a further improved, nanaomycin production was enhanced by sodium urate and reached a maximum of 750 μg/ml in three days. In the production of nanaomycin as well as other antibiotics, the ammonium ion-trapping agent to be added depended on the medium employed. Since it is difficult at present to predict the most effective combination of trapping agent and medium, selection must be made by trial-and-error. However, once a suitable trapping agent is selected, the effect of the trapping agent is comparable to, or rather more significant than, that of the medium.

Nanaomycin production appears to be unique in that it is promoted by amounts of ammonium ion up to about 10 mM, quantities at which the production of leucomycin (Omura *et al.*, 1980d) or of cephamycin (Aharonowitz and

TABLE III. Effects of Magnesium Phosphate (MgP), Calcium Phosphate (CaP), Sodium Urate (NaU), and a Natural Zeolite (ZL) on Nanaomycin Production and Ammonium Ion Concentration

Basal medium	NH_4^+ trapping agent	Amount added (%)	pH at peak nanaomycin titer	Peak nanaomycin E produced (μg/ml)	NH_4^+ (mM) at peak nanaomycin titer
A[a]	None		6.2	38	1.0
	MgP	1.0	6.8	170	1.1
	CaP	1.0	6.8	63	ND[d]
	NaU	0.5	6.8	100	9.5
	ZL	1.0	6.2	30	0.8
B[b]	None	-	6.6	130	2.1
	MgP	1.0	6.2	<10	ND
	NaU	1.0	7.0	560	9.8
	ZL	1.0	6.4	90	0.8
C[c]	None		6.2	340	5.1
	MgP	1.0	6.2	<10	ND
	NaU	0.5	7.0	750	12.5

[a] *2% glycerol, 2% soybean meal, 0.3% NaCl; pH 7.0; 27°C; 3 days.*
[b] *2% glucose, 1% meat extract, 0.5% NaCl, 0.3% $CaCO_3$; pH 6.0; 37°C; 36 hours.*
[c] *2% glycerol, 1% Bacto-soytone, 0.3% NaCl; pH 5.5; 27°C; 3 days.*
[d] *Not done.*

Demain, 1979) is inhibited. As shown in Table III, the improvement of medium A for higher nanaomycin production resulted in the selection of media B in which ammonium ion levels were higher and C with still higher levels. The addition of sodium urate caused further increases in the concentration of the ion, which reached nearly 10 mM. It is thus assumed that the enhancement of nanaomycin production by sodium urate resulted from the increase in ammonium ion levels in the media.

On the other hand, magnesium phosphate did not affect the ammonium level in medium A although it promoted nanaomycin production. The mechanism of this effect is now under study with reference to phosphate regulation of nanaomycin biosynthesis in *S. rosa* subsp. *notoensis*.

Various compounds such as ammonium sulfate, urea, and sodium urate served as ammonium donors and promoted nanaomycin production. Sodium or potassium nitrate were poor ammonium donors for this process. However, zeolite was found to be a good ammonium carrier. An ammonium sulfate-washed preparation of a natural zeolite (NH_4-zeolite) was prepared, and added to medium B as an ammonium donor. A small amount (2 mg/ml) of NH_4-zeolite yielded ammonium ion at about 10 mM, and the highest nanaomycin

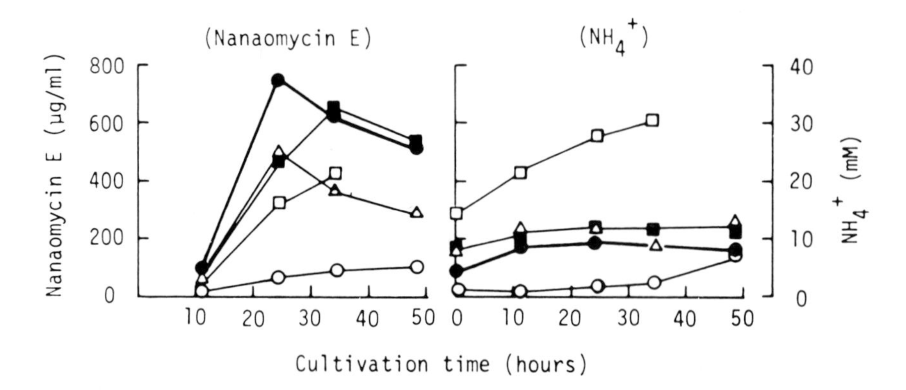

FIGURE 7. Effects of ammonium donors on nanaomycin production by Streptomyces rosa *subsp.* notoensis *OS-3966 and ammonium ion level in the medium. To medium B (see Table III), were added the following ammonium donors: NH_4-zeolite, 2 mg/ml (●); 5 mg/ml (■); $(NH_4)_2SO_4$, 0.5 mg/ml (△);* urea, 2 mg/ml (□); and none (O).

titer (750 μg/ml in 24 hours) (Fig. 7). Notably, the extent of enhancement was less than that when 0.5 mg/ml of ammonium sulfate was used to give nearly the same level of ammonium ion.

IV. MECHANISM OF THE NITROGEN REGULATION OF TYLOSIN BIOSYNTHESIS IN *S. FRADIAE*

Biosynthesis of tylosin is *S. fradiae* has been reported by Seno *et al.* (1977) and Omura *et al.* (1977, 1978, 1980a, 1982). Baltz and Seno (1982) studied the genetics of this process. The aglycone moiety of tylosin is derived from two acetates, five propionates, and one butyrate (Omura *et al.*, 1977). The precursors are condensed and then subjected to oxidation and reduction to give a lactonic intermediate, protylonolide. To the protylonolide, three sugars, mycaminose, mycinose, and mycarose or their equivalents, are then attach in that order; this is accompanied by oxidation and methylation of the glycosidic intermediates, leading to the final antibiotic product, tylosin (Omura *et al.*, 1982).

We isolated an *S. fradiae* mutant, strain 261, which produced protylonolide (Omura *et al.*, 1980a,b). Protylonolide was converted to tylosin by a second mutant, strain NP-10, by using the cosynthesis technique (Fig. 8), or by the wild-type tylosin producer strain KA-427, with the aid of cerulenin, an inhibitor of polyketide biosynthesis.

Biosynthetic regulation was also reported by Mardry *et al.* (1979) and Vu-

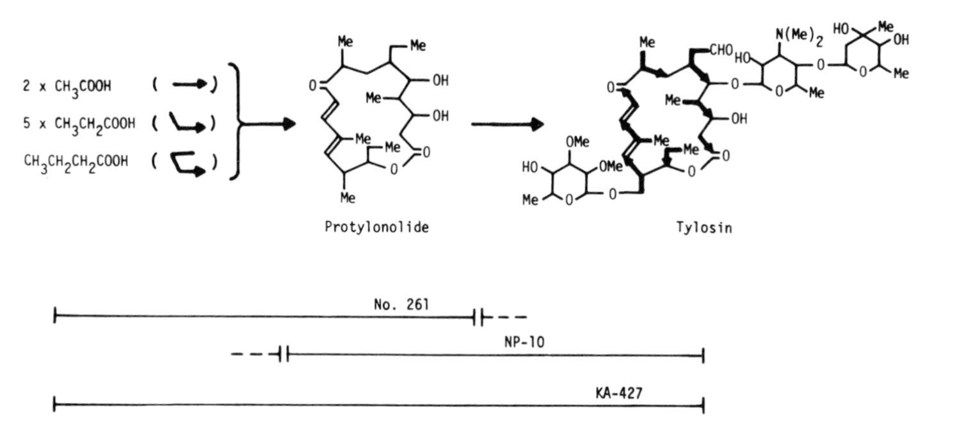

FIGURE 8. Biosynthesis of tylosin in Streptomyces fradiae *KA-427, and co-synthesis of tylosin by the mutant strains, No. 261 and NP-10.*

Trong *et al.* (1980, 1981) who indicated that tylosin biosynthesis is subject to regulation by carbon and phosphate. We found that tylosin production is enhanced three-fold by a natural zeolite in *S. fradiae* with a concomitant decrease in ammonium ion (Fig. 9). Based on this finding, we studied the mechanism of the nitrogen regulation of tylosin biosynthesis by using the three strains, KA-427, 261, and NP-10. One complex medium and one defined medium were used for the production of tylosin and protylonolide as well as for the conversion of protylonolide to tylosin by these three strains.

In a complex medium, a natural zeolite promoted protylonolide production by strain 261. This contrasted to a minimal change in the conversion rate of protylonolide to tylosin by strain NP-10 under similar conditions. The addition of ammonium chloride or methylamine, an unmetabolizable ammonium analog, resulted in considerable decreases in the titers of protylonolide. These results lead one to speculate that the nitrogen regulation of tylosin biosynthesis operates before protylonolide formation.

In a defined medium, when used as sole nitrogen sources or additions, L-alanine, L-asparagine, L-valine, and some other amino acids considerably stimulated protylonolide formation, suggesting that these amino acids are closely related to the precursors of protylonolide biosynthesis. It is probable that these amino acids are subjected to catabolic reactions to give acetate, propionate, and butyrate which serve as direct precursors. Ammonium ion is generated in the catabolic steps.

Ammonium ion-generating reactions are suppressed by exogenous ammonium ions. According to Magasanik *et al.* (1974), the enzyme protein of glutamine synthetase acts as a positive regulator in the synthesis of enzymes

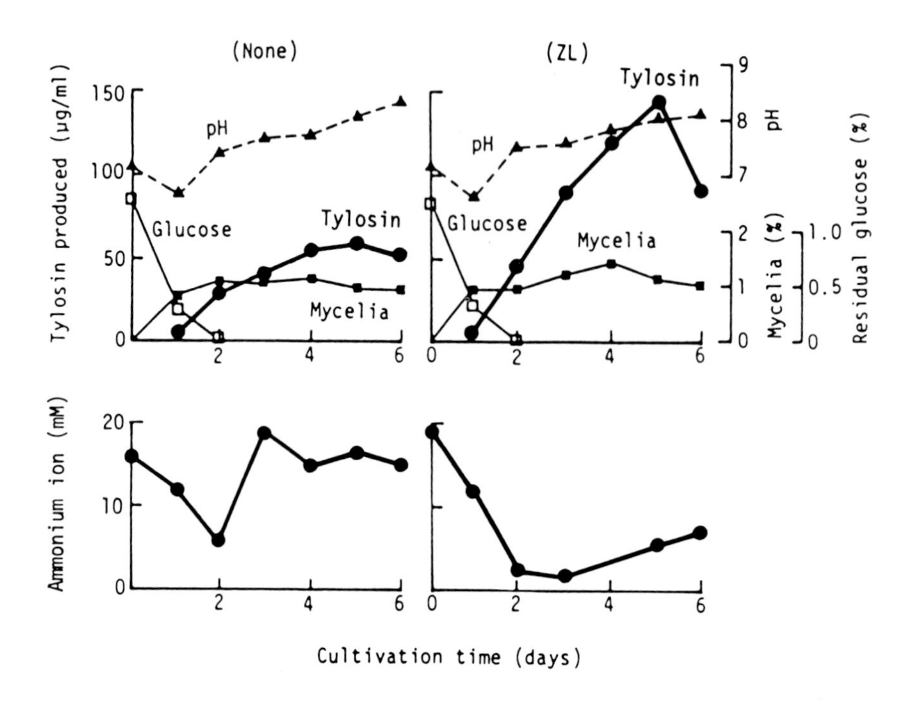

FIGURE 9. Time courses of tylosin production by S. fradiae *KA-427 (A) and ammonium ion concentration in the medium (B) in the presence or absence of a natural zeolite. Medium: 2% potato starch, 1% glucose, 0.5% peptone, 0.5% yeast extract, 0.4% $CaCO_3$, 0.3% L-asparagine, and 0 or 1% natural zeolite.*

responsible for the degradation of amino acids such as histidine, asparagine, and proline and of urea in enteric bacteria. A similar mechanism of nitrogen metabolism was proposed by Streicher and Tyler (1981) in *S. cattleya*, a thienamycin producer. The present authors (Tanaka *et al.*, 1981a) showed that, when magnesium phosphate was added to the medium, the synthesis of a glycine cleavage enzyme and glutamine synthetase remained high in another actinomycete, *Nocardia butanica*.

Based on the above information, it is reasonable to postulate that the catabolic steps of precursor amino acids leading to acetate, propionate, and butyrate are possible targets for nitrogen regulation of tylosin biosynthesis in *S. fradiae*.

V. CONCLUSION

Studies on ammonium ion-trapping agents for use in antibiotic fermentation are few. The usefulness of these agents in antibiotic production was recognized for the first time from our findings described in this paper. It was demonstrated that ammonium ion-trapping agents were useful not only for efficient production of many antibiotics but also for studying regulation mechanisms of antibiotic biosynthesis. Recently, the use of magnesium phosphate and magnesium salt in combination with phosphate salt was reported in the production of the antitumor antibiotics, gilvocarcin (Wei *et al.*, 1982) and tetrocarcin A (Tamaoki and Tomita, 1982).

Antibiotic fermentation media containing high concentrations of inorganic phosphate have been successfully used in combination with β-lactam-hypersensitive mutants in antibiotic screening studies, which led to the discovery of pyrrolnitrin, bicyclomycin, nocardicin, and FR-900148 (Imanaka, 1982). The antibiotic A-60 is produced only in the presence of high concentrations of magnesium sulfate (Ogata *et al.*, 1977). Likewise, magnesium phosphate and other ammonium ion-trapping agents may be of potential use as new tools for screening newer antibiotics (Iwai and Omura, 1982).

REFERENCES

Aharonowitz, Y., and Demain, A. L., *Can. J. Microbiol. 25:*61 (1979).
Baltz, R. H., and Seno, E. T., *Antimicrob. Agents Chemother. 20:*214 (1981).
Chang, L. T., and Elander, R. P., *Dev. Ind. Microbiol. 20:*367 (1979).
Drew, S. W., and Demain, A. L., *Annu. Rev. Microbiol. 31:*343 (1977).
Gray, P. P., and Bhuwapathanapun, S., *Biotechnol. Bioeng. 22:*1785 (1980).
Iwai, Y, and Omura, S., *J. Antibiot. 35:*123 (1982).
Imanaka, H., *J. Agric. Chem. Soc. Jpn. 56:*319 (1982). (in Japanese)
Kitao, C., Miyazawa, J. and Omura, S., *Agric. Biol. Chem. 43:*833 (1979).
Magasanik, B., Prival, M. J., Brenchley, J. E., Tyler, B. M., Deleo, A. B., Streicher, S. L., Bender, R. A., and Paris, C. G., *Curr. Top. Cell. Regul. 8:*119 (1974).
Mardry, N., Sprinkmeyer, R., and Pape, H., *Eur. J. Appl. Microbiol. Biotechnol. 7:365 (1979).*
Masuma, R., Tanaka, Y. and Omura, S., *J. Antibiot. 35:*1184 (1982).
Ogata, K., Osawa, H., and Tani, Y., *J. Ferment. Technol. 55:*285 (1977).
Omura, S., *in* "Methods in Enzymology", Vol. 72 (J. M. Lowenstein, ed.), p. 520. Academic Press, New York (1981).
Omura, S., Tanaka, H., Koyama, Y., Oiwa, R., Katagiri, M., Awaya, J., Nagai, T., and Hata, T., *J. Antibiot. 27:*363 (1974).
Omura, S., Takeshima, H., Nakagawa, A., Miyazawa, J., Piriou, F., and Lukacs, G., *Biochemistry 16:*2860 (1977).
Omura, S., Kitao, C., Miyazawa, J., Imai, H., and Takeshima, H., *J. Antibiot. 31:*254 (1978).
Omura, S., Kitao, C., and Matsubara, H., *Chem. Pharm. Bull. 29:*1963 (1980a).
Omura, S., Matsubara, H., Nakagawa, A., Furusaki, A., and Matsumoto, T., *J. Antibiot. 33:*915 (1980b).
Omura, S., Tanaka, Y., Tanaka, H., Takahashi, Y., and Iwai, Y., *J. Antibiot. 33:*1568 (1980c).
Omura, S., Tanaka, Y., Kitao, C., Tanaka, H., and Iwai, Y., *Antimicrob. Agents Chemother. 18:*691 (1980d).
Omura, S., Sadakane, N., and Matsubara, H., *Chem. Pharm. Bull. 30:*223 (1982).
Seno, E. T., Peter, R. L., and Hüber, F. M., *Antimicrob. Agents Chemother. 11:*455 (1977).
Streicher, S., and Tyler, B., *Proc. Nat. Acad. Sci. U.S.A. 78:*229 (1981).
Tamaoki, T., and Tomita, F., *Agric. Biol. Chem. 46:*1021 (1982).
Tanaka, Y., Omura, S., Araki, K., and Nakayama, K., *Agric. Biol. Chem. 45:*2661 (1981a).
Tanaka, Y., Takahashi, Y., Masuma, R., Iwai, Y., Tanaka, H., and Omura, S., *Agric. Biol. Chem. 45:*2475 (1981b).

Tanaka, Y., Tanaka, H., Omura, S., Araki, K., and Nakayama, K., *J. Ferment. Technol. 59:*447 (1981c).
Vu-Trong, K., Bhuwapathanapun, S., and Gray, P. P., *Antimicrob. Agents Chemother. 17:*519 (1980).
Vu-Trong, K., Bhuwapathanapun, S., and Gray, P. P., *Antimicrob. Agents Chemother. 19:*209 (1981).
Wei, T. T., Chain, J. A., Roller, P. P., Weiss, U., Stroshane, R. M., White, R. J., and Byrne, K. M., *J. Antibiot. 35:*529 (1982).
Weinberg, E. D., *Dev. Ind. Microbiol. 15:*70 (1973).

REGULATION OF MANNITOL DEHYDROGENASE AND PROTEASES IN *RHODOCOCCUS ERYTHROPOLIS*[1]

J. S. Bond
S. M. Sutherland
J. D. Shannon[2]
S. G. Bradley

Department of Biochemistry
Department of Microbiology and Immunology
Virginia Commonwealth University
Richmond, Virginia, U. S. A

I. INTRODUCTION

Genetic processes operative in the nocardioform bacteria, including members of the genus *Nocardia* and *Rhodococcus*, have been studied extensively since Adams and Bradley (1963) demonstrated gene transfer between two strains originally identified as *Nocardia erythropolis* and *Jensenia canicruria*. Subsequently Brownell, Adams, and Bradley (1967) described the actinophages ϕC and ϕEC which are relatively specific for species now know as *Rhodococcus erythropolis*. Crockett and Brownell (1972) were successful in lysogenizing *R. erythropolis* with actinophage ϕEC. More recently, Brownell, Enquist, and Denniston-Thompson (1980) have prepared deletion mutants of actinophage ϕEC and have identified potential cloning sites and transfected protoplasts of *R. erythropolis*. Similarly Kasweck, Little, and Alvardo (1981) have demonstrated gene transfer between strains of *N. asteroides*, and Kasweck, Little, and Bradley (1982) have detected and characterized cryptic

[1] *This work was supported by a grant (PCM-7807868) from the National Science Foundation. J. S. B. is a recipient of a Research Career Development Award from the National Institutes of Arthritis, Diabetes, Digestive and Kidney Diseases.*
[2] *Present address: Muscle Biology Group, University of Arizona, Tucson, Arizona.*

ISBN 0-12-528620-1

plasmids in this species and have accomplished genetic exchange by using protoplast fusion and regeneration. Clearly, nocardioform bacteria are amenable for genetic analysis by a wide variety of experimental techniques. While the producers for genetic manipulation of nocardioform bacteria are being refined, we have also examined the regulation of enzyme activities in nocardioform bacteria, particularly *R. erythropolis* (Bradley, 1978). The regulation of intracellular enzyme concentrations in nocardioform bacteria by transcriptional control of the rate of enzyme synthesis has been demonstrated for a variety of biosynthetic and catabolic reactions (Pang and Bradley, 1974a,b).

In this report, the role of selective inactivation and degradation of proteins in regulating intracellular enzyme concentration has been emphasized. The earlier views that bacterial proteins are completely stable in growing cells and that inactivation and degradation of enzymes either do not occur or are of little consequence can no longer be sustained (Bradley, 1981). Although there is an increasing amount of information on degradation of intracellular bacterial proteins in general (Goldberg and St. John, 1976), there is only a limited number of reports on the catabolism of specific enzymes in growing bacteria (Larrabee *et al.*, 1980; St. John *et al.*, 1979). Mannitol dehydrogenase (EC 1.1.1.67) of *R. erythropolis* has been selected as a good model for studying selective degradation of an intracellular bacterial enzyme. Mannitol dehydrogenase activity in *R. erythropolis* is inducible and its activity is lost rapidly when cells grown in mannitol are subcultured into succinate medium (Bond and Bradley, 1978). To determine whether the loss of activity is due to inactivation or degradation, antiserum was prepared to purified preparations of the enzyme. Antiserum, prepared in rabbits, to purified preparations of mannitol dehydrogenase inhibited the dehydrogenase activity. Using the antiserum in an inhibition assay, it was demonstrated that the decrease in mannitol dehydrogenase activity in cells subcultured into succinate medium was accompanied by a loss in immunoreactive protein. Our data, presented in this report, confirmed that the loss of mannitol dehydrogenase activity in cells growing on succinate is due to inactivation and degradation of the enzyme.

Proteolytic inactivation of desired proteins and polypeptides produced by genetically engineered bacteria is a potentially limiting factor in achieving maximum yields. Two intracellular protease activities were detected in *R. erythropolis* extracts, a sulfhydryl protease and a serine protease. The serine protease was purified 160-fold by ammonium sulfate fractionation and by chromatography on DEAE-cellulose, Sephadex G-150, hydroxyapatite, and benzamidine-glycylglycine-Sepharose. The molecular weight of the serine protease was approximately 90,000. It was monomeric and had a pH optimum of 7.0 to 7.2. The enzyme hydrolyzed benzoyl-arginine-4-methyl-7-coumarylamide, benzyloxycarbonyl-phenylalanyl-arginine-4-methyl-7-coumarylamide, hemoglobin, casein, and azocasein. The protease was inhibited by diisopropyl fluorophosphate, tosyl-lysine chloromethyl ketone, antipain, leupeptin, 4-amino-benzamidine, ovomucoid, and gramicidin

S. Magnesium salts stimulated activity. Intracellular protein degradation appears to be an important process in regulating enzyme concentration in nocardioform bacteria.

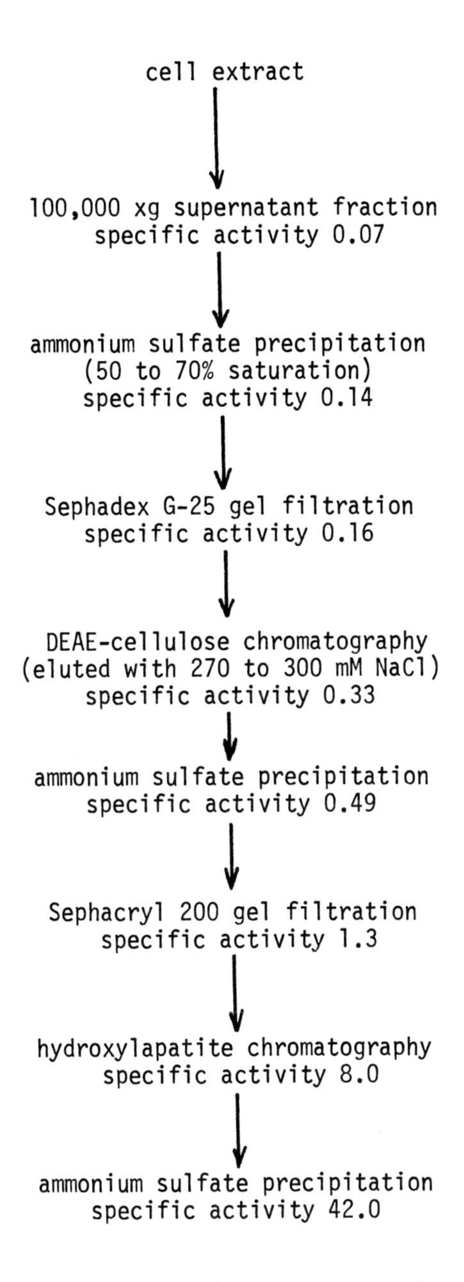

FIGURE 1. Purification of mannitol dehydrogenase from R. erythropolis.

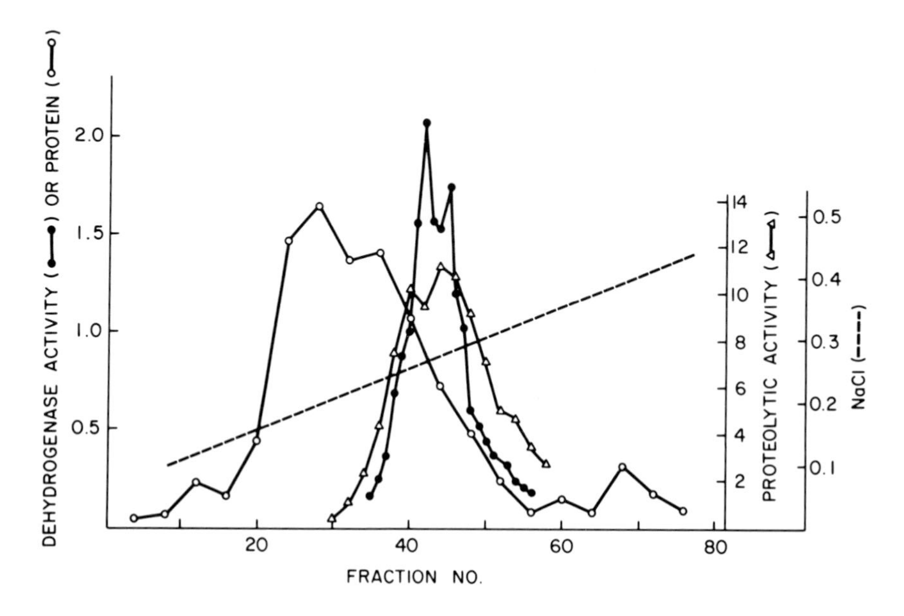

FIGURE 2. DEAE-cellulose chromatography of R. erythropolis *cell extracts. Twenty ml (5 mg/ml) of the 50-70% saturated* $(NH_4)_2SO_4$ *precipitated protein from cell extracts was applied to a DEAE-cellulose column. Eluted fractions contained 3.9 ml. Protein (O-O) is expressed as mg/ml; mannitol dehydrogenase activity (●-●) is expressed as units/ml; proteolytic activity (△-△) was measured with benzoyl-arginine-2-naphthylamide used as substrate and is expressed as nmoles substrate hydrolyzed/min/ml. Conductivity (M) (---) is expressed in terms of concentrations of standard sodium chloride solutions.*

II. MANNITOL DEHYDROGENASE

A. Purification

Mannitol dehydrogenase is difficult to purify. Under the best of conditions, the enzyme constitutes only 0.1% of the total cell protein. Moreover, it is a relatively unstable enzyme *in vitro*. Mannitol dehydrogenase was purified by combinations of centrifugation, precipitation, gel filtration, and ion exchange chromatography (Fig. 1). After centrifugation of disrupted *R. erythropolis* cells at 100,000 X *g* for 1 h, the supernatant fluid was treated with saturated $(NH_4)_2SO_4$ (50% final concentration), the mixture was centrifuged, and this supernatant was treated with saturated $(NH_4)_2SO_4$ (70% final concentration). This second precipitate containing protein which had been precipitated by $(NH_4)_2SO_4$ at 70% saturation but not at 50% saturation was dissolved in 20 mM tris (hydroxymethy)-aminomethane HC1 (Tris HC1), pH 7.5, and then sieved through a Sephadex G-25 column to remove salts. The protein sample was subjected to DEAE-cellulose chromatography within 1 h. The major protein peak was eluted from the DEAE- cellulose column between 200 and 275 mM NaCl. Mannitol dehydrogenase activity was eluted between 270 and 300 mM NaCl (Fig. 2). Fractions with more than 0.9 units mannitol dehydrogenase activity per ml were pooled and were precipitated with saturated $(NH_4)_2SO_4$ (80% final concentration). Upon subsequent gel filtration through Sephacryl 200, mannitol dehydrogenase was separated from the bulk of the remaining protein. Further purification of mannitol dehydrogenase was obtained by using hydroxyapatite chromatography. Overall purification was 300- to 600-fold.

B. Properties of Mannitol Dehydrogenase

The molecular weight of mannitol dehydrogenase, determined by Sephacryl 200 or Sephadex G-75 chromatography, was 37,000 (Fig. 3). When subjected to sodium dodecyl sulfate polyacrylamide gel electrophoresis in the presence of 2-mercaptoethanol, the enzyme showed a major protein band with a

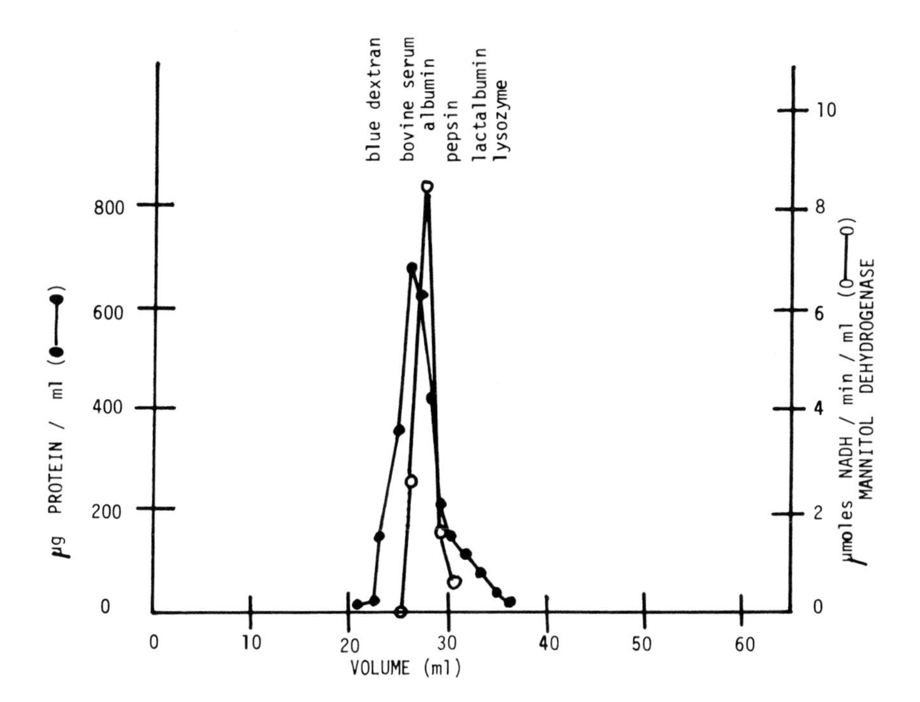

FIGURE 3. Gel filtration of fractions containing mannitol dehydrogenase eluted from the hydroxyapatite column. Proteins precipitated with 80% saturated $(NH_4)_2SO_4$ after hydroxyapatite chromatography were sedimented by centrifugation, dissolved in 20 mM Tris-HCl buffer containing 0.2 M NaCl, and applied to a Sephadex G-75 Column. Protein (●-●) is expressed as µg/ml; mannitol dehydrogenase (O-O) is expressed as units/ml.

molecular weight of 35,000. Mannitol dehydrogenase therefore appeared to be a monomeric enzyme. The purified fraction after hydroxyapatite chromatography contained small amounts of three other molecular weight bands (68,000, 84,000, and 22,000). These extraneous proteins could not be removed by gel filtration on Sephadex G-75 of by affinity chromatography using NAD^+-Sepharose, blue dextran-Sepharose, or by ion exchange columns.

Mannitol dehydrogenase acted on mannitol optimally at pH 9 to 10. With fructose as substrate, the pH optimum was 6 to 7 (Fig. 4). The enzyme did not catalyze the oxidation of mannitol-l-phosphate or the reduction of fructose-l-phosphate or fructose-6-phosphate. NADP was completely inactive in the oxidation of mannitol. At their optimal pH, mannitol appeared to be a better substrate for the enzyme than did fructose. The apparent K_M for mannitol was 3.1 mM and for NAD^+ was 0.1 mM. The apparent K_M for fructose was 55 mM. Sulfhydryl reagents such as iodoacetate or *p*-hydroxymercuribenzoate inhibited the dehydrogenase, indicating that reduced sulfhydryl groups on the enzyme were required for activity. Dithioerythritol and 2-mercaptoethanol did not markedly affect enzyme activity, indicating that the sulfhydryl groups

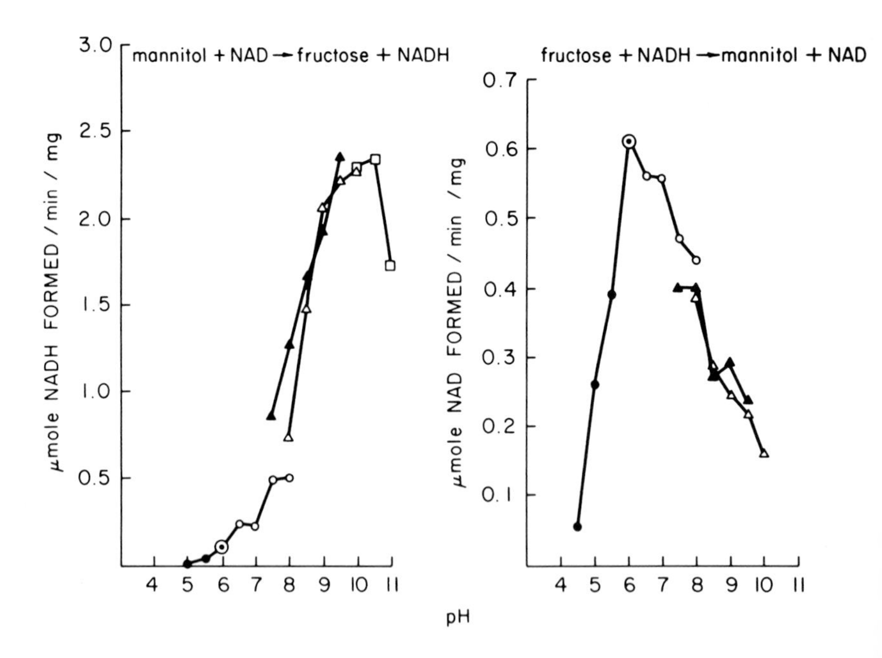

FIGURE 4. Effect of pH on mannitol dehydrogenase activity. The oxidation of mannitol to fructose was measured using 10 μg protein/ml, 100 mM mannitol, and 1 mM NAD^+; reduction of fructose to mannitol was measured using 20 μg protein/ml, 200 mM fructose, and 1 mM NADH. The buffers used were sodium acetate (●-●), sodium phosphate (O-O), Tris-HCl (▲-▲), diethanolamine (△-△), and glycine-NaOH (□-□), all at 20 mM.

were fully reduced or that oxidized groups were not readily reduced. Ethylenediaminetetraacetate, a metal chelating agent, and phenylmethylsulfonyl fluoride, a chemical that reacts with seryl groups at the active sites of many enzymes, had no effect on the dehydrogenase.

C. Immunologic Measurements

Antiserum from rabbits that had been immunized with purified mannitol dehydrogenase reacted with soluble extracts from cells grown in mannitol and with purified preparations of mannitol dehydrogenase to give two to four precipitin bands on Ouchterlony plates or after immunoelectrophoresis of the antigen preparations. At least two precipitin bands were observed when the antiserum was tested against soluble extracts from cells grown in succinate. Extracts from these cells were used to absorb antibodies to proteins other than mannitol dehydrogenase. A precipitin band unique to cells grown in mannitol was then observed. When immunoelectrophoresis plates were stained for mannitol dehydrogenase activity, the precipitin lines and enzyme activity lines were coincident and unique to samples from cells grown in mannitol.

The antiserum inhibited mannitol dehydrogenase activity, whereas control rabbit sera had no effect on the dehydrogenase activity. The amount of antiserum required to decrease mannitol dehydrogenase activity by 50% was proportional to the amount of enzyme in extracts of cells grown in mannitol. This antiserum was used to determine whether immunoreactive material disappeared in parallel with loss of mannitol dehydrogenase activity in extracts from cultures transferred to succinate medium. Within 10 h, the cell mass had increased 1.6-fold, whereas mannitol dehydrogenase activity had fallen by 80% and immunoreactive material had decreased by 70% (Table I). Loss of enzyme activity exceeded dilution through growth and paralleled loss of immunoreactive material.

These results indicate that mannitol dehydrogenase was degraded in growing *R. erythropolis*. The demonstration that loss of an inducible enzyme activity after subculture to medium lacking the nutritional inducer was associated with loss of immunoreactive protein provides strong support for the important role that protein degradation can play in the regulation of enzyme concentrations in growing bacteria. Slowly growing organisms, similar to non-dividing cells of constant size, may rely more on processes such as protein catabolism to decrease the concentrations of selected enzymes than do rapidly proliferating cells which can dilute out unnecessary enzymes once synthesis is stopped (Bond and Bradley, 1978).

Our data indicate that loss of activity precedes loss of immunoreactivity. It is possible that degradation of the enzyme involves two steps: (a) a modification or conformational change that inactivates the enzyme and (b) extensive proteolysis that destroys immunoreactivity. Aggregation or dimer formation may be the initial step leading to inactivation of the enzyme (Ueng *et al.*, 1976).

TABLE I. Changes in Mannitol Dehydrogenase Activity and Immunoreactivity in Soluble Extracts of Cells Grown in Mannitol and Subcultured in Succinate

	Mannitol dehydrogenase activity		*Immunoreactive material*[a]	
Hours of subculturing	*units/ml*	*% activity remaining*	*μl antiserum causing 50% inhibition*	*% active material*
0	0.38	100	0.40	100
2	0.21	55	0.29	73
4	0.23	60	0.29	73
6	0.23	60	0.30	75
8	0.14	36	0.20	50
10	0.08	21	0.13	33
12	0.07	18	0.13	32

[a] *Soluble extracts (100 μl) were incubated with either 100 υl antiserum or 100 μl antiserum diluted with normal rabbit sera for 10 min at 23°C. Sera from normal control rabbits did not inhibit mannitol dehydrogenase activity. Portions of the mixture of soluble extracts and antiserum, with or without control rabbit sera, were then assayed for mannitol dehydrogenase activity. For each soluble extract, six dilutions of antiserum were analyzed in duplicate to determine the amount of antiserum that produced 50% inhibition of activity.*

III. *RHODOCOCCUS* TRYPSIN

A. Purification

The proteases responsible for intracellular protein degradation in bacteria have not been previously identified and described in detail (Mount, 1980). *R. erythropolis* is a good model for such studies because it is generally considered non-proteolytic, in that it does not grow on media containing proteins as the sole carbon and nitrogen source. Two intracellular proteases have been detected in *R. erythropolis* (Bradley, 1981). One of these proteases has been partially purified and characterized.

When *R. erythropolis* cells were disrupted, most of the proteolytic activity was found in the supernatant fraction after centrifugation at 100,000 X *g* for 1h. The protease capable of hydrolyzing benzoyl-arginine-4-methyl-7-coumarylamide was purified by combinations of centrifugation, precipitation, ion-exchange chromatography, gel filtration, and affinity chromatography (Fig. 5). After centrifugation at 100,000 X *g* for 1 h, the supernatant fluid was first treated with saturated $(NH_4)_2SO_4$ (40% final concentration), the mixture was centrifuged, and this supernatant was treated with saturated $(NH_4)_2SO_4$ (65% final saturation). This second precipitate containing protein which had been precipitated by $(NH_4)_2SO_4$ at 65% saturation but not at 40% saturation was dissolved in Tris-HCl and then dialyzed against Tris-HCl to remove the salt. The protein sample was applied to a DEAE-cellulose column and was eluted with a salt gradient. The protease activity eluted between 275 and 325 mM NaCl. Fractions with more than 1.2 units of protease activity per ml were pooled and concentrated by ultrafiltration

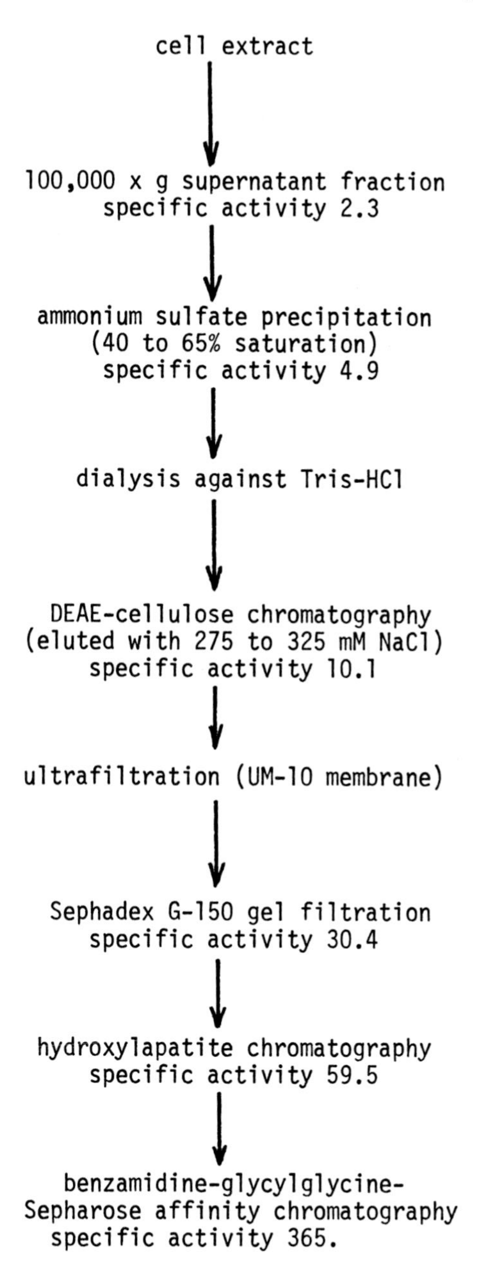

FIGURE 5. Purification of Rhodococcus *trypsin.*

(UM-10 membrane, Amicon Corp., Lexington, MA). The concentrated protein fraction was sieved through a Sephadex G-150 column. Further purification was achieved by hydroxyapatite chromatography. Several affinity chromatography systems were tested; the protease did not bind to a boronate column or to an ovomucoid-Sepharose column. The protease bound with high affinity to a gramicidin S-Sepharose column, but most enzyme activity was lost. Good purification was achieved with a benzamidine-glycylglycine-Sepharose column. Overall purification was 160-fold.

B. Properties of Rhodococcus *Trypsin*

The molecular weight of *Rhodococcus* trypsin as determined by Sephadex G-150 chromatography was 82,000. When subjected to sodium dodecyl sulfate polyacrylamide gel electrophoresis in the presence of 2-mercaptoethanol, the purified protease showed three bands which stained with Coomassie blue. To identify the band corresponding to the protease, the preparation was incubated with radiolabeled diisopropyl fluorophosphate, a reagent that reacts with serine residues in the active sites of enzymes. The radiolabeled protease was identified by fluorography after gel electrophoresis. The protease corresponded to the band with a molecular weight of 90,000. The protease appeared to be active as a monomer.

The pH optimum of the protease was 7.0 to 7.2. The apparent K_M of the protease was 42 μM for benzoyl-arginine-4-methyl-7-coumarylamide. The enzyme was stable at neutral and alkaline pH values (7 to 9) for at least 3 h at 25 °C, but 80% of the activity was lost within 3h at pH 6. Within 20 min, the protease lost 44% of its activity at 50 °C and 98% of its activity at 60 °C.

The most potent inhibitors of the enzyme were diisopropyl fluorophosphate, tosyl-lysine chloromethyl ketone, antipain, leupeptin, 4-aminobenzamidine, ovomucoid, and gramicidin S. All of these inhibitors also inhibit mammalian trypsin. Several other trypsin inhibitors (*e. g.*, phenvlmethylsulfonyl fluoride, lima bean trypsin inhibitor, aprotinin, and α-1-antitrypsin) were weakly or not inhibitory for *Rhodococcus* trypsin. The *Rhodococcus* trypsin was also not affected by inhibitors of cysteine proteinases (iodoacetate), aspartic proteinases (pepstatin), or metalloproteinases (ethylenediaminetetraacetate).

The purified protease preparation hydrolyzed several synthetic substrates and proteins. The best substrate of those tested was benzyloxycarbonyl-phenylalanyl-arginine-4-methyl-7-coumarylamide, a good substrate for cathepsin B and metalloproteinase from mouse kidney. Casein, hemoglobin, and azocasein were hydrolyzed, indicating proteinase activity. *Rhodococcus* trypsin was unable to hydrolyze the carboxypeptidase substrates benzyloxycarbonyl-glutamyl-tyrosine or benzyloxycarbonyl-phenylalanyl-alanine.

TABLE II. Effect of Mg^{+2} and Ca^{+2} on the Activity of Rhodococcus *Trypsin*

	Protease activity[a]	
Ionic Strength	Mg^{+2} *added*	Ca^{+2} *added*
none	2.04	2.08
0.02	8.05	3.11
0.04	7.85	3.46
0.06	8.92	2.83
0.10	8.96	2.49
0.20	7.46	1.58
0.30	5.17	1.06
0.40	3.92	0.76
0.50	2.84	0.57

[a]Protease activity was assayed using benzoyl-arginine-4-methyl-7-coumarylamide as the substrate. After incubation at 37°C for 5 to 10 min, the reaction was stopped with chloroacetate and the fluorescence measured in a spectrofluorometer with an excitation wavelength of 345 nm and an emission wavelength of 440 nm. Activity is expressed as nmoles substrate hydrolyzed per min per ml.

Salts markedly affected *Rhodococcus* trypsin activity (Table II). $MgCl_2$ was particularly effective in stimulating activity at ionic strengths of 0.02 to 0.2. Several other salts (*e. g.*, NaCl, KCl, NaF, and sodium acetate) activated the protease 2-to 3-fold at ionic strengths of 0.1 to 0.5. LiCl and Na_2SO_4 were less stimulatory at the same ionic strengths and $CaCl_2$ was inhibitory at ionic strengths greater than 0.15. Addition of 5 mM adenosine triphosphate, with or without 5 mM $MgCl_2$ and with 500 mM NaCl, did not alter the protease activity. *Rhodococcus* trypsin is not an ATP-dependent serine endoprotease.

IV. PROTEIN DEGRADATION

A. Physiological Role

Intracellular protein degradation appears to be an important process in regulating enzyme concentration in growing bacteria (Bond *et al.*, 1977). Growing bacteria must be able to adapt to new biosynthetic needs either because of individual cell cycle requirements or because of changing environmental conditions in batch cultures. Growing bacteria may degrade some of their proteins in order to synthesize other proteins if exogenous precursors are limiting.

Protein degradation appears to be important in determining both the rate of change of enzyme concentration from one steady state level to another, and the steady state level *per se* (Kenney and Lee, 1982). The steady state level of an enzyme is dependent upon both its rate of synthesis and its rate of elimination. If two proteins are synthesized at the same rate, the one that is degraded more slowly will have a higher steady state level (Bradley and Bond, 1974). Inducible enzymes that are degraded rapidly will disappear rapidly after removing the nutritional inducer or after inhibition of protein synthesis with inhibitors. Enzymes that are degraded more slowly respond more slowly to changes in the rate of protein synthesis (Bond and Bradley, 1978).

Protein degradation appears to serve as a surveillance system to eliminate abnormal proteins and salvage needed precursors. Abnormal proteins may arise as a result of premature chain termination, denaturation, or incorporation of analogues. Recently, bacteria have been genetically engineered to make proteins that are foreign to the producing organism; these proteins are recognized as abnormal or non-self by the producing bacterium. Proteolysis of desired enzymes or polypeptide products has become a serious limiting factor in commercial development of large scale processes. Nonfunctional proteins not only drain the cellular amino acid supply but may retard optimum cellular metabolism.

B. *Mechanism of Protein Degradation*

A substantial body of information now exists about the processes of protein degradation. Proteins that are degraded slowly possess a number of characteristics that distinguish them from rapidly degraded proteins, indicating that the information determining the stability of enzymes *in vivo* resides in the enzyme structure. There seems to be several variations on the general theme, but covalent modification appears to be the initial event leading to denaturation and subsequent degradation (Bond and Offermann, 1981; Levine *et al.*, 1981). A number of initial covalent modification reactions have been recognized: oxidation of metallic coenzymes, oxygenation of tyrosine residues, oxidation of sulfhydryl residues by oxidized glutathione, cleavage of polypeptides at particular sites, and deamidation of asparagine or glutamine. The denatured protein may form an aggregate which is more susceptible to proteolysis than is the native enzyme. Degradation of mannitol dehydrogenase by *R. erythropolis* cells appeared to follow these general steps: (a) the enzyme is inherently unstable; (b) inactive aggregates of the enzyme appear to be formed spontaneously; and (c) the denatured protein is degraded by general proteases such as serine proteases (Fig. 6). Mannitol dehydrogenase is inactivated by trypsin, and ligands of the dehydrogenase (*e. g*, mannitol, NAD^+) protect the dehydrogenase from inactivation by the protease. It is not yet known which proteases are responsible for degradation in *R. erythropolis*. In other systems, specific proteases have been shown to be responsible for the degradation of particular proteins, whereas ATP-dependent proteases have been implicated in the degradation of other proteins (Swamy and Goldberg, 1982). According to our scheme for protein degradation, the process can be impaired by (a) inhibiting the initial covalent modification, (b) impairing proteolysis with

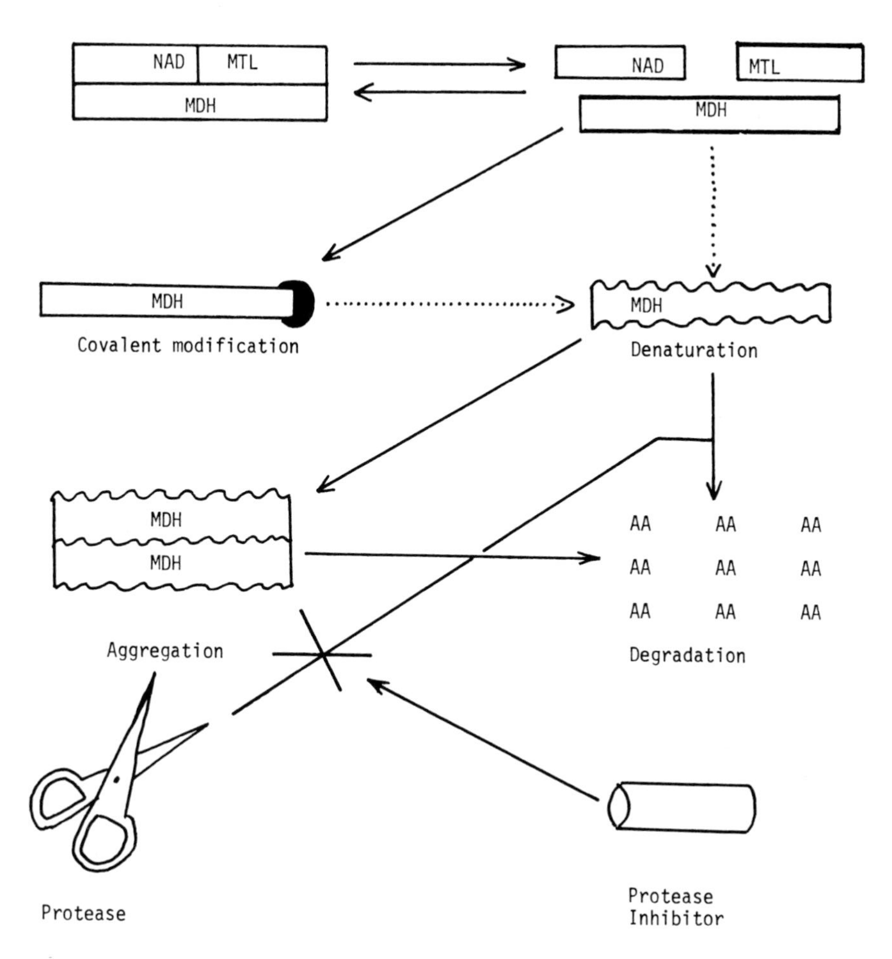

FIGURE 6. Scheme for degradation of mannitol dehydrogenase (MDH). NAD: nicotinamide dinucleotide; MTL: mannitol; AA: amino acid.

appropriate protease inhibitors, and (c) protecting the protein substrate by adding appropriate cofactors.

V. CONCLUDING REMARKS

A substantial body of data has accumulated documenting that selective protein degradation occurs in growing bacteria and is an important control mechanism. Examples are now being recognized in which transcriptional control is minimal and protein degradation is the primary means for controlling enzyme concentration and expression. The advances of genetic engineering in the development of microorganisms capable of producing theretofore "foreign" proteins and polypeptides has dramatically illustrated that there is a sophisticated surveillance system interlocked with the control systems of an organism. Not only do actinomycetes produce a full array of proteinases, they also produce an array of rather specific proteinase inhibitors. A battery of proteinases (Strongin *et al.*, 1979) in concert with differential synthesis, elimination, and compartmentalization of proteinase inhibitors, constitutes a powerful counterpoised mechanism for regulating intracellular protein concentration and enzyme activities (Bradley, 1981).

REFERENCES

Adams, J. N., and Bradley, S. G., *Science 140:*1392 (1963).

Bond, J. S., and Bradley, S. G., *in* "*Nocardia* and *Streptomyces*" (M. Mordarski, W. Kurylowicz, and J. Jeljaszewicz, eds.), p. 389. Gustav Fischer Verlag, Stuttgart (1978).

Bond, J. S., and Offermann, M. K., *Acta Biol. Med. Ger. 40:*1365 (1981).

Bond, J. S., Pang, R. H. L., and Bradley, S. G., *Dev. Ind. Microbiol. 18:*419 (1977).

Bradley, S. G., *in* "*Nocardia* and *Streptomyces*" (M. Mordarski, W. Kurylowicz and J. Jeljaszewicz, eds.), p. 287. Gustav Fischer Verlag, Stuttgart (1978).

Bradley, S. G., *in* "Actinomycetes" (K. Schaal and G. Pulverer, eds.), p. 367. Gustav Fischer Verlag, Stuttgart (1981).

Bradley, S. G., and Bond, J. S., *Adv. Appl. Microbiol. 18:*131 (1974).

Brownell, G. H., Adams, J. N., and Bradley, S. G., *J. Gen. Microbiol. 47:*247 (1967).

Brownell, G. H., Enquist, L. W., and Denniston-Thompson, K., *Gene 12:*311 (1980).

Crockett, J. K., and Brownell, G. H., *J. Virol. 10:*727 (1972).

Goldberg, A. L., and St. John, A. C., *Annu. Rev. Biochem. 45:*747 (1976).

Kasweck, K. L., Little, M. L., and Alvardo, F., *in* "Actinomycetes" (K. Schaal and G. Pulverer, eds.), p. 585. Gustav Fischer Verlag, Stuttgart (1981).

Kasweck, K. L, Little, M. L., and Bradley, S. G., *Dev. Ind. Microbiol. 23:*279 (1982).

Kenney, F. T., and Lee, K.-L., *Bioscience 32:*181 (1982).

Larrabee, K. L., Phillips, J. O., Williams, G. J., and Larrabee, A. R., *J. Biol. Chem. 255:*4125 (1980).

Levine, R. L., Oliver, C. N., Fulks, R. M., and Stadtman, E. R., *Proc. Nat. Acad. Sci. U.S.A. 78:*2120 (1981).

Mount, D. W., *Annu. Rev. Genet. 14:*279 (1980).

Pang, R. H. L., and Bradley, S. G., *Mycologia 66:*48 (1974a).

Pang, R. H. L., and Bradley, S. G., *J. Bacteriol. 118:*400 (1974b).

St. John, A. C., Jakubas, K., and Beim, D., *Biochim. Biophys. Acta 586:*537 (1979).

Strongin, A. Y., Gorodetsky, D. L., and Stepanov, V. M., *J. Gen. Microbiol. 100:*443 (1979).

Swamy, K. H. S., and Goldberg, A. L., *J. Bacteriol. 149:*1027 (1982).

Ueng, S. T. H., Hartanowicz, P., Lewandoski, C., Keller, J., Holick, M., and McGuinness, E. T., *Biochemistry 15:*1743 (1976).

ACTINOMYCETES AS MODELS OF BACTERIAL MORPHOGENESIS

R. Locci

Department of Mycology
University of Milan
Milan, Italy

I. INTRODUCTION

Systematic investigations of actinomycete morphogenesis (Baldacci *et al.*, 1971a,b; Cross *et al.*, 1973; Locci, 1971, 1975, 1976, 1978, 1980a,b,c, 1981; Locci and Petrolini, 1970, 1971; Locci and Rogers, 1975; Locci and Schaal, 1980; Locci and Schofield, 1981; Locci and Sharples, 1983; Schofield and Locci, 1981; Locci, Goodfellow, and Pulverer, unpublished data; Ridell and Locci, unpublished data) have revealed a series of developmental cycles which propose these organisms as models of bacterial diversification.

Micromorphological diversity in procaryotes is expressed in various phenotypic ways, as recently summarized by Starr and Schmidt (1981). In the present chapter, the characteristics and the significance of actinomycete diversification are reviewed and discussed.

II. METHODOLOGY

The methodology employed in the study of actinomycete morphogenesis includes light (LM), transmission (TEM), and scanning electron microscopy (SEM). The techniques are complementary and relative advantages of the different observation media are summarized in Table I.

Details of specific techniques employed can be found in previous works by the author (*loc. cit.*). Concerning the potentials of the various observation media, some comments are necessary. The term "sequential determination" means that it is possible to follow the development of the same propagule in

ISBN 0-12-528620-1

TABLE I. Relative Merits of Different Techniques of Observation in the Study of Actinomycete Morphogenesis

	Comparison of observation media			
Observation technique	*Definition capabilities*	*Sequential determination*	*Overall determination*	*Ultrastructure determination*
LM[a]	+[b]	+ + +	+	-
TEM	+ + +	-	+	+ + +
SEM	+ +	-	+ + +	+ +

[a]*LM = light microscopy; TEM = transmission electron microscopy; SEM = scanning electron microscopy.*
[b]*-: impossible; +: fair; + +: good; + + +: excellent.*

time. Overall determination involves examination of developing organisms at different time intervals, with the consequent possibility of reconstructing developmental patterns.

Naturally, the way to follow development of a single cell is time-lapse observation. The main obstacle at the moment is that this can only be done by light microscopy which has very poor definition. TEM is incomparable in ultrastructural investigations. SEM in part shares such advantages but is best in overall developmental determinations because of both the size of the sample that can be examined and the three dimensional capabilities of the instrument.

Actinomycete morphogenesis has been examined by time-lapse microphotography, immunofluorescent labeling (Locci and Schaal, 1980), and sequential SEM observations. An indirect verification consisted in following the effects of adverse conditions on development. Considering the fragility of the organisms involved, an adequate methodology, the membrane transfer-technique (MTT), was developed (Locci, 1980a) to investigate the immediate effects on micromorphology of sudden exposure to unfavorable environments (Locci, 1980a,b,c, 1981; Schofield and Locci, 1981).

III. RESULTS

A. Progress in Actinomycete Diversification

Actinomycetes are emblematic of procaryote diversification. Different aspects of the phenomenon may be tentatively summarized (Table II). The progression develops through a series of steps, analyzed in order below, from the coccus-rod induction up to the complexity of a sporulating mycelium.

B. Significance of Actinomycete Diversity

According to Brock (1971), the morphogenetic aspects of microbial differentiation should be viewed as being determined by ecological and evolu-

TABLE II. Aspects of Micromorphological Diversification in Actinomycetes

Progress in Actinomycete Diversification
—Coccus to rod induction
—Filament formation
—Elementary branching
—Mycelium formation
—Sporulation
—Secondary mycelium formation
—Sporulation of secondary mycelium

tionary factors. In Table III an attempt is made to correlate stages of development, as found in actinomycetes, with ecological consequences.

Filament formation leads to a larger size of the individual cell. Some consequences of mycelium formation are close adherence to the substrate and a coordinated exploitation of the same. Mycelial organization permits translocation of nutrients and provides additional mechanisms of variability. With aerial growth, the restriction of aquatic environments is overcome, with the consequent possibility of propagule dispersal by air. The role of antibiosis with reference to actinomycete survival was investigated.

1. Filament Formation. At the onset of actinomycete differentiation filament formation provides a way of increasing size. Since, for intrinsic reasons, this cannot be achieved by larger diameters (Henning, 1975; Pirie, 1973; Thompson, 1979), the obvious solution is cell elongation. Increase in size can be achieved in several ways (Fig. 1).

With reference to the cell wall, incorporation of new material is accomplished by intussusception which can be either diffuse or localized. In the case of typical actinomycetes, it is clearly polar (Locci and Sharples, 1983). This is also evident in less organized members of the group.

TABLE III. Significance of Morphological Characteristics in Terms of Ecological Adaptation

	Significance of actinomycete diversity
Filament formation:	*—Large size of individual cells*
Mycelium formation:	*—Close adherence to substrate*
	—Coordination in substrate utilization
	—Translocation of nutrients
	—Additional variability mechanisms
Aerial growth:	*—Liberation from aquatic environments*
	—Formation of aerial propagules
Antibiosis:	*—Competitive survival in microhabitats*

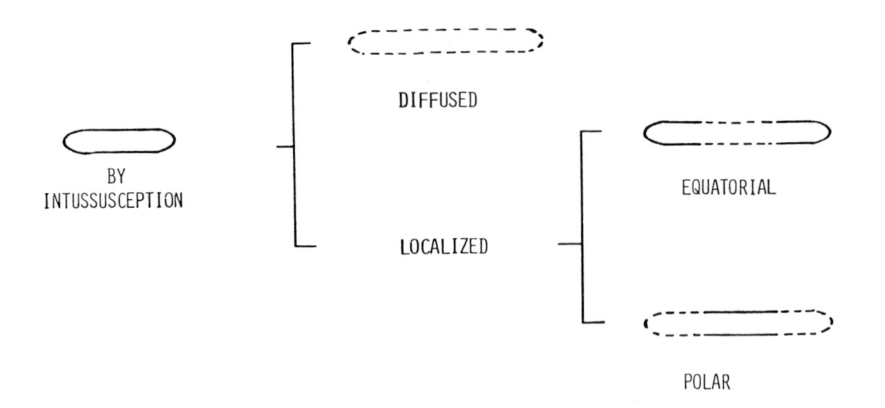

Concerning the significance of the phenomenon, Antonov, Ivanov, and Markov (1977) have shown a high kinetic complexity of actinomycete DNA. This suggests not only a genome which is two or three times larger than that of *Escherichia coli* but also the presence of repeated sequences.

To summarize, the diameter of an actinomycete is comparable to that of non-differentiated procaryotes; however quite early, there is a diversification toward elongation of the basic cylinder with a consequent larger mass of cytoplasm contained in the unit element. In practical terms, filament formation (and branching) may be regarded as a means of overcoming size limitations characteristic of procaryotes.

2. *Mycelial Organization.* Filament formation represents the first step toward branching. Advantages of mycelial organization are summarized in Table IV. Firm attachment to the substrate is beneficial in terms of its exploitation. The ability of actinomycetes to break down and utilize relatively "hard" organic compounds should be taken into account. At parity of cytoplasmic volumes, a branched organism has the advantage over a population of independent cells in that it can coordinate the utilization of broken down byproducts. In view of the fierce competition which exists in habitats such as soil, this is fundamental in order to avoid providing nutrients for other microorganisms.

The possibility of translocating nutrients is also pregnant with ecological implications, particularly in connection with the spreading of the organisms through areas void of nutrients, a common situation in soil. The spanning of barren areas would be a problem for "classical" procaryotes.

The colony conformation under conditions of nutrient restraint represents a further example of polarized growth. The margin continues to advance while filaments at the center are emptied of protoplasm. This type of development, in addition to the classical properties of growth (*e. g.*, the biosynthesis of wall material), also has some characteristics of motility, *i. e.*, some protoplasm is moved from its site of synthesis (Carlile, 1980).

The multinuclearity of actinomycetes is yet another advantage of mycelial organization. Additional variability mechanisms are available, ranging from heterokaryosis to parasexuality processes.

The disadvantage of a branching system is the severe impediment to dispersal of the organisms (Chater and Merrick, 1979). To overcome the difficulty, mycelia fragment (Fig. 2) into rods and/or cocci at the end of the developmental cycle (Locci, 1976); the mycelium may then form some *ad hoc* structures devoted to dispersal. The ultimate goal is achieved by the formation of spore-bearing aerial hyphae.

TABLE IV. Advantages, Disadvantages, and Ways of Overcoming the Latter of Mycelial Organization in Actinomycetes

Mycelial organization		
Advantages	*Disadvantages*	*Disaavantages overcome by*
—Close adherence to substrate *—Coordinated substrate utilization* *—Translocation of nutrients* *—Additional variability mechanisms*	*—Propagule dispersal*	*—Fragmentation* *—Sporulation* *—Aerial mycelium devoted to sporulation*

3. Aerial Growth. Vertically developing filaments may be observed in some actinomycetes as soon as a few hours after germ tube extrusion (Fig. 3). With further development, a network of aerial hyphae covers the colony surface of actinomycetes.

Reasons for discriminating between substrate and aerial growth are more profound than the mere location on solid media or on liquid surfaces. As reviewed by Ensign (1971), the two mycelia are different ontogenetically, morphologically, structurally, and physiologically. Cytological, cytochemical, and physiological differences are discussed in detail by Kalakoutskii and Agre (1976). In general terms, aerial growth appears less branched than does the substrate counterpart although some streptomycetes and streptoverticillia may show quite a complex organization. In contrast to the primary mycelium, the aerial mycelium is hydrophobic (Higgins and Silvey, 1966) most probably because of the presence of an "outer envelope" or "fibrous sheath" (Hopwood and Glauert, 1961; Wildermuth, *et al.*, 1971). Francisco and Silvey (1971) found that the substrate mycelium of a *Streptomyces* sp. was facultatively aerobic, whereas aerial growth was obligately so.

Aerial growth is an almost unique phenomenon among procaryotes and is related to their emergence from aquatic to terrestrial environments. Orientation of hyphae is possibly aided by the hydrophobic nature of the outer

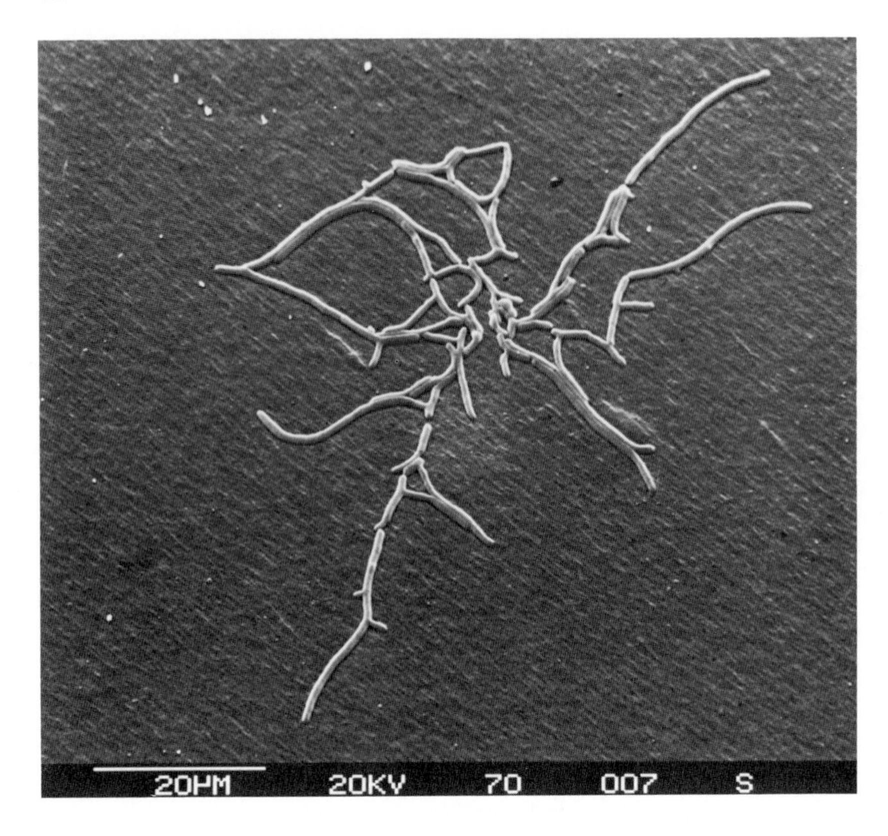

FIGURE 2. Initial fragmentation of a Rhodococcus rhodnii *microcolony.*

envelope, which also appears to protect cytoplasm from rapid desiccation. As pointed out by Gregory (1961), aerial hyphae bearing dry, powdery spores represent the first example of a spore elevation device, common in more elaborate organisms, which raises the spore-producing organ above the substratum and towards the moving layers of the atmosphere. It may be noted in this connection that typical aquatic actinomycetes are usually devoid of aerial growth (Kalakoutskii and Agre, 1976).

Aerial growth is not necessarily limited to actinomycetes showing well-developed substrate mycelia. In the genus *Sporichthya*, only aerial growth is present, consisting of filaments adhering through holdfast to the medium surface. Progress in aerial development may be summarized as shown in Table V.

Single, mostly unbranched, vertical hyphae which frequently coalesce into synnemata can often be observed in rhodococci (Locci, unpublished data), mycobacteria (Ridell and Locci, unpublished data), and sometimes even in organisms such as *Actinomyces bovis, A. naeslundii*, and *A. viscosus* (Locci and Schofield, 1981). In the latter case, their appearance is frequently ephimeral and may

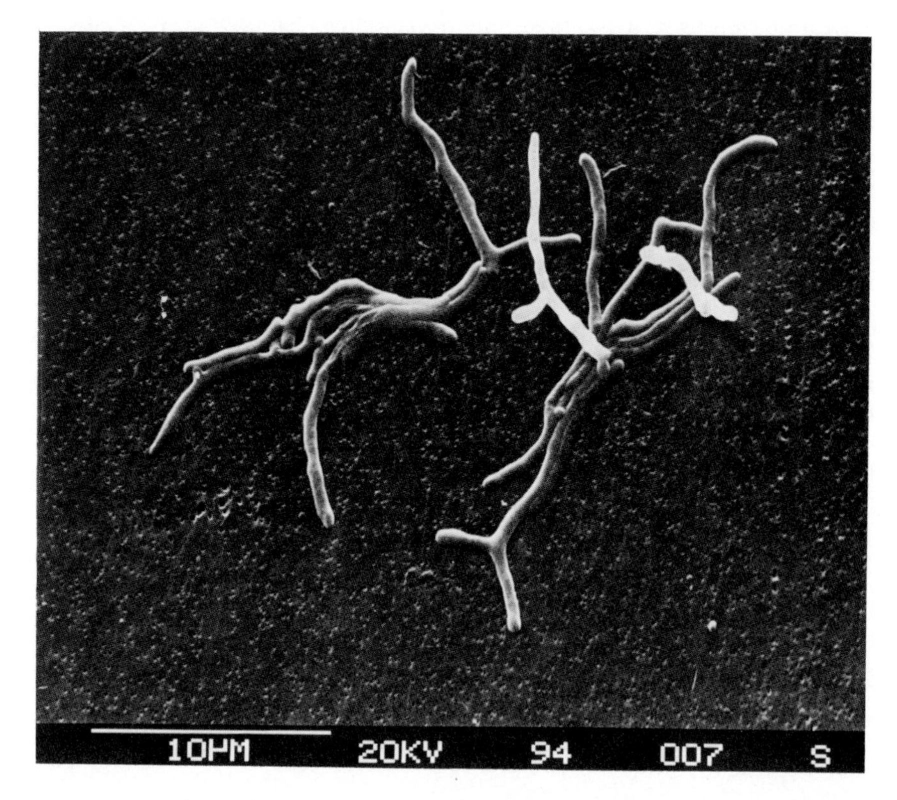

FIGURE 3. Early aerial filaments in Mycobacterium farcinogenes.

possibly represent an ancestral relic of little functional and ecological importance.

In some nocardiae, aerial growth may be quite abundant although hyphal differentiation is rather elementary. If sporulation takes place, the entire filament is involved in the process, with poor differentiation into sporophore and spore chain (Locci, 1976).

TABLE V. Stages of Aerial Mycelium Development

Progress in aerial development
—Aerial projection of filaments
—Aerial development of synnemata
—Non-branching filaments
—Aerial branched filaments
—Sporogenous filaments
—More sophisticated fruiting structures

In streptomycetes, aerial mycelium development ceases with sporulation, an indispensable dispersion mechanism for organisms otherwise closely anchored to their substrate during vegetative growth. Here too, lytic processes take place in aging colonies and may play a role in spore liberation, as in oligosporic actinomycetes (Locci, 1971, 1976). Some authors have stressed the parasitic nature of aerial on substrate growth. According to Wildermuth (1970), the latter undergoes extensive breakdown during aerial mycelium development. A general role of antibiotics has been proposed in this connection. Production of antibiotics in natural habitats has long been a subject of controversy (Williams and Khan, 1974). According to Chater and Merrick (1979), some facts ought to be kept in mind in this connection: a) the onset of antibiotic formation is relatively late with regard to the producer's development and b) antibiotic biosynthetic pathways demand energy and therefore their presence should be beneficial to the producer. The natural role of antibiotics is consequently thought to be a protective mechanism. If it is accepted that aerial growth develops as a sort of saprophyte at the expense of primary

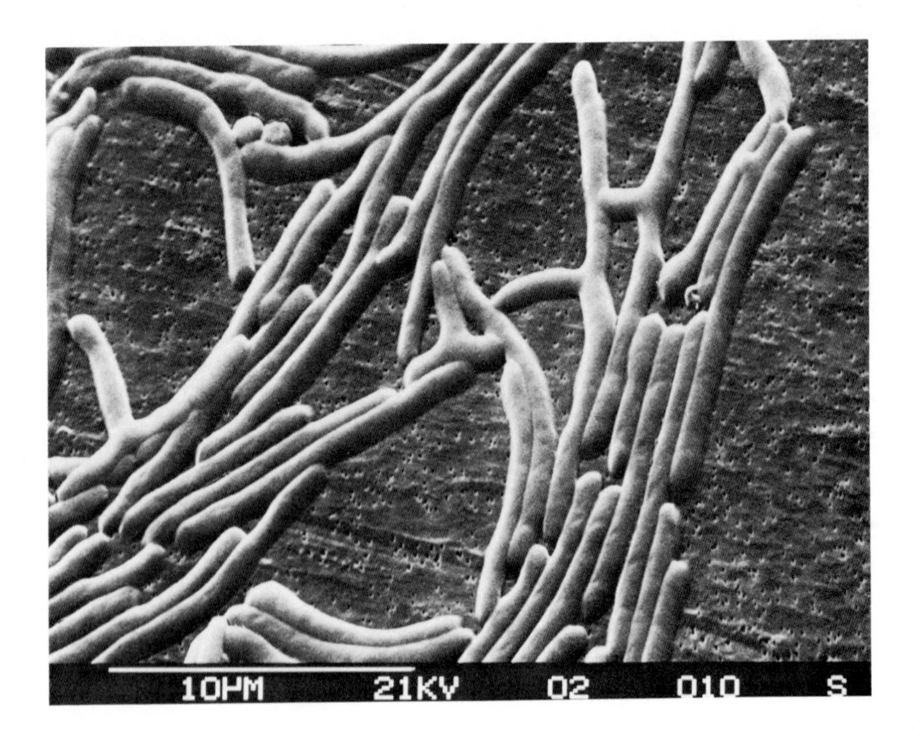

FIGURE 4. Elementary branching in Brevibacterium paraffionolyticum.

mycelium, then in natural environments, some protection against possible profiteers is required. In addition, the role of antibiotics at the colony level would explain the extremely low amounts detectable in nature.

IV. EXAMPLES OF MORPHOGENESIS

It is worthwhile considering some aspects of morphogenesis in "border" procaryotes in order to detect the initial symptoms of the diversification that characterizes actinomycete morphology. The developmental cycle has been, or is, being investigated in a number of representatives of the genera *Actinomyces, Arachnia, Bacterionema* (Locci, 1981), *Brevibacterium* (Fig. 4), *Cellulomonas* (Locci, unpublished results), *Corynebacterium* (Locci and Sar-

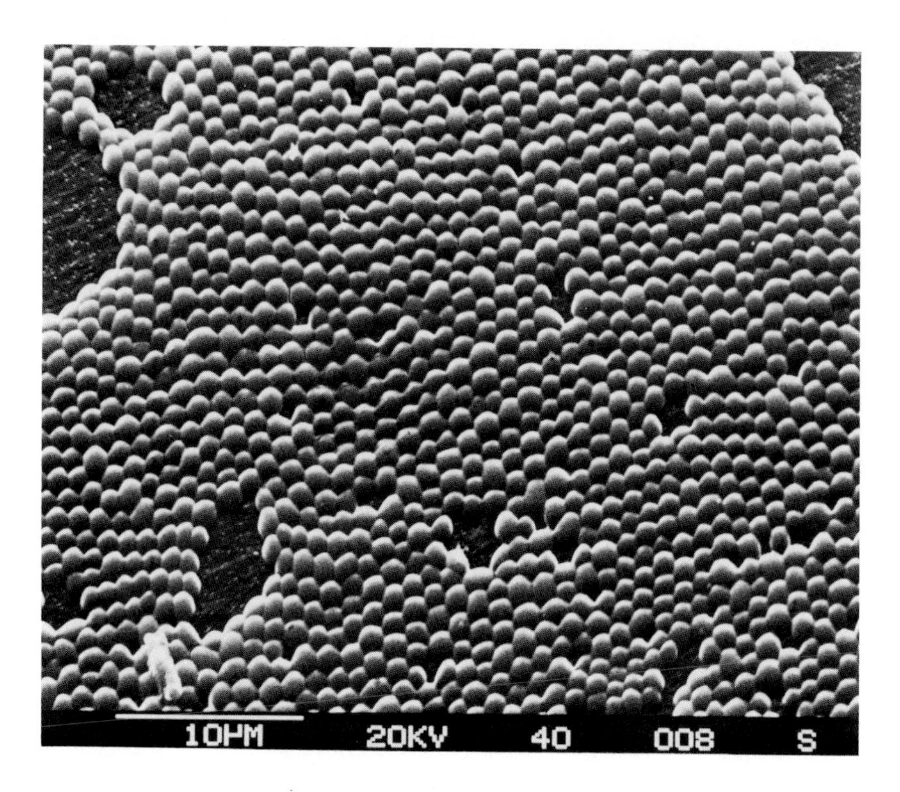

FIGURE 5. Arthrobacter globiformis *coccoid cells.*

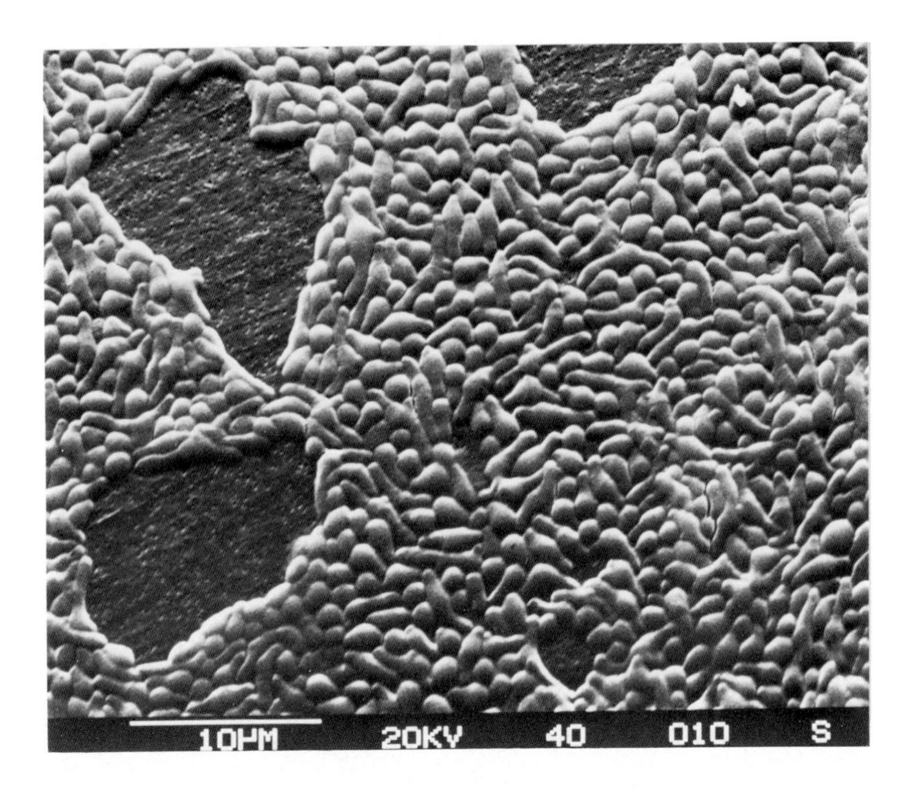

FIGURE 6. Elongation of Arthrobacter globiformis *cocci.*

di, unpublished results), *Microbacterium* (Locci, unpublished results), *Mycobacterium* (Ridell and Locci, unpublished data), *Nocardia* (Locci, 1976), *Rhodococcus (Locci et al.*, unpublished data), and *Rothia* (Locci, 1981). Most of the taxa investigated show complex morphogenetic patterns of extreme interest, which allow the various trends of diversification characteristic of the organisms to be followed.

Morphogenesis of arthrobacteria is rather emblematic in that it illustrates the initial stages of actinomycete development. Coccoid cells (Fig. 5) may develop, under particular conditions, as such or undergo induction, *i. e.*, elongate into rods (Fig. 6). Rods may sometimes (Locci, 1981) show elementary branching (Fig. 7) and at the end of the cycle revert to cocci (Clark, 1979). Reversion may be gradual by successive reduction in cell length or take place by simultaneous formation of cross septa along the rod (fragmentation).

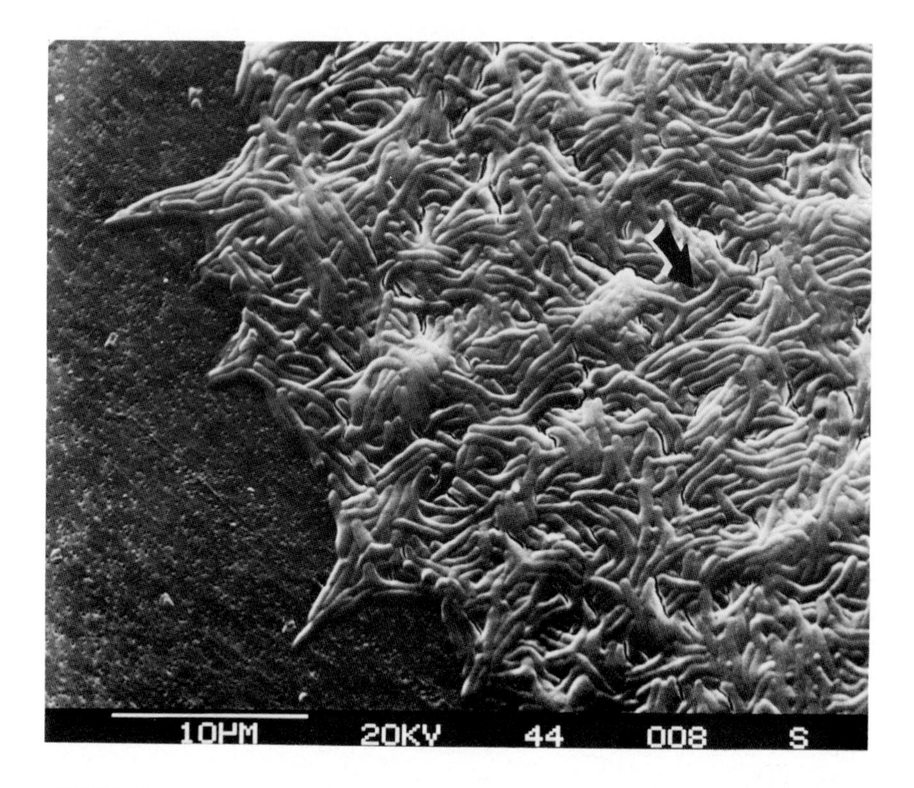

FIGURE 7. Arthrobacter globiformis *rods. Signs of elementary branching* (arrow) *are present.*

Already in the "no induction" microcycle, some interesting phenomena can be detected. Taking advantage of the presence of "scars" following cell division, the first signs of localized cell wall synthesis are observed.

As demonstrated by TEM (Kolenbrander and Hohman, 1977), new cell wall occurs in the plane of cell contact. As additional material is deposited, the cell pairs undergo snapping movements with the resulting sphere pairs exhibiting an "end-to-end orientation of surface rings". Each cell surface is them composed of part old and part new cell wall. Further divisions take place in a plane perpendicular to the previous one.

The process of localized growth is more evident when the coccus-rod transition takes place. Rod growth is clearly apical.

Here again, observation is facilitated by the reference points represented by the scars ("raised bands", "rings") following fission (Kolenbrander and Hohman, 1977). Cell surface elongation is unidirectional and the new cell wall is synthesized only at the poles encircled by the surface rings. Snapping post-division movements follow cell elongation with resulting V-shaped cell pairs (Figs. 8 and 9).

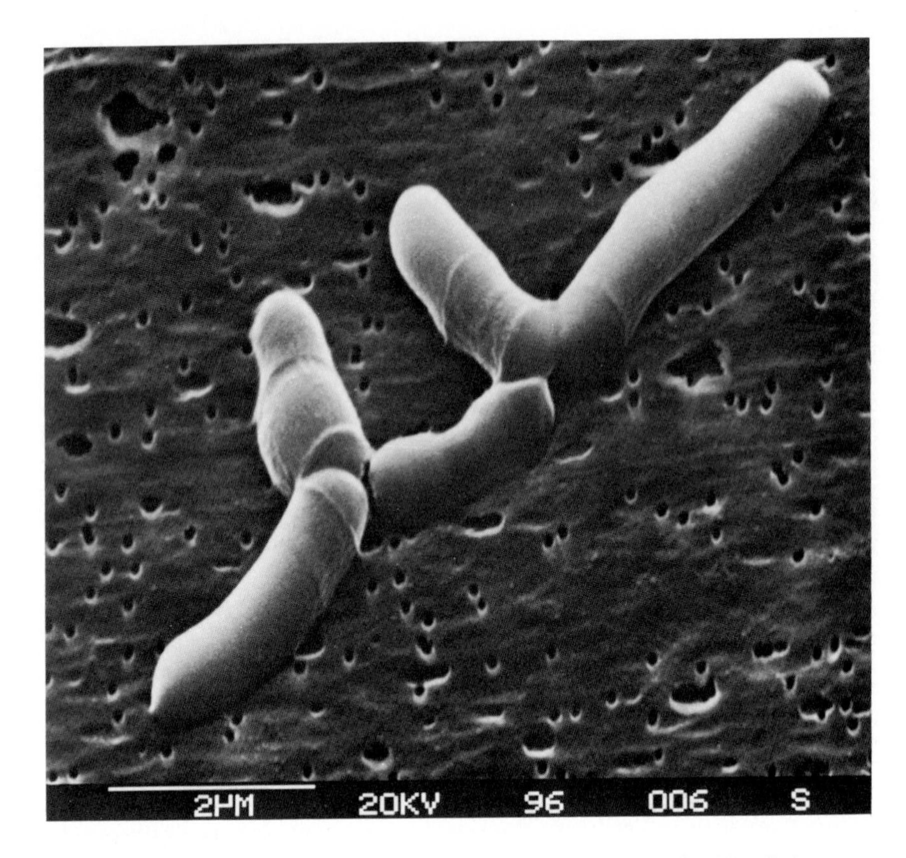

FIGURE 8. Appearance of Arthrobacter variabilis *rods after snapping with evident fission scars.*

V. CONCLUSIONS

Morphology is commonly regarded as responsible for the state of confusion which exists in actinomycete taxonomy. However, it should be stressed that, due to the size of the organisms in question, the instrumentation used for observation in the past was often inadequate. It is true that technical insufficiency caused misunderstanding and confusion. However, even at present, the term "morphological investigation" is often misused and taken as the equivalent of a quick bacterium-like examination at an unspecified moment of growth. It is superfluous to add that the description of an actinomycete smear is *not* morphology.

Often at the end of the growth cycle, actinomycetes show a similar ap-

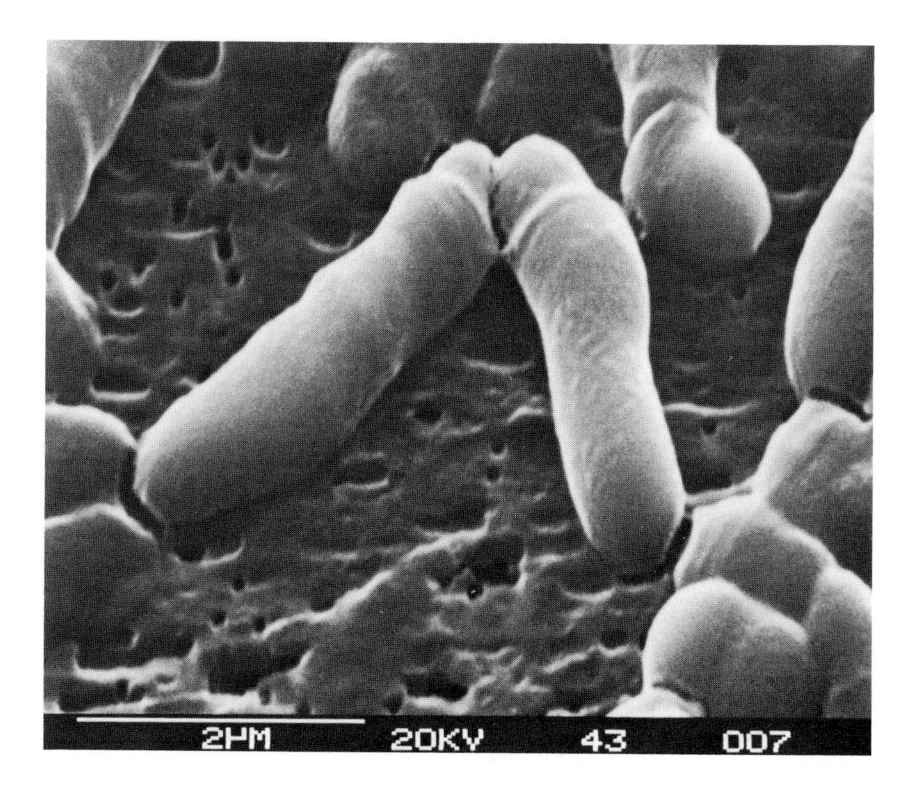

FIGURE 9. Unidirectional growth of Arthrobacter variabilis *rods following fission.*

pearance (Locci, unpublished data) which can hide clear-cut differences characterizing the morphogenetic cycle. The use of more powerful techniques of observation and a careful examination of ontogenesis should be highly rewarding, considering that size and shape which are admittedly the final result in living beings combine both genetic potential and environmental interferences.

Accordingly, an ecological and evolutionary appraisal of morphology, as suggested by Brock (1971), has been attempted. In this discourse on actinomycete morphology, evidence has been presented to show that shape and size are essentially the result of environmental adaptation and have a strong ecological significance. Apparent analogies should always be considered with extreme care, keeping in mind both the patterns leading to them (morphogenesis) and the teleological reasons for their appearance.

To conclude, it seems proper to quote T. D. Brock (1971): "Most studies on shape have had taxonomic or molecular orientation and one might almost assume that microorganisms produce such interesting and often intricate shapes solely to provide scientists with amusement". It seems that it is time to add a new dimension.

REFERENCES

Antonov, P. P., Ivanov, I. G., and Markov, G. G., *FEBS Lett. 79:*151 (1977).

Baldacci, E., Locci, R., and Petrolini, B. B., *Riv. Patol. Veg. 7* suppl.:21 (1971a).

Baldacci, E., Locci, R., and Petrolini, B. B., *Riv. Patol. Veg. 7* suppl.:45 (1971b).

Brock, T. D. *in* "Recent Advances in Microbiology" (A. Perez-Miravete and D. Pelaez, eds.), p. 77. Asoc. Mex. Microbiol., México (1971).

Carlile, M. J., *in* "The Eukaryotic Microbial Cell" (G. W. Gooday, D. Lloyd, and A. P. J. Trinci, eds.), p.1. Cambridge Univ. Press, Cambridge (1980).

Chater, K. F., and Merrick, M. J., *in* "Developmental Biology of Prokaryotes" (J. H. Parish, ed.), p. 93 Blackwell, Oxford (1979).

Clark, J. B. *in* "Developmental Biology of Prokaryotes" (J. H. Parish, ed.), p. 73. Blackwell, Oxford (1979)

Cross, T., Attwell, R. W., and Locci, R., *J. Gen. Microbiol. 75:*431 (1973).

Ensign, J. C., *in* "Recent Advances in Microbiology" (A. Perez-Miravete and D. Pelaez, eds.), p.91 Asoc. Mex. Microbiol., Mexico (1971).

Francisco, D. E., and Silvey J. K. G., *Can. J. Microbiol.17:*347 (1971).

Gregory, P. H., "The Microbiology of the Atmosphere". Leonard Hill Ltd., London (1961).

Henning, V., *Annu. Rev. Microbiol. 29:*45 (1975).

Higgins, M. L., and Silvey, I. K., *Trans. Am. Microsc. Soc. 85:*390 (1966).

Hopwood, D. A., and Glauert, A. M., *J. Gen. Microbiol. 26:*325 (1961).

Kalakoutskii, L. V., and Agre, N. S., *Bacteriol. Rev. 40:*469 (1976).

Kolenbrander, P. E., and Hohman, R. J., *J. Bacteriol. 130:*1345 (1977).

Locci. R., *Riv. Patol. Veg.* 7 suppl.:63 (1971).

Locci, R., *in* "Proc. I Intersectional Congr. IAMS" (T. Hasegawa, ed.), p. 77. Science Co., Tokyo (1975).

Locci, R., *in* "Actinomycetes: The Boundary Microorganisms" (T. Arai, ed.), p. 249. Toppan Co. Ltd., Tokyo (1976)

Locci, R., *Zentralbl. Bakteriol. 6* suppl.:173 (1978).

Locci, R., *Zentralbl. Bakteriol. I. Abt. Orig. A. 246:*98 (1980a).

Locci, R., *Zentralbl. Bakteriol. I. Abt. Orig. A. 249:*374 (1980b).

Locci, R., *Annu. Microbiol. 28:*63 (1980c).

Locci, R., *Zentralbl. Bakteriol. 11* suppl.:119 (1981).

Locci, R., and Petrolini, B. B., *G. Microbiol. 18:*69 (1970).

Locci, R., and Petrolini, B. B., *Riv. Patol. Veg., 7* suppl.:81 (1971).

Locci, R., and Rogers, J. L., *Annu. Microbiol. 25:*47 (1975).

Locci, R., and Schaal, K. P., *Zentralbl. Bakteriol. I. Abt. Orig. A. 246:*112 (1980).

Locci, R., and Schofield, G. M., *Annu. Microbiol. 31:*(1981). To be published.

Locci, R., and Sharples, G. P., *in* "The Biology of Actinomycetes"(M. Goodfellow, M. Mordarski and S. T. Williams, eds), (1983). (In press).

Pirie, N. W., *Annu. Rev. Microbiol. 27:*119 (1973).

Schofield, G. M., and Locci, R., *Annu. Microbiol. 31:* (1981) To be published.

Starr, M. P., and Schmidt, J. M., *in* "The Prokaryotes" (M. P. Starr, H. Stolp, H. G,Trüper, A. Balows, and H. G. Schlegel, eds.) p. 3, Springer Verlag, Berlin (1981).

Thompson, D. W., "On Growth and Form". Cambridge Univ. Press, Cambridge (1979).

Wildermuth, H., *J. Gen. Microbiol. 60:*43 (1970).

Wildermuth, H., Wehrli, E., and Horne, R. W., *J. Ultrastruct. Res. 35:*168 (1971).

Williams, S. T., and Khan, M. R., *Postepy. Hig. Med. Dosw. 28:*395 (1974).

STREPTOMYCES BAMBERGIENSIS SPORE ENVELOPE ULTRASTRUCTURE[1]

R. A. Smucker

Chesapeake Biological Laboratory
University of Maryland
Solomons, Maryland, U. S. A.

I. INTRODUCTION

Streptomyces sp. play an important role in modern societies in terms of antibiotic production and also have a significant role in nutrient recycling in natural processes. Because of pressures to produce maximum antibiotic yield, genetic studies have gained a degree of sophistication far beyond that of the basic biological studies of this important group. Classification has been dominated by morphological characters, of which the aerial spore ornamentation of *Streptomyces* has been used. Distinction among the aerial spore surfaces of *Streptomyces* has been made on the basis of smooth, warty, spiny, or hairy surface ornamentation which are discernible only with the electron microscope (Dietz and Mathews, 1962; Tresner *et al.*, 1961). These ornaments have been reported to be derived from the sheath (Williams *et al.*, 1973). The term "sheath" was also used by the same authors for a "sack" around a group of associated hyphae of *Microellobospora flavea*.

The *Streptomyces* sheath has been reported to be composed of elongated, hollow, or grooved elements, fine fibrillar elements, and amorphous material (Matselyukh, 1978; Wildermuth, 1972; Wildermuth *et al.*, 1971; Williams *et al.*, 1973). The spines of *S. viridochromogenes* are produced by the surface membrane ("fibrous sheath" of other authors) and not by the cell wall (Rancourt and Lechevalier, 1964). Lechevalier and Tikhonenko (1960) suggested that spine development in *S. viridochromogenes* was dependent on the culture

[1] *Supported in part (HVEM work) by Grant # RR 00570, Biotechnology Research Resource, NIH.*

ISBN 0-12-528620-1

medium. Furthermore, Coleman and Ensign (1982) showed that 0.5% casein hydrolysate completely inhibited secondary mycelia and spore development when added to phosphate buffered-glycerol-NH_4NO_3 solid medium.

Based on thin-sectioning studies of S. *violaceoruber* (S. coelicolor A3(2)), Hopwood and Glauert (1961) suggested that the fibrillar surface structures break when spores separate. In contrast, Vernon (1955) showed that *S. flaveoulus* spores, along with their spines, are liberated by a longitudinal splitting of the sheath. Spines only occasionally break off when the spores are released. These contrasting results are a function of the use of different species and of different methods of spore preparation.

The consensus in the *Streptomyces* literature regards the mature spore sheath as the "fibrous sheath" and, in thin sections, the loosely associated material, usually called the "fibrous sheath", appears to be detaching from maturing spores. However, Smucker and Pfister (1978) reported that *S. coelicolor* A3(2) has two layers outside the fibrous (rodlet mosaic) layer and concluded that the fibrous component remains with the mature spore.

The objective of this work is to present a spore envelope model for the hairy-spored *S. bambergiensis* and to test the hypothesis that the hairy-spore envelope follows a pattern similar to that of the smooth-spored *S. coelicolor* A3(2) (Smucker and Pfister, 1978).

II. MATERIALS AND METHODS

Streptomyces bambergiensis ISP # 5590 was obtained from Elwood B. Shirling, Ohio Wesleyan University. The culture was maintained in liquid nitrogen for long-term storage and on glycerol-asparagine medium for short-term storage (Shirling and Gottlieb, 1966). Spore production: *S. bambergiensis* was transferred by loop from glycerol-asparagine slants to either glycerol-asparagine agar (Shirling and Gottlieb, 1966), oatmeal agar (Shirling and Gottlieb, 1966), or chitin medium (Walker and Colwell, 1975), Spore and pigment production occurred most readily on chitin medium. All results reported in this paper were of aerial spores produced on chitin medium. Scanning electron microscopy (SEM): Two week-old colonies were quenched in liquid Freon 12 and then were maintained in liquid nitrogen until they were freeze-dried. After 4 h of lyophilization, intact colonies were mounted with adhesive on SEM stubs and coated with Au:Pd in a Hummer IV sputter coater (Technics, Alexandria, VA). Photographs were taken at 15kV accelerating voltage in a JEOL SEM U-3.

For transmission electron microscopy (TEM), two week-old colonies were fixed with a modification of the ruthenium red (RR) fixation scheme of Luft (1971). Colonies were immersed for 2 h in a mixture of equal volumes of 3.5% glutaraldehyde, 0.2 M sodium cacodylate, pH 7.2, and 1500 ppm (RR). Triton X-100 was added to the mixture to cause wetting of the colonies. Samples were washed three times with a mixture (1:2) of stock RR and 0.1 M sodium cacodylate, pH 7.2 (wash buffer) then post-fixed with a mixture of equal volumes of 5% osmium tetraoxide, 0.2 M sodium cacodylate, pH 7.2, and 1500 ppm RR for 16 h, the first 4 h at room temperature and the remaining 12 h at 5°C. Samples were rinsed free of unreacted OsO_4 with three changes of wash buffer and then dehydrated with ethanol (35, 50, 70, 90, 95 (2x), and 100% (2x)). Several drops of stock RR were added to each 5 ml alcohol through the 70% ethanol step; succeeding steps contained only ethanol. Colonies were transferred to propylene oxide (PO) (2x), then to a mixture of PO/Epon 812 (3:1) (Luft, 1961) for 8 h; a 1:1 mixture for 12 h; a 1:3 mixture for 12 h; then finally in 100% Epon 812 for 12 h. All of the impregnation steps were carried out at room temperature. Completed resin was polymerized first at 37°C (12h) and then at 60 °C 2 days.

Thin sections (300 to 700Å) for conventional TEM were doubly post-stained, first with 2% aqueous uranyl acetate (30 min) and then with Reynold's lead citrate (Reynold, 1963) (5 min). Sections were examined and photographed at 100 kV in a RCA-4A transmission electron microscope.

Thick sections (0.25 μm) for high voltage TEM (HVEM) were also doubly post-stained, first with 3% aqueous uranyl acetate (1 h) and then in Reynold's lead citrate (30 min). Sections were examined and photographed at 100 kV in an AEI EM-7 Million Volt electron microscope.

For freeze-fracture replicas, free spores were pelleted in distilled water containing Triton X-100, frozen in liquid Freon 12, fractured at -195°C, shadowed, and immediately coated with carbon. Some samples were etched at -100°C for 2 min prior to shadowing.

III. RESULTS

The dominating feature of *S. bambergiensis* spores was the conspicuous presence of the hair-like ornamentation covering mature spores (Fig. 1). Aerial spores in intact colonies were often covered with a membranous material (Fig. 1A). Many spores chains, however, were exposed and the hair-like appendages appeared on all visible sides of the spores (Fig. 2B).

TEM of thin sections demonstrated two wall layers. The inner spore wall was contiguous with the germ tube cell wall (unpublished data). The inner spore wall (SW_1) (Fig. 2) did not stain very well, and therefore was clearly delineated from the outer wall. There were, however, some cases in which the spore wall layers were stained in reverse, *i.e.*, the outer wall stained lightly and the inner wall intensely. No explanation is available for this phenomenon.

The spore envelope component lying outside the outer spore wall was the rodlet mosaic which was composed of individual fibers. These fibers and associated material comprised the spore hairs (Figs. 3-5).

Stained hairs (Fig. 3) showed the presence of two components lying outside the fibers. The outside component appeared peeled away from the fibers (Figure 3A) and, as is clearly shown in Figure 3B, hairs broken off during staining and air drying were divided into an inner fibrillar and an outer matrix. The lightly stained hair (Fig. 3A) clearly showed two components outside the

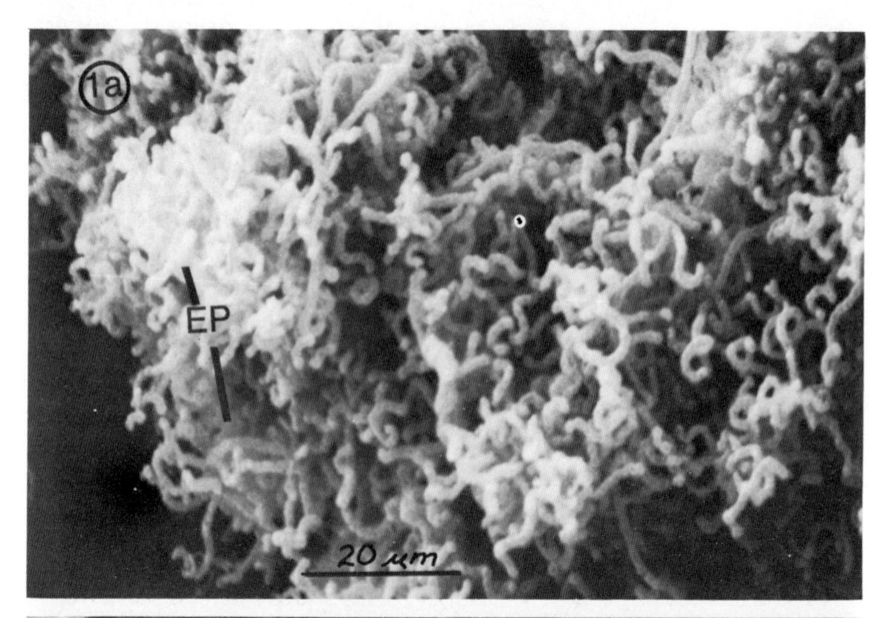

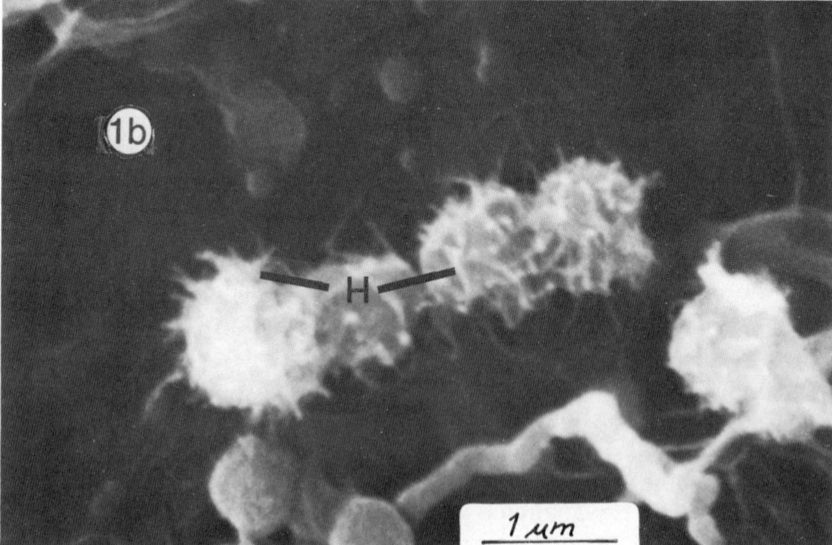

FIGURE 1. Scanning electron micrograph of a mature S. bambergiensis *colony. Note the extracellular coat surrounding groups of cells in A. In B, the long hairs are seen on all sides of spores.*

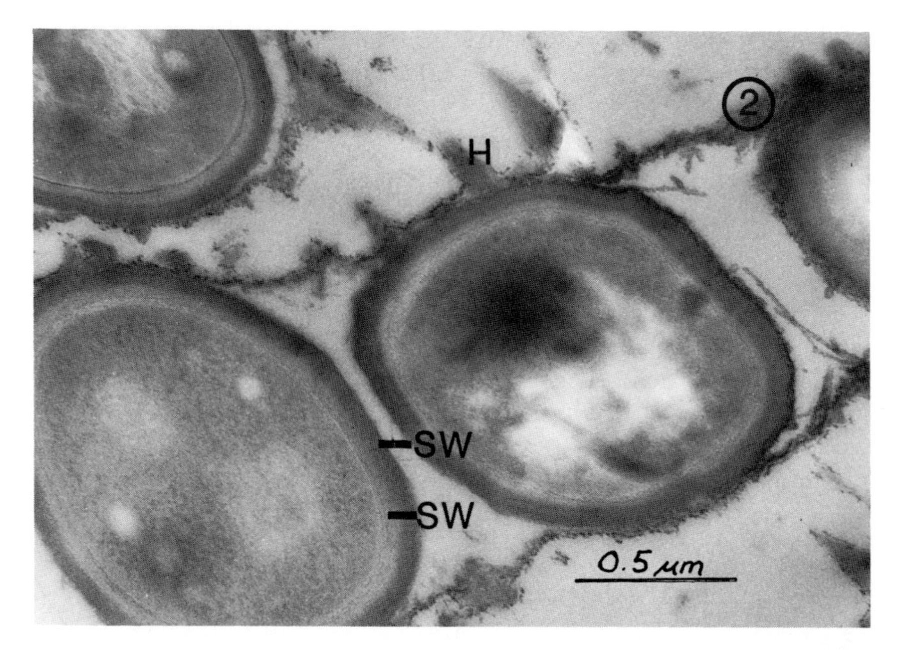

FIGURE 2. TEM micrograph of thin-sectioned S. bambergiensis *spores showing two spore wall layers.*

fibers: an electron-lucent layer and a thin, electron-dense layer. If conclusions were taken only from the negatively stained preparations (Fig. 3B), the thin outermost layer would have been visually "lost" in the phosphotungstic acid (PTA) stain.

Fracture planes obtained by freeze-etch replication generally occurred so that replicas were most often made of the rodlet mosaic fibrils and fibrils of the hair (Fig. 4). The basketwork pattern of intersecting fibrils in *S. violaceoruber* (*S. coelicolor* A3(2)) smooth spores (Hopwood and Glauert, 1961) was also found between the hairs of *S. bambergiensis*. These fibrils at the base of the hair were arranged in a spiral throughout the length of the hair. Deep etching and occasional fracture planes reveal the outer hair layers.

HVEM of 0.25 μm sections showed the complex structural system of the hairs (Fig. 5); the typical spore hair had three components. This was consistent with the results obtained by positively PTA stained hairs (Fig. 3A). Fibers appeared in the negatively stained hairs (Fig. 3B), and there was also an indication of striation patterns in the thick-sectioned hair (Fig. 5). Preparations showed the electron-lucent zone and the thin, outermost component.

In thick sections of hundreds of samples of single spores and of mature spores released from the sporophore (Fig. 6), hairs were seen to be associated all around the spore perimeters. This indicates that spore hairs are intimately associated with free spores and spore chains.

Figures 7 and 8 illustrate HVEM information regarding colony relationships between spore chains. Low magnification views of segments of aerial growth indicated a polymeric material in section, which appeared to closely appose some spores and to bridge spore chains. This corroborates the image seen

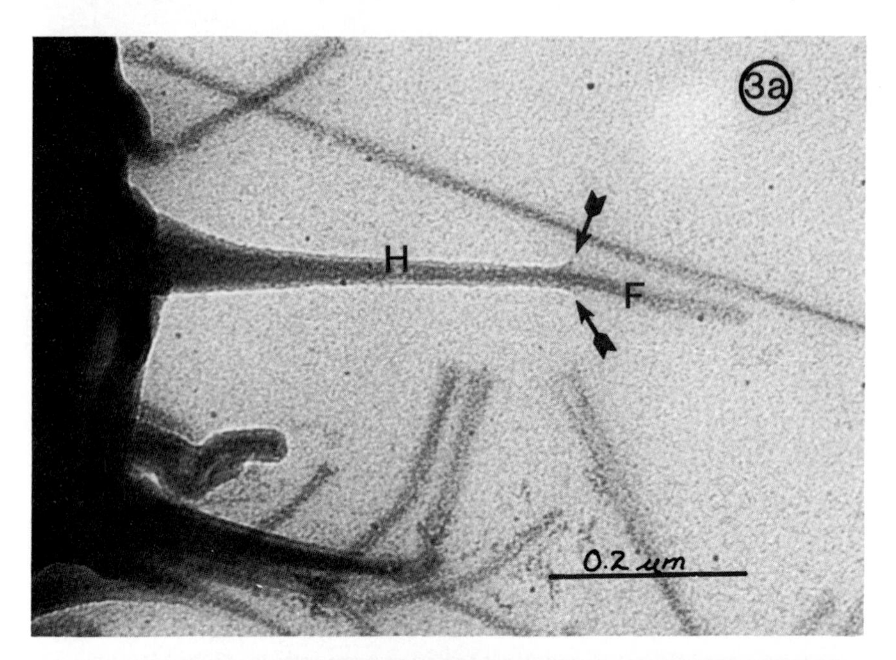

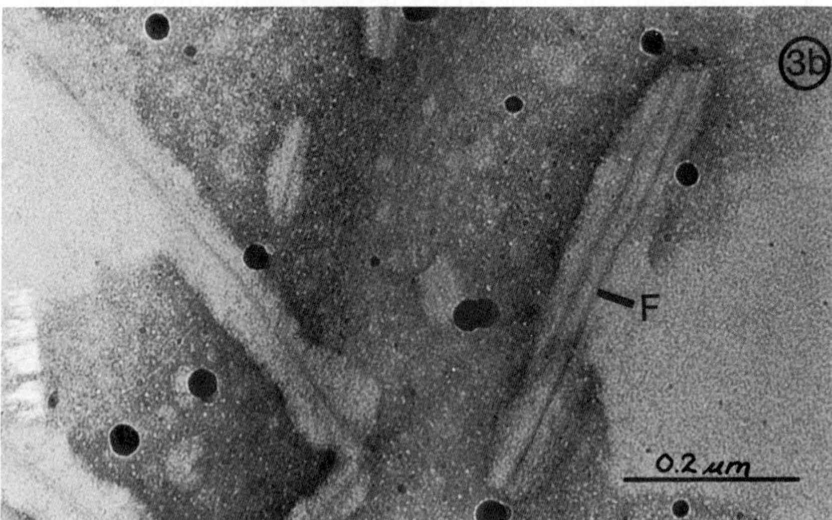

FIGURE 3. TEM of spore hairs in whole mounts. A: Lightly stained with PTA. These hairs took just enough stain to reveal the exterior layers (arrows) peeled away from the central part of the hair. B: PTA negatively stained hairs that broke away from spores during staining and air drying. The central fibers stained while the surrounding layers remained unstained.

in SEM of whole colonies. Higher magnification examination of sites of contact between the spores and the extraspore polymer (EP) (Figs. 7 and 8) showed the spore hairs penetrating the EP.

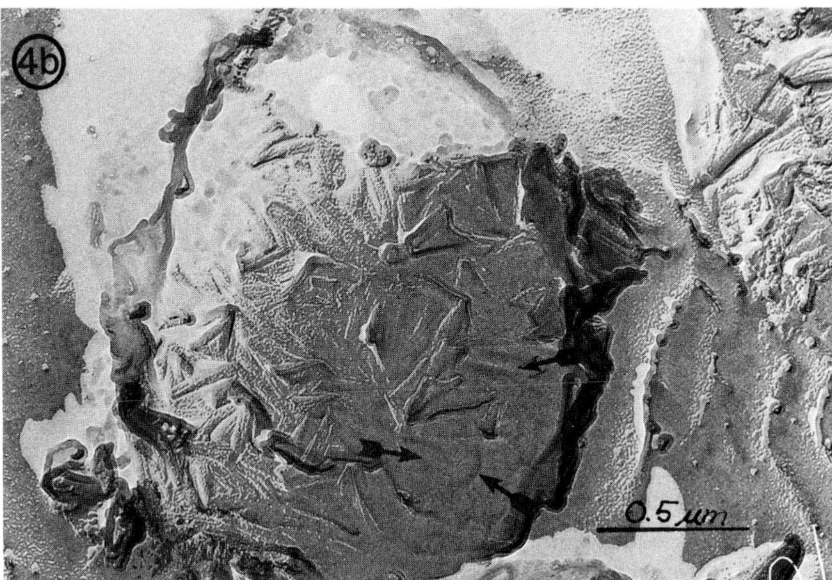

FIGURE 4. TEM of freeze-etch replica of aerial spores. A: Fibrils of the rodlet mosaic between the base of the hairs spiral up into and throughout the hairs. B: Fracture revealed a layer surrounding the hair exterior, thereby masking details of the fibers (arrows).

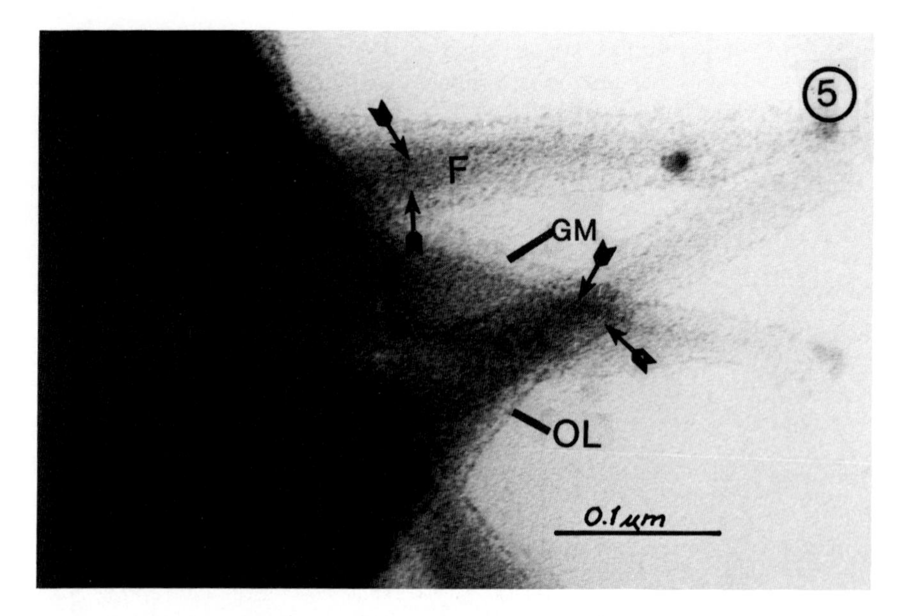

FIGURE 5. HVEM micrograph of thick sections (0.25 μm). These micrographs clearly demonstrate three components as indicated by staining patterns. In the distal region of the hair, the inner fibers, an electron-lucent zone, and a thin outer layer are clearly delineated. Arrows indicate the area where striation patterns are evident.

IV. DISCUSSION

The results presented here regarding the ultrastructural components and location of hair in *S. bambergiensis* are summarized in Figure 9. This schematic used information obtained by TEM of stained whole mounts, of freeze-etch replicas, and of thin sections and by HVEM of thick sections. In corroboration of the work of Matselyukh on *S. olivaceous* (1978) and in contrast to many reports based on thin section and freeze-etch replication, the present work demonstrated that spore hairs had a fibrillar center and an amorphous layer surrounding these fibrils. These results contrast with those of several authors who have indicated that the fibrillar component (also called fibrous sheath) is the outermost *Streptomyces* spore layer; Bradley and Ritzi (1968) indicated this for smooth spore *S. venezuelae*; Hopwood and Glauert (1961), for smooth spored S. *violaceoruber*; Rancourt and Lechevalier (1964), for spiny spored *S. viridochromogenes*; and Wildermuth *et al.* (1971), for smooth spored *S. coelicolor*.

The *S. bambergiensis* spore envelope model (Fig. 9) corroborated ultrastructural evidence presented for *S. coelicolor* A3(2) smooth spores (Smucker and Pfister, 1978).

Williams, Sharples, and Bradshaw (1973) summarized the trend in the

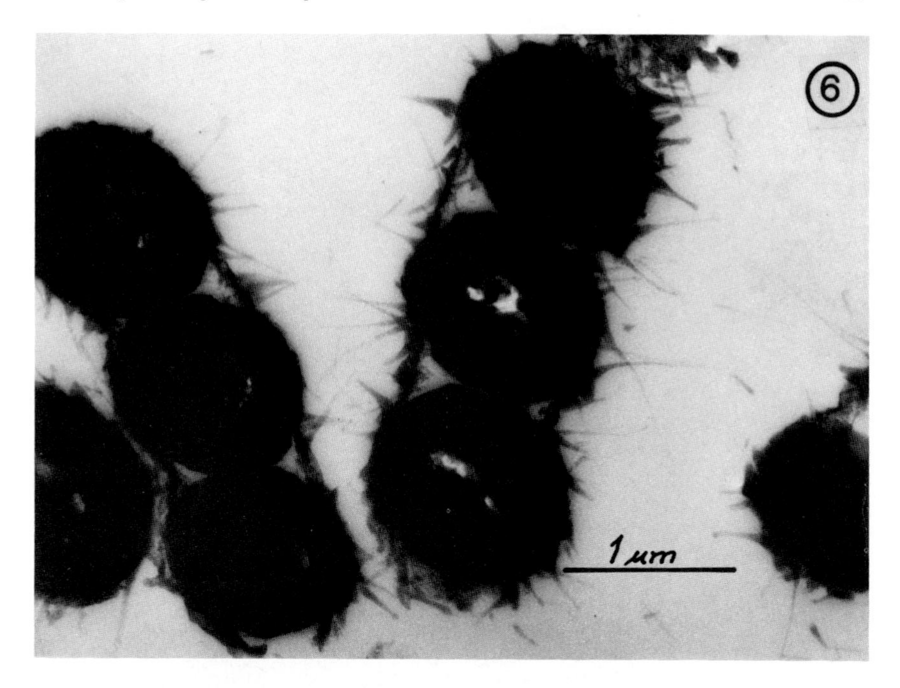

FIGURE 6. HVEM of a released spore chain showing that mature spores of S. bambergiensis *are surrounded with intact hairs. Even within the chain, hair components are seen inside the apparently loosely associated material which is usually called the "fibrous sheath".*

literature "to equate the rodlet mosaic (made of fibrous components) with the "fibrous sheath" which generally has a tenuous relationship with mature spores". Our work with S. *coelicolor* A3(2) (Smucker and Pfister, 1978) and with *S. bambergiensis*, presented herein, suggests that the rodlet mosaic fibrous layer is not on the outer surface, but appears covered with two spore coat layers. The stained preparations of spore hairs presented by Matselyukh (1978) and by the present work (Fig. 3) support the fibrillar component placement suggested by thick section (Fig. 5). Williams *et al.* (1972) previously suggested that an amorphous external layer, as seen by carbon replication, masked the details of the fibrous elements on *S. thermoflavus* smooth spores. Evidence presented in this paper from HVEM of thick (0.25 μm) sections is the first published indication of fibers in *Streptomyces* spore sections.

With the currently available data, it is clear that what is often called the "loosely associated sheath of fibrous sheath" in thin sections of *Streptomyces* spores may not in fact be fibrous. What has been called "the fibrous sheath" is in fact composed of at least three distinct layers. At present, only one of the hair components, the fibrillar component, has been chemically evaluated (Smucker, elsewhere in this volume)

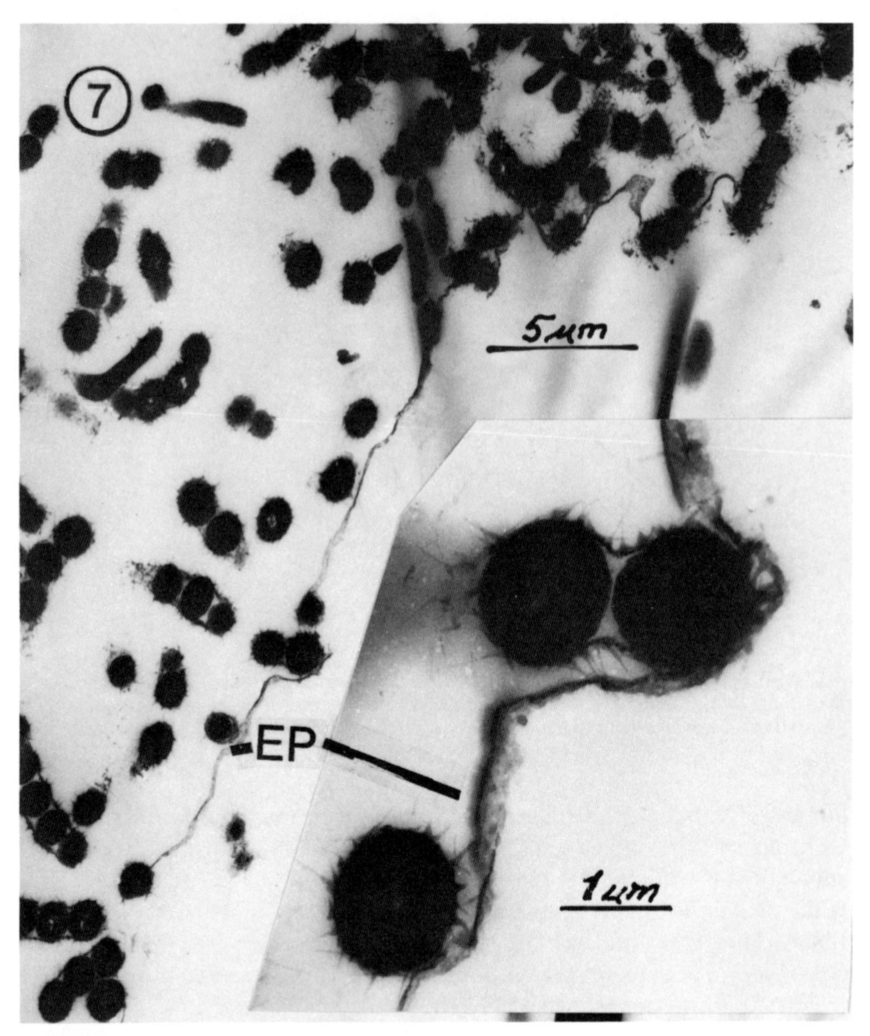

FIGURE 7. Low magnification HVEM of a thick section of aerial growth. The EP is seen bridging cells at the colony periphery. The inset is an enlargement of the boxed area and shows details of the spore and EP association.

V. SUMMARY

Previous reports using freeze-etch replication, whole cell replication, and thin sectioning have suggested that *Streptomyces* spores have an external ornamentation, called the rodlet mosaic, which may be responsible for the hydrophobic nature of *Streptomyces* spores. Previous work has shown that the rodlet mosaic in *S. coelicolor* A3(2) smooth spores is the third component

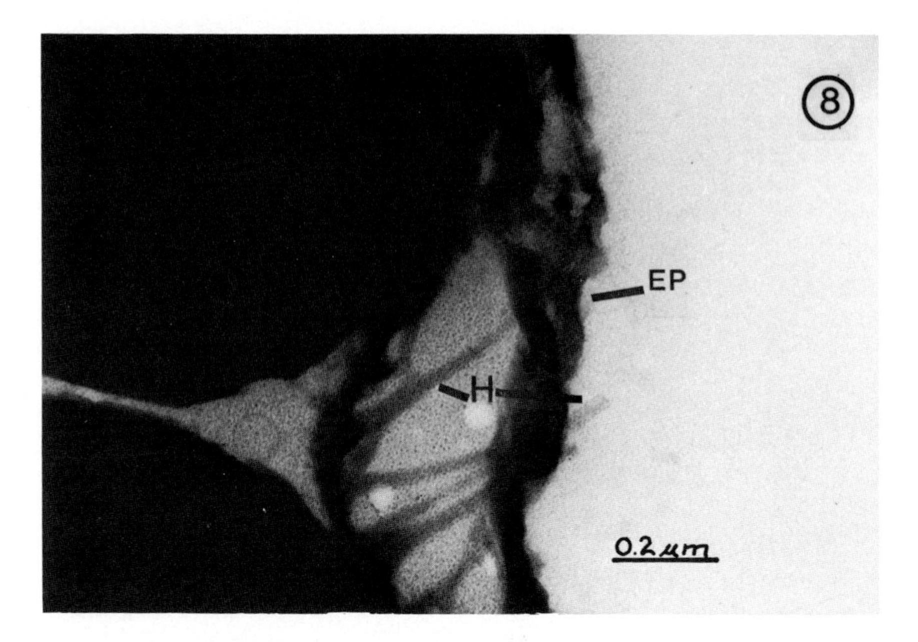

FIGURE 8. Higher magnification of the Figure 7 inset which shows the hairs penetrating the EP at several sites.

in from the external spore surface. In this study, *S. bambergiensis* aerial spores (hairy spores) which were harvested from chitin-agar lawns were prepared for transmission electron microscopy by sectioning, by freeze-etch replication, and by negative staining. Observations of hundreds of cells of each type of preparation led to the conclusion that the double-layered spore wall is covered with three morphologically distinct layers. The layers are, beginning with the one closest to the outer wall layer; the rodlet mosaic, the granular matrix, and the outermost layer. The fibrils of the layer which is called rodlet mosaic extended from the spore surface into a spiral which often appeared throughout the hair-like appendage. This layer of fibrils was covered with two layers. The central fibrils were covered with a negatively stained layer. High voltage electron microscopy of 0.25 μm sections revealed two layers outside the central fibers.

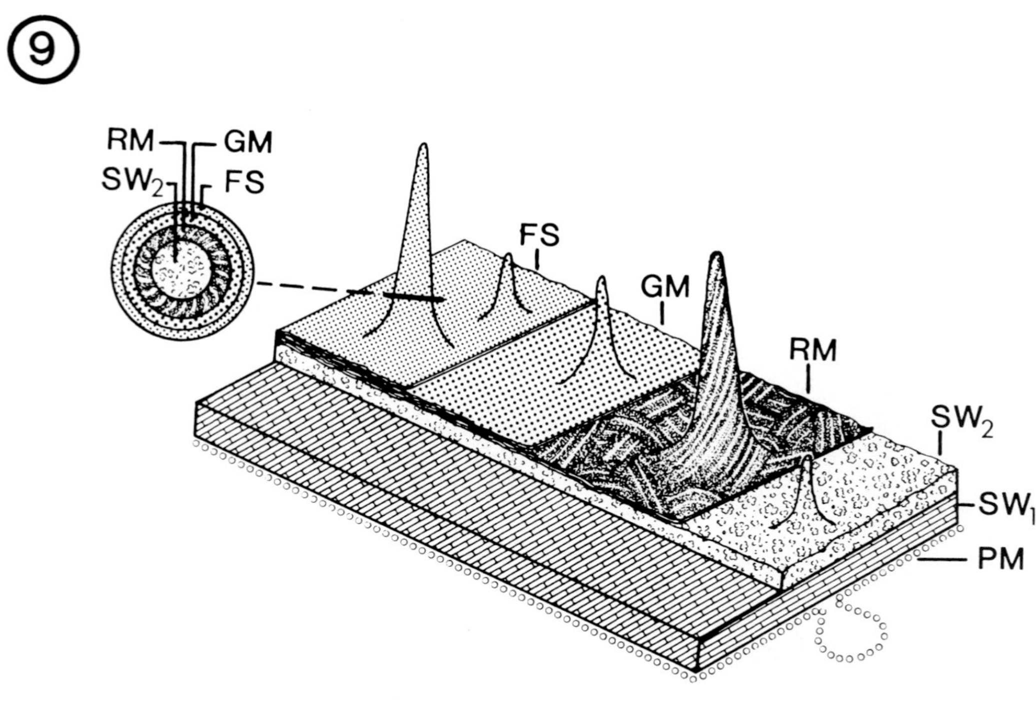

FIGURE 9. Schematic of S. bambergiensis *spore envelopes summarizing observations from conventional TEM of negatively stained spores, freeze-etch replicas, and thin sections and from HVEM of thick sectioned spores. Previous work suggests that the outer spore wall occasionally is found in the hair bases (Smucker and Simon, 1981). Legend; ESP, extra spore polymer; F, fibrils; GM, granular matrix, H, hair; SW_1, inner spore wall; SW_2, outer spore wall; PM, plasma membrane; RM, rodlet mosaic; S, spore.*

ACKNOWLEDGMENTS

The staff of the HVEM lab at the University of Wisconsin, Madison, Wisconsin, provided invaluable assistance in the HVEM analysis. Susanne Simon (University of Maryland, CEES) is also acknowledged for technical assistance with the 100 kV EM.
Freeze-etch replication was done by Mr. Eric Erbe and Dr. Russell R. Steere, USDA, Beltsville, Maryland. Fran Younger, Chesapeake Biological Laboratory, Center for Environmental and Estuarine Studies, University of Maryland, composed and drew the spore envelope schematic.
This paper is contribution No.1348 of the Center for Environmental and Estuarine Studies, University of Maryland, U.S.A.

REFERENCES

Bradley, S.G., and Ritzi, D., *J. Bacteriol. 95:*2358 (1968).
Coleman, R. H., and Ensign, J. C., *J. Bacteriol. 149:*1102 (1982).
Dietz, A., and Mathews, J., *Appl. Microbiol. 10:*258 (1962).
Hopwood, D. A., and Glauert, A. M., *J. Gen. Microbiol. 26:*325 (1961).
Lechevalier, H., and Tikhonenko, A. S., *Mikrobiologiya 29:*44 (1960).
Luft, J. H., *J. Biophys. Biochem. Cytol. 9:*409 (1961).
Luft, J. H., *Anat. Rec. 171:*347 (1971).
Matselyukh, B. P., *in* "*Nocardia* and *Streptomyces*" (M. Mordarski, W. Kurylowicz, and J. Jeljaszewicz, eds.), p. 440. Gustav Fischer Verlag, Stuttgart (1978).
Rancourt, M., and Lechevalier, H., *Can. J. Microbiol. 10:*311 (1964).
Reynold, D. M., *J. Cell. Biol. 17:*208 (1963).
Shirling, E. B., and Gottlieb, D., *Int. J. Syst. Bacteriol. 16:*313 (1966).
Smucker, R. A., and Pfister, R. M., *Can. J. Microbiol. 24:*397 (1978).
Smucker, R. A., and Simon, S. L., *in* "Sporulation and Germination" (H. S. Levinson, A. L. Sonenshein and D. J. Tipper, eds.), p. 317. American Society for Microbiology, Washington (1981).
Tresner, H. D., Davies, M. C., and Backus, E. J., *J. Bacteriol. 81:*70 (1961).
Vernon, T. R., *Nature (Lond.) 176:*935 (1955).
Walker, J., and Colwell, R. R., *Mar. Biol. 30:*193 (1975).
Wildermuth, H., *Arch. Mikrobiol. 81:*321 (1972).
Wildermuth, H., Wehrli, E., and Horne, R. W., *J. Ultrastruct. Res. 35:*168 (1971).
Williams, S. T., Bradshaw, R. M., Costerton, J. W., and Forge, A., *J. Gen. Microbiol. 72:*249 (1972).
Williams, S. T., Sharples, G. P., and Bradshaw, R. M., *in* "The Actinomycetales: Characteristics and Practical Importance" (G. Sykes and F. A. Skinner, eds.), p. 113. Academic Press, London (1973)

SPOROGENESIS IN THE *PILIMELIA* SPECIES[1]

Gernot Vobis

Fachbereich Biologie-Botanik
Philipps-Universität
Lahnberge, Marburg, F. R. G.

I. INTRODUCTION

The genus *Pilimelia* was classified by Kane (1966) as a member of *Actinoplanaceae*, having rod-shaped motile spores arranged in parallel chains within a sporangium. (In the present paper, the term sporangium is used in a functional sense (Ensign, 1978). It is a structure containing a mass of spores which are enveloped by its distinct wall layer. The spores escape only after the sporangial wall has ruptured; the sporangium, recognizable as an empty sac, is left behind. The sporangial wall is not a structural component of the spores.) The two described species, *Pilimelia anulata* and *P. terevasa*, are highly specialized in decomposing keratin. However, after less than a year in culture, they lost the ability to grow on keratin as a natural substrate (Kane-Hanton, 1974) and, consequently, also lost their typical morphological features. Presumably due to the general difficulties in culturing these spores and to the complicated baiting technique for their isolation, *Pilimelia* have not been examined as frequently as have related genera. Because of the high morphological similarity to *Ampullariella*, Cross and Goodfellow (1973) decided that the *Pilimelia* strains might be keratinophilic members of this genus.

A promising method of investigating the morphology and ultrastructure of *Pilimelia* is the culturing of these organisms on their natural keratinic substrate (Bland, 1968). We have now successfully cultured them both on mouse hair and on agar medium and have observed the stages of sporangial development in detail. We also examined the mechanism of spore discharge and the type of flagellation of the spores.

[1]*This work was supported by the grants PTB 8315/BCT 313 A of the "Bundesminister für Forschung und Technologie" of the Federal Republic of Germany.*

ISBN 0-12-528620-1

II. MATERIALS AND METHODS

Organisms and culture methods. The strains SK-6, SK-8, VK-114, VK-116, VK-118, VK-119, VK-121, VK-122, and VK-125 from our spore collection in Marburg (MB) were studied. All strains were isolated with the baiting technique described by Schäfer (1973) and were identified as *Pilimelia* species by morphological and physiological criteria. Good sporangial development was obtained on highly diluted (1:20) skim milk-agar (Gordon and Smith, 1955) which contained per liter of water: cattle horn meal (10 g); $Ca(NO_3)_2 \cdot 4H_2O$ (0.5 g); $MgSO_4 \cdot 7H_2O$ (0.7 g); K_2HPO_4 (0.005 g); $NaHCO_3$ (0.2 g) and trace $FeCl_3$ (Schäfer, personal communication).

Electron microscopy. For viewing thin sections in the transmission electron microscope (TEM), colonies were fixed, after 4, 6, or 21 days of incubation, at room temperature with glutaraldehyde and osmium tetroxide as previously described for *Streptomyces ramulosus* (Vobis, 1981). After dehydration in ethanol, they were embedded in Spurr low-viscosity embedding medium (Spurr, 1969). Sections were cut on a Reichert Ultracut 42 with a diamond knife and were post-stained with uranyl acetate (Watson, 1958) and lead citrate (Reynolds, 1963). For examination of flagellation, the extended spores were allowed to dry without fixation and were negatively stained with 2% uranyl acetate (Wildermuth, 1970). The preparations were examined with a Siemens Elmiskop I A.

For scanning electron microscopy (SEM), a very simple method of preparation was used (von Stosch, personal communication). Colonies grown on agar medium or on hair were fixed overnight with 1% osmium tetroxide, washed with distilled water, placed on cover slides, and immediately deep frozen on a cooled massive metal block. The samples were then dried in a closed plastic box containing phosphorus pentoxide. After several days, the cover slides were attached to the stubs and coated with a thin gold layer by using a sputter coater from Balzers Union. A Leitz AMR 1200 B was used to observe the samples.

FIGURES 1 to 4. Colonies on hair and agar medium. FIG. 1. Cylindrical to bell-shaped sporangia on mouse hair (VK-118, LM, 260 X). FIG. 2. Spherical sporangia on largely decomposed hair (SK-6, SEM, 860 X). FIG. 3. Sporangia on hair; damaged parts densely stained with cotton blue (enrichment culture, LM, 200 X). FIG. 4. Colony on agar medium with uneven surface and numerous sporangia (VK-122, SEM, 90 X).

Abbreviations in the figures
LM: light microscope; SEM: scanning electron microscope; TEM: transmission electron microscope; cl: columella; hw: hyphal wall; ma: matrix; me: mesosome; nu: nucleotide; ow: outer wall layer; pl: plasma membrane; ri: ribosomes; se: septum; sg: sporogenous hypha; so: spore; sp: sporangiophore; and sw: sporangial wall.

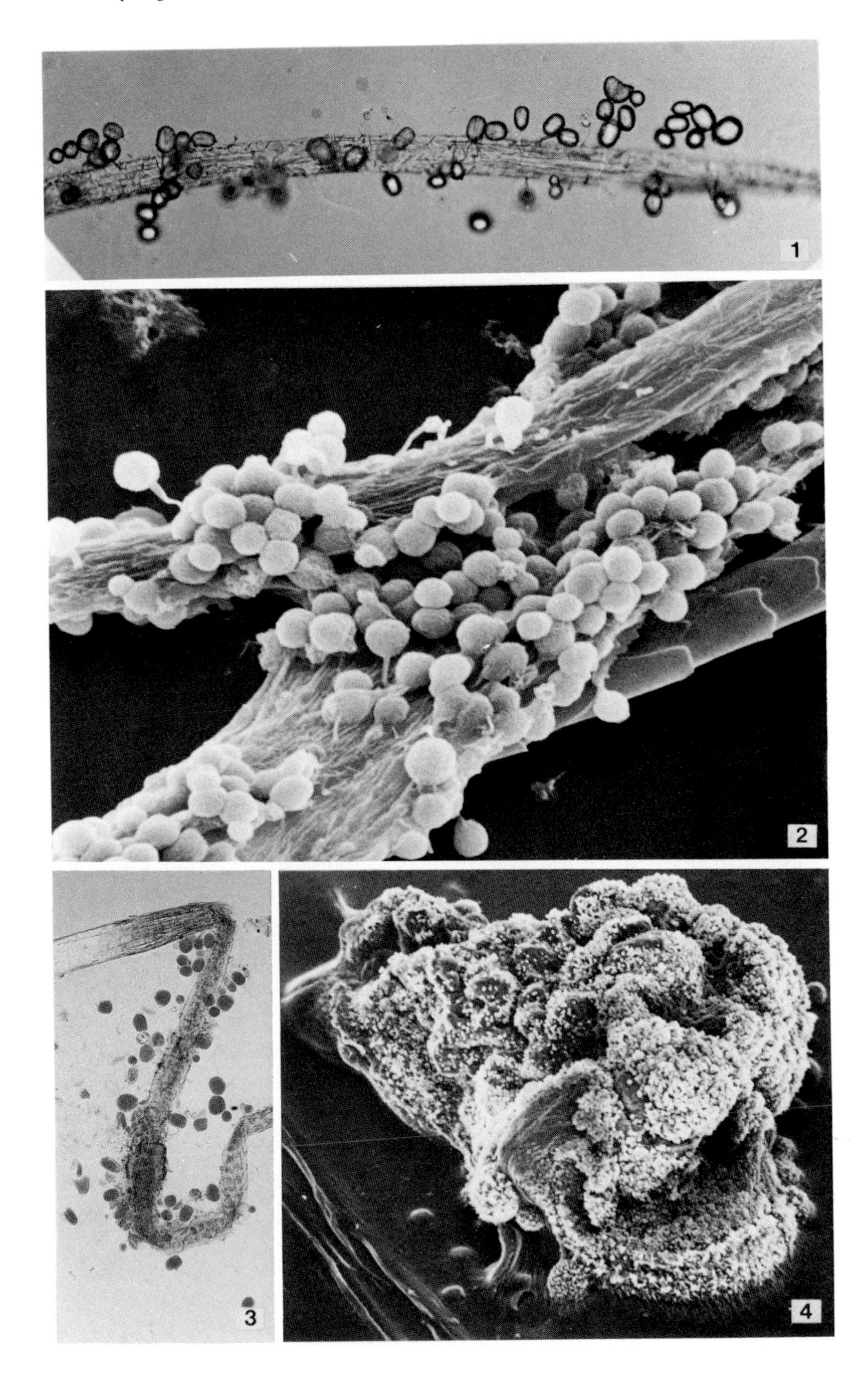
1
2
3
4

III. RESULTS

A. Morphology of Sporangia Producing Colonies on Hair and Agar Medium

For morphological study on the natural keratinic substrate, old colonies grown in Petri dishes were flooded with distilled water, and the mouse hair was placed on the surface of the water. A short time later, the hair was infested with released motile spores. After three weeks, tufts of sporangia were visible (Fig. 1). The hyphae not only covered the surfaces but also penetrated the hair. As a result of the decomposition of the keratinic substrate, the surface structure of the hair was completely destroyed (Fig. 2). As a result, the hair lost stability and became extremely fragile. The decomposed parts of the hair were stainable with cotton blue, whereas the undamaged parts rejected the dye (Fig. 3).

Colonies grew slowly on the agar medium, and were easily removed from the medium. Their surfaces were uneven and randomly patched with sporangia (Fig. 4). The typical morphology of sporangia of many strains was bell-shaped to cylindrical, with parallel rows of spore chains (Figs. 1 and 3). The isolates SK-6 and SK-8, described as *Pilimelia columellifera* (Schäfer, 1973), had globose sporangia in which the arrangement of spore chains was not clearly visible. These morphologically different sporangia also developed distinctively and, in following the text, they are called sporangial type I and type II.

B. Sporogenesis in Sporangia of Type I

On eight day-old colonies, finger-like aerial hyphae with rough exteriors grew above the surface of the colony (Fig. 5). Their tips thickened (Fig. 6) to form irregular mushroom-like heads (Fig. 7). These increased in size to become reverse bell-shaped, young sporangia which were carried by a large sporangiophore (Fig. 8). The ultrastructure of this development is demonstrated in Figures 10 to 13. The tip of the aerial hypha was covered by a loose, fibrous material that can be interpreted as a sheath or outer wall layer (Fig. 10). In the next stage, the sheath was extended at the end of the hypha and, below the tip, diminished to a small residue at the outside of the real hyphal wall of the young sporangiophore (Fig. 11). This hyphal wall layer thickened at the end of the hypha to a laminiform apex. All longitudinal growth then stopped. Further growth was possible only by extension of smaller hyphae in subter-

FIGURES 5 to 13. Development of sporangial type I. (FIGS. 7 and 9: VK-114; FIGS. 5, 6, 8, 10-13: VK-122) FIGS. 5-9. Single stages of sporangial development from aerial hypha to mature sporangium in SEM (FIG. 5: 400 X, FIG. 6: 450 X, FIG. 7: 460 X, FIG. 8: 270 X, FIG. 9: 310 X). FIG. 10. Tip of aerial hypha covered with a fibrous outer wall layer (TEM, 24,600 X). FIG. 11. Sporangial initium with laminiform apex and first extensions of sporogenous hyphae (arrowheads) (TEM, 34,500 X). FIG. 12. Sporangial initium with an increased network of fibrous material (TEM, 38,000 X). FIG. 13. Young sporangium with outgrowing sporogenous hyphae, embedded in intrasporangial matrix (TEM 17.500 X).

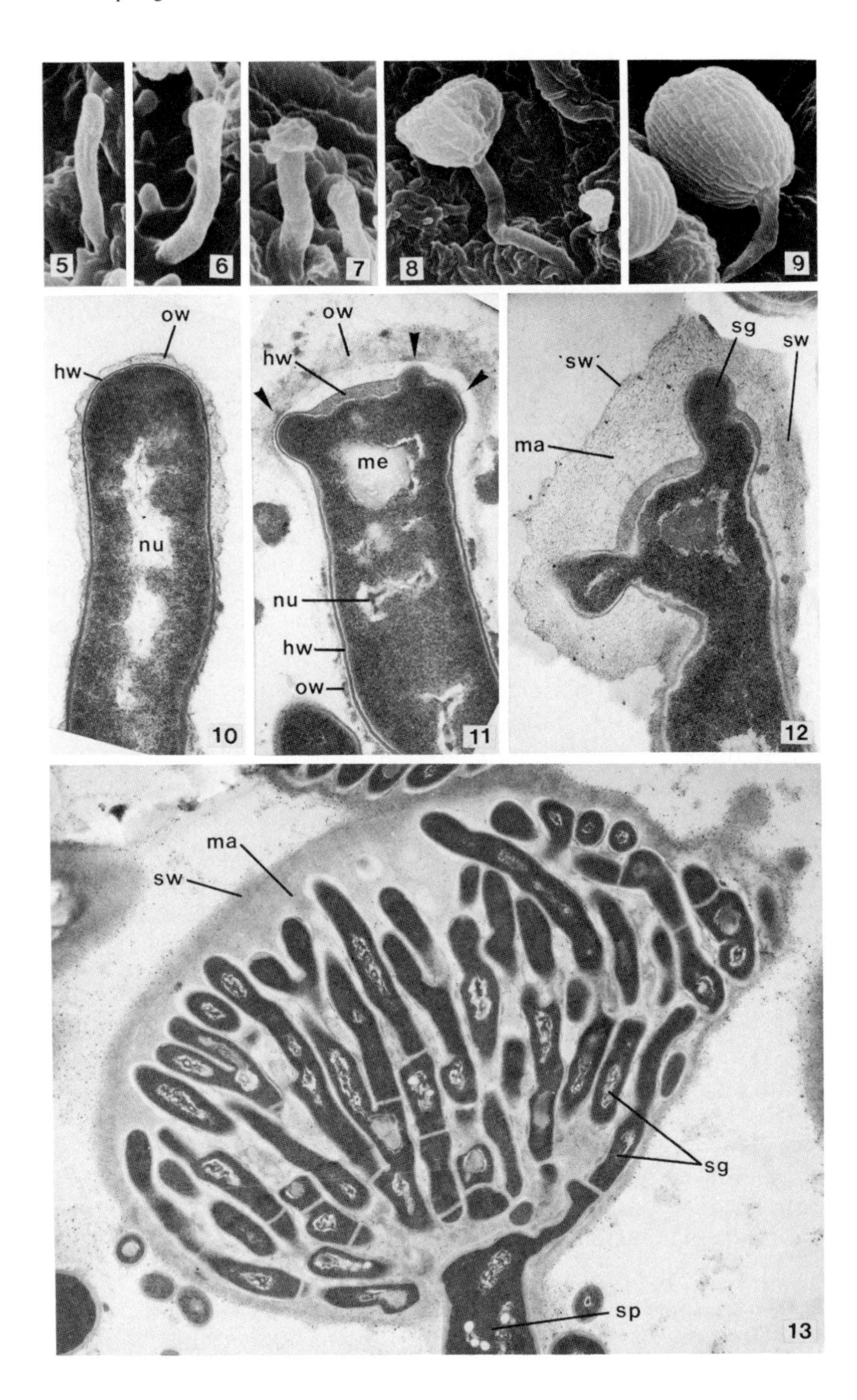
5
6
7
8
9
ow
hw
nu
10
ow
hw
me
nu
hw
ow
11
sg
sw
'sw'
ma
12
ma
sw
sg
sp
13

minal positions. The sporangial initium first enlarged through an increase of the fibrous material, which condensed on the surface to a more compact layer (Fig. 12). The small hyphae branched out immediately to become the initia of the sporogenous hyphae. Sporangiophore and sporogenous hyphae had a common cell wall layer. In the cytoplasm DNA-containing material and mesosomes were found (Figs. 10, 11, 12). Beside branching out, the sporogenous hyphae elongated, although always embedded in the fibrous material (Fig. 13). Growth of the hyphae caused enlargement of the young sporangium; the sporangiophore was separated from the sporogenous hyphae by cross walls at the base of the sporangium. The whole formation was enveloped by an outer wall layer, originating from the fibrous material named intrasporangial matrix.

At the tip of the sporogenous hyphae, the hyphal wall layer occasionally developed as a dispersed, lamellated network (Fig. 14). Apparently, the dichotomous ramifications were caused by these structures because they remained in the wedge areas of the branches as an integrated part of the cell wall (Fig. 15), until their eventual detachment from the cell wall. The division of sporogenous hyphae to rods of spore size resulted from formation of septa. From the inner part of hyphal wall a double-layered cross wall grew inward (Fig. 16). During spore maturation, the cross walls and longitudinal walls thickened. In the cytoplasm, ribosomes, mesosomes, and nucleotide-containing material were visible (Fig. 17). The walls of sporangiophorous hyphae were composed by the outer layer of amorphous material and the inner, real hyphal wall with small deposits of wall material in the cytoplasm. Nucleotide-material could be organized in a cylindrical form (Fig. 18). The mature sporangium was completely filled with dichotomously branched, parallel rows of rod-like spores embedded in the intrasporangial matrix (Fig. 19). The number of spores produced by one sporangium was about 20,000. The forms of sporangia were either bell-shaped (Fig. 19) or cylindrical (Fig. 9).

C. Sporogenesis in Sporangia of Type II

The sporangia arose only on the aerial hyphae that had extended above the surface of the colony (Figs. 20-23). Unlike the substrate hyphae, they had two distinct wall layers (Fig. 24). The inner, true hyphal wall was thin, but electron-dense in TEM sections; the outer layer was thicker, but more transparent (Fig. 25). At the tip of the aerial hyphae, the formation of sporangia began and was

FIGURES 14 to 19. Development of sporangial type I. (VK-122; TEM). FIG. 14. Lamellated hyphal wall (arrowheads) at the tip of sporogenous hyphae (55,000 X). FIG. 15. Lamellated hyphal wall (arrowhead) at the base of ramification (55,000 X). FIG. 16. Sporogenous hypha with ingrowing interspace septum (49,000 X). FIG. 17. Dichotomously branched chains of rod-like spores, lamellated parts of the wall separated (arrowheads) (40,500 X). FIG. 18. Cross section of sporangiophore with irregular deposits of wall material (arrowheads) (38,000 X). FIG. 19. Mature sporangium with numerous spores in parallel rows (4,400 X).

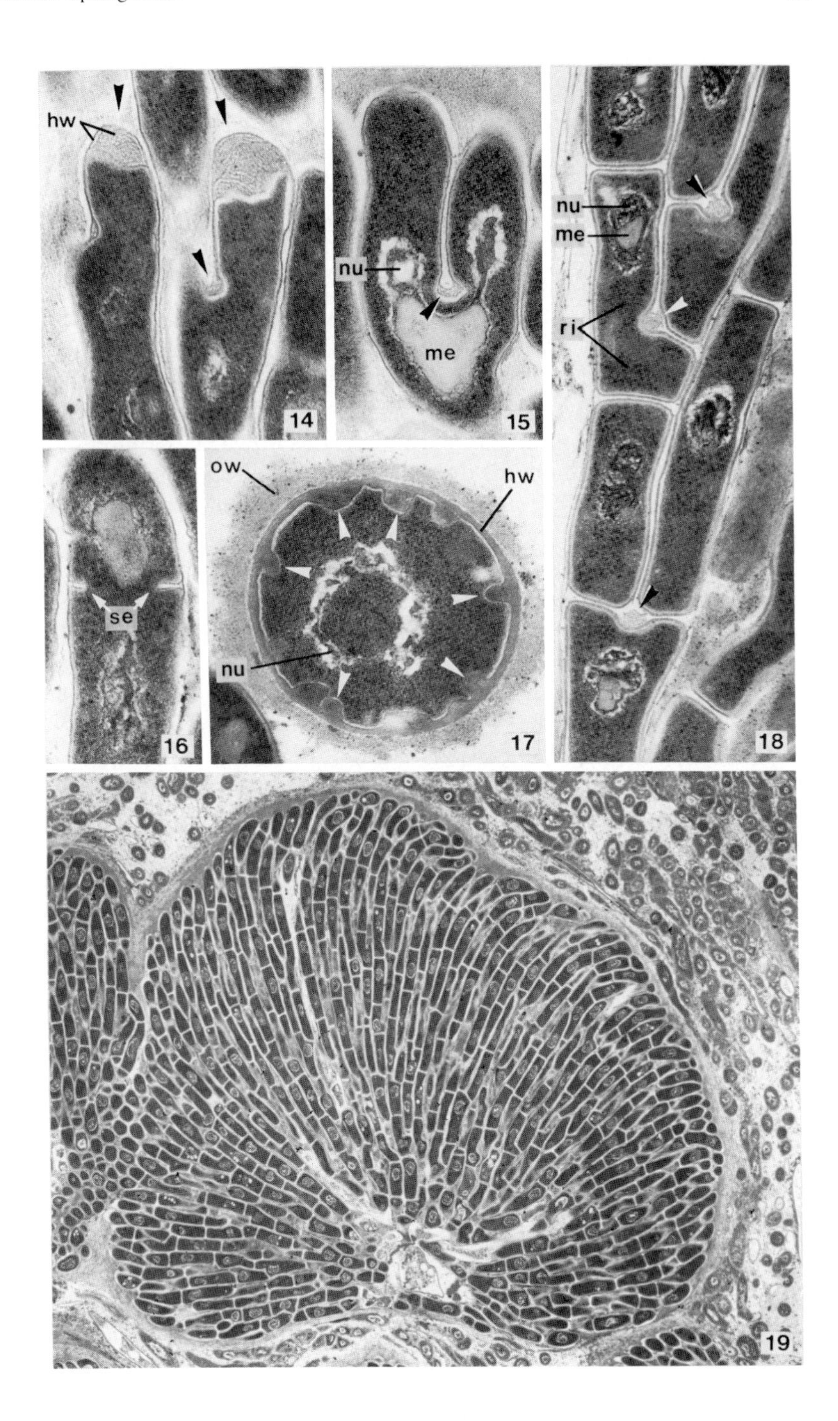
hw
nu
me
14
15
nu
me
ri
18
se
16
ow
hw
nu
17
19

first visible as a small vesicle (Fig. 22). The upper part of the aerial hypha extended into the sporangial initium, supported by the sporangiophore (Fig. 26). Small hyphae grew out from the uppermost tip of the sporangiophore. These young sporogenous hyphae branched out irregularly in all directions and also enclosed the upper part of the sporangiophore. This structure remained as a small column called columella (Fig. 26). The whole sporangial initium was filled with an amorphous material. The proportions of sporangial elements are shown at higher magnification in Figure 27. The columella and sporogenous hyphae had distinct, continuous hyphal walls. They were embedded in an amorphous matrix which could be separated from the hyphal wall. This matrix was closely connected to the outer wall layer that enveloped the sporangial initium and the sporangiophore. The common sheath can be designated as sporangial wall at the sporangial area and, further down, as the outer wall layer of the sporangiophore. On the inner, or true, wall layer, thickened ring-like deposits appeared on different segments of the sporangiophore (Fig. 28), but total formations of septa were never observed.

As the sporogenous hyphae increased, they lost their cytoplasmic contact with the sporangiophore through the formation of septa. Sporogenesis was introduced by septation of the sporogenous hyphae. Thin, ingrowing double cross walls divided the cytoplasm and mesosomes were observed in contact with these septa (Fig. 29). After formation of the typical interspace septa (Fig. 30), the sporogenous hyphae separated into single cells of spore size. The arrangement of the spore chains were visible only on sections of young sporangia in favorable positions (Fig. 31). The starting point or base of the sporogenous hyphae was at the tip of the columella. The arrangement of a spore chain is more or less comparable to a swivel, and in mature sporangia, these were organized into parallel rows visible even on the surface (Fig. 23). The interior ultrastructure did not change during sporangial maturation (Fig. 32) with the exception of the columella. The cytoplasmic content became increasingly transparent (Fig. 31) and totally disappeared in the ripe sporangium (Fig. 33). At the same time the cytoplasm of substrate hyphae also autolyzed and empty hyphal walls remained in the colony (Fig. 34).

D. Spore Discharge in Strain SK-6

Spore discharge was studied by using 0.4 mm thin sections of colonies on agar medium with abundant sporangia. These were observed under a cover

FIGURES 20 to 28. Development of sporangial type II. (SK-6). FIG. 20. Aerial hypha on surface of the colony (SEM, 2,500 X). FIG. 21. Transverse section of the margin of colony with aerial hypha (TEM, 2,700 X). FIG. 22. Sporangial initium at the tip of aerial hypha (SEM, 2,500 X). FIG. 23. Mature sporangium (SEM, 2100 X). FIG. 24. Aerial and substrate hyphae (TEM, 27, 000 X). FIG. 25. Tip of aerial hypha showing two distinct wall layers (TEM, 59,000 X). FIG. 26. Sporangial initium (TEM, 19,000 X). FIG. 27. Section of FIG. 26, showing the transition from sporangiophore to sporangium (TEM, 56,000 X). FIG. 28. Transverse section of sporangiophore with two wall layers (TEM, 56,000 X).

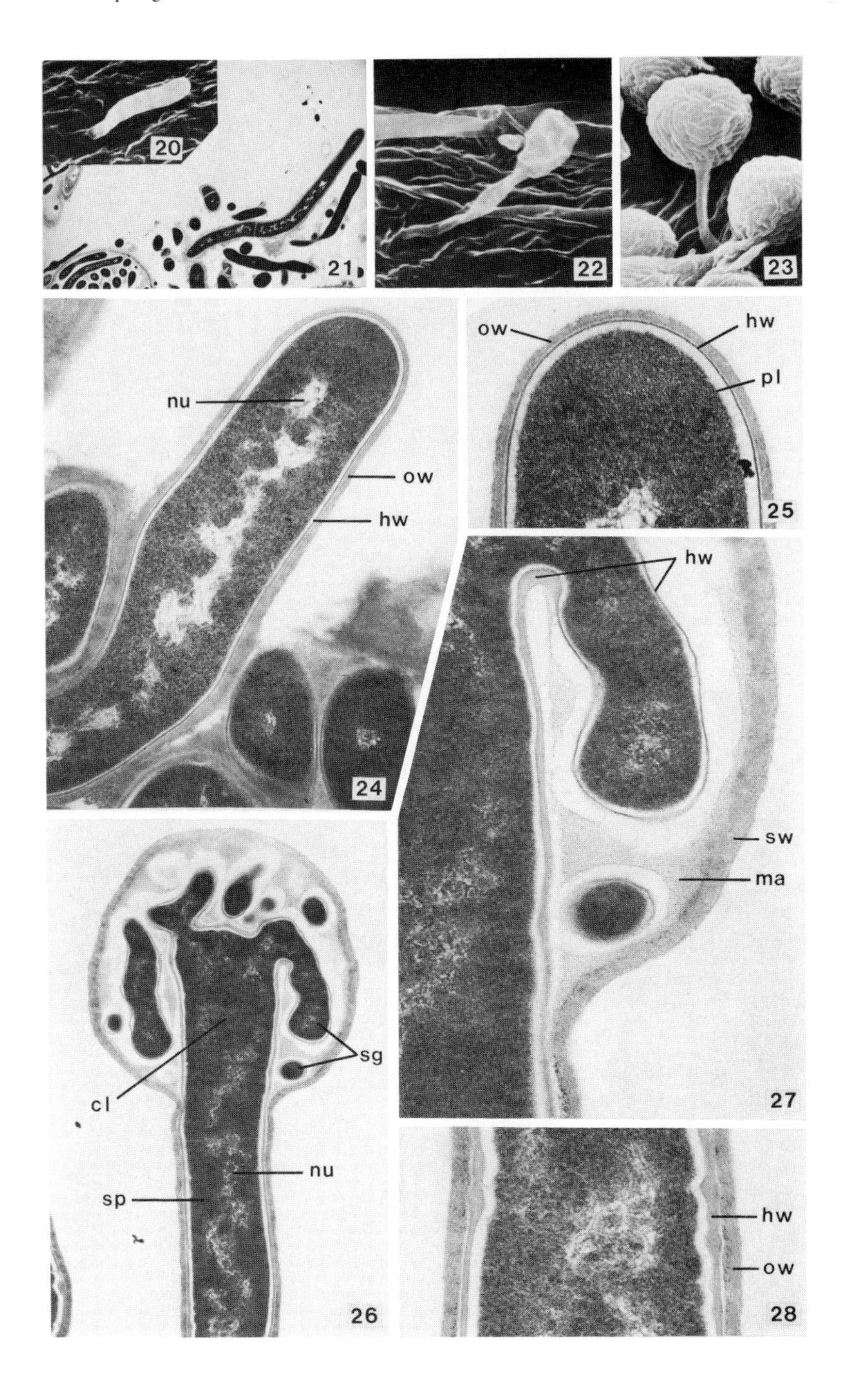
20
21
22
23
nu
ow
hw
24
ow
hw
pl
25
hw
sw
ma
27
sg
cl
nu
sp
26
hw
ow
28

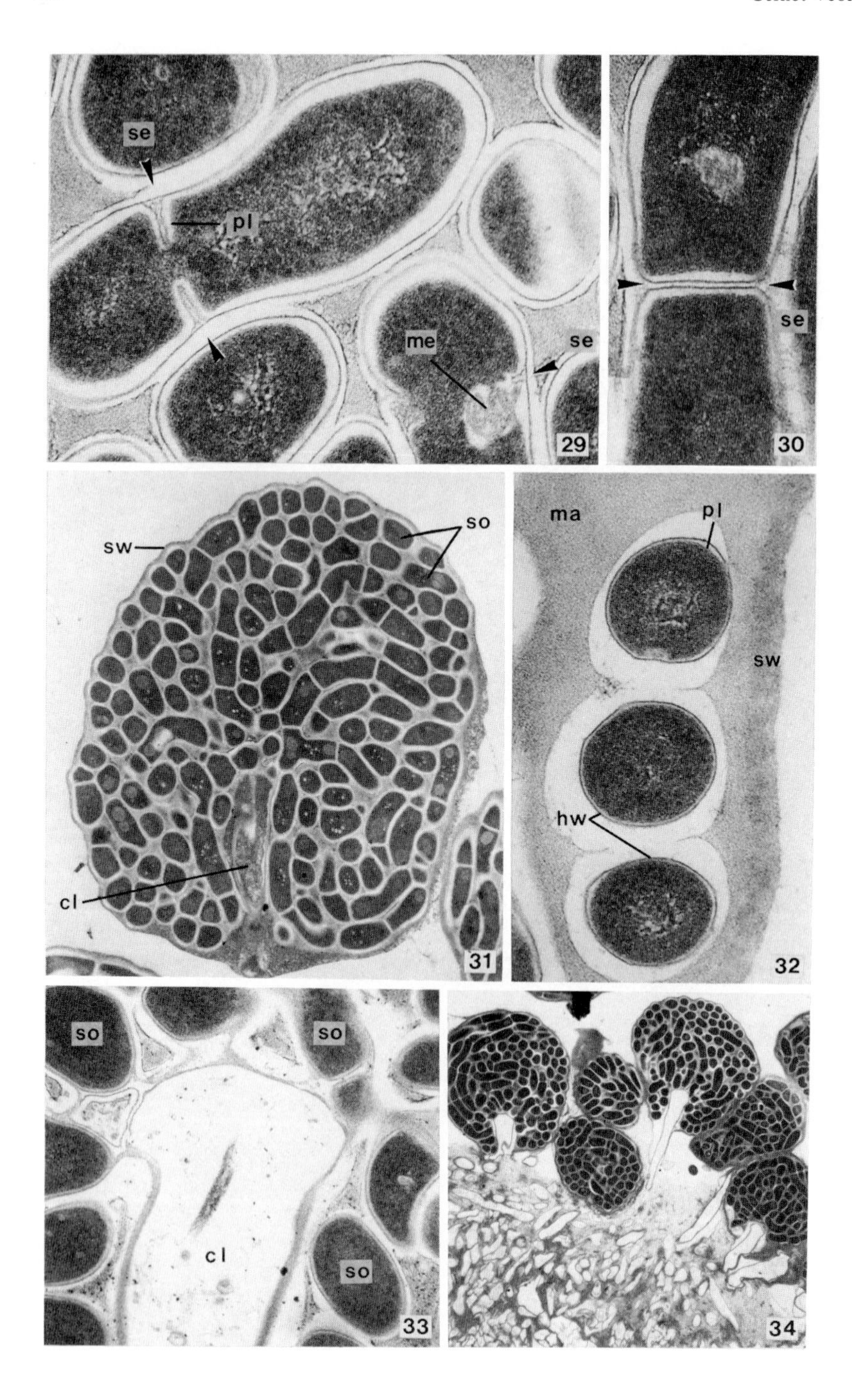
se
pl
me
se
29
se
30
so
sw
cl
31
ma
pl
sw
hw
32
so
so
so
cl
33
34

slide mounted in distilled water. After less than one hour, the spores were released. They were collected with a pipette and transferred onto grids for the negative staining procedure. The same sections of colonies were fixed for TEM ultrathin sections during spore discharge (Fig. 36). For SEM preparations, spores were allowed to discharge from sporangia which had been grown on hair (Fig. 37).

When not treated with water, the spores of mature sporangia remained embedded in the matrix and were held together by the sporangial wall (Fig. 35). Before movement was possible, disruption or even dissolution of the fibrous network of the matrix was necessary, after which the spores showed vigorous movement inside the sporangium before discharging. The upper part of the sporangial wall was partially perforated and the rest was recognizable as an empty sac (Fig. 36, 37). Peripherally situated spores seemed to escape first, followed by the spores near the sporangial base.

Dependent upon the capacity of swelling in water, the rod-like spores had different sizes; the average size was 0.35 to 0.45 by 0.8 to 1.5 μm. All spores had a laterally inserted tuft of flagella. The number of flagella varied from three to five (Fig. 38). A single flagellum might reach 4 or 5 μm in length and have a diameter of about 11 nm. In the most cases, the flagella of one spore lay together and functioned as one unit (Fig. 39). The spores swam indeterminately, without distinct direction. Not only single spores but also chains of two to three spores were observed.

IV. DISCUSSION

The family of *Actinoplanaceae* are distinguished from all other actinomycetes by the ability to form their spores within sporangia. With the exception of *Streptosporangium*, all genera are equipped with motile zoospores. Morphologically and developmentally, the *Actinoplanaceae* can be divided into two groups (Bland and Couch, 1981). The first group contains genera that have finger-like to pyriform sporangia with one of few spores. The members of the second group have spherical or cylindrical to irregular, multispored sporangia: *Actinoplanes, Amorphosporangium, Ampullariella, Pilimelia* and *Streptosporangium*. The morphology of sporangia, the arrangement of the spores, and their shapes are outlined in Figure 40. The spore chains can be organized in parallel rows, coils, or irregular patterns. No direct correlation between arrangement and shape of spores is evident. In general, combinations

FIGURES 29 to 34. Development of sporangial type II. (SK-6, TEM). FIG. 29. Formation of interspace septa within the sporogenous hyphae (70,000 X). FIG. 30. Completed interspace septum (70,000 X). FIG. 31. Young sporangium, spore chains arranged in form of a swivel (9,000 X), FIG. 32. Outer region of mature sporangium (70,000 X). FIG. 33. Upper part of the columella surrounded by sporogenous hyphae (36,400 X). FIG. 34. Mature sporangia on the surface of colony, plasma-like content of sporangiophores and substrate hyphae autolyzed (4,200 X).

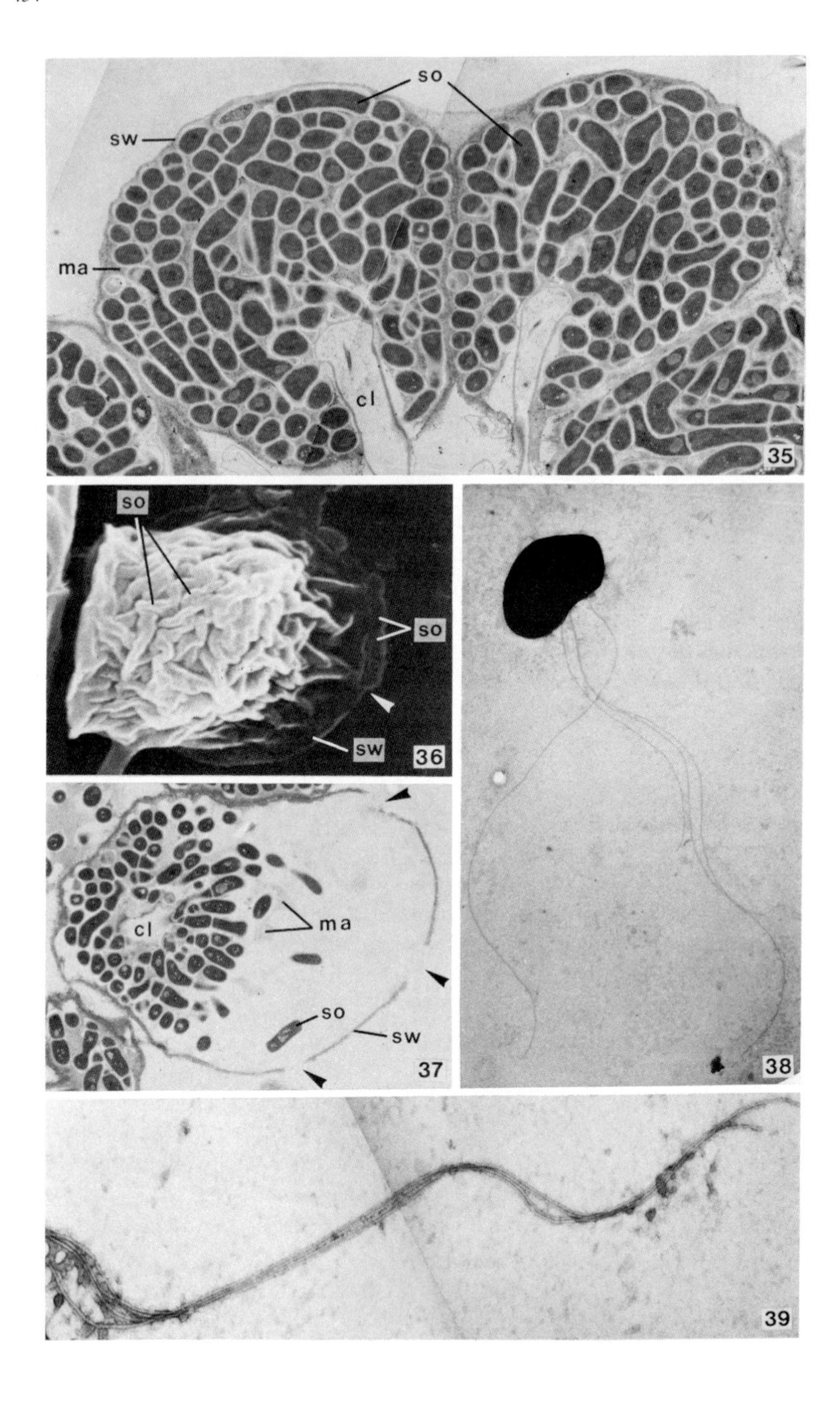
so
sw
ma
cl
35
so
so
sw
36
ma
cl
so
sw
37
38
39

of all the morphological features are sufficient to distinguish each genus; however, the genus *Pilimelia* is not clearly distinguishable from *Ampullariella* because both have parallel rows of spore chains and rod-shaped motile spores. Kane (1966) described the only two *Pilimelia*-species with a single polar flagellum that are accepted today (Skerman *et al.*, 1980). However, these observations are not verifiable because the type-strains lost their ability to form sporangia with intact spores in laboratory cultures. Other isolates of *Pilimelia* have spores with a laterally inserted tuft of flagella (Kane-Hanton, 1974; Schäfer, 1973), as shown in Figure 38. The type of flagellation is a distinct morphological feature for the genus *Pilimelia*.

After studying many isolates of keratinophilic members of *Actinoplanaceae*, we could demonstrate two different types of sporangial development. The bell-shaped or cylindrical type I has parallel rows of spore chains (Fig. 41) and resembles *Ampullariella* species (Couch, 1963, 1964) or *Actinoplanes rectilineatus* (Lechevalier and Lechevalier, 1975). All sporangia develop here with the same basic processes. The ultrastructural studies of *Pilimelia anulata* resulted in the same conclusion (Bland, 1968), but some remarkable differences exist. The tip of the sporangial initium has a thickened wall layer (Fig. 11); this laminiform apex was not visible in the *Actinoplanes* strains studied by Lechevalier and Holbert (1965) and Lechevalier, Lechevalier, and Holbert (1966). Additionally, the lamellated wall layers also appear at the tip of the sporogenous hyphae. A reference to similar structures in *Actinoplanes* is the striated cross wall at the base of the sporangium, which was observed by Sharples, Williams, and Bradshaw (1974). In *Acp. rectilineatus*, the formation of spores results from septation of the sporogenous hyphae, which occurs at widely separated intervals and is followed by intermediate septation (Lechevalier *et al.*, 1966). In *Pilimelia*-type I, the septation runs in basipetal direction in spore size intervals as the sporangium enlarges (Fig. 13). The septa correspond to the widely distributed cross wall type 2 described by Williams, Sharples, and Bradshaw (1973) and named "interspace septa" by Henssen *et al.* (1981).

The sporangial development of *Pilimelia*-type II (Fig. 41) is based on keratinophilic organisms with spherical sporangia penetrated by the column-like sporangiophore. The strains were described as *Pilimelia columellifera* in a thesis by Schäfer (1973). The spore size and the arrangement of the spore chains resemble those of the genus *Spirillospora* (Couch, 1963). Based only on these morphological characteristics and without studying any pure cultures, Tribe and Abu El-Suoud (1979) suggested transferring this species to

---

FIGURES 35 to 39. Spore discharge. (SK-6). FIG. 35. Sporangia with numerous spores, still embedded in matrix and held together by the sporangial wall (TEM, 8,400 X). FIG. 36. Nearly empty sporangium with ruptured wall, remnant of matrix and free spores visible (TEM, 6,200 X). FIG. 37. Partially emptied sporangium with perforated sporangial wall (arrowhead) and released spores (SEM, 4,500 X). FIG. 38. Spore with laterally inserted tuft of flagella (negatively stained, TEM, 19,000 X). FIG. 39. Tuft of four flagella as functional unit (Negatively stained, TEM, 47,500 X).

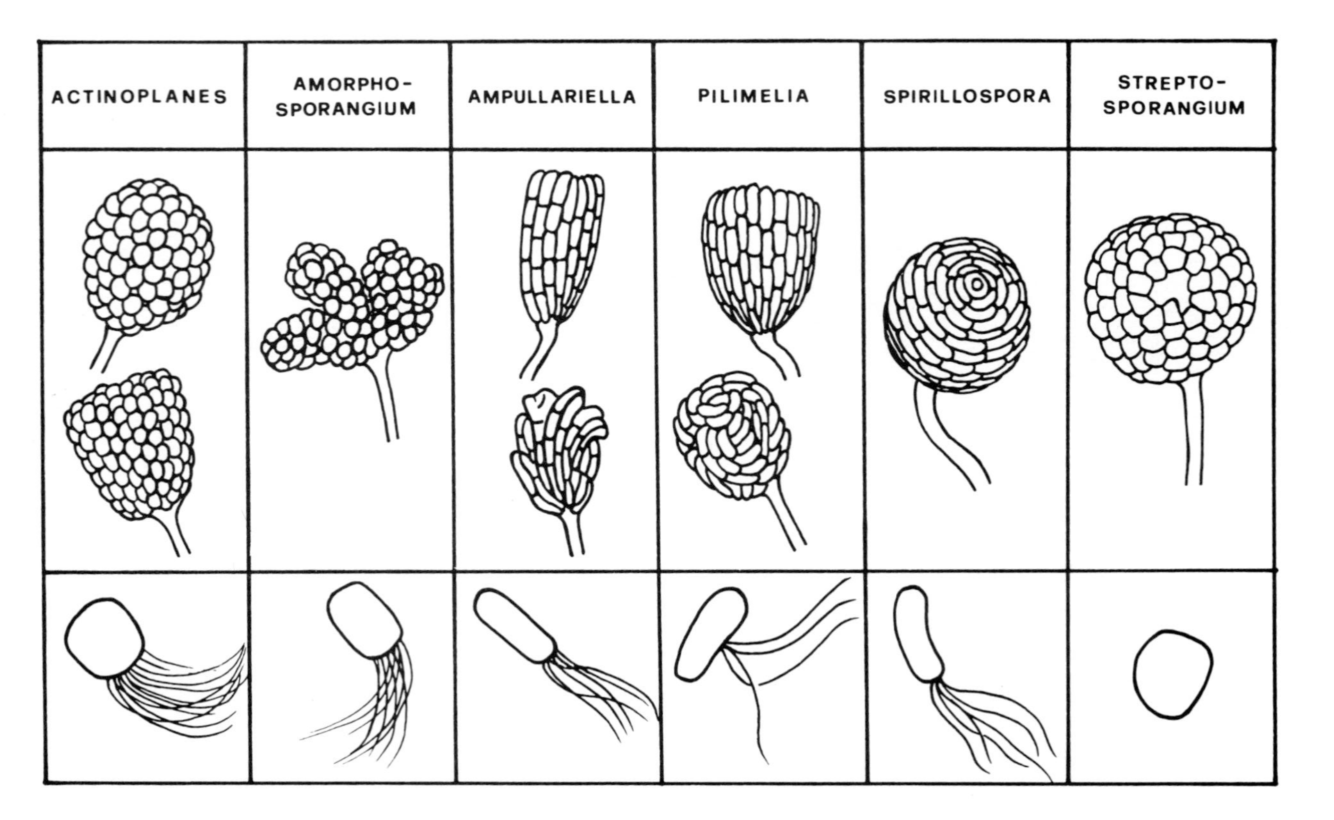

FIGURE 40. Comparison of the morphology of sporangia and spores of Actinoplanaceae *with multispored sporangia; compiled from Higgins* et al., *(1967), Kane-Hanton (1968, 1974), Lechevalier* et al., *(1966), Schäfer (1973) and personal data (Explanations see "Discussion").*

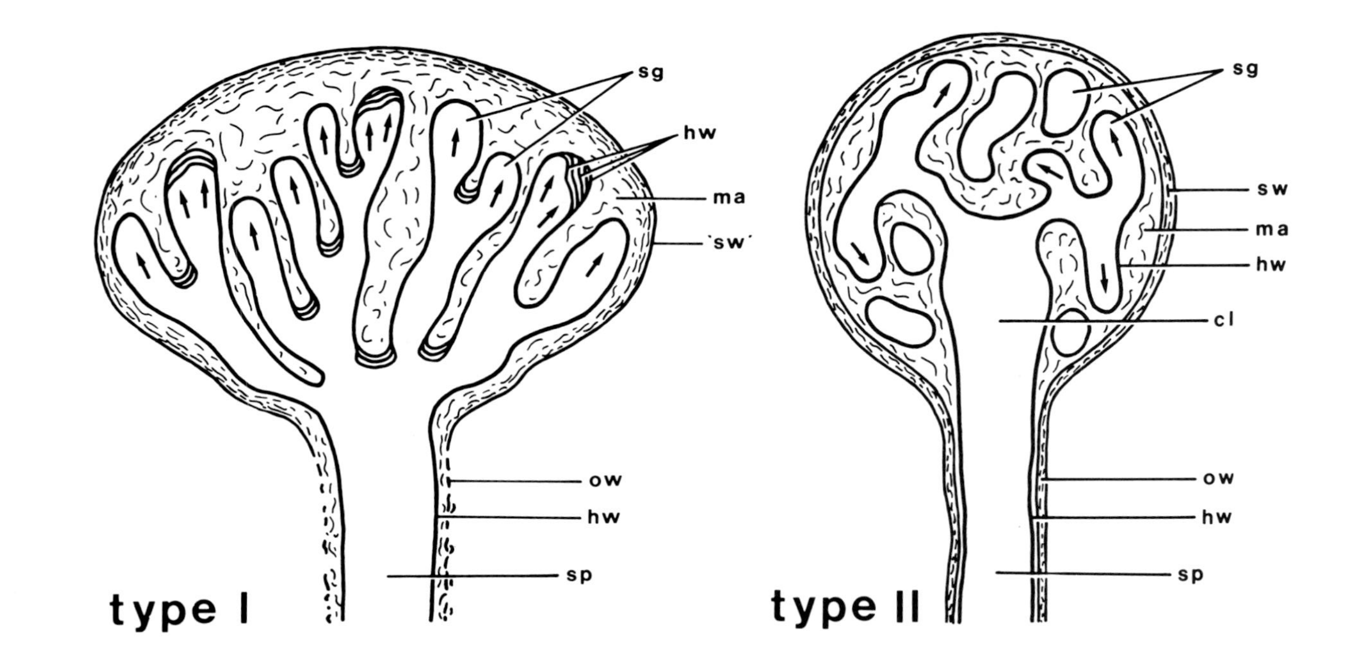

FIGURE 41. Schematic comparison of the two sporangial types of Pilimelia *at a young stage of development. Type I: Sporogenous hyphae grow out at the tip of sporangiophore and branch dichotomously (arrows), later forming parallel rows of spore chains. Type II: Sporogenous hyphae grow out at the tip of the sporangiophore and branch in all directions (arrows). The upper part of sporangiophore is also enclosed during growth and remains as a columella inside the sporangium.*

Spirillospora. The detailed study of the sporangial development allows a more exact interpretation. In *Spirillospora*, each branch of sporogenous hyphae grows in coils turning in the same direction (Bland and Couch, 1981; Lechevalier *et al.*, 1966). *Pilimelia*-type II also has branched sporogenous hyphae, but the coils twist in different directions. These coils even enclose the upper part of the sporangiophore. Therefore, the arrangement of the spore chains of mature sporangia is not as constant as in *Spirillospora*.

When the colonies of *Pilimelia* have reached maximal extension, the sporangia remain on the surface as multispored propagules. The spores are still embedded in the intrasporangial matrix and are additionally protected by the sporangial envelope (Fig. 35). In other genera of *Actinoplanaceae* with multispored sporangia, this envelope is interpreted to be analogous to the streptomycetous sheath (Cross, 1970; Lechevalier and Holbert, 1965; Williams *et al.*, 1973). In *Streptomyces*, the sheath surrounding the spore chains is composed of amorphous material and very small rod-like elements. When the spores disperse, the sheath remains as a structural component, forming the surface ornamentation of these spores (Wildermuth, 1970; Williams *et al.*, 1972). The sporangial envelope of *Actinoplanes* lacks the ultrastructural pattern of the streptomycetous sheath (Sharples *et al.*, 1974). In *Pilimelia*, the envelope is a thick wall layer distinguishable from the matrix, sometimes only by the degree of compactness.

ACKNOWLEDGMENTS

I am greatly indebted to Dr. D. Schäfer for his useful comments and to Mrs. H. Kisselbach-Heckmann, Mrs. E. Möller, and Mrs. C. Zimmermann for skillful technical help. Miss G. Traxler kindly reviewed the English text.

REFERENCES

Bland, C. E., *J. Elisha Mitchell Sci. Soc. 84:*8 (1968).
Bland, C. E. and Couch, J. N., *in* "The Prokaryotes, a Handbook on Habitats, Isolation, and Identification of Bacteria", Vol. 2. (M. P. Starr, H. Stolp, H. G. Trüper, A. Balows, and H. G. Schlegel, eds.), p. 2004. Springer, Berlin (1981).
Couch, J. N., *J. Elisha Mitchell Sci. Soc. 79:*53 (1963).
Couch, J. N., *Int. Bull. Bacteriol. Nomencl. Taxon. 14:*137 (1964).
Cross, T., *J. Appl. Bacteriol. 33:*95 (1970).
Cross, T. and Goodfellow, M., *in* "*Actinomycetales*: Characteristics and Practical Importance" (G. Sykes and F. A. Skinner, eds.), p. 11. Academic Press, London (1973).
Ensign, J. C., *Annu. Rev. Microbiol. 32:*185 (1978).
Gordon, R. E., and Smith, M. M., *J. Bacteriol. 69:*147 (1955).
Henssen, A., Weise, E., Vobis, G., and Renner, B., *in* "Actinomycetes" (K. P. Schaal and G. Pulverer, eds.), p. 137. G. Fischer, Stuttgart (1981).
Higgins, M. L., Lechevalier, M. P., and Lechevalier, H. A., *J. Bacteriol. 93:*1446 (1967).
Kane, W. D., *J. Elisha Mitchell Sci. Soc. 82:*220 (1966).
Kane-Hanton, W., *J. Gen. Microbiol. 53:*317 (1968)

Kane-Hanton, W., *in* "Manual of Determinative Bacteriology", 8th ed. (R. E. Buchanan and N. E. Gibbons, eds.), p. 718. Williams & Wilkins, Baltimore (1974).
Lechevalier, H. A., and Holbert, P. E., *J. Bacteriol. 89:*217 (1965).
Lechevalier, M. P., and Lechevalier, H. A., Int. J. Syst. Bacteriol. *25:*371 (1975).
Lechevalier, H. A., Lechevalier, M. P., and Holbert, P. E., *J. Bacteriol. 92:*1228 (1966).
Reynolds, E. S., *J. Cell Biol. 17:*208 (1963).
Schäfer, D., Dissertation, Marburg, (1973).
Sharples, G. P., Williams, S. T., and Bradshaw, R. M., Arch. Microbiol. *101:*9 (1974).
Skerman, V. B. D., McGowan, V., and Sneath, P. H. A., *Int. J. Syst. Bacteriol. 30:*225 (1980).
Spurr, A. R., *J. Ultrastruct. Res. 26:*31 (1969).
Tribe, H. T., and Abu El-Souod, S. M., *Nova Hedwigia 31:*789 (1979).
Vobis, G., *Zentralbl. Bakteriol. Hyg., I Abt. Orig. C. 2:*269 (1981).
Watson, M. L., *J. Biophys. Biochem. Cytol. 4:*475 (1958).
Wildermuth, H., *J. Bacteriol. 101:*318 (1970).
Williams, S. T., Sharples, G. P., and Bradshaw, R. M., *in* "*Actinomycetales*: Characteristics and Practical Importance" (G. Sykes and F. A. Skinner, eds.), p. 113. Academic Press, London (1973).
Williams, S. T., Bradshaw, R. M., Costerton, J. W., and Forge, A., *J. Gen. Microbiol. 72:*249 (1972).

THERMOACTINOMYCETES AS TERRESTRIAL INDICATORS FOR ESTUARINE AND MARINE WATERS[1]

Richard W. Attwell[2]
Rita R. Colwell

Department of Microbiology
University of Maryland
College Park, Maryland

I. INTRODUCTION

Actinomycetes have frequently been isolated from marine waters and sediments. They have been recovered from estuarine and coastal areas (Attwell *et al.*, 1981; Okazaki and Okami, 1976) and also from open ocean (Walker and Colwell, 1975; Weyland, 1969). But whether actinomycetes are active and autochthonous in saline natural waters has yet to be firmly established.

Many workers regard actinomycete marine isolates as terrigenous forms rather than as active members of the marine community. Grein and Meyers (1958) recovered actinomycetes from several sampling stations around the Canadian and North American coasts. They found no great difference in halotolerance between isolates from soil and those from the sea and concluded that the latter were wash-in forms. Kriss, Mishorstina, Mitskerich, and Zemksova (1967) reached similar conclusions since they isolated actinomycetes only infrequently during the course of extensive marine surveys. Other workers have provided evidence for at least some form of actinomycete activity in estuarine and marine environments.

Nocardia and *Streptomyces* species have often been isolated from decaying seaweed (Siebert and Schwartz, 1956). Weyland (1969) obtained large numbers of actinomycetes, generally in the range 1.0 X 10^2 to 1.5 X $10^3/cm^3$ sediment, from material obtained during extensive sampling in the North

[1]*Part of this work was performed under National Oceanic and Atmospheric Administration Grant 04-8-MO1-71 "Microbial Hazards Associated with Diving in Polluted Waters" and National Science Foundation Grant No. DEB 77-14646 "Systematics of the Genus* Vibrio".
[2]*Present Address: Department of Biological Sciences, Manchester Polytechnic, Chester Street, Manchester, U.K.*

ISBN 0-12-528620-1

Atlantic. He concluded that, since recoveries were frequent and in some cases were made more than one hundred miles from shore, the organisms obtained could not simply be wash-in forms. Subsequent work (Weyland, 1981) has substantiated this view, particularly with respect to nocardioforms isolates. Walker and Colwell (1975) and Okami and Okazaki (1978) observed a seasonal periodicity in the population size of some actinomycetes in coastal waters. The latter authors concluded that the organisms in question were active for at least some parts of the year.

II. THERMOACTINOMYCETES AS INDICATORS OF TERRIGENOUS ORIGIN

Members of the genus *Thermoactinomyces* have an optimum growth temperature of 50 °C although some strains are capable of growth at temperatures between 30° and 60°C. They grow in soils and decaying organic matter, particularly plant material, if suitably high temperatures are achieved. Their spores, which are resistant to a range of adverse conditions (Ensign, 1978), display considerable longevity and are found in large numbers in soil from which they are washed into aquatic systems (Cross and Johnston, 1972). Because these thermophilic organisms are unable to grow in cool waters, they are useful indicators of microbial contribution from surrounding land to aquatic environments (Al-Diwany and Cross, 1978).

Thermoactinomycetes have been isolated from water and sediments of streams, rivers (Al-Diwany and Cross, 1978; Cross and Johnston, 1972), estuaries, and coastal marine locations (Attwell *et al.*, 1981) by means of a very discriminating selective system. The recovery efficiency of the selective medium used is variable and often low when applied to natural populations, such found in aquatic environments (Al-Diwany *et al.*, 1978).

We therefore applied direct observation by means of epifluorescence microscopy combined with membrane filtration (Hobbie *et al.*, 1977) to the study of thermoactinomycetes in natural substrates (Attwell and Colwell, 1982). Water or sediment samples were mixed with the standard isolation medium, in liquid form, and then incubated at 50 °C for 3 h. The sample was stained with acridine orange and collected on the surface of a membrane filter which was then examined by means of epifluorescence microscopy. Spores with clearly developed germ tubes were counted and expressed as total viable count. The system provides an alternative to plate counts for estimating the number of viable spores in a substrate and is particularly useful when counts of thermoactinomycetes are low or not detectable by standard plate count.

TABLE I. Genera of Actinomycetales Isolated from Sediment Samples Collected in New York Harbor and from the New York Bight

	Thermoactino-myces	Dactylos-porangium	Microbi-spora	Micromo-nospora	*Nocardio-forms*	Saccharo-monospora	Strepto-myces	Strepto-sporangium
New York Harbor	+ [a]	+	+	+	+	*0* [b]	+	+
New York Bight "A"	+	*0*	+	+	+	+	+	+
New York Bight "B"								
capping material	+	*0*	*0*	+	+	*0*	+	+
dredging spoil	+	*0*	*0*	+	+	*0*	+	+

[a] *Isolated*
[b] *Not isolated*

III. THERMOACTINOMYCETES AS MARKER ORGANISMS IN DREDGING SPOIL

Since 1927, dredge spoils from New York Harbor have been barged approximately 12 miles offshore into the New York Bight area and dumped within designated sites. Between July and November 1980, sand was deposited over a previously used dump site area in the New York Bight in order to cap off and contain the dredging spoil. In the course of our studies on the distribution of actinomycetes in saline natural waters, we sampled sediments in New York Harbor and in the region of the New York Bight dredging spoil dump, *i. e.*, dumping reference point 73° 50.2' W, 40° 22.1' N (Attwell and Colwell, 1981).

Data presented in Tables I and II demonstrate that a range of actinomycetes were present in sediments of both New York Harbor and the New York Bight station 'A' which was not in vicinity of dump site. The range and numbers recovered from the harbor sediment are similar to those recovered from the Anacostia River and Chesapeake Bay (Attwell *et al.*, 1981) and are considerably greater than those found in the surface sediment of the New York Bight sampling site 'A', 73° 52' W, 40° 24' N. The dredging spoil contained a mesophilic aerobic actinomycete population similar to that of the harbor sediments, although two genera found in the latter were not represented in the spoil. Strains of *Micromonospora* constituted over 70% of the mesophilic actinomycetes recovered from the Harbor and Bight samples. This predominance may reflect longevity and consequent accumulation of *Micromonospora* spores, although there is evidence for growth in fresh water sediments under certain conditions (Johnston and Cross, 1976).

Those members of the genus *Thermoactinomyces* which were recovered would not have been capable of growth under the conditions prevailing in the sediments examined. When sediment samples from the New York Bight were held at 50 °C, the counts of thermoactinomycetes dropped and none of the bacteria could be recovered after 12 days. Elevated temperatures may promote germination of the highly resistant endospores, which are then unable to produce vegetative growth in sediment, with subsequent loss of viability. Counts in sediments held at 5 °C remained unchanged after one month. Thermoactinomycetes were present in similar numbers in the harbor sediment and in the capped dredge spoil in which they served as microbial markers. Counts were significantly lower in the capping material and Bight sediments from stations removed from the dump site. One species, *Thermoactinomyces sacchari*, was isolated only from harbor sediment and from dredging spoil.

Microbiological effects of the capping operation were easily detected, namely that bacteria were contained below the capping material and that large-scale mixing of spoil and capping sand did not occur. Detection of spoil-associated bacteria, *viz.*, *T. sacchari*, in upper layers of capping material at a future date would provide evidence of upward carriage of material through the capping layer, or of incomplete covering if erratic recoveries were obtained.

TABLE II. Actinomycete Counts Obtained for Sediment Samples Collected in New York Harbor and from the New York Bight

	Thermoactinomyces *(Total)*	*Mesophilic Actinomycetes*	Thermoactinomyces thalpophilis	Thermoactinomyces vulgaris	Thermoactinomyces sacchari
New York Harbor	9.7×10^{3} [a]	1.5×10^{4}	3.5×10^{2}	8.0×10^{3}	1.4×10^{3}
New York Bight "A"	3.3×10^{1}	7.3×10^{1}	1.5×10^{1}	1.8×10^{1}	*ND* [b]
New York Bight "B"					
capping layer	4.9×10^{1}	1.8×10^{2}	*ND*	4.9×10^{1}	*ND*
dredging spoil	1.3×10^{3}	1.1×10^{4}	1.2×10^{1}	1.0×10^{3}	3.3×10^{2}

[a] *Reported as colony forming units, (cfu)/g*
[b] *ND, Not detected*

Since the last spoil dump was in June 1980 and sampling at station 'B' was carried out in November 1980, the thermoactinomycete spores survived at least six months after dumping.

IV. THERMOACTINOMYCETES AS INDICATOR ORGANISMS IN OCEAN SEDIMENTS

Several sediment samples were obtained from a depth of over 4400 m in the Demarara Abyssal Plain by means of a box corer (Attwell *et al.*, 1981). Material was examined from two regions of the plain: region 'A', 49° 6' W, 8° 6' N, an area known to be influenced by outfall from the Amazon River; and region 'B', 45° 48' W, 10° 24' N, in an area remote from river or coastal influence.

Mesophilic aerobic actinomycetes were not recovered from any of the cores examined. Thermoactinomycete species were found in all cores from region 'A', giving a mean count of 1.5×10^1/g sediment, but were not isolated from those obtained in region 'B'. Counts were low, but consistent, showing no significant difference from core to core. It is possible that mesophilic actinomycetes may have been present in numbers too small to be detected by our methods. Those that were present may not have been able to develop colonies in the media and under the incubation conditions used, although these had proved successful in making isolations from other marine locations. An alternative but clear possibility is that organisms of this type cannot grow in such deep marine sediments. Furthermore, they do not have sufficient longevity to survive the period required for carriage from centers of dispersal and for deposition in such remote areas.

Thermoactinomycete species would not grow in the abyssal plain sediments, but did display sufficient longevity to retain viability during transfer to this region. That they are indicators of terrigenous inputs into the aquatic environments is emphasized by the fact that they were not recovered from that part of the abyssal plain remote from terrestrial influence. It is also significant that counts for aerobic heterotrophic marine bacteria made on 2216 marine agar (Difco), and not dependent on terrestrial input, were between 1.0×10^2 and 1.5×10^2/g sediment at both locations 'A' and 'B'.

V. THERMOACTINOMYCETES AS INDICATORS FOR THE INPUT OF TERRIGENOUS ACTINOMYCETES INTO AN ESTUARINE SYSTEM

The Chesapeake Bay forms a vast estuarine water system with many river tributaries. One such tributary, the Severn River, opens into the bay at Annapolis, Maryland.

TABLE III. Actinomycete Types Isolated from Sediment Samples Collected in the Severn River and Chesapeake Bay

Sampling station	Thermo-actinomyces	Micromono-spora	Strepto-myces	Strepto-sporangium	Microbi-spora	*Nocardio-forms*	Strepto-verticillium	Saccharo-monospora
1	+[a]	+	+	+	+	+	+	+
2	+	+	+	+	+	+	*0*[b]	+
3	+	+	+	+	+	+	*0*	*0*
4	+	+	+	+	+	+	*0*	*0*
5	+	+	+	+	+	+	*0*	*0*
6	+	+	+	*0*	*0*	*0*	*0*	+

[a] *Isolated*
[b] *Not isolated*

TABLE IV. Actinomycete Counts Obtained from Sediment Samples Collected in the Severn River and Chesapeake Bay

Sampling station	Thermo-actinomyces	*Mesophilic Aerobic Actinomycetes*	Thermoactinomyces *sp*	*Mesophilic Aerobic Actinomycetes*	Streptomyces *sp.* [a]	Micromonospora *sp.* [a]	Streptosporangtum *sp.* [b]
1	3.1×10^{3} [b]	3.6×10^{4}	7.9 [c]	92.1	26.6	51.3	13.9
2	1.6×10^{3}	1.9×10^{4}	8.0	92.0	45.0	40.0	5.0
3	1.2×10^{3}	3.0×10^{4}	3.9	96.1	32.2	33.0	2.0
4	1.4×10^{3}	2.0×10^{4}	6.7	93.3	38.0	47.6	4.7
5	6.4×10^{3}	8.2×10^{3}	43.8	56.2	6.8	34.0	6.1
6	9.5×10^{3}	6.0×10^{3}	62.3	38.7	7.7	25.8	ND

[a] *Most prevalent mesophiles isolated*
[b] *Counts expressed as colony forming units/g sediment*
[c] *Numbers expressed as % of total actinomycetes isolated*

Actinomycetes were isolated from stations on the Severn River and the Chesapeake Bay (Fig. 1). The stations selected covered a range of salinities from fresh water to marine levels. There was little difference between the range of genera isolated at station '1', a freshwater location far from the mouth of the Severn, and that at stations out in the Chesapeake Bay (Table III). However, the proportion of mesophilic isolates in the actinomycete population was significantly smaller at stations beyond Tolley Point (Table IV). Members of the genera *Micromonospora, Streptomyces*, and *Streptosporangium* were the

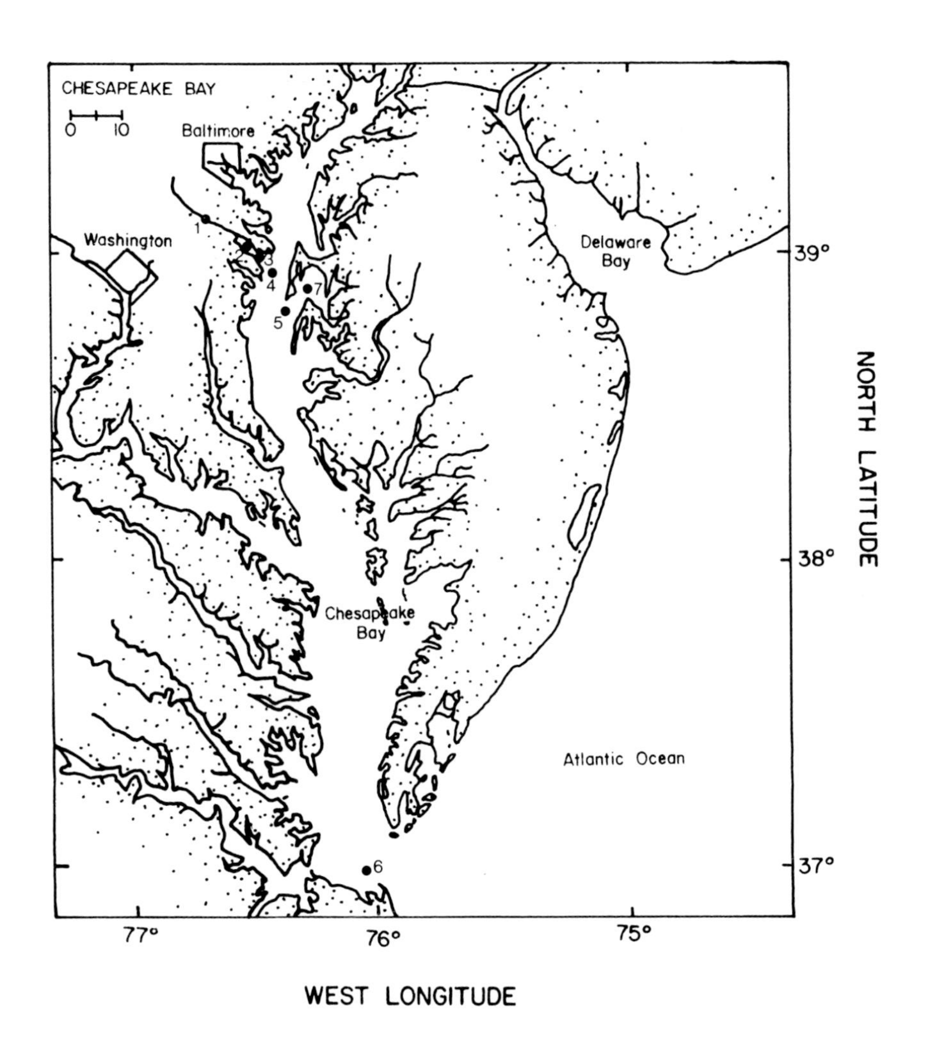

FIGURE 1. Location of sampling stations on the Severn River and Chesapeake Bay. 1, *Indian Landing;* 2, *Dreams Landing;* 3, *Annapolis Naval Academy;* 4, *Tolley Point;* 5, *Bloody Point;* 6, *Norfolk; and* 7, *Eastern Bay.*

most frequently isolated at all stations, but formed a much reduced fraction of the actinomycete population at stations '5' and '6'. This is particularly true of the latter two genera. Thermoactinomycetes, in contrast, accounted for a significantly greater proportion of the actinomycetes recovered from stations '5' and '6'. This increase cannot be accounted for by growth and must, therefore, be due to survival and accumulation of terrigenous thermoactinomycetes deposited in the sediments. None of the actinomycete genera detected increased their numbers relative to the thermoactinomycete indicator organism. Their distribution would thus appear to be governed by factors such as inactivation and dilution rates rather than by growth.

This study was restricted to a relatively short period of time (3 months). Nevertheless, the results demonstrate the value of thermoactinomycetes as markers for microbial input into an estuarine system. Numbers of mesophilic actinomycetes can be monitored throughout the year and changes in their numbers, rather than being expressed in absolute terms, can be correlated with those of the marker organisms, thus serving to demonstrate whether periodicity, such as that observed by Walker and Colwell (1975), is due to cycles of growth or to seasonal variation in the rates of microbial input into the estuary. In the latter case, a positive and significant correlation between the two groups should be evident.

VI. CONCLUSIONS

Thermoactinomycetes can clearly be used as indicator organisms in estuarine and marine situations, as they have been in fresh water (Al-Diwany and Cross, 1978; Cross and Attwell, 1974). They provide useful markers for the input of terrigenous microbes into natural waters and also for the movement or accumulation of microorganisms in man-made situations, such as the dredge spoil dump described above. We have also detected them in pharmaceutical wastes dumped into the ocean north of Puerto Rico, where they can be useful in monitoring the movement and dilution of microorganisms associated with the waste.

The use of *Thermoactinomyces* species as indicators is not restricted to inshore areas, since they can be isolated from deep marine locations, such as the Demarara Abyssal Plain as described earlier and the North Sea (Cross and Johnston, 1972), a finding attributable to the extreme longevity of thermoactinomycete endospores (Unsworth *et al.*, 1977), which enables these organisms to survive the long periods required for dispersal from land to open ocean locations. Studies to date, however, confirm that the presence of thermoactinomycetes is always associated with some form of terrestrial influence, even when remote from the sampling site.

Thermoactinomycetes have not yet been exploited as indicators in studies designed to resolve the question of whether selected actinomycete species are actively growing in the marine habitat. Marine actinomycetes have been shown to be capable of degrading complex organic compounds, such as cellulose, lignin, and starch, in oxygenated seawater (Willingham *et al.*, 1966). The isolates could have been wash-in forms capable of growth in saline water under laboratory conditions. It is quite possible that seawater acts selectively on terrestrial wash-in forms (Okami and Okazaki, 1974), with halotolerant survivors becoming active in coastal or estuarine waters where conditions and substrates are suitable. Changes in the size of actinomycete populations relative to thermophilic species known to be terrigenous and allochthonous thus can aid in the interpretation of results from samples taken *in situ*. Such an approach has been applied to bacterial populations in river water by Al-Diwany and Cross (1978).

Small numbers of thermoactinomycetes can be detected by means of epifluorescence technique (Attwell and Colwell, 1982). More useful information may also be gained if the source of individual thermoactinomycete species were known precisely. The ratios of the various species vary in different soil, water, and sediment samples for reasons which are not understood but presumably are associated with source and longevity of particular strains. *T. thalpophilus*, for example, can be isolated at all stations referred to in Figure 1, except from the waters and sediment of Eastern Bay.

In conclusion, members of the genus *Thermoactinomyces* have been found to be useful markers, yielding information about the distribution and survival of microorganisms in estuarine and marine environments. They offer the prospect of even more important application in pollution studies as more is learned of their physiology and ecology.

REFERENCES

Al-Diwany, L. J., and Cross, T., *in* "*Nocardia* and *Streptomyces*" (M. Mordarski, W. Kurylowicz, and J. Jeljaszewicz, eds.), p. 153. Gustav Fischer Verlag, Stuttgart (1978).
Al-Diwany, L. J., Unsworth, B. A., and Cross, T., *J. Appl. Bacteriol. 45:*249 (1978).
Attwell, R. W., and Colwell, R. R., *Mar. Pollut. Bull. 12:*351 (1981).
Attwell, R. W., Colwell, R. R., and Coolbaugh. *J. Mar. Sci. Technol. 15:*36 (1981).
Attwell, R. W., and Colwell, R. R., *Appl. Environ. Microbiol. 43:*478 (1982).
Cross, T., and Attwell, R. W., *in* "Spore Research 1973" (A. N. Barker, G. W. Gould and J. Wolf, eds.), p. 11. Academic Press, London (1974).
Cross, T., and Johnston, D. W., *in* "Spore Research 1971" (A. M. Barker, G. W. Gould and J. Wolf, eds.), p. 315. Academic Press, London (1972).
Ensign, J. C., *Annu. Rev. Microbiol. 32:*185 (1978).
Grein, A., and Meyers, S. P., *J. Bacteriol. 76:*457 (1958).
Hobbie, J. E., Daley, R. J., and Jasper, S., *Appl. Environ. Microbiol. 33:*1225 (1977).
Johnston, D. W., and Cross, T., *Freshw. Biol. 6:*464 (1976).
Kriss, A. E., Mishosstina, I. E., Mitskevich, N., and Zemtsova, E. V., "Microbial Populations of Oceans and Seas". Arnold, London (1967).
Okami, Y., and Okazaki, T., *J. Antibiot. 27:*240 (1974).
Okami, Y., and Okazaki, T., *in* "*Nocardia* and *Streptomyces*" (M. Mordarski, W. Kurylowicz, and J. Jeljaszewicz, eds.), p. 145. Gustav Fischer Verlag, Stuttgart (1978).

Okazaki, T., and Okami, Y., *in* "Actinomycetes: The Boundary Micro-organisms" (T. Arai, ed.), p. 125. Toppan Co. Ltd, Tokyo (1976).
Siebert, G., and Schwartz, W., *Arch. Hydrobiol. 52:*331 (1956).
Unsworth, B. A., and Cross, T., Seaward, M. R. O., and Sims, R. E., *J. Appl. Bacteriol. 42:*45 (1977).
Walker, J. D., and Colwell, R. R., *Mar. Biol. (Berl.) 30:*193 (1975).
Weyland, H., *Nature, (Lond.) 223:*858 (1969).
Weyland, H., *in* "Actinomycetes" (K.P. Schaal and G. Pulverer, eds.), p. 185. Gustav Fischer Verlag, Stuttgart (1981).
Wellingham, C. A., Roach, A. W., and Silvey, J. K. G., *Am. Midl. Nat. 75:*232 (1966).

ACTINOMYCETES IN MARINE SEDIMENTS

M. Goodfellow
J. A. Haynes[1]

Department of Microbiology
The University
Newcastle upon Tyne, U. K.

I. INTRODUCTION

Actinomycetes are a widely distributed and successful group of bacteria which have a number of properties that favor them in competition with other saprophytic microorganisms and ensure their survival under unfavorable environmental conditions. They are nutritionally versatile, produce several kinds of spores that serve as agents of dispersal and survival, and most form a radiating mycelium which facilitates the colonization of substrates distant from initial centers of growth. Actinomycetes are generally considered to contribute to the breakdown and recycling of the more recalcitrant naturally occurring organic compounds, but surprisingly little is known about the distribution, growth, activities, survival, and dissemination of actinomycete species in natural habitats. Until recently, ecological studies were severely hampered by the lack of suitable selective isolation techniques and the parlous state of much actinomycete systematics made it difficult to identify isolates even to the genus level.

Actinomycetes form an integral part of any balanced microbial community in soil, the majority of isolates being streptomycetes which mainly exist in the form of dormant spores (Cross, 1981; Williams 1978). The spores germinate in the presence of suitable plant and animal remains to form a limited branching mycelium bearing short chains of spores, the growth phase being restricted and discontinuous. The spores are continuously being washed into aquatic habitats where they accumulate in sediments. The widespread distribution and potential importance of streptomycetes in soil is indisputable, but it is becom-

[1] *J. H. gratefully acknowledges a studentship from the Natural Environmental Research Council.*

ISBN 0-12-528620-1

ing abundantly clear that actinomycete taxa, once considered rare, are in fact relatively common in terrestrial habitats. Thus, the development of specific selective isolation procedures have shown that large populations of *Actinomadura* (Athalye *et al.*, 1981), *Actinoplanes* (Makkar and Cross, 1982), *Micromonospora* (Cross and Attwell, 1974; Orchard, 1980), *Nocardia asteroides* (Orchard and Goodfellow, 1974), and *Rhodococcus coprophilus* (Rowbotham and Cross, 1977a) are present in terrestrial habitats.

Significant improvements have recently been made in the classification of actinomycetes by applying modern taxonomic methods (see Goodfellow and Board, 1980) to the revision of poorly described taxa. The application of numerical phenetic and chemotaxonomic techniques have been instrumental in establishing a framework for the current classification of the genera *Actinomadura* (Williams and Wellington, 1981), *Mycobacterium* (Goodfellow and Wayne, 1982), *Nocardia* (Goodfellow and Minnikin, 1981; elsewhere in this volume), and *Streptomyces* (Williams *et al.*, 1983a). Good classification is important as it provides an essential base for accurate identification. Microbial ecologists are not necessarily concerned with classification, but they do require simple, accurate diagnostic tests if they are to identify large numbers of isolates quickly. A combination of morphological, chemical, and spore characters can be recommended for the recognition of many actinomycete genera, but serious problems remain in assigning isolates to species (Goodfellow and Cross, 1974). Diagnostic tables have been recommended for the identification of actinomadurae, nocardiae, and rhodococci to the species level (Goodfellow and Schaal, 1979), but schemes such as these are no substitute for numerical methods of identification based upon diagnostic tests known to be reproducible (Wayne *et al.*, 1980). A probability matrix has recently been constructed and successfully used to identify streptomycetes from a variety of habitats (Williams *et al.*, 1983b; elsewhere in this volume).

The combination of isolation procedures selective for specific taxa and improved systematics makes the prospect of ecological studies on actinomycetes an attractive proposition. Considerable progress has already been made in unravelling the ecology of *Rhodococcus coprophilus* and *Nocardia asteroides* (Orchard, 1980, 1981; Orchard and Goodfellow, 1980; Rowbotham and Cross, 1977a,b). Strains of the former grow in herbivore dung, but the coccal survival stage contaminates grass and hay and remains viable after ingestion and passage through the rumen and intestines of ruminants, thereby contaminating the voided excrement. Dung from fields and farm effluents are washed into streams so that high numbers of *R. coprophilus* propagules can be found in rivers and lake muds where they remain inactive. *R. coprophilus* is a good example of a wash-in terrestrial organism and may prove to be a reliable indicator of farm animal pollution (Al-Diwany and Cross, 1978; Cross, 1981a). *N. asteroides* is widely distributed in soil and there is evidence that it is an active member of the autochthonous flora (Orchard, 1980, 1981). Nocardial populations in soil are influenced by a number of factors which include vegetation cover, soil type, and the presence of grazing animals.

The improved selective isolation and identification methods have also shed some light on the ecology of actinomycetes occurring in freshwater habitats (Cross, 1981a; Cross *et al.*, 1976). Large populations of *Micromonospora, Streptomyces*, and *R. coprophilus* have frequently been found in both water and sediment samples from streams, rivers, and lakes, but it seems likely that the majority of these actinomycetes are wash-in forms that lie dormant in such environments. Al-Diwany and Cross (1978) found a positive and significant correlation between the numbers of micromonosporas and thermophilic thermoactinomycetes isolated from the River Wharfe in West Yorkshire. This was a particularly interesting result for it is now known that species of *Thermoactinomyces* can grown rapidly in high temperature environments such as composts, overheated fodders, and isolated surface soil in which they produce endospores that can be washed into aquatic habitats where they can remain dormant but viable for hundreds of years (Cross 1981b, 1982; Cross and Attwell, 1974; Cross and Johnson, 1971). It seems most unlikely, therefore, that micromonosporas are an integral part of the freshwater microflora. On the other hand, it is increasingly being shown that thermoactinomycetes can serve as a useful indicator of the soil component of actinomycete propagules in water and sediments (Attwell and Colwell, 1981; elsewhere in this volume).

There is some evidence that actinomycetes can become active in freshwater habitats should they encounter suitable substrates for growth (Cross, 1981a). Thus, *Actinoplanes* can grow on vegetable matter in rivers, *Micromonospora* in timber foundation piles, *Streptomyces* on strips of chitin submerged in streams, *N. amarae* in activated sludge, and *N. asteroides* on rubber rings in water and sewage pipes.

II. ACTINOMYCETES IN MARINE HABITATS

Relatively little attention has been paid to actinomycetes found in the marine ecosystem. In early surveys of microorganisms from marine habitats, actinomycetes were mentioned incidentally if at all (Kriss, 1952; Kriss *et al.*, 1951; Wood, 1953; ZoBell, 1946; Zobell and Upham, 1944) and in later studies the emphasis was on "census taking" and the characterization of a small number of isolates (Table I and II). Since few of these studies were based on recognized selective isolation procedures and diagnostic tests, they provided little useful information on the occurrence, distribution, numbers, or types of actinomycetes occurring in the marine environment as a whole. The most comprehensive investigations have been carried out by Weyland (1969, 1970, 1981a,b) who found that actinomycete counts from marine sites were low compared with those from terrestrial habitats. The number of bacteria, as opposed to the corresponding actinomycete count, was found to decrease sharply with depth so that the ratio of actinomycetes to other bacteria was generally higher

TABLE I. Viable Counts (Mean Values of Colony Forming Units/ml Wet Sediment) and Types of Actinomycetes in Sediments Collected at Different Depths of the Seafloor [a]

Site	Depth (meters)	Number of sampling sites	Number of actinomycetes (mean values)	Actinomycetes Total bacteria (%)	Mean number of actinomycetes identified at each site: Number of sampling sites	Micromonospora		Streptomyces		Nocardioforms	
						Total	Number per site	Total	Number per site	Total	Number per site
Biscay	0-200	12	180	5.98	47	337	7.2	25	0.5	99	2.1
	200-1000	14	150	4.64							
	1000-2000	6	180	7.11							
	> 2000	20	100	25.64							
Iberian Sea	0-200	1	483	0.64	27	98	3.6	48	1.8	17	0.6
	200-1000	11	233	1.55							
	1000-2000	10	186	2.40							
	> 2000	13	131	4.68							
North Atlantic	0-200	5	100	0.17	21	21	1.0	4	0.2	240	11.4
	200-1000	6	1540	60.63							
	1000-2000	9	160	21.62							
	> 2000	10	16	9.41							

North Sea	0-200	38	480	0.29	106	485	4.6	465	4.4	74	0.7
	200-1000	5	790	11.88							
Off Faroes	0-200	21	178	0.06							
	200-1000	13	433	16.52	45	86	1.9	72	1.6	320	7.1
	1000-2000	11	569	51.26							
	> 2000	1	69	13.80							
Off North West Africa	0-200	10	46	0.19							
	200-1000	8	85	0.41	22	6.8	3.1	13	0.6	14	0.6
	1000-2000	7	85	6.16							
	> 2000	8	46	9.58							
Off Spitzbergen	0-200	16	160	0.18							
	200-1000	41	298	0.09	83	283	3.4	75	0.9	454	5.5
	1000-2000	14	164	1.54							
	> 2000	16	63	13.69							

[a] Data from Weyland (1981a).

TABLE II. Source and Types of Actinomycetes Isolated from Marine Habitats

Source	Organisms	Reference
Seawater and sediment, California, USA	Micromonospora, Mycobacterium, Nocardia, *and* Streptomyces *spp.*	*ZoBell* et al. *1943*
Sediment and kelp, California, USA	Streptomyces *spp.*	*ZoBell and Upham, 1944*
Intertidal sediment, Atlantic Coast, USA	Streptomyces *spp.*	*Hum and Shepard, 1946*
Seawater and sediment, Black Sea, USSR	Streptomyces *spp.*	*Kriss* et al. *1951; Kriss, 1963*
Sediment, Chukchi Sea, USSR	Nocardia *and* Streptomyces *spp.*	*Kriss, 1952*
Intertidal sediment, Scotland, UK	Streptomyces *spp.*	*Webley* et al. *1952*
Estuarine sediment, seawater and teleost fish, Eastern Australia	Nocardia *and* Streptomyces *spp.*	*Wood, 1953*
Fish-nets and cordage, Bombay, India	Nocardia *and* Streptomyces *spp.*	*Freitas and Bhat, 1954*
Intertidal sediment, Japan	Streptomyces *spp.*	*Saito, 1955*
Decaying seaweed, England, UK	Streptomyces *spp.*	*Chesters* et al. *1956*
Decaying seaweed, Germany	Nocardia *and* Streptomyces *spp.*	*Siebert and Schwartz, 1956*
Littoral sediments and seawater, Canada and USA	Micromonospora, Nocardia, *and* Streptomyces *spp.*	*Grein and Meyer, 1958*
Sediment, English Channel, Skagerrak and North Sea	Microbispora, Micromonospora, Nocardia, *and* Streptomyces *spp.*	*Weyland, 1969*
Sediment, Skagerrak, Barents and Norwegian Seas	Actinoplanes, Micromonospora, Nocardia, *and* Streptomyces *spp*	*Weyland, 1970*
Seawater, Baltic Sea	Geodermatophilus, *sp.* (Blastococcus aggregatus)	*Ahrens and Moll, 1970*
Canvas cloth in seawater, Arabian Sea, India	Nocardia *and* Streptomyces *spp.*	*Betrabet and Kasturi, 1971*
Sediment, Bay of Bengal, India	Streptomyces *spp.*	*Chandramohan* et al. *1972*
Intertidal sediment, seawater and algae, White Sea, USSR	Micromonospora, Nocardia, and Streptomyces *spp.*	*Solovieva, 1972*
Sediment, Pacific Ocean and Sagami Bay, Japan	Actinoplanes, Micromonospora, Nocardia, Streptomyces, *and* Streptoverticillium *spp.*	*Okami and Okazaki 1972; Okazaki and Okami, 1976*
Sediment, North Sea	Micromonospora *and* Nocardia *spp.*	*Boeye* et al. *1975*
Coastal sand, Lancashire, England, UK	Micromonospora, Nocardia, Streptomyces, *and* Streptosporangium *spp.*	*Watson and Williams, 1974*
Sediment, Chesapeake Bay and San Juan Harbor, USA	Actinoplanes, Micromonospora, Nocardia, *and* Streptomyces *spp.*	*Walker and Colwell, 1975; Austin* et al. *1977; Mallory* et al. *1977*
Sediment, Baltic Sea	Nocardia *and* Streptomyces *spp.*	*Steinmann, 1976*
Sediment, New York Bight and Harbor, USA	Dactylosporangium, Microbispora, Micromonospora, Nocardia, Saccharomonospora, Saccharopolyspora, Streptomyces, *and* Streptosporangium *spp.*	*Attwell and Colwell, 1981; Attwell* et al. *1981*

(continued)

Table II (continued)

Source	*Organisms*	*Reference*
Salt marsh soil, New Jersey, USA	Actinomadura, Microbispora, Micromonospora, Nocardia, Oerskovia, Streptomyces, *and* Thermomonospora *spp.*	*Hunter* et al. *1981*

in samples from the deeper sampling sites (Table I). Samples without actinomycetes were found from both shallow and deep sampling stations although about half of the samples from deep sites in the North Atlantic were free of actinomycetes (Weyland, 1981a).

It is still not clear whether actinomycetes are active in marine habitats. Some workers see them as part of an indigenous marine microflora (Okami and Okazaki, 1978; Okazaki and Okami, 1976; Weyland, 1981a,b; ZoBell, 1946), whereas others consider them as wash-in components from land, that merely survive as spores (Kriss *et al.*, 1967; Rubentschik, 1928). This latter proposition receives some support from the observation that the number of actinomycetes in marine habitats decreases with increasing distance from land (Attwell and Colwell, 1981; Attwell *et al.*, 1981; Okami and Okazaki, 1972; Weyland, 1969; 1981a); populations of up to 10^5 per ml have been reported from highly polluted inshore sediments (Walker and Colwell, 1975). It has been suggested (ZoBell, 1946) that the isolation of organisms from habitats in areas far removed from the possibilities of terrestrial contamination might be used as evidence of a marine origin. Actinomycetes have been recovered from sediments collected many miles from land (Okami and Okazaki, 1972, 1974; Okazaki and Okami, 1972, 1975, 1976; Weyland, 1981a). They have, for example, been isolated from samples taken at 7790 meters in the Puerto Rican Trench, 60 miles offshore (Walker and Colwell, 1975), and from sediments collected at depths of 3362 meters in the Atlantic Ocean, 175 miles from the coast of West Africa (Weyland, 1969). It is unfortunate that these latter workers were unable to examine their samples for thermoactinomycetes, as Attwell and Colwell (1981) have recovered viable thermoactinomycete endospores, but not actinomycetes, from a marine sediment taken 250 miles from land. There is some preliminary evidence that actinomycete counts in marine habitats may vary on a seasonal basis (Okami and Okazaki, 1974; Walker and Colwell, 1975).

Few attempts have been made to identify actinomycetes from marine sources to other than the genus level. Strains identified as *Micromonospora*, *Nocardia*, or *Streptomyces* seem to have a worldwide distribution (Table II) although most of the isolates included in these studies were from inshore sites and littoral zones where the possibility of contamination from the adjoining landmass was high. Weyland (1981a), however, found that the relative numbers of organisms identified as micromonosporas, streptomycetes, and nocardioforms were influenced by the location of the sampling sites. The nocardioforms formed the major part of the actinomycete flora in offshore sediments, streptomycetes were mainly distributed in the continental shelf and shallow sites, and the micromonosporas predominated in the deep sea sediments. The properties of selected nocardioforms (Weyland, 1981b) indicate that they belong to the genus *Rhodococcus* (Goodfellow and Alderson, 1977; Goodfellow and Minnikin, 1981).

Actinomycetes have been isolated from decaying marine algae and implicated in the degradation of substrates found in marine habitats. Strains from marine sources have been shown to degrade agar, alginates and laminarin (Chesters *et al.*, 1956; Humm and Shepard, 1946), cellulose (Betrabet and Kasturi, 1971; Chandramohan *et al.*, 1972; Rubentschik, 1928; Willingham *et al.*, 1966), chitin (Humm and Shepard, 1946), and oil and other hydrocarbons (Austin *et al.*, 1977; Walker and Colwell, 1975; ZoBell *et al.*, 1943). Such strains have been implicated in the degradation of wood submerged in seawater (Cavalcante and Eaton, 1980; Eaton and Dickinson, 1976). Many of the strains examined by these workers came from inshore areas where there could have been contamination by soil actinomycetes. Again, utilization of compounds such as alginates, cellulose, chitin, and laminarin is common among actinomycetes isolated from soil; therefore, spores from terrestrial habitats could contaminate marine algae in the sea and grow should suitable conditions arise. Similarly, many terrestrial actinomycetes can tolerate high salinities and hydrostatic pressures so that the mere ability of marine isolates to grow on media prepared from seawater, or under pressure, does not prove that they are part of the marine flora (Helmke, 1981; Kayamura and Takada, 1971; Okazaki and Okami, 1975; Watson and Williams, 1974). Indeed, salt-sensitive streptomycetes can adapt to high salt concentrations by culturing in media with a step-wise increase in salt (Okami and Okazaki, 1978). A similar observation was made for *Escherichia coli* by Doudoroff in 1940.

Given the increasing interest in the microbial ecology of aquatic environments, it is important to have an understanding of the natural microbial flora, not only in the context of numbers and potential activity, but also in terms of community structure and species composition. Since there are serious gaps in our information on the numbers and types of actinomycetes occurring in marine habitats, it was decided to determine the occurrence, numbers, and kinds of actinomycetes occurring in a diverse sample of marine sediments. Isolation procedures selective for specific fractions of the actinomycete community were used and an attempt made to identify representative strains to the species level.

III. ISOLATION, ENUMERATION, AND IDENTIFICATION METHODS

Sediment samples were obtained using gravity and piston cores taken from the sites shown in Table III. Samples were stored at either 4 °C or -10 °C until use. The top 6 cm of the cores taken from sediments 16, 17, and 18 were divided into 1 cm portions which were aseptically transferred to sterile universal bottles prior to storage. Only the top 1 to 2 cm of the remaining cores were examined. The pH of samples was determined by the method of Reed and Cummings (1945); percentage moisture content, by drying known weights of sediment at 105 °C to constant weight. After pretreatment to remove carbonate, a "loss on ignition" method was used to determine the amount of organic matter in samples (Piper, 1944). Air-dried samples wetted with deionized water were filtered, the conductivity was measured using a linear conductivity meter, and salinity (‰) was calculated by the method of Richards (1967).

Initially, experiments were designed to develop procedures for the selective isolation of micromonosporas, streptomycetes, and rhodococci, strains of which had been found to be relatively numerous in sediments included in a pilot study. Samples were also examined for actinomadurae, actinoplanetes, nocardiae, thermoactinomycetes, and mesophilic bacteria using the procedures outlined in Table IV. In all cases samples were diluted in the proportion, x g fresh weight of sediment to $2x$ ml artificial seawater (American National Standard 21169). Initial suspensions were shaken on a Griffin shaker (Griffin and George Ltd., Manchester) at setting 8 for 30 minutes, were allowed to settle for 5 minutes, and were pretreated where necessary. Ten-fold dilutions of treated and untreated suspensions were prepared and 0.2 ml aliquots were used to surface inoculate each of the selective media. All media were supplemented with sterile actidione (50 μg/ml) prior to pouring the plates. Inoculated media were incubated at a number of temperatures for varied periods of time (Table IV).

After incubation, plates were examined both by eye and by using a binocular microscope. The numbers and types of actinomycetes, thermoactinomycetes, and the total number of bacteria on the plates at the appropriate dilutions were counted and recorded. The average number of actinomycetes and thermoactinomycetes on the appropriate set of replica plates were calculated and the numbers present were expressed as colony forming units (cfu) per gram dry weight of sediment.

Colonies of actinomycetes and thermoactinomycetes growing on isolation plates were selected using random number tables and purified by streaking onto nutrient-rich media. Streptomycetes, micromonosporas, and rhodococci were cultured on Bennet's seawater agar, *Micromonospora* maintenance agar (Luedemann, 1971) supplemented with artificial seawater, and glucose-yeast extract-seawater agar, respectively, and incubated at 25 °C for 2 weeks. Thermoactinomycetes were cultured on Czapek dox yeast extract + casamino acids agar plates and incubated at 55 °C for 3 days (Cross and Attwell, 1974). A total

TABLE III. Source and Properties of the Marine Sediment Samples

Sediment sampling	Source	Distance from land (miles)	Depth (meters)	Moisture content (%)	Organic matter content (%)	pH	Salinity (%)
1	Morecambe Bay, UK	22	> 46	18.5	3.1	7.8	3.5
2	North Irish Sea, UK	24	183	15.6	4.7	7.9	3.4
3	Liverpool Bay, UK	15	> 46	12.1	9.6	8.2	3.2
4	Queens Channel, UK	6	> 46	28.7	6.0	7.5	3.4
6	Wylfa Head, UK	5	> 46	17.3	10.1	8.0	3.0
8	Carmarthen Bay, UK	5	> 46	28.0	9.3	8.1	3.5
9	Falmouth Bay, UK	5	> 46	41.4	20.4	7.8	3.2
11	South of Needles, UK	10	> 46	4.1	6.7	7.9	3.3
12	Off Isle of Wight, UK	10	> 46	10.7	4.9	8.0	3.3
14	South Falls, UK	10	> 46	8.5	4.2	8.0	3.2
16[a]	North Atlantic Ocean	500	4,920	52.2	16.7	7.2	3.2
17[a]	North Atlantic Ocean	112	2,880	54.7	12.6	7.2	3.2
18*	North Atlantic Ocean	45	158	26.5	7.1	7.8	3.0
19	North Atlantic Ocean	870	5,200	57.0	8.1	7.8	3.4
20	North-west of Puerto Rico	24	7,790	48.9	6.5	7.9	3.4
23	North-west of Puerto Rico	12	450	27.3	17.2	7.8	3.6

[a] Values are the mean of six samples.

TABLE IV. Summary of Methods Used for Isolating and Counting the Numbers of Actinomycetes, Thermoactinomycetes, and Mesophilic Bacteria in Marine Sediments

Organisms	*Media*	*Pretreatment of sediment suspension*	*Incubation*	*Method based*
Actinomadura	*Glucose yeast extract agar + rifampicin*	*Air-dried sediment heated at 100°C for min*	*30 °C; 3 weeks*	*Athalye* et al. *1981*
Actinoplanes	*Colloidal chitin seawater agar* [a]	*Sterile tap water added to air-dried sediment and incubated at 20 °C for 55 minutes*	*25 °C; 4 weeks*	*Makkar and Cross, 1982*
Micromonospora	*Cellulose asparagine seawater agar* [a] *+ novobiocin*	*60 °C for 40 minutes*	*18 °C; 10 weeks* [b]	*Goodfellow and Haynes, unpublished*
Nocardia	*Diagnostic sensitivity agar + methacycline*	*55 °C for 6 minutes*	*25 °C; 4 weeks*	*Orchard, 1978*
Rhodococcus	*M3 seawater agar* [a]	*55 °C for 6 minutes*	*18 °C; 10 weeks* [b]	*Rowbotham and Cross, 1977a*
Streptomyces	*Starch casein seawater agar* [a]	*50 °C for 10 minutes*	*18 °C; 10 weeks* [b]	*Goodfellow and Haynes, unpublished data*
Thermoactinomyces	*Czapek dox yeast extract + casamino acids, tyrosine and novobiocin*	*None*	*55 °C; 3 days*	*Cross, 1981b*
Heterotrophic bacteria	*Difco marine agar 2216*	*None*	*25 °C; 6 weeks*	*ZoBell, 1946*

[a] *Artificial seawater (American National Standard 21169).*
[b] *Duplicate plates also incubated at 4 °C for 6 months.*

of 250 *Micromonospora*, 140 *Rhodococcus*, 250 *Streptomyces*, and 92 *Thermoactinomyces* strains were obtained in pure culture. Glycerol suspensions of all of the purified strains were prepared and stored at -25 °C (Williams *et al.*, 1983a). All of the test strains were examined for their ability to grow on the five media described by Hidaka (1964).

The streptomycetes were assigned to 26 color groups by using media and methods recommended by Williams *et al.* (1969). Strains from each of the color groups, chosen on a proportional basis, were selected using random number tables. The 55 test strains were then examined for 41 diagnostic tests and identification achieved using the MATIDEN program and the *Streptomyces* probability matrix (for details see Williams *et al.*, 1983b; elsewhere in this volume). Print-out included information on the Willcox probability, taxonomic distance, and standard error of distance of the unknown strain against the three most likely groups of *Streptomyces*. The rhodococci were identified by using diagnostic tests highlighted by Goodfellow and Schaal (1979) and additional tests of value in characterization of *R. coprophilus* (Rowbotham and Cross, 1977b). It is well known that micromonosporas are difficult to identify to the species level, but an attempt was made to assign them to artificial groups by using a number of conventional morphological, physiological, and degradation tests. Thus, all of the isolates and 16 marker strains representing nine *Micromonospora* species were examined to determine their pH and temperature range, their ability to degrade cellulose, chitin, elastin, gelatin, nitrate, starch, tributyrin, tyrosine, and xylan, and their ability to grow in the presence of various concentrations of sodium chloride. The 92 thermoactinomycetes were identified using the methods recommended by Unsworth and Cross (1978).

IV. NUMBER OF TYPES OF ACTINOMYCETES IN MARINE SEDIMENTS

Actinomycetes, thermoactinomycetes, and mesophilic bacteria were recovered from all inshore sediments irrespective of depth (Table V). No correlation was found between the actinomycete counts and the salinity, pH, organic matter content or moisture content of the samples. In all inshore sediments, however, the actinomycetes only formed a small fraction of the total bacterial count. Actinomycetes were not found in deep sea samples from sites 16, 17, and 19 which were collected 500, 112 and 870 miles from land, respectively. Small numbers of thermoactinomycetes were recovered from two of twelve samples taken from sites 16 and 17 but not from the single sample from station 19 (Table V). The number of actinomycetes recorded from plates incubated at 4 °C were ten- to a hundred-fold fewer than those recorded on corresponding media incubated at the higher temperatures. All strains isolated at the lower temperature grew well at 25 °C.

TABLE V. Actinomycetes and Thermoactinomycetes in Marine Sediments

Sediment sampling site	*cfu/g dry weight* [a]				*Actinomycetes / Total bacteria*
	Micromonospora	Rhodococcus	Streptomyces	Thermoactinomyces	(%)
1	492	0	6	928	0.38
2	160	0	11	150	0.17
3	226	0	0	394	0.42
4	18,819	917	541	4,285	2.21
6	255	11	11	2,215	0.05
8	2,590	259	91	4,334	0.27
9	3,444	504	257	7,784	0.39
11	0	0	5	15	0.10
12	152	0	0	26	0.24
14	310	0	0	457	0.65
16 [b]	0	0	0	3 [c]	-
17 [b]	0	0	0	2 [c]	-
18 [b]	207	4	13	75	0.04
19	0	0	0	0	-
20	5,942	584	944	6,979	0.76
23	5,621	24	241	4,251	0.01

[a] *Actinomycete counts based on numbers growing on replica plates incubated at 18 °C.*
[b] *Six, as opposed to one, samples examined;*
[c] *Thermoactinomycetes present in only a single sample.*

The actinomycetes growing on the isolation plates were provisionally assigned to the genera *Micromonospora, Rhodococcus*, and *Streptomyces* on the basis of their colony morphology on the selective media. The identity of all 640 randomly chosen representatives of these taxa was confirmed in the characterization studies. However, despite the use of selective isolation procedures, *Actinomadura, Actinoplanes*, and *Nocardia* strains were not recovered from any of the samples. It was interesting that similar numbers of rhodococci and streptomycetes were found in six of the 13 samples containing actinomycetes although small numbers of the latter were also detected in an additional four sites. Micromonosporas were the predominant actinomycetes in all of the samples studied, but the numbers of these organisms were less than those recorded for the thermoactinomycete component. All 640 isolates included in the characterization studies grew on the media described by Hidaka (1964).

Only three of the 26 *Streptomyces* color groups contained more than four isolates although nearly half of the strains were classified in color group 6 (Table VI). Nearly 70% of the color group representatives were identified on the basis of the identification scores obtained using all three identification coefficients. Strains were assigned to named cluster-groups when they showed a Willcox probability greater than 0.85, low scores for taxonomic distance and its standard error, and the first group scores were significantly better than those against the next best two alternatives (Table VI; Williams *et al.*, 1983b). Most of the representative isolates were identified as *S. albidoflavus*, the name given to the largest cluster-group recovered by Williams *et al.* (1983a). Nineteen of the 21 representatives of color group 6 were identified as *S. albidoflavus* with all but two of them having Willcox probabilities above 0.98. In contrast, color group 1 does not appear to be homogeneous as it was not possible to identify half of the representative isolates, the remainder being assigned to either the *S. albidoflavus* or *S. atroolivaceus* color-groups. The representative strains of color group 8, the third largest grouping, were identified as *S. chromofuscus* and showed Willcox probabilities to the cluster-group bearing this name of 0.96 or greater. Representatives of the remaining color groups were assigned to a number of different cluster-groups (Table VI).

It proved difficult to identify the rhodococci to species by using the diagnostic tests described by Goodfellow and Schaal (1979), but it was possible to assign all 140 isolates to one of three artificial groups based on pigment production and morphological characters. Eighty percent of the isolates were placed in group 1. The strains in this group produced a well-developed primary mycelium, reddish-pink matt colonies, used testosterone and sebacic acid as sole carbon sources, grew at 40 °C and in the presence of crystal violet (0.0001%, w/v) and phenol (0.1%, w/v) but did not degrade adenine or tyrosine or use glycerol, inositol, maltose, or trehalose as sole sources of carbon for energy and growth. These properties are consistent with the isolates being identified as *R. coprophilus*. Further studies are required to determine the status of

TABLE VI. Identification of Streptomyces *Color Groups Based upon Identification Scores Achieved Using the Willcox Probability, Taxonomic Distance, and Standard Error of Taxonomic Distance Coefficients*

Color group	Color characteristics on oatmeal agar		Soluble pigment	No. of strains in color group	No. of strains		Cluster group identification
	Spore mass	Reverse substrate mycelium			Tested	Identified	
1	white	yellow-brown	-	44	8	4	S. albidoflavus/S. atroolivaceus
2	yellow	yellow-brown	-	3	1	1	S. violaceus
5	green	cream	pink	3	1	1	S. griseoflavus
6	white-yellow	yellow-brown	-	121	21	19	S. albidoflavus
7	cream	yellow-brown	-	4	1	1	S. albidoflavus
8	white	yellow-brown	-	19	3	3	S. chromofuscus
9	grey-white	yellow-brown	-	1	1	1	S. violaceniger
11	green-grey	yellow-brown	-	1	1	1	S. lydicus
15	grey	yellow-brown	-	4	1	1	S. atroolivaceus
16	blue	green-blue	blue	1	1	1	S. cyaneus
17	grey	yellow-brown	-	1	1	1	S. atroolivaceus
20	creamy-green	yellow-brown	-	3	1	1	S. albidoflavus
22	dark grey	dark grey	-	3	1	1	S. lydicus
24	grey	yellow-brown	-	4	1	1	S. diastaticus
25	grey-white	brown	-	4	1	1	S. rochei

the isolates classified in groups 2 and 3.

The *Micromonospora* isolates formed a remarkably homogeneous group. They formed orange colonies which bore dark brown or black spores, degraded cellulose, chitin, elastin, gelatin, starch, and xylan but not tributyrin, and grew between pH 6 and 8, between 20 and 40 °C, and in the presence of 3% (w/v) sodium chloride. Approximately half of the micromonosporas reduced nitrate, degraded tyrosine, and grew at pH 9, at 10 °C, and in the presence of 4% (w/v) sodium chloride. These tests did not allow the representatives of the nine *Micromonospora* species to be distinguished from one another. In complete contrast, all of the *Thermoactinomyces* isolates were identified to the species level by using the diagnostic table of Unsworth and Cross (1978). Seventy-seven strains (84%) were identified as *T. candidus*, one as *T. dichotomica*, and the remainder were equally distributed to *T. sacchari* and *T. vulgaris*. All 92 isolates were assigned to the same species on the basis of properties exhibited on Czapek dox yeast extract agar supplemented with tyrosine (Cross, 1981a), a result which allowed confidence to be placed in the results based solely on the latter method (Table VII).

V. BEHAVIOR OF ACTINOMYCETES IN MARINE HABITATS

Actinomycetes are widely distributed in marine and littoral environments, but this does not necessarily mean that they are able to multiply and grow significantly in such habitats. Indeed, actinomycetes normally form only a small fraction of the bacterial community in marine sediments and rigorous

TABLE VII. Thermoactinomyces *Species in Marine Sediments*

Sediment sampling site	T. candidus	T. dichotomica	T. sacchari	T. vulgaris
1	*835*	*0*	*28*	*65*
2	*128*	*0*	*7*	*15*
3	*299*	*0*	*83*	*12*
4	*3,819*	*0*	*42*	*424*
6	*2,082*	*0*	*22*	*111*
8	*4,031*	*0*	*43*	*260*
9	*7,161*	*78*	*0*	*545*
11	*14*	*0*	*1*	*0*
12	*23*	*0*	*0*	*3*
14	*157*	*0*	*94*	*206*
16	*20* [a]	*0*	*0*	*0*
17	*11* [a]	*0*	*0*	*0*
20	*6,827*	*69*	*83*	*0*
23	*4,165*	*0*	*43*	*43*

[a] *Isolated from one of six samples studied*

pretreatment regimes and selective media are needed if isolation plates are not to be overrun by unwanted fast-growing Gram-negative bacteria. It is also clear that actinomycete counts from marine habitats are much lower than are those reported for terrestrial and freshwater habitats (Cross, 1981a; Cross *et al.*, 1976). These relatively low counts are consistent with the idea that actinomycetes are simply washed into the marine ecosystem, a view supported by the observation that the number of actinomycetes decrease with increasing distance from land. In the present study, a count of over 7 x 10^3 cfu/g dry weight were obtained for sediment 20 which had been collected at a depth of 7,790 meters, 24 miles from land, whereas at the other extreme, sediment 16, collected 500 miles from land, did not yield any actinomycetes. The isolation of small numbers of thermoactinomycetes from sediment 16 serves to underline the value of these organisms as indicators of terrestrial pollution and shows, once again, that the endospores of *Thermoactinomyces* can be transported long distances by ocean currents before coming to rest (Attwell and Colwell, 1981; Cross and Johnson, 1971). The presence of thermoactinomycetes in all of the marine sediments found to contain actinomycetes is a further indication that many of the latter are wash-in components.

It is unquestionable that large numbers of *Micromonospora, Rhodococcus*, and *Streptomyces* can reside and remain viable in marine sediments since they can readily be cultivated using the appropriate selective isolation procedures. The failure of most early workers to detect rhodococci is probably because the latter were mistaken for nocardiae or included under the term nocardioform (Weyland, 1981a,b). In the present study most of the rhodococci were identified as *R. coprophilus*, the coprophilic organism which is being increasingly used as an indicator of farm animal pollution in aquatic habitats (Al-Diwany and Cross, 1978; Cross, 1981a). It also seems likely that most of the streptomycetes are wash-in components as over half of the streptomycete component were assigned to the *S. albidoflavus* cluster group, a taxon known to accommodate organisms common in soil. Micromonosporas are also numerous in soils and freshwater habitats and their recovery as the predominant actinomycete in the deeper marine sediments is in good agreement with the results of Weyland (1981a). The ability of *Micromonospora* spores to tolerate reduced oxygen tensions (Watson and Williams, 1974) may help to account for their ability to remain viable in marine sediments.

It is generally well known that marine bacteria consist mainly of Gram-negative rods with specific requirements for sodium and other ions; these are provided by seawater. Hidaka (1964) suggested that true marine bacteria required seawater in media for optimal growth. On this basis, all the micromonosporas, rhodococci, and streptomycetes examined in the present study were terrestrial isolates as they grew just as well on media prepared with distilled water as on that with seawater.

It is evident both from the present and earlier studies that actinomycete are not uncommon in marine habitats. It seems likely, however, that most actinomycetes have been washed into the sea and collect in marine sediments where they can survive for long periods of time as spores or resting propagules. The possibility cannot be ruled out that, under certain exceptional conditions, some of the spores may germinate and grow; experimental evidence on this point is urgently needed. Actinomycetes in marine habitats may, however, be seen as a selected gene pool that might contain organisms capable of producing useful metabolites. The availability of methods which permit the preferential isolation of specific actinomycete taxa (Cross, 1982) make the exploitation of the actinomycete community in marine habitats an attractive prospect.

ACKNOWLEDGMENTS

We wish to thank Drs. M. Rolfes, C. M. Brown, B. Austin, and R. R. Colwell for samples of marine sediments and Dr. S. T. Williams and Mrs. J. Vickers for their help and encouragement with the computation.

REFERENCES

Ahrens, R., and Moll, G., *Arch. Mikrobiol. 70:*243 (1970).
Al-Diwany, L. J., and Cross, T., *Zentralbl. Bakteriol. Parasitenk. Infektionskr. Hyg. 1. Abt. Orig. 6*, suppl.:153 (1978).
Athalye, M., Lacey, J., and Goodfellow, M., *J. Appl. Bacteriol. 51:*289 (1981).
Attwell, R. W., and Colwell, R. R., *Mar. Pollut. Bull. 12:*351 (1981).
Attwell, R. W., Colwell, R. R. and Coolbaugh, J., *Mar. Technol. Soc. J. 15:*36 (1981).
Austin, B., Colwell, R. R., Walker, J. D. and Calomiris, J. J., *Appl. Environ. Microbiol. 34:*60 (1977).
Betrabet, S. M., and Kasturi, K., *J. Sci. Technol. Life Sci. (India) 98:*180 (1971).
Boeye, A. Wayenbergh, M., and Aerts, M., *Mar. Biol. 32:*263 (1975).
Cavalcante, M. S., and Eaton, R. A., "The International Research Group of Wood Preservation. Working Group 1. Biological Problems". Document No. IRG/WP/1110 (1980).
Chandramohan, D., Ramu, S., and Natarajan, R., *Curr. Sci. (Bangalore) 41:*245 (1972).
Chesters, C. G. C., Apinis, A., and Turner, M., *Proc. Linn. Soc. Lond. 166:*87 (1956).
Cross, T., *J. Appl. Bacteriol. 50:*397 (1981a).
Cross, T., *in* "The Prokaryotes: a Handbook on Habitats, Isolation and Identification of Bacteria" (M. P. Starr, H. Stolp, H. G. Trüper, A. Balows and H. G. Schlegel, eds.), p. 2091. Springer Verlag, Berlin (1981b).
Cross, T., *Dev. Ind. Microbiol. 23:*1 (1982).
Cross, T., and Attwell, R. W., *in* "Spore Research" 1973 (A. N. Barker, G. W. Gould and J. Wolf, eds.), p. 11. Academic Press, London (1974).
Cross, T., and Johnson, D. W., *in "Spore Research" 1971* (A. N. Barker, G.W. Gould and J. Wolf, eds.), p. 315 Academic Press, London (1971).
Cross, T., Rowbotham, T. J., Mishustin, E. N., Tepper, E. Z., Portaels, F., Schaal, K.P., and Bickenbach, H., "The Biology of the Nocardiae" (M. Goodfellow, G. H. Brownell, and J. A. Serrano, eds.), p. 337. Academic Press, London (1976).
Doudoroff, M., *J. Gen. Phys. 23:*585 (1940).
Eaton, R. A., and Dickinson, D. J., *Mater. Org. 11:*521 (1976).
Freitas, Y. M., and Bhat, J. V., *J. Univ. Bombay 23:*53 (1954).
Goodfellow, M., and Alderson, G., *J. Gen. Microbiol 11:*99 (1977)
Goodfellow, M., and Board, R. G. (Eds.), "Microbiological Classification and Identification". Academic Press, London (1980).

Goodfellow, M., and Cross, T., *in* "Biology of Plant Litter Decomposition" (C. H. Dickinson and G. J. F. Pugh, eds.), p. 269. Academic Press, London (1974).
Goodfellow, M., and Minnikin, D. E., *in* "The Prokaryotes: A Handbook on Habitats, Isolation and Identification of Bacteria" (M. P. Starr, H. Stolp, H. R. Trüper, A. Balows and H. G. Schlegel, eds.), p. 2016. Springer Verlag, Berlin (1981).
Goodfellow, M., and Schaal, K. P., *in* "Identification Methods for Microbiologists" (D. W. Lovelock and F. A. Skinner, eds.), p. 261. Academic Press, London (1979).
Goodfellow, M., and Wayne, L. G., *in* "The Biology of the Mycobacteria". *Volume 1* (C. Ratledge and J. L. Stanford, eds.), p. 472. Academic Press, London (1982).
Grein, A., and Meyers, S. P., *J. Bacteriol. 76:*457 (1958).
Helmke, E., *Zentralbl. Bakteriol. Mikrobiol. Hyg. 1. Abt. Orig. 11*, suppl:321 (1981).
Hidaka, T., *Mem. Fac. Fish. Kagoshima Univ. 12:*135 (1964).
Humm, H. J., and Shepard, K. S., *Duke Univ. Mar. St. Bull. 3:*76 (1946).
Hunter, J. C., Eveleigh, D. E., and Casella, G., *Zentralbl. Bakteriol. Mikrobiol. Hyg. 1. Abt. Orig. 11*, suppl.:195 (1981).
Kayamura, Y., and Takada, H., *Trans. Mycol. Soc. Jpn. 12:*161 (1971).
Kriss, A. E., *K. Sev. Vost. Soyuza SSR 2:*336 (1952).
Kriss, A. E., "Marine Microbiology" Oliver and Boyd, London (1963).
Kriss, A. E., Rukina, Y., and Markianovich, Y., *Tr. Sevastop. Ser. Biol. 8:*220 (1951).
Kriss, A. E., Mishostina, I. E., Mitskevich, N., and Zemtskova, E. V., "Microbial Populations of Oceans and Seas". Arnold, London (1967).
Luedemann, G. M., *Int. J. Syst. Bacteriol. 21:*240 (1971).
Makkar, N. A., and Cross, T., *J. Appl. Bacteriol. 52:*209 (1982).
Mallory, L. M., Austin, B., and Colwell, R. R., *Can. J. Microbiol. 23:*733 (1977).
Okami, Y., and Okazaki, T., *J. Antibiot. 25:*456 (1972).
Okami, Y., and Okazaki, T., *J. Antibiot. 27:*240 (1974).
Okami, Y., and Okazaki, T., *Zentralbl. Bakteriol. Parasitenk. Infektionskr. Hyg. 1. Abt. Orig. 6*, suppl. 145 (1978).
Okazaki, T., and Okami, Y., *J. Antibiot. 25:*461 (1972).
Okazaki, T., and Okami, Y., *J. Ferment. Technol. 53:*833 (1975).
Okazaki, T., and Okami, Y., *in* "Actinomycetes: The Boundary Microorganisms" (T. Arai, ed.), p. 125. Toppan Co. Ltd., Tokyo (1976).
Orchard, V. A., *N. Z. J. Agric. Res. 21:*21 (1978).
Orchard, V. A., *Soil Biol. Biochem. 12:*477 (1980).
Orchard, V. A., *Zentralbl. Bakteriol. Mikrobiol. Hyg. 1. Abt. Orig. 11*, suppl.:167 (1981).
Orchard, V. A., and Goodfellow, M., *J. Gen. Microbiol. 85:*160 (1974).
Orchard, V. A., and Goodfellow, M., *J. Gen. Microbiol. 118:*295 (1980).
Piper, C. S., "Soil and Plant Analysis". Adelaide University (1944).
Reed, J. F., and Cummings, R. W., *Soil Sci. 59:*97 (1945).
Richards, L.A., "Diagnosis and Improvement of Saline and Alkaline Soils". Agricultural Handbook Number 6. United States Department of Agriculture (1967).
Rowbotham, T. J., and Cross, T., *J. Gen. Microbiol. 100:*231 (1977a).
Rowbotham, T. J., and Cross, T., *J. Gen. Microbiol. 100:*123 (1977b).
Rubentschik, L., *Zentralbl. Bakteriol. Parasitenk. Infektionskr. Hyg. II Abt. 76:*305 (1928).
Saito, T., *Tohoku Univ. Fourth Ser. (Biol.) 21:*145 (1955).
Siebert, G., and Schwartz, W., *Arch. Hydrobiol. 52:*321 (1956).
Solovieva, N. K., *Antibiotiki (Mosc.) 17:*387 (1972).
Steinmann, J., *Bot. Mar. 19:*47 (1976).
Unsworth, B. A., and Cross, T., *J. Appl. Bacteriol. 45:*16 (1978).
Walker, J. D., and Colwell, R. R., *Mar. Biol. 30:*193 (1975).
Watson, E. T., and Williams, S. T., *Soil Biol. Biochem. 6:*43 (1974).
Wayne, L. G., Krichevsky, E. J., Love, L. L., Johnson, R., and Krichevsky, M. I., *Int. J. Sys. Bacteriol. 30:*528 (1980).
Webley, D. M., Eastwood, D. J., and Gimingham, C. H., *J. Ecol. 40:*169 (1952).
Weyland, H., *Nature (Lond.) 223:*858 (1969).
Weyland, H., "The Ocean World", p. 13. Joint Oceanographic Assembly, Tokyo (1970).
Weyland, H., *Zentralbl. Bakteriol. Mikrobiol. Hyg. 1. Abt. Orig. 11*, suppl.:185 (1981a).
Weyland, H., *Zentralbl. Bakteriol. Mikrobiol. Hyg. 1. Abt. 11*, suppl.:309 (1981b).

Williams, S. T., *Zentralbl. Bakteriol. Parasitenk. Infektionskr. Hyg. 1. Abt. 6*, suppl.:137 (1978).
Williams, S. T., and Wellington, E. M. H., *in* "The Prokaryotes: a Handbook on Habitats, Isolation and identification of bacteria" (M. P. Starr, H. Stolp, H. G. Trüper, A. Balows and H. G. Schlegel, eds.), p. 2103. Springer-Verlag, Berlin (1981).
Williams, S. T., Davies, F.L., and Hall, D.M., *in* "The Soil Ecosystem" (J. G. Sheals ed.), p. 107. The Systematics Association, London (1969).
Williams, S. T., Goodfellow, M., Alderson, G., Wellington, E. M. H., Sneath, P. H. A., and Sackin, M. J., *J. Gen. Microbiol. 129:*1743 (1983a).
Williams, S. T., Goodfellow, M., Wellington, E. M. H., Vickers, J. C., Alderson, G., Sneath, P. H. A., Sackin, M. J., and Mortimer, A. M., *J. Gen. Microbiol. 129:*1815 (1983b).
Willingham, C. A. Roach, A. W., and Silvey, J. K. G., *Am. Midl. Nat. 75:*231 (1966).
Wood, E. J. F., *Aust. J. Mar. Freshw. Res. 4:*160 (1953).
ZoBell, C. E., "Marine Microbiology". Chronica Botanica, Waltham, Mass (1946).
ZoBell, C. E., and Upham, H. C., *Bull. Scripps Inst. Oceanogr. Univ. Calif. 5:*239 (1944).
ZoBell, C. E., Grant, C. W., and Haas, H. F., *Bull. Am. Assoc. Petrol. Geol. 27:*1175 (1943).

STUDIES OF THE ECOLOGY OF STREPTOMYCETE PHAGE IN SOIL[1]

S. T. Williams
S. Lanning

Botany Department
University of Liverpool
Liverpool, U. K.

I. INTRODUCTION

Actinophage may be readily detected in most soils by using specific enrichment procedures in which a concentrated inoculum of the prospective host is incubated with soil in a suitable liquid medium. Phage active against *Streptomyces* spp., the predominant actinomycete genus in soil, are most readily detected, but phage to other soil genera have also been isolated (Table I). It appears that that actinophage, and particularly streptomycete phage, are widespread in the soil environment. There is, however, little information on the numbers, distribution, survival, and significance of any phage in soil. Reanney and Marsh (1973) suggested that, if phage occurred in soil at 0.1% of the titer obtained in the laboratory, phage would be the most numerous "genetic objects" in that habitat. Thus, phage are potentially of prime ecological importance.

Ecological studies of microbes in soil and in other environments are based on a knowledge of their distribution, frequency, and response to environmental factors. While such information is available for streptomycetes in soil (Williams, 1978), it is lacking for their phage. Therefore, we have assessed and developed methods for the direct counting of streptomycete phage in soil and have studied the effect of various environmental factors on the stability of these phage in soil.

[1] *This work was supported by a grant from the N.E.R.C.*

ISBN 0-12-528620-1

TABLE I. Examples of Actinophage Isolated from Soils by Specific Enrichment

Genus of propagation strain	*Authors*
Amorphosporangium	*Wellington and Williams (1981)*
Chainia	*Prauser (1976); Wellington and Williams (1981)*
Micromonospora	*Wellington and Williams (1981)*
Nocardia	*Prauser (1976); Williams* et al. *(1980); Wellington and Williams (1981)*
Nocardioides	*Prauser (1976)*
Oerskovia	*Prauser (1976)*
Promicromonospora	*Prauser (1976)*
Rhodococcus	*Prauser (1976); Williams* et al.. *(1980); Wellington and Williams (1981)*
Streptomyces	*Prauser (1976); Wellington and Williams (1981)*
Streptoverticillium	*Prauser (1976); Wellington and Williams (1981)*

II. METHODS FOR DIRECT ISOLATION AND ENUMERATION OF STREPTOMYCETE PHAGE

While enrichment procedures can be used to readily detect many actinophage in soil, such techniques provide no indication of the numbers of phage present in the soil sample. The relatively few attempts at direct isolation of bacteriophage from soil have yielded very low titers (Casida and Liu, 1974; Collard, 1970; Khavina, 1954; Reanney, 1968; Tan and Reanney, 1976; Welsch *et al.*, 1955). This may be due to any of several factors, including: (i) inactivity of hosts in the nutrient-poor soil mass (Gray and Williams, 1971; Williams, 1978); (ii) non-detection of phage adsorbed either to resting cells (Casida and Liu, 1974) or to soil colloids (Robinson and Corke, 1959; Sykes and Williams, 1978); or (iii) inactivation or loss of phage during the extraction procedures. In trying to improve yield of phage, we have paid particular attention to the methods of extraction. A summary of the results is presented here; details were given by Lanning and Williams (1982).

The efficiency of various procedures at each stage of extraction was assessed by determining the recovery of a streptomycete phage (f069) which had been added at known titers to sterile soils. Choice of the agitation procedure to prepare soil extracts was important; whereas virtually total recovery of phage was achieved by reciprocal shaking, only 4% recovery was obtained by magnetic stirring. Nutrient broth containing 0.1% (w/v) egg albumin at pH 8.0 was selected as the best eluent. Centrifugation of soil extracts at 1,200 X *g* for 15 min produced little or no reduction in numbers of phage recovered. Filter sterilization of soil suspensions resulted in losses of 33 to 45% of the titer obtained in unfiltered suspensions. Filtration losses could be avoided by substituting chloroform-sterilization, if chloroform-resistant phage were present.

The efficiency of the new procedures devised as a result of these test was compared with those of two earlier methods for direct isolation of actinophage (Table II). The efficiency of recovery of an actinophage (fMx2) added to sterile soils was clearly improved with chloroform sterilization (Method D), giving almost total recovery (Table III). Counts of streptomycete phage in various natural soils were also increased significantly by the new procedures (Table IV). The numbers of phage in the nutrient-rich compost were high, presumably due to the increased activity of streptomycetes, and more phage were propagated on the recently collected soil isolates than on the three type cultures. Considering that only six host strains were used, the results suggest that the total numbers of streptomycete phage in these soils were in excess of 10^4 g^{-1} soil. Numbers of phage in soil have clearly been grossly underestimated in the past.

TABLE II. Methods for Direct Counts of Actinophage in Soil

Method	*Agitation procedure*	*Eluent*	*Sterilization procedure*
A. Welsch et al. *(1955)*	*Frequent shaking by hand over 2 to 3 h*	*Distilled water*	*Membrane filtration*
B. Sykes (1977)	*Rotary shaking for 1h at 25 °C*	*Nutrient broth (pH 7.0)*	*Membrane filtration*
C. Lanning and Williams (1982)	*Reciprocal shaking for 30 min at 4 °C*	*Nutrient broth + egg albumin (pH 8.0)*	*Membrane filtration*
D. Lanning and Williams (1982)	”	”	*2% chloroform*

(Reproduced from Lanning, S., and Williams, S. T., J. Gen. Microbiol. 128:*2063 (1982) by copyright permission of the Society for General Microbiology.)*

TABLE III. Recovery of Phage fMx2 from Sterile Soils by Four Methods

Soils	*% phage recovered*			
	Method A[a]	*Method B*	*Method C*	*Method D*
Sand	18	25	87	96
Arable	13	18	74	93
Garden	32	30	34	118

[a]*For details, see Table II.*
(Reproduced from Lanning, S., and Williams, S. T., J. Gen. Microbiol. 128:*2063 (1982) by copyright permission of the Society for General Microbiology.)*

III. THE INFLUENCE OF ENVIRONMENTAL FACTORS ON STREPTOMYCETE PHAGE

It seems likely and logical that the tolerance of a phage to environmental conditions is at least as great as that of its host. Thus, for example, phage for the extreme thermophile *Thermus thermophilus* (Sakaki and Oshima, 1975) and for extreme halophiles (Torsvik and Dundas, 1974, 1980; Wais *et al.*, 1975) have been isolated. In the actinomycetes, phage attacking thermophiles in the genera *Thermoactinomyces* (Kurup and Heinzen, 1978; Patel, 1969b; Sarfert *et al.*, 1979), *Thermomonospora* (Patel, 1969a), and *Micropolyspora* (Kurup and Heinzen, 1978) have been detected. Since there is little information on the effects of environmental factors on other actinophage, it was decided to study the influence of environmental factors on streptomycete phage in soil.

Initial studies were carried out by adding known titers of a streptomycete phage (f85) to sterile or non-sterile soils under laboratory conditions. The effects of various environmental factors were assessed by measuring changes in phage titers after a period of two weeks. Each factor was studied separately, with other conditions being kept, as far as possible, at levels optimum for phage survival. The results are summarized in Table V. The phage titer generally decreased more in non-sterile soil, but patterns of response to environmental variables were similar to those in sterile soil. Although soil type and moisture content generally had little effect on phage stability, high moisture levels in non-sterile soil resulted in approximately ten-fold reductions in phage titers. Stability was greatest at low temperatures and this mesophilic phage was completely inactivated at 40 °C. Large scale inactivation also occurred at pH 3.6.

TABLE IV. Comparison of Four Methods for Direct Enumeration of Actinophage in Natural Soils.

Soil	*Streptomycete host*	*Number of phage recovered (p.f.u. g^{-1} soil)*			
		Method[a]			
		A	*B*	*C*	*D*
Compost	*Soil isolate Mx1*	*515*	*2,173*	*2,917*	*12,387*
	Soil isolate Mx2	*669*	*5,820*	*3,916*	*12,008*
	Soil isolate Mx3	*785*	*6,885*	*7,263*	*22,950*
	S. lavendulae *ISP 5069*	*0*	*0*	*0*	*0*
	S. michiganensis *ISP 5015*	*0*	*0*	*0*	*0*
	S. griseus *ISP 5236*	*87*	*566*	*397*	*806*
Arable	*Soil isolate Mx1*	*17*	*82*	*225*	*275*
	Soil isolate Mx2	*50*	*117*	*642*	*1,283*
	Soil isolate Mx3	*56*	*181*	*667*	*1,858*
	S. lavendulae *ISP 5069*	*0*	*0*	*0*	*0*
	S. michiganensis *ISP 5015*	*0*	*0*	*0*	*0*
	S. griseus *ISP 5236*	*16*	*107*	*277*	*660*
Garden	*Soil isolate Mx1*	*13*	*87*	*233*	*1,517*
	Soil isolate Mx2	*16*	*35*	*207*	*558*
	Soil isolate Mx3	*41*	*98*	*558*	*2,868*
	S. lavendulae *ISP 5069*	*0*	*0*	*0*	*0*
	S. michiganensis *ISP 5015*	*0*	*0*	*0*	*0*
	S. griseus *ISP 5236*	*0*	*0*	*0*	*0*

[a]*For details, see Table II*
(Reproduced from Lanning, S., and Williams, S. T., J. Gen. Microbiol. 128:*2063 (1982) by copyright permission of the Society for General Microbiology.)*

These results and some of our previous work suggested that pH is a major factor influencing phage stability and activity in soil. It is therefore relevant to consider this environmental factor in more detail.

IV. THE INFLUENCE OF pH ON STREPTOMYCETE PHAGE

Acidity is an important factor governing the activity and survival of streptomycetes in soil, most isolates being either neutrophilic or acidophilic in their pH requirements and tolerances (Williams *et al.*, 1971). Neutrophilic streptomycetes grow between about pH 5.5 to 8.5 with an optimum around pH 7.0, whereas acidophiles grow between about pH 3.5 to 6.5 with optimum growth around pH 4.5. Acidoduric strains, with responses overlapping these ranges, have also been isolated (Hagedorn, 1977; Williams and Robinson, 1981). The presence of acidophilic streptomycetes

TABLE V. The Effect of Various Factors on a Streptomycete Phage (f85) Added to Soil

Factors	Changes in numbers (log_{10} p.f.u. g^{-1} air-dried soil) after 14 days	
	Sterile soil	Non-sterile soil
Soil type [a]		
Sand	-0.8	-2.1
Loam	-0.8	-2.1
Clay	-1.1	-2.4
Moisture content (% moisture-holding capacity)		
15	-0.30	-0.90
25	-0.10	-0.70
50	-0.10	-1.60
75	-0.10	-2.00
100	-0.04	-2.00
Temperature (°C)		
5	+0.1	-0.3
12	-0.5	-1.4
25	-0.9	-2.6
40	-6.0	-6.0
pH		
7.7	+0.1	-1.1
7.0	-1.3	-0.9
5.4	-1.2	-1.1
4.2	-1.2	-1.5
3.6	-5.6	-4.2

[a] *Values corrected for recovery efficiency.*

in 17 acidic soils has been demonstrated (Khan and Williams, 1975). Studies on the effect of pH on growth rate and on enzyme activity under laboratory conditions have emphasized the differential responses of streptomycetes (Flowers and Williams, 1977; Williams and Flowers, 1978). The roles of both neutrophiles and acidophiles in the decomposition of chitin in acidic soils have been elucidated (Williams and Robinson, 1981). It is therefore clear that potential hosts for streptomycete phage exist and are active in acidic soils.

A. The Effect of pH on Replication of Phage in Laboratory Cultures

Two streptomycete phage (f6 and f13) isolated from neutral soil and initially propagated on neutrophilic hosts were used in a study by Sykes *et al.* (1981). However, some acidophilic isolates were also susceptible to these phage. Thus, the responses of the same phage to pH at various stages of the replication cycle can be studied by using hosts with different pH requirements. These results will be summarized here; details were given by Sykes *et al.* (1981).

Inactivation of the phage in broths of different pH follows simple first-order kinetics; both phage are rapidly inactivated at pH levels below 5.0 or above 9.0. They are also rapidly inactivated when introduced into sterile soils with a pH <4.2. Rates of adsorption to neutrophilic and acidophilic hosts are not significantly affected by pH. Likewise, efficiency of infection is influenced more by the host than by pH. The efficiency of intracellular replication of phage is greatest when the host, either neutrophile or acidophile, is growing at its optimum pH. These results show that acidophilic streptomycetes are susceptible to these phage, providing that the free phage particles are not subjected to levels of acidity (below pH 5.0) which inactivate them.

The presence of clay minerals can also influence the response of phage to acidity (Sykes and Williams, 1978). It has been shown that phage f6 readily adsorbs to kaolin in nutrient broth. Most of the adsorbed phage remains infective, but the pH stabilities of adsorbed and of free phage differ: for adsorbed phage 50% inactivation occurs at pH 6.1; for free phage, at pH 4.9. Since hydrogen ion concentrations are increased at the surface of negatively charged colloidal particles, adsorbed phage experience a pH lower than that measured in the bulk suspension, the difference (△ pH) being apparently 1 to 2 units. This suggests that phage in clay soils are more susceptible to pH levels below neutrality than those in soils with less colloidal particles.

TABLE VI. The Effect of Acidity on Streptomycete Phage Growth in Nutrient Broth

pH	*% original titer recovered after 16h*				
			Phage		
	fO15	*fO69*	*fMx1*	*fMx2*	*fMx3*
2.0	*0.0*	*0.0*	*0.00*	*0.00*	*0.00*
2.5	*0.0*	*0.0*	*0.01*	*0.08*	*0.03*
3.0	*0.0*	*0.0*	*15.30*	*11.70*	*13.30*
3.5	*0.0*	*0.0*	*23.80*	*34.00*	*89.90*
4.0	*1.0*	*0.4*	*53.50*	*42.00*	*109.00*
4.5	*18.2*	*0.7*	*97.60*	*54.80*	*92.40*
5.0	*62.5*	*16.2*	*88.80*	*57.20*	*117.00*
6.0	*85.4*	*115.0*	*100.40*	*92.50*	*114.00*
7.0	*100.0*	*100.0*	*100.00*	*100.00*	*100.00*

B. Further Studies of the Effects of Acidity on Free Phage

These results and those obtained with phage f85 (Table V) indicated that soil acidity was a major cause of inactivation of free phage. It was therefore decided to conduct further studies with a range of phage isolated from soils by the improved, direct isolation procedures described above. The phage and their propagation hosts were f015 (*S. michiganensis*, ISP 5015), f069 (*S. lavendulae*, ISP 5069), fMx1 (*Streptomyces* sp., Mx1), fMx2 (*Streptomyces* sp. Mx2), and fMx3 (*Streptomyces* sp., Mx3).

The stabilities of these phage at different levels of acidity in broth and sterile soils were compared (Tables VI and VII). Whereas phage f015 and f069 were largely or completely inactivated at or below pH 4.5, a significant proportion of Mx phage remained active at pH 3.0, with a small number withstanding pH 2.5. A similar pattern occurred in sterile soil, although the proportions recovered were lower than those from broth; some Mx phage remained stable at pH 4.6, with fMx3 showing the most resistance to acidity, as it did in broth. The Mx phage also differed from f015 and f069 in their considerably greater tolerances to chloroform (Table VIII); indeed, they were isolated using the chloroform sterilization technique. When negatively stained preparations were examined by electron microscopy, phage f015 and f069 were found to have hexagonal heads with long non-contractile tails. This places them into the morphological group B of Bradley (1967) to which most previously examined streptomycete phage have been placed (Bradley and Ritzi, 1967; Korn *et al.*, 1978). In contrast, the Mx phage had hexagonal heads with short tails, thereby resembling Bradley's group D.

Thus the Mx phage, unlike the other streptomycete phage which we have studied, were not rapidly inactivated at the pH levels which commonly occur in acidic soils. However, they were originally isolated from neutral soils and propagated on neutrophilic hosts at pH 7.0. Attempts to propagate these three phage on 19 acidophilic streptomycetes which had been isolated from various acidic soils were unsuccessful. It was therefore decided to attempt to isolate streptomycete phage from acidic soils.

TABLE VII. The Effect of Acidity on Streptomycete Phage Growth in Sterile Soil

	% added phage recovered after 16h				
			Phage		
Soil pH	*f015*	*f069*	*fMx1*	*fMx2*	*fMx3*
3.6	*0*	*0*	*0*	*0*	*0.002*
4.6	*0*	*0*	*4.4*	*4.8*	*32.600*
8.0	*119.0*	*94.5*	*46.7*	*69.8*	*108.300*

TABLE VIII. The Effect of 2% (v/v) Chloroform on Streptomycete Phage Growth in Nutrient Broth

Phage	*% original titer recovered after 7 min*
f015	0.00080
f069	0.00001
fMx1	1.00000
fMx2	92.80000
fMx3	92.20000

C. Attempts to Isolate Phage from Acidic Soils

Studies were carried out on a range of acidic podzols (Table IX). As the sole aim was to detect the presence of phage in these soils, a specific enrichment technique was used, each soil sample being enriched in broth culture with spores of a streptomycete. Both neutrophilic and acidophilic streptomycetes were used as prospective hosts, a proportion of the strains used having originated from the soils under study. Enrichment broths were adjusted to pH 5.0 for acidophiles and to pH 7.0 for neutrophiles. After incubation, one sample of the supernatant was sterilized by filtration and the other by 2% (v/v) chloroform. Sterilized extracts were spotted onto plates

TABLE IX. Summary of Attempts to Isolate Phage from Acidic Soils by Specific Enrichment with Streptomycetes

Soils	*pH*	*No. isolates tested*		*No. positive results*	
		Acidophiles	*Neutrophiles*	*Acidophiles*	*Neutrophiles*
Podzol (F_2/H horizon) (Freshfield, Merseyside, U.K.)	4.2	38	26	1	0
Podzol (A_2 horizon) (Storeton, Merseyside, U.K.)	4.0	38	26	1	0
Podzol (A_2 horizon) (Thurstaston, Merseyside, U.K.)	3.6	38	26	0	0
Podzol (F,H,A1,A2 horizons) (Northumberland, U.K.)	4.0-4.5	19	26	0	0

which had been seeded with the appropriate streptomycete. Plates at both pH 5.0 and 7.0 were prepared for all strains used.

No phage were isolated using neutrophilic streptomycetes and, for two of the soils, only one acidophilic strain gave a positive reaction (Table IX). These plates showed typical areas of confluent lysis; the presence of phage particles (Bradley group B) was demonstrated by electron microscopy. However, despite repeated attempts using various media and pH protocols, it proved impossible to propagate this phage.

V. CONCLUSIONS

Our studies showed that streptomycete phage were widespread and more numerous in soil than previously had been realized. Survival and replication of many of these phage in soil will be aided by their demonstrated polyvalency (Korn *et al.*, 1978; Prauser, 1976; Wellington and Williams, 1981). Those with more restricted host ranges may well be in lysogenic association in their natural habitat and may be isolated when indicator strains are inadvertently used. Free phage particles in soil tolerated a wide range of environmental conditions, but most were inactivated below pH 5.0. Acidophilic streptomycetes were susceptible to phage isolated from neutral soils and propagated on neutrophilic hosts, providing that the free phage were not subjected to low pH. It is also clear that relatively acid-tolerant phage occurred in neutral soils. Although there is therefore no reason *a priori* why acidophilic or acidoduric streptomycetes in acidic soils should not be infected by phage, efforts to detect such phage have met with only limited success so far.

Phage are a neglected ecological entity. Our knowledge of the ecology of streptomycete phage and other bacteriophage clearly lags behind that of their hosts. There are several reasons why this should be rectified:

(i) Phage probably play a significant role in regulating host activities in soil.
(ii) Direct counts of phage in soil may prove to be a sensitive indicator of host activity.
(iii) Phage may be important agents of genetic exchange between organisms in soil by contributing to the natural variation among isolates.
(iv) Improved isolation techniques, by providing more information about the distribution of phage in soil, would provide a wide range of new material for genetical studies.

REFERENCES

Bradley, D. E., *Bacteriol. Rev. 31:*230 (1967).
Bradley, S. G., and Ritzi, D., *Dev. Ind. Microbiol. 8:*206 (1967).
Casida, L. E., and Liu, L., *Appl. Microbiol. 28:*951 (1974).
Collard, C. A., *C. R. Seances Soc. Biol. 164:*465 (1970).
Flowers, T. H., and Williams, S. T., *Microbios 18:*223 (1977).
Gray, T. R. G., and Williams, S. T., *in* "Microbes and Biological Productivity" (E.D. Hughes and A. Rose, eds.), p. 255. Cambridge University Press, Cambridge (1971).
Hagedorn, C., *Appl. Environ. Microbiol. 32:*368 (1977).
Khan, M. R., and Williams, S. T., *Soil Biol. Biochem. 7:*345 (1975).
Khavina, E. S., *Tr. Inst. Mikrobiol. Akad. Nauk, SSSR3:*224 (1954).
Korn, F., Weingartner, B., and Kutzner, H. J., *in* "Genetics of the Actinomycetales" (E. Freerksen, I. Tarnok, and T. H. Thumin, eds.), p. 251. Gustav Fischer Verlag, Stuttgart (1978).
Kurup, V. P., and Heinzen, R. J., *Can. J. Microbiol. 24:*794 (1978).
Lanning, S., and Williams, S. T., *J. Gen. Microbiol. 128:*2063 (1982).
Patel, J. J., *Arch. Mikrobiol. 65:*601 (1969a).
Patel, J. J., *Arch. Mikrobiol. 69:*294 (1969b).
Prauser, H., *in* "The Biology of the Nocardiae" (M. Goodfellow, G. H. Brownell, and J. A. Serrano, eds.), p. 266. Academic Press, London (1976).
Reanney, D. C., *N. Z. J. Agric. Res. 11:*763 (1968).
Reanney, D. C., and Marsh, S. C. N., *Soil Biol. Biochem. 5:*399 (1973).
Robinson, J. B., and Corke, C. T., *Can. J. Microbiol. 5:*479 (1959).
Sakaki, Y., and Oshima, T., *J. Virol. 15:*1449 (1975).
Sarfert, E., Kretschmer, S., Triebel, H., and Luck, G., *Z. All. Mikrobiol. 19:*203 (1979).
Sykes, K. I., Ph. D. thesis, University of Liverpool, (1977).
Sykes, K. I., and Williams, S. T. *J. Gen. Microbiol. 108:*97 (1978).
Sykes, K. I., Lanning, S., and Williams, S. T., *J. Gen. Microbiol. 122:*271 (1981).
Tan, J. S. H., and Reanney, D. C., *Soil Biol. Biochem. 8:*145 (1976).
Torsvik, T., and Dundas, I. D., *Nature (Lond.) 248:*680 (1974).
Torsvik, T., and Dundas, I. D., *J. Gen. Virol. 47:*29 (1980).
Wais, A.C., Kon, M., MacDonald, R. E., and Stollar, B. D., *Nature (Lond.) 256:*314 (1975).
Wellington, E. N. H., and Williams, S. T., *Zentralbl., Bakteriol. Mikrobiol. Hyg. I Abt. 11* suppl., 93 (1981).
Welsch, M., Minon, A., and Schonfield, J. K., *Experientia 11:*24 (1955).
Williams, S. T., *in "Nocardia* and *Streptomyces*" (M. Mordarski, W. Kurylowicz, and J. Jeljaszewicz, eds.), p. 137. Gustav Fischer Verlag, Stuttgart (1978).
Williams, S. T., and Flowers, T. H., *Microbios 20:*99 (1978).
Williams, S. T., and Robinson, C. S., *J. Gen. Microbiol. 127:*55 (1981).
Williams, S. T., Wellington, E. M. H., and Tipler, L. S., *J. Gen. Microbiol. 119:*173 (1980).
Williams, S. T., Davies, F. L., Mayfield, C., and Khan, M. R., *Soil Biol. Biochem. 3:*187 (1971).

MOLECULAR SYSTEMATICS OF ACTINOMYCETES AND RELATED ORGANISMS

Erko Stackebrandt
Karl-Heinz Schleifer

Lehrstuhl für Mikrobiologie
Institut Botanik und Mikrobiologie
Technischen Universität München
München 2, Federal Republic of Germany

I. INTRODUCTION

Significant progress in genetics and biochemistry has made it possible to determine changes in the primary structure of macromolecules, either directly by sequencing or indirectly by reassociation techniques. As a result of comparative studies, the phylogenetic structure of the procaryotic kingdoms is emerging (Woese and Fox, 1977), allowing detection of the various lines of descent leading to recent organisms (Stackebrandt and Woese, 1981a). One major aspect of the phylogenetic tree refers to the ultimate goal of taxonomy (Stanier and van Niel, 1941), in that it is now possible to classify bacteria in a hierarchic system on the basis of their natural relationships. Through the combining of different approaches, a meaningful taxonomy can be achieved. By starting with methods which measure large phylogenetic distances by the comparison of highly conserved and universally distributed molecules, such as 5S or the 16S rRNA, the main groups can be detected. At the family level, the studies may be enlarged by the use of RNA cistron similarity and by immunological comparison of conserved proteins, allowing allocation of a large number of organisms within a rather short period. The fine structure of the resulting higher taxa then may be detected by DNA-DNA reassociation studies, by phage typing experiments, and by immunological studies on less conserved proteins. These genetically defined taxa may be described with key characters from physiological, biochemical, and morphological studies.

This strategy has been applied to two groups of organisms, the coryneform

ISBN 0-12-528620-1

bacteria and the actinomycetes, whose classification into families and genera has been established mainly on the basis of morphological characteristics (Cross and Goodfellow, 1973; Gottlieb, 1974; Rogosa *et al.*, 1974).

II. METHODS

A. Comparative Analysis of Oligonucleotides of 16S Ribosomal (r) RNA

Taxonomists may choose between two methods to determine sequences of oligonucleotides of the 16S rRNA. The traditional method, originally developed by Sanger, Brownlee, and Barrell (1965) and later improved and introduced into phylogenetic studies by Woese and colleagues (Uchida *et al.*, 1974), will not be discussed in further detail. The second method (Stackebrandt *et al.*, 1981a) circumvents some of the disadvantages of the traditional method: (i) Labelling *in vivo* or rRNA is replaced by labelling techniques *in vitro* which allow the investigation of slow-growing bacteria, as well as of symbionts and parasites, and which permit the investigation of the 18S rRNA of eucaryotic cytoplasmatic ribosomes; (ii) The two dimensional separation of oligonucleotides by high voltage electrophoresis is replaced by a combination of high-voltage electrophoresis and chromatography which allow not only a reduction in the number of high voltage electrophoresis apparatus, but also the use of DEAE thin-layer plates instead of DEAE paper; (iii) The sequences of oligonucleotides (hexamers or larger), instead of being deduced from the electrophoretic mobility of enzymic digest products of oligonucleotides as in the traditional method, are determined by a two-dimensional separation of alkaline cleavage-products of oligonucleotides which allows for rapid and easy sequencing. Figure 1 shows the individual steps of the method which has been described in detail by Stackebrandt *et al.* (1981a).

It is worth noting that these two sequencing techniques differ only in the methodical steps leading to the sequencing of oligonucleotides. Once these sequences have been determined, the results from both approaches can be directly compared and used for the generation of a common phylogenetic tree of the 16S rRNA. The dendrograms of relationship, presented in Figures 2 to 7 are such composite trees.

The degree of relatedness among two organisms is expressed as a similarity coefficient (S_{AB} value). These values range between 0.03 (randomly related rRNA's) and 1.0 (complete identity of rRNA's). The S_{AB} value is defined as follows: (number of bases in sequences common to a given pair of catalogues) X 2, divided by the total number of bases present in the pair of catalogues (Fox *et al.*, 1977). Average linkage clustering among the merged groups allows a graphic presentation of the degree of relatedness.

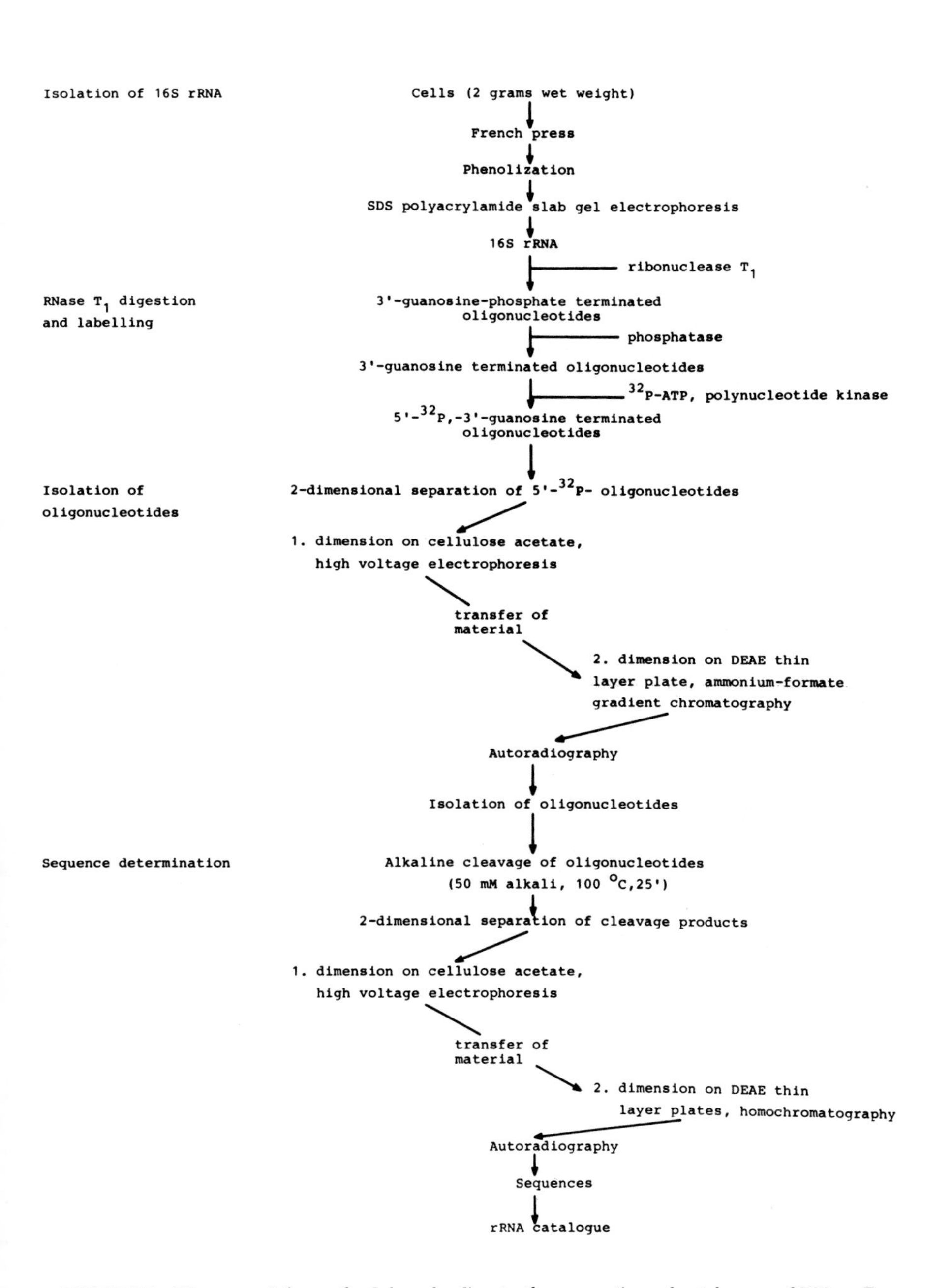

FIGURE 1. Diagram of the methodology leading to the generation of catalogues of RNase T_1 resistant oligonucleotides of 16S ribosomal RNA's.

B. Nucleic Acid Hybridization Procedures

1. Isolation of Nucleic Acids. Cells from coryneform bacteria and micrococci were lysed with lysozyme (Back and Stackebrandt, 1978; Meyer and Schleifer, 1978), whereas those from actinomycetes were sheared by passage through a French pressure cell at 8,000 psi (Stackebrandt *et al.*, 1981b). The procedures used for the purification of high-molecular weight DNA, as well as for tritium-labelled DNA, have been described (Meyer and Schleifer, 1978). For the isolation of tritium-labelled rRNA's, cells were sheared by passage through a French press at 20,000 psi. Phenolization of the cells, separation of the 16S and 23S rRNA by one-dimensional sodium dodecyl sulfate-polyacrylamide slab gel electrophoretic elution were performed by using described procedures (Stackebrandt *et al.*, 1981a,b).

2. DNA-DNA Hybridization. DNA-DNA homology studies were performed by using the membrane-filter technique (Denhardt, 1966; Gillespie and Spiegelman, 1965; McConaughy *et al.*, 1969). Seven to 10 μg of single-stranded DNA, fixed to a cellulose acetate filter (5.5 mm in diameter), was hybridized to 0.1 to 0.5 μg ^{3}H-DNA (depending on the specific activity) in 200 μg of 3X SSC (1 X SSC: 0.15 M NaCl-0.015 M trisodium citrate, pH 7.0) which contained formamide in a concentration which allowed hybridization at 60°C and which fulfilled conditions described to be optimal (25°C below the melting point of the DNA). Hybridization was carried out for 24 h.

3. DNA-rRNA Cistron Similarity Studies. In DNA-rRNA non-competition reassociation experiments, the method of De Ley and De Smedt (1975) was basically followed. Modifications included the use of rRNA with a high specific activity (3.5×10^4 to 1.1×10^5 cpm/μg rRNA per experiment) and, as a consequence, of a small amount of filter-bound DNA (7 to 10 μg/filter) Döpfer *et al.*, 1982; Stackebrandt *et al.*, 1981b). Hybridization was carried out in 200 μl of 5 X SSC containing 25% formamide for 18 h at 60°C. Enzymic digestion of incompletely reassociated rRNA and determination of the thermal stability of the heteroduplexes ($T_{m(e)}$) between 60 and 95°C were done as described by De Ley and De Smedt (1975).

C. Determination of Peptidoglycan Types

The greatest diversity of peptidoglycan types can be found in coryneform bacteria and actinomycetes. Among these organisms, strains are found in which the peptidoglycan have the same amino acid composition but show differences in their amino acid sequences and in the mode of their crosslinkage. Hence, it is necessary not only to determine the main amino acids, *e.g.*, lysine, ornithine, or diaminopimelic acid (Dpm), and their configuration, *e.g.*, *meso-* or L L-Dpm, but in addition, the amino acid sequence of both the peptide subunit and the interpeptide bridge.

The procedures used in the determination of the amino acid composition and sequence, together with the peptidoglycan types and their abbreviations, have been described in detail by Schleifer and Kandler (1972).

III. RESULTS

A. Comparison of the Results of Oligonucleotide Sequencing, DNA-rRNA Cistron Similarity, and DNA Homology Studies

The application of these three methods in phylogenetic/taxonomic studies is suitable to cover the whole range of taxonomic categories. Though the comparative analysis of the 16S rRNA permits detection of the highest ranks (urkingdoms) (Woese and Fox, 1977) up to the species level, this method is too time consuming and too expensive to be routinely used in taxonomic studies at the intergeneric level. Relationships between closely related species can be detected most efficiently in DNA homology studies (Stackebrandt and Fiedler, 1979; Stackebrandt and Kandler, 1979; Steigerwalt *et al.*, 1975), whereas the degree of relatedness of more remotely related species up to the intrafamily level can be detected in DNA-rRNA cistron similarity studies (De Smedt and De Ley, 1977; De Ley *et al.*, 1978; Stackebrandt *et al.*, 1981b).

Many representatives of the coryneform group of organisms and of the order *Actinomycetales* have been included in studies using all three methods, and a high correlation of the degree of relatedness could be detected. The branching pattern of the phylogenetic tree of representatives of various genera of spore-forming actinomycetes included in sequencing studies (III B) and in DNA-rRNA cistron similarity studies (Stackebrandt *et al.*, 1981b) was found to be almost identical. S_{AB} values of greater than 0.55, separating genera related to *Streptomyces* from those related to *Actinoplanes*, correlate with $\Delta T_{m(e)}$ values lower than 10 °C. ($\Delta T_{m(e)}$ is the difference in the $T_{m(e)}$ values of the homologous and heterologous DNA-rRNA duplexes (De Smedt and De Ley, 1977).)

More data are available for the comparison of S_{AB} values and DNA-DNA homology values. Investigations on *Arthrobacter* (Stackebrandt and Fiedler, 1979; Stackebrandt *et al.*, 1980b), *Micrococcus* (Schleifer *et al.*, 1979; Stackebrandt *et al.*, 1980b), *Cellulomonas* (Stackebrandt *et al.*, 1980a,b), *Streptomyces* as well as *Ampullariella* and *Actinoplanes* (Stackebrandt *et al.*, 1981b) have shown that S_{AB} values greater than 0.75 correlate well with DNA homology values greater than 20%. A similarly good correlation was found between DNA homology and DNA-rRNA cistron similarity studies on spore-forming members of genera of the order *Actinomycetales* (Stackebrandt *et al.*, 1981b) and on representatives of *Microbacterium, Curtobacterium*, and plant pathogenic corynebacteria (Döpfer *et al.*, 1982). In these investigations, DNA-DNA homology values greater than 20% correlated with $\Delta T_{m(e)}$ values less than 6 to 7°C.

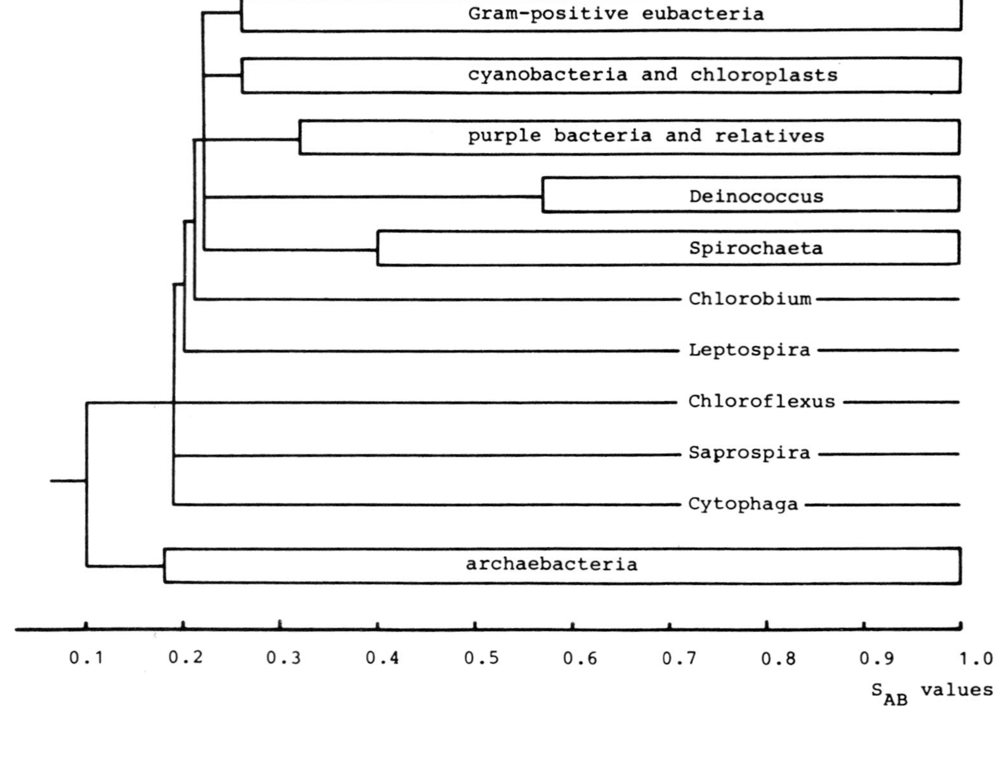

FIGURE 2. Dendrogram of relationships among procaryotes. A more detailed picture of the clusters defined by cyanobacteria and chloroplasts, purple bacteria, and archaebacteria has been presented by Fox et al. *(1980) and Stackebrandt and Woese (1981c).*

As a consequence, the phylogenetic tree of the 16S rRNA may be extended by the results of nucleic acid homology studies, as is shown in the dendrograms of relationships of some of the groups mentioned below (Figs. 5-7). It should also be mentioned that immunological studies on coryneform bacteria, in which the 50S subunit of the ribosomes was used as a phylogenetic probe, revealed that the phylogenetic branching pattern of representatives of *Cellulomonas, Oerskovia, Arthrobacter, Micrococcus, Brevibacterium,* and *Microbacterium* was almost identical to that derived from the 16S rRNA (Seeberger, 1982).

B. The Phylogeny of the Order Actinomycetales

1. The Main Lines of Descent of Procaryotes. All actinomycetes and coryneform bacteria included in comparative studies on the 16S rRNA have been found to be members of one of the several major lines of descent of eubacteria, namely the one comprising Gram-positive eubacteria. Figure 2 presents the dendrogram of relationship showing the main groupings of procaryotes recognized so far. Some of the lines are represented by a single or a few organisms only, *e. g., Deinococcus, Spirochaeta, Chloroflexus, Leptospira, Chlorobium, Cytophaga,* and *Saprospira*; other are better defined, *e. g.*, cyanobacteria and chloroplasts, the group of purple sulfur and non-sulfur bacteria, and their non-phototrophic relatives, as well as the Gram-positive eubacteria. A detailed presentation of the eubacteria have been published by Fox *et al.* (1980) and Stackebrandt and Woese (1981a).

2. The Phylogeny of Gram-positive Eubacteria. Within the line of the Gram-positive eubacteria, organisms with a DNA G + C content of less than about 55 mole% (*Clostridium-Lactobacillus-Bacillus* branch) are clearly separated phylogenetically from those organisms exhibiting a higher G + C content (*Bifidobacterium-Propionibacterium-Actinomyces* branch). In the current systematics of bacteria (Gottlieb, 1974), *Kurthia zopfii, Eubacterium tenue, E. limosum,* and *Thermoactinomyces vulgaris* are classified within the coryneform group of organisms, *Propionibacterium* and *Actinomycetales*, respectively. These organisms, showing a DNA G + C content of lower than 55 mole%, were shown to be members of the *Clostridium-Lactobacillus-Bacillus* branch (Fig. 3).

K. zopfii shows a remote relationship to the lactic acid bacteria, *Aerococcus, Gemella* and *Bacillus* (Ludwig *et al.*, 1981b; unpublished results). *E. limosum*, together with *Clostridium barkeri* and *Acetobacterium wodii* (Tanner *et al.*, *1981), and E. tenue,* together with *Clostridium lituseburense* and *Peptostreptococcus anaerobius*, form two distinct groups (Stackebrandt and Woese, 1981a). *Thermoactinomyces vulgaris* is specifically related to bacilli (Cross, 1982; Collins *et al.*, 1982b; Stackebrandt and Woese, 1981a,b). The high DNA G + C containing members of *Micrococcus*, on the other hand, traditionally classified with other Gram-positive, spherical bacteria, *e. g., Staphylococcus*

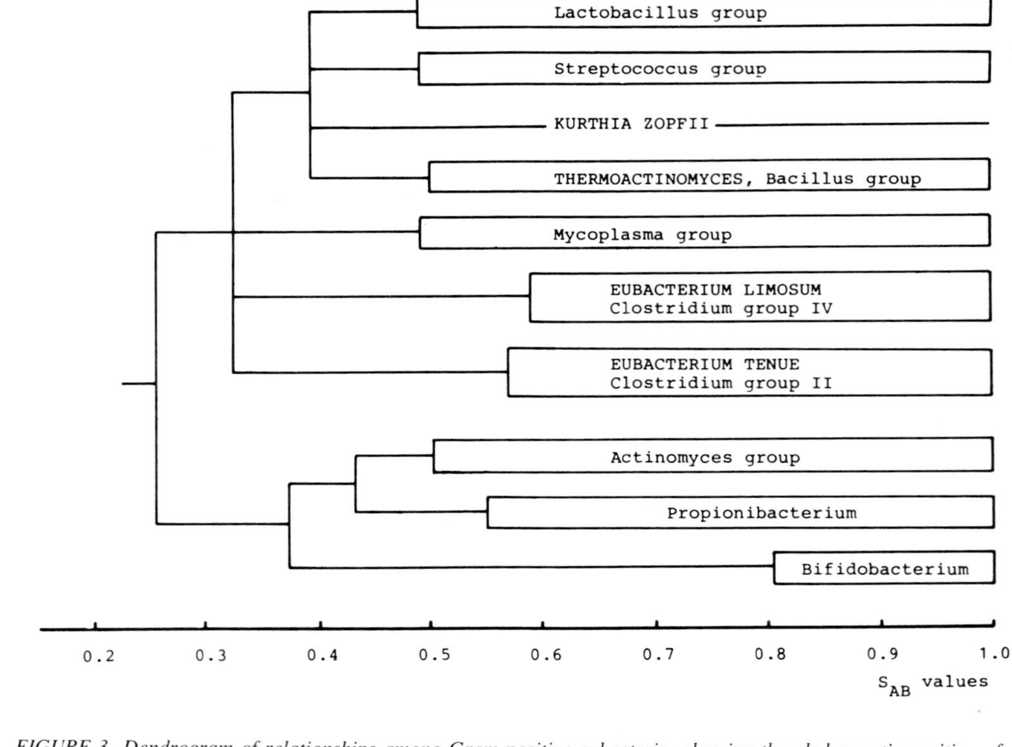

FIGURE 3. Dendrogram of relationships among Gram-positive eubacteria, showing the phylogenetic position of Kurthia, Thermoactinomyces, *and* Eubacterium. *A more detailed picture of the Gram-positive eubacteria of the* Clostridium-Bacillus-Lactobacillus *branch has been published (Fox* et al., *1980; Ludwig* et al., *1981b; Tanner* et al., *1981; Stackebrandt and Woese, 1981b; Woese et al.,* 1980).

and *Planococcus*, are actually members of the *Bifidobacterium-Propionibacterium-Actinomyces* branch. It is certainly not premature to postulate that the determination of the DNA G + C content of a Gram-positive eubacterium will immediately allow it to be allocated into one of these two major branches.

3. The Phylogeny of Gram-positive Eubacteria with a High DNA G + C Content. Within the *Bifidobacterium-Propionibacterium-Actinomyces* branch (Fig. 4), the organisms are related phylogenetically (Fox *et al.*, 1980; Stackebrandt and Woese, 1981 a,b). The line leading to *Bifidobacterium* branched off from the main stem earlier in evolution than do those leading to the other organisms of this branch (low S_{AB} values), indicating that the bifidobacteria are phylogenetically older than propionibacteria, coryneform bacteria, and actinomycetes. The phylogenetic depth of the genus *Bifidobacterium* has not been discovered because only two representatives have been analyzed.

A similar situation holds for the propionibacteria which branch off later (high S_{AB} values) than do the bifidobacteria. Investigation of more representatives of both genera is certainly needed; the results will reveal valuable information about their phylogeny and about the early evolution of the Gram-positive eubacteria with a high DNA G + C content.

All the remaining organisms of this branch, which have been investigated so far, are more closely related to each other. The phylogenetic coherency of this group of organisms, which includes the genus *Actinomyces*, has led to the informal proposal of giving this group the rank of an order (order *Actinomycetales*) (Stackebrandt and Woese, 1981b).

Several sub-lines of descent branched off at about the same time in evolution (same S_{AB} values) for which the rank of families can be informally proposed: *Arthrobacteriaceae, Actinoplanaceae, Streptomycetaceae, Streptosporangiaceae, Actinomycetaceae, Corynebacteriaceae, Mycobacteriaceae*, and *Thermomonosporaceae. Geodermatophilus obscurus, Micrococcus mucilaginosus*, and *Arthrobacter simplex* show similar deep branching points as do the eight families and may therefore constitute families of their own; however, more organisms related to these three species have to be investigated before a definite statement can be made.

4. The Phylogeny of the Order Actinomycetales. The information available concerning the phylogenetic structure of some of the various families of the order *Actinomycetales* has reached such an extent that only a few representatives can be included in the phylogenetic trees. We will restrict the information on results from our own laboratory because the same species have been included in comparative studies (see III A) and all nucleic acid hybridization studies have been carried out under the same defined conditions.

Other recent investigations include DNA-rRNA cistron similarity studies on spore-forming actinomycetes (Mordarski *et al.*, 1980b) and DNA-DNA homology studies on corynebacteria (Denhaive *et al.*, 1982; Suzuki *et al.*, 1981)

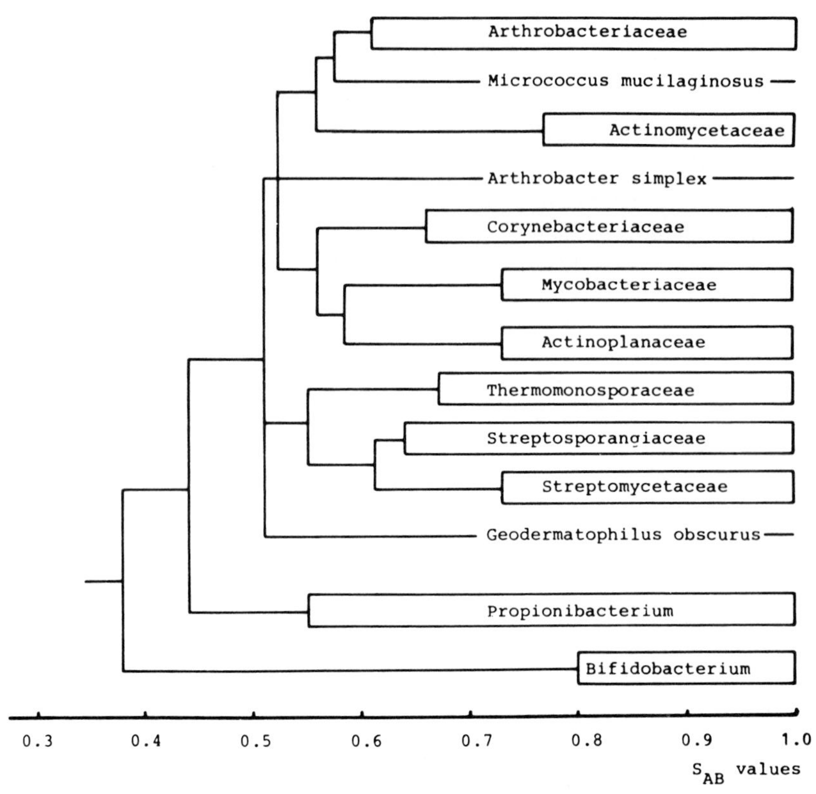

FIGURE 4. Dendrogram of relationships among Gram-positive eubacteria with a high DNA G+C content. A detailed analysis of the phylogeny of actinomycetes and related organisms is presented in TablesI-III.

on *Rhodococcus* (Mordarski *et al.*, 1980a), and on *Actinomyces* (Coykendall and Munzenmaier, 1979).

The designation of genera presented below is not exclusively based on phylogenetic data; in addition, chemotaxonomic characteristics of proven taxonomic value have been taken into account (Figs. 5-7 and Table I). Another point to be mentioned concerns the disregard of nomenclatural rules. We are aware of the fact that the union of *Arthrobacter* and *Micrococcus* and of *Actinoplanes* and *Micromonospora* has to be named *Micrococcus* and *Micromonospora*, respectively, and that the respective families have consequently to be named *Micrococcaceae* and *Micromonosporaceae*. We use the illegitimate terms *Arthrobacter (Arthrobacterieaceae)* and *Actinoplanes (Actinoplanaceae)* since *Arthrobacter* and *Actinoplanes* are more typical representatives of their families.

a. The family Arthrobacteriaceae. Figure 5 shows the phylogenetic structure of the genera constituting the family *Arthrobacteriaceae*. Within the family, the genera *Cellulomonas/Oerskovia, Arthrobacter/Micrococcus*, and *Pro-*

micromonospora are more closely related to each other than they are to *Brevibacterium, Microbacterium, Curtobacterium*, and related organisms.

The genus *Arthrobacter* is not restricted phylogenetically to coryneform bacteria which are characterized by morphological life cycle and strictly aerobic growth since, in addition, it contains spherical bacteria of the genus *Micrococcus* (Stackebrandt and Woese, 1979; Stackebrandt *et al.*, 1980b). The close relationship among representatives of these two genera was also confirmed by DNA-homology studies (Schleifer and Lang, 1980). The type of strain of *Arthrobacter, A. globiformis*, is more closely related to various species of *Micrococcus, e. g., M. luteus, M. lylae, M. roseus*, and *M. varians*, than to *A. atrocyaneus*. It seems that micrococci must be regarded as morphologically degenerated forms of arthrobacteria (Stackebrandt and Woese, 1979; Stackebrandt *et al.*, 1980b). In addition to micrococci, this phylogenetically defined genus also contains a number of misclassified species, *e. g., Brevibacterium sulfureum* and *B. protoformiae* (Stackebrandt and Fiedler, 1979), as well as *Corynebacterium uratoxidans, C. aurantiacum, Brevibacterium liquefaciens*, and *B. fuscum* (unpublished results). The only micrococci groupings outside the *Arthrobacter* group are *M. sedentarius* and *M. nishinomiyaensis*. Even more distantly related is *Micrococcus mucilaginosus* (Ludwig *et al.*, 1981).

A situation similar to that in *Arthrobacter* is seen in the relationship of *Cellulomonas* and *Oerskovia* although in this case mycelium-producing strains are closely related genetically to rod-shaped strains (Stackebrandt *et al.*, 1980a). The high genotypic and phenotypic similarity found between organisms of two genera led to the union of *Cellulomonas* and *Oerskovia* in a redefined genus *Cellulomonas*. This emended genus in addition contains *Arthrobacter luteus, Corynebacterium manihot, Nocardia cellulans, Brevibacterium fermentans*, and *B. liticum* which are, together with *Oerskovia xanthineolytica*, synonymous with *Cellulomonas cartae* (Stackebrandt *et al.*, 1982).

Two monotypic genera are represented by *Promicromonospora* (unpublished) and *Brevibacterium (Stackebrandt et al.*, 1980b). Genotypically more complex is a cluster comprising *Microbacterium, Curtobacterium*, and *Agromyces*, together with misclassified strains of *Corynebacterium, Arthrobacter*, and *Brevibacterium*. Extensive nucleic acid hybridization experiments revealed that the genera *Microbacterium, Curtobacterium*, and *Agromyces*, cannot be clearly distinguished from each other (Döpfer *et al.*, 1982). This finding, together with the distinct positions of *Microbacterium lacticum, Corynebacterium betae*, and *C. mediolanum* in the dendrogram of the 16S rRNA (Stackebrandt *et al.*, 1980b) and the possession of a unique peptidoglycan structure (group B, according to Schleifer and Kandler, 1972), was taken as an indication that the strains forming this coherent cluster should be described as members of one genus, namely *Microbacterium* (Döpfer *et al.*, 1982).

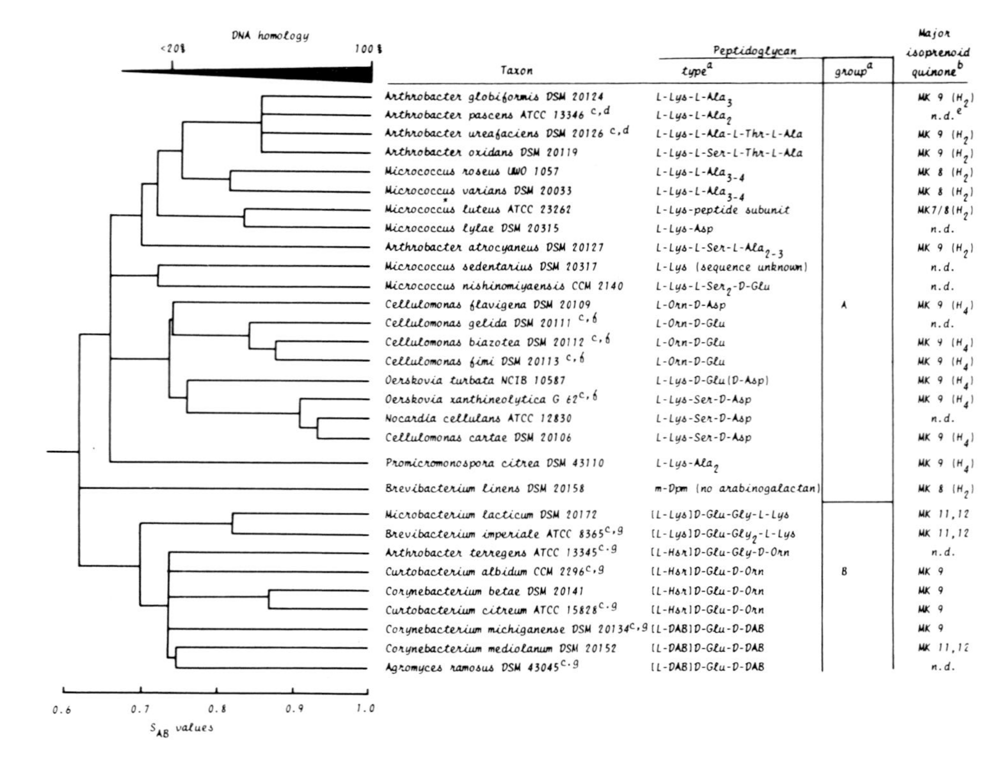

FIGURE 5.–Dendrogram of relationships of the family Arthrobacteriacea *with additional information on peptidoglycan types and major isoprenoid quinones.* [a]*Schleifer and Kandler (1972).* [b]*Collins and Jones (1981).* [c]*Organisms are assigned to the 16SrRNA clusters on the basis of DNA homologies and rRNA cistron similarities.* [d]*Stackebrandt and Fiedler (1979).* [e]*n.d., not determined.* [f]*Stackebrandt* et al., *(1980a).* [g]*Döpfer* et al. *(1982).*

b. The family Actinomycetaceae. This family contains a single genus, *Actinomyces*. The two representatives, *A. viscosus* and *A. bovis*, included in phylogenetic studies were found to be highly related (S_{AB}: 0.78) (Stackebrandt and Woese, 1981b). Phylogenetically, *Actinomyces* is more closely related to the family *Arthrobacteriaceae*, composed almost exclusively of non-mycelium-forming organisms, than it is to the other mycelium-producing organisms with which *Actinomyces* constitutes the order *Actinomycetales* in the present phenetic classification (Gottlieb, 1974).

c. The family Corynebacteriacea. Little phylogentic work has been carried out on those members of *Corynebacterium* which are pathogenic or parasitic for animals and humans. However, phylogenetic analysis of *C. diphtheriae* (Stackebrandt *et al.*, 1980b) revealed that this species, together with strains related to *C. diphtheriae* on the basis of similarities in chemotaxonomic markers (Collins *et al.*, 1982a; Schleifer and Kandler, 1972) and numerical analysis (Jones, 1975, 1978), constitutes a genus which so far is the only one in the family *Corynebacteriaceae*. *Arthrobacter variabilis* and *Corynebacterium glutamicum* have also been regarded as members of *Corynebacterium* (Fig. 6).

d. The family Mycobacteriaceae. A situation similar to that in *Corynebacteriaceae* is also seen in *Mycobacteriaceae* in that too few representatives have been included in phylogenetic studies (Fig. 6). Until a wider selection of species is characterized, it will not be possible to propose a generic structure for this family. Representatives of *Mycobacterium, Rhodococcus*, and *Nocardia* are more closely related than might be expected from their allocation into different genera (Stackebrandt and Woese, 1981b). The degree of relatedness is comparable to that found among species of the genus *Arthrobacter* (Fig. 5).

e. The family Actinoplanaceae. This family contains the mainly morphologically defined genera *Actinoplanes, Ampullariella, Amorphosporangium, Dactylosporangium*, and *Micromonospora* (Fig. 6). Within this family, strains of *Actinoplanes* and *Ampullariella* are closely related, whereas *Amorphosporangium auranticolor, Dactylosporangium aurantiacum, Micromonospora chalcea* (Stackebrandt and Woese, 1981b), as well as *M. echinospora* (Stackebrandt *et al.*, 1981b) are somewhat more distantly related to *Actinoplanes*. Although it does not form distinct spore vesicles, *Micromonospora* resembles the other genera in many chemotaxonomic and physiological properties (Jones and Bradley, 1964; Stackebrandt *et al.*, 1981b). The high phylogenetic similarity, supported by biochemical features strongly favors the union of these genera into one genus.

f. The family Thermomonosporaceae. To date, the family *Thermomonosporaceae* is composed of three organisms, *Thermomonospora curvata* and two representatives of *Actinomadura*, the types species *A. madurae*, and *A. verrucospora*. The specific relationship of the two *Actinomadura* species and their remote relatedness to *T. curvata* indicate the presence of two

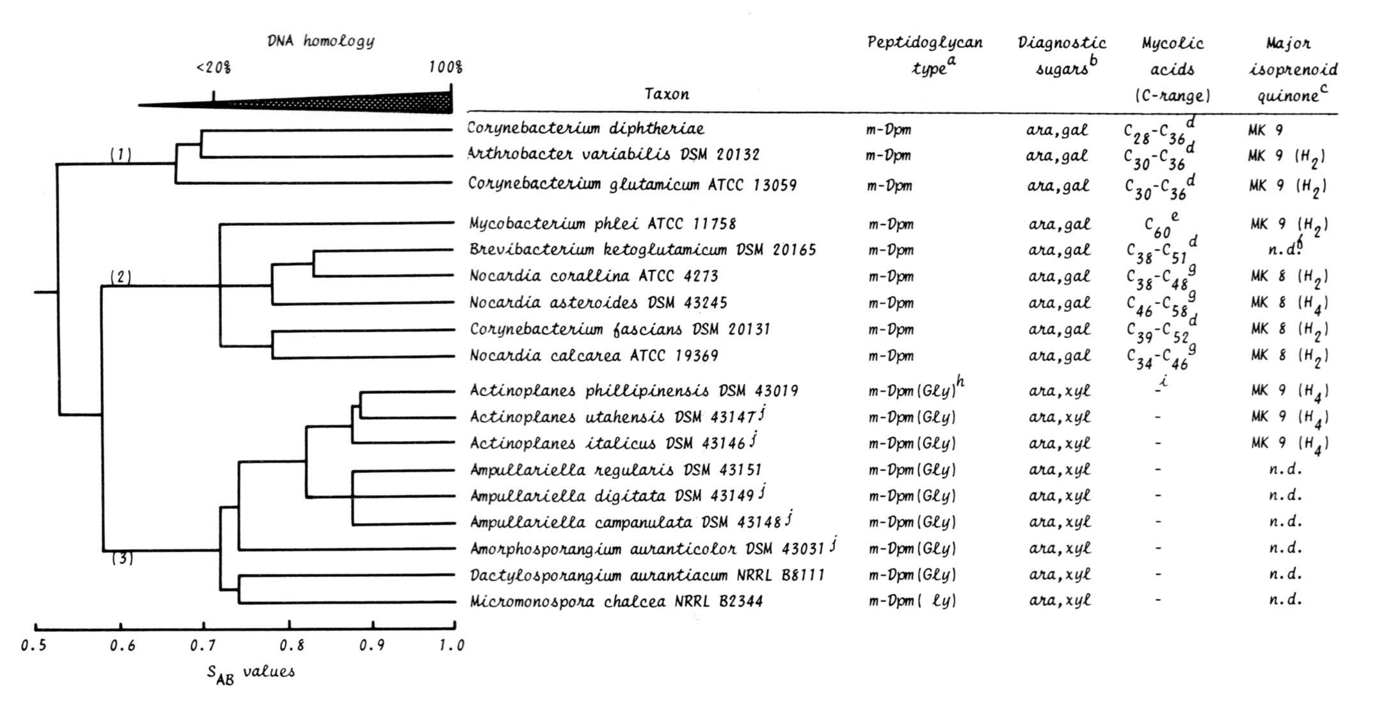

Taxon	Peptidoglycan type[a]	Diagnostic sugars[b]	Mycolic acids (C-range)	Major isoprenoid quinone[c]
Corynebacterium diphtheriae	m-Dpm	ara,gal	C_{28}-C_{36} [d]	MK 9
Arthrobacter variabilis DSM 20132	m-Dpm	ara,gal	C_{30}-C_{36} [d]	MK 9 (H_2)
Corynebacterium glutamicum ATCC 13059	m-Dpm	ara,gal	C_{30}-C_{36} [d]	MK 9 (H_2)
Mycobacterium phlei ATCC 11758	m-Dpm	ara,gal	C_{60} [e]	MK 9 (H_2)
Brevibacterium ketoglutamicum DSM 20165	m-Dpm	ara,gal	C_{38}-C_{51} [d]	n.d.[f]
Nocardia corallina ATCC 4273	m-Dpm	ara,gal	C_{38}-C_{48} [g]	MK 8 (H_2)
Nocardia asteroides DSM 43245	m-Dpm	ara,gal	C_{46}-C_{58} [g]	MK 8 (H_4)
Corynebacterium fascians DSM 20131	m-Dpm	ara,gal	C_{39}-C_{52} [d]	MK 8 (H_2)
Nocardia calcarea ATCC 19369	m-Dpm	ara,gal	C_{34}-C_{46} [g]	MK 8 (H_2)
Actinoplanes phillipinensis DSM 43019	m-Dpm(Gly)[h]	ara,xyl	-[i]	MK 9 (H_4)
Actinoplanes utahensis DSM 43147[j]	m-Dpm(Gly)	ara,xyl	-	MK 9 (H_4)
Actinoplanes italicus DSM 43146[j]	m-Dpm(Gly)	ara,xyl	-	MK 9 (H_4)
Ampullariella regularis DSM 43151	m-Dpm(Gly)	ara,xyl	-	n.d.
Ampullariella digitata DSM 43149[j]	m-Dpm(Gly)	ara,xyl	-	n.d.
Ampullariella campanulata DSM 43148[j]	m-Dpm(Gly)	ara,xyl	-	n.d.
Amorphosporangium auranticolor DSM 43031[j]	m-Dpm(Gly)	ara,xyl	-	n.d.
Dactylosporangium aurantiacum NRRL B8111	m-Dpm(Gly)	ara,xyl	-	n.d.
Micromonospora chalcea NRRL B2344	m-Dpm(ly)	ara,xyl	-	n.d.

FIGURE 6. Dendrogram of relationships of the families Corynebacteriaceae *(1),* Mycobacteriaceae *(2), and* Actinoplanaceae *(3), with additional information on chemotaxonomic markers.* [a]*Schleifer and Kandler (1972).* [b]*Lechevalier and Lechevalier (1981).* [c]*Collins and Jones (1981).* [d]*Collins* et al. *(1982a).* [e]*Minnikin and Goodfellow (1980).* [f]*n.d., not determined.* [g]*Minnikin and Goodfellow (1976).* [h]m-*Dpm may be hydroxylated (*m-*Hy-Dpm) (Schleifer and Kandler, 1972); Gly in position 1 of the peptide subunit (Kawamoto* et al., *1981).* [i] *(–), not present.* [j]*Organisms are assigned to the 16S rRNA clusters on the basis of DNA homologies and rRNA cistron similarities (Stackebrandt* et al., *1981b).*

genera. *Thermomonosporaceae* shows a specific, but not close, relationship to *Streptomycetaceae*.

g. The family Streptosporangiaceae. Three groups worthy of genus status are found within the phylogenetically defined family *Streptosporangiaceae: Streptosporangium, Nocardiopsis*, and a separate cluster containing two strains of *Actinomadura. Planomonospora parontospora* and *Planobispora longispora* are members of *Streptosporangium* (Stackebrandt *et al.*, 1981b). *Nocardiopsis dassonvillei*, formerly *Actinomadura dassonvillei*, is clearly separated phylogenetically from *A. madurae*, justifying its description as the type species of the new genus *Nocardiopsis* (Meyer, 1976). The phylogenetic position *of Actinomadura pusilla* and *A. roseoviolaceae* reveals, that the morphologically defined genus *Actinomadura* is genetically heterogenous. This has already been recognized by analysis of lipid composition (Agre *et al.*, 1975) and by numerical analysis (Goodfellow *et al.*, 1979).

h. The family Streptomycetaceae. This family is composed of a number of representatives of different genera which are almost indistinguishable from *Streptomyces* in respect to nucleic acid homologies: strains of *Chainia, Kitasatoa, Streptoverticillium, Elytroporangium, Microellobosporia, Actinosporangium* (Stackebrandt *et al.*, 1981b), "*Actinoplanes*" *armeniacus* (Kroppenstedt *et al.*, 1981), and *Actinopygnidium caeruleum*. Cross and Goodfellow (1973) transferred *Kitasatoa* into the family *Streptomycetaceae*, and *Actinopygnidium* and *Actinosporangium* were considered synonymous with *Streptomyces*. Further indication for a close relationship among the taxa listed above come from phage typing experiments (Kroppenstedt *et al.*, 1981; Prauser, 1976) and from high similarity in biochemical features (Lechevalier and Lechevalier, 1981; Stackebrandt *et al.*, 1981b).

IV. DISCUSSION AND TAXONOMIC IMPLICATIONS

By subjecting a representative selection of organisms which are included in Part 17 of "Bergey's Manual" (Gottlieb, 1974; Rogosa *et al.*, 1974) to phylogenetic analyses it could be shown that, with certain exceptions, all these organisms form a coherent group which constitutes one of the several branches of the phylogenetic tree of the eubacteria. The exceptions are the genera *Eubacterium, Kurthia*, and *Thermoactinomyces* which are members of the branch that is mainly defined by clostridia, bacilli, and lactic acid bacteria. If one intends to maintain the order *Actinomycetales*, the order as previously defined (Gottlieb, 1974) has to be emended to exclude *Bifidobacterium* and *Thermoactinomyces* and to include those organisms classified in the "coryneform group" of bacteria (Rogosa *et al.*, 1974) as well as *Micrococcus*.

Phylogenetically, an actinomycete can be described as a Gram-positive eubacterium showing a DNA G + C content of more than 55 mole%, which cannot be allocated to either *Bifidobacterium* or *Propionibacterium* (Stackebrandt,

1982). These genera can be distinguished from each other as well as from members of *Actinomycetales* by characteristics easy to determine, such as relationship to oxygen, end-products of glucose fermentation, and cell wall composition (Table I).

The traditional separation of the coryneform organisms from the mycelium- and spore-forming actinomycetes is not justified from a phylogenetic point of view. Neither the coryneform, nor the nocardioform organisms are phylogenetically more ancient than are the actinomycetes and therefore cannot be considered the progenitors of the actinomycetes. Moreover, the mycelium-producing organisms do not form a coherent cluster exclusively separate from non-mycelium forming eubacteria. There are several examples of mycelium-forming actinomycetes being more closely related to rod-shaped or even spherical bacteria, *e. g., Oerskovia* to *Cellulomonas, Promicromonospora* to the *Arthrobacter/Micrococcus/Cellulomonas* cluster, and at a higher level, *Actinoplanaceae* to *Mycobacteriaceae* and *Corynebacteriaceae*, as well as *Actinomycetaceae* to *Arthrobacteriaceae*.

A comparison of the phenetic system (Gottlieb, 1974; Rogosa *et al.*, 1974) with the phylogenetic system, presented in Fig. 5 and 6, shows little correspondence at the family level. *Streptomycetaceae* and *Actinomycetaceae* (excluding *Bifidobacterium*) are the only phenetically defined families which are phylogenetically coherent. (The status of *Frankia* has not yet been elucidated.)

All other phenetically defined families must be emended or they are not valid in that the type genus has to be transferred to another family. *Actinoplanaceae* and *Micromonosporaceae* may serve as impressive examples. So far, only three of the ten genera of *Actinoplanaceae* (Couch and Bland, 1974) remain in the phylogenetically defined family *Actinoplanaceae* which in addition contains *Micromonospora*. *Planomonospora, Planobispora*, and *Streptosporangium*, forming one taxon of genus status, are members of *Streptosporangiaceae*, whereas members of *Kitasatoa* should be reclassified as streptomycetes (Cross and Goodfellow, 1973). The transfer of *Micromonospora* into *Actinoplanaceae* removes the type genus of *Micromonosporaceae* (Küster, 1974) in which *Thermoactinomyces* is not an actinomycete at all (Cross, 1982; Stackebrandt *et al.*, 1981b). The phylogenetic analysis of *Actinobifida, Micropolyspora*, and *Microbispora* will show whether or not these remaining members of *Micromonosporaceae* form the nucleus of a new family. The family *Dermatophilaceae* (Gordon, 1974), of which only *Geodermatophilus* has been investigated so far, is probably not coherent either (Goodfellow and Pirouz, 1982).

Only a small number of genera have been investigated to such an extent that a general statement can be made about the correspondence of the phenetic and the phylogenetic system. Especially within *Arthrobacteriaceae*, phylogenetic analysis, together with numerical studies and investigations on biochemical

TABLE I. Chemotaxonomic Markers of Bifidobacterium, Propionibacterium, *and Phylogenetically Defined Families of the Order* Actinomycetales

Taxon	*Oxygen utilization*	*Peptidoglycan type* [a]	*Cell wall type* [b]	*Diagnostic sugar* [b]	*Mycolic acid* [c]	*Major fermentation end product*
Bifidobacterium	- (catalase -)	*variable diamino acid: Lys and/or Orn*	*not defined*	-	-	*lactic acid, acetic acid*
Propionibacterium	-(+) (catalase +)	*m-Dpm-direct/ L,L-Dpm-Gly*	*II/I*	-	-	*propionic acid*
Actinomycetaceae	+(-)	*unique: Orn-Lys-Glu or Lys-Lys-Asp*	*not defined*	-	-	*succinic acid, lactic acid, acetic acid*
Arthrobacteriaceae	+	*highly variable: no L,L-Dpm, no m-Dpm-arabino-galactan*	*not character-istic*	-	-	-
Corynebacteriaceae	+(-)	*m-Dpm-direct*	*IV*	*arabinose, galactose*	C_{22}-C_{36}	-
Mycobacteriaceae	+	*m-Dpm-direct*	*IV*	*arabinose, galactose*	C_{36}-C_{66}/ C_{60}-C_{90}	-
Actinoplanaceae	+	*m-Dpm-direct Gly in position 1*	*II*	*arabinose, xylose*	-	-
Streptomycetaceae	+	*L,L-Dpm-Gly*	*I*	-	-	-
Streptosporangiacea *(not yet well defined)*	+	*m-Dpm*	*III*	*madurose may be present*	-	-
Thermomonosporaceae *(not yet well defined)*	+	*m-Dpm*	*III*	*madurose may be present*	-	-

[a] *Schleifer and Kandler (1972); see Figures 5 to 7*
[b] *Lechevalier and Lechevalier (1981); see Figures 6 to 7*
[c] *See Figure 6*

characteristics, has offered a solution to taxonomic problems which have existed for decades. The nucleus of *Arthrobacter* (Keddie and Cure, 1978; Keddie and Jones, 1981; Stackebrandt and Fiedler *et al.*, 1979), of *Brevibacterium* (Fiedler *et al.*, 1981; Keddie and Jones, 1981), and that of *Cellulomonas* (Stackebrandt and Kandler, 1979; Stackebrandt *et al.*, 1982) are now recognized, and many misclassified strains of the individual genera found their place in other genera. The generic status of *Microbacterium, Curtobacterium*, and *Agromyces* which form a tight cluster in phylogenetic studies is not as clearcut when the differences in biochemical characteristics (Collins and Jones, 1980; Collins *et al.*, 1980; Keddie and Jones, 1981; Schleifer and Kandler, 1972) found among members of these genera are considered. A number of genera of the descriptive order *Actinomycetales* have been established by an overestimation of their complex morphologies, *e. g., Actinoplanes* and *Ampullariella; Streptosporangium, Planomonospora,* and *Planobispora;* and representatives of various genera which are actually species of *Streptomyces.*

To use the phylogeny of the actinomycetes as a basis for a future classification, it is necessary to point out biochemical and other phenotypic similarities which characterize the emerging taxa. The high degree of similarity in biochemical features (Figs. 5, 6), which in many examples extends to the family level, is obvious. It can therefore be concluded that the phylogenetic grouping of organisms reflects not only similarities in ribosomal ribonucleic acids and deoxyribonucleic acids as detected by sequencing analysis and hybridization studies but also by the overall genetic similarity. Table I presents a diagnostic key based on biochemical and morphological features which can be used for identification and classification at the family level. Future studies on taxonomically valuable characteristics, including those on lipids, polysaccharides, teichoic acids, cytochromes and other cell constituents, combined with features already proved to be reliable will extend the key to the genus level as is partially outlined in Figures 5 and 6.

REFERENCES

Agre, N. S., Efimova, T. P., and Guzeva, L. N., *Microbiol. 44:*220 (1975).
Back, W., and Stackebrandt, E., *Arch. Microbiol. 118:*79 (1978).
Collins, M. D., and Jones, D., *J. Appl. Bacteriol. 48:*459 (1980).
Collins, M. D., and Jones, D., *Microbiol. Rev. 45:*316 (1981).
Collins, M. D., Goodfellow, M., and Minnikin, D. E., *J. Gen. Microbiol. 118:*29 (1980).
Collins, M. D., Goodfellow, M., and Minnikin, D. E., *J. Gen. Microbiol. 128:*129 (1982a).
Collins, M. D., Mckillop, G. C., and Cross, T., *Microbiol. Lett. 13:*151 (1982b).
Couch, N. J., and Bland, C. E., *in* "Bergey's Manual of Determinative Bacteriology" (R. E. Buchanan and N. E. Gibbons, eds.), 8th ed., p. 706. Williams and Wilkins, Baltimore (1974).
Coykendall, A. L., and Munzenmaier, A. J., *Int. J. Syst. Bacteriol. 29:*234 (1979).
Cross, T., *Actinomyces 16:*77 (1982).
Cross, T., and Goodfellow, M., *in* "Actinomycetes: Characteristics and Practical Importance" (F. A. Skinner, and G. Sykes, eds.), p. 11. Academic press, New York (1973).
De Ley, J., and De Smedt, J., *Antonie Leewenhoek J. Microbiol. Serol. 41:*287 (1975).

De Ley, J., Segers, P., and Gillis, M., *Int. J. Syst. Bacteriol. 28:*154 (1978).
Denhaive, P., Hoet, P., and Cocito, C., *Int. J. Syst. Bacteriol. 32:*70 (1982).
Denhardt, D. T., *Biochem. Biophys. Res. Commun. 23:*641 (1966).
De Smedt, J., and De Ley, J., *Int. J. Syst. Bacteriol. 27:*222 (1977).
Döpfer, H., Stackebrandt, E., and Fiedler, F., *J. Gen. Microbiol. 128:*1697 (1982).
Fiedler, F., Schäffler, M. J., and Stackebrandt, E., *Arch. Microbiol. 129:*85 (1981).
Fox, G. E., Pechman, K. R., and Woese, C. R., *Int. J. Syst. Bacteriol. 27:*44 (1977).
Fox, G. E., Stackebrandt, E., Hespell, R. B., Gibson, J., Maniloff, J., Dyer, T., Wolfe, R. S., Balch, W., Tanner, R., Magrum, L., Zablen, L. B., Blakemore, R., Gupta, R., Luehrsen, K. R., Bonen, L., Lewis, B. J., Chen, K. N., and Woese, C. R., *Science 209:*457 (1980).
Gillespie, D., and Spiegelman, S., *J. Mol. Biol. 12:*829 (1965).
Goodfellow, M., and Pirouz, T., *J. Gen. Microbiol. 128:*503 (1982).
Goodfellow, M., Alderson, G., and Lacey, J., *J. Gen. Microbiol. 112:*95 (1979).
Gordon, M. A., *in* "Bergey's Manual of Determinative Bacteriology" (R. E. Buchanan and N. E. Gibbons, eds.), 8th ed., p. 723. Williams & Wilkins, Baltimore (1974).
Gottlieb, D., *in* "Bergey's Manual of Determinative Bacteriology" (R. E. Buchanan and N. E. Gibbons, eds.), 8th ed., p. 657. Williams & Wilkins, Baltimore (1974).
Jones, D., *J. Gen. Bacteriol. 87:*52 (1975).
Jones, D., *in* "Coryneform Bacteria" (I. J. Bousfield and A. G. Callely, eds.), p. 13. Academic Press, New York (1978).
Jones, L. A., and Bradley, S. G., *Mycologia 56:*505 (1964).
Kawamoto, J., Tetsuo, O., and Takahashi, N., *J. Bacteriol. 146:*527 (1981).
Keddie, R. M., and Cure, G. L., *in* "Coryneform Bacteria" (I. J. Bousfield, and A. G. Callely, eds.), p. 47. Academic Press, New York (1978).
Keddie, R. M., and Jones, D., *in* "The Prokaryotes" (M. P. Starr, H. Stolp, H. G. Trüper, A. Balows, and H. G. Schlegel, eds.), p. 1838. Springer, Berlin (1981).
Kroppenstedt, R. M., Korn-Wendisch, F., Fowler, V. J., and Stackebrandt, E., *Zentralbl. Bakteriol. I. Abt. Orig. C2:*254 (1981).
Küster, E., *in* "Bergey's Manual of Determinative Bacteriology" (R. E. Buchanan and N. E. Gibbons, eds.), 8th ed., p. 855. Williams & Wilkins, Baltimore (1974).
Kutzner, H. J., *in* "The Prokaryotes" (M. P. Starr, H. Stolp, H. G. Trüper, A. Balows, and H. G. Schlegel, eds.), p. 2028. Springer, Berlin (1981).
Lechevalier, H. A. and Lechevalier, M. P., *in* "The Prokaryotes" (M. P. Starr, H. Stolp, H. G. Trüper, A. Balows, and H. G. Schlegel, eds.), p. 1915. Springer, Berlin (1981).
Ludwig, W., Schleifer, K. H., Fox, G. E., Seewaldt, E., and Stackebrandt, E., *J. Gen. Microbiol. 125:*357 (1981a).
Ludwig, W., Seewaldt, E., Schleifer, K. H., and Stackebrandt, E., *FEMS Microbiol. Lett. 10:*193 (1981b).
McConaughy, B. L., Laird, C. D., and McCarthy, B. J., *Biochemistry 8:*3289 (1969).
Meyer, J., *Int. J. Syst. Bacteriol. 26:*487 (1976).
Meyer, S. A., and Schleifer, K. H., *Arch. Microbiol. 117:*183 (1978).
Minnikin, D. E., and Goodfellow, M., *in* "The Biology of Nocardiae" (M. Goodfellow, G. H. Brownell, and J. A. Serrano, eds.), p. 160. Academic Press, London (1976).
Minnikin, D. E., and Goodfellow, M., *in* "Microbiological Classification and Identification: (M. Goodfellow, and R. G. Board, eds.), p. 189. Academic Press, London (1980).
Mordarski, M., Goodfellow, M., Kaszen, J., Tkacz, A, Pulverer, G., and Schaal, K.P., *Int. J. Syst. Bacteriol. 30:*521 (1980a).
Mordarski, M., Goodfellow, M., Tkacz, A., Pulverer, G., and Schaal, K. P., *J. Gen. Microbiol. 118:*313 (1980b).
Prauser, H., *in* "The Biology of Nocardiae" (M. Goodfellow, G. H. Brownell, and J. A. Serrano, eds.), p. 266. Academic Press, New York (1976).
Rogosa, M., Cummins, C. S., Lelliott, R. A., and Keddie, R. M., *in* "Bergey's Manual of Determinative Bacteriology" (R. E. Buchanan, and N. E. Gibbons, eds.), 8th ed., p. 599. Williams & Wilkins, Baltimore (1974).
Sanger, F. J., Brownlee, G. G., and Barrell, B. G., *J. Mol. Biol. 13:*373 (1965).
Schleifer, K. H., and Kandler, O., *Bacteriol. Rev. 36:*407 (1972).
Schleifer, K. H., and Lang, K., *FEMS Microbiol. Lett. 9:*223 (1980).
Schleifer, K. H., Heise, W., and Meyer, S. A., *FEMS Microbiol. Lett. 6:*33 (1979).
Seeberger, L., Ph. D. Thesis, Technical University, Munich (1982).

Stackebrandt, E., *Actinomycetes 16:*132 (1982).
Stackebrandt, E., and Fiedler, F., *Arch. Microbiol. 120:*289 (1979).
Stackebrandt, E., and Kandler, O., *Int. J. Syst. Bacteriol. 29:*273 (1979).
Stackebrandt, E., and Woese, C. R., *Curr. Microbiol. 2:*317 (1979).
Stackebrandt, E., and Woese, C. R., *in* "Molecular and Cellular Aspects of Microbial Evolution" (M. J. Carlile, J. F. Collins, and B. E. B. Moseley, eds.), p. 1. University Press, Cambridge (1981a).
Stackebrandt, E. and Woese, C. R., *Curr. Microbiol. 5:*131 (1981b).
Stackebrandt, E., Häringer, M., and Schleifer, K. H., *Arch. Microbiol. 127:*179 (1980a).
Stackebrandt, E., Lewis, B. J., and Woese, C. R., *Zentralbl. Bakteriol. I. Abt. Orig. C2:*137 (1980b).
Stackebrandt, E., Ludwig, W., Schleifer, K. H., and Gross, H. J., *J. Mol. Evol. 17:*227 (1981a).
Stackebrandt, E., Wunner-Füssl, B., Fowler, V. J., and Schleifer, K. H., *Int. J. Syst. Bacteriol. 31:*420 (1981b).
Stackebrandt, E., Seiler, H., and Schleifer, K. H., *Zentralbl. Bakteriol. I. Abt. Orig. C3:*401 (1982).
Stanier, R. Y., and van Niel, C. B., *J. Bacteriol. 42:*437 (1941).
Steigerwalt, A. G., Fanning, G. R., Fife-Ashbury, M. A., and Brenner, D. J., *Can. J. Microbiol. 22:*441 (1975).
Suzuki, K., Kaneko, T., and Komagata, K., *Int. J. Syst. Bacteriol. 31:*131 (1981).
Tanner, R. S., Stackebrandt, E., Fox, G. E., and Woese, C. R., *Curr. Microbiol. 5:*35 (1981).
Uchida, T., Bonen, L., Schaup. H. W., Lewis, B. J., Zablen, L., and Woese, C. R., *J. Mol. Evol. 3:*63 (1974).
Woese, C. R., and Fox, G. E., *Proc. Nat. Acad. Sci. U.S.A. 74:*5088 (1977).
Woese, C. R., Maniloff, J., and Zablen, L. B., *Proc. Nat. Acad. Sci. U.S.A. 77:*494 (1980).

CLASSIFICATION AND IDENTIFICATION OF CLINICALLY SIGNIFICANT *ACTINOMYCETACEAE*[1]

Klaus P. Schaal
Geraldine M. Schofield[2]

Institute of Hygiene
University of Cologne
Federal Republic of Germany

I. INTRODUCTION

The family *Actinomycetaceae* as defined in the 8th edition of Bergey's Manual of Determinative Bacteriology (Slack, 1974) is composed of Gram-positive, non-spore-forming bacteria which are predominantly diphtheroid in shape but tend to form branched filaments and which possess a fermentative carbohydrate metabolism. Compared with previous descriptions of the family, this definition has greatly improved classification as well as identification of the facultatively anaerobic, filamentous bacteria. However, after recent numerical phenetic (Fillery *et al.*, 1978; Holmberg and Hallander, 1973; Holmberg and Nord, 1975; Melville, 1965; Schaal and Schofield, 1981; Schofield and Schaal, 1981), chemotaxonomic (Alshamaony *et al.*, 1977; Collins *et al.*, 1979; Minnikin *et al.*, 1978), and genetic (Coykendall and Munzenmaier, 1979; Johnson and Cummins, 1972) analyses, the *Actinomycetaceae* remain a heterogeneous, partly ill-defined, and partly unrelated collection of microorganisms. Furthermore, in light of numerical phenetic and chemotaxonomic evidence, bacteria which previously were classified with other taxa now are being considered for classification with the actinomycetes. It is therefore not surprising that the composition and the taxonomic justification of the family *Actinomycetaceae* as currently recognized have been increasingly questioned.

[1] *Supported in part by a grant from the Deutsche Forschungsgemeinschaft.*
[2] *Present address: PHLS Centre for Applied Microbiology and Research, Microbiological Safety Reference Laboratory, Salisbury, Wiltshire, U. K.*

ISBN 0-12-528620-1

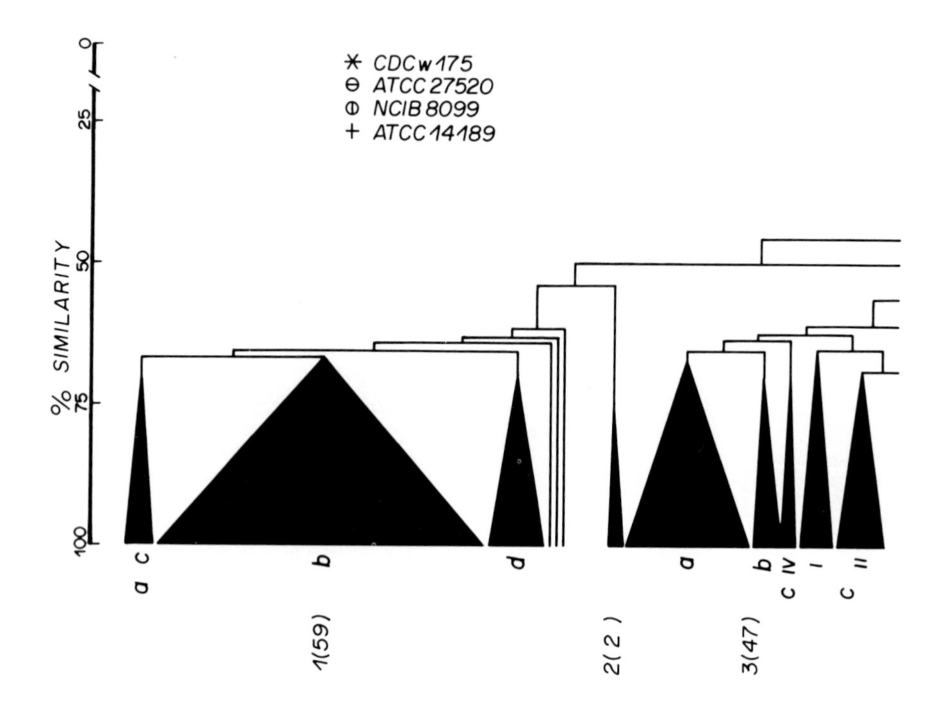

II. CLASSIFICATION

Genetic and chemotaxonomic data have clarified certain problem areas in actinomycete taxonomy. However, these studies have as yet had little impact on the classification of the fermentative actinomycetes in general because the tests were usually performed on only a few strains and their results therefore are highly dependent on the quality of the representative cultures used. In contrast, numerical taxonomic techniques can simultaneously handle a large number of test organisms and a large number of characters. Therefore, such techniques can provide basic information on systematic relationships in broader terms, which can then be verified in detail by more elaborate methods.

With the fermentative actinomycetes, several such numerical analyses have been performed, the most comprehensive one being that of Schofield and Schaal (1981). In this study, most of the recognized species of the *Actinomycetaceae* were recovered as well-defined and clearly separated phena (Fig. 1.). This is especially true for the major human pathogen, *Actinomyces israelii*, which forms a tight cluster with many differential characters (Fig. 1, cluster 1). However, as the abbreviated dendrogram (Fig. 1) based upon the S_J coefficient shows, this cluster could be subdivided into three to four subgroups delimited at 84.5% S_{SM} and linked only at rather low similarity levels with other species of the genus *Actinomyces*.

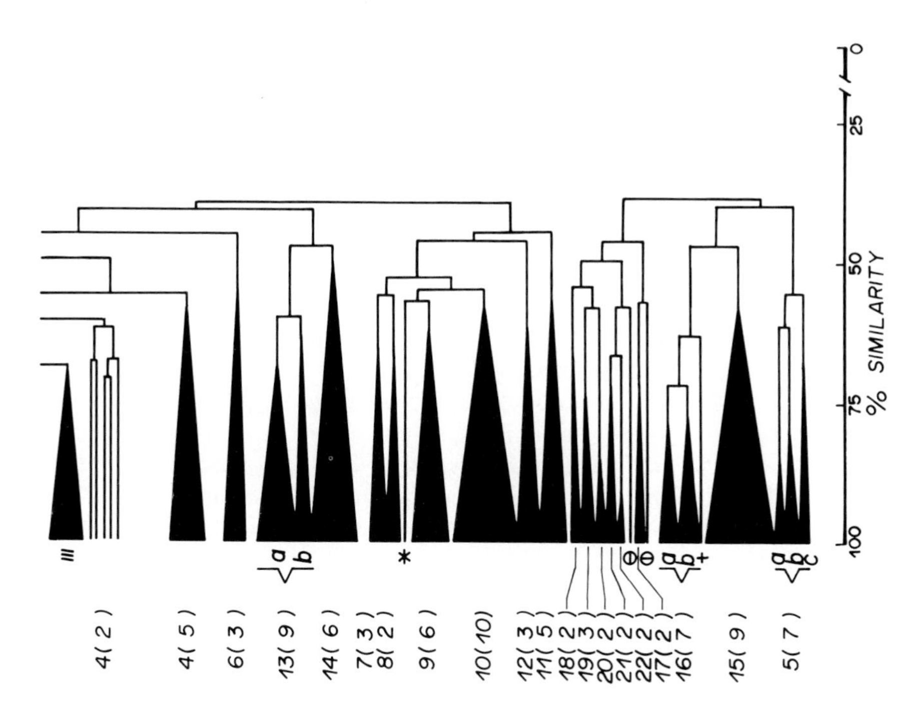

FIGURE 1. Abbreviated dendrogram showing similarity between strains based upon the S_J coefficient and average linkage.
(Reproduced from Schofield and Schaal, J. Gen. Microbiol. 127*:237 (1981) by copyright permission from the Society for General Microbiology.)*

The second major phenon (cluster 3) contained strains designated either *A. naeslundii* or *A. viscosus*. These two species were previously thought to be so similar that their combination into one taxon has been suggested (Holmberg and Hallander, 1973; Slack and Gerencser, 1975). The data of Schofield and Schaal (1981) also indicate a close relationship between both species. However, with all of the coefficients used, several subclusters were delineated which corresponded well with formerly established species and serovars.

Subcluster 3a appeared to be especially compact and was solely composed of *A. naeslundii* reference strains (serovars I and II) and of oral and clinical isolates which had previously been identified as *A. naeslundii*. The strains which formed subcluster 3b were all recent oral isolates and possibly constitute a new center of variation. The reference cultures of *A. viscosus* were recovered in subcluster 3c, together with serovar III strains of *A. naeslundii*. This subcluster was further subdivided into four subgroups which corresponded to the serovars I and II of *A. viscosus*, to the so-called atypical viscosus strains, and to *A. naeslundii* serovar III strains, respectively.

Cluster 4 comprised only clinical isolates without any reference culture. Although these organisms were linked at comparatively low similarity levels, they seem to represent another new center of variation and might well form the nucleus of a new species.

Cluster 7 to 12 formed an aggregate group of species which appeared quite heterogeneous although the single phena were individually well defined. The named taxa included were *A. bovis*, the type species of the genus (cluster 7); *Bifidobacterium bifidum* (cluster 8); *Erysipelothrix rhusiopathiae* (cluster 9); *Corynebacterium pyogenes* (cluster 10); *A. odontolyticus (cluster 11); "Actinobacterium" meyerii* (single strain) and *C. haemolyticum* (cluster 12).

A. bovis, C. pyogenes, and *E. rhusiopathiae* were grouped together by all statistical methods employed; in fact, only *A. odontolyticus* changed its position when the pattern coefficient was used, joining the naeslundii/viscosus group at a low similarity level. Despite this constant grouping, the taxonomic relevance of the aggregate cluster is doubtful. At least *E. rhusiopathiae* and possibly also *B. bifidum* and *C. haemolyticum* would appear not to be related to each other or to the *Actinomyces* species. Obviously, a large number of negative characters shared by these organisms was responsible for the aggregation of the phena.

However, an affinity between *A. bovis, A. odontolyticus*, and *C. pyogenes* has already been considered by Slack and Gerencser (1975), and recently has been supported by chemical data (Collins *et al.*, 1982a). On the other hand, *C. pyogenes* and *C. haemolyticum* were clearly separated from each other, indicating that both organisms were not as closely related as Barksdale *et al.* (1957) suggested some years ago. On the basis of chemical data, Collins *et al.* (1982a) came to the same conclusion.

In cluster 13, all the cultures labelled *Arachnia propionica* were combined. Although *Arachnia* is currently recognized as a single species genus, the strains formed two distinct subclusters which separated the two serovars of the species. No relation was observed with the *Propionibacterium* clusters (17 to 22). This is in good agreement with the results of Johnson and Cummins (1972) who found low DNA homologies between various *Propionibacterium* and *Arachnia* strains. Representatives of the genera *Bacterionema* and *Rothia* formed the distinct clusters 15 and 16 which showed little affinity with each other or with the other *Actinomycetaceae*. For *Rothia*, no other suitable taxonomic niche has yet been found, but it has been suggested by Minnikin *et al.* (1978) that *Bacterionema* be transferred to the genus *Corynebacterium*. It should also be mentioned that *Rothia* strains fell into two distinct subclusters which might be considered two new species, but which do not coincide with the two recognized serovars.

As some interesting correlations had been observed between serovars and cluster or subclusters, we decided to check the validity of the phena by serological methods. As yet, only a few antisera against center strains of selected clusters have been prepared. Their preliminary reactivity patterns as demonstrated by an indirect immunofluorescence assay are shown in Table I. Cross-reactions between different phena were seen, but were usually weak.

TABLE I. Serological Cross-Reactivity between Selected Strains of Fermentative Actinomycetes as Determined by Indirect Immunofluorescence

Strains	*Antisera against*							
	A. israelii *I* *(1 a)*	A. israelii *I* *(1 b)*	A. naeslundii *I* *(3 a)*	Actinomyces *sp.* *(3 b)*	A. viscosus *(3 c)*	A. odontolyticus *I* *(11)*	Ar. propionica *I* *(13 a)*	R. dentocariosa *I* *(16 a)*
A. israelii *I (1 a)*	+ + +[a]	+ + +	-	-	-	+	+	+
A. israelii *II (1 b)*	+ + +	+ + +	+	+	+	-	+	+
A. israelii (1 b)	+ + +	+ + +	+	+	+	-	+	+
A. israelli *(1 c)*	+	+ +	+	-	+	-	-	+
A. israelli *(1 d)*	+	+	-	-	-	-	-	-
A. naeslundii *I (3 a)*	+	+	+ + +	+	+	-	-	+
Actinomyces *sp. (3 b)*	+	-	+	+ + +	+	-	-	-
A. viscosus *(3 c)*	+	-	+	+ + +	+ + +	+	-	-
A. bovis *I (7)*	-	-	-	+	+	-	-	-
A. bovis *II (9)*	-	-	-	-	-	-	-	-
A. odontolyticus *I (11)*	-	-	-	-	-	+ + +	-	-
A. propionica *I (13 a)*	-	+	-	-	-	-	+ + +	-
Ar. propionica *II (13 b)*	+	+ +	-	-	-	-	+ + +	-
B. matruchotii *(15)*	-	-	-	-	-	-	-	+
R. dentocariosa *I (16 a)*	-	-	-	-	-	-	-	+ + +
R. dentocariosa *(16 b)*	-	-	-	-	-	-	-	+ + +
Act. meyerii	-	-	-	-	-	-	-	-

[a] *(+ + +) = Titer > 1/5 of the homologous reaction; (+ +) = titer ≦ 1/5 to > 1/10 of the homologous reaction; (+) = Titer ≦ 1/10 to 1/100 of the homologous reaction; (-) = Titer < 1/100 of the homologous reaction.*

However, marked cross-reactivity was observed between members of the various subclusters in the israelii, naeslundii/viscosus, *Arachnia*, and *Rothia* phena indicating that these strains are also antigenically related. The apparent serological differences noted in the israelii and naeslundii/viscosus clusters require further detailed analysis but generally support the subdivisions made.

As so much new numerical phenetic, chemotaxonomic, and genetic data have been accumulated, it may be appropriate to reconsider the taxonomic structure of the genus *Actinomyces* and of the whole family *Actinomycetaceae*. Previously, the family *Actinomycetaceae* was composed of the five genera *Actinomyces, Bifidobacterium, Arachnia, Bacterionema*, and *Rothia* (Table II) which contained the established species previously mentioned. Similar organisms were also found outside the family, in the taxon *Corynebacterium* as *C. pyogenes* and *C. haemolyticum*, or as *"Actinobacterium" meyerii* which, although it is an illegitimate designation, stands for an existing bacterial species.

Whether *C. pyogenes* and *C. haemolyticum* are members of the genus *Corynebacterium* has long been questioned, and recently it was formally proposed that both organisms be reclassified. *C. pyogenes* was transferred to the genus *Actinomyces* as *A. pyogenes* comb. nov. (Collins and Jones, 1982); for *C. haemolyticum*, the new genus *Arcanobacterium* was created (Collins *et al.*, 1982b) on the basis of phenetic and chemotaxonomic evidence. The reclassification of *Bacterionema matruchotii* in the genus *Corynebacterium* had already been suggested by Goodfellow and his group.

Future changes will chiefly involve the genus *Actinomyces*. According to the numerical taxonomic results obtained by Schofield and Schaal (1981), this taxon may be subdivided into three subgroups which may be considered either as subgenera or as new independent genera. If we visualize the taxonomic

TABLE II. Previous Taxonomic Structure of the Actinomycetaceae *and Related Organisms*

Family	*Genus*	*Species*
Actinomycetaceae	Actinomyces	A. bovis A. israelii A. naeslundii A. viscosus A. odontolyticus
	Bifidobacterium	*Several species*
	Arachnia	A. propionica
	Bacterionema	B. matruchotii
	Rothia	R. dentocariosa
	Corynebacterium	C. pyogenes C. haemolyticum
	Actinobacterium	A. meyerii

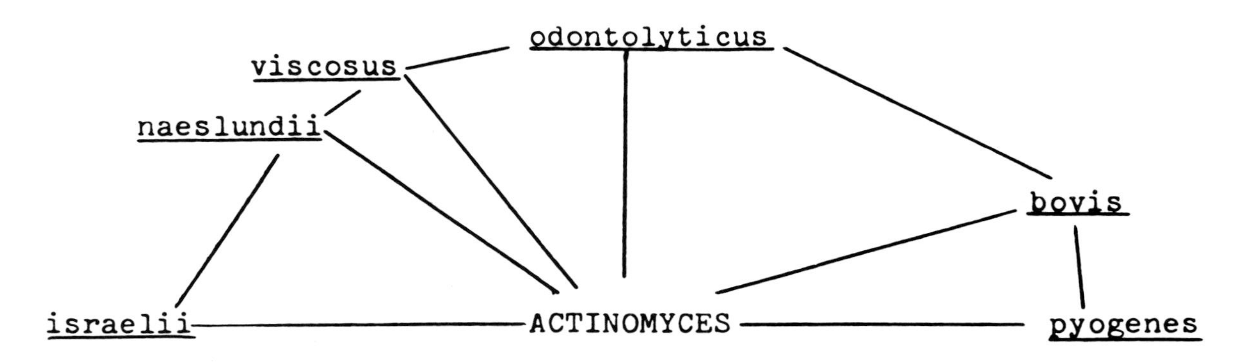

FIGURE 2. Diagrammatic representation of the taxonomic distances within the genus Actinomyces.

distances within the genus as it is currently recognized, this subdivision becomes immediately apparent (Fig. 2). *A. israelii* is situated at one end of the taxonomic spectrum and *A. bovis* and *A. pyogenes* are found at the other end. *A. naeslundii* and *A. viscosus* occupy an intermediate position, thereby forming the third subgenus. At present, it is difficult to decide whether *A. odontolyticus* would better be included in the naeslundii/viscosus group or in the bovis/pyogenes aggregate. More strains of *"Actinobacterium" meyerii* must be investigated before the species can be reclassified and possibly allocated to one of these groups.

The formation of two or three new genera from members of the old genus *Actinomyces* would create considerable nomenclatural problems. The type species of the genus is *A. bovis* which therefore would retain the designation *Actinomyces* as would *A. pyogenes*, thereby forming a new taxon which contains only typical animal pathogens. *A. israelii* as well as the naeslundii/viscosus group would need new genus names. This undoubtedly would cause confusion in the medical world where the designations actinomycosis for a well-known disease and *Actinomyces* for its causative agents have been in use for about one hundred years.

Clearly, more work is needed before all these problems can be solved. However, one may speculate a little on the future taxonomic structure of the filamentous bacteria (Table III). It is difficult to say whether the family *Actinomycetaceae* should be retained and, if it is, which genera it should comprise. Besides the subdivision of the genus *Actinomyces*, several new groups of fermentative, diphtheroid to filamentous bacteria await nomination and further detailed description. These include the subclusters of the viscosus group, strains of clusters 4 and 14, and the subclusters in the taxa *Arachnia* and *Rothia*. As yet, only the complex *Actinomyces* genus and *Arachnia* would appear suitable members of a redefined family *Actinomycetaceae*. The systematic position of the other genera mentioned in Table III and of *Bifidobacterium* is uncertain and may be clarified by techniques such as 16S rRNA sequencing (Stackebrandt and Woese, 1981).

III. IDENTIFICATION

Numerical phenetic analyses not only have provided new insight into the taxonomic relationships of the fermentative actinomycetes, but also have produced a large set of differential characters and diagnostic methods. In this respect, the study of Schofield and Schaal (1981) is especially valuable since these authors developed a series of new techniques or adjusted old ones for application to the actinomycetes. The major advantage of these tests seems to be that they were miniaturized thereby allowing a favorable correlation between inoculum and substrate which appears to be especially important for slow growing and metabolically less active organisms such as the actinomycetes.

TABLE III. Proposed New Structure

Genus	*Subgenus*	*Species*	*Subspecies*
Actinomyces	*I*	A. israelii	*4 subspecies (?)*
	II	*1.* A. naeslundii	
		2. A. viscosus	*1 = serovar I* *2 = serovar II* *3 = "atypical" viscosus strains* *4 =* A. naeslundii, *serovar III*
		3. Actinomyces *sp* *= subcluster B (recent oral isolates)*	
		4. Actinomyces *sp.* *= cluster 4*	
	III	*1.* A. bovis	
		2. A. pyogenes (= C. pyogenes)	
		3. A. odontolyticus	
		4. "A. meyerii"	
Arachnia		*1.* A. propionica	*1 = serovar I* *2 = serovar II*
		2. Arachnia *sp.* *(= cluster 14)*	
Corynebacterium		C. matruchotii (= B. matruchotii)	
Arcanobacterium		A. haemolyticum (= C. haemolyticum)	
Rothia		R. dentocariosa	*2 subspecies (?)*

For identification of the *Actinomycetaceae* at the genus level and for differentiating them from morphologically related Gram-positive bacteria, a set of comparatively simple and inexpensive physiological, chemotaxonomic, and end-product analysis tests are recommended (Table IV). If the examination of cell wall constituents and wall lipids be restricted to the demonstration of diaminopimelic acid (DAP) and mycolic acids, simple techniques with whole organisms can be employed which are suitable for routine application. At present, end-product analyses by gas-liquid chromatography (GLC) are widely used in the anaerobe laboratory and are applicable under routine conditions (Schaal, 1982). Typical differential characters derived from these tests are the presence of LL-DAP in *Arachnia* and *Propionibacterium* and of DL-DAP in *Bacterionema* and *Eubacterium* cell walls, the mycolic acids in *Bacterionema*, the positive catalase test in *Bacterionema, Rothia*, and *Propionibacterium*, and the patterns of acid end-products as shown in Table IV

TABLE IV. Differentiation of the Genera of the Actinomycetaceae *from Each Other and from Related Taxa*[a]

Genus	*Physiological characters*		*Chemotaxonomic characters*			*Major acid end products from glucose fermentation*[d]										
	Catalase	*Nitrate reduction*	*DL-DAP*[b]	*LL-DAP*[b]	*Mycolic acid*[c]	*Acetic*	*Propionic*	*Iso-butyric*	*N-butyric*	*Iso-valeric*	*N-valeric*	*Iso-caproic*	*N-caproic*	*Pyruvic*	*Lactic*	*Succinic acids*
Actinomyces	-[e,f]	*D*	-	-	-	+	-	-	-	-	-	-	-	-	+	+
Arachnia	-	+	-	+	-	+	+	-	-	-	-	-	-	-	(+)	(+)
Bifidobacterium	-	-	-	-	-	+	-	-	-	-	-	-	-	-	+	(+)
Bacterionema	+	+	+	-	+	+	*D*	-	-	-	-	-	-	*D*	*D*	-
Rothia	+	+	-	-	-	+	-	-	-	-	-	-	-	(+)	+	(+)
Propionibacterium	*D*	*D*	(+)	+[g]	-	+	+	-	-	*D*	-	-	-	-	(+)	(+)
Eubacterium	-	*D*	*D*	-	-	+	(+)	-	+	-	-	-	*D*	-	*D*	(+)
Lactobacillus	-	-	-	-	-	(+)	-	-	-	-	-	-	-	-	+	(+)
Erysipelothrix	-	-	-	-	-	+	-	-	-	-	-	-	-	-	+	+
Arcanobacterium (C. haemolyticum)	-	*D*	-	-	-	+	-	-	-	-	-	-	-	-	+	*D*

[a]*Data compiled from Slack and Gerencser, 1975; Holmberg and Nord, 1975; Holdeman* et al., *1977; Schaal* et al., *1980; Schaal and Pulverer, 1981; Schofield and Schaal, 1981; Schaal and Schofield, 1981a, b*
[b]*DAP* = meso- *or LL-diaminopimelic acid in whole-cell hydrolysates*
[c]*In whole-cell methanolysates*
[d]*In PYG medium (Holdeman* et al., *1977)*
[e]*+ = present; - = absent; D = species or type differences; (+) = present in small amounts or absent*
[f]A. viscosus *is catalase-positive*
[g]P. lymphophilum *does not contain DAP.*

(Schaal, 1982; Schaal and Pulverer, 1981).

For reliable identification at the species level, a comparatively large number of characters may be required (Table V) because pronounced strain variation within the various taxa tends to render shorter schemes equivocal. Nevertheless, certain key reactions can be selected which are of especially high resolving power. These include, besides culture characteristics, catalase reaction, nitrite reduction, aesculin hydrolysis, acid formation from selected carbohydrates, deamination of arginine and urea, and end-product analysis.

We have tried such an abbreviated scheme under routine conditions using the commercial Minitek differentiation system for all tests excluding catalase and end-product analysis (Table VI). The results appeared to be very reliable as confirmed by immunofluorescence; the percentages of positive reactions differed only slightly from the data obtained in the numerical phenetic study.

If strains of the naeslundii/viscosus group are to be identified at the subcluster level, certain additional tests must be performed and evaluated together with the results of the basis set of tests (Table VII). High differential values can be expected from the following reactions: nitrite reduction, catalase, aesculin hydrolysis, acid from glycerol, meso-inositol and ribose, resistant to 20% bile and 0.2% sodium taurocholate, deamination of arginine and urea, and esterase (C4) activity. The differentiation of the two subclusters of *Arachnia* may be achieved by the tests for acid production from glycerol and sorbitol and for deamination of ornithine and serine; in these tests only the subcluster corresponding to serovar I is positive. As yet, no good criteria for distinguishing the four subgroups of the israelii phenon have been found.

In clinical laboratories, direct or indirect immunofluorescence techniques have widely been used for the identification of actinomycete isolates chiefly because these procedures are simple, rapid, and usually more reliable than are conventional physiological tests (Slack and Gerencser, 1975). However, since antisera to only *A. israelii* and *A. naeslundii* are commercially available and since the heterogeneity of the actinomycetes has been shown by numerical phenetic evidence, it may be predicted that a considerable number of actinomycetes will be missed when a small spectrum of labelled antisera is used as the sole means of detecting and identifying the organisms. In our view, carefully standardized, miniaturized physiological tests are at present the most dependable way of recognizing fermentative actinomycetes and similar bacteria.

IV. CONCLUSION

It may be stated that the classification and identification of fermentative actinomycetes have undergone substantial changes in recent years. This not only has improved the knowledge of the systematic structure of these bacteria, but also has created additional problems, especially with respect to the taxonomy and nomenclature of the genus *Actinomyces*. The authors would, therefore,

TABLE V. Characters of Use in the Identification of Actinomycetaceae *and Related Taxa*

Character	A. israelii	A. naeslundii A. viscosus		A. bovis	A. odontolyticus	C. pyogenes	C. haemolyticum	Arachnia		Bacterionema	Rothia	"Actinobacterium" meyerii
	(59)[a] (1)[b]	(46) (3)	(4) (4)	(3) (7)	(5) (11)	(10) (10)	(3) (12)	(9) (13)	(6) (14)	(9) (15)	(7) (16)	(1) -
Colony rough	+[c]	v	-	-	v	-	-	+	v	+	+	-
Aerobic growth	-	v	v	-	v	-	-	-	v	+	+	-
Catalase	-	v	-	-	-	-	-	-	-	+	+	-
NO_2-reduction	-	v	v	-	-	-	-	-	v	-	+	-
NO_3-reduction	v	+	v	-	+	v	v	+	+	+	+	+
DNAase	-	-	v	+	v	+	+	-	v	v	v	-
Hydrolysis tests:												
Aesculin	+	v	+	-	v	-	-	-	v	-	+	-
Starch	-	v	v	+	v	v	v	-	-	+	-	+
Alkali produced in peptone-containing media	-	-	-	-	-	-	+	-	-	+	+	-
Acid from:												
Amygdalin	+	-	v	-	-	-	-	+	-	-	-	+
Cellobiose	+	v	v	-	-	-	-	-	-	-	-	-

Glycerol	-	v	-	+	-	-	-	v	v	-	-	-
Meso-inositol	+	v	-	v	-	-	-	v	-	-	-	-
Mannitol	v	-	-	-	-	-	-	+	v	v	-	-
Raffinose	+	+	v	-	-	-	-	+	v	-	-	-
Xylose	+	-	v	-	v	v	-	-	-	-	-	+
Growth in the presence of:												
Bile (10% (w/v)	-	+	+	v	-	v	-	+	-	-	+	-
Ammonia from:												
Arginine	+	v	-	-	-	-	-	-	v	-	-	-
Serine	-	-	-	-	-	-	-	v	v	-	+	-
Alanine	-	-	-	-	-	-	-	v	-	-	-	-
Urea	-	v	v	-	-	-	-	-	v	v	-	-
Decarboxylation of:												
Lysine	-	-	-	-	v	-	-	-	-	v	-	-
Ornithine	-	-	-	-	-	-	-	v	-	v	-	-
Lysis by lysozyme + SDS:												
Strong clearing, 4 h	-	v	-	+	+	v	-	-	v	v	+	+
End-products of glucose fermentation:												
Propionic acid	-	-	-	-	-	-	-	+	v	v	-	-
Succinic acid	+	v	v	v	-	v	v	-	-	-	v	+
API enzyme tests:												
4. Esterase lipase (C8)	-	-	-	-	-	-	-	+	v	+	v	-
18. N-acetyl-β-glucosaminidase	-	-	-	+	-	+	+	-	-	-	-	-

[a]*No. of strains.*
[b]*Cluster number.*
[c]*(+), positive; (-), negative; (v), variable.*

TABLE VI. Results of Selected Tests Under Routine Conditions

Character	Percentages of positive reactions for							
	A. israelii	A. naeslundii	A. viscosus	A. bovis	A. odonto-lyticus	Ar. propio-nica	Act. meyerii	B. matru-chotii
	(177)[a]	(41)	(44)	(4)	(4)	(10)	(1)	(1)
Colony smooth	0[b]	96[b]	98[b]	100[b]	100[b]	0[b]	+[b]	-[b]
Catalase	0	0	61[b]	0	0	0	-	+[b]
NO_3-reduction	49	100[b]	57	0[b]	100[b]	100[b]	+	+[b]
Indole production	0	0	0	0	0	0	-	-
Aesculin hydrol.	98[b]	93	52	75	75	0[b]	-[b]	-[b]
Acid from:								
Arabinose	58[b]	2	0	0	50[b]	0	-	-
Cellobiose	90[b]	25	7	0	0	0	-	-
Glucose	100	100	100	100	100	100	+	+
Glycerol	9	41	34	50	0	30	-	-
Meso-*inositol*	95[b]	92[b]	65	75	0[b]	10	-	-
Lactose	89	56	68	100	0[b]	80	-	-
Mannitol	57[b]	0	0	0	0	100[b]	-	-
Maltose	94	85	91	100	75	100	+	+
Raffinose	90[b]	95[b]	93[b]	0[b]	0[b]	100[b]	-[b]	-[b]
Rhamnose	10	0	0	0	50[b]	0	-	-
Salicin	81	56	23	25	25	20	-	-
Sorbitol	80[b]	2	3	0	0	80[b]	-	-
Sucrose	98	100	98	100	100	100	+	+
Trehalose	81	98	79	0[b]	25	60	-[b]	-[b]
Xylose	95[b]	4	0	25	75[b]	0	+[b]	-
NH_3 from:								
Arginine	95[b]	12	3	0	0	0	-	-
Urea	0	92[b]	9	0	0	0	-	-
Propionic acid as end product	0	0	0	0	0	100[b]	-	+[b]

[a]*No. of strains.*
[b]*For each organism, the minimum characters required for identification are indicated.*

TABLE VII. Characters Which May Be of Use in the Differentiation of Strains of A. naeslundii *and* A. viscosus

Character	A. naeslundii *(subcluster a)* *(17)*[b]	*Recent oral isolates (subcluster b)* *(4)*	A. viscosus *(subcluster c)*[a] *I* *(4)*	*II* *(4)*	*III* *(5)*	*IV* *(2)*	*Total* *(16)*
Catalase	*0*[c]	*0*	*100*	*100*	*80*	*0*	*81*
NO_2*-reduction*	*88*	*100*	*0*	*0*	*40*	*0*	*19*
NO_3*-reduction*	*100*	*100*	*75*	*75*	*80*	*100*	*81*
Hydrolysis of:							
Aesculin	*94*	*100*	*0*	*0*	*20*	*100*	*19*
Tween 40	*35*	*0*	*0*	*50*	*0*	*0*	*19*
Tween 60	*12*	*0*	*0*	*50*	*40*	*0*	*25*
Acid from:							
Cellobiose	*35*	*75*	*0*	*0*	*40*	*0*	*13*
Glycerol	*18*	*0*	*0*	*25*	*100*	*50*	*50*
Meso-*inositol*	*94*	*100*	*0*	*75*	*100*	*100*	*63*
Mannose	*100*	*50*	*50*	*75*	*100*	*100*	*75*
Melibiose	*100*	*75*	*100*	*100*	*20*	*100*	*69*
Ribose	*35*	*25*	*0*	*100*	*100*	*100*	*69*
Trehalose	*100*	*75*	*25*	*100*	*100*	*100*	*81*
Xylose	*35*	*0*	*0*	*0*	*0*	*0*	*0*
Growth in the presence of:							
NaCl 2% (w/v)	*77*	*100*	*100*	*100*	*100*	*100*	*100*
Bile 20% (w/v)	*77*	*25*	*100*	*0*	*40*	*0*	*38*
Sodium selenite 0.01 % (w/v)	*65*	*0*	*0*	*75*	*40*	*50*	*44*
Sodium taurocholate 0.2 % (w/v)	*65*	*100*	*50*	*0*	*0*	*0*	*13*
Ammonia from:							
Arginine	*29*	*100*	*0*	*25*	*0*	*0*	*6*
Urea	*94*	*75*	*25*	*25*	*0*	*0*	*13*
Inhibition by nalidixic acid (30 μg per disc)	*71*	*75*	*100*	*0*	*80*	*100*	*69*
Lysis by lysozyme + SDS:							
Weak clearing, 4 h	*0*	*100*	*50*	*0*	*0*	*50*	*19*
Strong clearing, 4 h	*100*	*0*	*25*	*75*	*60*	*50*	*56*
API enzyme tests:							
3. Esterase (C4)	*0*	*50*	*50*	*0*	*20*	*100*	*31*
17. β-Glucosidase	*53*	*0*	*0*	*25*	*0*	*0*	*6*

[a]*Subgroups: I,* A. viscosus *serotype 1; II,* A. viscosus *serotype 2; III, 'Atypical'* A. viscosus *strains; IV,* A. naeslundii *serotype 3 strains (B120, B102).*
[b]*No. of strains.*
[c]*Percentage of positive reactions.*

welcome any comment or proposal on this matter, for instance, on the possibilities of transferring the type species of *Actinomyces* from *A. bovis* to *A. israelii* in the old or in a new sense.

ACKNOWLEDGMENTS

The technical assistance of Monika Pinkwart is gratefully acknowledged. The authors also thank Sigrid Glanschneider for photographic services and Evelyn Heidermann for typing the manuscript.

REFERENCES

Alshamaony, L., Goodfellow, M., Minnikin, D. E., Bowden, G. H., and Hardie, J. M., *J. Gen. Microbiol. 98:*205 (1977).
Barksdale, W. L., Li, K., Cummins C. S., and Harris, H., *J. Gen Microbiol. 16:*749 (1957).
Collins, M. D. and Jones, D., *J. Gen. Microbiol. 128:*901 (1982).
Collins, M. D., Goodfellow, M., and Minnikin, D. E., *J. Gen. Microbiol. 110:*127 (1979).
Collins, M. D., Jones, D., Kroppenstedt, R. M., and Schleifer, K. H., *J. Gen. Microbiol. 128:*335 (1982a).
Collins, M. D., Jones D., and Schofield, G. M., *J. Gen. Microbiol. 128:*1279 (1982b).
Coykendall, A. L., and Munzenmaier, A. J., *Int. J. System. Bacteriol. 29:* 234 (1979).
Fillery, E. D., Bowden, G. H., and Hardie, J. M., *Caries Res. 12:*299 (1978).
Holmberg, K. and Hallander, H. O., *J. Gen. Microbiol. 76:*43 (1973).
Holmberg, K., and Nord, C. E., *J. Gen. Microbiol. 91:*17 (1975).
Johnson, J. L., and Cummins, C. S., *J. Bacteriol. 109:*1047 (1972).
Melville, T. H., *J. Gen. Microbiol. 40:*309 (1965).
Minnikin, D. E., Goodfellow, M., and Collins, M. D., *in* "Coryneform Bacteria" (I. J. Bonsfield, and A. G. Callely, eds.), p. 85. Academic Press, London (1978).
Schaal, K. P., *in* "Biology of the Actinomycetes" (M. Goodfellow, M. Mordarski, and S. T. Williams, eds.). Academic Press, London (1982). (In press).
Schaal, K. P., and Pulverer, G., *in* "The Prokaryotes: A Handbook on Habitats, Isolation and Identification of Bacteria" (M. P. Starr, H. Stolp, H. G. Trüper, A. Balows, and H. G. Schlegel, eds.), p. 1923. Springer, Heidelberg (1981).
Schaal, K. P., and Schofield, G. M., *in* "Actinomycetes" (K. P. Schaal, and G. Pulverer, eds.), p. 67. Gustav Fischer, Stuttgart (1981).
Schofield, G. M., and Schaal, K. P., *J. Gen. Microbiol. 127:*237 (1981).
Slack, J. M., *in* Bergey's "Manual of Determinative Bacteriology" 8th edition. (R. E. Buchanan, and N. E. Gibbons, eds.), p. 659. Williams and Wilkins Co., Baltimore (1974).
Slack, J. M., and Gerencser, M. A., "Actinomyces, Filamentous Bacteria". Burgess Publishing Co., Minneapolis (1975).
Stackebrandt, E., and Woese, C. R., *in* "Molecular and Cellular Aspects of Microbial Evolution" (M. J. Carlile, J. F. Collins, and B. E. B. Moseley, eds.), p. 1. Cambridge University Press, Cambridge (1981).

TAXONOMY OF *THERMOMONOSPORA* AND RELATED OLIGOSPORIC ACTINOMYCETES[1]

A. J. McCarthy[2]
T. Cross

Postgraduate School of Biological Sciences
University of Bradford
Bradford, England

I. INTRODUCTION

The application of numerical and chemical taxonomic methods has significantly improved actinomycete classification, but there remain many groups which have not been studied systematically and are therefore relatively poorly classified. The genus *Thermomonospora* is one such group in which an overreliance on morphology has resulted in an unsatisfactory classification. We therefore decided to apply numerical taxonomic methods to the classification of *Thermomonospora* and related organisms, the primary objective being to provide a sound basis for species identification. Representatives of *Thermoactinomyces, Saccharomonospora,* and *Micropolyspora* were also studied to yield further information on the intra- and intergeneric relationships between oligosporic actinomycetes.

The generic name *Thermomonospora* was originally proposed by Henssen (1957) for a group of monosporic, thermophilic actinomycetes found in rotted cow and sheep dung. Although two mesophilic species have since been described (Nonomura and Ohara, 1971, 1974), *Thermomonospora* strains are most commonly found on isolation plates incubated at 50 to 55 °C. Thermophilic actinomycete spores have been isolated from a wide range of habitats (for review see Cross, 1981), but it is mainly in overheated substrates, such as moldy fodders and composts, that these organisms are both active and numerous. We have found that thermophilic actinomycetes are most efficiently isolated by

[1] *This work was supported by a grant from the Natural Environment Research Council.*
[2] *Present address: Dept. of Biochemistry and Applied Molecular Biology, University of Manchester, Institute of Science and Technology, Manchester, England.*

ISBN 0-12-528620-1

using a sedimentation chamber and Andersen sampler (Lacey and Dutkiewicz, 1976) to reduce the growth of thermophilic *Bacillus* spp. This technique has enabled us to confirm that, while *Thermomonospora* spp. are heavily outnumbered by thermoactinomycetes and *Micropolyspora faeni* in moldy hay, they dominate the actinomycete population in composts prepared for cultivation of the edible mushroom, *Agaricus bisporus* (Lacey, 1973; McCarthy and Cross, 1981). The prevalence of thermophilic actinomycetes in moldy fodders and related substrates constitutes, in many cases, a serious health hazard. *Thermoactinomyces* spp., *Mip. faeni*, and *Sam. viridis* are all causative agents of hypersensitivity pneumonitis (Lacey, 1971; Pepys *et al.*, 1963; Treuhaft *et al.*, 1980) but, as yet, there is no firm evidence for the involvement of *Thermomonospora* spp. in respiratory disease. Indeed, members of the genus *Thermomonospora* have a more beneficial role in natural substrates, since by virtue of their ability to degrade cellulose, hemicellulose, and pectin, they contribute to the nutrient recycling process. Consequently, *Thermomonospora* strains have been the subject of intense study in relation to single-cell protein and ethanol production from cellulosic wastes (Bellamy, 1977; Crawford *et al.*, 1973; Ferchak *et al.*, 1980; Ginnivan *et al.*, 1977; Humphrey *et al.*, 1977; Stutzenberger, 1979).

II. A REVIEW OF *THERMOMONOSPORA* TAXONOMY

A. The status of Thermomonospora *and Related Genera*

The rejection of "*Actinobifida*" as an invalid generic name (Cross and Goodfellow, 1973; Skerman, *et al.*, 1980) and the improvement in the definition of *Thermoactinomyces* (Cross and Goodfellow, 1973) have simplified the classification of *Thermomonospora* at the genus level. However, the relationships between monosporic actinomycete genera have often been confused and controversial. The concept of one genus for monosporic actinomycetes was instigated by Waksman, Umbreit, and Cordon (1939) and Krasilnikov (1941), who classified *Thermoactinomyces* (Tsiklinsky, 1899) within the genus *Micromonospora* (Orskov, 1923). Unlike *Micromonospora* spp. which are devoid of aerial growth, thermoactinomycetes produce a true, well-developed aerial mycelium, and it was this fundamental difference which later led to the re-instatement of *Thermoactinomyces* as valid genus (Waksman and Corke, 1953). Subsequent to the original description of *Thermomonospora* (Henssen, 1957), the argument that these organisms did not differ significantly from *Micromonospora* was raised again. Krasilnikov (1964) regarded *Thermomonospora* as a questionable genus which differed from *Micromonospora* only in its thermophilic character and he later proposed that two of Henssen's species, *Thm. curvata* and "*Thm. lineata*", should be included in *Micromonospora*

(Krasilnikov and Agre, 1964). However, this ignored the formation of a sporulating aerial mycelium by *Thermomonospora* spp., even though it was the characteristic which prohibited Henssen (1957) from initially identifying her isolates within the genus *Micromonospora*. Studies on the wall composition of actinomycetes (Becker *et al.*, 1965; Lechevalier and Lechevalier, 1970) provided an additional criterion for the separation of genera. Using this criterion, *Micromonospora* (wall chemotype II) was clearly distinguished from other monosporic actinomycete genera (Table I).

There remained, however, the problem of distinguishing *Thermomonospora* from both *Thermoactinomyces* and "*Actinobifida*" (Krasilnikov and Agre, 1964). Substrate mycelium sporulation and wall composition initially were used to differentiate between *Thermoactinomyces* and *Thermomonospora*, but were later found to be of no value in the separation of these two genera. Henssen (1957) described *Thermomonospora* spp. as forming single spores only on the aerial mycelium; whereas, thermoactinomycetes formed spores also on the substrate mycelium (Küster and Locci, 1963a). In accordance with these observations, Küster and Locci (1963b) reclassified "*Tha. viridis*" (Schuurmans *et al.*, 1956) as "*Thm. viridis*". Since a later detailed morphological study of monosporic actinomycetes (Cross and Lacey, 1970) revealed that many *Thermomonospora* strains produce substrate mycelium spores, this character was no longer a valid indicator of generic identity. In addition, it was unfortunate that *Thermomonospora* was represented by "*Thm. viridis*" in actinomycete wall composition studies (Becker *et al.*, 1965; Lechevalier and Lechevalier, 1970), for unlike other *Thermomonospora* spp. which have a chemotype III wall, "*Thm. viridis*" has a wall chemotype IV. Heterogeneity in wall composition within *Thermomonospora* led to the creation of *Saccharomonospora* (Nonomura and Ohara, 1971) for monosporic, wall type IV actinomycetes. At present, this genus contains only one species, *Sam. viridis* (synonym "*Thm. viridis*"), but it has been suggested that "*Mip. caesia*" and *Mip. internatus*, which form mainly single spores and have a type IV wall composition, could be accommodated within *Saccharomonospora* (Kurup, 1981). Hence, *Thermomonospora* comprised wall type III actinomycetes which form single spores on aerial or on both aerial and substrate

TABLE I. Classification of Monosporic Actinomycete Genera

	Aerial mycelium	*Endospores*	*Wall chemotype* [a]
Micromonospora	-	-	*II*
Thermoactinomyces	+	+	*III*
Thermomonospora	+	-	*III*
Saccharomonospora	+	-	*IV*

[a] *Lechevalier and Lechevalier (1970).*

hyphae, a genus circumscription which applied equally to *Thermoactinomyces*. It was with some relief, therefore, that *Thermoactinomyces* was redefined as comprising actinomycetes whose endogenously formed spores have the structure and properties of true bacterial endospores (Cross and Goodfellow, 1973). Evidence for a close relationship between thermoactinomycetes and endospore-forming bacilli has been provided by data on menaquinone composition (Collins *et al.*, 1982), DNA base ratios (Craveri *et al.*, 1973), and 16S ribosomal RNA sequences (Stackebrandt and Woese, 1981a,b). The probable reclassification of *Thermoactinomyces* as a member of the family *Bacillaceae* (Cross, 1982) has far-reaching implications for actinomycete taxonomy in general.

The genus "*Actinobifida*" was proposed by Krasilnikov and Agre (1964) to describe a group of thermophilic actinomycete isolates which are characterized by the formation of single spores on dichotomously branched sporophores. Differences in colony pigmentation permitted the recognition of two species: "*Acb. dichotomica*" (yellow) (Krasilnikov and Agre, 1964) and "*Acb. chromogena*" (brown) (Krasilnikov and Agre, 1965). Locci, Baldacci, and Petrolini (1967) described a new "*Actinobifida*" species, "*Acb. alba*", which in producing a colorless substrate mycelium and white aerial mycelium is distinct from the above species. One of these species, "*Acb. dichotomica*", was shown to produce heat-resistant bacterial endospores (Cross *et al.*, 1968) and was therefore renamed *Tha. dichotomicus* (Cross and Goodfellow, 1973). With the placing of "*Thm. viridis*" in the genus *Saccharomonospora* (Nonomura and Ohara, 1971), strains of "*Actinobifida*" and *Thermomonospora* could be included in one, non-endospore-forming, wall type III genus. Since the definitive characteristic of dichotomously branched sporophores had been described first in "*Thm. fusca*" (Henssen, 1957), the generic name *Thermomonospora* was retained and "*Actinobifida*" rejected (Cross and Goodfellow, 1973; Skerman *et al.*, 1980).

Actinomycetes which produce single, non-motile spores are presently classified in four genera; Table I lists their major differentiating characteristics. Determination of wall chemotype can now be routinely applied for the identification of actinomycetes and may be used in addition to morphological characteristics to separate *Thermomonospora* spp. from members of the genera *Micromonospora* and *Saccharomonospora*. However, wall analysis cannot be used to differentiate *Thermomonospora* and *Thermoactinomyces*; conclusive evidence of endospore formation can only be obtained by the application of electron microscopy. Similarities in the morphology of *Thermomonospora* and *Thermoactinomyces* strains having a white aerial mycelium increase the possibility of misidentification which is a particularly important problem in view of the role of *Thermoactinomyces* spp. in respiratory disease. The application of heat-resistance properties of spores in the identification of thermophilic actinomycetes has been complicated by the demonstration that spores which were not formed endogenously possess limited heat resistance (Dorokhova *et al.*, 1970). However, Attwell (1973) and Nonomura and Ohara (1974) found significant differences in the ability of the spores of *Thermoactinomyces* and *Thermomonospora* to resist heat, and our own studies confirmed that, unlike sporulating cultures of *Thermoactinomyces* spp., those of *Saccharomonospora*, *Thermomonospora*, and *Micropolyspora* strains do not survive wet-heat treatment at 90°C for 30 minutes.

B. Thermomonospora *Species*

Morphology has been used as the principal basis for species classification in *Thermomonospora*, in which spore arrangement, their presence on substrate mycelium, and their colony pigmentation have assumed particular significance. Reliance on a few morphological characters inevitably led to a classification which was neither stable nor predictive. In their comparative study on the morphology of thermophilic actinomycetes, Cross and Lacey (1970) found an almost continuous range of variation among the samples strains, which included types that had been described as distinct species of *Thermomonospora* or of "*Actinobifida*". Their study clearly demonstrated the inadequacy of species descriptions based on morphological examination of only a few strains. Against this background, with the prevalent nomenclatural confusion and in the absence of type cultures for many species, Skerman *et al.* (1980) selected only four of the ten strains previously described as *Thermomonospora* spp. for inclusion in the *Approved Lists of Bacterial Names*.

Thermomonospora was one of three new thermophilic actinomycete genera described by Henssen in 1957. Although that study made an important contribution to thermophilic actinomycete classification, a number of species unfortunately were not isolated in pure culture and, therefore, were unavailable for comparative study. Since, of the three *Thermomonospora* spp. described (Henssen, 1957), only *Thm. curvata* has been obtained in pure culture, it was later named as the type species of the genus (Henssen and Schnepf, 1967). The other species described by Henssen (1957) are now nomenclaturally invalid (Skerman *et al.*, 1980). Although there is certainly insufficient information to permit the identification of isolates with "*Thm. lineata*", the name "*Thm. fusca*", whose members are characterized by the production of single spores on repeatedly branched sporophores, has been used frequently in the literature. Waksman and coworkers (1939) previously had observed, but not isolated, a thermophilic actinomycete which exhibited this characteristic morphology, and had identified it with "*Mim. fusca*" (Jensen, 1932). Henssen (1957) believed that these authors had, in fact, observed "*Thm. fusca*", a thermophilic actinomycete with a sporulating aerial mycelium and, therefore, clearly distinct from "*Mim. fusca*" Jensen. Waksman (1961) compounded the confusion by re-describing "*Thm. fusca*" Henssen as an organism which formed a brown colored colony on agar medium, in sharp contrast to Henssen's earlier statement that members of the genus *Thermomonospora* formed either colorless or yellow colonies with a white aerial mycelium. Similarities between descriptions of "*Thm. fusca*" (Waksman, 1961) and "*Acb. chromogena*" (Krasilnikov and Agre, 1965) resulted in further confusion, since microbiologists who were later involved in the isolation and identification of actinomycetes referred to the work of Waksman (1961), rather than to that of Henssen (1957), for a description of "*Thm. fusca*". One component of this complex nomenclatural situation was removed by Luedemann (1971) who reduced "*Mim. fusca*" (Jensen, 1932) to a synonym for *Mim. purpureochromogenes* (Waksman and Curtis, 1916); Cross and Goodfellow (1973), in their treatise on actinomycete taxonomy, stated that this left "the specific name *fusca* for a species in the genus *Thermomonospora* forming a white aerial mycelium and no diffusible pigments". Such a strain, whose morphological description complied with that given for "*Thm. fusca*" by Henssen (1957), was subsequently isolated, described, and deposited in several culture collections under that name (Crawford, 1975).

In addition to "*Thm. fusca*" and *Thm. curvata*, two other species names can be associated with the white *Thermomonospora* group. *Thm. alba* (Locci *et al.*, 1967 *emend*; Cross and Goodfellow, 1973) is morphologically similar to "*Thm. fusca*" and differs from it only in the ability to produce substrate mycelium spores, a character of doubtful validity for species identification within this group (Cross and Lacey, 1970). *Thm. mesouviformis* (Nonomura and Ohara 1974) is likewise morphologically similar, but is described as a mesophilic organism, incapable of growth at 50°C. Detailed examination of other characteristics, in addition to morphology and growth temperature range, should reveal the taxonomic structure of the white *Thermomonospora* group and this has been one of our main objectives. A study of this nature has already been attempted to some extent by Kurup (1979) who examined morphological, biochemical, and serological data obtained for twenty strains representing five *Thermomonospora* spp. In relation to the white *Thermomonospora* group, Kurup (1979) suggested that *Thm. curvata* and *Thm. mesouviformis* were distinct species, while "*Thm. fusca*" and *Thm. alba* were synonymous. However, Kurup recognized the need for a more detailed study on a comprehensive collection of strains and, indeed, our results (section III D) do not support his tentative proposals.

The *Thermomonospora* population of overheated natural substrates comprises two major subgroups: the white *Thermomonospora* group discussed above, and "*Thm. chromogena*" which is recognizable as discrete, brown colonies on isolation plates incubated at 50 to 55°C (McCarthy and Cross, 1981). The taxonomic position of the latter species is not well defined and it has not been included in the *Approved Lists of Bacterial Names* (Skerman *et al.*, 1980). When the genus "*Actinobifida*" was rejected, it remained to be determined whether or not "*Acb. chromogena*" (Krasilnikov and Agre, 1965) formed heat-resistant endospores and could accordingly be classified in either *Thermoactinomyces* or *Thermomonospora*. Evidence that "*Acb. chromogena*" produced thermoactinomycete-like endospores (Dorokhova *et al.*, 1970; Mach and Agre, 1970) was later refuted by Attwell (1973) who conclusively demonstrated that the spores neither were formed endogenously nor exhibited resistance to heat comparable to that of thermoactinomycete endospores. This species could therefore be accommodated within *Thermomonospora*, and we have endeavored to determine its relationship to other members of the genus.

Two new species of *Thermomonospora*, "*Thm. spiralis*" and "*Thm. falcata*", were described by Henssen in 1970. Since the descriptions contained information only on the method of spore formation, both species names were cited by Henssen as *nominanuda*. "*Thm. spiralis*" produced spore chains in addition to single spores and, therefore, would not conform to the genus definition of *Thermomonospora* (Cross and Goodfellow, 1973). The taxonomic status of this species cannot be evaluated because all the original strains died and no further isolations have been reported. "*Thm. falcata*" strains are, however, available and their sporulation pattern—single spores on a falcate or sickle-shaped sporophore leading to the formation of a spore cluster— would not, as Kurup (1979) suggested, exclude them from the genus *Thermomonospora*. In fact, this spore arrangement could also be interpreted as irregular sporophore branching, previously described in "*Thm. chromogena*" (Krasilnikov and Agre, 1965); therefore, we were particularly interested in the relationship between these two species.

The two mesophilic "*Thermomonospora*" spp. described by Nonomura and Ohara (1971, 1974) were correctly assigned to the genus according to their wall composition and spore heat resistance properties. The possibility that *Thm. mesouviformis* is closely related to other white *Thermomonospora* spp. has already been discussed; however, *Thm. mesophila* would appear to be a distinct species (Nonomura and Ohara, 1971; Kurup, 1979). The most recently described member of the genus, "*Thm. galeriensis*" (Szabó *et al.*, 1976), is important as a producer of the antibiotic primycin, but due to the absence of wall composition data, its generic affiliation has remained unconfirmed.

III. NUMERICAL CLASSIFICATION

A. Collection and Analysis of Data

This study was based on a detailed examination of 113 strains which included fresh isolates, type or named cultures of available *Thermomonospora* spp., and representatives of the genera *Thermoactinomyces*, *Saccharomonospora*, and *Micropolyspora*. Strains were grown routinely at 50°C, with the exception of the type strain of *Thm. mesophila* which was grown at 40°C. Mor-

TABLE II. Tests Selected for Use in the Numerical Taxonomic Study

Test category		*Unit characters*
Morphology		*18*
Growth temperature		*6*
pH tolerance		*4*
Sensitivity to growth inhibitors		*18*
Sensitivity to antibiotics		*11*
Degradative tests		*19*
Specific enzyme tests		*8*
Carbon source utilization		*17*
	Total	*101*

phological, physiological, and biochemical tests provided 101 discriminatory unit characters of acceptable reproducibility (Table II), and test error, estimated at 1.1%, had no deleterious effect on the numerical classification. Similarity matrices were computed with the Simple Matching (S_{SM}), Jaccard (S_J), and Pattern Difference (D_p) coefficients. For the D_p analysis, 15 morphological characters were excluded and the S_{SM} matrix was computed with, and without, these characters. Clustering was achieved using both single and average (UPGMA) linkage algorithms (Sneath and Sokal, 1973).

A simplified dendrogram derived from the S_{SM} UPGMA analysis on the full complement of strains and characters is presented in Figure 1. Both clustering methods recovered the same four major clusters for each of the three coefficients. Cluster composition and arrangement were unaffected when S_J was used instead of S_{SM}, thus confirming that none of the groupings were based primarily on negative correlations. Strain vigor also had no effect on the classification, since analysis with the D_p coefficient yielded a similar dendrogram to that presented in Figure 1. The classification therefore consisted of a number of stable and distinct clusters which could be equated with taxa.

B. "Thm. chromogena"

The type strain of "*Acb. chromogena*", two reference cultures of "*Thm. falcata*", and isolates provisionally identified as "*Thm. chromogena*" were all recovered in cluster 1, defined at the 88% similarity level (Fig. 1). Two subclusters, 1a and 1b, were consistently recovered in all the analyses, but are not considered sufficiently distinct to merit separate species status in the classification proposed here. In addition to the diagnostic characteristics of "*Thm. chromogena*", listed in Table III, two features of this species are worthy of note. First, the sporulation process, as revealed by light microscopy, corresponded to that described for "*Thm. falcata*" by Henssen (1970) (see section II B), in which spore clusters are produced by incurving hyphae bearing single spores rather than by repeated sporophore branching (Krasilnikov and

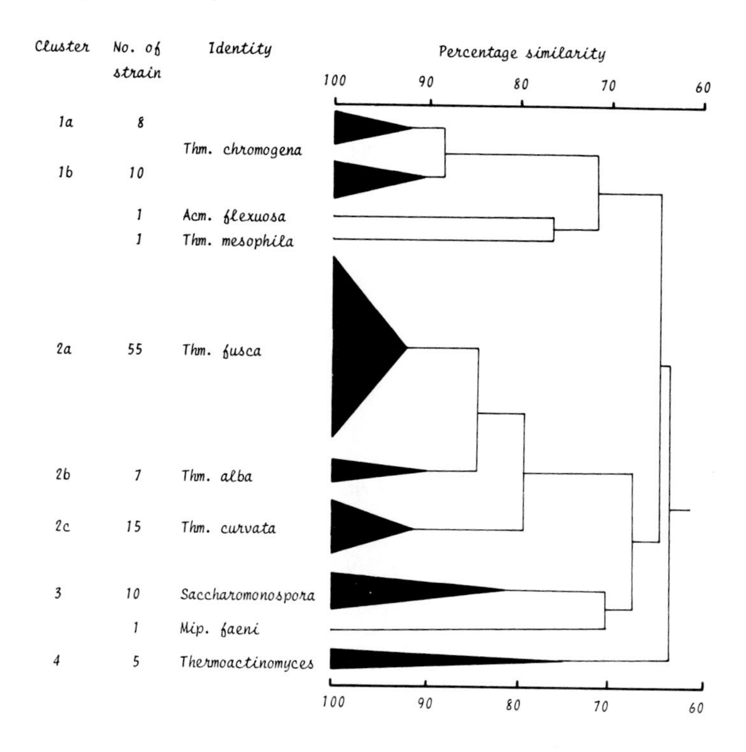

FIGURE 1. Simplified dendrogram derived from the S_{SM} UPGMA analysis of data for 113 strains.

Agre, 1965). Second, "*Thm. chromogena*" strains, unlike those of other *Thermomonospora* spp., exhibited resistance to the aminoglycoside antibiotics tobramycin, gentamicin, and kanamycin. More information on the occurrence and distribution of "*Thm. chromogena*" probably will result from the use of an improved isolation medium which contains kanamycin as the selective agent (McCarthy and Cross, 1981).

"*Thm. chromogena*" strains had little in common with the other *Thermomonospora* strains located in cluster 2 (Fig. 1) and this tends to suggest that a generic separation is justified. The fact that representatives of clusters 1 and 2 had fundamentally different menaquinone patterns (Collins *et al.*, 1982) could be used to support such a proposal; however, in the absence of further chemical and genetic data, the formation of a new genus for "*Thm. chromogena*" is premature. "*Thm. chromogena*" conformed to the genus definition of *Thermomonospora* (section II A) and, for the present, should be re-instated as a valid species.

TABLE III. A Proposed Identification Scheme for Thermomonospora *spp.*

Character [a]	Thm. chromogena	Thm. mesophila	Thm. fusca	Thm. alba	Thm. curvata
Colony reverse color [b]	*brown*	*brown*	*yellow*	*yellow*	*yellow orange*
Growth at [b]					
53 °C	+ [c]	-	+	-	+
pH 11.0	-	-	+	+	+
Growth in: [b]					
0.00002% (w/v) crystal violet	+	+	+	-	+
0.002% (w/v) tetrazolium chloride	+	-	+	-	+
Degradation of:					
tyrosine					
xanthine	+	+	-	-	-
hypoxanthine					
starch	-	+	+	+	+
pectin	+	+	+	+	-
elastin	+	+	+	*v*	-
cellulose MN 300	-	-	+	+	+
Nitrate reduction	+	+	-	*v*	+
Oxidase	+	+	-	-	-
Utilization of:					
D-galactose	+	+	+	*v*	-
D-ribose	*v*	-	-	-	+
D-xylose	-	+	-	-	+
sucrose	-	-	+	+	+
lactose	-	-	+	*v*	-
melezitose	-	-	+	-	*v*

[a] *Cultures incubated at 50°C, except* Thm. mesophila *(40°C).*
[b] *CYC agar (Cross and Attwell, 1974) used as growth medium.*
[c] *(+):80 to 100% positive; (v): 21 to 79% positive; (-): 0 to 20% positive.*

C. Thm. mesophila

The type strain of *Thm. mesophila* did not cluster with the other *Thermomonospora* strains included in the study and, in support of the original description (Nonomura and Ohara, 1971), clearly represented a distinct species. In gross colony morphology and biochemical properties, *Thm. mesophila* had more in common with "*Thm. chromogena*" than with the *Thermomonospora* strains recovered in cluster 2, but could be differentiated from the former by its spore arrangement, mesophilic growth temperature requirement, antibiotic sensitivity pattern, and ability to degrade starch, agar, and DNA. More strains are required to circumscribe *Thm. mesophila* and, as with "*Thm. chromogena*", further chemical and genetic studies may lead to relocation of this species outside *Thermomonospora*. The tenuous relationship between *Thm. mesophila* and the type strain of "*Actinomadura flexuosa*" (Fig. 1) was probably not significant, since the latter forms chains of spores, contains madurose (Guzeva *et al.*, 1972) and is therefore more appropriately classified in either *Actinomadura* or *Excellospora* (Lacey *et al.*, 1978).

D. The White Thermomonospora *Group*

Cluster 2, formed at 79% similarity (Fig. 1), represents what is often referred to as the white *Thermomonospora* group on which Henssen (1957) based her original description of the genus. Previous studies on *Thermomonospora* taxonomy (Cross and Lacey, 1970; Kurup, 1979) demonstrated the range of variation within this group, but failed to adequately define its internal structure. In this study, three subclusters were revealed and, although a number of characteristics served to distinguish them, there were some common features. All strains formed colonies with a white aerial mycelium which bore single spores on sporophores. Optimum growth and sporulation occurred in the range pH 8.0 to 11.0, and few strains were capable of growth below pH 6.5. The white *Thermomonospora* group is also a potential source of commercially useful strains, since all of those examined thus far can degrade both cellulose and hemicellulose.

1. "Thm. fusca". Subcluster 2a comprised the majority of white *Thermomonospora* isolates in addition to strains received as "*Thm. fusca*", *Thm. curvata* and "*Thm. lineata*". Morphologically, these strains conformed to the original description of "*Thm. fusca*" (Henssen, 1957), but not to that later cited by Waksman (1961). Kurup (1979) identified all his *Thermomonospora* isolates as "*Thm. fusca*" (Henssen, 1957), and this supports the suggestion that the strains recovered in cluster 2a represent the most numerous group of white thermomonosporas present in high-temperature environments.

Morphological variation and instability was a feature of these strains and, in common with Cross and Lacey (1970), we found substrate mycelium sporulation to be a particularly unsuitable taxonomic character. However, the production of single spores on repeatedly branched sporophores borne on the aerial mycelium was consistent and, together with physiological and biochemical characteristics (Table III), permited the delimitation of "*Thm. fusca*". The suggestion that this species name should be revived (Crawford, 1975) is now supported by data on a representative collection of strains.

2. Thm. curvata. Previous authors have maintained that the type species,

Thm. curvata, is morphologically distinct. *Thm. curvata* is characterized by the production of spores on branched and unbranched sporophores borne along the aerial hyphae, rather than on repeatedly branched sporophores leading to the formation of dense spore clusters as observed in strains of "*Thm. fusca*" (Cross and Goodfellow, 1973; Cross and Lacey, 1970; Henssen, 1957; Henssen and Schnepf, 1967; Kurup, 1979). While we generally agree with this separation, the fact that several strains received as *Thm. curvata* were not located with the type strain in subcluster 2c suggests that reliance on morphology alone has led to mis-identification. Numerical analysis of morphological, physiological, and biochemical data has confirmed the separate identity of *Thm. curvata* and "*Thm. fusca*" (Fig. 1), and application of diagnostic tests (Table III) should permit a more reliable identification of these two species.

3. Thm. alba. In contrast to the two white *Thermomonospora* spp. discussed above, the taxonomic position of *Thm. alba* is not well defined. The type strain was recovered in cluster 2b together with the type strain of *Thm. mesouviformis* and five white *Thermomonospora* isolates. In the S_{SM} UPGMA dendrogram (Fig. 1), this subcluster was closely related to subcluster 2a but, when the data were examined more closely, it was apparent that two of the seven strains had more in common with members of subcluster 2c. The most prominent feature of the subcluster 2b strains was their low optimum growth temperature — aerial mycelium production and sporulation were much improved when cultures were incubated at 40°C rather than at 50°C which was close to the upper growth temperature limit of most strains. This leads to the suggestion that subcluster 2b is a collection of low temperature variants of "*Thm. fusca*" and *Thm. curvata*, and it may be significant that the only three white *Thermomonospora* strains which were originally isolated from soil were all recovered in this subcluster. However, subcluster 2b was identified by all four numerical analyses (section III A) and, therefore, could be a center of variation which was inadequately represented in this study.

Until more data on a larger group of strains suggest otherwise, the name *Thm. alba* should be retained for white thermomonosporas which grow optimally at 40 to 45°C and which cannot be readily identified with either "*Thm. fusca*" or *Thm. curvata* (Table III). Nonomura and Ohara (1974) and Kurup (1979) stated that *Thm. mesouviformis* was a mesophilic organism distinct from other white *Thermomonospora* spp. because it did not grow at 50°C and 55°C, respectively. But, mesophily and thermophily are not easily definable terms. Since we found that the *Thm. mesouviformis* type strain grew at 50°C and had an S_{SM} value >0.9 with the type strain of *Thm. alba*, it is proposed that *Thm. mesouviformis* (Nonomura and Ohara, 1974) be regarded as a later synonym of *Thm. alba* (Locci *et al.*, 1967 *emend*; Cross and Goodfellow, 1973).

E. The Genus Saccharomonospora

All the monosporic, wall type IV strains were recovered in cluster 3, defined at 81% similarity (Fig. 1), which can therefore be considered as representing the genus *Saccharomonospora* in this study. Since considerable heterogeneity was observed within this cluster, studies on a larger number of strains are required to determine the number of valid and distinct *Saccharomonospora* spp. Strains received as "*Mip. caesia*" produced single spores and their inclusion in cluster 3 supports the proposal that they be transferred to *Saccharomonospora* (Kurup, 1981). Recommendations can also be made with respect to the delimitation of *Saccharomonospora* and its single species, *Sam. viridis*. As presently defined (Cross, 1981), *Saccharomonospora* excludes actinomycetes which exhibit sporulation on the substrate mycelium. Two of the cluster 3 strains produced single spores on a fragmenting substrate mycelium and, although this feature is not considered sufficiently stable and unequivocal to merit particular taxonomic significance, a degree of flexibility should be introduced into the genus definition. The definition of *Sam. viridis* should also be widened to include pigmented variants. One strain, received as a *Thermomonospora* sp., was very closely related to the type strain of *Sam. viridis*, but produced a lilac/purple, rather than the characteristic dark green pigment.

The type strain of *Thm. galeriensis* (Szabó *et al.*, 1976) was received too late for inclusion in the numerical taxonomic study, but a number of its properties were determined. The following characteristics demonstrated that *Thm. galeriensis* had more in common with members of cluster 3 than with *Thermomonospora* strains recovered in cluster 2: wall chemotype IV; growth at pH 6.0; tolerance to 5% (w/v) sodium chloride; tyrosine degradation; glycerol utilization; no activity against carboxymethylcellulose. Reference to an earlier description of *Thm. (Thermopolyspora) galeriensis* (Vályi-Nagy *et al.*, 1970) indicates that this is a *Saccharomonospora* strain which has lost the ability to produce aerial mycelium, spores, and a dark green pigment.

The tenuous relationship between the type strain of *Mip. faeni* and cluster 3 (Fig. 1) was not apparent when the data were analyzed with the D_p coefficient. Nevertheless, the classification of genera within the whole group of wall type IV actinomycetes which lack mycolic acids is in need of further study.

F. The Genus Thermoactinomyces

The recovery of five *Thermoactinomyces* strains, representing three species, in a cluster which showed little similarity to other groups in the study emphasizes the superficial nature of the resemblance between thermoactinomycetes and thermomonosporas (section II A). Although the diversity within *Thermoactinomyces* was not well represented, comparison of our results with those of a detailed numerical taxonomic study on this genus (Unsworth, 1978) yields a number of characters which could be used to discriminate between *Thermoactinomyces* and the white *Thermomonospora* group. Thus, cellulose degradation, β-galactosidase activity, and sensitivity to 25 μg ml^{-1} novobiocin are characters for which white *Thermomonospora* spp. are positive and thermophilic thermoactinomycetes, negative. With respect to species identification within *Thermoactinomyces*, the identification scheme proposed by Unsworth and Cross (1978) was generally supported by the data obtained in this study.

IV. CONCLUSIONS

Numerical phenetic classifications were derived from the examination of a large number of strains for a range of characters, and were therefore stable, could accommodate a degree of error, and provided a sound basis for further taxonomic studies. The morphological classification of *Thermomonospora* into valid species clearly required substantiation, and the application of numerical taxonomic methods solved some of the species identification problems within this genus. Five *Thermomonospora* spp. were recognized and it is hoped that the identification scheme presented in Table III will prove to be reliable when applied in other laboratories. Perhaps more significantly, numerical classification of *Thermomonospora* and related organisms has also highlighted specific areas where further investigation is needed. These include elucidation of the internal taxonomic structure of *Saccharomonospora*, its relationship to other wall type IV actinomycete genera, and the possible relocation of *Thm. chromogena* outside the genus *Thermomonospora*.

ACKNOWLEDGMENTS

We are grateful to all of those colleagues who donated strains.

REFERENCES

Attwell, R. W., Ph. D. thesis, University of Bradford U. K. (1973).
Becker, B., Lechevalier, M. P., and Lechevalier, H. A., *Appl. microbiol. 13:*236 (1965).
Bellamy, W. D., *Dev. Ind. Microbiol. 18:*249 (1977).
Collins, M. D., Mackillop, G. C., and Cross, T., *FEMS Microbiol. Lett. 13:*151 (1982).
Collins, M. D., McCarthy, A. J., and Cross, T., *Zentralbl. Bakteriol. I Abt.* (1982). In press.
Craveri, R., Manachini, P. L., Aragozzini, F., and Merendi, C., *J. Gen. Microbiol. 74:*201 (1973).
Crawford, D. L., *Can. J. Microbiol. 21:*1842 (1975).
Crawford, D. L., McCoy, E., Harkin, J. M., and Jones, P., *Biotechnol. Bioeng. 15:*833 (1973).
Cross, T., *in* "The Prokaryotes, a Handbook on Habitats, Isolation and Identification of Bacteria" Vol. II, (M. P. Starr, H. Stolp, H. G. Trüper, A. Balows, and H. G. Schlegel, eds), p. 2091. Springer-Verlag, Berlin (1981).
Cross, T., *Actinomycetes 16:*77 (1982).
Cross, T., and Attwell, R. W., *in* "Spore Research 1973" (A.N. Barker, G.W. Gould, and J. Wolf, eds.), p. 11. Academic Press, London (1974).

Cross, T., and Goodfellow, M., *in* "Actinomycetales: Characteristics and Practical Importance" (G. Sykes, and F. A. Skinner, eds.), p. 11. Academic Press, London (1973).
Cross, T., and Lacey, J., *in* "The Actinomycetales" (H. Prauser, ed.), p. 211. Gustav Fischer Verlag, Jena (1970).
Cross, T., Walker, P. D., and Gould, G. W., *Nature (Lond.) 220:*352 (1968).
Dorokhova, L. A., Agre, N. S., Kalakoutskii, L. V., and Krasilnikov, N. A., *in* "The Actinomycetales" (H. Prauser, ed.), p. 227. Gustav Fischer Verlag, Jena (1970).
Ferchak, J. D., Hägerdal, B., and Pye, E. K., *Biotechnol. Bioeng. 22:*1527 (1980).
Ginnivan, M. J., Woods, J. L., and O'Callaghan, J. R., *J. Appl. Bacteriol. 43:*231 (1977).
Guzeva, L. N., Agre, N. S., and Sokolov, A. A., *Mikrobiologiya 41:*1080 (1972). (In Russian)
Henssen, A., *Arch. Mikrobiol. 26*:373 (1957).
Henssen, A., *in* "The Actinomycetales" (H. Prauser, ed.), p. 205. Gustav Fischer Verlag, Jena (1970).
Henssen, A., and Schnepf, E., *Arch. Mikrobiol. 57:*214 (1967).
Humphrey, A. E., Moreira, A., Armiger, W., and Zabriskie, D., *Biotechnol. Bioeng. Symp. 7:*45 (1977).
Jensen, H. L., *Proc. Linn. Soc. N. S. W. 57:*173 (1932).
Krasilnikov, N. A., *in* "Guide to the Actinomycetales" Izdatel'stvo Akademii Nauk, USSR, Moskva-Leningrad (1941). (In Russian)
Krasilnikov, N. A., *Hind. Antibiot. Bull. 7:*1 (1964).
Krasilnikov, N. A., and Agre, N. S., *Mikrobiologiya 33:*935 (1964). (In Russian)
Krasilnikov, N. A., and Agre, N. S., *Mikrobiologiya 34:*284 (1965).
Kurup, V. P., *Curr. Microbiol. 2:*267 (1979).
Kurup, V. P., *Microbiologica (Bologna) 4:*249 (1981).
Küster, E., and Locci, R., *Arch. Mikrobiol. 45:*188 (1963a).
Küster, E., and Locci, R., *Int. Bull. Bacteriol. Nomencl. Taxon. 13:*213 (1963b).
Lacey, J., *J. Gen. Microbiol. 66:*327 (1971).
Lacey, J., *in* "Actinomycetales: Characteristics and Practical Importance" (G. Sykes, and F. A. Skinner, eds.), p. 231. Academic Press, London (1973).
Lacey, J., and Dutkiewicz, J., *J. Appl. Bacteriol. 41:*315 (1976).
Lacey, J., Goodfellow, M., and Alderson, G., *in* "Nocardia and Streptomyces" (M. Mordarski, W. Kurylowicz, and J. Jeljaszewicz, eds.), p. 107. Gustav Fischer Verlag, Stuttgart (1978).
Lechevalier, M. P., and Lechevalier, H. A., *Int. J. Syst. Bacteriol. 20:*435 (1970).
Locci, R., Baldacci, E., and Petrolini, B., *G. Microbiol. 15:*79 (1967).
Luedemann, G. M., *Int. J. Syst. Bacteriol. 21:*240 (1971).
Mach, F., and Agre, N. S., *in* "The Actinomycetales" (H. Prauser, ed.), p. 221 Gustav Fischer Verlag, Jena (1970).
McCarthy, A. J., and Cross, T., *J. Appl. Bacteriol. 51:*299 (1981).
Nonomura, H., and Ohara, Y., *J. Ferment. Technol. 49:*895 (1971).
Nonomura, H., and Ohara, Y., *J. Ferment. Technol. 52:*10 (1974).
Orskow, J., *in* "Investigations into the Morphology of the Ray Fungi". Levin and Munksgaard, Copenhagen (1923).
Pepys, J., Jenkins, P. A., Festenstein, G. N., Gregory, P. H., Skinner, F. A., and Lacey, M. E., Lancet *ii:*607 (1963).
Schuurmans, D. M., Olson, B. H., and San Clemente, C. L., *Appl. Microbiol. 4:*61 (1956).
Skerman, V. B. D., McGowan, V., and Sneath, P. H. A. *Int. J. Syst. Bacteriol. 30:*225 (1980).
Sneath, P. H. A., and Sokal, R. R., *Numerical Taxonomy.* W. H. Freeman and Co, San Francisco (1973).
Stackebrandt, E., and Woese, C. R., *in* "Molecular and Cellular Aspects of Microbial Evolution" (M. J. Carlile, J. F. Collins, and B. E. B. Moseley, eds.), p. 1. University Press, Cambridge (1981a).
Stackebrandt, E., and Woese, C. R., *Curr. Microbiol. 5:*197 (1981b).
Stutzenberger, F. J., *Biotechnol. Bioeng. 21:*909 (1979).
Szabo, I. M., Marton, M., Kulcsar, G., and Buti, I., *Acta Microbiol. Acad. Sci. Hung. 23:*371 (1976).
Treuhaft, M. W., Green, J. G., Arusel, R., and Borge, A., *Am. Rev. Respir. Dis. 121:*100 (1980).
Tsiklinsky, P., *Ann. Inst. Pasteur (Paris) 13:*500 (1899).
Unsworth, B. A., Ph.D. thesis, University of Bradford, U. K. (1978).
Unsworth, B. A., and Cross, T., *J. Appl. Bacteriol. 45:*16 (1978).

Valyi-Nagy, T., Kulcsar, G., Szilagyi, I., Valu, G., Magyar, K., Kiss, G. H., and Horvath, I., U.S. Patent Office, Patent No. 3,498,884 (1970).
Waksman, S. A., "The Actinomycetes Volume 2. Classification, Identification and Descriptions of Genera and Species." Williams and Wilkins, Baltimore (1961).
Waksman, S. A., and Corke, C. T., *J. Bacteriol. 66:*377 (1953).
Waksman, S. A., and Curtis, R. E., *Soil Science, 1:*99 (1916).
Waksman, S. A., Umbreit, W. W., and Cordon, T. C., *Soil Sci. 47:*37 (1939).

NUMERICAL CLASSIFICATION AND IDENTIFICATION OF STREPTOMYCETES[1]

S. T. Williams
J. C. Vickers

Department of Botany
University of Liverpool
Liverpool, U. K.

M. Goodfellow

Department of Microbiology
University of Newcastle
Newcastle upon Tyne, U. K.

G. Alderson

School of Medical Sciences
University of Bradford
Bradford, U. K.

E. M. H. Wellington

Department of Biology
Liverpool Polytechnic
Liverpool, U. K.

P. H. A. Sneath
M. J. Sackin

Department of Microbiology
University of Leicester
Leicester, U. K.

A. M. Mortimer

Department of Botany
University of Liverpool
Liverpool, U. K.

I. INTRODUCTION

The definition and recognition of *Streptomyces* species have provided taxonomists with a major problem for many years. Hundreds of species have been legitimately described (Pridham and Tresner, 1974; Shirling and Gottlieb, 1967, 1968a,b, 1969, 1972) and included in the "Approved List of Bacterial Names" (Skerman *et al.*, 1980). Still more have merely been cited in the patent literature (Trejo, 1970). Many attempts have been made over the

[1] *We gratefully acknowledge research grants from the Science Research Council which supported this work.*

ISBN 0-12-528620-1

past 30 years to allocate the numerous species to groups (or series), thereby facilitating their identification (see Williams *et al.*, 1981, 1983a,b for detailed discussion). The vast majority of such groupings were based on a few subjectively chosen morphological and pigmentation properties; hence, the species groups were artificial and inclusion could be ruled out by one aberrant character state. Biochemical, nutritional, and physiological characters have been used in many species descriptions, but there has been little standardization in test selection, test procedures, or the range of species to which they have been applied. One notable exception was the use of standardized carbon source utilization tests in the International *Streptomyces* Project (I.S.P.) (Shirling and Gottlieb, 1966) and, hence, in the last edition of Bergey's "Manual" (Pridham and Tresner, 1974).

The first and most comprehensive attempt to construct a numerical classification of streptomycetes by using a wide range of characters was made by Silvestri *et al.* (1962). Twenty-five centers of variation were recognized and the data were used to construct a probabilistic identification key (Hill and Silvestri, 1962; Möller, 1962). Subsequently, there were several attempts to construct both numerical classification and identification systems, but they were based on a relatively small number of characters (Gyllenberg, 1970; Gyllenberg *et al.*, 1967, 1975; Kurylowicz *et al.*, 1975). Results of factor analysis also suggested that many characters used to describe *Streptomyces* species were highly variable and prone to errors in interpretation (Gyllenberg, 1970).

The taxonomic relationships between *Streptomyces* and several other genera in the family *Streptomycetaceae* are also unclear. Thus, for example, the genera *Chainia* and *Streptoverticillium* may be regarded as members of *Streptomyces*, which from sclerotia and verticillate sporophores respectively, rather than as separate genera. In the current edition of Bergey's "Manual", *Chainia* is included in *Streptomyces*, while *Streptoverticillium* is not (Pridham and Tresner, 1974).

The work reported here summarizes our attempts over the past six years to improve the taxonomy of streptomycetes by construction of polythetic classification and identification systems. Full details, including test methods and strain histories, were given by Williams *et al.* (1981, 1983a,b).

II. CONSTRUCTION OF NUMERICAL CLASSIFICATION SCHEME

A. Strains and Characters

Four hundred and seventy five strains and 44 duplicate cultures were included, with type cultures being selected whenever possible. Emphasis was placed on *Streptomyces* species which included 394 I. S. P. cultures (Shirling and Gottlieb, 1967, 1968a,b, 1969, 1972); marker strains of 14 other genera were also studied. After preliminary evaluation of the reproducibility of tests, 162 unit characters were determined for all strains. The characters included those used traditionally for streptomycetes as well as newly applied tests. They were categorized as morphological traits, pigmentation, antimicrobial activity, biochemical properties, degradative ability, antibiotic resistance, growth requirements, and utilization of carbon and nitrogen sources.

B. Computation and Analysis

Most characters existed in one of two mutually exclusive states and were scored plus or minus. Qualitative multistate characters, such as pigmentation and spore chain morphology, were coded as several independent characters; each was scored plus for the character state shown and minus for the alternatives. Quantitative multistate characters, such as tolerance to inhibitors, were coded by the additive method (Sneath and Sokal, 1973). Data were computed to determine both the simple matching coefficient (S_{SM}) (Sokal and Michener, 1958), which includes both positive and negative matches, and the Jaccard coefficient S_J) (Sneath, 1957) which includes only positive matches. Clustering was achieved using the unweighted average linkage algorithm of Sneath and Sokal (1973).

The distinctness of the major clusters so defined was assessed by calculation of the degree of overlap and the overlap statistics between all cluster pairs by using the OVCLUST program (Sneath, 1979a). Test reproducibility was assessed by examining the determinations of the 139 (see below) unit characters for each pair of the 44 duplicate cultures. Test variance (S_i^2) was calculated (Sneath and Johnson, 1972) and the average test variance (S^2) was used to calculate the average probability (p) of an erroneous test result (Sneath and Johnson, 1972). Similarity between duplicates was calculated by computation of the S_{SM} coefficient.

C. Results of Numerical Classification

1. Test Error. Analysis of the results obtained with the duplicate cultures showed that the average probability (p) of an erroneous test result was 3.36%, which was well within the limit of 10% suggested by Sneath and Johnson (1972). The 44 pairs of duplicate cultures showed a mean similarity of 93.1% S_{SM}. Thus, test error was clearly within acceptable limits. The majority of tests gave S_i^2 values <0.05. However, since seven tests showed a variance >0.1 and another 16 provided little or no separation value, these tests were deleted from the data matrix before computation of overall similarities. The final matrix therefore contained 139 unit characters.

2. Composition of Cluster Groups. Of the cluster groups defined at 70.1% S_{SM}, by far the largest was group A which contained 218 (73%) of the *Streptomyces* strains (Table I). Several other genera sharing wall chemotype I (Lechevalier and Lechevalier, 1970) with *Streptomyces* were also included in this group. *Actinopycnidium* and *Actinosporangium* were clearly synonyms of *Streptomyces; Chainia, Elytrosporangium*, and *Microellobosporia* showed morphological differences to *Streptomyces*, but fell within the genus on the basis of overall similarity. *Nocardioides albus* was on the fringe of this cluster, but was not included in it by the S_J coeffi-

TABLE I. Composition of Cluster Groups Defined at 70.1% S_{SM}

Group No.	*No. of strains*	*No. clusters defined at 77.5% S_{SM}* [a]	*Major components*	*No. of* Streptomyces *strains*
A	*340*	*48*	Streptomyces *spp.* Actinopycnidium caeruleum Actinosporangium violaceum Chainia *spp.* Elytrosporangium *spp.* Microellobosporia *spp.* Nocardiopsis dassonvillei Saccharopolyspora hirsuta Nocardioides albus	*218*
B	*8*	*2*	Streptomyces rimosus	*8*
C	*16*	*8*	Streptomyces *spp.*	*16*
D	*2*	*2*	Microellobosporia flavea Streptomyces massasporeus	*1*
E	*25*	*8*	Actinomadura *spp.* Nocardia mediterranea Streptomyces *spp.*	*8*
F	*60*	*19*	Kitasatoa *spp.* Streptoverticillium *spp.* Streptomyces lavendulae	*32*
G	*3*	*2*	Streptomyces fradiae	*3*
H	*2*	*2*	Streptomyces *spp.*	*2*
I	*3*	*2*	Nocardia *spp.*	*1*
J	*15*	*7*	Actinomadura pelletieri Intrasporangium calvum *Acidophilic* Streptomyces *spp.*	*11*

[a] *Includes single member clusters.*

(Reproduced from Williams et al., J. Gen. Microbiol. 129:*1743 (1983) by copyright permission of the Society for General Microbiology.)*

cient analysis; this also applied to *Saccharopolyspora hirsuta* (wall chemotype IV). The main anomaly was *Nocardiopsis dassonvillei* (wall chemotype III) which was grouped into this cluster by both coefficients. Excluded from cluster group A were *Intrasporangium, Kitasatoa* and *Streptoverticillium*, genera with wall chemotype I; *Actinomadura* spp. and *Microtetraspora glauca* (wall chemotype III); and *Nocardia asteroides* and *N. mediterranea* (wall chemotype IV).

Some *Streptomyces* species (27%) clearly fell outside the "*Streptomyces*" cluster group A, including those in the major clusters *S. rimosus* and *S. lavendulae*. The former constituted cluster group B and the latter joined with *Kitasatoa* and *Streptoverticillium* species to form cluster group F. Thus, the overall generic status of the cluster groups is uncertain.

3. The Major Clusters. Groups defined at the 77.5% S_{SM} level consisted of 22 major clusters containing six or more strains (Table II) and 51 minor clusters. Where possible, clusters were named after the earliest validly described species which they contained. The major clusters contained 306 strains (64.4%) and the minor ones, 140 (29.5%); 28 strains (5.9%) remained unclustered.

Some details of the major clusters are given in Table II. These fell into the "*Streptomyces*" cluster group A, with the exception of *S. rimosus, S. lavendulae, Streptoverticillium griseocarneum, Actinomadura* spp., and *Nocardia mediterranea*. As four characters were used to group *Streptomyces* species in the latest edition of Bergey's "Manual" (Pridham and Tresner, 1974), the predominant states of these characters within each cluster are given (Table II). Some clusters, such as *S. albidoflavus* (which approximated the "Griseus" group of Hütter (1963) and other workers), *S. albus* and *S. violaceoniger*, were reasonably consistent in their character states. Others, such as *S. chromofuscus* and *S. diastaticus*, showed considerable variation in these characters. It is not surprising that polythetic groups defined by using 139 characters did not always show concordance with those groups constructed with four subjectively chosen characters. A detailed discussion of the relationships of these clusters to previous groupings of streptomycetes was given by Williams *et al.* (1983a). Until current studies provide information on the genetic relationships and chemotaxonomy of these clusters, it is most appropriate to regard them as species groups.

All of the major clusters and most of the minor ones defined by the S_{SM} coefficient were also defined using the S_J coefficient, although in some cases cluster composition was changed. Of the 22 major clusters, 14 remained intact and eight were split into two or more sub-clusters (Table II). This suggested that the classification system was quite robust. It was also encouraging that most of the major clusters defined in the S_{SM} analysis showed little significant overlap (taking 5% as expected overlap), especially in view of the difficulties experienced in distinguishing streptomycete taxa in earlier studies (Gyllenberg,

1970; Gyllenberg *et al.*, 1967). The value of 5% is not stringent, but is less than that of about 8.3% which corresponds to continuous variation (Sneath, 1977). The pattern of groups defined may represent some overlapping variation rather than entirely well-separated and sharply defined groups.

III. CONSTRUCTION OF A PROBABILISTIC IDENTIFICATION MATRIX

Numerical classification results in the definition of phena at selected levels of similarity. It also provides quantitative data on the test reactions within each defined group, these being expressed as the percentage of

TABLE II. Major Clusters Defined at 77.5% by S_{SM} Coefficient and at 63.0% by S_J Coefficient

Cluster name	No. of strains		Predominant characteristic features [a]			
	77.5% S_{SM}	63% S_J	Spore surface [b]	Spore chain[c]	Spore color[d]	Melanin pigment
Streptomyces albidoflavus	71	72	Sm	RF	Y-Gy	-
S. atroolivaceus	9	6	Sm	RF	W-Gy	-
S. exofoliatus	18	16 (2)[e]	Sm	RF	R-Gy	+/-
S. violaceus	8	9	Sm/Spy	RF/RA/S	V-Y	+
S. fulvissimus	9	9	Sm	RF/RA/S	R	+/-
S. rochei	26	23	Sm/Spy/Hy	S/RA	Gy	-
S. chromofuscus	9	7 (3)	Sm/Spy/Hy	RF/S	Y-W-Gy	+/-
S. albus	6	6	Sm	S	W	-
S. griseoviridis	6	6	Sm	S	R	-
S. cyaneus	38	37	Sm/Spy	S	B-R-Gy	+
S. diastaticus	20	22 (5)	Sm	RF/RA/S	W-R-Gy	+/-
S. olivaceoviridis	7	8 (2)	Sm	S	Gy	-
S. griseoruber	8	9	Sm	S	Gy	-
S. lydicus	11	10 (2)	Sm	S	Gy	-
S. violaceoniger	6	6	Rug	S	Gy	-
S. griseoflavus	6	5	Spy/Hy	RA	Gn	-
S. phaeochromogenes	6	6	Sm	RA/S	?	+/-
S. rimosus	7	7	Sm	RF/S	W-Y	-
Actinomadura *spp.*	6	6 (3)	Sm	RF/RA/S	?	-
Nocardia mediterranea	8	8	Sm	RF	?	-
Streptoverticillium griseocarneum	9	9 (3)	Sm	V	R-Y-Gy	+
Streptomyces lavendulae	12	11 (4)	Sm	RF	R	+

[a] *Features used to define species groups in Bergey's "Manual" (1974)*
[b] *Sm, smooth; Spy, spiny; Hy, Hairy; and Rug, rugose*
[c] *RF,* Rectus Flexibilis; *RA,* Retinaculum-Apertum; *and S,* Spira
[d] *Y, yellow; Gy, gray; W, white; R, red; V, violet; B, blue; and Gn, green.*
[e] *Figures in parentheses are the numbers of clusters into which a group splits with S_J coefficient. (Reproduced from Williams* et al. J. Gen. Microbiol. 129: *1743 (1983) by copyright permission of the Society for General Microbiology.)*

strains which show a positive state for each character studied. Such data are in a form which is ideal for the construction of an identification matrix (Hill, 1974; Sneath, 1978). Such a matrix contains the minimum number of selected characters required for discrimination between the groups previously defined by numerical classification. The matrix can then be used for the probabilistic identification of unknown strains. Surprisingly few numerical classifications of actinomycetes or other bacteria have been accompanied by such an identification system, although it would seem to be the logical end-product of a numerical taxonomic study. Therefore, we used the classification test data to construct an identification matrix for the major clusters defined by numerical classification (Williams *et al.* 1983b).

A. Selection of Tests

Twenty three clusters consisting of all the major clusters (Table II) and *Streptomyces fradiae*, a well-known source of antibiotics, were selected for the matrix. The characters most diagnostic for these clusters were selected from the 139 tests used in the classification matrix.

The first step in the selection procedure was the determination of the number of clusters in which each test was predominantly positive or negative, a good test showing a consistent state within as many clusters as possible and, ideally, giving a good balance between positive and negative reactions between clusters. The product of these values gives the separation index, S_i, of Gyllenberg (1963). A further selection of tests was achieved using the CHARSEP program (Sneath, 1979c) which includes five different separation indices for assessing the diagnostic value of characters, including the V. S. P. index which gives higher scores for the more useful characters. The next step was to apply the DIACHAR program (Sneath, 1980a) which selects the most diagnostic tests for each group in an identification matrix. In a well-constructed matrix, there should be several strongly diagnostic characters for each group. Results were satisfactory for the majority of clusters in the matrix. Simple inspection showed that a few characters were strongly diagnostic for one particular cluster (*e. g.*, rugose spore surface), and these also were included in the final matrix, although their overall separation values (S_i and V. S. P.) were low (Table III). The final matrix therefore consisted of 23 clusters X 41 tests, the latter covering a wide range of characters (Table III). The percentage positives for each test and cluster were stored in a computer for subsequent testing and use. Some examples of tests with a poor diagnostic value are given in Table IV.

B. Theoretical Evaluation of the Matrix

The importance of evaluating identification matrices has been stressed by Sneath and Sokal (1973) and Sneath (1978). The quality of our matrix was

TABLE III. Diagnostic Value of 41 Characters Selected for the Identification Matrix

Characters	*No. clusters in which character is predominantly:* +	-	S_i *index* [a]	*V. S. P. index* [b]
Morphology [c]				
1. Spore surface smooth	15	2	30	27.0
2. Spore surface rugose	1	22	22	3.5
3. Spore chain B.V.	2	21	42	10.5
4. Spore chain R.A.	2	18	36	16.8
5. Spore chain R.F.	3	14	42	38.3
6. Spore chain S.	6	7	42	54.9
7. Fragmentation of mycelium	1	22	22	3.6
Pigmentation				
8. Melanin	3	11	33	38.6
9. Substrate yellow-brown	14	2	28	22.1
10. Substrate red-orange	1	18	18	12.1
11. Spore mass grey	3	12	36	40.5
12. Spore mass red	2	13	26	34.2
13. Spore mass green	0	21	0	2.4
Carbon source utilization				
14. Adonitol	4	10	40	47.3
15. Cellobiose	18	2	36	19.7
16. D-fructose	15	1	15	10.1
17. Meso-inositol	9	3	27	39.0
18. Inulin	1	13	13	18.6
19. D-mannitol	17	4	68	42.1
20. Raffinose	6	2	12	39.3
21. L-rhamnose	8	2	16	43.0
22. D-xylose	14	2	28	26.5
Nitrogen source utilization				
23. α-aminobutyric acid	3	8	24	33.4
24. L-histidine	8	2	16	33.0
25. L-hydroxyproline	2	6	12	35.1
Degradation				
26. Allantoin	3	5	15	39.9
27. Arbutin	13	4	52	40.6
28. Xanthine	9	5	45	53.8
Enzyme production				
29. Lecithinase	4	13	52	48.6
30. Pectinase	3	10	30	38.5
31. H_2S *production*	13	3	39	35.7
32. NO_3 *reduction*	4	6	24	42.6
Antibiosis				
33. Aspergillus niger	3	11	33	39.7
34. Bacillus subtilis	5	4	20	44.4
35. Streptomyces murinus	6	5	30	45.8

(continued)

Table III (continued)

Antibiotic resistance				
36. Neomycin (50 μg ml $^{-1}$)	*4*	*14*	*56*	*42.0*
37. Rifampicin (50 μg ml $^{-1}$)	*8*	*4*	*32*	*43.0*
Growth				
38. 45°C	*4*	*8*	*32*	*48.1*
39. Sodium azide (0.01%)	*2*	*9*	*18*	*36.9*
40. Sodium chloride (7.0%)	*3*	*5*	*15*	*44.6*
41. Phenol (0.1%)	*8*	*6*	*48*	*55.8*

[a]*Gyllenberg, 1963*
[b]*Sneath, 1979c*
[c]*See Table II for explanation of abbreviations*
(Reproduced from Williams et al., J. Gen. Microbiol. *129:1815 (1983) by copyright permission of the Society for General Microbiology.)*

therefore assessed both theoretically and practically.

A program (OVERMAT, Sneath, 1980c) for determining overlap between groups in an identification matrix was applied to the percent positive values for characters in the matrix; it is not possible to assess overlap in a large matrix by simple inspection. If there is much overlap between groups, unknowns may not identify well to any one of them. OVERMAT determines the disjunction index (W) for each pair of groups and the corresponding nominal overlap (V_G); the significance of the determined overlap can also be assessed against a selected critical overlap value (V_o). In this case the

TABLE IV. Some Examples of Tests with Poor Diagnostic Value for the Major Clusters

Characters	*No. clusters in which character is predominantly:* +	-	*S_i index* [a]	*V. S. P. index* [b]
1. Spore surface hairy	*0*	*21*	*0*	*0.3*
2. Spore mass blue	*0*	*22*	*0*	*0.2*
3. Utilization of nitrate	*20*	*0*	*0*	*2.8*
4. Utilization of L-arginine	*19*	*0*	*0*	*1.3*
5. Utilization of D-mannose	*21*	*0*	*0*	*1.3*
6. Proteolysis	*14*	*0*	*0*	*3.0*
7. Degradation of R.N.A.	*20*	*0*	*0*	*0.9*
8. Degradation of aesculin	*20*	*0*	*0*	*0.9*
9. Resistance to cephaloridine (100 μg ml $^{-1}$)	*23*	*0*	*0*	*0.02*
10. Resistance to phenyl ethanol (0.3% w/v)	*17*	*0*	*0*	*1.1*

[a] *Gyllenberg (1963)*
[b] *Sneath (1979c)*
(Reproduced from Williams et al., J. Gen. Microbiol. 129:*1815 (1983) by copyright permission of the Society for General Microbiology.)*

chosen critical value was 5% (see OVCLUST program for cluster overlap). No significant overlap between any of the clusters in the matrix was detected.

All subsequent assessments of the matrix (both theoretical and practical) involved use of the MATIDEN program (Sneath, 1979b) to obtain the best identification scores for known or unknown strains against the groups in the matrix. Of the identification coefficients included in this program, the three used were (i) Willcox probability (Willcox *et al.*, 1973). This is the likelihood of unknown character state values against a particular group divided by the sum of the likelihoods against all groups; the closer the score is to 1.0, the better is the fit. (ii) Taxonomic distance. This expresses the distance of an unknown from the centroid of the group with which it is being compared; a low score indicates relatedness to the group and ideally is less than about 0.15. (iii) Standard error of the taxonomic distance. This assumes that the groups are in hyperspherical normal clusters. An acceptable score is less than about 2.0 to 3.0; about half the members of a taxon will have negative scores, *i. e.*, they are closer to the centroid than average.

Identification is achieved if the best scores are good *and* sufficiently better than the next best two alternatives against other groups. The output of the program also lists atypical properties of the unknown against its best group, which should be few for a good identification.

The first identification scores obtained by using the matrix were determined by the MOSTTYP program (Sneath, 1980b) which evaluates matrices by calculating the best scores which the Hypothetical Median Organism (H.M.O.) of each group could achieve. Results obtained with nine of the 23 clusters, selected to illustrate the range of response, are given in Table V. It

TABLE V. Examples of Identification Scores for Hypothetical Median Organisms (MOSTTYP Program, Sneath, 1980b).

Cluster name	*No. strains in cluster*	*Identification scores*		
		Willcox probability	*Taxonomic distance*	*Standard error of taxonomic distance*
S. lavendulae	*12*	*1.000*	*0.23*	*-2.76*
S. fulvissimus	*9*	*1.000*	*0.19*	*-3.62*
S. griseoflavus	*6*	*1.000*	*0.18*	*-3.35*
S. atroolivaceus	*9*	*0.999*	*0.20*	*-3.17*
S. albus	*6*	*1.000*	*0.14*	*-3.97*
Stv. griseocarneum	*9*	*1.000*	*0.15*	*-4.02*
S. albidoflavus	*71*	*0.999*	*0.24*	*-2.70*
S. exofoliatus	*18*	*0.999*	*0.24*	*-2.74*
S. olivaceoviridis	*7*	*0.999*	*0.19*	*-3.25*

(Reproduced from Williams et al., J. Gen. Microbiol. 129:*1815 (1983) by copyright permission of the Society for General Microbiology.)*

is clear that the matrix withstood this theoretical test. All Willcox probabilities were either 0.999 or 1.0, taxonomic distances were low, ranging from 0.14 to 0.24, and standard errors of taxonomic distance were negative.

The next step in evaluation of the matrix was to feed in the test results (obtained from the classification data) of a randomly chosen strain from each cluster. Results obtained for the nine representative clusters are given in Table VI. All strains identified to the correct cluster. Willcox probabilities were high, but those for the representative strains of the *S. exofoliatus* and *S. olivaceoviridis* clusters were lower than the values for the H.M.O. Taxonomic distances were all somewhat higher than that for the H.M.O.; standard errors had low or negative values, with that for the *S. olivaceoviridis* cluster being highest. These results were regarded as very satisfactory, since some deterioration of the scores compared with those the H.M.O. was inevitable.

C. Practical Evaluation of Matrix

The identification matrix appeared to be theoretically sound, so the next logical step was to assess it by feeding in data obtained from the independent

TABLE VI. Examples of Identification Scores for Cluster Representatives Using Classification Test Data

Cluster Name	*Cluster representative*	*Identification scores*		
		Willcox probability	*Taxonomic distance*	*Standard error of taxonomic distance*
S. lavendulae	S. lavendulae	*1.000*	*0.35*	*0.75*
S. fulvissimus	S. spectabilis	*1.000*	*0.29*	*-0.82*
S. griseoflavus	S. hirsutus	*1.000*	*0.28*	*0.04*
S. atroolivaceus	S. scabies	*0.999*	*0.29*	*-0.76*
S. albus	S. albus	*0.999*	*0.24*	*-0.06*
Stv. griseocarneum	Stv. cinnamomeum	*0.998*	*0.29*	*0.55*
S. albidoflavus	S. griseus	*0.988*	*0.30*	*-0.91*
S. exofoliatus	S. umbrinus	*0.924*	*0.35*	*-0.03*
S. olivaceoviridis	Elytrosporangium brasiliense	*0.879*	*0.34*	*1.27*

(Reproduced from Williams et al., J. Gen. Microbiol. 129: *1815 (1983) by copyright permission of the Society for General Microbiology.)*

determination of the character states of both known and unknown strains.

The character states of the same cluster representatives used in the theoretical evaluation (examples given in Table VI) were independently re-determined, and identification scores were obtained (Table VII). Generally, there was little deterioration in the scores compared with those obtained using the original classification data (Table VI), the notable exceptions being the reduced Willcox probabilities for the strains representing clusters *S. albidoflavus, S. exofoliatus* and *S. olivaceoviridis*. Such changes were clearly due to discrepancies between some of the classification test results and their re-determinations. However, overall test agreement for all tests and clusters was high; total discrepancies were 55 of 943 (5.8%). The average test variance (S_i^2) was 0.029. From the formula of Sneath and Johnson (1972), this is equivalent to a probability of error of 3% and is well below the acceptable limit of 5% for test error within the same laboratory.

Finally, the 41 characters were determined for unknown isolates from soil and water, and their identification scores calculated. The criteria for a successful identification, based on the output from the MATIDEN program, were (i) A Willcox probability greater than 0.850, with low scores for tax-

TABLE VII. Examples of Identification Scores for Cluster Representatives from the Independent Re-determination of Character States

Cluster name	*Cluster representative*	*Identification scores*		
		Willcox probability	*Taxonomic distance*	*Standard error of taxonomic distance*
S. lavendulae	S. lavendulae	*1.000*	*0.37*	*1.20*
S. fulvissimus	S. spectabilis	*1.000*	*0.27*	*-1.39*
S. griseoflavus	S. hirsutus	*0.999*	*0.31*	*0.92*
S. atroolivaceus	S. scabies	*0.997*	*0.27*	*-1.33*
S. albus	S. albus	*0.999*	*0.29*	*1.59*
Stv. griseocarneum	Stv. cinnamomeum	*0.999*	*0.29*	*0.84*
S. albidoflavus	S. griseus	*0.856*	*0.31*	*-0.67*
S. exofoliatus	S. umbrinus	*0.641*	*0.36*	*0.36*
S. olivaceoviridis	Elytrosporangium brasiliense	*0.240*	*0.35*	*1.43*

(Reproduced from Williams et al., J. Gen. Microbiol. 129:*1815 (1983) by copyright permission of the Society for General Microbiology.)*

onomic distance and its standard error; (ii) All first scores significantly better than those for the next best two alternative groups; and (iii) A small number of characters of the unknown listed as being atypical of those of the group in which it is placed.

A diversity of information is provided in the output and the user must decide, from experience, what scores are acceptable (Sneath, 1979b). Examples of scores for identified and non-identified isolates are given in Table VIII. As values for the Willcox probability decreased, those for taxonomic distance and its standard error generally increased, the latter's increase being more marked. Of 64 isolates, obtained from a variety of natural sources, 52 (81.3%) were identified. The clusters to which the isolates identified were *S. albidoflavus* (60% of isolates), *S. rochei* (32%), *S. diastaticus* (14%), *S. chromofuscus, S. griseoruber*, and *S. exofoliatus* (each 4%). Of the strains which identified, 60% did so at Willcox probability levels of 0.990 or above.

IV. CONCLUSIONS

Numerical taxonomy is of value for both classification and identification of bacteria. However, most numerical classification schemes have not been supported by probalistic identification systems, one of the few exceptions being the probability matrix for identification of slowly growing mycobacteria (Wayne *et al.*, 1980). Conversely, most probabilistic identification schemes (*e.g*, Gyllenberg *et al.*, 1975; Hill *et al.*, 1978; Lapage *et*

TABLE VIII. Examples of Identification Scores for Unknown Isolates

Isolate number	*Cluster identification*	*Identification scores*		
		Willcox probability	*Taxonomic distance*	*Standard error of taxonomic distance*
1	S. albidoflavus	*0.999*	*0.34*	*-0.04*
2	S. albidoflavus	*0.999*	*0.32*	*-0.58*
3	S. rochei	*0.992*	*0.35*	*0.27*
4	S. diastaticus	*0.986*	*0.40*	*0.93*
5	S. griseoruber	*0.913*	*0.32*	*1.90*
6	S. chromofuscus	*0.907*	*0.35*	*0.45*
7	*Not identified*	*0.840*	*0.41*	*3.35*
8	*Not identified*	*0.790*	*0.39*	*0.78*
9	*Not identified*	*0.640*	*0.41*	*1.80*
10	*Not identified*	*0.520*	*0.40*	*1.41*
11	*Not identified*	*0.480*	*0.43*	*3.54*

(Reproduced from Williams et al., J. Gen. Microbiol. 129:*1815 (1983) by copyright permission of the Society for General Microbiology.)*

al., 1973) have been constructed using data less comprehensive than that provided by numerical classification. Here, we have attempted to produce a more objective classification of streptomycetes and a probabilistic identification scheme for the major clusters.

The classification results illustrated the wide range of variation in *Streptomyces*; while 22 major clusters were defined, 35.4% of the type strains fell into minor clusters or remained unclustered at the 77.5% S_{SM} level. This should represent real variation as a sufficient number of tests were used to avoid creation of artificial discontinuities. Some clusters were homogeneous with respect to the "traditional" morphological and pigmentation characters and, thus, were comparable with species groups of earlier workers (*e.g.*, Hütter, 1967; Pridham and Tresner 1974). However, these characters were not generally cluster-specific, showing that a meaningful sub-generic classification of streptomycetes cannot be based on a few subjectively chosen characters. The results vindicate Sneath's (1970) view that numerical analysis was the only practical way of dealing with the overspeciation in this genus. It is not yet clear whether the clusters defined represent species or species groups. Further studies which use DNA:DNA and DNA:RNA pairing and chemotaxonomy and which assess capacities for genetic exchange are underway.

The identification matrix used in this work is the most comprehensive and fully tested of any published to date. Nevertheless it has its imperfections. Practically, it was not feasible to include all clusters, and a few of the major clusters included were not as distinct as they might have been. Also, the minimum number (41) of tests needed to distinguish between the clusters is quite large. This is a reflection of the variation within clusters and the necessity of having at least as many tests as taxa in a matrix (Sneath and Chater, 1978). The matrix is now in use for the identification of isolates from a variety of habitats. Initial testing with unknown strains gave identification frequencies (81%) which compare favorably with those of other matrices applied to field strains (Hill *et al.*, 1978; Lapage *et al.*, 1973; Willcox *et al.*, 1980). We applied a less stringent level of the Willcox probability than those workers, but this can be justified by the use of the additional identification data provided by the MATIDEN program and by the likelihood that our clusters are species groups.

Therefore, numerical taxonomy has provided a more objective means of dealing with the genus *Streptomyces*. The results should serve as a sound basis for further improvements in the taxonomy of this difficult, but important group of actinomycetes.

REFERENCES

Gyllenberg, H. G., *Ann. Acad. Sci. Fenn. Ser. A IV Biol. 69:*1 (1963).
Gyllenberg, H. G., *in* "The Actinomycetales" (H. Prauser, ed.), p. 101. Gustav Fischer Verlag, Jena (1970).
Gyllenberg, H. G., Woznicka, W., and Kurylowicz, W., *Ann. Sci. Fenn. Ser. A IV Biol. 114:*3 (1967).
Gyllenberg, H. G., Niemelä, T. K., and Niemi, J. S., *Postepy Hig. Med. Dosw. 29:*357 (1975).
Hill, L. R., *Int. J. Syst. Bacteriol. 24:*494 (1974).
Hill, L. R., and Silvestri, L. G., *G. Microbiol. 10:*1 (1962).
Hill, L. R., Lapage, S. P., and Bowie, I. S., *in* "Coryneform Bacteria" (I. G. Bousefield, and A. G. Callely, eds.), p. 181. Academic Press, London (1978).
Hütter, R., *G. Microbiol. 11:*191 (1963).
Hütter, R., "Systematik der Streptomyceten". Karger, Basel (1967).
Kurylowicz, W., Paszkiewicz, A., Woznicka, W., Kurzatkowski, W., and Szulga, T., *Postepy. Hig. Med. Dosw. 29:*281 (1975).
Lapage, S. P., Bascomb, S., Willcox, W. R., and Curtis, M. A., *J. Gen. Microbiol. 77:*273 (1973).
Lechevalier, M. P., and Lechevalier, H., *Int. J. Syst. Bacteriol. 20:*435 (1970).
Möller, F., *G. Microbiol. 10:*29 (1962).
Pridham, T. G., and Tresner, H. D., *in* Bergey's "Manual of Determinative Bacteriology" (R. E. Buchanan, and N. E. Gibbons, eds.), 8th ed., p. 747. Williams and Wilkins, Baltimore (1974).
Shirling, E. B., and Gottlieb, D., *Int. J. Syst. Bacteriol. 16:*313 (1966).
Shirling, E. B., and Gottlieb, D., *Int. J. Syst. Bacteriol. 17:*315 (1967).
Shirling, E. B., and Gottlieb, D., *Int. J. Syst. Bacteriol. 18:*69 (1968a).
Shirling, E. B., and Gottlieb, D., *Int. J. Syst. Bacteriol. 18:*279 (1968b).
Shirling, E. B., and Gottlieb, D., *Int. J. Syst. Bacteriol. 19:*391 (1969).
Shirling, E. B., and Gottlieb, D., *Int. J. Syst. Bacteriol. 22:*265 (1972).
Silvestri, L. G., Turri, M., Hill, L. R., and Gilardi, E., *Symp. Soc. Gen. Microbiol. 12:*333 (1962).
Skerman, V. B. D., McGowan, V., and Sneath, P. H. A., *Int. J. Syst. Bacteriol. 30:*225 (1980).
Sneath, P. H. A., *J. Gen. Microbiol. 17:*201 (1957).
Sneath, P. H. A., *in* "The Actinomycetales" (H. Prauser, ed.), p. 371. Gustav Fischer Verlag, Jena (1970).
Sneath, P. H. A., *J. Math. Geol. 9:*123 (1977).
Sneath, P. H. A., *in* "Essays in Microbiology", (J. R. Norris, and M. H. Richmond, eds.), p. 10/1. J. Wiley, Chichester (1978).
Sneath, P. H. A., *Comput. Geosci. 5:*143 (1979a).
Sneath, P. H. A., *Comput. Geosci. 5:*195 (1979b).
Sneath, P. H. A., *Comput. Geosci. 5:*349 (1979c).
Sneath, P. H. A., *Comput. Geosci. 6:*21 (1980a).
Sneath, P. H. A., *Comput. Geosci. 6:*27 (1980b).
Sneath, P. H. A., *Comput. Geosci. 6:*267 (1980c).
Sneath, P. H. A., and Johnson, R., *J. Gen. Microbiol. 72:*377 (1972).
Sneath, P. H. A., and Sokal, R. R., "Numerical Taxonomy. The Principles and Practice of Numerical Classification". W. H. Freeman, San Francisco (1973).
Sneath, P. H. A., and Chater, A. O., *in* "Essays in Plant Taxonomy" (H. E. Street, ed.), p. 79. Academic Press, London (1978).
Sokal, R. R., and Michener, C. D., *Kans. Univ. Sc. Bull. 38:*1409 (1958).
Trejo, W. H., *Trans. N. Y. Acad. Sci. 32:*989 (1970).
Wayne, L. G., Krichevsky, E. J., Love, L. L., Johnson, R., and Krichevsky, M. I., *Int. J. Syst. Bacteriol. 30:*528 (1980).
Willcox, W. B., Lapage, S. P., Bascomb, S., and Curtis, M. A., *J. Gen. Microbiol. 77:*317 (1973).
Willcox, W. R., Lapage, S. P., and Holmes, B., *Antonie van Leeuwenhoek 46:*233 (1980).
Williams, S. T., Wellington, E. M. H., Goodfellow, M., Alderson, G., Sackin, M., and Sneath, P. H. A., *Zentralbl. Bakteriol. Mikrobiol. Hyg. Abt. I. 11*, suppl.:45 (1981).
Williams, S. T., Goodfellow, M., Alderson, G., Wellington, E.M.H., Sneath, P.H.A., and Sackin, M. J., J. *Gen. Microbiol. 129:*1743 (1983a).
Williams, S. T., Goodfellow, M., Wellington, E. M. H., Vickers, J. C., Alderson, G., Sneath, P.H.A., Sackin, M.J., and Mortimer, A.M., *J. Gen. Microbiol. 129:*1815 (1983b).

A TAXONOMIC APPROACH TO SELECTIVE ISOLATION OF STREPTOMYCETES FROM SOIL

J. C. Vickers
S. T. Williams

Department of Botany
University of Liverpool
Liverpool, U. K.

G. W. Ross

Glaxo Group Research Ltd.
Greenford, Middlesex, U. K.

I. INTRODUCTION

The discovery of the antibiotic actinomycin by Waksman and Woodruff (1940) stimulated interest in the isolation and screening of actinomycetes for the production of novel secondary metabolites. The vast majority of antibiotics subsequently discovered are produced by members of the genus *Streptomyces* (Kurylowicz, 1976) and it is likely that the rarer species of this group contain as yet undiscovered isolates and metabolites. However, the search for "novelty" is becoming increasingly difficult and new developments in both isolation and screening methods are necessary to exploit the streptomycetes to their full potential.

The selective isolation methods employed at present are rather subjective. Media have often been developed on an empirical basis without regard to the nutritional requirements and growth responses of streptomycetes (Williams and Wellington, 1982). Recent numerical taxonomic studies of 7500 streptomycetes and related organisms (Williams *et al.*, 1983a,b) have provided extensive data on physiological and growth requirements of streptomycetes belonging to 23 major species-groups. These results have been summarized by

ISBN 0-12-528620-1

Williams *et al.* (elsewhere in this volume). This information also has provided a more logical basis for the formulation of new isolation media for selected species-groups.

Lack of objective methods for qualitative assessment of streptomycete isolates also has been a problem for many years. We tried to solve this by using a computer-assisted identification system for the major streptomycete groups (Williams *et al.*, 1983b), based on data obtained from a numerical classification study (Williams *et al.*, 1983a).

The results of these studies have made it possible to prepare selective isolation media based on sound objective criteria and to accurately identify the strains so isolated. The principles and preliminary results of this approach are outlined here.

II. COMPUTER-ASSISTED IDENTIFICATION OF STREPTOMYCETES

The computer identification system was developed for the 23 major cluster groups of streptomycetes that had been defined by numerical classification (Williams *et al.*, 1983a). A computer matrix containing the positive frequencies of 41 selected identification tests against the major clusters (Williams *et al.*, 1983b) was constructed. Identification of an unknown which had been scored for the presence or absence of properties was made by using both this matrix and a computer program MATIDEN (Sneath, 1979). The properties of the unknown were entered into a computer and were compared with each group, in turn, of the identification matrix. The program calculated the identification scores of the unknown against each group and the best identifica-

TABLE I. Summary of Identification Scores Obtained for Unknown Isolates from Soil and Water

Willcox probability level [a]	*Number of isolates*
0.999	16
0.998	6
0.997	1
0.996	3
0.990 - 0.995	5
0.970 - 0.989	4
0.950 - 0.969	3
0.930 - 0.949	3
0.910 - 0.929	4
0.850 - 0.909	7
<0.850 (unidentified)	12

[a] *Willcox* et al. *(1973).*

tions were printed. The identification scores consisted of three coefficients: (1) Willcox probability (Willcox *et al.*, 1973), (2) taxonomic distance, and (3) standard error of this distance. Good scores were indicated by high values of coefficient (1) and low values of coefficients (2) and (3), (Sneath, 1979). Lapage *et al.* (1973), in a study of the probabilistic identification of Gram-negative, aerobic, rod-shaped bacteria, showed that a Willcox probability $\geqq 0.999$ was the most favorable level for accurate identification. However, this level is too stringent for identification of a group of organisms as large and heterogeneous as the streptomycetes. We therefore adopted a Willcox probability of 0.850 as the minimum score for identification (Table I), with additional reference to coefficients (2) and (3) (Williams *et al.*, 1983b and elsewhere in this volume).

III. COMPUTER SELECTION OF THE MOST DIAGNOSTIC CHARACTERS FOR MAJOR STREPTOMYCETE CLUSTERS

A. Theoretical Aspects

An important result of the identification studies was the acquisition of several other computer programs which can gain access to the data matrix. A program DIACHAR (Sneath, 1980), in which the most diagnostic characters for individual groups within the matrix are selected, was used. With this program, high diagnostic scores were given by those characters which were either consistently positive or negative for species of one group when compared with those of all other groups. Since many of the identification tests used were those for nutritional or growth-affecting characters, the program highlighted those constituents which could be used in a medium to selectively isolate a particular major species-group(s) of streptomycetes.
high diagnostic scores were given by those characters which were either consistently positive or negative for species of one group when compared with those of all other groups. Since many of the identification tests used were those for nutritional or growth-affecting characters, the program highlighted those constituents which could be used in a medium to selectively isolate a particular major species-group(s) of streptomycetes.

B. Practical Application

The aim of this study was to select against strains of the frequently isolated species-group, *Streptomyces albidoflavus*. It was hoped that, by selectively reducing numbers of this group on soil plates, rarer species of streptomycetes would develop more readily.

The DIACHAR program was run with the identification matrix and a printout of the most diagnostic characters for the *S. albidoflavus* cluster was obtained (Table II). The program indicated that the addition of neomycin (50 μg

TABLE II. Use of the DIACHAR Program[a] to Determine the Most Diagnostic Character States of the Streptomyces albidoflavus *Cluster*

Character	*Character state*	*Score*
Degradation of xanthine	+	*0.501*
Spore chain spirales	-	*0.474*
Spore chain rectiflexibiles	+	*0.446*
Resistance to neomycin (50 μg ml^{-1})	-	*0.445*
Degradation of arbutin	+	*0.441*
Growth on 0.1% phenol	+	*0.426*
Lecithinase production	-	*0.421*
Growth at 45 °C	-	*0.420*
Spore mass: red	-	*0.392*
Mannitol utilization	+	*0.383*
Spore surface: smooth	+	*0.380*
Melanin production on PYIA	-	*0.361*
Growth on 7% NaCl	+	*0.357*
Raffinose utilization	-	*0.340*

[a]*Sneath, 1980.*

ml^{-1}) to medium was the character with the highest diagnostic score for selection against development of the *S. albidoflavus* group. However, by examining the data matrix (Table III), it was apparent that only three of the other 22 major species-groups grew well in the presence of neomycin at 50 μg ml^{-1}. Thus, since the specific aim was not to isolate only these three groups, it was more appropriate to use a character (*e. g.*, raffinose utilization) that gave a slightly lower diagnostic score but that allowed broader development of the other groups in the matrix. The data matrix showed that histidine was the best choice of nitrogen source to complement raffinose to reduce numbers of *S. albidoflavus* strains, while allowing growth of a variety of other streptomycete groups. Therefore, a selective medium was made, which contained the following per liter: raffinose, the major carbon source, (10 g); L-histidine, the major nitrogen source, (1 g); $MgSO_4 \cdot 7H_2O$ (0.5 g); $FeSO_4 \cdot 7H_2O$ (0.01 g); K_2HPO_4 (1.0 g); and bacto-agar (Difco) (12.0 g), adjusted to pH 7.0 to 7.4. A second selective medium consisted of starch-casein agar (Küster and Williams, 1964) containing rifampicin (50 μg ml^{-1}). The only antibiotic resistance tests included in the identification matrix were those for rifampicin and neomycin. Since the incorporation of antibiotics into isolation media had been used successfully for other actinomycete genera, we decided to test the effects of rifampicin which, at 50 μg ml^{-1}, was shown to inhibit 46% of the *S. albidoflavus* strains tested in the classification study but to allow greater development of several other major groups. As discussed previously, neomycin inhibited the growth of most of the groups in the matrix and, therefore, was not considered useful for this particular study.

TABLE III. Examples of Scores Included in the Identification Matrix (Expressed as Frequency of Positive Reactions)

Characters [b]	Species groups [a]					
	S. albidoflavus	S. cyaneus	S. chromofuscus	S. platensis	S. rochei	S. diastaticus
Raffinose utilization	*0.17*	*0.99*	*0.22*	*0.82*	*0.69*	*0.84*
Histidine utilization	*0.65*	*0.85*	*0.78*	*0.36*	*0.77*	*0.68*
Resistance to neomycin ($50\ \mu g\ ml^{-1}$)	*0.01*	*0.01*	*0.01*	*0.18*	*0.08*	*0.01*
Resistance to rifampicin ($50\ \mu g\ ml^{-1}$)	*0.54*	*9.46*	*0.33*	*0.09*	*0.89*	*0.68*

[a] *Only six of the 23 species groups studied are listed in this table.*
[b] *Only four of the 41 characters determined are listed in this table.*

The three media finally chosen for the selective isolation study were:

i) starch-casein (Küster and Williams, 1964) — the control medium;
ii) starch-casein medium + rifampicin (50 μg ml^{-1});
iii) raffinose-histidine medium.

IV. PROCEDURES FOR SELECTIVE ISOLATION

A. Preliminary Grouping

The soils chosen for this study were from a mature sand dune, a garden rosebed, and a grassland posture. For each soil, three dilution plates were prepared by using the two new selective media and the starch-casein (Küster and Williams, 1964) control medium which represented a so-called "general" isolation medium for streptomycetes.

Due to the large numbers of colonies developing on the isolation plates, it was impractical to identify all isolates. Therefore, the preliminary grouping system of Williams *et al.* (1969) was used for isolates prior to their identification by computer. With this system, the isolates from each soil were grouped on the basis of their pigmentation on oatmeal agar and peptone-iron agar slants. Isolates with *identical* color characteristics and melanin reaction were grouped together, and representatives from each group were randomly selected for identification as were any isolates which did not group.

A preliminary study showed this system to be successful in reducing repetition in the analysis of major groups present in a soil. Unknown isolates from a variety of habitats were grouped and then *all* of the isolates were put through the computer identification system. Eighty percent of the strains previously grouped together remained so when identified by computer. Prior to computer identification, this preliminary grouping system successfully defined the dominant streptomycetes and the rarer isolates isolated from a soil and, therefore, provided a means to deal with the large numbers of isolates required for a statistically sound isolation study.

B. Computer Identification of Isolates

Isolates (preliminary group representatives and ungrouped strains) to be identified were run through the 41 tests for identification (Section II). This allowed the precise analysis of the selective effects of the new media and a comparison of these effects with those of the control medium (Table IV). When these results had been examined together with the percent positive data contained in the matrix (Table III), it was possible to assess if an observed increase or decrease in numbers of a species-group on the new media, compared with that on the control, was mainly due to selection by media constituents or to

TABLE IV. Computer Identification of Isolates Obtained from Different Soils Using Three Media

Source	*Species group*	*Percentage of total isolates on:*		
		Starch-casein	*Starch-casein + rifampicin (50 μg ml^{-1})*	*Raffinose-histidine*
Sand dune	Streptomyces albidoflavus	2.9	2.0	0.0
	S. chromofuscus	1.0	0.0	0.0
	S. cyaneus	14.7	0.0	29.4
	S. diastaticus	0.0	10.8	0.0
	S. platensis	12.8	0.0	2.0
	S. rochei	3.9	1.0	4.9
	Unidentified isolates	3.9	1.0	4.9
Rosebed soil	S. albidoflavus	2.1	0.0	8.5
	S. chromofuscus	2.1	0.0	6.4
	S. cyaneus	0.0	0.0	8.5
	S. rochei	29.8	0.0	21.3
	Unidentified isolates	4.3	13.0	4.3
Grassland pasture soil	S. albidoflavus	6.6	16.6	0.0
	S. atroolivaceus	0.0	16.6	0.0
	S. chromofuscus	20.0	0.0	22.2
	S. cyaneus	6.6	0.0	11.1
	S. diastaticus	0.0	0.0	11.1
	Unidentified isolates	66.6	66.6	55.5

competition effects. For example, the sand dune soil (Table IV) showed both a reduction in numbers of *S. albidoflavus* strains and a marked increase in members of the *S. cyaneus* group on the medium containing raffinose and histidine. From the data matrix, it is clear that 100% of the *S. cyaneus* group were positive for raffinose utilization and 85% were positive for histidine utilization, whereas the *S. albidoflavus* group was only 17% and 65% positive for raffinose and histidine utilization, respectively. The selective effect of the medium on the isolation of these two groups is thus explicable. The reactions of the other groups present in the sand dune soil to the raffinose-histidine medium were less predictable. Competition effects on isolation plates seemed likely to be important, since the data matrix showed that these groups were neither significantly inhibited nor enhanced by the main constituents of the medium. The reduction in the number of strains of the group *S. platensis* was particularly marked and there seemed to be strong competition from the *S. cyaneus* group under conditions which enhanced the growth of the latter. The addition of rifampicin to starch-casein medium caused a selective reduction in numbers of all groups previously isolated; *S. albidoflavus* and *S. rochei* strains were the least affected, as might be predicted, since the data matrix showed that these two groups were generally more resistant to rifampicin at 50 μg ml^{-1}. The isolation of an additional group, *S. diastaticus*, on the rifampicin-containing medium indicated that competition had previously excluded this group from both the control and raffinose-histidine media.

The garden rosebed soil produced a high proportion of *S. rochei* strains and low numbers of other species-groups on the control medium; *S. cyaneus* species were not detected (Table IV). On the raffinose-histidine medium, *S. cyaneus* was detected and the proportions of *S. albidoflavus* and *S. chromofuscus* strains were increased, whereas *S. rochei* strains were decreased in number. Reduction of competition from *S. rochei* spp., in conjunction with the selective effects of the raffinose-histidine medium, would account for the observed differences. The proportion of *S. albidoflavus* strains from the rosebed soil was not reduced on the selective medium. This was probably due to strain differences which would be expected in members of such a large species-group (Williams *et al.* 1983a). None of the isolates obtained from the starch-casein + rifampicin selective medium could be identified to a group in the matrix. It is possible, therefore, that the isolates obtained by this isolation medium belong to rarer streptomycete groups or are 'outliers' of the major species-groups.

Although the grassland pasture soil yielded a large proportion of unidentified isolates on all three media, some major groups were detected (Table IV). It thus appeared that this soil contained a more diverse population of streptomycetes than did the other soils studied. The results of the raffinose-histidine medium on this soil were predictable. Numbers of *S. cyaneus* isolates were increased, while those of *S. albidoflavus* decreased. Strains of the *S. diasaticus* group, not detected on starch-casein, were also isolated. An additional group, *S. atroolivaceus*, was isolated on the starch-casein + rifampicin medium. Eighty-nine percent of the strains in this group were resistant to rifampicin (50 μg ml^{-1}). Strains of *S. albidoflavus*, resistant to this concentration of rifampicin, were also isolated.

V. CONCLUSIONS

The two newly devised media had clear selective effects which could generally be predicted from physiological and growth requirements provided by the numerical taxonomic study.

Competition also played a role. Elimination of one group by selection sometimes led to the isolation of increased numbers of another group.

Starch-casein is not a suitable selective isolation medium for all streptomycetes.

The principles of this approach to selective isolation have been justified in practice. There is no such thing as a "general" isolation medium for streptomycetes; a battery of selective media must be used to give a more accurate picture of the qualitative nature of populations in soil. Analysis of appropriate growth requirements of different species-groups by the DIACHAR program (Sneath, 1980) provides an objective basis for devising selective isolation media.

Further work will include the use of a program which selects more positively for particular species-groups of streptomycetes, with the aim of developing a range of selective isolation media. This should prove valuable to microbial ecologists and taxonomists, and for the isolation of both rarer species-groups and those with known industrial potential.

ACKNOWLEDGMENT

This work was supported by the Science and Engineering Research Council and Glaxo Group Research Ltd.

REFERENCES

Kurylowicz, W., "Antibiotics, a Critical Review". Polish Medical Publishers, Warsaw (1976).
Küster, E. and Williams, S. T., *Nature (Lond.) 202:*928 (1964).
Lapage, S. P., Bascomb, S., Willcox, W. R., and Curtis, M. A., *J. Gen. Microbiol. 77:*273 (1973).
Sneath, P.H.A., *Comput. Geosci. 5:*195, (1979).
Sneath, P.H.A., *Comput. Geosci. 6:*21, (1980).
Waksman, S. A., and Woodruff, H. B., *J. Bacteriol. 40:*581, (1940).
Willcox, W. B., Lapage, S. P., Bascomb, S., and Curtis, M., *J. Gen. Microbiol. 77:*317 (1973).
Williams, S. T. and Wellington, E. M. H., *in* "Methods of Soil Analysis", Vol. II, p. 969. Agronomy Monograph No. 9 (2nd ed.). American Society of Agronomy, (1982).
Williams, S. T., Davies, F. L., and Hall, D. M., 1969 A *in "The Soil Ecosystem"* J. G. Sheals, ed. p 107.
Williams, S. T., Goodfellow, M., Alderson, G., Wellington, E. M. H., Sneath, P. H. A., and Sackin, M. J., *J. Gen. Microbiol. 129:*1743 (1983a).
Williams, S. T., Goodfellow, M., Wellington, E. M. H., Vickers, J. C., Alderson, G., Sneath, P. H. A., Sackin, M. J., and Mortimer, A. M., *J. Gen. Microbiol. 129:*1815 (1983b).

FRANKIA: NEW LIGHT ON AN ACTINOMYCETE SYMBIONT

Dwight Baker[1]

Department of Biology
Middlebury College
Middlebury, Vermont, U. S. A.

Ed Seling

Museum of Comparative Zoology
Harvard University
Cambridge, Massachusetts, U. S. A.

I. INTRODUCTION

For almost a century, the symbiotic relationship between certain angiosperm plants and the actinomycete *Frankia* has been described and studied. Brunchorst (1886) was the first to describe the gall-like growths on plant roots that would later be identified as the site of a nitrogen-fixing, mutualistic symbiosis. During the past hundred years, a great deal of conflicting information has been accumulated concerning these symbioses which are now named actinorhizae. The confusion in this field arises from the fact that, until the late 1970's, no scientist could reproducibly isolate and cultivate *in vitro* the actinomycetes involved in the symbioses. Progress in that aspect has been made during the past few years and we shall summarize the new information obtained, or more appropriately, the information which has not been obtained, during the recent revival of interest in actinorhizal symbioses.

[1]*Present address: Charles F. Kettering Research Laboratory, Yellow Springs, Ohio.*

ISBN 0-12-528620-1

II. ISOLATION OF FRANKIAE

Attempts to isolate endophytes from actinorhizal plants were begun in the late 1800's and have been continued through the present. Contrary to reports in the scientific literature, no reproducible success had been achieved until Callaham, Del Tredici, and Torrey (1978) isolated an infective actinomycete from *Comptonia peregrina*. Since Baker and Torrey (1979) extensively reviewed the previous isolation research, no additional discussion will be included here.

Actinorhizal microsymbionts are particularly difficult to isolate because of their very slow growth rate and somewhat fastidious nature. Frankiae compete very poorly with other soil microorganisms in rich nutrient media. However, isolations currently are being performed by using a number of traditional and non-traditional isolation techniques, all of which are reproducible. Baker and Torrey (1979) reviewed the merits of each of the isolation procedures. More recently Lalonde, Calvert, and Pine (1981) and Burggraaf, Quispel, Tak, and Valstar (1981) suggested modifications to the serial dilution techniques and sucrose-density fractionation (Baker *et al.*, 1979). No isolation procedure has ever been shown to be more than 50% effective in isolating frankiae from a number of different genera of plant hosts. No successful isolations have ever been achieved from soil suspensions. Nevertheless, given the inherent difficulties involved (Baker, 1982), a respectable number of bacterial strains have been isolated from plant root nodules.

III. DESCRIPTION OF THE FRANKIAE

Frankia, the only mutualistically symbiotic member of the *Actinomycetales*, is currently the sole genus in the family *Frankiaceae* (Lechevalier and Lechevalier, 1979). Taxonomic characters and current phylogenetic classification of the frankiae are discussed by Lechevalier and Lechevalier (elsewhere in this volume). Definitive taxonomic classification of the frankiae will depend on future successes in isolating more broadly representative strains from natural environments.

Frankia are symbiotically associated with a diverse number of angiosperm plants belonging to 20 genera of eight plant families (Table I). This is in stark contrast to the specificity shown by the nitrogen-fixing rhizobia for the legume family. As scientists achieve a greater understanding of nodulated plants, additional actinorhizal plant genera, if not families, may be identified, while others, previously described as being actinorhizal, may be removed as was recently shown by Tiffney *et al.*, (1978). Phylogenetic speculation on the host plants (Bond and Becking, 1982) has shown that the large plant families, *Rhamnaceae* and *Rosaceae*, are good starting points for searching for new ac-

TABLE I. Actinorhizal Plant Genera. Families and Genera of Plants Reported to be Nodulated by the Actinomycete, Frankia[a]

Family	*Genera*	*Isolated endophytes*
Betulaceae	Alnus	+
Casuarinaceae	Casuarina	+ *(?)*
Coriariaceae	Coriaria	
Datiscaceae	Datisca	
Elaeagnaceae	Elaeagnus	+
	Hippophae	+
	Shepherdia	
Myricaceae	Comptonia	+
	Myrica	+
Rhamnaceae	Ceanothus	+ *(?)*
	Colletia	
	Discaria	
	Kentrothamnus	
	Trevoa	
Rosaceae	Cerococarpus	
	Chaemabatia	
	Cowania	
	Dryas	
	Purshia	+
	Rubus	

[a] *The right hand column indicates genera from which endophytes have been isolated and cultured* in vitro. *Question marks denote* Frankia *strains unable to reinfect their source host genus.*

tinorhizal genera. Little information exists on the evolution of the actinorhizal symbioses although fossil evidence of actinorhizae has been reported (Baker and Miller, 1980).

In symbiotic association with its host plant, *Frankia* causes a nodule to form on the plant root. Developmentally, the nodule is a determinate, modified lateral root (Torrey, 1978). Within this nodule, the actinomycete occupies enlarged cortical cells and is surrounded by an encapsulation material (Lalonde and Devoe, 1976) to separate it from the host cell cytoplasm. Within the cortical cells, the actinomycete actively reduces atmospheric dinitrogen to a combined form which can be transported and used by the host plant. In return, the plant provides an energy supply in the form of carbohydrates derived from photosynthesis. A direct correlation between amount of photosynthesis and the amount of bacterial nitrogen fixation has been shown conclusively (Gordon and Wheeler, 1978; Wheeler, 1971).

Nitrogen fixation by *Frankia* cells *in vitro* has been demonstrated by using both the indirect acetylene reduction assay (Tjepkema *et al.*, 1980) and the direct $^{15}N_2$ assay (Torrey *et al.*, 1981). Interestingly, nitrogen fixation *in vitro* is always associated with the presence of vesicles, specialized heavy-walled cells borne on side branches of the vegetative mycelium (Tjepkema *et al.*, 1981).

IV. CHARACTERISTICS OF ACTINORHIZAE

A. Distribution

Actinorhizal plants and therefore the microsymbiotic frankiae are distributed worldwide. Silvester (1977) describes extensively the specific distribution and the natural histories of actinorhizae genera. The common ecological features of all actinorhizae plants are that they are predominantly temperate, not tropical, are perennial, and are generally pioneer species, *i. e.*, they are the first plants to colonize a novel or disturbed ecosystem.

Casuarina (Casuarinaceae) is probably the only truly tropical actinorhizal genus although other genera may be distributed throughout the tropics in montane environments. Klemmedson (1979) has pointed out that in warm southwestern North America, actinorhizal plants are rarely found below 900 m, whereas leguminous plants are distributed widely in the lowlands. Whether this suggests that the microsymbiont likewise is restricted to temperate or montane tropical regions is not known although this seems highly unlikely.

B. Morphology

All actinorhizal plant genera are woody shrubs or trees or have a woody perennial rootstock. For this reason, the root nodules themselves are perennial and increase in size each year due to secondary growth. This is in contrast to many legume nodules which are commonly small and ephemeral.

Actinorhizal root nodules can be classified into two major morphological types (Torrey, 1978). The *Alnus*-type has a coralloid structure in which the nodule meristem remains predominantly quiescent during much of the year. The *Myrica*-type also has a basic coralloid structure, but this is usually obscured by a mass of roots growing from the nodule lobes. In this case the nodule meristem does not become quiescent and nodule roots, ageotropic determinant roots, grow out of the tips of the nodule lobes.

C. Ultrastructure

1. Vesicles. Much work has been undertaken concerning the ultrastructure of the actinomycete endophyte within the root nodule. From this work, it has been shown that the morphologies of the endophytes differ from plant genus to plant genus. Figure 1 shows the range of endophyte morphologies observed in actinorhizae. The most prominent feature of *Frankia* within host cells is the

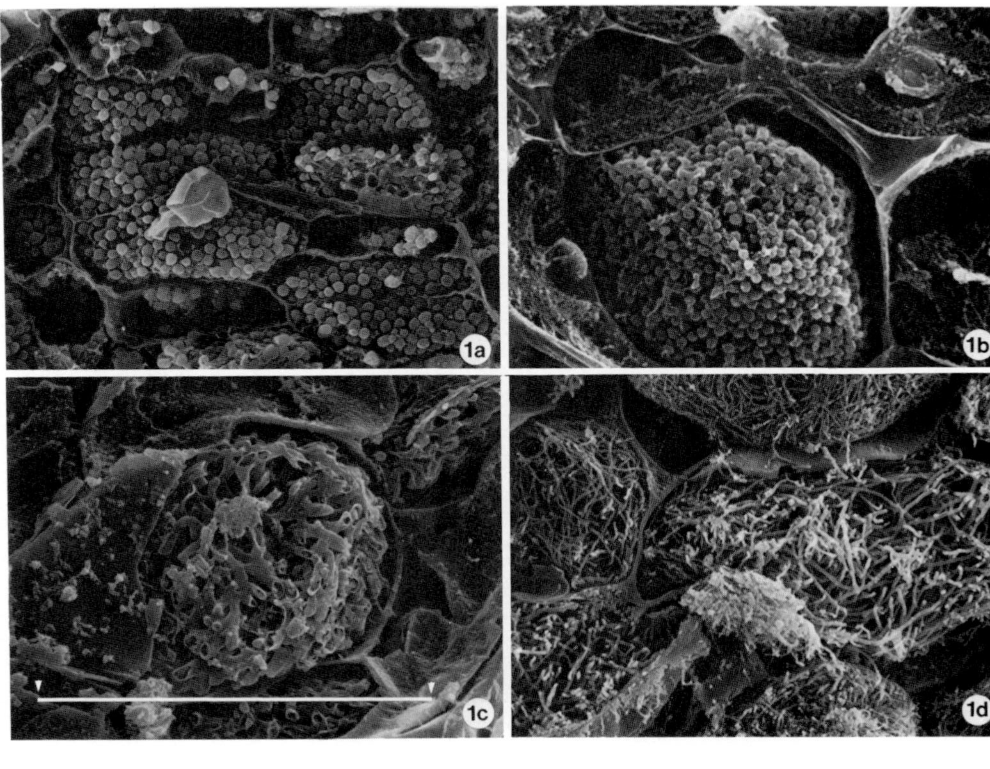

FIGURE 1. Infected host cells of actinorhizal plants: a) Alnus rubra; b) Ceanothus americanus; c)Myrica gale; d) *ineffective endophyte within* Shepherdia canadensis. *Bar = 100* μm.

presence of the endophytic vesicles which vary in both shape and size from host to host. With few exceptions, vesicles can be recognized in root nodule cells. One major exception is the absence of any vesicular structure in ineffective (*i.e.*, non-nitrogen-fixing) symbioses (Fig. 1d). In such a symbiosis, the cells are filled with endophytic hyphae, but neither structures recognizable as vesicles can be distinguished ultrastructurally, nor can any nitrogenase activity be detected (Baker *et al.*, 1980). Similar to the research performed on frankiae *in vitro*, this suggests that the vesicle is essential to the process of nitrogen fixation and is the site of active nitrogenase.

As mentioned above, vesicle shapes differ in different host plants. Even though the vesicle is a bacterial structure, its morphology within the host cell is determined by the host plant. How the host regulates vesicle morphology is unknown. Vesicle morphologies can be classified into three (possibly four) major categories (Fig. 2). The first category is the spherical vesicle (Fig. 2a) which is characteristic of such plants as *Alnus* and *Elaeagnus*. The vesicles of this type are not strictly spherical but may appear lumpy or deformed due to compaction within the host cell. The second category is the pear of flask-shaped vesicle (Fig. 2b) which is characteristic of *Ceanothus* and *Dryas*. These vesicles are generally smaller than are spherical vesicles. The third category is the club-shaped vesicle (Fig. 2c) characteristic of *Myrica* and *Coriaria*. Club-shaped vesicles are sometimes difficult to distinguish because they look much like terminal hyphae and may not be any larger than the undifferentiated mycelium. This is probably the case with *Casuarina* (Fig. 2d) in which vesicular structures have not been identified ultrastructurally or histochemically. Some researchers feel that *Casuarina* should be classified in a fourth category in which vesicles are truly absent. However, the endophyte within *Casuarina* (Fig. 2d) is distinguishable from ineffective nodule endophytes (Fig. 1d) which for some reason lack vesicles.

That vesicles, in general, are located at the periphery of the host cells (*Alnus, Ceanothus*) or surround a large central vacuole (*Coriaria, Datisca*) probably reflects the necessity of their being in proximity to a host cell membrane for transport of nutrients, N_2, and the products of the nitrogenase reaction.

Vesicles produced *in vitro* by *Frankia* are always spherical and are always borne on mycelial side branches. Structurally, the vesicle is a multilamellar, heavy-walled cell which sometimes contains random cross-walls. The function or importance of the random cross-walls is not known, but the multilamellar outer wall is important to the nitrogenase reaction. O_2 which is deleterious to nitrogenase is thought to be excluded from the vesicle by these walls (Tjepkema *et al.*, 1981) in a system analogous to that of a cyanobacterial heterocyst.

2. Sporangia. Cultures of *Frankia* characteristically produce large sporangia

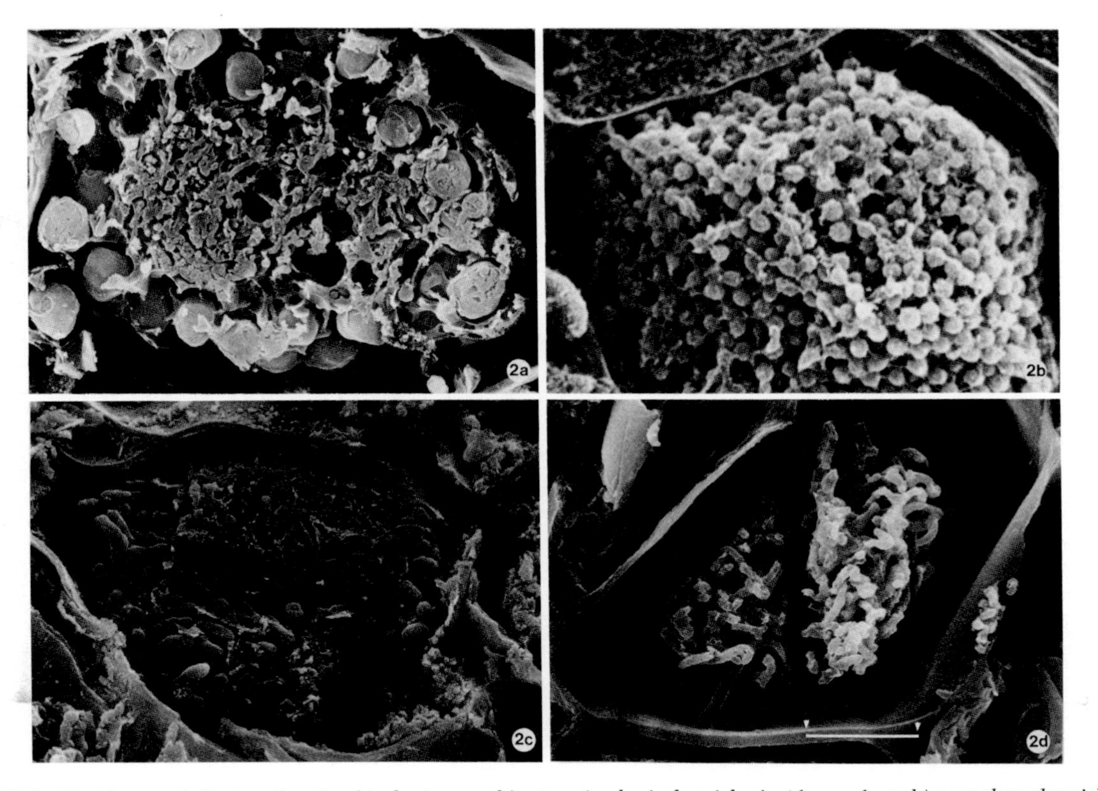

FIGURE 2. Vesicle morphologies of actinorhizal microsymbionts: a) spherical vesicles in Alnus rubra; *b) pear-shaped vesicles in* Ceanothus americanus; *c) club shaped vesicles in* Myrica gale, *d) undifferentiated (?) mycelium in* Casuarina glauca. *Bar = 10 μm.*

on the vegetative mycelium *in vitro* (Fig. 3a). No aerial mycelia have been observed. The spores which are produced are non-motile and not ornamented. Occasionally sporangia are observed within root nodules (Fig. 3b, 3c). This occurs frequently in the host plant *Alnus*; work on this genus suggests that there are differences between bacterial strains which produce many sporangia *in vivo*, Sp(+), and strains which produce few or none, Sp(-) (van Dijk, 1978; van Dijk and Merkus, 1976). No correlation between production of sporangia *in vivo* and production *in vitro* can be made since all isolated *Frankia* strains produce numerous sporangia in laboratory culture. However, some researchers feel that there are nutritional differences between the two types, with Sp(+) strains requiring a root lipid extract for growth (Burggraaf *et al.*, 1981). Sp(+) strains have been shown to be more infective than are Sp(-) strains (Houwers and Akkermans, 1981), but there is evidence to suggest that Sp(+) strains may be less efficient in fixing nitrogen (van Dijk, 1978). A good deal of confusion about this aspect still exists and, until further studies are undertaken, generalizations cannot be made. Sporangia are occasionally observed in other plant genera although they may be associated with ineffective symbioses (Baker *et al.*, 1980).

V. ECOLOGY OF SYMBIOTIC ACTINORHIZAL PLANTS

A. Nodulation Parameters

In a survey of nodulation of actinorhizal plants (Bond, 1976), it was determined that some genera, particularly those of the families *Rosaceae* and *Rhamnaceae*, are irregularly observed to be nodulated in native sites. Conversely, some genera like *Alnus* are rarely observed *not* to be nodulated. Although this may reflect some difference in the capabilities of the host plants to nodulate, it

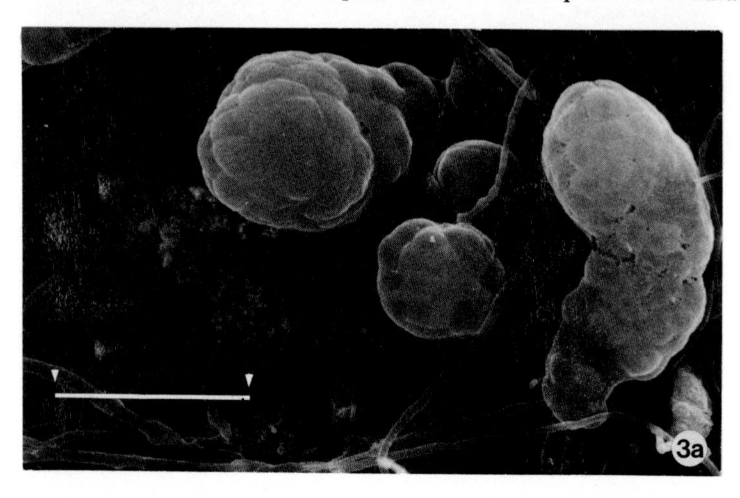

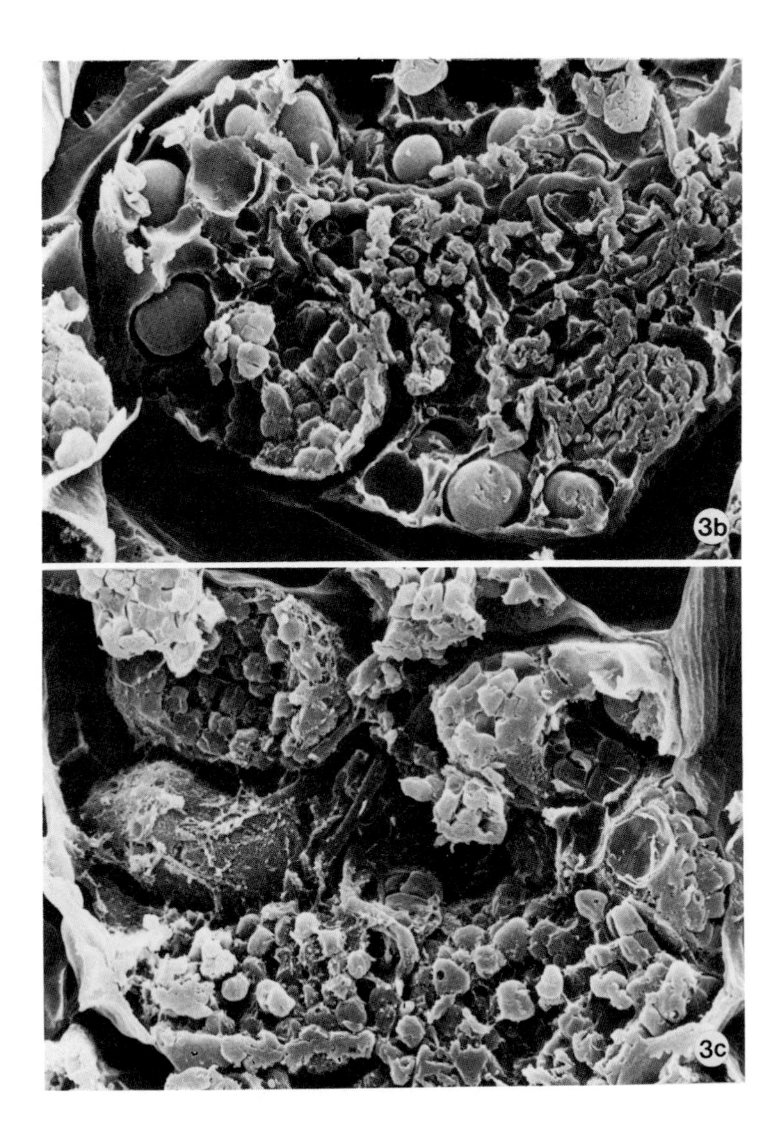

FIGURE 3. Sporangia of Frankia*: a) Sporangia of* Frankia *sp. Cp I1* in vitro*; b,c) sporangia and spores within infected cells of* Alnus incana *ssp.* rugosa; *vesicles are also visible. Bar* = *10* μm.

may be a problem of observation or of collection. In our own experience, nodules of plants of the *Rosaceae* in western North America were easily found if a systematic effort and total excavation were undertaken. This is difficult with large plants because the below-ground biomass is considerable. Plants existing in semiarid regions will bear nodules only at some considerable depth from the surface of the soil. All these problems contribute to a rather poor knowledge of nodulation *in situ*. Nonetheless, a few generalizations about nodulation in a field environment can be made.

The conditions of the soil substrate are very important to the success of nodulation and the performance of the symbioses. Moisture content is probably the most important factor affecting the nodulation and effectivity of the actinorhizal plants. This is particularly so in semiarid climates (Klemmedson, 1979); there is a strong correlation between presence of soil moisture and incidence of root nodules. The concentration of free combined nitrogen is another important factor in the occurrence of nodulation. *Alnus* is relatively insensitive to combined nitrogen in the substratum, but in other genera such as *Hippophaë*, nodulation is strongly inhibited by free combined nitrogen (Oremus, 1980).

Carpenter and Hensley (1979) demonstrated that increased calcium in the form of $CaCO_3$ improved nodulation and performance of symbiotic nitrogen-fixing plants in all cases. They suggested that this resulted from an increase in substrate pH, but further work on this subject needs to be performed to prove this.

A very important parameter in the nodulation of actinorhizal plants is the presence of the actinomycete in the soil. Unfortunately, very little information is available on this topic. van Dijk (1979) reviewed a number of factors dealing with *Frankia* distribution in the soil and determined that confusion exists because little is known of the physiological status of the bacterium in the soil. Compounding this problem is the fact that attempts to isolate frankiae directly from soil have not been successful (Baker and Torrey, 1979). An added problem is the possibility that different species exist together and that ecotypes of the actinomycetes may exist (*e. g.*, Sp(+) or Sp(-)). It has not yet been determined whether *Frankia* are able to grow asymbiotically in soil or exist as non-growing propagules until developing a new symbiotic association. By whatever means, in a natural environment, *Frankia* can remain viable for long periods of time, with some reports claiming at least 100 years (Wollum *et al.*, 1968).

B. Compatibility

Similar to the situation in the legume symbioses, *Frankia* strains belong to host compatibility groups (Baker, 1982). In general the bacteria of one compatibility group are unable to infect host of another, and *vice versa*. It has been shown, however, that within these groups there are degrees of compatibility among hosts and microsymbionts. Wheeler, McLaughlin, and Steele (1981) demonstrated differences between nitrogenase activities of the natural endophytes of *Alnus glutinosa* and *A. rubra* in Scotland when cross-inoculation trials were performed. Dillon and Baker (1982) undertook a more thorough investigation of this question by analyzing the performance of five isolated *Frankia* strains on five host species. In their study, they discovered that certain *Frankia* strains were consistently good or consistently poor symbionts, whereas others might be superior with one host genus but inferior with another. This proves that symbiotic collaboration between host and microsymbiont is extremely important to the efficient performance of the association. Additional studies of these phenomena are warranted as more pure-cultured strains become available.

C. Practical Importance

Because symbiotic actinorhizal plants contribute a valuable commodity, fixed nitrogen, to ecosystems, they are prime candidates for exploitation by mankind. Since they are evolutionarily adapted to disturbed sites and to pioneer roles, they are well suited for land reclamation projects. Actinorhizal plants have already been extensively distributed and used in North America for this purpose (Fessenden, 1979). Utilization in the forestry industry will occur because of their woody and perennial nature (DeBell, 1979; Gordon and Dawson, 1979). Use of actinorhizal plants for biomass and energy production has also been suggested (Zavitkovski *et al.*, 1979). In any event, the scientific community will continually become aware of the importance of actinomycete-nodulated plants.

VI. CONCLUSIONS

Because of recent successes in the isolation of frankiae from actinorhizal plants, knowledge of these nitrogen-fixing symbioses is increasing. The knowledge available is limited, however, because of the following deficiencies: 1) An insufficient number of frankiae representative of the many reported actinorhizal symbioses have been isolated. 2) The biology of *Frankia* as an aerobic actinomycete is impeded by the slow growth of the bacterium. 3) The natural history of *Frankia* in soil and other natural environments is not understood. 4) The dynamics of the symbiotic relationship between this actinomycete and its host plants is poorly understood. For these reasons much research must be undertaken to obtain even a limited understanding of this symbiotic actinomycete.

Fortunately the practical utilization of actinorhizal plants has already been started. The fields of forestry, land reclamation, and biomass production will increase their use of these symbiotic plants in the near future. This impetus has and will provide the momentum to promote and encourage *Frankia* research.

REFERENCES

Baker, D., *The Actinomycetes* (1982). To be published.

Baker, D., and Miller, G., *Can. J. Bot. 58:*1612 (1980).

Baker, D., and Torrey, J. G., *in* "Symbiotic Nitrogen Fixation in the Management of Temperate Forest" (J. Gordon, C. Wheeler, and D. Perry, eds.), p. 38. Oregon State Univ. Forestry Res. Lab., Corvallis (1979).

Baker, D., Torrey, J. G., and Kidd, G. H., *Nature (Lond.) 281:*76 (1979).

Baker, D., Newcomb, W., and Torrey, J. G., *Can. J. Bot. 26:*1072 (1980).

Bond, G., *in* "Symbiotic Nitrogen Fixation in Plants" (P. Nutman, ed.), p. 443. Cambridge Univ. Press, Cambridge (1976).

Bond, G., and Becking, J. H., *New Phytol. 90:*57 (1982).

Brunchorst, J., *Untersuchung. Bot. Inst. Tübigen 2:*151 (1886).

Burggraaf, A., Quispel, A., Tak, T., and Valstar, J., *Plant Soil 61:*157 (1981).

Callaham, D., Del Tredici, P., and Torrey, J. G., *Science 199:*899 (1978).

Carpenter, P. L., and Hensley, D. L., *Bot. Gaz. 140:*S76 (1979).

DeBell, D., *in* "Symbiotic Nitrogen Fixation in the Management of Temperate Forests" (J. Gordon, C. Wheeler, and D. Perry, eds.), p. 451. Oregon State Univ. Forestry Res. Lab., Corvallis (1979).

Dillon, J. T., and Baker, D., *New Phytol.* (1982). (In Press)

Fessenden, R. J., *in* "Symbiotic Nitrogen Fixation in the Management of Temperate Forests" (J. Gordon, C. Wheeler, and D. Perry, eds.), p. 403. Oregon State Univ. Forestry Res. Lab., Corvallis (1979).

Gordon, J. C., and Dawson, J. O., *Bot. Gaz. 140:*S88 (1979).

Gordon, J. C., and Wheeler, C. T., *New Phytol. 80:*179 (1978).

Houwers, A., and Akkermans, A. D. L., *Plant Soil 61:*189 (1981).

Klemmedson, J. O., *Bot. Gaz. 140:*S91 (1979).

Lalonde, M., and Devoe, I. W., *Physiol. Plant Pathol. 8:*123 (1976).

Lalonde, M., Calvert, H. E., and Pine, S., *in* "Current Perspectives in Nitrogen Fixation" (A. H. Gibson and W. E. Newton, eds.), p. 296. Australian Academy of Science, Canberra (1981).

Lechevalier, M. P., and Lechevalier, H. A., *in* "Symbiotic Nitrogen Fixation in the Management of Temperate Forests" (J. Gordon, C. Wheeler, and D. Perry, eds.), p. 111. Oregon State Univ. Forestry Res. Lab., Corvallis (1979).

Oremus, P. A. I., *Plant Soil 56:*123 (1980).

Silvester, W. B., *in* "A Treatise on Dinitrogen Fixation", Vol. III (R. Hardy and W. Silver, eds.), p. 141. John Wiley & Sons, New York (1977).

Tiffney, W. N., Jr., Benson, D. R., and Eveleigh, D. E., *Am. J. Bot. 65:*625 (1978).

Tjepkema, J. D., Ormerod, W., and Torrey, J. G., *Nature (Lond.) 287:*633 (1980).

Tjepkema, J. D., Ormerod, W., and Torrey, J. G., *Can. J. Microbiol. 27:*815 (1981).

Torrey, J. G., *Bioscience 28:*586 (1978).

Torrey, J. G., Tjepkema, J. D., Turner, G. L., Bergersen, F. J., and Gibson, A. H., *Plant Physiol. 68:*983 (1981).

van Dijk, C., *New Phytol. 81:*601 (1978).

van Dijk, C., *in* "Symbiotic Nitrogen Fixation in the Management of Temperate Forests" (J. Gordon, C. Wheeeler, and D. Perry, eds.), p. 84. Oregon State Univ. Forestry Res. Lab., Corvallis (1979).

van Dijk, C., and Merkus, E., *New Phytol. 77:*73 (1976).

Wheeler, C. T., *New Phytol. 70:*487 (1971).

Wheeler, C. T., McLaughlin, M. C., and Steele, P., *Plant Soil 61:*169 (1981).

Wollum, A. G. II, Youngberg, C. T., and Chichester, F. W., *For. Sci. 14:*114 (1968).

Zavitkovski, J., Hansen, E. A., and McNeel, H. A., *in* "Symbiotic Nitrogen Fixation in the Management of Temperate Forests" (J. Gordon, C. Wheeler, and D. Perry, eds.), p. 388. Oregon State Univ. Forestry Res. Lab., Corvallis (1979).

TAXONOMY OF *FRANKIA*[1]

Mary P. Lechevalier
Hubert A. Lechevalier

Waksman Institute of Microbiology
Rutgers, The State University of New Jersey
Piscataway, New Jersey. U. S. A.

I. INTRODUCTION

Species of *Frankia* are actinomycetes which form nitrogen-fixing nodules on the roots of some woody shrubs and trees. It has been proposed that the term "actinorhizal" be used to describe this association (Torrey and Tjepkema, 1979). Since 1978 (Callaham *et al.*, 1978), frankiae have been cultivated *in vitro*. Presently, an estimated 50 strains have been isolated from about 20 different species of plants (Baker, 1982). Grown in liquid culture, these organisms are characterized by the production of hyphae that bear sporangia in which the spores are produced by division of the hyphae in several planes. In this respect, frankiae differ from the *Actinoplanaceae* which form sporangiospores through division of ingrowing hyphae in the plane perpendicular to their long axis (Lechevalier and Holbert, 1965; Lechevalier *et al.*, 1966). Rather, frankiae resemble species of *Dermatophilus* in which masses of spores are produced by multiplanar division of hyphae (Gordon, 1964). However, unlike the latter, frankiae do not form motile spores (Lechevalier and Lechevalier, 1979). In some *Frankia* strains, in addition to sporangia, small vesicles may be routinely formed (M.P. Lechevalier, unpublished); in other strains, vesicles may be induced only under special culture conditions. It has been shown that, *in vitro* the vesicles are associated with the capacity to fix nitrogen; they are also believed to be the site of this function in the host plant (Tjepkema *et al.*, 1980).

Strains of *Frankia* grow slowly, but all may be grown on relatively simple

[1]*This work was supported in part by the U. S. Department of Agriculture Grant No. 59-2341-0-1-439-0 and by the Charles and Johanna Busch Fund.*

ISBN 0-12-528620-1

media. Like the *Dermatophilaceae*, frankiae are aerobic to microaerophilic. Since they cause the formation of nodules on the roots of susceptible plants, one might assume that soil is their natural reservoir; however, although a variety of plating techniques has been used, frankiae have not yet been detected in soil in general, nor in the rhizosphere in particular (Lechevalier *et al.*, 1982). In any case, plants may be caused to nodulate by planting them in appropriate soils (Becking, 1977). In addition, seeds of susceptible plants have been shown to be free internally of the endophyte, and non-nodulated plants may be routinely obtained following seed-coat sterilization (Knowlton *et al.*, 1980). The aim of the present paper is to review the morphological, chemical, physiological, and serological properties of frankiae and to compare these properties with those of species of *Dermatophilus* and related organisms.

II. MORPHOLOGY

We will not be concerned here with the morphology of the frankiae in the host plants, but will limit our discussion to their growth behavior in culture media. In general, frankiae exhibit the same morphological features whether grown in the natural habitat offered by the host plant or in our man-made culture media. These features are branching procaryotic hyphae bearing small (5 μm in diameter) vesicles which are usually much more numerous when frankiae are grown in the plant than in cultures and small (10 μm) to large (up to 100 μm in length) sporangia containing masses of spores measuring 1 μm to 3 μm in diameter. The mature spores have very thick walls (up to 0.1 μm) with an unusual triple-layered membrane-like outer envelope. They are heat labile at

TABLE I. Frankia *spp. Used in Taxonomic Studies*

Strain	*Host plant*	*Isolated by*	*Group*
ACN1	Alnus viridis *var.* crispa	*Lalonde (1979)*	*B*
AirI1	Alnus incana *ssp.* rugosa	*Lechevalier (1981)*	*B*
AirI2	Alnus incana *ssp.* rugosa	*Lechevalier* et al. *(1983)*	*A*
ArI3	Alnus rubra	*Berry and Torrey, 1979*	*B*
ArI5	Alnus rubra	*Baker (1982)*	*B*
AvsI2	Alnus viridis *spp.* sinuata	*Baker (1982)*	*B*
CaI1	Ceanothus americanus	*Horriere (to be published)*	*A*
CpI1	Comptonia peregrina	*Callaham* et al., *(1978)*	*B*
DII	Casuarina equisetifolia	*Dommergues (Gautier* et al., *1981)*	*A*
EuI1 (b)	Elaeagnus umbellata	*Baker* et al., *(1980)*	*A*
G2	Casuarina equisetifolia	*Dommergues (Gautier* et al., *1981)*	*A*
McI1	Myrica cerifera	*Baker (1982)*	*B*
MgI5	M. gale	*Baker (1982)*	*A*
MpI1	M. pensylvanica	*Lechevalier and Lechevalier (1979)*	*B*
PtI1	Purshia tridentata	*Baker (1982)*	*A*

100°C for 10 min. Motility has not been observed.

We currently maintain in pure culture more than 30 frankiae strains and about 15 co-isolates (presumably replicate isolates from the same plant). Some of the isolates we have studied are listed in Table I, as are their plant hosts. Of these, we have studied 14 in detail. The color of the vegetative mycelia of these various strains is quite varied: white, tan to dark brown, or light pink to red. The sporangiospores are usually the same color as the vegetative mycelium. Strain AirI2, isolated from *Alnus incana* spp. *rugosa*, is an especially striking exception in that it has on off-white to orange-tan vegetative mycelium, but forms black spores (Lechevalier *et al.*, 1982). The same authors noted that about one half of the strains of *Frankia* grown *in vitro* produce no soluble pigments, whereas other strains produced yellow, green, brown, or red pigments. The chemical nature of these pigments is not known.

III. CHEMICAL PROPERTIES

All the strains of *Frankia* analyzed thus far have a cell wall of Type III (Lechevalier and Lechevalier, 1980). This cell wall type is that most commonly encountered in actinomycetes (Lechevalier and Lechevalier, 1981) and is characterized by the presence of *meso*-diaminopimelic acid (DAP), as the only diagnostic marker. The other actinomycetes with multiplanar sporangia, namely, the members of the genera *Dermatophilus* and *Geodermatophilus*, also belong to this cell wall group.

The various genera with cell walls of Type III can be separated into two groups on the basis of the presence or absence of a sugar in the whole cell hydrolyzates of their members. One group contains 3-O-methyl-D-galactose (madurose); the other lacks this sugar. A third group of possible importance, which also contains galactose and rhamnose, has been described (Labeda *et al.*, elsewhere in this volume) . We were surprised to discover that the actinomycetic endophytes not only lacked homogeneity in their whole-cell sugar composition, but most of them had whole-cell sugar patterns not previously found among organisms with a cell wall of type III. The only frankiae falling into one of the classical patterns are strains G2, DI, and DII isolated from *Casuarina equisetifolia* by Gautier *et al.* (1981). These strains have a Type B whole-cell pattern which is characterized by the presence of madurose. In contrast, most of the other *Frankia* isolates have a Type D whole-cell sugar pattern which means that they contain xylose and arabinose as diagnostic sugars. Previously, the D pattern had always been associated with a cell wall of Type II (*meso*-DAP plus glycine as diagnostic constituents), not with type III as we found here. We do not yet understand the significance of this variation. In addition, four other *Frankia* strains could not be assigned to one of the previously recognized sugar patterns. Of these, three (EuI1b, CaI1, and MgI5) contained fucose, and one (PtI1) contained large amounts of glucose (Tables II and III).

TABLE II. Chemistry and Serology of Frankiae Group A

Culture no.	*Whole-cell sugar pattern* [a]	*Cell wall type*	*Phospholipid type*	*Serogroup*
AirI2	*Xylose (D)*	*III* [b]	*I* [c]	*II*
CaI1	*Fucose (X)* [d]	*III*	*I*	*II*
DII	*Madurose (B)*	*III*	*I*	*II*
EuI1b	*Fucose (X)*	*III*	*I*	*II*
G2	*Madurose (B)*	*III*	*I*	*II*
MgI5	*Fucose (X)*			*II*
PtI1	*Glucose (X)*	*III*	*I*	*II*

[a]*All strains contain 2-O-methyl-D-mannose.*
[b]*Contain* meso-*diaminopimelic acid, glutamic acid, alanine, glucosamine, and muramic acid.*
[c]*Contain phosphatidyl inositol mannosides (variable), phosphatidyl inositol, and diphosphatidylglycerol.*
[d]*No whole-cell sugar pattern letter assigned.*

Fucose has not been found in *meso*-DAP-containing aerobic actinomycetes. It occurs occasionally in streptomycetes (M.P. Lechevalier, unpublished), in the anaerobes *Actinomyces bovis* and *A. naeslundii*, and in plant pathogens such as *Corynebacterium poinsettiae* and *C. betae* (Cummins, 1973).

In addition to the chemotaxonomic markers discussed above, the undetermined hexose which we previously reported to be associated with cells of *Frankia* (Lechevalier and Lechevalier, 1979) was found in all strains examined. Unfortunately, a chromatographically identical sugar occurs in the hydrolysates of some members of the taxa *Actinoplanes, Ampullariella,*

TABLE III. Chemistry and Serology of Frankiae Group B

Culture no.	*Whole cell sugar pattern* [a]	*Cell wall type*	*Phospholipid type*	*Serotype*
AirI1	*D* [b]	*III* [c]	*I* [d]	*I*
ArI3	*D*	*III*	*I*	*I*
ArI4	*D*	*III*		*I*
ArI5	*D*	*III*	*I*	*I*
AvcI1	*D*	*III*	*I*	*I*
AvsI1	*D*	*III*	*I*	*I*
CpI1	*D*	*III*	*I*	*I*
CpI3	*D*	*III*	*I*	*I*
MpI1	*D*	*III*	*I*	*I*

[a]*All strains also contain 2-O-methyl-D-mannose.*
[b]*Major amounts of xylose; minor amounts of arabinose.*
[c]*Contain* meso-*diaminopimelic acid, glutamic acid, alanine, glucosamine, and muramic acid.*
[d]*Contain phosphatidyl inositol mannosides (variable), phosphatidyl inositol, and diphosphatidylglycerol.*

Micromonospora, Dactylosporangium, and *Actinomadura*; thus this sugar probably is not suitable as a marker for the genus. Lalonde (1981) reported that the hexose from *Frankia* is 2-O-methyl D-mannose. Since the unknown polar amino acid which gives a blue reaction with ninhydrin and which we reported to occur in cells of *Frankia* (Lechevalier and Lechevalier, 1979) was absent from some strains (CaI1 and AirI2), it also does not appear to be a reliable marker for the genus.

All the frankiae that have been analyzed for phospholipids have been found to have a phospholipid pattern of Type PI (Lechevalier *et al.*, 1977). This is the type found in the genus *Dermatophilus*, but not in *Geodermatophilus*. Phospholipid pattern PI is characterized by the presence of phosphatidyl inositol mannosides, phosphatidyl inositol, and either disphosphatidyl glycerol, or acyl phosphatidyl glycerols. In *Frankia*, phosphatidyl glycerols are not present. No phosphatidyl ethanolamine, phosphatidyl methylethanolamine, phosphatidyl choline, or GluNU's (unknown glucosamine-containing phospholipids) are produced. Besides *Dermatophilus* and *Frankia* among the sporangiate actinomycetes, the phospholipid pattern PI has been found only in one strain of *Spirillospora*. This pattern is also found in many nonsporate actinomycetes (*Agromyces, Bacterionema, Nocardioides, Rothia*) and in a few strains of conidiate actinomycetes belonging to the genus *Actinomadura* and its close relative *Microtetraspora*.

IV. PHYSIOLOGY

One of the most striking features of frankiae is their slow rate of growth. It takes two to 12 weeks (doubling time: 2 to 5 days) for a culture of *Frankia* to reach stationary phase; the biomasses obtained *in vitro* are extremely small. Typically, 2 to 15 mg (wet weight) per ml culture fluid may be obtained; this is perhaps 1/100th of the biomass commonly obtained from a streptomycete.

Because of this slow and meager growth, the determination of the biochemical properties of strains of *Frankia* is difficult. So far, with the cultures that we tested, we found them all to be sensitive to lysozyme under growth conditions. This lysozyme sensitivity was characteristic of the vast majority of actinomycetes having a cell wall of type III. Only three strains (EuIlb, AirI2, and PtIl) produced a protease (caseinase); none produced cellulase. Starch was hydrolyzed by all strains examined. One strain (G2) hydrolyzed starch completely, but in most cases the hydrolysis was partial, giving a dextrinoid reaction with the iodine reagent. That starch was not fully broken down by most strains was surprising, since it is the principal storage carbohydrate in plants and might be expected to be an important source of carbon for the endophyte. The intervention of the plants appears necessary here. About one third of the strains tested could reduce nitrate to nitrite. Utilization of ammonia was variable.

At this moment, we feel that *Frankia* strains can be divided into at least two groups (Table I). Until we understand them more fully, we will refer to these as groups A and B. In general, members of group A produced pigmented cells and the majority of them produced, vesicles in culture media. They had very diverse whole-cell sugar patterns. They utilized and produced acid from a variety of sugars (0.5% w/v) in a casein-hydrolysate basal medium. They grew well in media containing ammonia or amino acids as a source of nitrogen and glucose, xylose, maltose, arabinose, or sucrose as carbohydrate sources. Addition of Tween usually reduced or inhibited their growth. Most produced proteases. Group A strains tended to be more aerobic than were group B strains and could be maintained in slant culture. All were either non-infective or non-effective in the plant host. In contrast, the members of group B were unpigmented, had a whole-cell sugar pattern of Type D (major amounts of xylose and minor quantities of arabinose) and required special culture conditions to induce production of vesicles *in vitro*. Most did not utilize sugars and none produced acid in the above-mentioned basal medium. They did not produce proteases. Although only some group B members utilized ammonia as a source of nitrogen, all grew well on amino acids. In the absence of Tween, growth of this group was slower than that of group A; with Tween, growth was more rapid. Because they were strictly microaerophilic, no members of group B could be maintained on slants. In the host plant, all but one strain were infective and effective. It is possible that members of Group B were metabolically crippled in some way and, thus, may more readily entered into a symbiotic association with the host. The lack of effectivity (capability of fixing nitrogen) of members of Group A in the host plant from which they were isolated may mean that they should be considered endoparasites, not endosymbionts. It is also possible that we do not yet understand fully the conditions necessary for producing effective nodules in all cases.

V. SEROLOGY

Baker *et al.* (1981a,b) have shown that members of the genus *Frankia* may be distinguished from other actinomycetes on the basis of serology. There are at least 2 serogroups among the isolates tested to date. One serogroup (I) corresponded to our physiological group B.

VI. CONCLUSIONS

The great variation in whole-cell sugar patterns found among the various isolates of *Frankia* is unique among species of actinomycetes which belong to what appears to be an otherwise chemically and morphologically homogeneous taxon. Whether these external polysaccharides serve as recognition signals for the host plant remains to be determined. The phylogenetic implications of this variability remain speculative.

If we were only to consider cell wall composition and phospholipid pattern in conjunction with morphology, we would conclude that species of *Frankia* are related to species of *Dermatophilus*. In both cases we are dealing with organisms that are detected by our methods only in association with tissues of higher organisms, plant roots in the case of *Frankia* and mammalian skin in the case of *Dermatophilus*. In both groups of organisms, soil is suspected to be the vehicle of transmission of the parasite from one individual to another, but the soil has not been demonstrated to be the reservoir of the pathogens.

In summary, we can conclude that frankiae and dermatophili have much in common: certain aspects of morphology (mode of sporangium formation), ecology (found in living host cells; apparently do not multiply in soil), chemistry (cell wall and phospholipid composition), and physiology (tendency to microaerophily). However, they are sufficiently different (animals vs. plant cells as host, whole-cell sugars, serology (Baker *et al.*, 1981a)), that we feel that grouping them together taxonomically would be premature. Thus, we still retain a slightly modified version of our 1979 definition of the genus *Frankia*:

1) Actinomycetic, nitrogen-fixing, nodule-forming endophytes or endoparasites which have been grown in pure culture *in vitro* and which:
   (a) induce effective or ineffective nodules in a host plant and may be reisolated from within the nodules of that plant, and
   (b) produce sporangia-containing non-motile spores in submerged liquid culture and may also form vesicles.
2) Free-living actinomycetes having no known nodule-forming or nitrogen-fixing capacity, but which show the morphology described in 1(b) above.

With regard to species, Becking (1970) redefined the genus *Frankia* prior to the *in vitro* cultivation of the endophytes and proposed ten species based on host-plant compatability. Recent data using the endophytes cultivated *in vitro* have shown that his species are not valid. Any species to be proposed in the future will have to be based on more complete physiological, serological, and chemical data than are available at the moment.

REFERENCES

Baker, D., *The Actinomycetes 17:*35 (1982).
Baker, D., Lechevalier, M. P., and Dillon, J. T., *in* "Current Perspectives in Nitrogen Fixation" (A. Gibson, and W. Newton, eds.), p. 479. Australian Academy of Science, Canberra (1981a).
Baker, D., Newcomb, W., and Torrey, J. G., *Can. J. Microbiol. 26:*1072 (1980).
Baker, D., Pengelly, W. L., and Torrey, J. G., *Int. J. Syst. Bacteriol. 31:*148 (1981b).
Becking, J. H., *Int. J. Syst. Bacteriol. 20:*201 (1970).
Becking, J. H., in "Treatise of Dinitrogen Fixation" (R. W. F. Hardy, and W. S. Silver, eds.), p. 215. Sect. III. Biol., John Wiley, New York (1977)
Berry, A., and Torrey, J. G., *in* "Symbiotic Nitrogen Fixation in the Management of Temperate Forests" (J. C. Gordon, C. T. Wheeler, and D. A. Perry, eds.), p. 69. Oregon State Univ., Corvallis (1979).

Callaham, D., Del Tredici, P., and Torrey, J. G., *Science 199:*899 (1978).
Cummins, C. S., *in* "Handbook of Microbiology" (A. I. Laskin, and H. A. Lechevalier, eds.), vol. II, p. 167. CRC Press, Cleveland (1973).
Gautier, D., Diem, H. G., and Dommergues, Y., *Appl. Environ. Microbiol. 41:*306 (1981).
Gordon, M. A., *J. Bacteriol. 88:*509 (1964).
Knowlton, S., Berry A., and Torrey, J. G., *Can. J. Microbiol. 26:*971 (1980).
Lalonde, M., *in* "Symbiotic Nitrogen Fixation in the Management of Temperate Forests" (J. C. Gordon, C. T. Wheeler, and D.A. Perry, eds.), p. 480. Oregon State Univ., Corvallis (1979).
Lalonde, M., Cited in Wheeler, C. T., *in* "Current Perspectives in Nitrogen Fixation" (A. Gibson, and W. Newton, eds.), p. 254. Australian Academy of Science, Canberra (1981).
Lechevalier, H. and Holbert, P. E., *J. Bacteriol. 89:*217 (1965).
Lechevalier, H. A., and Lechevalier, M. P., *in* "The Prokaryotes" (M. P. Starr, H. Stolp, H. G. Truper, A. Balows, and H. G. Schleger, eds.), p. 1915. Springer-Verlag, Berlin (1981).
Lechevalier, H. A., Lechevalier, M. P., and Holbert, P. E., *J. Bacteriol. 92:*1228 (1966).
Lechevalier, M. P., *in "Actinomycetes"* (K. P. Schaal, and G. Pulverer, eds.), p. 159. Gustav Fischer Verlag, Stuttgart (1981).
Lechevalier, M. P., Baker, D., and Horrière, F., *Can. J. Bot. 61* (1983). (In press).
Lechevalier, M. P., deBièvre, C., and Lechevalier, H. A., *Biochem. Syst. Ecol. 5:*249 (1977).
Lechevalier, M. P., Horrière, F., and Lechevalier, H., *Dev. Ind. Microbiol. 23:*51 (1982).
Lechevalier, M. P., and Lechevalier, H. A., *in* "Symbiotic Nitrogen Fixation in the Management of Temperate Forests" (J. C. Gordon, C. T. Wheeler, and D. A. Perry, eds.), p. 111. Forest Research Laboratory, Oregon State University, Corvallis (1979).
Lechevalier, M. P., and Lechevalier, H. A., *in* "Actinomycete Taxonomy" (A. Dietz, and D. W. Thayer, eds.), SIM Special Publication No. 6, p. 225. Soc. for Industrial Microbiology, Arlington (1980).
Tjepkema, J. D., Ormerod, W., and Torrey, J. G., Nature (Lond.) *287:*633 (1980).
Torrey, J. G., and Tjepkema, J. D., *Bot. Gaz. 140*, suppl.: i (1979).

A CRITICAL EVALUATION OF *NOCARDIA* AND RELATED TAXA

M. Goodfellow
Department of Microbiology

D. E. Minnikin
Department of Organic Chemistry
The University
Newcastle upon Tyne, U. K.

I. INTRODUCTION

The family *Nocardiaceae* (Castellani and Chalmers, 1919) currently contains actinomycetes that are aerobic, Gram-positive, have *meso*-diaminopimelic acid, arabinose, and galactose in the wall (wall chemotype IV *sensu* Lechevalier and Lechevalier, 1970), and form a rudimentary-to-extensive substrate mycelium that usually fragments to give bacillary and coccoid elements (Cross and Goodfellow, 1973; Goodfellow and Minnikin, 1981a). *Nocardia* is the type genus of the family. It is becoming increasingly evident that the family is markedly heterogeneous (Goodfellow and Minnikin, 1981b) and can be divided into two aggregate groups bases on *Nocardia* and *Micropolyspora*, respectively. Actinomycetes with a wall chemotype IV can be assigned to one or other of these groups on the basis of morphology and fatty acid composition (Table I).

Strains classified in the group based on *Nocardia* contain mycolic acids, high molecular weight 3-hydroxy fatty acids with a long alkyl branch in the two position; whereas, those in the second aggregate taxon lack mycolic acids and contain major amounts of branched chain *iso* and *anteiso* acids (Goodfellow and Minnikin, 1981a,b). Several easy and reliable methods are available for the detection of mycolic acids (see Minnikin and Goodfellow, 1980). Identification to the two aggregate taxa can also be achieved using morphological properties. Thus, the mycolic acid-containing strains either lack mycelia or reproduce by fragmentation of hyphae into bacillary or coccoid elements

ISBN 0-12-528620-1

TABLE I. Morphology and Fatty Acid and DNA Base Composition of Actinomycetes with Wall Chemotype IV sensu *Lechevalier and Lechevalier (1970)*[a]

	Morphology		*Fatty Acid Profiles*					
Genus	*Substrate mycelium*	*Aerial mycelium spores*	*Mycolic*	*Straight chain*	*Unsat-urated*	*Tuberculo-stearic*	*Iso and anteiso*	*G + C (mole %)*
Caseobacter	*Rods and cocci*	-	+	+	+	+	-	*65 - 67*
Corynebacterium	*Straight, curved and pleomorphic rods*	-	+	+	+	±	-	*51 - 59*
Mycobacterium	*Rods or hyphae frag-menting to make rods and cocci*	-	+	+	+	+	-	*62 - 70*
Nocardia	*Fragmentation to rods and cocci*	*Fragmentation spores may be present*	+	+	+	+	-	*64 - 72*
Rhodococcus	*Rods or hyphae frag-menting into rods and cocci*	-	+	+	+	+	-	*63 - 73*

Actinopolyspora	*Fragmentation may occur*	*Long chains of smooth spores*	-	+	-	-	+	*64*
Micropolyspora	*Fragmentation occurs, short chains of spores produced*	*Single to short chains of smooth, lumpy or spiny spores*	-	+	-	-	+	*ND*
Pseudonocardia	*Stable mycelium with characteristic zig-zag pattern; fragmentation spores produced*	*Chains of smooth or spiny spores*	-	+	±	±	+	*79*
Saccharomonospora	*Stable mycelium*	*Single rough to warty spores*	-	+	+	-	+	*74 - 75*
Saccharopolyspora	*Fragmentation into rod-shaped elements*	*Long chains of hairy spores*	-	+	+	-	+	*77*

[a] +, *Feature present; -, Feature absent; ND, not determined. Data from Collins* et al. *(1982a,b); Goodfellow and Minnikin (1981a); Minnikin and Goodfellow (1980); Minnikin* et al. *(unpublished data); Nesterenko* et al. *(1982); Williams and Wellington (1980, 1981); Williams* et al. *(1976).*

(Locci, 1976, 1981); whereas, members of the *Micropolyspora* group show a more distinctive morphology that includes the formation of spores on aerial hyphae or on both substrate and aerial mycelium (Lechevalier *et al.*, 1961).

This review is concerned with major trends in the systematics of actinomycetes with a wall chemotype IV and more specifically with some of the recent developments in the taxonomy of the genera *Nocardia* and *Rhodococcus*. Comprehensive reviews are available on the systematics of *Corynebacterium* (Barksdale, 1970, 1981) and *Mycobacterium* (Barksdale and Kim, 1977; Goodfellow and Minnikin, 1983; Goodfellow and Wayne, 1982).

II. MYCOLIC ACID-CONTAINING TAXA

Actinomycetes assigned to this group are currently classified in the genera *Caseobacter, Corynebacterium, Mycobacterium, Nocardia*, or in the *'aurantiaca'* taxon and have many properties in common (Barksdale, 1970, 1981; Barksdale and Kim, 1977; Bradley and Bond, 1974; Goodfellow and Minnikin, 1977, 1981b; Goodfellow and Wayne, 1982; Minnikin and Goodfellow, 1980). Representative strains have also been examined using a variety of serological techniques and have been shown to be closely related (Lind and Ridell, 1976, 1983). It is also very encouraging that preliminary DNA:ribosomal RNA (rRNA) pairing (Mordarski *et al.*, 1980a, 1981a) and 16S rRNA cataloguing data (Stackebrandt and Woese, 1981) have shown that mycolic acid-containing bacteria form a recognizable phyletic line. These initial studies need to be extended to determine whether all of the mycolic acid-containing taxa should be classified in a single family, but it is now possible to describe the aggregate taxon in some detail.

In addition to having a wall chemotype IV and mycolic acids, the organisms in this aggregate group are aerobic, Gram-positive, catalase positive, and nonmotile, but they do show a wide range of morphological features. Caseobacters, corynebacteria, aurantiaca strains, some rhodococci (*R. bronchialis, R. equi, R. maris, R. rubropertinctus, R. terrae*), and most mycobacteria are amycelial; whereas, nocardiae, other rhodococci (*R. coprophilus, R. erythropolis, R. globerulus, R. luteus, R. rhodnii, R. rhodochrous, R. ruber*), and some mycobacteria (*M. farcinogenes, M. senegalense*) typically form a branched mycelium that undergoes fragmentation. The formation of aerial hyphae is restricted to nocardiae and a few mycobacteria (Portaels *et al.*, 1982). In some nocardiae, notably *N. brevicatena*, aerial hyphae may be differentiated into short chains of arthrospores (Locci, 1976; Williams *et al.*, 1976). Members of the group also contain major amounts of diphosphatidylglycerol, phosphatidylinositol, and phosphatidylinositol mannosides and have a DNA base composition within the range 51 to 73% guanine plus cytosine (G + C). They also contain major amounts of straight chain and unsaturated fatty acid (Table II).

TABLE II. Diagnostic Lipids of Mycolic Acid-containing Actinomycetes

Long chain [a]	*Mycolic acids*			*Major menaquinones* [c]	*Phospholipids* [d]	
	Ester released on pyrolysis [b]	*No. of double bonds*	*Overall size (no. of carbons)*			
S,U,T	*22:0-26:0*	*1,2*	*60 - 90*	*MK-9*(H_2)	*PE,PI,PIM*	Mycobacterium *spp.*
"	*20:0,20:1*	*1-5*	*68 - 74*	*MK-9*	*PE,PI,PIM*	'aurantiaca' *taxon*
"	*22:0,22:1*					
"	*16:1,18:1*	*0-3*	*46 - 54*	*MK-9*(H_2)	*ND* [e]	Nocardia amarae
"	*12:0-18:0*	*0-3*	*46 - 60*	*MK-8*(H_4)	*PE,PI,PIM*	Nocardia asteroides, N. brasiliensis, N. brevicatena, N. otitidis-caviarum
"		*0-3*	*34 - 52*	*MK-8*(H_2)	*PE,PI,PIM*	Rhodococcus coprophilus, R. equi R. erythropolis, R. rhodnii, R. rhodochrous, R. ruber
		1-4	*46 - 64*	*MK-9*(H_2)	*PE,PI,PIM*	R. bronchialis R. rubopertuictus R. terrae
"	*14:0,18:0*	*0,1,2*	*40 - 36*	*MK-9*(H_2)	*ND*	Caseobacter polymorphus
"	*8:0,10:0*	*0,1,2*	*22 - 32*	*MK-9*(H_2)	*PI,PIM*	Corynebacterium bovis
S,U	*14:0,18:0 (14:1-18:1)*	*0,1,2*	*26 - 38*	*MK-8,9*(H_2)	*PI,PIM*	Corynebacterium matruchotii, C. callunae, C. diphtheriae, C. glutamicum, C. kutscheri, C. lilium, C. pseudodiphtheriticum, C. pseudotuberculosis, C. renale, C. xerosis

[a] *Abbreviations: S, straight chain; U, monounsaturated; T, tuberculostearic (10-methyloctadecanoic)*
[b] *Abbreviations exemplified by 14:0, tetradecanoate; 14:1, tetradecanoate*
[c] *Abbreviations exemplified by MK-9*(H_2)*: dihydrogenated menaquinone with nine isoprene units*
[d] *Abbreviations: PE, phosphatidylethanolamine; PI, phosphatidylinositol; and PIM, phosphatidylinositol mannosides.*
[e] ND, not determined.
Data taken from Collins et al. *(1982a,b); Goodfellow and Minnikin (1983) and Minnikin and Goodfellow (1980, 1981).*

Mycolic acid-containing actinomycetes have been extensively studied using conventional numerical taxonomic techniques (Goodfellow and Minnikin, 1978, 1983; Jones, 1975); in broad-ranging surveys, corynebacteria, mycobacteria, nocardiae, rhodococci, and aurantiaca strains formed aggregate clusters considered to merit generic rank (Goodfellow and Wayne, 1982; Goodfellow *et al.*, 1982b,c). The numerical phenetic surveys have provided few good characters for separating the genera, although tests of presumptive diagnostic value have been used, after suitable reproducibility studies, to construct a computer probability matrix for the identification of slow growing mycobacteria (Wayne *et al.*, 1980). Although chemical characters have generally been found useful in assigning unknown actinomycetes to genera, the analyses of wall amino acids, sugars, and peptidoglycan structure have merely underlined the affinity between the mycolic acid-containing taxa (Goodfellow and Minnikin, 1981c, 1983; Goodfellow and Wayne, 1982). However, the presence of N-glycolylmuramic acid in the peptidoglycan of mycobacteria, nocardiae, and rhodococci provides a means of distinguishing them from corynebacteria which contain the more usual N-acetylmuramic acid (Uchida and Aida, 1977, 1979).

The rich pool of lipids occurring in mycolic acid-containing bacteria has provided the most valuable chemical markers for the recognition of *Nocardia* and related taxa. The structural discontinuities that exist between the various kinds of mycolic acids, menaquinones, and polar lipids have provided the most promising results so far (Table II) (Minnikin and Goodfellow, 1980). The structural complexity of mycobacterial mycolic acids, their high molecular weight (C_{60} to C_{90}), and characteristic components having more than two points of unsaturation in the molecule serve to distinguish mycobacteria from related strains with smaller (C_{22} to C_{64}) and more highly unsaturated mycolic acids. On pyrolysis gas chromatography, the latter release C_8 to C_{18} esters, whereas mycobacterial mycolic acids yield C_{22} to C_{26} components. The *'aurantiaca'* taxon is especially well defined by its lipid composition, as it contains fully unsaturated menaquinones with nine isoprene units (abbreviated as MK-9), and characteristic, highly unsaturated mycolic acids intermediate in size (C_{68} to C_{74}) between those of mycobacteria, nocardiae, or rhodococci (Goodfellow *et al.* 1978). With the exception of *N. amarae*, nocardiae can also be recognized on the basis of menaquinone composition (Table II).

Good characters are needed to distinguish between *Corynebacterium* and *Rhodococcus* strains which fall into separate aggregate clusters (Goodfellow *et al.* 1982a, b) and have different DNA base compositions (Table I). Representatives of these two genera are, however, difficult to delineate solely on the grounds of lipid analyses as they each contain dihydrogenated menaquinones with eight or nine isoprene units and relatively simple mycolic acids. Fatty acid and polar lipid analyses may prove to be of value in distinguishing between these taxa. Thus, corynebacteria usually contain high proportions of phosphatidylglycerol but lack phosphatidylethanolamine, whereas for rhodococci the reverse applies. Further, 10-methyloctadecanoic (tuberculostearic) acid is found in rhodococci but is absent from corynebacteria, except for *C. bovis* (Collins *et al.*, 1982b; Lechevalier *et al.*, 1977). In this context, it is notable that the proposal to transfer *C. equi* to the genus *Rhodococcus* as *R. equi* (Goodfellow and Alderson, 1977) is supported by numerical phenetic, genetic, and chemical data (Goodfellow *et al.*, 1982a,b; Mordarski *et al.*, 1980b). A strong case has also been argued to transfer *Bacterionema matruchotii* to the genus *Corynebacterium* as *C. matruchotii* comb. nov. (Goodfellow *et al.*, 1982b).

A. Nocardia *Trevisan 1889*

In light of recent developments, an improved definition of the genus can be given. Thus, nocardiae are aerobic, Gram-positive, non-motile, partially to completely acid alcohol fast actinomycetes that produce a primary mycelium that fragments into bacillary and coccoid elements which germinate to form branched hyphae 0.5 to 1.0 μm in diameter. Aerial hyphae are usually produced and may differentiate into arthrospores. The organism has an oxidative metabolism, is generally resistant to lysozyme, and can use a wide range of organic compounds as sole sources of carbon for energy and growth. Nocardiae have a wall chemotype IV, N-glycolyl muramic acid, a peptidoglycan of A1γ type, and, with the exception of *N. amarae*, contain tetrahydrogenated menaquinones with eight isoprene units as the predominant isoprenologue. They also have large proportions of straight-chain unsaturated and tuberculostearic acids, major amounts of diphosphatidylglycerol, phosphatidylethanolamine, phosphatidylinositol, and phosphatidylinositol mannosides, and mycolic acids with 46 to 60 carbons and up to three double bonds. The fatty acid esters released on pyrolysis gas chromatography of mycolic esters have 12 to 18 carbon atoms. The DNA composition is within the range 64 to 72 mole % G + C. The type species is *N. asteroides*, the type strain ATCC 19247.

The redefined genus *Nocardia* contains the well-established taxa, *N. amarae, N. asteroides, N. brasiliensis, N. farcinica* (group Kyoto-1 *sensu* Tsukamura, 1969), and *N. otitidis-caviarum* (formerly *N. caviae*), and the less well-studied *N. carnea, N. hydrocarbonoxydans, N. petroleophila, N. transvalensis* and, *N. vaccinii*. All of these species were cited on the *Approved Lists of Bacterial Names* (Skerman *et al.*, 1980). An eleventh species can be added, if the proposal to transfer *Micropolyspora brevicatena* to the genus *Nocardia* as *N. brevicatena* were accepted (Goodfellow and Pirouz, 1982). Conversely, strains currently classified as *N. aerocolonigenes, N. autotrophica* (formerly *coeliaca*), *N. mediterranea*, and *N. orientalis* should be excluded from the genus as they lack mycolic acids and have properties consistent with their inclusion in the *Mycropolyspora* aggregate taxon.

N. amarae, N. brasiliensis, N. farcinica (group Kyoto-1), and *N. otitidiscaviarum* are good taxospecies (Goodfellow, 1971; Goodfellow *et al.*, 1982c; Tsukamura, 1969), but the generic assignment of the former is not unequivocal. Thus, *N. amarae* strains form a homogeneous cluster on the edge of the *Nocardia* aggregate taxon, have dihydrogenated menaquinones with nine isoprene units as major isoprenologue, release the somewhat unusual C_{16} and C_{18} monounsaturated esters on pyrolysis of methyl mycolates, and are unable to grow in lysozyme broth (Goodfellow *et al.*, 1982c; Lechevalier and Lechevalier, 1974). It is also interesting that the type strain of *N. amarae* is not lysed by nocardiophages (Williams *et al.*, 1980). Further comparative studies are needed to determine whether this species should be retained in the genus *Nocardia*. There is no doubt about the generic position of *N. farcinica* (group Kyoto 1) (Ridell *et al.*, 1982; Tsukamura, 1969; Ridell and Goodfellow, unpublished data), but an acceptable name for this species awaits resolution of the confusion surrounding the epithet *N. farcinica* (Tsukamura, 1982).

It is widely acknowledged that *N. asteroides* is a markedly heterogeneous species (Goodfellow and Minnikin, 1978, 1981a; Orchard and Goodfellow, 1980). Strains labelled *N. asteroides* have been assigned to subgroups, but the results of the various studies are difficult to compare as they are invariably based on different strains and tests. Schaal and Reutersberg (1978) did, however, recognize two well-defined taxa in the *N. asteroides* complex, *N. asteroides* subgroup A and subgroup B. These taxa appear to merit species status, a view endorsed by preliminary DNA homology data (Mordarski *et al.*, 1977). *N. brevicatena, N. carnea, N. transvalensis*, and *N. vaccinii* should be included in any future comparative studies as they appear to be closely related to strains classified in the *N. asteroides* complex. Improved descriptions are also required for *N. hydrocarbonoxydans* and *N. petroleophila*, both of which can grow autotrophically.

The thrust of much recent work in nocardial taxonomy has been focussed on improving the classification of these organisms. This emphasis can undoubtedly be defended on the thesis that good classification is a prerequisite for accurate identification. However, most species of *Nocardia* are still difficult to identify, although several diagnostic schemes have been recommended (Goodfellow and Minnikin, 1981a; Goodfellow and Schaal, 1979; Gordon *et al.*, 1978). Some of the characters which may be of value in distinguishing between *Nocardia* species are shown in Table III. However, diagnostic tables such as these are no substitute for numerical methods of identification based upon diagnostic tests which are known to be reproducible (Wayne *et al.*, 1980).

B. Rhodococcus *(Zopf, 1981) Tsukamura 1974*

The genus *Rhodococcus* was re-introduced for actinomycetes which had been previously classified in a number of taxa including *Mycobacterium rhodochrous, Gordona, Jensenia*, and the *'rhodochrous'* complex

TABLE III. Characteristics of Nocardia *Species*

	N. amarae	N. asteroides [a]	subgroup A	subgroup B	N. brasiliensis	N. brevicatena	N. carnea	N. farcinica	N. hydrocarbonoxydans	N. otitidis-caviarum	N. petroleophila	N. transvalensis	N. vaccinii
Morphological and staining characters:													
Acid-fastness	- [b]		v	v	v	v	v	+	+	v	-	v	+
Substrate mycelium color [c]	Cr, W		Y,O	Y,O,R	C,O,R	O,R	Cr,Pe	Y,O	O,Br	W,C	W,Y,O	Cr,Pu	Cr,O,R
Decomposition of:													
Casein	-		-	-	+	-	-	-	-	-	-	-	-
Elastin	-		-	-	+	-	-	ND	-	-	-	-	-
Hypoxanthine	-		-	-	+	-	-	-	-	+	+	+	-
Testosterone	-		+	+	+	+	+	+	ND	-	ND	-	-
Tyrosine	-		-	-	+	-	-	-	-	-	-	+	-
Xanthine	-		-	-	-	-	-	-	-	+	+	v	-
Resistance to:													
Lysozyme	-		+	+	+	+	+	+	+	+	+	+	+
Rifampicin	v		+	+	+	ND	+	ND	ND	+	ND	+	v
Urease production	+		+	+	+	-	-	+	-	+	+	+	+

[a] N. asteroides *subgroups A and B can be distinguished by their sensitivity to gentamicin (10 μg/ml) and by their ability to grow on isoamyl alcohol, 2,3-butylene glycol, gluconate, 1,2-propylene glycol, or rhamnose as sole carbon sources (see Schaal, 1977)*

[b] *Symbols: +, positive; -, negative; v, variable; ND, not determined. Data taken from Goodfellow (1971); Goodfellow and Schaal (1979); Goodfellow* et al. *(1982c) Gordon* et al. *(1978); and Ridell and Goodfellow (unpublished data). Details on methods can be found in Goodfellow and Schaal (1979) and Gordon* et al. *(1978).*

[c] *B, brown; C, colorless; Cr, cream; O, orange/ pale orange; Pe, Peach; Pu, purple; R, red.*

(Goodfellow and Alderson, 1977; Tsukamura, 1974). Rhodococci can be defined as aerobic, Gram-positive actinomycetes which show considerable morphological diversity and can be acid-alcohol fast. In all cases, the life-cycle begins with the coccus or short rod stage, with differences between organisms related to the presence of a number of more or less complex morphological stages. Thus, cocci may germinate, into short rods, form branched filaments, or in the most differentiated types, produce an extensive primary mycelium. The next generation of cocci or short rods are formed by the fragmentation of the rods, branched filaments, or hyphae. Feeble aerial hyphae or aerial synnemata consisting of coalesced unbranched filaments may be formed by some strains.

Rhodococci also have an oxidative metabolism, reduce nitrate, are sensitive to lysozyme, and use a wide range of carbon compounds as sole carbon sources for energy and growth. They have a wall chemotype IV, a peptidoglycan containing N-glycolylmuramic acid and of the A1γ type, and dihydrogenated menaquinones with eight or nine isoprene units as the predominant isoprenologue. Strains also contain large proportions of straight-chain unsaturated and tuberculostearic acids, major amounts of diphosphatidylglycerol, phosphatidylethanolamine, phosphatidylinositol, and phosphatidylinositol mannosides, and mycolic acids with 32 to 66 carbons and up to four double bonds. The fatty acid esters released on pyrolysis gas chromatography of mycolic esters contain 12 to 18 carbon atoms. The DNA base composition is within the range 63 to 73 mole % G+C. The type species is *Rhodococcus rhodochrous*; the type strain, ATCC 13808.

TABLE IV. Morphological and Chemical Properties of Rhodococcus *Species*

	Life cycle [a,b]				
	coccus-rod	*coccus-rod -hyphae*	*Mycolic acids* [c] *(no. of carbons)*	*Predominant menaquinones* [d]	*G + C* [e] *mole %*
R. bronchialis	+	-	*54 - 66*	*MK-9(H$_2$)*	*63 - 65*
R. coprophilus	-	+	*38 - 48*	*MK-8(H$_2$)*	*67 - 70*
R. equi	+	-	*30 - 38*	*MK-8(H$_2$)*	*70 - 72*
R. erythropolis	-	+	*34 - 48*	*MK-8(H$_2$)*	*64 - 71*
R. globerulus	-	+	*ND*	*ND*	*63 - 67*
R. luteus	-	+	*ND*	*MK-8(H$_2$)*	*64*
R. maris	+	-	*ND*	*MK-8(H$_2$)*	*73*
R. rhodochrous	-	+	*36 - 50*	*MK-8(H$_2$)*	*67 - 70*
R. ruber	-	+	*40 — 50*	*MK-8(H$_2$)*	*69 - 73*
R. ruberpertinctus	+	-	*46 - 62*	*ND*	*67 - 69*
R. terrae	+	-	*52 - 64*	*MK-9(H$_2$)*	*64 - 69*

[a] *+, present; -, absent; ND, not determined.*
[b] *Locci* et al. *(unpublished data).*
[c] *Collins* et al. *(1982a), Minnikin and Goodfellow (1976); Minnikin* et al.. *(unpublished data).*
[d] *Minnikin and Goodfellow (1980)*
[e] *Goodfellow* et al. *(1982b,c); Mordarski* et al. *(1980b, 1981b), and Nesterenko* et al. *(1982).*

The genus *Rhodococcus* currently contains twelve species (Table IV), most of which were originally defined in numerical phenetic analyses (Goodfellow and Alderson, 1977; Goodfellow *et al.*, 1982b; Rowbotham and Cross, 1977; Tsukamura, 1974) and shown to be homogeneous in chemical and DNA:DNA pairing studies (Minnikin and Goodfellow, 1980; Mordarski *et al.*, 1980b, 1981b). The congruence between the three independent kinds of taxonomic data suggests that the classification is good, but simple and reliable microbiological tests are urgently needed for the differentiation of species of *Rhodococcus*.

III. TAXA LACKING MYCOLIC ACIDS

Much less attention has been devoted to this aggregate group which currently encompasses the genera *Actinopolyspora, Micropolyspora, Pseudonocardia, Saccharomonospora*, and *Saccharopolyspora*. It is not yet clear whether all of these taxa merit generic status, or if they collectively form a natural taxon. They do, however, have a number of properties in common. They are all aerobic, Gram-positive, non-acid-alcohol fast, non-motile, and catalase positive, but do encompass considerable morphological heterogeneity (Table

TABLE V. Lipids of Actinomycetes with Wall Chemotype IV but Lacking Mycolic Acids [a]

Long-chain fatty acids [b]	*Predominant menaquinones* [c]	*Diagnostic phospholipids* [d]	*Taxon*
S,I,A	*MK-9(H_4)*	*PC,PG*	Actinopolyspora halophila
	MK-9(H_4,H_6)	*PC,PG,PI,PME*	Micropolyspora faeni
	MK-9(H_4)	*PE,PG,PI,PIM*	"Nocardia" mediterranea
	MK-9(H_4)	*PE,PG,PI,PIM*	"Nocardia" orientalis
S,I,A(T)	*MK-8(H_4)*	*PC,PE,PI,PME(PIM)*	"Nocardia" autotrophica
S,I,A(U,T)	*MK-9(H_4)*	*PC,PE,PI,PME(PIM)*	Pseudonocardia thermophila
S,U,I,A	*MK-9(H_4)*	*PE,PG,PI,PIM*	'Nocardia' aerocolonigenes
	MK-9(H_4)	*PE,PI,PIM(APG)*	Saccharomonospora viridis
	MK-9(H_4)	*PC,PI,PME(APG) PE,PG,PIM)*	Saccharopolyspora hirsuta

[a] *Data taken from Collins* et al. *(1981); Goodfellow and Minnikin (1983); Lechevalier* et al. *(1977, 1981); Minnikin and Goodfellow (1980, 1981); and Athalye* et al. *(unpublished data).*
[b] *Abbreviations: S, straight-chain; U, mono-unsaturated; T, tuberculostearic (10-methyloctadecanoic); A,* anteiso*; I,* iso*; parenthesis indicates variable occurrence.*
[c] *Abbreviations exemplified by MK-9(H_4), tetrahydrogenated menaquinone with nine isoprene units*
[d] *Abbreviations: APG, acyl phosphatidylglycerol; PC phosphalidylcholine; PE, phosphatidylethanolamine; PG, phosphatidylglycerol; PI, phosphatidylinositol; PIM, phosphatidylinositol mannosides and PME, phosphatidylmonomethylethanolamine.*

I). Fragmentation is generally much less pronounced than in mycolic acid-containing actinomycetes and may be partly due to localized areas of autolysis (Williams *et al.*, 1976). Further, in addition to having a wall chemotype IV and a high content of branched chain *iso* and *anteiso* fatty acids, these organisms contained tetrahydrogenated menaquinones with nine isoprene units as major isoprenologue and, with a single exception, have DNA remarkably rich in G + C (Table I). They also show some variation in polar lipid composition; phosphatidylinositol is found in most strains but the occurrence of phosphatidylinositol is variable (Table V). An interesting point is the presence of phosphatidylcholine in *Actinopolyspora, Micropolyspora*, and *Saccharopolyspora*.

Micropolyspora, like most of the other genera lacking mycolic acids, is essentially morphological in concept. It was introduced to accommodate actinomycetes that carried short chains of spores on both substrate and aerial hyphae (Lechevalier *et al.*, 1981). It is, therefore, somewhat puzzling that the genus contains *M. caesia, M. internatus*, and *M. coerulea*, strains of which predominantly bear spores on both the substrate and aerial mycelium. The status of *Micropolyspora* is, however, in a state of flux as the reclassification of *M. brevicatena* as *N. brevicatena* comb. nov. (Goodfellow and Pirouz, 1982) leaves the genus without a type species and nomenclaturally invalid (Lapage *et al.*, 1975). A possible way out of this nomenclatural confusion is to conserve the name *Micropolyspora*, with *M. faeni* as the type species (McCarthy *et al.*, 1983).

In preliminary numerical phenetic studies *M. faeni, P. thermophila*, and *S. viridis* formed distinct clusters (Goodfellow and Minnikin, 1981b; Goodfellow and Pirouz, 1982). These initial surveys need to be extended to include representatives of other taxa with a wall chemotype IV, but lacking mycolic acids. The latter should include isolates from fodder, which were recovered in two numerically defined clusters (Goodfellow *et al.*, 1978), and *N. aerocolonigenes, N. autotrophica, N. mediterranea*, and *N. orientalis* which have morphological and chemical properties in common with *Micropolyspora* and related taxa (Table V). *N. aerocolonigenes* was redescribed by Gordon *et al.*, (1978), but was not included on the *Approved Lists of Bacterial Names* (Skerman *et al.*, 1980).

REFERENCES

Barksdale, L., *Bacteriol. Rev. 34:*378 (1970).

Barksdale, L, *in* "The Prokaryotes: a Handbook on Habitats, Isolation and Identification of Bacteria" (M. P. Starr, H. Stolp, H. G. Trüper, A. Balows, and H. G. Schlegel, eds.), p. 1827. Springer Verlag, Berlin (1981).

Barksdale, L, and Kim, K. S., *Bacteriol. Rev. 41:*217 (1977).

Bradley, S. G., and Bond, J. S., *Adv. Appl. Microbiol. 18:*131 (1974).

Castellani, A., and Chalmers, A. J. "Manual of Tropical Medicine", 3rd ed. Williams, Wood and Co, New York, (1917).

Collins, M. D., Goodfellow, M., and Minnikin, D. E., *J. Gen. Microbiol. 128:*129 (1982a).

Collins, M. D., Goodfellow, M., and Minnikin, D. E., *J. Gen. Microbiol. 128:*2503 (1982b).

Collins, M. D., Ross, H. N. M., Tindall, B. J., and Grant, W. D., *J. Appl. Bacteriol. 59:*559 (1981).

Cross, T., and Goodfellow, M., *in* "Actinomycetales: Characteristics and Practical Importance" (G. Sykes, and F. A. Skinner, eds.), p. 11. Academic Press, London (1973).

Goodfellow, M., *J. Gen. Microbiol. 69:*33 (1971).

Goodfellow, M., and Alderson, G., *J. Gen. Microbiol. 100:*99 (1977).

Goodfellow, M., and Minnikin, D. E., *Annu. Rev. Microbiol. 31:*159 (1977).

Goodfellow, M., and Minnikin, D. E., *Zentralbl. Bakteriol. Parasitenkd., Infektionskr. Hyg. Abt. I 6*, suppl: 43 (1978).

Goodfellow, M., and Minnikin, D. E., *in* "The Prokaryotes: a Handbook on Habitats, Isolation and Identification of Bacteria" (M. P. Starr, H. Stolp, H. G. Trüper, A. Balows, and H. G. Schlegel, eds.), p. 2016. Springer Verlag, Berlin (1981a).

Goodfellow, M., and Minnikin, D. E., *Zentralbl. Bakteriol. Mikrobiol. Hyg. Abt. I 11,* suppl. 7 (1981b).

Goodfellow, M., and Minnikin, D. E., *in* "The Prokaryotes: a Handbook on Habitats, Isolation and Identification of Bacteria" (M. P. Starr, H. Stolp, H. g. Trüper, A. Balows, and H. G. Schlegel, eds.), p. 1811. Springer-Verlag, Berlin (1981c).

Goodfellow, M., and Minnikin, D. E., *in* "The Mycobacteria: a Sourcebook" (G. P. Kubica, and L. G. Wayne, eds.), *Volume I.* Marcel Dekker, New York (1983). (In press)

Goodfellow, M., and Pirouz, T., *J. Gen. Microbiol. 128:*503 (1982).

Goodfellow, M., and Schaal, K. P., *in* "Identification Methods for Microbiologists" (F. A. Skinner, and D. W. Lovelock, eds.), p. 261. Academic Press, London (1979).

Goodfellow, M., and Wayne, L. G., *in* "The Biology of the Mycobacteria" (C. Ratledge, and J. L. Stanford, eds.), *Volume I*, p. 472. Academic Press, London (1982).

Goodfellow, M., Beckham, A. R., and Barton, M. D., *J. Appl. Bacteriol.* (1982a). (In press)

Goodfellow, M., Weaver, C. R., and Minnikin, D. E., *J. Gen. Microbiol. 128:*731 (1982b).

Goodfellow, M., Minnikin, D. E., Todd, C., Alderson, G., Minnikin, S. M., and Collins, M. D., *J. Gen. Microbiol. 128:*1283 (1982c).

Goodfellow, M., Orlean, P. A. B., Collins, M. D., Alshamaony, L., and Minnikin, D. E., *J. Gen. Microbiol. 109:*57 (1978).

Gordon, R.E., Mishra, S. K., and Barnett, D.A., *J. Gen. Microbiol. 109:*69 (1978).

Jones, D., *in* "Coryneform Bacteria" (I. J. Bousfield, and A. G. Callely, eds.), p. 13. Academic Press, London (1978).

Lapage, S. P., Sneath, P. H. A., Lessel, E. F., Skerman, V. B. D., Seeliger, H. P. R., and Clark, W. A., *in* "International Code of Nomenclature of Bacteria. 1975 Revision". American Society for Microbiology, Washington (1975).

Lechevalier, M. P., and Lechevalier, H. A., *Int. J. Syst. Bacteriol. 20:*435 (1970).

Lechevalier, M. P., and Lechevalier, H. A., *Int. J. Syst. Bacteriol. 24:*278 (1974).

Lechevalier, M. P., De Bièvre, C., and Lechevalier, H. A., *Biochem. Syst. Ecol. 5:*249 (1977).

Lechevalier, H. A., Solotorovsky, M., and McDurmont, C. I., *J. Gen. Microbiol. 26:*11 (1961).

Lechevalier, M. P., Stern, A. E., and Lechevalier, H. A., *Zentralbl. Bakteriol. Mikrobiol. Hyg. Abt. I 11*, suppl.:111 (1981).

Lind, A., and Ridell, M., *in* "The Biology of the Nocardiae" (M. Goodfellow, G. H. Brownell, and J. A. Serrano, eds.), p. 220. Academic Press, London (1976).

Lind, A., and Ridell, M., *in* "The Mycobacteria: a Sourcebook" (G. P. Kubica, and L. G. Wayne, eds.), *Volume I.* Marcel Dekker, New York (1983). (In press)

Locci, R., *in* "Actinomycetes: The Boundary Microorganisms" (T. Arai, ed.), p. 249. Toppan Co. Ltd., Tokyo (1976).

Locci, R., *Zentralbl. Bakteriol. Mikrobiol. Hyg. Abt. I 11*, suppl.:119 (1981).

McCarthy, A. J., Cross, T., Lacey, J., and Goodfellow, M., *Int. J. Syst. Bacteriol.* (1983) (In press)

Minnikin, D. E., and Goodfellow, M., *in* "The Biology of the Nocardiae" (M. Goodfellow, G. H. Brownell, and J. A. Serrano, eds.), p. 160. Academic Press, London (1976).

Minnikin, D.E., and Goodfellow, M., *in* "Microbiological Classification and Identification" (M. Goodfellow and R. G. Board, eds.), p. 189. Academic Press, London (1980).

Minnikin, D. E., and Goodfellow, M., *Zentralbl. Bakteriol. Mikrobiol. Hyg. Abt. I 11*, suppl.:99 (1981).

Mordarski, M., Schaal, K. P., Szabo, K., Pulverer, G., and Tkacz , A., *Int. J. Syst. Bacteriol. 27:*66 (1977).

Mordarski, M., Goodfellow, M., Tkacz, A., Pulverer, G., and Schaal, K. P., *J. Gen. Microbiol. 118:*313 (1980a).
Mordarski, M., Goodfellow, M., Kaszen, I., Tkacz, A., Pulverer, G., and Schaal, K. P., *Int. J. Syst. Bacteriol. 30:*521 (1980b).
Mordarski, M., Tkacz, A., Goodfellow, M., Schaal, K. P., and Pulverer, G., *Zentralbl. Bakteriol. Mikrobiol. Hyg. Abt. I 11*, suppl.:79 (1981a).
Mordarski, M., Kaszen, I., Tkacz, A., Goodfellow, M., Alderson, G., Schaal, K. P., and Pulverer, G. *Zentralbl. Bakteriol. Mikrobiol. Hyg. Abt I 11*, suppl.:25 (1981b).
Nesterenko, O. A., Nogina, T. M., Kasumova, S. A., Kvasnikov, E. I., and Batrakov, S. G., *Int. J. Syst. Bacteriol. 32:*1 (1982).
Orchard, V. A., and Goodfellow, M., *J. Gen. Microbiol. 118:*295 (1980).
Portaels, F., Goodfellow, M., Minnikin, D. E., Minnikin, S. M., and Hutchinson, I. G., *Ann. Soc. Belge Med. Trop. 61:*477 (1982).
Ridell, M., Goodfellow, M., Minnikin, D. E., Minnikin, S. M., and Hutchinson, I. G., *J. Gen. Microbiol. 128:*1299 (1982).
Rowbotham, T. J., and Cross, T., *J. Gen. Microbiol. 100:*123 (1977).
Schaal, K. P., and Reutersberg, H., *Zentralbl. Bakteriol. Parasitenkd. Infektionskr. Hyg. Abt. I 6*, suppl.:53 (1978).
Skerman, V. B. D., McGowan, V., and Sneath, P. H. A., *Int. J. Syst. Bacteriol. 30:*225 (1980).
Stackebrandt, E., and Woese, C. R., *in* "Molecular and Cellular Aspects of Microbial Evolution" (M. J., Carlile, J. F. Collins, and B. E. B. Moseley, eds.), p. 1. University Press, Cambridge (1981).
Trevisan, V., "I Generie e le Specie delle *Battieriaceae*" Zanaboni and Gabussi, Milan (1889).
Tsukamura, M., *J. Gen. Microbiol. 56:*265 (1969).
Tsukamura. M., *Jpn. J. Microbiol. 18*:37 (1974).
Tsukamura, M., *Int. J. Syst. Bacteriol. 32:*235 (1982).
Uchida, K., and Aida, K., *J. Gen. Appl. Microbiol. 23:*249 (1977).
Uchida, K., and Aida, K., *J. Gen. Appl. Microbiol. 25:*169 (1979).
Wayne, L. G., Krichevsky, E. J., Love, L. L., Johnson, R., and Krichevsky, M. I., *Int. J. Syst. Bacteriol. 30:*528 (1980).
Williams, S. T., and Wellington, E. M. H., *in* "Microbiological Classification and Identification" (M. Goodfellow, and R. G. Board, eds.), p. 139. Academic Press, London (1980).
Williams, S.T., and Wellington, E.M.H., *in* "The Prokaryotes: A Handbook on Habitats, Isolation and Identification of Bacteria", (M. P. Starr, H. Stolp, H. G. Trüper, A. Balows, and H. G. Schlegel, eds.), p. 2103. Springer-Verlag, Berlin (1981).
Williams, S. T., Wellington, E. M. H., and Tipler, L. S., *J. Gen. Microbiol. 119:*173 (1980).
Williams, S. T., Sharples, G. P., Serrano, J. A., Serrano, A. A., and Lacey, L., "The Biology of the Nocardiae" (M. Goodfellow, G. H. Brownell, and J. A. Serrano, eds.), p. 102. Academic Press, London (1976).
Zopf, W., *Ber. Dtsch. Bot. Ges. 9:*22 (1981).

NUMERICAL METHODS IN THE TAXONOMY OF SPOROACTINOMYCETES

G. Alderson

School of Studies in Medical Sciences
University of Bradford

M. Athalye

Department of Bacteriology
Institute of Dermatology
London

R. P. White

Department of Statistics
Rothamsted Experimental Station
Harpenden, U. K.

I. GENERAL INTRODUCTION

Microbiologists have only recently become aware of statistical techniques which have been used by behavioral scientists since the 1930's and which were pioneered by Hotelling (1933a,b). The application of these techniques to microbiology was first visualized by Sneath (1957b) in his classic monograph, but it has taken the microbiologist at the bench many years to realize the potential of his insight.

Computer-assisted numerical methods can handle large amounts of data and have been used to produce logical schemes representing the phenetic relationships between groups of microorganisms. Such schemes, providing they are reproducible, can then be used to identify fresh isolates.

A thorough literature survey would yield more than 500 scientific publications which have applied numerical methods to bacterial systematics (for some

ISBN 0-12-528620-1

references to this literature, see Colwell, 1970; Jones and Sackin, 1980; Sneath, 1962, 1972, 1976; Sneath and Sokal, 1973). The vast majority of these publications have concentrated on the application of hierarchical techniques of analysis to the classification of bacteria. Ordination techniques have not been taken up so enthusiastically by microbial taxonomists, but their use is slowly becoming more acceptable. When Sokal and Sneath (1963) first introduced ordination methods to microbiologists, such methods were not considered to be of value in bacterial taxonomy. However, opinions have since been revised (Sneath, 1978; Sneath and Sokal, 1973) and Marriott (1974) considered that clustering and ordination should not be considered as alternatives, but as complementary procedures. A number of recently reported numerical taxonomic studies of bacteria have successfully used ordination methods (Gutteridge *et al.*, 1979; Logan and Berkeley, 1981; Rosswall and Kvillner, 1978; Sneath, 1978; Szulga, 1978) and these studies have confirmed the usefulness of the methods which were first indicated by a few pioneering workers, notably Gyllenberg (Gyllenberg, 1970; Gyllenberg and Eklund 1967; Gyllenberg *et al.*, 1967; Quadling and Hopkins, 1967; Sundman, 1968, 1970; Sundman and Carlberg, 1967; Sundman and Gyllenberg, 1967).

The aims of this chapter are three-fold: (a) to clarify terminology and applications of ordination methods in the field of bacterial classification; (b) to review some past applications of ordination techniques to sporoactinomycete systematics; and (c) to report the results of a new study of some sporoactinomycetes utilizing ordination techniques. Since we will not deal with the mathematics of ordination, we refer the reader to papers by Gower (1966, 1969, 1975), Boyce (1969), and Marriott (1974).

II. NUMERICAL METHODS

Many microbial taxa are still described on the basis of single characteristics, or a series of single characters, so that the possession of a small complete set of such features is both necessary and sufficient for inclusion in these groups. Such artificial classifications, termed monothetic (Sneath, 1962), are notoriously unreliable as they have a low information content and cannot usually accommodate strain variation (Sneath, 1957a,b; 1962; 1972; 1978). Conversely, general purpose classifications or natural classifications (*sensu* Gilmour, 1937, 1951) are based on groups which have a high information content and which are composed of organisms that have the greatest number of shared characters. Such groupings, termed polythetic (Sneath, 1962), permit a limited number of exceptional characters in individual strains.

A reliable and relatively quick way of establishing centers of variation in a collection of organisms is to examine many strains for a large number of tests (characters). Equal weight is given to every character in the construction of polythetic groups. The eighteenth century biologist, Adanson, was an early proponent of such principles; thus, numerical taxonomy is often called Adansonian, but it was Sneath (1957a,b) who introduced the numerical taxonomic method to bacterial taxonomists. It should be noted that the relationships studied are phenetic; they are based on observed characters of microorganisms and no phylogenetic inferences are claimed.

The theoretical basis of numerical taxonomy is well documented (Goodfellow, 1977; Jones and Sackin, 1980; Sneath, 1971, 1972, 1978; Sneath and Sokal, 1973). Thus, only the broad outline of the procedures involved in undertaking a numerical phenetic study is presented here, before the relative merits of different ordination methods for taxonomy is discussed.

A. Planning a Numerical Taxonomic Study

Planning a numerical classification requires an understanding of the logical steps involved in any classification exercise. The computer is simply a tool which enables the microbiologist to perform otherwise lengthy calculations. It has been shown that the final classification achieved by numerical methods may be influenced by the selection and the number of strains and characters, as well as by the test reproducibility and the statistics employed (Austin and Colwell, 1977; Goodfellow *et al.*, 1979; Jones and Sackin, 1980; Sneath, 1972, 1974, 1978; Sneath and Johnson, 1972; Sneath and Sokal, 1973). For these reasons, it is wise to compare the groupings from a numerical classification with those from other kinds of investigations such as chemotaxonomy and immunotaxonomy in a polyphasic taxonomic study (Goodfellow and Wayne, 1982; Sneath, 1976, 1978).

It is clear from what has gone before that the choice of strains and tests will depend on a thorough examination of the relevant literature. At this point, discussions with statisticians are also essential. There are often limitations in the capacity of the central processing unit of the computer or in computer programs which will influence the numbers of organisms and of tests that can be included. The type of tests may also be restricted due to absence of certain coding facilities.

1. Microbiological Aspects

a. Strains (OTU's). Entities to be classified may be strains, species, or genera, all of which can be collectively called operational taxonomic units (OTU's). The test strains selected for a numerical phenetic classification should include in addition to many representatives of the taxa under study marker cultures of possible related species or genera and duplicate cultures to provide an internal check on test error (for details of test error, see Sneath 1978; Sneath and Johnson, 1972). Special effort should be made to use authentic type strains as well as new isolates, whenever available. A lower limit of about 150 strains has been recommended by Goodfellow (1977), but the final number should be chosen in consultation with the statistician.

b. Tests (characters). A taxonomic character is any property that may vary between OTU's and that is useful in taxonomy. Each independent character used to construct a numerical classification provides one piece of phenetic information on an OTU and may be the expression of one or more genes. OTU's are examined for a large number of properties or "bits of information", each of which is termed a unit of character (Goodfellow, 1977; Jones and Sackin, 1980; Sneath, 1957b, 1962, 1970). The number and choice of unit characters can be critical; a minimum of 50 is advised but, if possible, 100 to several hundred should be studied (Goodfellow, 1977; Jones and Sackin, 1980; Sneath, 1972, 1978). Unit characters should be taken to represent a spread of taxonomic criteria such as biochemical, chemical, morphological, and physiological properties, although an attempt should be made to randomize the selection deliberately. The tests should represent the entire genome, *i.e.*, the genotype and the phenotype (Sneath, 1978). Microbiological tests are particularly susceptible to experimental error due to difficulties in standardization of test conditions as well as to problems arising from contamination or mishandled cultures. Only "good" unit characters should therefore be used for numerical phenetic classifications, namely, ones which exhibit genetic stability and are not susceptible to minor experimental modification or subject to observational uncertainties. Test procedures should be standardized as far as possible and it is useful to repeat all tests on approximately 5% of the test strains to estimate test reproducibility (Jones and Sackin, 1980). Numerical taxonomists did not, however, invent test error, they merely have utilized methods for quantifying it and its effects.

When the test data have been obtained for all of the OTU's, an evaluation should be made of quality by making use of repeat tests and/or of results from duplicate cultures. Unreliable information may be removed from the data matrix, but in any culling exercises, a balance must be struck between the effects of poor test reproducibility on the clustering of strains (Sneath and Johnson, 1972) and information lost by deletion of data (Wilkinson and Jones, 1977).

2. Computational Aspects

a. Coding and preparation of raw data matrix. The microbiological data which is often in the form of words must be coded into a form suitable for transferring onto computer punch cards or into some other form suitable for input into a computer. Characters may be quantitative, binary, or qualitative multistate and the coding method employed largely depends on the kinds of tests used and the computer programs available (see Sneath, 1978; Sneath and Sokal, 1973). The statistician should be consulted again at this stage, before a final $n \times t$ table is produced (n OTU's by t tests).

b. Analysis of data. The origin of all numerical analyses in this $n \times t$ table which can be analyzed in one of two ways: either by forming correlations between all pairs of tests (t), or by calculating the associations between all pairs of OTU's (n). These two matrices have been referred to as the **R** and **Q** matrices, respectively. Both the **R** and the **Q** matrices can be analyzed using a variety of hierarchical and ordination techniques. Table I lists some of those currently used in microbial systematics. In bacteriology, however, the term *numerical taxonomy* is usually associated with cluster analysis of **Q** matrix data (Goodfellow, 1977; Gower, 1969; Jones and Sackin, 1980; Marriott, 1974; Sneath, 1978). This area of application has been very well reviewed (Goodfellow, 1977; Jones and Sackin, 1980; Sneath, 1971, 1972, 1978; Sneath and Sokal, 1973) and only the commonly used association coefficients and clustering methods will be dealt with here. Ordination methods are discussed in detail in Section III.

Hierarchical classifications in microbiology have usually involved calculation of a **Q** matrix giving the association between every pair of OTU's. Although a wide range of coefficients of association are available, only a few have found favor with bacterial taxonomists (Sneath and Sokal, 1973). The two most commonly used are (1) simple matching coefficient, S_{SM}, (Sokal and Michener, 1958), and (2) the coefficient of Jaccard, S_J (Sneath, 1975b), which excludes negative matches. Thus,

1. $S_{SM} = (a + d)/(a + b + c + d)$
2. $S_J = a/(a + b + c)$

where a (+ +) and d (--) are the numbers of shared positive and negative unit characters in a pair of OTU's, and b(+ -) and c(- +) are the numbers of differences. It has been found useful to use these coefficients in conjunction to test robustness of the clustering (Goodfellow *et al.*, 1978, 1979).

The coefficients, S_J and S_{SM}, use binary data only and a more recent addition to useful coefficients has been Gower's coefficient, S_G (Gower, 1971), which is a generalization of S_J and S_{SM} to allow for multistate characters.

TABLE I. Numerical Methods Applied to Bacterial Systematics

Methods	***Q** Matrix Analyses*	***R** Matrix Analyses*
Hierarchical	*Average linkage* *Furthest neighbor (complete linkage)* [a] *Minimum spanning tree /single linkage/ Wroclaw taxonomy method* [b]	*Centrifugal correlation (sequential dendrite)*
Ordination	*Principal coordinate analysis*	*Principal component/ factor analysis; canonical variate/ discriminant analysis*

[a] *Term in parenthesis is synonymous to the foregoing term.*
[b] *Terms separated by slash give very similar results in practice.*

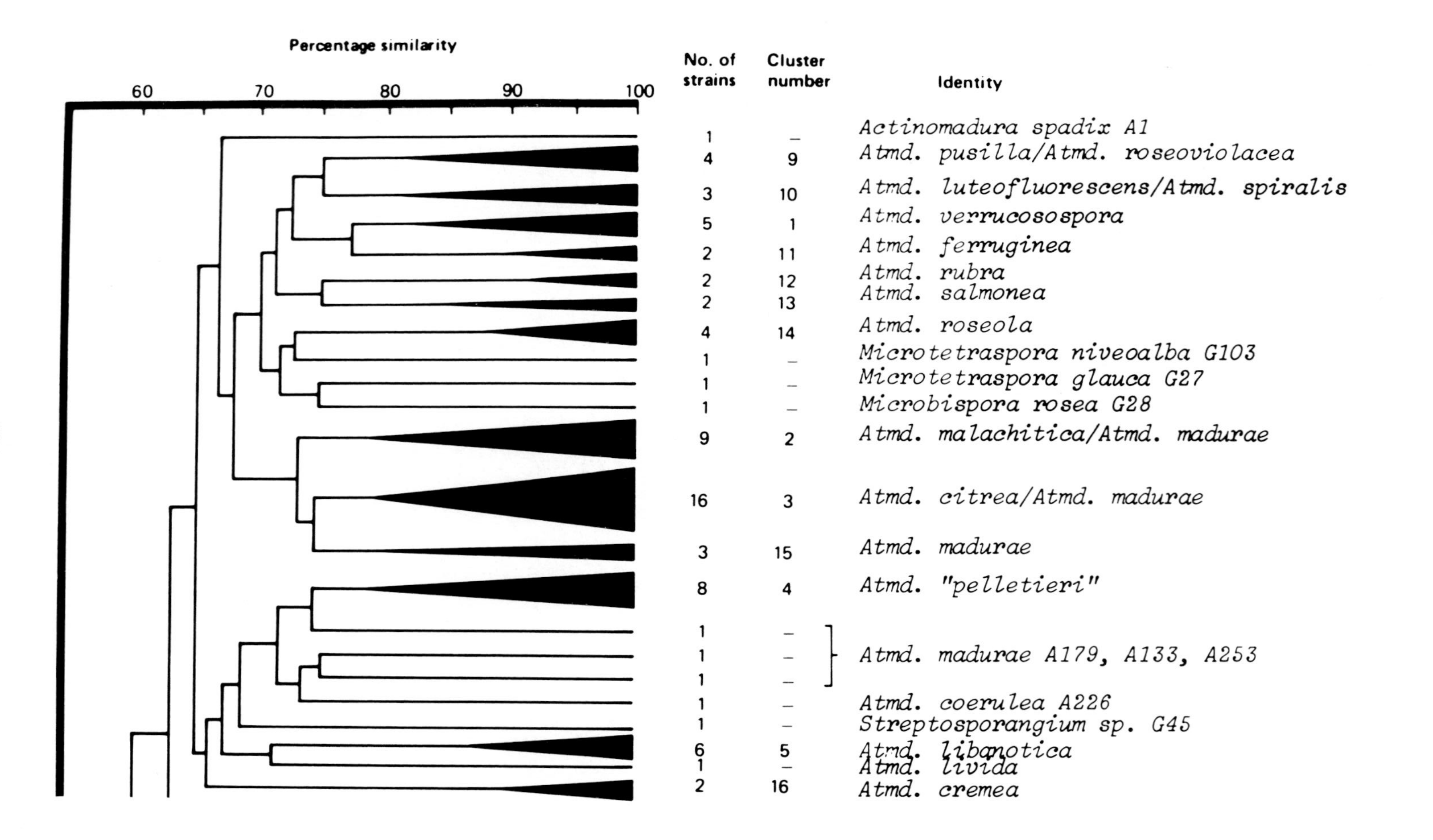

Percentage similarity
60
70
80
90
100
No. of strains
Cluster number
Identity
1 – Actinomadura spadix A1
4 9 Atmd. pusilla/Atmd. roseoviolacea
3 10 Atmd. luteofluorescens/Atmd. spiralis
5 1 Atmd. verrucosospora
2 11 Atmd. ferruginea
2 12 Atmd. rubra
2 13 Atmd. salmonea
4 14 Atmd. roseola
1 – Microtetraspora niveoalba G103
1 – Microtetraspora glauca G27
1 – Microbispora rosea G28
9 2 Atmd. malachitica/Atmd. madurae
16 3 Atmd. citrea/Atmd. madurae
3 15 Atmd. madurae
8 4 Atmd. "pelletieri"
1 –
1 – Atmd. madurae A179, A133, A253
1 –
1 – Atmd. coerulea A226
1 – Streptosporangium sp. G45
6 5 Atmd. libanotica
1 – Atmd. livida
2 16 Atmd. cremea

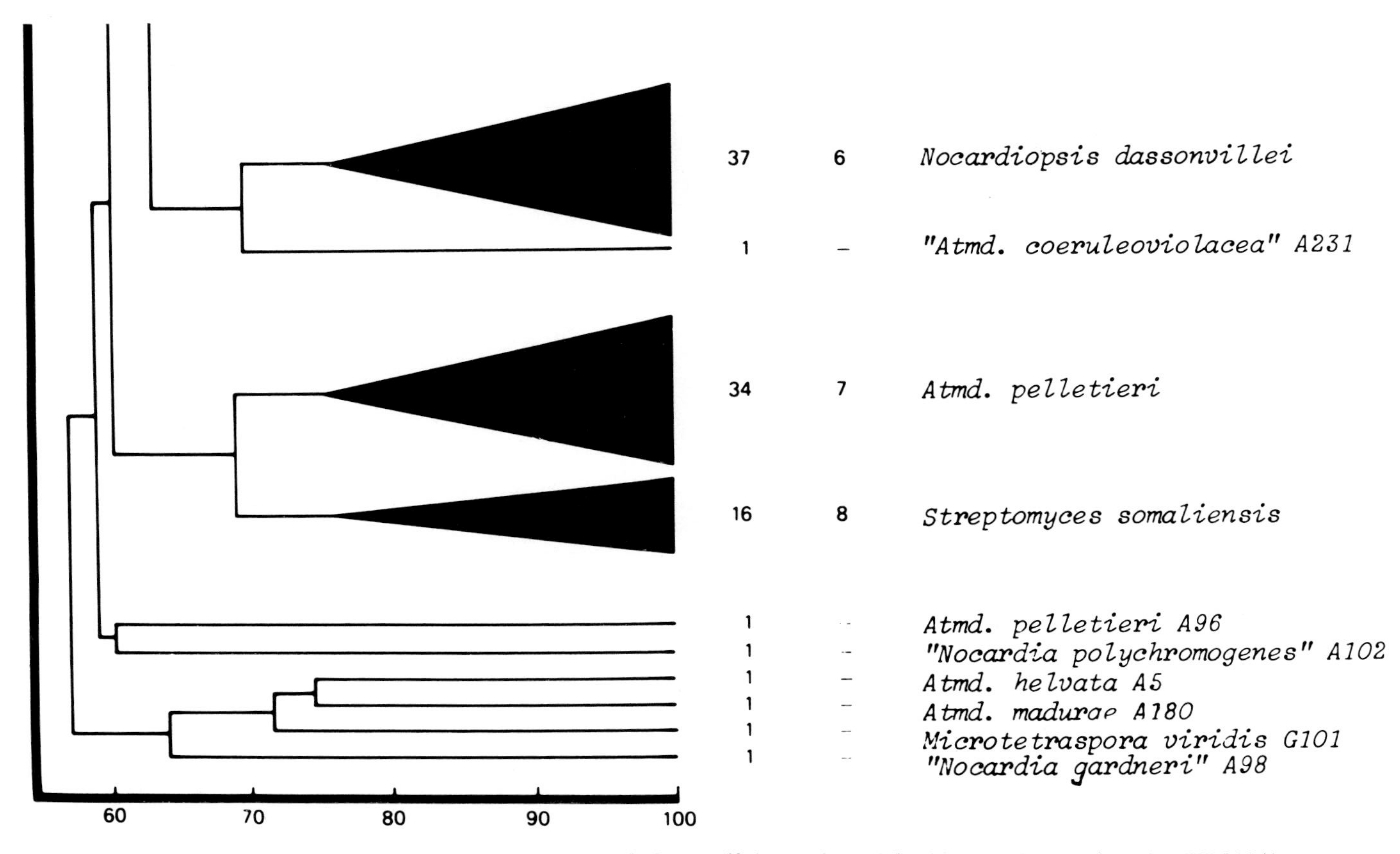

FIGURE I. Abridged dendrogram showing clusters based on the S_{SM} coefficient and unweighted average linkage clustering (UPGMA).

Gower (1971) also discusses ways by which missing values can be handled by all coefficients. More commonly used in bacteriology than S_G is the pattern difference coefficient, D_P (Sneath, 1968). The latter, like S_J and S_{SM}, is defined for two state characters but, unlike these two, is a dissimilarity coefficient which measures the differences between pairs of OTU's, leaving out the component due to strain vigor.

Agglomerative clustering algorithms have been used commonly in bacterial taxonomy. Single linkage tends to produce long straggly clusters, if they are present, whereas the more popular average linkage methods (*e. g.*, UPGMA; unweighted pair group method (Sneath, 1978)) give more compact clusters.

The results of such clustering are usually visualized as a tree diagram or dendrogram (sometimes termed phenogram if based on phenetic evidence). An abridged example of a dendrogram is given in Figure 1. Shaded diagrams are also used, with the similarity matrix plotted out and shaded in with OTU's in the same order as along the top of the corresponding phenogram. The densest shading usually represent high similarity values. Minimum spanning trees have also been used to show *how* clusters form during single linkage analyses.

It must be emphasized that not all data sets lend themselves to hierarchical, *i. e.*, nested, non-overlapping arrangements. Goodness of clustering should always be tested using, for example, co-phenetic correlation or the **W** test (Sneath, 1977, 1978).

Visualization of results of clustering using an ordination method can be an excellent way of confirming that the suggested groupings are really indicated by the observations (Gower, 1969; Marriott, 1974).

III. ORDINATION METHODS

Hierarchical clustering techniques are generally reliable when examining the affinities between close neighbors but have been shown to give a poor representation of relationships between major clusters (Sneath, 1978; Sneath and Sokal, 1973). Ordination methods, on the other hand, are good at depicting relationships between major clusters but have been shown to falsify distances between close neighbors (Gower, 1966; Sneath, 1978; Sneath and Sokal, 1973). Thus, to get an overall picture of the taxonomic structure in a chosen set of OTU's, it is recommended that they be examined by cluster analysis followed by ordination (Marriott, 1974; Sneath, 1978).

Ordination may be derived from either the **Q** or **R** matrix (Table II), but a common principle underlies all methods. Mathematical techniques are used to find the best representation of distances in multidimensional phenetic space using a reduced number of dimensions. The first dimension is that which expresses the greatest spread or scatter of the OTU's which are arranged along this dimension; the next dimension, *i. e.*, that which shows the next greatest scatter of the OTU's, is then sought and the OTU's are arranged along this dimension. This procedure is repeated until a high proportion of the variance is accounted for. Large capacity computers are required to carry out these analyses which are based on eigenvalues (variance) and eigenvectors (new coordinates). The results of such analyses may be visualized as taxonomic maps in two dimensions or as three-dimensional taxometric models (fish-tank models). Clusters are *not* formed; they must be formed later by eye using guidelines from an earlier cluster analysis.

A. Methods Based on the Correlation (R) Matrix

1. Principal Component/Factor Analysis. The method of principal component analysis consists of finding the transformation of a set of variables or test results in which the first few principal components account for most of the variability in the original data (Gower, 1969; Marriott, 1974; Sneath, 1978). Each principal component is uncorrelated with preceding components. Gower (1966) showed that factor analysis gave results similar to those of principal component analysis but dismissed factor analysis as an inappropriate model for taxonomic purposes. A disadvantage of both these methods is that missing values in the raw data are usually not allowed.

2. Canonical Variate/Discriminant Analysis. For both these techniques, OTU's are known to belong to already specified groups. From the taxonomists' point of view, these methods do not seem relevant for classification, but they have found favor with some bacteriologists for discriminating between "genus groups" (Gutteridge *et al.*, 1980; Gutteridge and Norris, 1979; Macfie *et al.*, 1978).

B. Methods Based on the Association (Q) Matrix

In the analysis of the **Q** matrix, the associations are assumed to be related to distances in multidimensional phenetic space. Under specific conditions, using certain distances, the **Q** matrix analysis gives equivalent results to principal component analysis or canonical variate/discriminant analysis.

1. Principal Coordinate Analysis. In principal coordinate analysis (Gower, 1966) eigenvalues and eigenvectors are calculated directly from the **Q** matrix. Since the **Q** matrix has already been formed for clustering purposes and there may be *no* corresponding **R** matrix analysis, it is more convenient to use principal coordinate analysis. This is in contrast with the fact that, in general, **R** matrices are smaller than **Q** matrices, and thus the **R** matrix analysis would be preferred where large numbers of OTU's are involved. In recent years, the advent of large capacity, high speed computers than can rapidly perform many highly iterative computations have removed most restrictions on the size of both data sets and **Q** matrices. Thus, when there are missing test results—a common phenomenon in bacteriology, similarity measures remain reliable and

TABLE II. Numerical Taxonomic Methods Applied in Sporoactinomycete Classification

Study	*Major taxa studied*	*No. of strains*	*No. of characters*	*Methods used*	*No. of major clusters recovered*
Gilardi et al., *1960* [a]	Actinomadura, Nocardia, Streptomyces	69	91	S_J *correlated features*	5
Hill et al., *1961*	Actinomadura, Nocardia, Streptomyces, Streptosporangium	69	91	S_{SM} *; single linkage*	21
Silvestri et al., *1962*	Micromonospora, Mycobacterium, Nocardia, Streptomyces, Streptoverticillium	190	105	S_{SM} *; taxonomic distance*	28
Gyllenberg et al., *1967* [b]	*Streptomyces, ("Yellow" series)*	60	33	*"Phenetic" classification and factor analysis*	5
Woznicka, 1967a,b	Streptomyces, *("Yellow" series)*	55	74	S-*index; Wroclaw taxonomy method*	4
Kurylowicz et al., *1969*	Streptomyces	150	35	*Similarity measure NS; taxonomic tree*	NS
Williams et al., *1969*	Streptomyces	18	46	S_{SM} *; single linkage*	5
Gyllenberg, 1970 [b]	Streptomyces	174	37	*Factor analysis*	NS
Kurylowicz et al., *1970*	Streptomyces, Streptoverticillium	150	143	S_{SM} *; taxonomic tree*	NS
Goodfellow, 1971	Actinomadura, Dermatophilus, Geodermatophilus, Mycobacterium, Nocardia, Orskovia, Rhodococcus	283	241	S_{SM} *; average and single linkage*	15
Paszkiewicz, 1972	Streptomyces, Streptoverticillium	300	200	*Wroclaw taxonomy method*	3
		300	7 [a]	*Centrifugal correlation*	12

Goodfellow et al., *1974*	Actinomadura, Mycobacterium, Nocardia, Rhodococcus	*98*	*180*	S_J, S_{SM}; *single linkage*	*4*
Kurylowicz et al., 1975	Streptomyces, Streptoverticillium	*448*	*31*	*Wroclaw taxonomy method*	*14*
		448	*23*	*Centrifugal correlation*	*21*
		448	*31*	S_{SM}; *single linkage*	*14*
Gyllenburg et al., *1975*	Streptomyces, Streptoverticillium	*559*	*24*	**Q** *index; non-hierarchical clustering*	*15*
Szulga, 1978	Streptomyces	*33*	*11*	*Similarity coefficient not stated*	
				Average linkage	*4*
				Single linkage	*3*
				Principal components	*NS*
				Sequential dendrite	*NS*
Alderson and Goodfellow, 1979	Actinomadura	*157*	*90*	S_J; *average and single linkage*	*4*
Goodfellow et al., *1979*	Actinomadura	*156*	*90*	D_P; S_J; S_{SM}; *average and single linkage*	*13*
Konev and Mitnetskii, 1980	Streptoverticillium	*42*	*44*	*Total similarity*	*NS*
Williams et al., *1981*	Streptomyces, Streptoverticillium	*519*	*139*	S_J	*10*
				S_{SM}; *average linkage*	*56*
Goodfellow	Actinomadura *and strains representative of 18 other genera* [c]	*130*	*108*	D_P; S_J; S_{SM}; *average and single linkage*	*21*

[a] *Correlated features.*
[b] *Study used ordination techniques.*
[c] *With* meso-*diaminopimelic acid in the cell wall.*
D_P: *pattern difference coefficient (Sneath, 1968);* S_J: *Jaccard coefficient (Sneath, 1957b);* S_{SM}: *simple matching coefficient (Sokal and Michener, 1958);* **Q** *index (Gyllenberg and Niemelä, 1975);* NS: *not stated.*

robust, whereas replacing missing values by guesses is not satisfactory (Marriott, 1974). Since principal coordinate analyses are based on such similarity measures, we suggest that principal coordinate analysis be the method of choice to complement cluster analysis in bacterial classification.

IV. APPLICATION OF NUMERICAL METHODS TO SPOROACTINOMYCETE SYSTEMATICS

Numerical methods have been used extensively in the revision of some actinomycete taxa, and the genera *Mycobacterium* and *Nocardia* in particular have benefited from this application (Goodfellow and Minnikin, 1977, 1978; Goodfellow and Wayne, 1982; Sneath, 1976). In contrast with these and other studies on nocardioform actinomycetes, the sporoactinomycetes have received comparatively little attention. Thus, the current weakness in the classification of these organisms is well reflected in the latest edition of Bergey's "Manual of Determinative Bacteriology" (Buchanan and Gibbons, 1974). Indeed some taxa, such as those classified in the families *Actinoplanaceae* (Couch and Bland, 1974) and *Micromonosporaceae* (Küster, 1974) may still be described as strictly monothetic. The various major numerical studies on sporoactinomycetes are summarized in Table II. In general, hierarchical techniques have been the sole analyses of choice; only a few workers have used ordination methods (see footnote b, Table II).

A. Ordination Methods

Gyllenberg, Woźnicka, and Kuryłowicz (1967) were the first to apply ordination methods in sporactinomycete taxonomy. These workers applied factor analysis as well as single linkage analysis to the data obtained in earlier studies (Woźnicka, 1967a,b; see Table II). Woźnicka, in her numerical analyses of the "Yellow" series of the genus *Streptomyces* (equivalent to the "Griseus" series), revealed anomalies in the classical morphological groupings. Thus, strains classified in the same series on the basis of subjective morphological characters such as the color of the vegetative mycelium had similarity values ranging from a low of 0.37 up to 0.81. Gyllenberg *et al.* (1967) applied factor analysis to Woźnicka's data and showed the occurrence of significantly correlated characteristics. The pattern of factors indicated that the 60 strains were a rather heterogeneous group; the four groups found in Woźnicka's studies were rearranged on the basis of the factor analysis and a new fifth group was defined. Gyllenberg and his co-workers concluded that factor analysis showed that antibiotic and physiological activities contributed more to the common factors than did traditional morphology. The authors emphasized that their data had to be interpreted in the context of the restricted number of strains studied but nevertheless felt that the study had shown the need for further work to improve tests employed in streptomycete taxonomy.

In a subsequent study, Gyllenberg (1970) again employed factor analysis, but he increased both the number and range of strains. One hundred and seventy four streptomycetes were studied for 37 features, thus the character data base was only marginally increased from the previous study. Clusters were not defined, but once again, Gyllenberg recognized several significant character correlations (factors) and stressed that these covered only a minor part of the total variance of tests. He dismissed variable tests and random interpretational errors as explanations for the latter and suggested that the test used were not sufficiently sensitive to define the structure of the genus *Streptomyces*. Gyllenberg suggested that new approaches such as analytical chemistry and molecular biology were needed to unravel the sub-generic classification of this complex group of organisms. This suggestion nevertheless begs the question as to how a manageable number of *representatives* of the genus *Streptomyces* may be chosen for examination by the more analytical but time-consuming methods.

Szulga (1978) used a selection of strains (33) from the 21 groups which Kuryłowicz *et al.* (1975) had recovered in their centrifugal correlation analysis of 448 streptomycetes (Table II). Szulga attempted to evaluate numerical methodology in the classification of *Streptomyces*; both **Q** and **R** techniques were applied, and hierarchical and ordination methods were represented. Only 11 characters were used and the majority of these were morphological in nature. Szulga concluded that average linkage was preferable to single linkage, but that principal components provided an easier identification of individuals. The sequential dendrite method segregated groups and allowed the taxonomic evaluation of characters. The study of Szulga was interesting to numerical taxonomists, but the minimal data base limited the usefulness of the exercise. Sneath, as early as 1970, noted that a more extensive and varied character set than that employed by collaborators (Gottlieb and Shirling, 1967) on the International *Streptomyces* Project was required to reveal natural phenetic groups within the genus *Streptomyces*. Szulga, however, utilized only 11 of the 31 ISP characters.

These few pioneering papers which have utilized ordination methods in sporoactinomycete taxonomy have not as yet shown other bacteriologists the full potential of such methods for taxonomy. This is understandable given their narrow data base and also given the problems in the taxonomy of the organisms they selected for study (see Williams *et al.*, elsewhere in this volume).

B. The Application of Numerical Methods to the Genus Actinomadura *with Special Emphasis on Principal Coordinate Analysis*

1. Introduction. Current thought on the classification of the sporoactinomycete genus *Actinomadura* has been especially influenced by the application of numerical taxonomic methods. Lechevalier and Lechevalier (1970) originally proposed the genus for three species of *Nocardia* with a wall chemotype III (*sensu* Lechevalier and Lechevalier, 1970); in the current edition of Bergey's "Manual of Determinative Bacteriology" (McClung, 1974), *Actinomadura* is a genus *incertae sedis*. To date, however, over 31 new species have been described (Horan and Brodsky, 1982; Huang, 1980; Meyer, 1979; Preobrazenskaya *et al.*, 1979; Skerman *et al.*, 1980). The application of cluster analysis (Alderson and Goodfellow, 1979; Goodfellow and Pirouz, 1982; Goodfellow *et al.*, 1979) has shown that the genus *Actinomadura* is a heterogeneous group consisting of at least three major centers of variation, one of which is equated with *Actinomadura sensu stricto*. In 1976, Meyer proposed that strains of *Actinomadura dassonvillei* were sufficiently different from other actinomadurae to warrant their removal to the new genus *Nocardiopsis*. Numerical taxonomic studies have supported this proposal; thus, this group of strains represents the second center of variation. The third area of variation can be loosely equated with *Atmd. pelletieri*. Strains of this latter species have shown an unexplained, high overall similarity with strains of *Streptomyces somaliensis* (Goodfellow *et al.*, 1979; Kurup and Schmitt, 1973) although chemotaxonomic criteria may be employed to distinguish easily between the two groups (Collins *et al.*, 1977; Lechevalier *et al.*, 1977).

A comprehensive numerical phenetic study was undertaken to investigate this problem and also to examine the relationship of the newly described *Actinomadura* species with the three original species in the genus and with each other. It was decided to embark on a bi-phasic study by first examining a large number of strains for a range of tests with analysis of the resulting data set by clustering and subsequently by examining the results of clustering by ordination. The results of this latter study will be discussed in detail in an attempt to illustrate the principles of application and with the hope of generating interest so that other workers will apply such methods to new (or old) problems.

2. Methods. One hundred and seventy strains representing 25 species of the genera *Actinomadura, Microbispora, Microtetraspora, Nocardiopsis*, and *Streptomyces* were examined for 120 unit characters. The characters included a range of biochemical, morphological, nutritional and physiological tests. The data were analyzed using the CLASP package of programs (written by Ross, Laukner, and Hawkins; Rothamsted Experimental Station, Harpenden, Herst, England). Both the S_{SM} and D_P coefficients of association were used to calculate the **Q** matrix which was clustered using UPGMA and single linkage analyses. Only the S_{SM}, UPGMA analysis is considered here (see Athalye *et al.*, 1983 for details of the other analyses). Ordination by using principal coordinate analysis was based on results from the S_{SM}, UPGMA analysis.

3. *Results and Conclusions.* Results of the clustering analysis are shown in the abridged dendrogram (Fig. 1). Strain of *Nocardiopsis dassonvillei* were well separated from all actinomadurae, again providing confirmation of the validity of this recently described genus. *Atmd. pelletieri* strains were recovered as two distinct clusters, clusters 4 and 7 (Fig. 1). Cluster 4 consisted of a few fast growing strains which showed a high average similarity to several newly described *Actinomadura* species as well as to clusters 2, 3, and 15 (*Atmd. madurae* clusters). All of the strains in clusters 2, 3, 4, and 15 and in the newly described strains were found in an aggregate cluster-group defined at the 65% similarity level (S_{SM}). This cluster-group can be equated with the genus *Actinomadura sensu stricto*. The majority of *Atmd. pelletieri* strains, including the type strain, were recovered in a second well-defined cluster, cluster 7, which fell outside *Actinomadura sensu stricto* (Fig. 1). This *Atmd. pelletieri* cluster once again was very closely linked to the *S. somaliensis* strains in cluster 8.

The results of the present ordination analysis based on principal coordinates are visualized as a taxonomic map in two dimensions and as a three-dimensional taxometric (fish tank) model (Figs. 2 and 3, respectively). Cluster

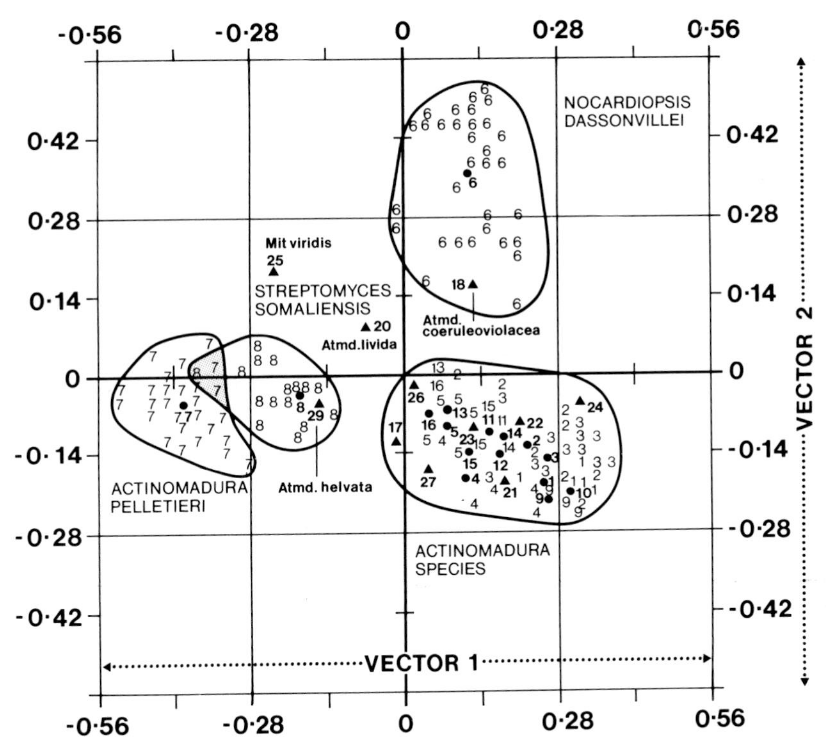

FIGURE 2. Principal coordinate analysis. Two dimensional taxonomic map based on the S_{SM} (UPGMA) analysis: (●), mean of each cluster; (▲17, etc.), positions of unclustered type strain of Actinomadura, Microbispora, *and* Microtetraspora *species; enclosed areas represent scatter about the mean; 6, 7, etc., positions of individual strains within each cluster.*

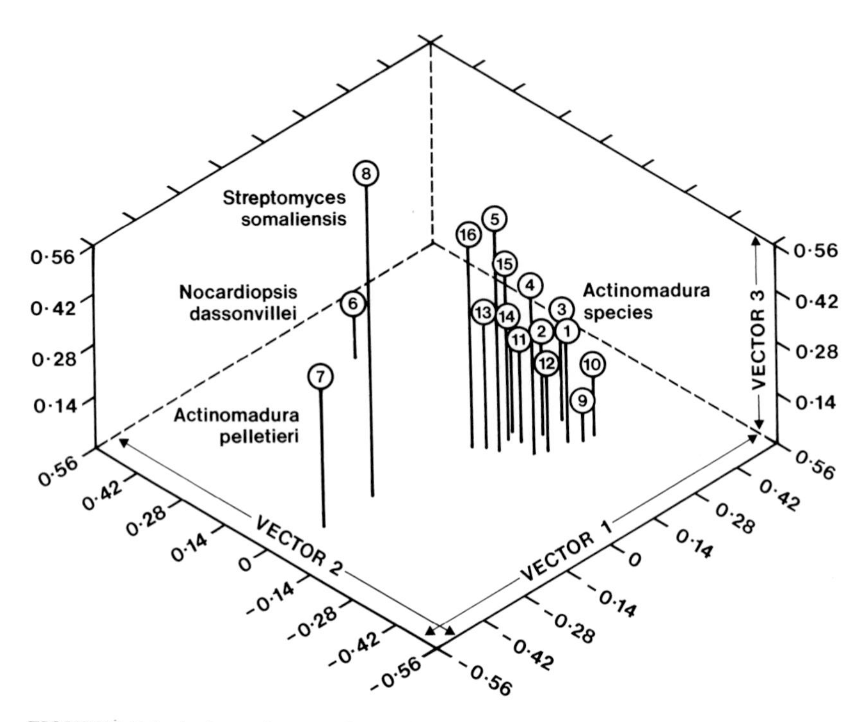

FIGURE 3. Principal coordinate analysis. Three dimensional taxometric model based on the S_{SM} (UPGMA) analysis.

(8), etc represents the positions of the hypothetical mean organisms of the clusters.

numbers in Figures 2 and 3 correspond to those in Figure 1. Figure 2 once again shows the distinct separation of strains in the genus *Nocardiopsis* from all other strains in the study. The aggregation of the *Actinomadura* species as a cluster-group is also well depicted as is the separation of the two *Atmd. pelletieri* clusters. Cluster 4 was recovered in *Actinomadura sensu stricto*, whereas cluster 7 is found overlapping cluster 8 which contains the *S. somaliensis* strains.

Thus, in this case, two-dimensional ordination only serves to confirm the results of the hierarchical analysis. A consideration of the third dimension, however, provides new information on the taxonomic structure inherent in these groups of closely related organisms (Fig. 3).

Atmd. pelletieri strains in cluster 7 are shown here to be quite distinct from *S. somaliensis* strains in cluster 8; a result which is in agreement with the available chemical data (Collins *et al.*, 1977; Lechevalier *et al.*, 1977). Yet again *Nocardiopsis* strains are well separated from the other organisms in all three dimensions. This third dimension, however, does not differentiate between members within the aggregate *Actinomadura* cluster-group to any significant extent. Since over 40% of the variance in the data set is explained by the three dimensions, it can be said that this aggregation of clusters may truly represent the variation within the species of the genus *Actinomadura sensu stricto*.

V. CONCLUSIONS

In past years the application of conventional numerical taxonomy of a hierarchical nature has led to serious and rapid taxonomic changes in the groups of organisms to which it has been applied. Numerical analyses, however, have also produced anomalous results. Thus, a high level of similarity has been indicated between strains of *Atmd. pelletieri* and those of *S. somaliensis* by using numerical methods, whereas chemical analyses have shown them to be quite different. The application of principal coordinates to data based on the S_{SM}, UPMGA analysis clearly confirmed the chemical data and showed the two groups to be well separated (Fig. 3).

Marriott (1974) concluded that ordination methods and hierarchical methods should be considered as complementary. Our results certainly show that ordination can provide important new insights into phenetic relationships between microorganisms, and we suggest that other microbiologists may find advantages in using the two approaches in conjunction.

REFERENCES

Alderson, G., and Goodfellow, M., *Postepy Hig. Med. Dosw. 33:*109 (1979).

Athalye, M., Goodfellow, M., Lacey, J., and White, R.P., *Int. J. Syst. Bacteriol.* 1984 (To appear)

Austin, B., and Colwell, R. R., *Int. J. Syst. Bacteriol. 27:*204 (1977).

Boyce, A. J., *in* "Numerical Taxonomy" (A. J. Cole, ed.), p. 1. Academic Press, London (1969).

Buchanan, R. E., and Gibbons, N. E. (eds.), Bergey's "Manual of Determinative Bacteriology", 8th ed. The Williams and Wilkins Co., Baltimore (1974).

Collins, M. D., Pirouz, T., Goodfellow, M., and Minnikin, D. E., *J. Gen. Microbiol. 100:*221 (1977).

Colwell, R. R., *Dev. Ind. Microbiol. 11:*154 (1970).

Couch, J. N., and Bland, C. E., *in* Bergey's "Manual of Determinative Bacteriology" (R. E. Buchanan and N.E. Gibbons, eds.) 8th ed., p. 706. The Williams and Wilkins Co., Baltimore (1974).

Gilardi, E., Hill, L. R., Turri, M., and Silvestri, L. G., *G. Microbiol. 8:*203 (1960).

Gilmour, J. S. L., *Nature (Lond.) 139:*1040 (1937).

Gilmour, J. S. L., *Nature (Lond.) 168:*400 (1951).

Goodfellow, M., *J. Gen. Microbiol. 69:*33 (1971).

Goodfellow, M., *in* "CRC Handbook of Microbiology" (A. T. Laskin and H. A. Lechevalier, eds.), 2nd ed., Vol 1, p. 579. CRC Press, Ohio (1977).

Goodfellow, M., Alderson, G., and Lacey, J., *J. Gen. Microbiol. 112:*95 (1979).

Goodfellow, M., Lind, A., Mordarska, H., Pattyn, S., and Tsukamura, M., *J. Gen. Microbiol. 85:*291 (1974).

Goodfellow, M., and Minnikin, D. E., *Annu. Rev. Microbiol. 31:*159 (1977).

Goodfellow, M., and Minnikin, D. E., *Zentralbl. Backteriol. Parasitenkd. Infektionskr. Hyg. Abt. I. 6* suppl:43 (1978).

Goodfellow, M., Orlean, P. A. B., Collins, M. D., and Alshamaony, L., *J. Gen. Microbiol. 109:*57 (1978).

Goodfellow, M., and Pirouz, T., *J. Gen. Microbiol. 128:*503 (1982).

Goodfellow, M., and Wayne, L. G., *in* "The Biology of Mycobacteria" (C. Ratledge and J. L. Stanford, eds.), p. 472. Academic Press, London (1982).

Gottlieb, D., and Shirling, E. B., *Int. J. Syst. Bacteriol. 17:*315 (1967).
Gower, J. C., *Biometrika 53:*325 (1966).
Gower, J. C., *Acarologia 11:*357 (1969).
Gower, J. C., *Biometrics 27:*857 (1971).
Gower, J. C., *in* "Biological Identification with Computers" (R. J. Pankhurst, ed.), p. 251. Academic Press, London (1975).
Gutteridge, C. S., Macfie, H. J. H., and Norris, J. R., *J. Anal. Appl. Pyrolysis 1:*67 (1979).
Gutteridge, C. S., Mackey, B. M., and Norris, J. R. *J. Appl. Bacteriol. 49:*165 (1980).
Gutteridge, C. S., and Norris, J. R. *J. Appl. Bacteriol. 47:*5 (1979).
Gyllenberg, H. G., *in* "The Actinomycetales" (H. Prauser ed.), p. 101. Gustav Fischer Verlag, Jena (1970).
Gyllenberg, H. G., and Eklund, E., Ann. Acad. *Sci. Fenn. Ser. A IV Biol. 113:*1 (1967).
Gyllenberg, H. G., Niemelä, T. K., and Niemi, J. S., *Postepy Hig. Med. Dosw. 29:*357 (1975).
Gyllenberg, J. G., Wőznicka, W., and Kuryłowicz, W., *Ann. Acad. Sci. Fenn. Ser. A IV Biol. 114:*3 (1967).
Hill, L. R., Turri, M., Gilardi, E., and Silvestri, L. G., *G. Microbiol. 9:*56 (1961).
Horan, A. C., and Brodsky, B. C., *Int. J. Syst. Bacteriol. 32:*195 (1982).
Hotelling, H., *J. Educ. Psychol. 24:*417 (1933a).
Hotelling, H., *J. Educ. Psychol. 24:*498 (1933b).
Huang, L. H., *Int. J. Syst. Bacteriol. 30:*565 (1980).
Jones, D., and Sackin, M. J., *in* "Microbiological Classification and Identification" (M. Goodfellow and R. G. Board, eds.), p. 73. Academic Press, London (1980).
Konev, Y. F., and Mitnetskii, A. B. Mıkrobiologiya *49:*110 (1980).
Kurup, P. V., and Schmitt, J. A., *Can. J. Microbiol. 19:*1035 (1973).
Kuryłowicz, W., Paszkiewicz, A., Woźnicka, W., Kurzatkowski, W., and Szulga, T. *Postepy Hig. Med. Dosw. 29:*281 (1975).
Kuryłowicz, W., Woźnicka, W., Malinowski, K., and Paszkiewicz, A. Exp. Med. Microbiol. 21:143 (1969).
Kuryłowicz, W., Woźnicka, W., Paszkiewicz, A., and Malinowski, K. *in* "The Actinomycetales" (H. Prauser, ed.), p. 107. Gustav Fischer Verlag, Jena (1970).
Küster, E., *in* Bergey's "Manual of Determinative Bacteriology" (R. E. Buchanan and N. E. Gibbons, eds.), 8th ed., p. 846. The Williams and Wilkins Co., Baltimore (1974).
Lechevalier, H. A., and Lechevalier, M. P. *in* "The Actinomycetales" (H. Prauser, ed.), p. 393. Gustav Fischer Verlag, Jena (1970).
Lechevalier, M. P., De Bievre, C., and Lechevalier, H. *Biochem. Syst. Ecol. 5:*249 (1977).
Logan, N. A., and Berkeley, R. C. W. *in* "The Aerobic Endospore Forming Bacteria" (R. C. W. Berkeley and M. Goodfellow, eds.), p. 105. Academic Press, London (1981).
Macfie, H. J. H., Gutteridge, C. S., and Norris, J. R., *J. Gen. Microbiol. 104:*67 (1978).
Marriott, F. H. C. "The Interpretation of Multiple Observations". Academic Press, London (1974).
McClung, N. M. *in* Bergey's "Manual of Determinative Bacteriology" (R. E. Buchanan and N. E. Gibbons, eds.), p. 726. The Williams and Wilkins Co., Baltimore (1974).
Meyer, J., *Int. J. Syst. Bacteriol. 26:*487 (1976).
Meyer, J., *Z. Allg. Mikrobiol. 19:*37 (1979).
Paszkiewicz, A., *Arch. Immunol. Ther. Exp. 20:*307 (1972).
Preobrazenskaya, T. P., Sveshnikova, M. A., Maksimova, T. S., Ol'Khovatova, O. L., Chormonova, N. T., and Terekhova, L. P., *Actinomycetes 14:*21 (1979).
Quadling, C., and Hopkins, J. W., *Can. J. Microbiol. 13:*1379 (1967).
Rosswall, T., and Kvillner, E., *in* "Advances in Microbial Ecology" (M. Alexander, ed.), Vol2, p. 1. Plenum Press, New York (1978).
Silvestri, L., Turri, M., Hill, L. R. and Gilardi, E., *in* "Microbial Classification" (G. E. Ainsworth and P. H. A. Sneath, eds.), p. 333. Cambridge University Press, Cambridge (1962).
Skerman, V. B. D., McGowan, V., and Sneath, P. H. A., *Int. J. Syst. Bacteriol. 30:*225 (1980).
Sneath, P. H. A., *J. Gen. Microbiol. 17:*184 (1957a).
Sneath, P. H. A., *J. Gen. Microbiol. 17:*201 (1957b).
Sneath, P. H. A., *in* "Microbial Classification" (G. E. Ainsworth and P. H. A. Sneath, eds.), p. 289. Cambridge University Press, Cambridge (1962).
Sneath, P. H. A., *J. Gen. Microbiol. 54:*1 (1968).

Sneath, P. H. A., *in* "The Actinomycetales" (H. Prauser, ed.), p. 371. Gustav Fischer Verlag, Jena (1970).
Sneath, P. H. A., *in* "Recent Advances in Microbiology" (A. Perez-Miravete and D. Pelaez, eds.), p. 581. Asociación Mexicana de Microbiología, Mexico City (1971).
Sneath, P. H. A. *in* "Methods in Microbiology" (J. R. Norris and D. W. Ribbons, eds.), Vol. 4, p. 29. Academic Press, London (1972).
Sneath, P. H. A., *Int. J. Syst. Bacteriol. 24:*508 (1974).
Sneath, P. H. A., *in* "The Biology of the Nocardiae" (M. Goodfellow, G. H. Brownell, and J. A. Serrano, eds.), p. 74. Academic Press, London (1976).
Sneath, P. H. A., *Math. Geol. 9:*123 (1977).
Sneath, P. H. A., *in* "Essays in Microbiology" (J. R. Norris and M. R. Richmond, eds.), Vol. 9, p. 1. John Wiley and Sons, Chichester (1978).
Sneath, P. H. A., and Johnson, R., *J. Gen. Microbiol. 72:*377 (1972).
Sneath, P. H. A., and Sokal, R. R. "Numerical Taxonomy: the Principles and Practice of Numerical Classification" W. H. Freeman, San Francisco (1973).
Sokal, R. R., and Michener, C. D., *Kans. Univ. Sci. Bull. 38:*1409 (1958).
Sokal, R. R., and Sneath, P. H. A. "Principles of Numerical Taxonomy". W. H. Freeman, San Francisco (1962).
Sundman, V., *Acta Agric. Scand. 18:*22 (1968).
Sundman, V., *Can. J. Microbiol. 16:*455 (1970).
Sundman, V., and Carlberg, G., Ann. *Acad. Sci. Fenn. A IV Biol. 115:*1 (1967).
Sundman, V., and Gyllenberg, H. V., *Ann. Acad. Sci. Fenn. A IV Biol. 112:*1 (1967).
Szulga, T., *Zentralbl. Backteriol. Parasitenkd. Infektionskrankheiten Hyg. Abt. I. 6*, suppl.:31 (1978).
Williams, S. T., Davies, F. L., and Hall, D. M. *in* "The Soil Ecosystem" (J. G. Sheals, ed.), p. 107. Systematics Association Publication no. 8 (1969).
Williams, S. T., Wellington, E. M. H., Goodfellow, M., Alderson, G., Sackin, M., and Sneath, P. H. A., *Zentralbl. Backteriol. Midrobiol. Hyg. Abt. I. 11, Suppl.: 47 (1981).*
Wilkinson, B. J., and Jones, D., *J. Gen. Microbiol. 98:*399 (1977).
Woźnicka, W., *Med. Dosw. Mikrobiol. 19:*9 (1967a).
Woźnicka, W., *Med. Dosw. Mikrobiol. 19:*21 (1967b).

PHAGE HOST RANGES IN THE CLASSIFICATION AND IDENTIFICATION OF GRAM-POSITIVE BRANCHED AND RELATED BACTERIA

H. Prauser

Zentralinstitut für Microbiologie und Experimentelle Therapie
Akademie der Wissenschaften der DDR
Jena, German Democratic Republic

I. INTRODUCTION

The term "Gram-positive branched bacteria" is used here to replace the terms *Actinomycetales* and the commonly used actinomycetes. Taxonomists should do their best to develop a new classification of the organisms under study by including results from biochemical assays, from DNA:DNA and DNA:RNA hybridizations, from comparative cataloging of 16S ribosomal RNA oligonucleotide sequences, and from actinophage host-range studies. This would reduce the chance of arriving at misleading conclusions in the diverse disciplines of microbiology currently based on artificial systems.

In this paper, the results of actinophage host-range studies are presented and discussed with the aim of contributing to the elucidation of natural interrelationships between groups of actinomycetes and a few related organisms. Information on this approach comes mainly from three groups: Bradley and Anderson, 1958; Prauser and Falta, 1968; Wellington and Williams, 1981. A review on host-phage relationships in nocardioform organisms was given by Prauser (1976a).

II. GENERAL CONDIDERATIONS

A. The Character of Actinophage Host-Range Studies

The molecular events of the vegetative multiplication of phage, involving adsorption, DNA injection, DNA replication, transcription and translation,

ISBN 0-12-528620-1

as well as lysogenization, establishment of repression, integration of phage, etc. are well known for particular bacteriophage and actinophage (Lomovskaya *et al.*, 1980). However, the large number of phage used and host-phage relationships involved in the extensive cross-infection experiments cannot be studied in detail. For example, the attempt to elucidate the background of one of the many observed clearing effects (Klein *et al.*, 1981) gives some idea of the extent of the work necessary in a single case. Thus, the results of phage host-range studies must be understood first of all as phenomenal ones.

Another drawback of phage-host range studies lies in the sequence of obtaining both phage and strains. From 1965 to the present, our laboratory has isolated 136 phage after elimination and loss of phage. Cross-infection experiments carried out with different sets of phage and bacterial strains over the course of the years are too extensive to be presented in a survey such as this.

B. Isolation and Application of Phage

For the purpose under consideration, phage were isolated mainly from soil and composts. Usually enrichment procedures involving the addition of fresh inocula of prospective host strains were used. The presence of phage was demonstrated by spotting or plating filtrates on or in soft agar overlays which had been seeded with prospective host strains. Purification was performed by three to seven serial transfers of material from single plaques to fresh overlays. Phage were propagated for enrichment by the confluent-plate lysate method or in submerged shaker culture. Titers ranged between 10^7 to 10^{10} p.f.u./ml. Media for all procedures were those usually applied for the strains involved, generally with the addition of Ca^{++} ions. Sterile filtrates were stored in ampoules at 4 °C for several months and even for several years. Storage in ampoules in liquid nitrogen was also successful.

Multiple-loop applicators usually were used in cross-infection experiments, allowing up to 64 transfers in one step. In order to demonstrate single plaques, several dilutions of the original phage suspension were used. In our laboratory, the phage suspension with the highest available concentration was used. To demonstrate phage propagation, material from the lytic zones was transferred by loops to fresh overlays. Repeated lysis there, whether clear or turbid, was taken as indicating phage propagation on the first plate.

C. The Lytic Effects

The appearance of lytic effects depends on the method used in the cross-infection experiments. When high-titer suspensions were applied, lytic zones in the bacterial lawn may be clear more or less turbid, *i. e.*, lysis usually is confluent. Within a turbid zone, less turbid or clear single plaques occasionally occur, indicating mutation or the phage tested.

As pointed out above, material is transferred by use of a loop from the lytic zone to a second plate freshly seeded with the strain in question. Lysis on the second place verifies phage propagation on the first plate. Absence of lysis on the second plate proves the lytic effect on the first plate to be that which we call a clearing effect: phage-dependent clearing of the bacterial lawn without phage propagation. Random tests showed that these effects were independent of inhibitory factors such as antibiotics in the phage suspension.

The purpose of the various methods for applying low-titer suspensions is the demonstration of single plaques and, consequently, of phage propagation in one single test. However, the capacity of a phage to produce clearing effects cannot be detected with this method.

D. Taxonomic Level and Phage Susceptibility

Statements on the taxonomic level of actinophage activity, *i. e.*, on the taxon specificity of phage attack, are influenced by several factors:

(i) The weight of statements depends on the number and taxa studied and the number and selection of phage applied.

(ii) Almost no phage is virulent for all strains of a taxon under study.

(iii) The concept of taxon specificity implies that a phage is active only against members of one taxon.

(iv) Taxa are man-made. Results of 16S RNA oligonucleotide cataloging (Fox *et al.*, 1980); Stackebrandt and Woese, 1981) clearly demonstrate that the common conception, composition, and degree of subclassification of taxonomic groups frequently diverge from the degree of similarities expressed by genetic sequences. Many taxa are overclassified, *e. g.*, the family *Streptomycetaceae*.

(v) The statement on specificity is preliminary in the case of a genus that is monotypic, *i. e.*, represented by one species only, at the date of the studies. The statement is also preliminary in those cases in which representatives of only one species of a genus were available.

E. Taxon Specificity of Clearing Effects

Clearing effects. *i. e.*, phage-dependent clearing of the bacterial lawn without phage propagation, were found to be taxon specific (Prauser, 1981b). Table III shows the frequency and distribution of clearing effects in the group of genera including *Nocardia, Rhodococcus*, and the "*aurantiaca* taxon". The N-phage applied did not give clearing effects with all tested strains of other taxa, which represented all other cell wall chemotypes. For all strains and taxa, except *Nocardia amarae*, clearing effects could be shown. On the other hand, for some strains and taxa, *i. e. Rhodococcus bronchiales, R. terrae, R. rubropertinctus*, and the "*aurantiaca* taxon", clearing effects only were found. In the taxonomic interpretation of phage-host relationships, this suggests that the clearing effects relate the four taxa in question to other members of the involved genera; this is in agreement with all other taxonomic data. Whether clearing effects, as the only response to phage, indicate a limited degree of relationship may be resolved by further comparative studies. The occurrence of clearing effects is not restricted to nocardiae and rhodococci. We found them with nearly all the actinomycete taxa studied (Prauser, 1981b).

F. Phage Susceptibility and Cell Wall Chemotype

When Bradley and his coworkers (Bradley and Anderson, 1958) started the actinophage host range studies, the importance of differences in the cell wall composition of actinomycetes had not yet been realized or adequately understood (Becker *et al.* 1964, 1965; Yamaguchi, 1965). Consequently, at that time, the taxonomic position of the strains used could not be checked with regard to this character, thereby implying the risk of misclassification of strains. Thus, phage were reported to be equally active against strains of the genera *Streptomyces* and *Nocardia, Streptomyces* and *Micromonospora, Streptomyces* and *Actonomadura* — pairs of genera differing in their cell wall chemotype (see review by Prauser, 1976a). Later, misclassification of strains included in this study were reported in the literature or, in most cases, by personal communication.

Kikuchi and Perlman (1977) reported that their soil-isolated phage ϕUM51 was active against strains of several species of *Micromonospora, Nocardia mediterranei* ATCC 13685, and *Streptomyces griseus* NRRL B-292. According to the names given, the cell wall chemotypes II, IV, and I appeared to be involved. In our hands, phage ϕUW 51 was active against several strains of *Micromonospora* but was ineffective not only against the other two species mentioned, but also against 29 additional international reference strains of *Streptomyces griseus*.

Kurup and Heinzen (1978) found that their temperate phage ϕ-150A, isolated from a strain of *Micropolyspora faeni* (cell wall chemotype IV), was lytic against most of the tested strains of this species as well as against four of their isolates of *Thermoactinomyces candidus* (cell wall chemotype III). We were unable to obtain strains and phage for re-evaluation. Recent suggestions for the classification of *Thermoactinomyces vulgaris* (Stackebrandt and Woese, 1981) would seem to make such re-evaluation highly necessary.

In this author's laboratory, since the start of our studies in which approximately 200 phage and hundreds of strains had been used, no phage was observed that cross-reacted with strains of a different cell wall chemotype, with the exception of the clearing effects of *Streptomyces* phage against *Nocardiopsis* strains (Prauser 1970a, b; Prauser 1976a,b; Prauser 1981a,b; Prauser and Falta, 1968; Prauser and Momirova, 1970; Prauser *et al.*, 1967a). Recently, Williams, Wellington, and Tipler (1980) and Wellington and Williams (1981) came to the same conclusion after studying a large number of phage, strains, and taxa.

Since cell wall composition is a generally accepted criterion in the classification of actinomycetes and since phage susceptibility in the sense pointed out is in full correspondence with cell wall composition, wall composition will be taken as a guideline for displaying the results of actinophage host range studies and for discussing their taxonomic implications.

III. ACTINOPHAGE SUSCEPTIBILITY IN SEVERAL GROUPS OF ACTINOPHAGES

Table I gives some general impression of the number and range of strains and phage that were included in three large-scale studies (Prauser and Falta, 1968; Wellington and Williams, 1981; Prauser and Rössler, unpublished data). Phage are available now for most of the actinomycete genera. Although intensive attempts at isolation, especially for actinomaduras, have been performed, the lack of phage for all sporoactinomycetes of cell wall chemotype III is striking.

A. Actinomycetes of Cell Wall Chemotype I

Strains of *Streptomyces;*, *Steptoverticillium*, *Chainia*, *''Actinopycnidium''*, *''Actinosporangium''*, *Microellobosporia*, *Elytrosporangium*, *''Microechinospora''*, and *Kitasatoa* are susceptible to sets of phage which are called family-specific for the *Streptomycetaceae* (Prauser and Falta, 1968; Wellington and Williams, 1981; Prauser and Rössler, unpublished data). As expected, there is no difference whether the phage were originally isolated for strains of the genus *Streptomyces* (Prauser and Falta, 1968) or for members of other genera of this family (Wellington and Williams, 1981) (Table I). Cross-infection experiments contributed to the transfer of the genus *Microellobosporia* from the family *Actinoplanaceae* to the family *Streptomycetaceae* (Prauser *et al.* 1967a) and to a similar transfer of the genera *Microechinospora* (Prauser *et al.* 1967b) and *Elytrosporangium* (Prauser, 1970). The inclusion of the genus *Kitasatoa* in the *Streptomycetaceae* was also supported by cross-infection experiments. Common phage susceptibility was understood as one of the distinctive characters of the family *Streptomycetaceae* (Prauser, 1970).

Recently we isolated phage for a strain of *Sporichthya polymorpha* which

TABLE I. Genera and Phage Included in Some Actinophage Host-Range Studies

Cell wall chemotype	Genus	Susceptibility [a]	Prauser and Falta (1968)		Prauser and Rößler (unpublished data)		Wellington and Williams (1981)		No. of phage available [b]
			No. of Strains	No. of phage	No. of Strains	No. of phage	No. of Strains	No. of phage	
I	Streptomyces	+	27	4	162	16	25	27	26
	Streptoverticillium	+	3	-		-	7	3	-
	Chainia	+	1	-	5	-	5	3	-
	"Actinopycnidium"	+	1	-	1	-	1	1	-
	"Actinosporangium"	+	1	-	1	-	1	1	-
	Microellobosporia	+	2	-	4	-	3	-	-
	Elytrosporangium	+	-	-	3	-	3	-	-
	"Microechinospora"	+	-	-	1	-	-	-	-
	Kitasatoa	+	-	-	-	-	5	-	-
	Nocardioides	+	5	3	32	4	2	2	24
	Arthrobacter simplex	+	1	1	-	1	-	-	1
	Arthrobacter tumescens	+	1	-	1	1	-	-	1
	Intrasporangium	- [c]	1	-	1	-	1	-	-
	Sporichtyha	+	-	-	-	-	-	-	1
II	Actinoplanes	+	2	-	10	-	-	-	4
	Ampullariella	+	1	-	1	-	-	-	-
	Dactylosporangium	- [c]	-	-	2	-	-	-	-
	Amorphosporangium	+	-	-	1	-	1	1	-
	Micromonospora	+	3	-	6	2	1	1	5

III	Nocardiopsis	+	-	-	-	-	1	-	4
	Actinomadura	- [c]	5	-	8	-	6	-	-
	Microtetraspora	-	-	-	5	-	1	-	-
	Microbispora	-	1	-	7	-	-	-	-
	Streptosporangium	- [c]	1	-	10	-	1	-	-
	Spirillospora	-	1	-	1	-	-	-	-
	Planomonospora	- [c]	-	-	1	-	-	-	-
	Dermatophilus	-	1	-	-	-	-	-	-
	Geodermatophilus	- [c]	-	-	-	-	-	-	-
	Thermoactinomyces	+	-	-	-	-	-	-	2
IV	Nocardia	+	2	1	{ 153 (Nocardia and Rhodococcus)	1	9	4	14 [d]
	Rhodococcus	+	8	4		6	3	2	19
	Mycobacterium	+	2	1	16	-	-	-	-
	Saccharomonospora	+	1	-	-	-	-	-	3
	Saccharopolyspora	- [c]	-	-	-	-	1	-	-
	Micropolyspora faeni	+	1	-	-	-	-	-	3
V	Actinomyces viscosus	+	-	-	-	-	-	-	-
VI	Oerskovia	+	4	3	7	3	1	1	11
	Promicromonospora	+	5	2	3	3	1	1	6
	Total	27	81	19	442	37	79	47	136

[a] *Present situation*
[b] *No. of phage available in author's laboratory.*
[c] *Attempts at isolating phage failed.*
[d] *In addition, 12 phage active against "nocardiae" lacking mycolic acids (see TABLE IV)*

had been obtained from M. Lechevalier. The phage was ineffective against several strains of *Streptomyces, Nocardioides*, and *Arthrobacter simplex*. Experiments are different because of the slow and limited growth of this organism.

The situation of *Nocardioides* needs to be discussed in some detail. *Nocardioides* phage were ineffective against all strains tested which belong to the family *Streptomycetaceae* and *Nocardioides* strains were not susceptible to *Streptomyces* phage, as far as lytic effects were concerned (Prauser and Falta, 1968; Wellington and Williams, 1981). However, clearing effects caused by some *Streptomyces* phage in some *Nocardioides albus* strains are well known (Prauser, 1976; Prauser, 1981b). This caused hesitation in regard to the family placement of *Nocardioides*. After DNA:DNA reassociation experiments (Tille *et al.* 1978), we consider *Nocardioides* to be a genus in search of a family, a view shared by H. and M. Lechevalier (1981). To our surprise, we recently found that *Nocardioides* strains IMET 7806 and IMET 7830 were susceptible to true lysis by the *Streptomyces* phage S6. This preliminary result needs to be confirmed and studied in detail.

On the other hand, the *Nocardioides* phage X6 was virulent against strains of *Arthrobacter simplex* which have cell wall chemotype I; *Nocardioides* strains were susceptible to the *Arthrobacter simplex* phage X4 (Prauser 1976b). The results are summarized in Table II.

At present, the taxonomic position of the genus *Nocardioides* is very interesting. Some data favor the placement of *Nocardioides* and *Arthrobacter simplex* in one genus: a) DNA:DNA homology around 16% (Tille and Prauser, unpublished data); b) very similar fatty acid, menaquinone, and polar lipid profiles (M. Goodfellow, personal communication); and c) cross-susceptibility to phage (Prauser, 1976b). However, the morphology and life

TABLE II. Actinophage Susceptibility of Streptomyces, Nocardioides, *and* Arthrobacter simplex

Phage	*S7*	*S6*	*X4*	*X6*
Taxa tested / *propagated on*	S. olivaceus *NCIB 8238* [a]	S. griseus *NCIB 8232* [a]	A. simplex *IMET 10283* [b]	N. albus *IMET 7810* [c]
Streptomyces	+ [d]	+	-	-
Nocardioides	*o*	*?*	+	+
A. simplex	-	-	+	+

[a] *W. Kurylowicz, Warsaw*
[b] *Sci. Counc. Bact. Div. Canada as* A. globiformis
[c] *Isolate from author's laboratory*
[d] *(-): no response; (+): lysis; (o): clearing effect; (?) indeterminate.*

cycle of both organisms are extremely different: a uni-cellular, motile bacterium *versus* a mycelial-, even aerial mycelium-producing actinomycete with arthrospores. Classification problems like this will be more and more frequent in the future. Unfortunately, there are no data available concerning the 16S rRNA oligonucleotide sequences of Nocardioides.

Arguments against a close relationship between *Nocardioides* and *Streptomyces* are a) lack of DNA:DNA homology (Tille and Prauser, unpublished data), b) different kind of phospholipids (M. Lechevalier *et al.*, 1977); and c) different profiles of fatty acids, menaquinones, and polar lipids (M. Goodfellow, personal communication). The morphologies are similar. Actinophage host ranges are conflicting (Table II) and seem to attribute *Nocardioides* to an intermediate position between *A. simplex* and *Streptomyces*. This, however, obviously would not be supported by comparative 16S and rRNA cataloging, since *Arthrobacter simplex* and *Streptomyces* occupy very different positions (Stackebrandt and Woese, 1981).

B. Actinomycetes of Cell Wall Chemotype II

In the beginning, a few *Micromonospora* phage were found to be genus-specific (Prauser and Rössler, unpublished data). After we recently succeeded in isolating four *Actinoplanes* phage for three strains of the genus *Actinoplanes* which had been isolated and identified by Regine Vettermann of our laboratory, we subjected 48 strains of 19 *Actinoplanes* species; two strains of two *Amorphosporangium* species; five strains of five *Ampullariella* species; two strains of two *Dactylosporangium* species; nine strains of six *Micromonospora* species, including three unidentified strains; one strain of *Planomonospora parontospora*; and three strains of two *Streptosporangium* species, including one unidentified strain to a cross-infected experiment which included the four *Actinoplanes* phage (Ap-phage) and five *Micromonospora* phage (Mm-phage). Species identification of 34 of the *Actinoplanes* strains is tentative, and all studies will have to be repeated. Cross-infection results of slow growing organisms like most of those used here are difficult to read and to reproduce. Nevertheless, the general situation seems to be clear. All *Micromonospora* strains, including *M. aurantiaca* ATCC 47029 (Table III, source 1), *M. echinospora* ATCC 15837 (1), *M. brunnea* INA 166 (1), and *M. peucetica* B 211 (4) were susceptible to at least three Mm-phage and one Ap-phage. The majority of the *Actinoplanes* strains was susceptible by clearing effects and/or lysis to at least one Ap-phage. A few strains likewise were susceptible to at least one Mm-phage. Among the latter strains, none of the 14 international reference strains was included. *Amorphosporangium auranticolor* RIA 819 (8) and *Amorphosporangium globisporum* KCC A-0186 (19) were susceptible to lysis by Ap- and Mm-phage. *Ampullariella campanulata* DSM 9244 (18) was lysed by phage Ap4.

The common susceptibility of strains *Actinoplanes, Amorphosporangium*, and *Ampullariella* to *Actinoplanes* phage is in good correspondence with the results of 16S rRNA oligonucleotide sequence cataloging (Stackebrandt and Woese, 1981), of DNA:DNA homology studies (Stackebrandt *et al.*, 1981[b]) and of DNA:RNA duplexing (Stackebrandt *et al.*, 1981). There is also correspondence concerning the common susceptibility to Mm- and Ap-phage of strains *Actinoplanes, Amorphosporangium* and *Micromonospora* (Stackebrandt and Woese, 1981; Stackebrandt *et al.*, 1981 a, b). The observations of Kikuchi and Perlman (1977) as a phage active against *Micromonospora* spp. were discussed above.

The *Micromonospora/Actinoplanes* phage are understood to be family-specific for the family *Micromonosporaceae* including the genera *Micromonospora, Actinoplanes, Amorphosporangium, Ampullariella*, and *Dactylosporangium, i. e.*, the actinomycetes of cell wall chemotype II.

C. Actinomycetes of Cell Wall Chemotype III

For nearly all the genera, phage could not or have not yet been isolated, though we are especially interested in obtaining phage for *Actinomadura* strains.

For *Nocardiopsis dassonvillei*, phage were briefly mentioned by Prauser (1981a). As far as we know, these phage are species-specific, since they did not attack strains previously thought to be members of the genus *Actinomadura*.

The clearing effects caused by two *Streptomyces* phage in the lawns of strains of *Nocardiopsis dassonvillei* (Prauser 1981b) are interesting, since this is the only case to date in which phage effect organisms of a different cell wall chemotype. Only a few arguments favor a relationship between *Nocardiopsis dassonvillei* and Streptomyces. *N. dassonvillei* was called a macroscopic replica of *S. griseus* (Gordon and Horan, 1968). After numerical taxonomic analysis, *N. dassonvillei* falls within the *Streptomyces* cluster (Williams *et al.*, 1981).

Strains of the genus *Thermoactinomyces*, which according to Stackebrandt and Woese (1980) should no longer be considered to be an actinomycete, are susceptible to taxon-specific phage that do not show any cross-reaction with strains of other genera (Prauser and Momirova, unpublished data).

D. Actinomycetes of Cell Wall Chemotype IV

Phage isolated for strains of *Nocardia* and *Rhodococcus* are cross-reactive to strains of these genera, but do not cause any response in strains of other taxa (Prauser, 1981a,b; Prauser and Falta, 1968; Williams *et al.*, 1980; Prauser and Rössler, unpublished data). Though mycobacteria were not broadly represented in these studies, it seems that they are not susceptible to *Nocar-*

dia/Rhodococcus phage (Prauser and Falta, 1968; Prauser and Rössler, unpublished data). I understand the *Nocardia/Rhodococcus* phage as family-specific for the family *Nocardiaceae*. Like the *Streptomyces* phage, they do not differentiate the genera.

Micropolyspora brevicatena IMRU 1086 is susceptible to *Nocardia/Rhodococcus* phage. This supports its transfer to the genus *Nocardia* (Goodfellow and Pirouz, 1982).

Table III and data from Prauser (1981a,b) contain only selections of the total strains tested. Nevertheless, the data show that there are quantitative differences between the strains and taxa concerning the number of lytic and clearing effects. These results correspond to some extent to findings of other authors, that cannot be discussed here in detail, *e. g.*, the separate position of the "aurantiaca taxon" (Goodfellow and Minnikin, 1981). Thus, the family *Nocardiaceae* is understood to harbor the genera *Nocardia*, *Rhodococcus*, and the genus to be established for the "aurantiaca taxon".

Table III represents the main part of the strains of *Nocardia* and *Rhodococcus* which were tested in a study that also included 16 strains of seven "*Nocardia*" species lacking mycolic acids as well as seven strains of *Saccharopolyspora hirsuta*.

In this study, 12 phage were included which could be isolated for five of the strains and species (Table IV). None of the 16 strains showed any lytic or clearing effect with any of the *Nocardia/Rhodococcus* phage. None of the W-phage caused any response in *Nocardia, Rhodococcus*, and "*aurantiaca*" strains given in Table IV. None of the W-phage lysed any strain other than its propagation host. Williams *et al.* (1980) already had found that 12 strains of *Nocardia aerocoloniegenes, N. autotrophica, N. orientalis, N. rugosa,* and *N. saturnea* were not susceptible to 23 *Nocardia/Rhodococcus* phage. Obviously the "Nocardia" species in question are in search of a new genus that has yet to be established.

Saccharomonospora viridis and *Micropolyspora faeni (= M. rectivirgula)* were found to be susceptible to two species-specific sets of three phage each (Prauser and Momirova, 1970). One of the Tm-phage was isolated later. In the author's opinion, both these genera are in search of a family. The studies of Kurup and Neinzen (1978) were discussed above.

E. Actinomycetes of Cell Wall Chemotype VI

Exclusive susceptibility of *Nocardia turbata* to a set of three phage (O-phage) (Prauser and Falta, 1968) was one of the reasons for the present author to support the transfer of the former *N. turbata* to the new genus *Oerskovia*. In the same study, the taxon-specific activity of two phage (P-phage) against *Promicromonospora* strains was shown. All 34 *Promicromonospora* strains tested were susceptible to at least one of six *Promicromonospora* phage but failed to show any response to 11 Oerskovia phage. It is concluded that

TABLE III. Host Range Pattern of Nocardia/Rhodococcus *phage*

Species	*Strain No.*	*Source* [a]	*N-phage* 2	4	10	13	17	18	21	22	24	25	26	28	30	32	34	36
N. amarae	*IMRU 1604*	6	- [b]	-	-	+	-	*n*	-	-	-	-	-	-	-	-	-	-
N. amarae	*IMRU 3960 (T)* [c]	6	-	-	-	-	-	*n*	-	-	-	-	-	-	-	-	-	-
N. asteroides	*ATCC 19247 (T)*	4	*o*	*o*	*o*	-	+	+	*o*	*o*	+	*o*	-	-	-	*o*	*o*	*o*
N. farcinica	*ATCC 3318 (T)*	11	*o*	*o*	+	-	*o*	+	+	+	+	+	+	-	-	+	+	+
N. otidis-caviarum	*IP 751*	13	*o*	+	+	-	-	*o*	-	+	-	-	+	-	-	*o*	-	+
N. brasiliensis	*ATCC 19295*	10	*o*	+	+	-	-	*o*	-	+	-	-	+	-	-	+	-	+
N. vaccinii	*ATCC 11092 (T)*	4	*o*	+	+	+	+	+	+	+	+	+	+	+	+	-	+	+
R. rhodochrous	*ATCC 13808 (T)*	18	+	+	-	*o*	+	+	+	*o*	+	+	+	+	-	-	+	*o*
R. bronchialis	*ATCC 25592 (T)*	18	-	-	-	*o*	-	-	-	-	-	-	-	-	-	-	-	-
R. corallinus	*ATCC 25593 (T)*	18	-	+	-	+	-	+	-	-	-	-	-	-	-	-	-	-
R. terrae	*ATCC 25594 (T)*	18	*o*	*o*	-	*o*	-	-	-	*o*	-	-	-	-	-	-	-	-
R. ruber	*KCC A-0205 (T)*	18	*o*	-	-	*o*	+	+	+	-	+	+	+	+	+	-	+	-
R. equi	*ATCC 25729 (T)*	5	*o*	+	+	*o*	-	+	+	+	-	-	+	-	-	-	+	+
R. rubropertinctus	*NCIB 9664 (T)*	14	-	*o*	*o*	*o*	-	*o*	-	*o*	-	-	*o*	-	-	-	-	*o*
R. erythropolis	*ATCC 4277 (T)*	5	*o*	+	+	*o*	+	+	*o*	+	*o*	+	+	*o*	+	-	+	+
R. rhodnii	*B/I*	5	*o*	+	*o*	*o*	-	+	-	+	-	-	+	-	-	-	-	+
R. coprophilus	*ATCC 29080 (T)*	5	-	+	+	*o*	*o*	+	*o*	+	*o*	*o*	+	*o*	-	+	*o*	+
"aurantiaca"-*taxon*	*80004*	18	-	-	-	-	-	-	-	-	-	-	-	-	-	-	-	-
"aurantiaca"-*taxon*	*M 296*	5	-	*o*	-	-	-	-	-	-	-	-	-	-	-	-	-	-

[a] *Source of strains (also of strains mentioned in the text in parenthesis)*

1) D. Claus, Göttingen; 2) I. Bousfield, Aberdeen; 3) T. Cross, Bradford, 4) G. F. Gause, Moscow; 5) M. Goodfellow, Newcastle upon Tyne; 6) R. Gordon, New Brunswick; 7) A. Grein, Milan; 8) P. Hirsch, Kiel; 9) R. Hütter, Zürich; 10) V. D. Kuznetsov, Moscow; 11) H. A. Lechevalier, New Brunswick; 12) M. P. Lechevalier, New Brunswick: 13) F. Mariat, Paris; 14) H. Mordarska, Wroclaw; 15) A. Seino, Tokyo; 16) R. Schweißfurth, Saarbrücken; 17) I. M. Szabó, Budapest; 18) M. Tsukamura, Aichi.

[b] *(-): no response; (O): clearing effect; (+): lysis; (n): not tested.*

[c] *(T): Type Strain*

TABLE IV. Strains and Phage of Nocardia *and* Saccharopolyspora *Lacking Mycolic Acids*

Species	*Strain [a] No.*	*Source [b]*	*Phage isolated*
N. "aerocolonigenes"	*IMRU 1663*	6	
N. "	*ISP 5034 [c]*	6	*W1*
N. autotrophica	*ATCC 19727 [c]*	8	
N. mediterranei	*ATCC 13685 [c]*	9	*W2, W4, W5, W6*
N. nitrificans	*Hirsch 379*	8	
N. "	*Hirsch 377*	8	*W3*
N. "	*Hirsch 211*	8	
N. "	*Hirsch 210*	8	
N. orientalis	*ATCC 19795 [c]*	6	*W5, W8, W10, W11*
N. "	*IMRU 1311*	6	
N. "	*IMRU 1149*	6	
N. petroleophila	*ATCC 15777*	17	*W12*
N. "	*ATCC 45*	16	
N. "	*Hirsch 102*	8	
N. rugosa	*Hirsch 675*	8	
N. "	*NCIB 8926*	14	
S. hirsuta	*ATCC 27875 [c]*	6	
S. "	*IMRU 1437*	6	
S. "	*IMRU 1560*	6	

[a] *All strains were included in the same study as the strains listed in Table III.*
[b] *Source of strains see Table III.*
[c] *Type strain.*

there is no close relationship between the genera *Oerskovia* and *Promicromonospora.*

After the introduction of an additional species named *Oerskovia xanthineolytica*, only clearing effects of *Oerskovia turbata* phage could be demonstrated against strains of the new species. One of them was studied in detail (Klein *et al.*, 1981). In Table V, the taxa are given that were interrelated by overlapping susceptibility to phage previously or more recently isolated and propagated on *Oerskovia turbata* IMRU 761 (Table III, source 3) (phage 02), *Oerskovia xanthineolytica* IMET 7075 (phage 05), *Nocardia cellulans* NCIB 8868 (20) (phage 011), *Brevibacterium fermentans* NCIB 9943 (2) (phage cf2), and *Cellulomonas fimi* NCIB 8980 (2) (phage Cl1). Among the additional strains which were included in the cross-infection experiments and that did not show any response to the phage under study were *Corynebacterium manihot* NCIB 9097 (20), *Cellulomonas flavigena* NCIB 8073 (20), *Cellulomonas biazotea* NCIB 8077 (20), and *Cellulomonas subalbus* NCIB 8075 (20). *Cellulomonas cartae* DSM 20106 (18) was susceptible to *Oerskovia* phage (Table V).

On the basis of cell wall chemotype (Keddie and Cure, 1977), isoprenoid

TABLE V. Phage Susceptibility of Oerskovia turbata, O. xanthineolytica, Nocardia cellulans, Brevibacterium fermentans, Cellulomonas cartae, *and* C. fimi

Phage	*02*	*05*	*N11*	*cf2*	*Cl1*
propagated on / *Taxa tested*	Oerskovia turbata *IMRU 761 (6)* [a]	Oerskovia xanthineo-lytica *IMET 7075* [b]	Nocardia cellulans *NCIB 8868 (2)* [a]	Brevibacterium fermentans *NCIB 994 (5)* [a]	Cellulomonas fimi *NCIN 8980 (5)* [a]
Oerskovia turbata	+ [d]	*o*	-	-	-
Oerskovia xanthineolytica	*o*	+	+	+	-
Nocardia cellulans	-	*o*	+	+	-
Brevibacterium fermentans	-	-	*o*	+	-
Cellulomonas cartae *DSM20106* [c]	-	-	+	*o*	-
Cellulomonas fimi	-	-	*o*	-	+

[a] *Source of strains see Table III.*
[b] *Isolate from anthor's laboratory.*
[c] *N. Agre. Moscow.*
[d] *(-): no response; (+) lysis; (o): clearing effect.*

quinone patterns (Collins *et al.*, 1979), DNA:DNA homology and 16S rRNA oligonucleotide sequences (Stackebrandt *et al.*, 1980) and numerical taxonomic analyses (Bousfield, 1972; Goodfellow, 1971), it was suggested by some of these authors to unite *Brevibacterium fermentans, Corynebacterium manihot*, and *Nocardia cellulans* with *Oerskovia xanthineolytica* which, moreover, was transferred to the genus *Cellulomonas* by Stackebrandt *et al.* (1980). The results of cross-infection experiments support the transfer of the above species to the genus *Oerskovia*.

I am undecided as to the inclusion of both the *Oerskovia* species in the genus *Cellulomonas*. I am well aware that, for example, in *Cellulomonas biazotea* small and transient mycelia may be formed under certain conditions. However, in my opinion, *Cellulomonas cartae*, with its typical *Oerskovia*-like life cycle on agar slide cultures, with its relatively low DNA:DNA homology to other *Cellulomonas* species but high homology to both the *Oerskovia* species (Stackebrandt *et al.*, 1980), with its L-lysine containing cell wall identical to that of *Oerskovia xanthineolytica* (Stackebrandt *et al.*, 1980) but different from the ornithine containing walls of *Cellulomonas* Fiedler and Kandler, (Prauser 1966; 1973), and with its actinophage susceptibility reported here, is a member of the genus *Oerskovia*. Of course, the genera *Cellulomonas* and *Oerskovia* obviously are closely related. In addition to the fundamental data of Stackebrandt *et al.* (1980), this is also slightly indicated by the clearing effect on *Cellulomonas fimi* caused by the phage 011 which must be understood as an *Oerskovia* phage (Table V).

IV. THE USE OF ACTINOPHAGE FOR IDENTIFICATION

Frequent attempts have been made to introduce the data with actinophage into the species identification of actinomycetes. The main disadvantage is the fact that there is almost no phage which specifically lyses all strains of a given species. For example, Rautenstein (1960) obtained three phage specific for *Streptomyces griseus*, but only the phage streptomycini III was reported to lyse all strains of the species. Knowing the taxonomic situation within this area of the genus *Streptomyces*, it is clear that a general agreement on thee specificity even of this phage will not be attainable. Phage S2 (Prauser and Falta, 1968) proved to be specific for *Streptomyces albus*. Of the 14 available reference strains, only seven displayed the morphology typical of *Streptomyces albus*. All seven of these strains were lysed by phage S2. A study by Korn *et al.*, (1978) involving more or less species-specific as well as polyvalent phage for the genus *Streptomyces* illustrates the general situation.

The chances for the genus and family identification of a given unknown strain are better. The availability of a set of genus- or family-specific phage is a prerequisite. The members of a set must be selected from a variety of phage after intensive cross-infection studies including as many strains and phage as is possible. Only by this procedure will it be possible to develop a set of phage which is broad enough to encompass all kinds of strains of the given genus, yet which is small enough to be applied in the laboratory. It is difficult to give the number of phage of an appropriate set. Four to eight phage may be a suitable size for practical purposes. In any case, devices such as multi-loop applicators are indispensable as it is impossible in large-scale studies to do all the work by hand. Also, the chance of misplacing suspensions is nearly excluded by the use of multi-loop applicators, especially if the phage suspensions are applied in parallel.

It is our practice to collect 20 to 25 strains of interest and subject them to two to 32 appropriate phage suspensions. In this way, we can relatively reliably pick up strains of interesting taxonomic position from a preselected collection of fresh isolates. Application of a few sets of phage is also the most rational way to check material which does not display, or no longer displays, the characters of sporulation.

V. CONCLUDING REMARKS

In our experience, actinophage host-range studies are above all a simple and quick method to determine natural relationships, *i. e.*, they are one of the tools for classification. The results from these studies have to be compared continually with the results from other methods, preferably with those with the greatest decisive power. In this way, phage host-range studies may contribute to a classification that reflects the natural history of the organisms under study.

ACKNOWLEDGMENT

The author thanks all those, listed in the footnotes, who donated strains.

REFERENCES

Becker, B., Lechevalier, M. P., Gordon, R. E., and Lechevalier, H. A., *Appl. Microbiol. 12*:421 (1964).
Becker, B., Lechevalier, M. P., and Lechevalier, H. A., *Appl. Microbiol. 13*:236 (1965).
Bousfield, I. J., *J. Gen. Microbiol. 71*:441 (1972).
Bousfield, I. J., and Goodfellow, M., *in* "The Biology of the Nocardiae" (M. Goodfellow, G. H. Brownell, and J. A. Serrano, eds.), p. 39. Academic Press, London (1976).
Bradley, S. G., and Anderson, D. L., *Science 128*:413 (1958).
Collins, M. D., Goodfellow, M., and Minnikin, D. E., *J. Gen. Microbiol. 110*:127 (1979).
Delisle, A. L., Nauman, R. K., and Minah, G. E., *Infect. Immun. 20*:303 (1978).
Fiedler, F., and Kandler, O., *Archiv. Mikrobiol. 89*:41 (1973).

Fox, G. E., Stackebrandt, E., Hespell, R. B., Gibson, J., Maniloff, J., Dyer, T. A., Wolfe, R. S., Balch, W. E., Tanner, R. S., Magrum, L. J., Zablen, L. B., Blakemore, R., Gupta, R., Bonen, L., Lewis, B. J., Stahl, D. A., Luehrsen, K. R., Chen, K. N., and Woese, C. R., *Science 209*:457 (1980).
Goodfellow, M., *J. Gen. Microbiol. 69*:33 (1971).
Goodfellow, M., and Minnikin, D. E., *in* "Actinomycetes", (K. P. Schaal, and G. Pulverer, eds.), p. 7 Gustav Fisher, Stuttgart (1981).
Goodfellow, M., and Pirouz, T., *J. Gen. Microbiol. 128*:503 (1982).
Goodfellow, M., Weaver, C. R., and Minnikin, D. E., *J. Gen. Microbiol. 128*:731 (1982).
Gordon, R. E., and Horan, A. C., *J. Gen. Microbiol. 50*:235 (1968).
Keddie, R. M. and Cure, G. L., *J. Appl. Bacteriol. 42*:229 (1977).
Kikuchi, M. and Perlman, D., *J. Antibiot. 30*:423 (1977).
Klein, I., Kittler, L., Kretschmer, S., Füss, F. and Taubeneck, U., *Zeitsch. Allgem. Mikrobiol. 21*:427 (1981).
Korn, F. Weingärtner, B., and Kutzner, H. J., *in* "Genetics of the Actinomycetales" (E. Freerksen, I. Tarnok, and J. H. Thumim, eds.), p. 251. Gustav Fischer, Stuttgart (1978).
Kurup, V. P., and Heinzen, R. J., *Can. J. Microbiol. 24*:794 (1978).
Lechevalier, M. P., De Bievre, C., and Lechevalier, H., *Biochem. System. Ecol. 5*:249 (1977).
Lechevalier, H. A., and Lechevalier, M. P., *in* "The Prokaryotes, a Handbook on Habitats, Isolation, and Identification of Bacteria" (M. P. Starr, H. Stolp, H. G. Trüper, A. Balows, and H. G. Schlegel, eds.), p. 2118. Springer, Berlin (1981).
Lomovskaya, N. D., Chater, K. F., and Mkrtumian, N. M., *Microbiol. Rev. 44*:206 (1980).
Prauser, H., Publications de la Faculté des Sciences de l'Université J. E. Purkyne Brno, *475*:268 (1966).
Prauser, H., *in* "The Actinomycetales" (H. Prauser, ed.), p. 407. Gustav Fischer, Jena (1970a).
Prauser, H., Publications de la Faculté des Sciences de l'Université J. E. Purkyne Brno, *123*:509 (1970b).
Prauser, H., *in* "The Biology of the Nocardiae" (M. Goodfellow, G. H. Brownell, and J. A. Serrano, eds.), p. 266. Academic Press, London (1976a).
Prauser, H., *Int. J. Syst. Bacteriol. 26*:58 (1976b).
Prauser, H., *in "Nocardia* and *Streptomyces"* (M. Mordarski, W. Kurylowicz, and J. Jeljaszewicz, eds.), p. 3. Gustav Fischer, Stuttgart (1978).
Prauser, H., *in* "Actinomycetes", (K. P. Schaal, and G. Pulverer, eds.) p. 17. Gustav Fischer, Stuttgart (1981a).
Prauser, H., *in* "Actinomycetes", (K. P. Schaal, and G. Pulverer, eds.) p. 87. Gustav Fischer, Stuttgart (1981b).
Prauser, H., and Falta, R., *Zeitschr. Allgem. Mikrobiol. 8*:39 (1968).
Prauser, H., and Momirova, S., *Zeitschr. Allgem. Bacteriol. 10*:219 (1970).
Prauser, H., Müller, L. and Falta, R., *Int. J. Syst. Bacteriol. 17*:361 (1967a).
Prauser, H., Müller, L. and Falta, R., *Monatsber. Deutsch. Akad. Wissensch. 9*:730 (1967b).
Rautenstein, J., *Proc. Inst. Microbiol. Acad. Sci. (USSR) 8*:29 (1960).
Stackebrandt, E., Häringer, M., and Schleifer, K. H., *Arch. Microbiol. 127*:179 (1980).
Stackebrandt, E., and Woese, C. R., *in* "Molecular and Cellular Aspects of Microbial Evolution" (Carlile, Collins, and Moseley, eds.), p. 1. Society of General Microbiology, Cambridge (1981a).
Stackebrandt, E., and Woese, C. R., *Curr. Microbiol. 5*:197 (1981b).
Stackebrandt, E., Wunner-Füssl, B., Fowler, V. J., and Schleifer, K. H., *Int. J. Syst. Bacteriol. 31*:420 (1981).
Tille, D., Prauser, H., Szyba, K., and Mordarski, M. *Zeitschr. Allgem. Mikrobiol. 18*:459 (1978).
Wellington, E. M. H., and Williams, S. T., *in* "Actinomycetes", (K. P. Schaal, and G. Pulverer, eds.), p. 93. Gustav Fischer, Stuttgart (1981).
Williams, S. T., Wellington, E. M. H., Goodfellow, M., Alderson, G., Sackin, M., and Sneath, P. H. A., *in* "Actinomycetes", (K. P. Schaal, and G. Pulverer, eds.), p. 47. Gustav Fischer, Stuttgart (1981).
Williams, S. T., Wellington, E. M. H., and Tipler, L. S., *J. Gen. Microbiol. 119*:173 (1980).
Yamaguchi, T., *J. Bacteriol. 89*:444 (1965).

INDEX

K

L

M

N

O

P

Q

R

S

T

U

V